Textbooks in Telecommunication Engineering

Series Editor

Tarek S. El-Bawab
Professor and Dean
School of Engineering
American University of Nigeria
Yola, Nigeria

The Textbooks in Telecommunications Series:

Telecommunications have evolved to embrace almost all aspects of our everyday life, including education, research, health care, business, banking, entertainment, space, remote sensing, meteorology, defense, homeland security, and social media, among others. With such progress in Telecom, it became evident that specialized telecommunication engineering education programs are necessary to accelerate the pace of advancement in this field. These programs will focus on network science and engineering; have curricula, labs, and textbooks of their own; and should prepare future engineers and researchers for several emerging challenges.

The IEEE Communications Society's Telecommunication Engineering Education (TEE) movement, led by Tarek S. El-Bawab, resulted in recognition of this field by the Accreditation Board for Engineering and Technology (ABET), November 1, 2014. The Springer's Series Textbooks in Telecommunication Engineering capitalizes on this milestone, and aims at designing, developing, and promoting high-quality textbooks to fulfill the teaching and research needs of this discipline, and those of related university curricula. The goal is to do so at both the undergraduate and graduate levels, and globally. The new series will supplement today's literature with modern and innovative telecommunication engineering textbooks and will make inroads in areas of network science and engineering where textbooks have been largely missing. The series aims at producing high-quality volumes featuring interactive content; innovative presentation media; classroom materials for students and professors; and dedicated websites.

Book proposals are solicited in all topics of telecommunication engineering including, but not limited to: network architecture and protocols; traffic engineering; telecommunication signaling and control; network availability, reliability, protection, and restoration; network management; network security; network design, measurements, and modeling; broadband access; MSO/cable networks; VoIP and IPTV; transmission media and systems; switching and routing (from legacy to next-generation paradigms); telecommunication software; wireless communication systems; wireless, cellular and personal networks; satellite and space communications and networks; optical communications and networks; free-space optical communications; cognitive communications and networks; green communications and networks; heterogeneous networks; dynamic networks; storage networks; ad hoc and sensor networks; social networks; software defined networks; interactive and multimedia communications and networks; network applications and services; e-health; e-business; big data; Internet of things; telecom economics and business; telecom regulation and standardization; and telecommunication labs of all kinds. Proposals of interest should suggest textbooks that can be used to design university courses, either in full or in part. They should focus on recent advances in the field while capturing legacy principles that are necessary for students to understand the bases of the discipline and appreciate its evolution trends. Books in this series will provide high-quality illustrations, examples, problems and case studies.

For further information, please contact: Dr. Tarek S. El-Bawab, Series Editor, Professor and Dean of Engineering, American University of Nigeria, telbawab@ieee.org; or Mary James, Senior Editor, Springer, mary.james@springer.com

More information about this series at https://link.springer.com/bookseries/13835

Krishnamurthy Raghunandan

Introduction to Wireless Communications and Networks

A Practical Perspective

 Springer

Krishnamurthy Raghunandan
Metropolitan Transit Authority
New York City Transit
New York, NY, USA

Additional material to this book can be downloaded from https://link.springer.com/book/978-3-030-92188-0

ISSN 2524-4345   ISSN 2524-4353   (electronic)
Textbooks in Telecommunication Engineering
ISBN 978-3-030-92190-3   ISBN 978-3-030-92188-0   (eBook)
https://doi.org/10.1007/978-3-030-92188-0

This Springer imprint is published by the registered company Springer Nature Switzerland AG
The registered company address is: Gewerbestrasse 11, 6330 Cham, Switzerland

Foreword

Krishnamurthy Raghunandan (Raghu) has delivered a comprehensive and timely book covering many aspects of wireless communications and networks. One can argue that wireless communication is the most important type of communication that exists today, and by extension, wireless technologies are the most important to understand. This is so because communication using wireless media is ubiquitous. It is the starting point for the vast majority of interactions people have over networks today, whether it is with another person, a machine, or a service.

Wireless network access is available in almost every location in developed territories. First, wireless communication was broadly used by consumers for voice services. With the advent of wireless local area networks, the benefits of wireless and mobile data access became apparent. As device capabilities improved, wireless data started to become available on smart phones, starting with email and then progressing to web browsing. Now wireless communication is used by consumers for entertainment, managing business affairs, healthcare, and many other purposes through a seeming endless supply of apps. Many businesses provide services through mobile apps.

This underlines the importance of understanding the uses, limitations, and applicability of the various wireless technologies that are available and emerging. This applies to people that use the technologies, plan and run business operations, those that invest in technology, and those that create it. This book is a great resource for this broad audience.

The first ten chapters of the book are widely accessible for business people, scientists, and engineers. They provide an introduction to wireless networks and the basics of wireless technologies. It is important to understand the basics behind the different wireless technologies to appreciate when different technologies can be used and to have a working vocabulary of the field. Important aspects of wireless systems are covered in Chaps. 3–6. Wireless communication does not occur in a vacuum so it is important to understand how wireless systems work. These chapters cover cellular systems, short-range wireless systems, and even tap-and-go systems that are being used at points-of-sale. While wireless communication is often used as an access technology, it is also used for backhaul in networks to reduce costs; this type of system is also reviewed. These chapters also cover very practical concerns regarding standards and licensing. These are two aspects of systems that must be well understood by businesses that will use or offer wireless services. Security, one of the most critical aspects of any wireless system is also reviewed.

Chapters 7–10 cover the wireless ecosystem. Wireless technologies can be expensive and require equipment from multiple vendors. Understanding the economics of such systems is important when deciding how to leverage wireless technologies. These chapters conclude with discussion on the impact wireless systems have on society with a special focus on healthcare.

Chapters 11–19 are targeted towards engineers and provide a deeper dive into technology. The first set of these chapters is organized by systems and provides technical details about these systems. These include cellular systems up to and including 5G networks. 5G networks are currently being deployed and will be used widely for many different types of applications. WiFi networks and satellite systems are also covered.

The last several chapters cover selected topics of interest to engineers. These include vehicular networks, SCADA networks which are widely used to control critical infrastructure, and details on wireless transmission and antennas and RADAR, with a separate chapter on measurements. These chapters will be useful to engineers working in these fields.

This book is ideal as a reference book for a broad audience, or as a textbook for engineers. For engineers in particular, it provides an excellent context for wireless networks before diving into technical details. For users of wireless networks, it provides an excellent primer on wireless technologies. I am sure you will enjoy it.

Thomas F. La Porta
Evan Pugh Professor
William E. Leonhard Endowed Chair
Director of Electrical Engineering and Computer Science
Pennsylvania State University
State College, USA

Preface

With the proliferation of wireless devices, offering an understanding of this technology to the younger generation of undergraduate students is felt by engineers, scientists, and mathematicians who have contributed to its growth. Having researched certain aspects of satellite systems and worked in the wireless industry for four decades, I felt it is time to share the immense knowledge gained over these years. I was fortunate to have worked with some of the brightest minds in industry and felt students the world over could benefit from it. This book was written to support and supplement some of the courses in communication technologies taught in universities today.

Many universities offer courses in wireless communication during postgraduate degree level. Some may offer a course or two in wireless as optional. But there is a dire need in industry for fresh graduates who have a broader understanding of this important area of knowledge. To fill this need, it was felt that an introductory course in wireless communication is appropriate at undergraduate level.

To address this need, this book is organized into two parts. The first ten chapters lay the basic foundation for wireless and also address social, government regulations, and commercial aspects of what it means to set up and operate a wireless network. While introducing these technologies, mathematical aspects are deliberately kept to the minimum. At the same time, the security aspects are highlighted as a separate chapter to provide evidence of how the industry made strides to ensure the smart phone became one of the most secure devices and even became the preferred device for financial transactions. Social aspects such as public health concerns about RF radiation are addressed in a separate chapter. It explores the impact of wireless devices on the society, based on a historic account of the technology and how in the past two decades it has changed social interactions.

The second half of the book from Chaps. 11 to 19, uses mathematical and design aspects based on Physics when needed. The objective is to keep it focused on how these systems evolved with pointers to design of algorithms that make wireless system secure and reliable. In some instances, references are made to material covered in other chapters, with the intent to avoid duplication. Last few chapters address technologies that were in use for about a century but now made new strides, such as radar coming into every car and into meteorology as Doppler, or SCADA coming into every home or hospital for monitoring or the autonomous car that is making major strides in transportation. The book concludes with the most important topic of metrology essential for the professional. It elaborates on RF measurements, setting up systems, that needs deeper understanding of Physics, at the same time use application software. This is addressed in depth, emphasizing how standards are built and operated by different countries. An important aspect is the phenomenal accuracy achieved in measuring time and how it affected fundamental understanding of the physical quantities—mass, length, and time but now all of them are referenced to time.

The author has some humble suggestions of how to use this material as a textbook. Since this book organized into two parts, but not two separate books, its use as a textbook can be evaluated by universities in different ways. One possibility is to use the first ten chapters as an introductory subject to students of science (including biology and medicine), engineering, and business since it supplements knowledge in their chosen topics. Specific chapters such as 7–10 could be useful to students majoring in social studies or liberal arts, which enhances their understanding of wireless technology—in terms of social aspects of wireless as a personal device, public health concerns, financial transactions, and regulatory aspects. Later in their career, while interacting with the public, it can help them address the psychological, social, and other aspects related to wireless, but based on scientific evidence—this will be something they would cherish. Similarly, students of biology and medicine may see patients, pets, and wild animals have wireless devices under their skin or within their vicinity and major allocation of frequency bands by regulatory authorities is a sign of major growth that they must become aware of.

The second part of the book from Chap. 11 onwards could be for students of electrical engineering, telecommunication engineering, and computer science/engineering, since it affects several aspects of their chosen topic as majors. It has been pointed out in industry that IT departments of most organizations do not have personnel with background in wireless (except for WiFi). In reality, cellular and other wireless technologies are markedly superior in terms of security and have a different approach to network set up and operation. They are making major strides into IT and the traditional approach of IP-based network is being challenged.

The number of references at the end of each chapter vary, are not only based on the chapter content but are also based on how reader may need to find extensive theoretical/practical supporting material. In rare instances, it also includes video links, to provide better understanding with 3D view, given the complexity of electromagnetic wave and its motion through space.

Each chapter of this book was reviewed by a group of reviewers who are subject matter experts from the academic and industry teams. Their review and comments immensely helped me in terms of adding topics or even add an additional chapter to enhance or support the material presented.

It has been pointed out that this material can be broadly used as a reference book by industry professionals. Personnel in industry could incorporate topics or chapters of this book as a formal training material. Many of them may want to keep a copy of this book for ready reference—which helps in perusal of what are the latest advances in a specific topic.

To me writing this book has been a major journey, often involving thoughts of what went through during product design, network implementation, regulatory, or other aspects that the author experienced in the industry or even while making presentations to students at various conferences. The rapid pace at which this technology advanced was envisioned by standards bodies such as the 3GPP or the IEEE, but it has certainly gone even beyond their own imagination.

To share such vision and allow a wider audience such as the student body to enjoy and enhance this book, is the hope that the author envisions. This also comes with a humble request to provide feedback that may get incorporated at a later date. All the best wishes for an enjoyable and thought-provoking journey called "wireless".

Holmdel, USA   Krishnamurthy Raghunandan

Acknowledgements

I am grateful to the IEEE Communication Society for giving me an opportunity to write this textbook.

I would like to express my gratitude to Prof. Tarek El-Bawab of Communications Society, who patiently went through multiple edits of my proposal on what the contents of this book should be. I am thankful to him for his thoughtful suggestions and supportive actions.

A few words about this book. This book is a tribute to the thousands of hardworking, thoughtful engineers, scientists, and professionals of the wireless communication industry. Their unrelenting efforts have made wireless products and services adored by the world population. I had the good fortune to meet and work with some of them in the Bell labs at Holmdel, and some others in standards body 3GPP during multiple meetings, as well as others while setting up examinations like WCET of the IEEE. Many chapters in this book bear testimony to the efforts they put in, to make wireless technology what it is today.

I would like to acknowledge the following reviewers:

1. Prof. Thomas La Porta
2. Prof. Mathini Sellathurai
3. Dr. Jeffrey H. Sinsky
4. Dr. Anjali Agarwal
5. Dr. Ashok Rudrapatna
6. Dr. Rez Karim
7. Eng. Ying Zhu
8. Dr. Mani Iyer
9. Prof. Manu Malek

These professionals are all subject matter experts, who were very supportive, and provided comments that helped enrich contents of this textbook. They all felt that this book would be very valuable to students and implying that the lessons run as a seamless learning.

In addition to the above experts, all members of my family have been very supportive, by reading specific chapters and providing helpful suggestions. I would like to thank my niece, Engr. Ashwini Gopinath, who reviewed the initial chapters and provided feedback on how to make this appealing to the younger generation. Her insight helped me to modify my style of writing.

I am thankful to my dear daughter, Engr. Gowri, who took time and interest in reviewing some of the chapters such as 7–9 related to her field of expertise of Pharmaceutical Engineering.

I would also like to take this opportunity to thank my wife, Ar. Samidheni, being a College professor of Mathematics herself, provided valuable comments on how people from other disciplines and faculties (other than engineering) would perceive it. She has supported me through the entire period of writing this book. I am immensely thankful to her for her insights and helpful suggestions.

It is my fervent hope that this book will open up the minds of many in the younger generation who have grown up with the cell phone and enjoyed the benefits of wireless communication.

Krishnamurthy Raghunandan

Contents

About the Author

Krishnamurthy Raghunandan has been a Technical Lead for the IEEE Communication Society's WCET exam. He is also a Professional Engineer licensed in the state of New York. He holds a Master's in Electrical Engineering from the Indian Institute of Technology, Kharagpur, and conducted research on satellite communication receiving M.Phil at the University of Surrey, England. Since 2004, he has been a Project Engineer responsible for introducing the latest wireless technologies into subway systems and buses at New York City Transit. From 1993 to 2004, he worked at AT&T/Lucent Bell Laboratory, Holmdel, New Jersey, developing and testing cellular handsets (evaluation in lab and field trials), was a Team Leader for 3GPP standards RAN (Radio Access Network) and was a Bell Labs field representative developing new products and networks for customers. During 1977–87 he worked in the Indian Space Research Organization working on launch vehicles and satellite checkout systems. He has published numerous research papers with IEEE.

Supportive Material

Supportive material at the end of each chapter is organized into six topics (see notes below):

1. End of chapter summary
2. Emphasis on participation
3. Knowledge gained
4. Homework problems
5. Project ideas
6. Out-of-the box concepts

Conclusion at the end of each chapter indicates author's remarks on how different aspects of that chapter have progressed and how the industry or topic may progress.

But "end of chapter summary" lists the topics actually covered in that chapter.

"Emphasis on participation" provides insights into what a student can do to gain a better understanding, it can be theoretical, hands-on effort as a team or even using an app or an algorithm.

"Knowledge gained" touches on new material addressed in that chapter that helps the student to recall, at times connecting such thoughts to earlier chapters.

"Homework problems" focus on eliciting work from the students, how well they understood the material in that chapter, can they supplement it with other material they have access to, including reference material in each chapter. Instead of just providing numbers that can be substituted in a formula given in that chapter, it focuses on whether the students understand deeper aspects.

"Project ideas" are meant to be pointers and indicate possibilities of what projects could be done to enhance understanding of the topic. Since each topic has considerable potential for growth, project ideas can evolve based on progress of technology at that time.

"Out-of-the-box concepts"—Perhaps the most important aspect of wireless technology is this idea of thinking about something totally off the traditional line. CDMA and OFDM are only two examples, but the topic in each chapter is expected to open up the minds to look at possibilities that may seem contrary to the common understanding.

Introduction

Abbreviations

1G	First-Generation cellular network
2G	Second-Generation cellular network
3G	Third-Generation cellular network (3GPP standard)
4G	Fourth-Generation cellular network (3GPP standard)
5G	Fifth-Generation cellular network (3GPP standard)
6G	Sixth-Generation Cellular network (3GPP standard—in planning stages)
3 GPP	Third-Generation Partnership Project (standard that governs all cellular systems)
3 GPP2	Third-Generation Partnership Project (for 2G and 3G CDMA)
NMT	Nordic Mobile Telecom
TACS	Total Access Communication System, UK first-generation network
AMPS	American Mobile Public System, North American first-generation network
C-450 or C-Netz	German first-generation network
TMA	1G cellular network of Spain
RTMI	1G cellular network of Italy
PDC	1G cellular network of Japan
GSM	Global System for Mobiles (2G network)
TDMA	Time Division Multiple Access
CDMA	Code Division Multiple Access
OFDMA	Orthogonal Frequency Division Multiple Access
LTE	Long-Term Evolution
PC	Personal Computer
RFID	Radio Frequency Identification (Tag—either active or passive)
Bluetooth	Low-throughput personal area technology
ZigBee	Low-throughput personal area technology
Beacon	Low-power broadcast transmitter
TV	Tele Vision (broadcast technology)
Doppler Radar	Wireless system to monitor atmosphere
HF	High Frequency (band in 3–30 MHz)
VHF	Very High Frequency (bands in 30–300 MHz)
UHF	Ultra High Frequency (band 300–3000 MHz)
Long Haul Microwave	Link distances of 10 km–180 km
First Net	Public safety network of USA
Pioneer	Space probe to study outer planets and escape beyond solar system
Mercury	First "man-in-space" program from USA
Voyager	Robotic interstellar probe to study solar system
RAN	Radio Access Network (part of network that addresses wireless and mobility aspect)
JPEG	Joint Photographic Experts Group (Standards body that defines all digital images)
MPEG	Moving Pictures Expert Group (Standards body governing all moving images with audio)

© Springer Nature Switzerland AG 2022
K. Raghunandan, *Introduction to Wireless Communications and Networks*,
Textbooks in Telecommunication Engineering, https://doi.org/10.1007/978-3-030-92188-0_1

XML	Extensible Mark-up Language (used to structure data for storage and transport)
IoT	Internet of Things (term used to describe machine-to-machine communication)
Podcast	Digital audio file made available on the Internet for downloading to PC or mobile device
IETF	Internet Engineering Task Force (standards body for IP networks

In communication engineering, "wireless" refers to communication without the use of wires over any distance—from a millimeter to vast deep space, beyond our solar system. To begin study of wireless as a communication technology, it is important to consider whether alternatives exist, if they did why are they being replaced by wireless. Is this concept being promoted by a certain industry, a set of powerful lobbyists, or even by some other force that we don't understand? Reasons for these questions arise from the fact that since the industrial revolution this is one technology that seems to be growing too fast. Therefore, we must begin by asking ourselves, whether we are caught in a short-term fad or does it really have long-term view?

1.1   Is Wireless a Game changer—Or is It just a Fad?

The word "wireless" is often used in the context of a smart phone. However, there is vast area of wireless beyond the smart phone. Even within the smart phone, there are two distinct segments that access the Internet using (a) the cellular network and (b) the WiFi network (which is part of LAN). Use of the word "wireless" to mean both networks and the assumption that these two are similar and interchangeable is quite a common notion. But the cellular network is much larger and independent of the wired LAN network, although it does provide interfaces to the IP network. In contrast, WiFi is a 100 m "over the air" extension for a wired LAN (IP network). For example, WiFi in a bus or a train can only work if it gets connected to Internet through a cellular or satellite network. Therefore, cellular is the essential wireless network and supports devices inside any building as well as outside such as on the street and beyond (land, water, and air).

To address the question of whether wireless is a game changer or not, let us review cellular systems with its brief history. It started as a fad, with many predicting its early demise.

Due to persistent efforts by some, it stayed on and continued in that state for a few decades; the journey started in the last century.

The concept of cellular started in the late 1940s when covering large geographic areas with cells was conceived [1]. Almost 30 years later, in the 1970s, the first generation or 1G was tried in various parts of the world and national standards emerged slowly in the 1980s.

With major improvements in technology, such as chip fabrication and dramatic improvements in reducing power consumption (size of battery), wireless bounced back as a game changer only in the past 20 years. It took center stage of communication networks in the beginning of twenty-first century. Everyone today, affirms the use of cell phone and its impact on society. Figure 1.1 shows a range of cell phones that started in 1973 and now all the way to 5G showing dramatic reduction in size, yet major improvement in performance. Early versions were bulky, consumed a lot of power. If used in cars, it was even necessary to use car battery and attach the portable unit next to the driver. Figure 1.2 shows a typical mobile unit during 1G which was so bulky that rested in a compartment within the vehicle. The antenna was long, and it was necessary to mount it on roof of the vehicle, for it to work properly.

There were 12 separate standards in different countries of Europe (NMT, TACS, C-450, TMA, RTMI, etc.), while Japan and North America had their own standards for cellular [2]. None of these standards worked with each other. Progress was slow and wireless was considered nothing more than a fad, reserved for those with a penchant to try something new. At that stage, it was easy to make counterfeits of cell phones and that was a major problem, perceived by some as the death knell to cellular technology.

The 1G phones typically weighed 5 kg, emitting 5 W but could just make a voice call. It took almost 10 h to charge the battery that supported only 30-min of voice call. In contrast, 4G smart phone today weighs about ¼ Kg and emits 1/5 W while performing innumerable tasks in addition to phone call and Internet access. The change in three decades is quite dramatic.

During early 1990s, second generation of cellular digital standards evolved in North America (D-AMPS or IS-136), Japan (PDC) and in Europe where most countries adopted a common standard called GSM (Global System for Mobiles) [3]. Europe abandoned all their old 12 standards. GSM or Global system for Mobile Communications quickly became a dominant standard. This move by most nations of Europe, made a significant global impact. Cellular became important for voice and low-speed data. The access technology supporting all digital systems was TDMA (Time Division Multiple Access). Since digital ushered an era quite distinct and separate from earlier analog, it was designated 2G or the second-generation standard.

Fig. 1.1 Evolution of cell phones from 1 to 4G

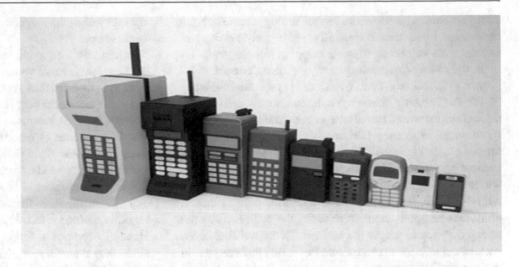

Fig. 1.2 Portable 1G cell phone used in cars

There is another technology called CDMA (Code Division Multiple Access) which became a 2G standard in North America (ANSI-41 or American National Standard Interim-41). These two standards extended to most countries of the globe as part of 2G initiative. These two technologies were considered competitors moving towards 3G that did not inter work with each other. During the initial deployment phase, fraud and counterfeits were possible, but this was aggressively addressed by building a robust security feature in CDMA, bringing a fresh lease of life to cellular. This security feature called the CAVE algorithm (Cellular Authentication and Voice Encryption), was invented by researchers with a well thought out plan to stop hacking and fraud [4]. It was very effective, continuing well into the third generation of cellular technology. We will learn more on this famous algorithm and its impact on cellular technology in Chap. 6 that elaborates on security. It was perhaps the biggest contributor to stop cellular fraud; that approach towards security is now firmly established in all cellular systems.

The late 1990s saw expansion of CDMA using larger bandwidth (5 MHz) that brought higher throughput in digital

cellular. While it started out in 2G as ANSI-41, the special features of CDMA technology [5] such as soft handoff and better security, made it the default technology of 3G. CDMA came to be known as 3G or third-generation standard, although two versions of CDMA persisted in various parts of the globe. An important standards body known as 3GPP (Third-Generation Partnership Project) was formed that became responsible for later generations of cellular technology. It established the relevance of cellular as a common global communication system with well-defined interfaces to all existing communication technologies such as satellite, IP network, and PSTN (Public Switched Telephone Network). Considerable details such as how pictures, videos are sent over the cellular system, how they are displayed depending on the screen size of various devices, were all addressed by this standards body (3GPP). Let us take a moment to visualize the importance of this effort. When users have cell phones of different sizes, different screen types, made by different vendors, used in different parts of the world—how does the system know what phone is used, where, and how does the system know how to fit the picture into this phone and a host of other features on my cell phone? Perhaps the most important contribution was how to make this technology secure to bring confidence in its user community, since early versions of cellular were vulnerable to eavesdropping [6].

In the twenty-first century, a truly global standard of 4G known as LTE (Long-Term Evolution) [7] was established by 3GPP that became mainstay common to all cell phones throughout the world. Based on access technology of OFDMA (Orthogonal Frequency Division Multiple Access), it is the fourth generation or 4G of cellular technology. LTE truly became one standard by which every cell phone works anywhere in the world. During 5G (which is backward compatible with 4G LTE), its popularity has increased even further. It reinforces all applications to think of the cell phone as the primary device to access the Internet. Every website, news, social media site has to reckon with this reality that everyone in the twenty-first century has access to cell phone, but not necessarily wired LAN or a PC with a large screen. There are separate test laboratories and detailed and well-established procedures to put a new smart phone through all the phases of testing before it gets released to the public.

Currently, the 3GPP standards group has a vision of 5G, which is being deployed in phases [8], and initial deployments started in 2019. Later, deployments are expected to continue well into late 2020s, and perhaps by then 6G will start.

The other popular term for wireless concerns WiFi (Wireless Fidelity) which started out as an IEEE standard in 1999. While this has a very different flavor and is mostly used at destination, its value as a broad band technology is widely recognized.

We shall re-visit these standards and technological impacts during reviews in Chap. 5.

1.2 What Are the Wireless Technologies?

Wireless technologies could be classified based on distance —they range from the closest (near field) that operates within a few centimeters, to the farthest which is into deep outer space—Voyager which is already beyond our solar system continues to send us signals. Wireless communication includes all those operating in air (earth's troposphere), to satellites and beyond that operate mostly in vacuum. We can broadly classify them based on the distance and capacity.

1. Near field and short distance: (100 m or less): RFID (Radio Frequency Identification), Blue tooth, WiFi, ZigBee, wireless microphones, Beacons/remote starters in cars (fixed and mobile), garage door openers and others. Most of these devices operate on low power— typically 250mW or less. Most short-distance technologies only support either fixed units or units at pedestrian speeds but can provide throughput that is quite high (several hundred Mbps in some cases). Near field devices operate within the magnetic field which is less than a few centimeters.

2. Medium distance with mobility (between 100 m and 10 km): Broadcast systems (Radio and TV), cellular systems, land mobile systems (public safety), police radar, airport instrument landing systems, port traffic and docking systems, marine radio (boat to boat and beacons), personal beacons in races (running, bicycle, cars, marathons), Amateur Band radio. Fixed Millimeter wave and microwave links. They typically operate using medium power varying from ½ Watt to almost 100 W (over the air). Most medium-distance technologies support full mobility defined at 220 kmph (high-speed rail) or higher. Handoff becomes an important feature in these systems since they cover larger distances. Throughput can be in several Mbps depending on features. Many cellular systems operate in the UHF band over similar distances.

3. Long-distance communication (between 10 and 100 km): Doppler radar, short wave radio (radio broadcast, Amateur Band international and others), marine ship-to-ship HF band (international waters). Long-haul microwave links. Although radiated power in these systems vary from 100 to 1000 W, many of them operate with short pulses and may not continuously radiate. In addition to mobility, several other features such as atmospheric conditions are important in the design of these systems. Throughput capacity could be a few Mbps, but number of users could be quite large, particularly in broadcast systems.

4. Very long distances and regional services (between 100 and 1000 km): Satellite services such as satellite digital radio broadcast, satellite communication links (regional spot beam within a region), emergency communication links to support disaster zones, public safety FirstNet, cellular systems within an entire region, ground radar, and navigation support for commercial aircraft. Instead of using high power, many of these rely on other techniques such as very high gain antennas, low noise amplifiers, etc. Speeds in these services are predominantly based on the speed of satellites in orbit (typically 4–5 times the speed of jet aircraft). Capacity can vary from a few Mbps to hundreds of Kbps. Geostationary satellites at 36000 km from earth is a good example.

5. Deep space/international communication (beyond 1000 km): Communication during space missions, Astro Physics and radio telescopes, missions such as Pioneer, Mercury, Voyager, and others. Intercontinental phone/data calls travelling through major satellite earth station links. These include calls to aircraft, ships, and remote regions. All such communications are concerned with non-real time links, typically taking several hundreds of milliseconds to several seconds of travel time for radio waves. To minimize loss over the air and vacuum, most of these rely on lower frequency bands such as HF

and VHF. Due to this limitation, applications that require large throughput (such as pictures, video) take considerable time to arrive. Generally, throughput is low (a few Kbps in many cases) due to distance involved. Note that delay could be considerable, running into several minutes or hours of deep space missions.

The general theme is that shorter distances allow more throughput, uses less power, and operates under limited atmospheric conditions. This is generally true for all wireless systems.

1.3   Commonality and Differences Between Wired and Wireless Systems

It is important to state that wired communication networks closely follow and are often laid close to or near power cables. For most of twentieth century, the twisted copper pair was commonly used as the primary method of reaching any telephone, in a residence, a business unit or industrial complex. But since late 1980s, fiber optic cable became a major part of wired network and is very widely used today.

Access and Backhaul

These are terms commonly used in both wired and wireless —"Access" and "Backhaul". If we perceive telecommunication network as a commercial building, then Backhaul refers to back entrance of shipping dock from where goods and services are received or sent out. Backhaul often has large capacity to allow goods and services to move from/to the building. It could receive and send trucks, for example, involving large cargo. Similarly, typical telecom provider would be interested in the backhaul. Others who visit the building may not even know about the backhaul. "Access" on the other hand can be perceived as the front door with its lock and key and other security features. It can also involve specific inner doors that lead to cordoned areas within the building. The telecom users (wireless and wired) look for controlling "access" which can be a smart phone, or a PC/laptop and desk telephone.

This concept of backhaul and access although not very accurate, describes some telecom network terms to the novice, in simple terms. The common standard between these two happens to be "Ethernet interface", which can be perceived as a "shipping container for data" that can be sent over rail, truck or even a ship. Similarly, Ethernet interface is common to wired, wireless and satellite links. We will describe Ethernet standard in later chapters which evolved from wired standards as a common interface.

While wireless and wired networks often complement each other, in recent years their roles became somewhat clear. Mostly access using wireless is now the preferred choice, since it offers flexibility in terms of mobility (not stuck in a room with a desktop PC or wired phone, for example). The wired network is most effective as a backhaul network. This is not a strict definition but depicts evolution of each technology and how the engineering profession perceives and deploys them. Most new deployments globally follow this strategy. Depending on the terrain and geographic location, combination of these two are always possible. The following table provides a cursory view of commonality and differences.

From the Table 1.1, a general pattern emerges—access mainly uses wireless, while backhaul is mainly through wired, but an interchange or combination may be needed in some cases.

Both wired and wireless systems provide consistent performance in terms of all the applications listed. There is obvious advantage for each—a fixed wire link telephone provides a known voice quality while wireless can range from extremely good to marginal, including dropped call in regions with poor coverage. On the other hand, fixed service limits the ability to move. Both cordless and WiFi links are only short-distance extensions of wired system and will not allow users to get on the road or move at vehicular speeds. There are important technical differences to consider, but they will be elaborated in later chapters.

1.4   Introduction to Wiretapping—Why is Wireless Tapping not Practical

With the progress in telecommunications, came concerns about privacy. The act of wiretapping is probably as old as telephone itself. The concern continued as systems became more complex. Therefore, wiretapping acts were enacted by regulatory bodies to address this issue. For example, the ECPA or Electronic Communication Privacy Act of 1986, is s federal law, aimed at protecting privacy in communications with other persons. While the act includes tapping of conversation on radio as well, it was a major challenge to implement this on a cellular system.

To illustrate how wiretapping is affected, consider the simple connection shown in Fig. 1.3. The "Act" not only concerns tapping a voice call but also data moving through a wired network. We could therefore perceive both voice and data connections from telephone as capable of being tapped (ISDN phones of 1970s already had a data port).

Unlike the scheme indicated in Fig. 1.3 (for wired), the cell phone does not have a fixed connection to one single base station. Different base stations it connects to (over the air), can change even during a single conversation without knowledge of the user.

For a potential hacker trying to tap the user conversation, this poses a major problem. Firstly, the conversation "over

Table 1.1 Comparison of wired and wireless systems

Application	Wired	Wireless	Comment
Public Switched Telephone Network (PSTN)	Most widely deployed, known as trunk circuit	Cellular networks interface to PSTN at the switch	Physical location of wired phone is known constant. For wireless, location is based on triangulation and algorithms
Voice coder (to synthesize voice to digital signals)	Originally analog trunk quality voice. Now G 711 coder, Voice over IP (VoIP)	Varies with technology. 2G and WiFi use ACELP (Algebraic Code Excited Linear Prediction), 3G EVRC (Enhanced Variable Rate Coder), and 4G VoLTE (Voice over LTE)	Each generation of wireless brings new versions of voice coder. However, ACELP is used as windows media player. Therefore, in terms of voice quality the differences between the two are no longer distinct
Data services— video/CCTV	Multiple devices and channels are well supported	Capacity based on link. 60 GHz links support up to 1Gbps. Lower frequency links support less. 300 Mbps at 23 GHz. 4G modem supports 50 Mbps	For fixed links (similar to wired), Millimeter wave links support the same throughput as fiber links. Link distances are limited based on line-of-sight up to distances of 1 km
Backhaul 10Gbps or more	Standard used for gateways, switches	Available on 70/80 Gbps Millimeter wave links with line of sight, up to 5 km	For short distances and semi-urban locations where other services are needed, wired links are cost-effective
Data services— web/interactive	Standard for PC, laptops in buildings	Smart phone, tablet, and others. WiFi in building, Cellular everywhere else	Access in offices becomes flexible by WiFi. For all other situations, cellular is common
Overseas/inter-continental links	Undersea cable/fiber optic cable	Satellite links are popular globally. Connected at international gateways	A combination of wired and satellite links is required in most cases
Broadcast/cable TV	Only cable extension to homes	Satellite TV and radio broadcast use standard wireless links	Cable TV is a wired extension from major broadcast receive stations to consumers

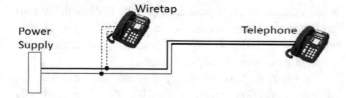

Fig. 1.3 Wiretapping—prohibited by law

the air" is encrypted using an elaborate scheme. This encryption process is quite complex as will be explained in later Chaps. 6 and 11 (CAVE algorithm indicated earlier in Sect. 1.1 is an example). In addition, since the user and the cell phone can move at any given time, there is no single base station that the hacker could tap into. Some service providers may turn off encryption of the air links, which can result in reduced security—allowing potential hackers [6].

Let us consider the next part where some governments regulated that with legal consent it should be possible to tap into conversations of certain criminals. The ECPA has provisions where "Some information can be obtained from providers with a subpoena; other information requires a special court order; and still other information requires a search warrant". If wiretapping with such legal consent were to be considered, how does one tap into a cell phone conversation? The standards body 3GPP was posed this challenge and had to come up with an elaborate scheme to conform. It was determined that interception would be possible at the Mobile Switching Center (MSC) [7]. This is contrary to popular thinking that tapping a wireless conversation is easy. During the analog cell phone era of 1G, yes, it was easy, which was the basis of popular thinking. But during the digital cell phone era of 2G and beyond, considerable effort was made to change this situation including specific algorithms. In fact, these changes made connections very secure which is why hacking cell phone became a major challenge. This will be explained in greater detail later in Chap. 6 on security. In conclusion, wiretapping is limited to wired networks, it is quite difficult to perform wireless tapping.

1.5   Why Are Wireless Devices in Widespread Use?

There are multiple reasons why wireless devices are in widespread use. Perhaps the most important and practical one relates to logistics of deploying the network (also relates to cost). It is not practical to lay cables (copper, fiber optic) over extended distances to establish connection. The second part relates to what is known as "cut the tether"—which means wiring is perceived as a limitation in the sense that it does not allow user device to move around over great distances.

To elaborate on these aspects, consider the example of connecting parts of a country that is made up of a group of islands. In the wired communication network, this situation is handled by a combination of satellite links along with local wired links known as Subscriber Loop Carrier (SLC) system. This example also applies to rural communities where a community of few hundred residents lives far away from the next town. Calls/data messages from this isolated community are grouped together and then sent via long distance fiber optic links or satellite links. A classic example of such communication can be seen in countries such as Indonesia or the Hawaiian Islands. How wireless network accomplishes the same task is to link them directly via satellite, by consolidating local traffic using base stations (instead of SLC). The standards body 3GPP offers a common Radio Access Network (RAN) which has commonality with satellite link. The terrestrial links use UTRAN (Universal Terrestrial Radio Access Network), the satellite links use URAN (Universal Radio Access Network). The same GSM cell phone accommodates both and works with either satellite or the terrestrial tower [9], depending on the availability at that location. This commonality of interface standard means users of a GSM phone to communicate with the satellite since they fully interwork with the GSM network.

This brings us to the question of whether the globe is really wired or is wiring even necessary. The answer is guarded "no" to both. In retrospect, major part of the developing world including parts of Europe, Africa, and Asia were not wired as extensively as North America. Many of these regions steadily evolved towards wireless networks since it offered effective solutions to their communication needs. This is the main reason why 3GPP standards body suggests that wireless Internet or the modified version of wired Internet of today, will become the future [9]. But in countries where extensive fiber-optic long-distance networks exist, situation is different. Fiber can carry considerably more traffic than their wireless counterparts. They will not only stay but will also get enhanced to carry even higher throughput. Several protocols that were originally written for the wired network, were gradually incorporated into wireless allowing co-existence of both wired and wireless. We will learn more on this in Chap. 5 on wireless standards.

1.6   Pictures and Videos

A major consequence of wireless taking center stage was to bring changes to incorporate co-existence. This was due to limited bandwidth over the air (wireless) and how it imposes efficiency on all transmission protocols. Since all long-distance communication (wired or wireless) requires data to move in a serial fashion (one single line or channel on which bits of data move). To send pictures or movies over long distances, standards are established, known as JPEG (Joint Photographic Experts Group) [10] and MPEG (Moving Pictures Experts Group) [11]. These standard bodies specify detailed protocols with the assumption that it is a wired network, and bandwidth is not much of concern. Large parts within the frame were reserved "for future use". When these standards were established during early 1990s, it was not perceived that the humble cell phone would become center piece of multimedia messages (picture and video).

However wireless imposed limitations on these standards —therefore picture was modified to become a standard called JPEG-2000 [10]. Some important features of JPEG 2000 are:

1 Multiple resolutions and quality layers in a single file.
2 Region extraction: Random access code stream allows fast extraction of sub-regions and quality layers.
3 Compression: Support for lossless and "lossy" compression.
4 "Self-Contained-ness": Support for embedded XML metadata in file.
5 Progressive Transmission: Stream information so that image quality improves progressively as the downloading proceeds.

Note that features 3 and 5 indicate JPEG2000 provides pictures lend themselves to transmission over a cell phone (wireless link) since it allows "lossy" medium. Similar improvements in MPEG allowed streaming using SP (Simple Profile). A careful look at these protocols reveals profiles and levels that allow "reasonable quality video" transmission using specific techniques. Figure 1.4 from the MPEG standard provides some details, showing when SP at level 1 is

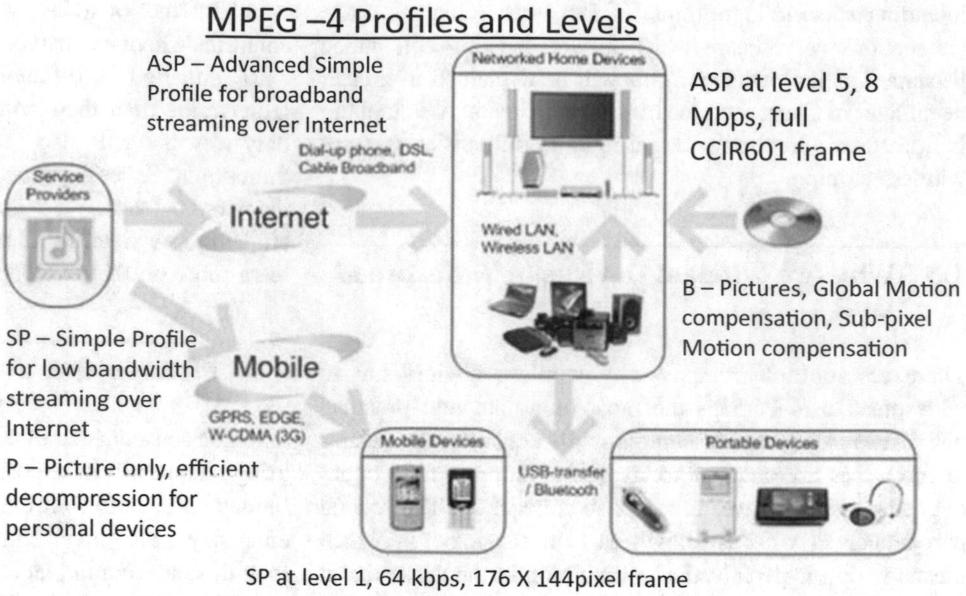

Fig. 1.4 Motion Picture MPEG
—basic profiles and levels

MPEG –4 Profiles and Levels

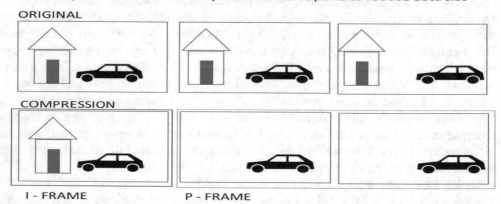

ASP – Advanced Simple
Profile for broadband
streaming over Internet

ASP at level 5, 8
Mbps, full
CCIR601 frame

SP – Simple Profile
for low bandwidth
streaming over
Internet

B – Pictures, Global Motion
compensation, Sub-pixel
Motion compensation

P – Picture only, efficient
decompression for
personal devices

SP at level 1, 64 kbps, 176 X 144pixel frame

Fig. 1.5 Concept of compression
and efficiency with MPEG-4
using H.264 standard

MPEG-4, H.264
• Capture one full image as a first reference
• Take next set of images – compare them with the reference – only
  compress and transmit only the difference parts to reduce data size

ORIGINAL

COMPRESSION

I - FRAME   P - FRAME

used only 64 kbps (low-throughput rate) basic wireless link can be used for streaming. This has become popular and widely used by all cell phones today.

Figure 1.5 provides more details of compression and frames. Using few hundred-kbps sent over a wireless link can provide video of decent quality sufficient for surveillance. High-Definition (HD) video could still be stored at site (locally using wired link), but transmission of decent quality video to command center in real-time uses wireless links. Figure 1.5 provides the process by which original picture is used as a reference—this is known as an I-Frame (highlighted in red), which is transmitted occasionally for reference. Only changes in moving object (car moving away from home) or movement in the background scene are captured through the P-Frame and regularly sent over the air. High-quality, fast-moving image with the compression technique on a CCTV camera produces still 2 Mbps at 30

FPS (frames per second). In wireless link, decent quality video using SP would require less than 500 Kpbs to send these motion pictures. But full-resolution picture for later analysis could be stored locally using wired links. Full-speed motion video of good quality with ASP (Advanced Simple Profile) for broadband streaming at Level 5 can be done over wired links, when needed.

The Moving Picture Experts Group (MPEG) introduced their first standard MPEG-1 in 1993. Over the years, it regularly released more sophisticated and efficient methods to not only build, but also transport motion pictures over the network. The importance of JPEG and MPEG must be stressed again since every cell phone, or device such as television, use these standards every day. For a list of their standards, visit [10, 11].

MPEG-4 offers technology that covers a large range of existing applications as well as new ones such as 3D sound.

The low-bit rate and error resilient coding allows for robust communication over limited rate wireless channels, useful, e.g., mobile videophones and space communication. During development of 3G standards, incorporation of the MPEG version in cell phones was reviewed. Table 1.2 shows a typical list of features currently available to developers who use software to support Audio services on a smart phone based on Android platform. Note that several formats conform to the 3GPP standard types as well as other popular formats that are supported. Mobile phones use 3GP, a simplified version of MPEG-4 Part 12 (MPEG-4/JPEG2000 ISO Base Media file format), with the 3gp and 3g2 file extensions. These files also store non-MPEG-4 data (H.263, AMR, TX3G). Android platform, which is very popular for cell phones provides guidelines to developers [12]. Both in audio and video formats, this platform for developers lists features and capability. Similarly, for video support by MPEG standard offers Android media framework video encoding profiles and parameters. Table 1.3 shows options recommended for playback using the H.264 Baseline Profile codec. The same recommendations apply to the Main Profile codec, which is available in Android 6.0 and later [12].

Table 1.2 Audio support by MPEG standard (for Android developers)

Format	Encoder	Decoder	Details	File Types container formats
AAC LC	YES	Yes	Support for mono/stereo/5.0/5.1 content with standard sampling rates from 8 to 48 kHz	• 3GPP (0.3gp) • 3GPP (0.3gp) • MPEG-4 (.mp4,.m4a) • ADTS raw AAC (.aac, decode in Android 3.1+, encode in Android 4.0+, ADIF not supported) • MPEG-TS (.ts, not seekable, Android 3.0+)
HE-AACv1 (AAC+)	Android 4.1+	Yes	
HE-AACv2 (enhanced AAC +)		Yes	Support for stereo/5.0/5.1 content with standard sampling rates from 8 to 48 kHz	
• MPEG-TS (.ts, not seek able, Android 3.0+)		Android 9+	Support for up to 8ch content with standard sampling rates from 8 to 48 kHz	
xHE-AAC	Android 4.1+	YES Android 4.1+	Support for mono/stereo content with standard sampling rates from 16 to 48 kHz	
AAC ELD (enhanced low delay AAC)	YES	Yes	4.75–12.2 kbps sampled @ 8 kHz	• 3GPP (0.3gp) • AMR (.amr)
AMR-NB	Yes	Yes	9 rates from 6.60 kbit/s to 23.85 kbit/s sampled @ 16 kHz	
AMR-WB	Yes	Yes	Mono/Stereo (no multichannel). Sample rates up to 48 kHz (but up to 44.1 kHz is recommended on devices with 44.1 kHz output, as the 48–44.1 kHz downsampler does not include a low-pass filter). 16-bit recommended; no dither applied for 24 bit	• FLAC (.flac) MPEG-4 (.mp4,.mp4a, Android 10 +)
FLAC	Android 4.1+	Android 3.1+	MIDI Type 0 and 1. DLS Version 1 and 2. XMF and Mobile XMF. Support for ringtone formats RTTTL/RTX, OTA, and iMelody	• Type 0 and 1 (.mid,.xmf, mxmf) • RTTTL/RTX (.rttl,.rtx) • OTA (.ota) • iMelogy (.imv)
MIDI		Yes	Mono/Stereo 8-320kbps constant (CBR) or variable bit-rate (VBR)	• MP3 (.mp3) • MPEG-4 (.mp4,.m4a, Android 10 +) • Matroska (.mkv, Android 10+)
MP 3	Android 10+	Yes		Ogg (.ogg) • Matroska (.mkv)
OPUS	Android 10+	Android 5.0+	8- and 16-bit linear PCM (rates up to limit of hardware). Sampling rates for raw PCM recordings at 8000, 16,000, and 44,100 Hz	WAVE (.wav)
Vorbis				Ogg (.ogg) • Matroska (.mkv, Android 4+) • MPEG-4 (.mp4,.mp4a, Android 10 +)

Table 1.3 Video encoding recommendations by MPEG standard

	SD (Low quality)	SD (High quality)	HD 720p (N/A on all devices)
Video resolution	176 × 144 px	480 × 360 px	1280 × 720 px
Video frame rate	12 fps	30 fps	30 fps
Video bitrate	56 Kbps	500 Kbps, 2 Mbps	2 Mbps, 4Mbps
Audio codec	AAC-LC	AAC-LC	AAC-LC
Audio channels	1 (mono)	2 (stereo)	2 (stereo)
Audio bitrate	24 Kbps	128 Kbps	192 Kbps

Depending on the size of phone (and size of its display) and throughput capability of the smart phone, different options exist. Obviously HD display is possible only on phones with large display and with high-throughput subscription (typically 4G or 5G service). 30 fps (frames per second) provide full motion video (for example a car moving at 120kmph) can be used for games or other types of recording using smart phone.

In Table 1.4, observe that the most popular picture formats of JPEG and PNG are supported by both encoder and decoder. What does this mean? It means picture sent or received can be clearly managed, edited, etc., much like the way it is done on a laptop or a desktop PC. Yet, BMP or GIF files sent from a PC can be decoded and received by the cell phone. Such details indicate how the major standard bodies related to pictures and motion pictures have quickly realized that the cell phone originates and receives images and movies. It is gradually becoming the primary means of multimedia communication. Detailed guidelines are provided by MPEG on what is supported in terms of streaming video, that has now led to variety of online services such as lecture presentations, news videos, other pre-arranged programmed streamed to audience around the globe. While laptop can be used if there is WiFi, the smart phone allows enjoying such programs wherever there is cellular service.

1.7   Internet of Things (IoT)

There is a whole world of wireless devices that can fill the void of sensor networks, in general referred to as "machine to machine communication". These are devices that work in locations like at home, outside, in industry and all around provide information based on context, just a few times a day. This area is considered important since wireless services now pervade most areas of globe, wherever sensor network are needed. Such a platform is a natural choice for wireless devices.

To standardize and visualize technology based on wireless sensor network and IP network is important to note that data sent by "Things" each time is no more than a few hundred bytes. Typical examples include credit card information, swipe in ID cards at entry points of buildings, or vital patient health parameters such as blood pressure or pulse rate. These are best served by a flexible, dynamic platform grouped as "IoT—or Internet of Things" that connects to Wireless.

Figure 1.6 shows the concept of splitting the total infrastructure into 4 layers (on the right) to receive context supported e-services out of raw data from the "Internet of Things" [13]. These layers only establish a general framework, but do not alter the current infrastructure. These layers

Table 1.4 Shows images supported by the MPEG and JPEG standards

Format	Encoder	Decoder	Details	File types container formats
BMP		Yes		BMP (.bmp)
GIF		Yes		GIF (.gif)
JPEG	Yes	Yes	Base + progressive	JPEG (.jpg)
PNG	Yes	Yes		PNG (.png)
WebP	Android 4.0 + Lossless: Android 10 + Transparency: Android 4.2.1+	Android 4.0 + Lossless: Android 4.2.1 + Transparency: Android 4.2.1 +	Lossless encoding can be achieved on Android 10 using a quality of 100	WebP (. webp)
HEIF		Android 8.0 +		HEIF (.heic;. heif)

Fig. 1.6 Four-layer context
aware conceptual framework

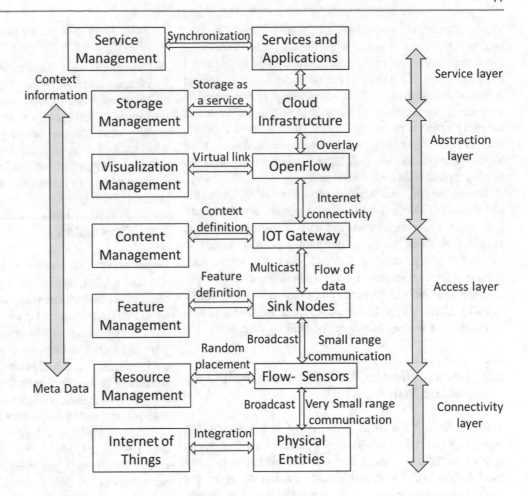

indicated on the right side in Fig. 1.6, help in creating interfaces that will interwork well and provide a generic architecture for all IoT. A brief description of each layer is described below which will clarify what was stated earlier.

Connectivity layer: This layer connects all physical devices involved in the framework and interconnection among them. Such devices should be uniquely identifiable. This layer assigns ranges to low power devices such as sensors, actuators, RFID tags, etc., and resource management checks that uses resources in terms of wireless infrastructure is available.

Access layer: Access layer consists of topology definition, network initiation, creation of domains. This layer includes connection setup, intra-inter domain communication, scheduling, packet transmissions between flow-sensors and IoT gateway. It also selects data based on context and rejects redundant data to help streamline relevant information and store it as needed.

Abstraction layer: One of the important functions of abstraction layer is to characteristic of OpenFlow by adding virtual layer, as shown in Fig. 1.6, which provides a common platform among various communication systems. This

layer becomes the central common platform from physical layer perspective but allows distribution of services. It helps control traffic, assign QoS (Quality of Service) to achieve efficient use of bandwidth, reliability.

Service Layer: Service layer is responsible for storage and management and works with future technologies that bring scalability and efficiency in the system. It involves data supervision, storage, and business management and operations. Although shown as a single layer, business support resides slightly above cloud computing service, whereas OpenFlow is placed below to include visualization and management.

The general concept moving forward is to have service providers offer such infrastructure layers to users or even to agencies that collect data from various organizations and provide the data they need using a meta data pipe to drop off information to organizations that need them. This generic concept may work very well since small amount of data that each organization receives from a large number of sensors, may not be cost-effective to invest in massive infrastructure. It may be better to use service of such "neutral IoT agency". Until the age of IP networks, such "things" did communicate using analog signals sent over wires (twisted copper pair).

This may soon become a thing of the past. This concept of analog signal and main control center at a central location to manage data received from various sensors will be discussed in Chap. 16 on SCADA. It is important to note that such an independent network and its monitoring is not practical anymore since the growing number of devices will make it far more efficient to use IoT network. For home security, such neutral providers already do provide monitoring your home while you are away. All that is needed is an access to the Internet with an IP connection.

It is no surprise that wireless service providers across the globe have made major plans to accommodate such IoT devices on a very large scale. The scale is so large that IoT is projected at about ten times the human population. IoT is covered in greater detail in Chap. 5, Sect. 5.6, where standards and devices relating to the expanding universe of IoT are reviewed. IoT will be explored further in Chap. 16 where SCADA (Supervisory Control and Data Acquisition) networks that are mainly based on IoT will be described.

1.8 Introduction to the Culture of Wireless— Tesla's Vision

It is not difficult for those growing up in the twenty-first century, to envision the culture of wireless. What this means is that it is the only device that you will ever have to carry around. It provides personal identity at airports, opens car doors, pays for purchases, can connect to social networks, television, radio, navigation, bank transactions, and everything else that needs Internet access, anything such as a chat or podcast.

However, at the end of twentieth century, there were skeptics who felt wireless technology would not come to fruition. Not so for others who had worked on this technology such as Nichola Tesla [14]. We know him as the forefather who envisioned all electrical power distribution systems of today. He was a treasure house of inventions although not widely recognized in his own lifetime. He also did envision a world where wireless would accomplish many things that have only become possible today (about one century later). In honor of this great electrical engineer and visionary, it is only appropriate to recall what he said at the beginning of the twentieth century:

"As soon as [the Wardenclyffe plant is] completed, it will be possible for a businessman in New York to dictate instructions and have them instantly appear at his office in London or elsewhere. He will be able to call up, from his desk, and talk to any telephone subscriber on the globe, without any change whatever in the existing equipment. An inexpensive instrument, not bigger than a watch, will enable the bearer to hear anywhere, on sea or land, music or song, the speech of a political leader, the address of an eminent man of science, or the sermon of an eloquent clergyman, delivered in some other place, however

distant. In the same manner any picture, character, drawing or print can be transferred from one to another place…"

Nichola Tesla, "the future of wireless art", Wireless Telegraphy and Telephony, 1908, pp. 67–71.

Have we accomplished everything Tesla forecast over a century ago? Perhaps—yes. We would conclude this chapter paying tribute to the genius who predicted the future of wireless as a culture, quite precisely. The words he spoke reflect the culture of society in his times, yet his vision of what wireless can do, is indeed, quite accurate.

1.9 Conclusion

Objective of this chapter was to introduce the reader to wireless communication as a growing technology. In this effort, the chapter started off with some of the known facts by classifying them and portrayed the reality that wireless made a major leap only in recent years but is here to stay. It provided useful examples of why this is so. In the global environment, communication was always perceived as something that required wires. The fact that these wires are not always needed, makes sense from a Physicist's perspective, but not perceived by the public.

There is also the widespread perception that everything over the air is freely accessible. In terms of security, this notion was quickly put to rest at the beginning of second generation of wireless network with digital technology and algorithms. It will be discussed later in Chaps. 6 and 11.

Summary

The chapter provides an overview of wireless technology using various examples in terms of distance, type of service, etc. It elaborates on some of the well-known wireless technologies. It compares the current wired and wireless technologies used in communication networks to provide a conclusion on their strengths. An important legal aspect of both wired and wireless networks relates to privacy. This was discussed briefly and the difficulty of tapping into a conversation/data over wireless network was elaborated. The chapter concludes with impacts such as Pictures, videos and IoT indicating some reasons why society has embraced wireless devices in a big way. Reasons for its popularity are highlighted indicating a major sea of change is forthcoming due to IoT (Internet of Things). The chapter concludes with how our society has come very close to the vision given by the great innovator Nichola Tesla over a century ago.

Emphasis on participation

(i)   The concept of cellular was first initiated in which year? Why is the word "cellular" used?—hint: look for cellular design in nature, by insects.

(ii) Why should governments mandate to tap details of conversations? Under what circumstances do these not conflict with privacy of electronic communication?

(iii) Why did the GSM network become popular in so many countries? Give at least two reasons.

(iv) Tables 1.5.1.1, 1.5.1.2, and 1.5.1.3 describe a variety of file formats. To familiarize, search some of the file types and formats. Which of them are used on your cell phone?

(v) The standards body 3GPP represents third generation of wireless. Were there other standard bodies during the 3G time frame? If so, what happened to them? (see Ref. [5]).

(vi) "Internet of Things" seems to be a whole world of its own. What was the predecessor to this? Does it still exist? What do some of the old power substations in your area use to monitor power equipment?

Knowledge gained

(a) Wireless and wired networks—both have their strengths and limitations. In general, wireless in moving into the area of "access to the network" with smart phone as an important device. Several other wireless devices have taken a similar approach.

(b) Contrary to popular thinking—it is not easy to hack into a cellular network. This has become true since the digital wireless systems introduced excellent security features.

(c) An important reason for popularity of wireless devices is based on the premise of "cut the tether" making the device free and independent to move around. The flexibility this concept provides was elaborated.

(d) From a historic perspective, some of the visions of engineers has taken almost a century to come to fruition. However, the journey along the way has brought forth devices that are very close to the source (near field) as well as to the edges of our known universe.

(e) The medium for all wireless communication is well established—air in the global terrestrial region and vacuum in the bigger space of our universe.

(f) Changes are being made to accommodate wireless devices in all major application areas of communication systems (JPEG and MPEG are examples).

Homework problems:

1. WiFi and Cellular are both wireless technologies. Which one of these supports mobility while you are "on the road"?

2. "Near-field" devices—require that the device must operate within the room. Is this statement accurate? Substantiate your answer.

3. What were the earliest wireless services offered to the public? Are these services still in use today?

4. Tapping into an electronic communication was prohibited by law in the year 1986? Why is this important? What are the challenges of tapping into a cellular link?

5. Are there commercially available devices to tap into a conversation? Do they work on the smart phone today? Provide possible answers of why and why not?

6. What are the advantages of having a common radio for terrestrial and satellite? Where are these useful?

7. Size of files such as JPEG and MPEG are based on picture resolution and quality. What are the practical limitations of sending them over wireless networks?

Project ideas

For backhaul network, both fiber-optic cable and microwave links have been used. Make a list of (at least five) countries that predominantly use microwave links? What are the reasons? Propose your concept of moving forward with (a) fiber optic links (b) microwave links for backhaul for a new urban development. What are the challenges?

Visit a museum or the history society of IEEE and make a list of devices that interest you. Perform more detailed search in terms of who built it, when, what did it do? Were there any predecessors, any successors?

Out-of-the-box concepts

Search for an integrated cellular network using natural power resources (such as solar panel, wind or geothermal) and where is it possible? Where are these useful—only in specific countries or globally? Provide a baseline model, with technology.

Security has always been a concern for communication networks. There are detailed security schemes for IP networks, available from IETF (Internet Engineering Task Force). Why did the cellular network choose to develop a new security scheme?

References

1. Ring DH (1947) Mobile telephony—wide area coverage, case 20564. Technical Memorandum, Bell Laboratories, 11 Dec 1947
2. A brief history of mobile communications. Richard Frenkiel, WINLAB, Rutgers University. http://www.winlab.rutgers.edu/~narayan/Course/Wireless_Revolution/vts%20article.pdf
3. GSM history. www.3gpp.org/specifications/gsm-history

4. Analysis of security based on CDMA Air interface system, by Chen Dan, Liu Yijun, Tang Jiali. Available on www.sciencedirect.com, Elsevier, Energy proceeding 16 (2012) 2003–2010
5. Network interworking between GSM-MAP and ANSI-41 MAP. https://www.3gpp2.org/Public_html/Specs/N.S0028-0_v1.0.pdf
6. Eavesdropping on GSM: State of affairs, by Fabian van den Broek, Radboud University, Nijmegen, Institute for Computing and Information Sciences (iCIS). http://www.cs.ru.nl/~fabianbr/pub/WISSec2010_GSM_Eavesdropping.pdf
7. Long term Evolution—LTE 3GPP. www.3gpp.org/technologies/keywords-acronyms/98-lte
8. Tentative timeframe for 5G. www.3gpp.org/news-enews/1674-timeline
9. UTRAN (UMTS terrestrial radio Access / USRAN (UMTS Satellite Radio Access). https://www.etsi.org/deliver/etsi_tr/123900_123999/...00.../tr_123930v040000p.pdf
10. Wireless JPEG2000. https://www.vocal.com/video-codecs/jpwl/
11. Moving Pictures Experts Group. https://mpeg.chiariglione.org/docs/full-list-mpeg-standards
12. Supported media format—for Android phones. https://developer.android.com/guide/topics/media/media-formats
13. Conceptual framework for Internet of Things' Visualization via OpenFlow in context-aware networks, by Theo Kanter, Rahim Rahmani, Arif Mahmoud, Jan 2014, arxiv. https://www.researchgate.net/publication/259953949_Conceptual_Framework_for_Internet_of_Things'_Visualization_via_OpenFlow_in_Context-aware_Networks
14. Ljiljana Trajkovic, Tesla's vision of global wireless communications. http://www.ensc.sfu.ca/research/cnl

Background to Wireless

Abbreviations

FSPL	Free Space Path Loss (minimum loss suffered by Radio signal depends on Frequency and distance)
dBm	Decibel Milliwatt (expression for RF power with 1 mW as reference)
EM Wave	Electro-Magnetic Wave (Electric and Magnetic waves orthogonal to each other and to direction of travel)
Fleming's right-hand rule	Practical way to visualize the Electric, Magnetic waves and direction of travel
Vertical/horizontal polarization	Orientation of the Electric wave as it leaves the transmit antenna
Short wave radio	Radio used for "beyond the horizon" coverage with the help of reflection from Ionosphere, for broadcast, Amateur and over-the-horizon applications (includes HF, MF, and VHF bands)
Ionosphere	The ionized part of the Earth's atmosphere is known as the ionosphere (ions created due to ultraviolet lights colliding with this layer, knocking out electrons creating ions)
ALE	Automatic Link Establishment (ALE) is the worldwide de facto standard for digitally initiating and sustaining HF (High Frequency) radio communications.
NVIS	Near Vertical Incident Sky wave, using reflection from the F layer of ionosphere to communicate over great distance of up to 650 km, marine and long-distance communication in the 1.8–8 MHz band.
FDM	Frequency Division Multiplex (use of separate frequency bands for transmit and receive—Duplex)
TDM	Time Division Multiplex (use of same frequency band for transmit and receive but at only separated by time—Simplex)
OFDM	Orthogonal Frequency Division Multiplex (method using sine and cosine sub channel waves to separate signals maintaining orthogonal signals to avoid interference)
Fourier Series	A series consisting of cosine and sine function sub channel components
Fourier Transform	Allows transformation of signal from time to frequency domain and vice versa frequency to time domain
DFT	Discrete Fourier Transform. A variant of Fast Fourier Transform, but uses discrete-time datasets are converted into a discrete-frequency representation
FFT	Fast Fourier Transform (based on the Forward and Inverse Fourier Transforms)
DSP	Digital Signal Processing—area of technology that processes signals in digital format, with umpteen number of applications
ADC	Analog to Digital Converter. A device that converts real life analog signal into digital signal (for use by DSP or similar processing)
DAC	Digital to Analog Converter: A device that converts digital stream

© Springer Nature Switzerland AG 2022
K. Raghunandan, *Introduction to Wireless Communications and Networks*,
Textbooks in Telecommunication Engineering, https://doi.org/10.1007/978-3-030-92188-0_2

LCD	into analog signal (usually output from DSP to the end device)
	Liquid Crystal Display. A display device that uses low energy to provide a display, often used in battery operated electronic devices
BBU	Base Band Unit is a device in telecom systems that transports a baseband frequency, usually from a remote radio unit, to which it may be tied through optical fiber.
POTS	Plain Old Telephone Service. Refers to the early analog voice transmission phone system implemented over copper twisted pair wires.
Vocoder	A vocoder is a category of voice codec that analyzes and synthesizes the human voice signal for audio data compression, multiplexing, voice encryption or voice

2.1 Introduction

It is important to have a basic understanding or background to study wireless technology. That background is mostly based on Physics and Mathematics. Radio Physics is a branch of Physics that deals with electromagnetic radiation which includes very low frequency to very high frequency (light and beyond). All of them have the same property and we will start with that assumption. Starting with the basic assumption of "waves that radiate are radio waves", there would be a whole range of frequency bands that would include every alternating wave including the electrical power grid (used at homes and offices) operating at 50 Hz/60 Hz, all the way to light and beyond (gamma rays, etc.). These are all electromagnetic waves that travel through different mediums (such as cable, air, vacuum). It is essential to understand their properties so that wireless technology becomes discernable.

2.2 Radio Wave Propagation—Basics

In practice, the term "radio wave" is applied to those used for wireless communication that quite often limits the range. Therefore, the term does not include the lowest frequencies, nor does it include light or highest frequencies. However, the term "radiation" applies to all of them. Generally, the accepted norm for radio waves is frequencies ranging from 300 kHz all the way up to the millimeter-wave frequencies of 30 GHz. The millimeter frequency bands extend up to

about 300 GHz. Although millimeter-wave band, is classified as radio waves, some of their features resemble light (but millimeter waves cannot be seen). Briefly, the accepted range of frequencies for radio wave include the range 300 kHz–300 GHz.

Basically, all electromagnetic waves are governed by the fundamental equation:

$$C = f \cdot \lambda \tag{2.1}$$

where C = Speed of light. At about 3×10^8 m/s, this is a well-known constant described in Physics; "f" is the frequency, and λ is the wavelength. Therefore, Eq. 2.1 can be summed up in words as "the product of frequency and wavelength is always constant and is expressed by the speed of light". This is perhaps the most important fundamental equation that applies to all electro-magnetic radiation including radio waves.

2.2.1 Free Space Path Loss

Propagation or travel of radio waves in free space (or through air) is generally described as "Free Space Path Loss (FSPL)" and it is governed by the fundamental equation:

$$FSPL(dB) = 10 log_{10}\left(\left(\frac{4\pi df}{C}\right)^2\right) \tag{2.2}$$

which can be simplified as

$$FSPL(dB) = 20 log_{10}\left(\frac{4\pi df}{C}\right) \tag{2.3}$$

This is re-written as

$$FSPL(dB) = 20 log_{10}d + 20 log_{10}(f) + FSPL(dB) + 20 log_{10}\left(\frac{4\pi}{C}\right) \tag{2.4}$$

Replacing the constants π and c with their value, we get

$$FSPL(dB) = 20 log_{10}d + 20 log_{10}(f) - 147.55 \tag{2.5}$$

In the above equations, "d" is the distance between transmitter and receiver, "f" is the frequency of the electromagnetic wave. In the original Eq. 2.2, the term $\left(\frac{4\pi df}{C}\right)^2$ represents surface area of a sphere which is $4\pi r^2$. IEEE defines FSPL as "The loss between two isotropic radiators in free space, expressed as a power ratio". There are two important factors that indicate how the signal attenuates (reduces in power) with (i) square of the distance, (ii) square of the frequency. This is the fundamental reason why distance plays a major role. Even 10 m away the signal would reduce to 1/100th of its value. By the same count as frequency increases 10 times, signal would reduce 100 times, for the same distance.

To account for such dramatic loss in signal, logarithmic scale becomes essential. In wireless communication, this results in power being measured in dBm. What is dBm scale and why is it so popular in wireless communication? dBm— is an abbreviation for "dB milliwatt". It is a scale that provides measurement of power, using 1 mW as the reference. RF power will be further elaborated in Sect. 2.2.2. To get an overview of RF power, Table 2.1 shows a practical power level to dBm conversion scale. It helps in understanding the range of power levels encountered in real life wireless systems. Main objective of this table is to view dBm as convenient and practical scale, instead of using conventional method of expressing power in Watts, mW, or kW.

The second and the last column give a clue as to why the dBm scale directly tracks as $(10)^x$—it can be readily observed. Since 1 mW is taken as "reference" for this scale, it is written as "0 dBm". All positive dBm numbers are always multiple of 10. For example, 30 dBm is $(10)^3$ mW (1W or 1000 mW). The largest power used in the table is 1 MW = $(10)^9$ mW. At the other end of scale is lowest level of 1 attowatt = $(10)^{-15}$, signal which is so small that it requires "carefully designed high gain, low noise receiver" used typically in deep space mission receiving stations. Normally, such low power signal would be buried deep in the noise floor.

There is another important thumb rule that wireless engineers regularly use. It is known as 3 dB rule, which helps in quick mental arithmetic. +3 dB means the signal is double. Since 20 dBm is 100 mW (see Table 2.1) 23 dBm will be 200 mW. Similarly, −3 dB is half the power 20–3 = 17, which is 50 mW. This thumb rule goes very well with Table 2.1 where to convert any RF power to dBm, simply to raise the power of mW by the exponent of 10, and then apply the 3 dB rule as and when needed.

Is there a difference between dB and dBm? Yes, there is. dB simply refers to a ratio and it has no dimension. Therefore, gain of an antenna in dB means its "output/input". An antenna with 6 dB or (3 dB + 3 dB) gain means signal is increased by four times (2 + 2). If its input is 20 dBm (100 mW) then output of antenna will be 26 dBm or 400 mW. Similarly, if there is loss of 3 dB in the connecting cable with 100 mW input, then 20–3 = 17 dBm which is 50 mW at its other end.

In contrast, dBm is a unit of power with dimension in milliwatts, with 1 mW as reference. The two should not be confused and always conversion must be kept by maintaining the same dimensional units but applying dB only as a ratio to compare. For comparison, +10 dBm and −10 dBm are 20 dB apart therefore their difference in power is 100. That means when compared to (−10 dBm), (+10 dBm) is 100 times larger. This is the reason why the dBm scale is popular with engineers, although in Physics, energy may still be expressed in microvolts/meter. Standard tables are readily available for conversion since laboratory and field measurements use both.

2.2.2 How Can we Perceive Radio Waves?

This brings us to the fundamental question of "what are these radio waves which cannot be seen, heard or felt?" While they are fundamentally the same as light, the fact that their wavelengths are much longer, it makes them impossible to visually observe. Since they are higher in frequency than the audio wave they cannot be heard even when air is used as a medium. Note that sound is not an electromagnetic wave, but a vibration that is carried through air. Unlike the electric current which can be felt by the skin (as a shock) radio wave is not felt because the equivalent power (therefore current) is too small to be felt by us. Alright, then how do we even recognize the presence of these radio waves, and where do they exist?

Physicists and mathematicians provide considerable analysis to prove that they exist. There are important equations by James Maxwell that provide mathematical methods to describe radio waves [1]. Before entering the world of mathematics and analysis in terms of how these equations relate to the electromagnetic waves, let us consider a more fundamental method of recognizing these waves. It is wisely said that "if something can be measured, it expands our understanding of that phenomenon". These electromagnetic waves at high frequencies do have something in common with our normal, daily use of electricity. One such important measurement of electricity is its power, used for heating which is described by the equation:

$$\text{Power } P = I^2 R \tag{2.6}$$

where P is the Power that results in heat, "I" the current flowing through the resistive element R (such as a heating coil). This is a basic experiment that we perform in school using a simple battery and a resistive coil or a bulb. The power delivered to a resistive coil or bulb heats up the coil (or lights up a bulb). It can be measured indirectly by measuring the current "I" through the wire, using an ammeter. Knowing the value of resistor R, we can calculate the power in watts. For example, if current is 2 Amps and resistor has a value 5 Ω, then we know the power is $(2)^2 \times 5 = 20$ W. How is this useful to measure radio waves? Fortunately, this power law applies to radio waves as well and therefore if we could use a suitably formed carbon cone

Table 2.1 Power levels in the dB milliwatt scale and applications

Power level expressed in Watts	Equivalent level in dBm	Scale (multiplier)	Application	Relation to 1 mW as reference)
1 MW	+90 dBm	1000,000,000	TV transmitter	$(10)^9$
100 KW	+80 dBm	100,000, 1000	TV transmitter	$(10)^8$
10 KW	+70 dBm	10,000,000	TV transmitter	$(10)^7$
1 KW	+60 dBm	1000,000	Radio transmitter	$(10)^6$
100 W	+50 dBm	100,000	Public safety radio transmitter	$(10)^5$
10 W	+40 dBm	10,000	Cellular base station transmitter	$(10)^4$
1 W	+0 dBm	1000	Public safety portable	$(10)^3$
100 mW	+20 dBm	100	Smart phone/WiFi transmitter	$(10)^2$
10 mW	+10 dBm	10	Bluetooth, IoT Zigbee transmitter	$(10)^1$
1 mW	**0**	**1**	**Cordless phone, IoT transmitter**	**Reference** $(10)^0$
0.1 mW	−10 dBm	1/10	Wrist radio transmitter	$(10)^{-1}$
0.01 mW	−20 dBm	1/100	Baby monitors	$(10)^{-2}$
1 μW	−30 dBm	1/1000	Receiver near Base station	$(10)^{-3}$
0.1 μW	−40 dBm	1/ 10,000	Receiver near Base station	$(10)^{-4}$
0.01 μW	−50 dBm	1/ 100,000	Smart phone—strong signal	$(10)^{-5}$
1 μW	−60 dBm	1/ 1000,000	WiFi—strong signal	$(10)^{-6}$
0.1 μW	−70 dBm	1/ 10,000,000	WiFi—very good signal	$(10)^{-7}$
0.01 nW	−80 dBm	1/100,000,000	Smart phone—very good signal	$(10)^{-8}$
1 pW	−90 dBm	1/ 1000,000,000	Smart phone—medium signal	$(10)^{-9}$
0.1 pW	−100 dBm	1/1000,000,000,000	Smart phone—weak signal	$(10)^{-10}$
0.01 pW	−110 dBm	1/10,000,000,000,000	Geostationary satellite receiver	$(10)^{-11}$
1 fW	−120 dBm	1/100,000,000,000,000	GPS satellite receiver	$(10)^{-12}$
0.1 fW	−130 dBm	1/1000,000,000,000,000	Signal from space stations	$(10)^{-13}$
0.01 fW	−140 dBm	1/10,000,000,000,000,000	Signals from Mariner or Lunar missions	$(10)^{-14}$
1 aW	−150 dBm	1/100,000,000,000,000,000	Signals from intergalactic probe	$(10)^{-15}$

that absorbs radio waves, it develops current that can be measured and therefore calibrated in Watts.

This fundamental measurement of electric power is used globally to recognize the existence of radio waves and to quantify radio waves. It is one of the fundamental methods used to measure the power of radio waves [2]. All of us can readily relate to infrared and ultraviolet radiation coming from the sun that warms up our skin. With extended exposure, it tans our skin. Ultraviolet rays are also electromagnetic waves but are at much higher frequencies compared to radio wave (well into the visible spectrum). The other practical device that most of us can relate to is the microwave oven which is used to warm up food. Here, the radio waves are carefully focused and concentrated towards a small area to effectively heat up food.

This is very much like focusing Sun rays with lens that we would have used as children—to focus sun rays on to a piece of paper to burn it. There are two more important parameters that describe the radio waves, indicated in Eq. 2.1. These two are frequency and wavelength. The advantage in this measurement is that, so long as we know either the frequency or the wavelength, it is always possible to calculate the other, because C—the speed of light, is a constant and well known.

2.2.3 Measuring Radio Waves—Imagine What They Would Look Like

To make measurements in the radio frequency bands we are concerned with, wavelength is something we could consider. At frequency of 300 kHz, the wavelength λ is about 1000 m. Although this may seem incredible, the wave is in fact 1 km long. At the upper end of radio band is 300 GHz which is only one mm long; again, this too seems incredibly small, but is true. How does this impact practical use of waves? We will learn in later chapters that the length of an antenna or the width of waveguide to support any frequency must be at least ½ of this wavelength, to be effective.

Therefore, we will see devices of different sizes depending on frequency band they operate. But there are other techniques to squeeze devices into chip sets.

Other than wavelength, what are the other properties of a radio wave? How can one imagine this wave moving all around us? Fundamentally electromagnetic waves are illustrated by the three-dimensional characteristic of the wave indicated in Fig. 2.1 which is carefully drawn to display waves on different planes that are perpendicular (orthogonal) to each other. This is in fact the simplest form in which it can be perceived. Let us consider the waves in Fig. 2.1—starting from the left. The current in the antenna (represented by a rod on the left) produces circular magnetic waves around the antenna—magnetic field/wave B is represented in the horizontal plane in red pulsates and radiates around the antenna rod. The electric field/wave E in the direction of the current is represented in blue by the vertical plane. That electric wave oscillates along the length of the rod. These two fields/waves always maintain orthogonal symmetry as shown. In the third dimension is the movement of the wave represented by direction "d". This is the direction of motion of the electromagnetic wave. Since both electric wave and magnetic waves are perpendicular to the direction of motion, these waves are known as transverse waves. These waves indicate the basic nature of electromagnetic waves that travel at the speed of light.

The two waves "electric and magnetic" are related and the ratio of electric to magnetic field is expressed by the Eq. 2.7

Fig. 2.1 Pictorial representation of an Electromagnetic (Radio) Wave

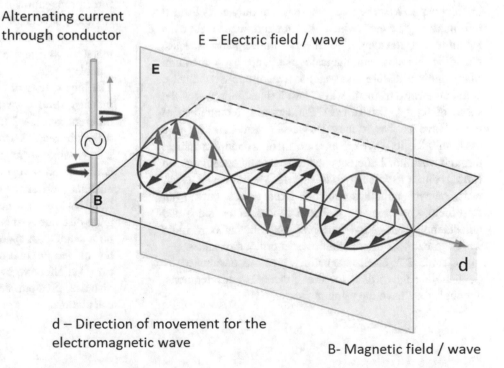

Alternating current through conductor

E – Electric field / wave

d – Direction of movement for the electromagnetic wave

B- Magnetic field / wave

$$\frac{E}{B} = c \qquad (2.7)$$

where E is the electric field vector, B the magnetic field vector, and C is the speed of light. In general, magnetic field is dominant within close vicinity of the antenna. Even in this proximity, it is considered weak compared to earth's magnetic field. As the wave moves away from antenna the electric field becomes dominant. This phenomenon is expressed by the terms "near field" and "far field". The accepted convention is that near field is up to a distance equal to one wavelength which is λ and beyond this distance far field is assumed. In far field, only electric field is considered (although magnetic wave is always present it is assumed to be very weak). In Chap. 3, we will review near-field communication in detail. Electric field is always used to describe orientation of the antenna. The antenna is described as vertically polarized, if antenna is physically held vertical (as shown in Fig. 2.1); the electric field is in fact physically oscillating in the vertical plane or the E-plane.

Antenna is said to be horizontally polarized, if the antenna is placed physically horizontal and the E field would then be in the horizontal plane. These two polarizations are always orthogonal to each other and the separation of these waves is useful in terms of carrying entirely independent information streams. Given the facts common to all electromagnetic waves, let us next consider the question *why is it that only the radio wave band is chosen as a means of communication?*.

Characteristics of lower frequencies (from tens of Hertz up to about 30 kHz) suggest that they can only carry limited information. "Sound" which is vibration, needs air as a medium and travels up to several miles. Most species, including humans communicate using air as a medium. Many marine creatures use water as a medium. "Sound wave is not an electromagnetic wave" and it does not travel at the speed of light (much slower). Fundamental requirement of radio wave is that it must propagate "independent of a medium". Radio waves require no medium and can radiate through vacuum. Different mediums that it encounters on earth, reduces its intensity (attenuates). Medium such as air, water or even buildings attenuate radio waves, but do not stop them. Only metal—an electrical conductor, stops and reflects the waves back. Therefore, most objects allow radio waves to carry of information over enormous distances.

In Table 2.2, electromagnetic waves are designated by frequency bands, due to the practical considerations, although they have the same property.

Direction of the three forces—electric, magnetic, and the direction of wave motion are visually expressed using Fleming's right-hand rule shown in Fig. 2.2. Fleming's right-hand rule applies to the entire range of electromagnetic waves listed in Table 2.2, irrespective of the frequency band separation shown in that table.

This is a practical method, is a quick check used by physicists and engineers, to recollect electromagnetic waves. In Fig. 2.2, the fore finger points to the "magnetic field", middle finger (coming out of the paper) points to the "electric current", and thumb points to the "direction of motion", or the direction in which the EM wave moves.

The signal voltage develops an electric field around the antenna elements. The current flow in the antenna produces a magnetic field. The electric and magnetic fields combine and regenerate one another according to Maxwell's famous equations (discussed later in Chap. 17), and the combined wave is launched from the antenna to travel through space. At the receiving site, the electromagnetic wave induces a voltage in the antenna, which converts the electromagnetic wave back into an electrical signal that can be further processed.

Table 2.2 classification of electromagnetic waves is based on frequency band. In turn, radio band is separated from other bands due to specific qualities of waves within this band such as.

(i) Radio waves can travel over great distances "without the need for a metallic wire".

(ii) Most objects *except electrical conductors* such as metals, carbon, etc., do not stop these waves. Therefore, propagation through natural impediments such as trees, layers of insulation, and other objects is possible (but signal strength reduces). Therefore, only metal as well as electrical conductors are considered "blockages".

(iii) The largest user of "reflected waves" is the broadcast industry. Short waves are radio waves reflected off different layers of ionosphere (brief description will follow in Sect. 2.3).

(iv) It is possible to retain orientation of the electric field waves based on the physical orientation of the source antenna (vertical or horizontal).

(v) Radio waves can carry information effectively over great distances—this allows a transmitter to be placed on a satellite, a space probe or even in Voyager that has moved beyond our solar system. It is possible to carry on effective communication independent of the distance. System design incorporates distance into calculations.

Table 2.2 Electromagnetic waves designated by different frequency bands

Lower bands (1 Hz–300 kHz)	Radio Bands (300–300 GHz)	Light bands (beyond 300 GHz)	Bands above light (X-Ray, gamma ray)
Use of metallic wire/cable as medium. Power lines are most common waves in lower frequencies. But radiation in this band is avoided by shielding	*Does not need metal as medium to propagate.* Radiates both in air and vacuum. No limit on distance but needs proper antenna for waves to radiate	Travels in air and vacuum, but only in straight lines. Cannot travel through opaque objects including thin paper or dense smoke	Can travel short distances through dense objects. Often used for analyzing failures (such as defects in material, bones, etc.)
Bandwidth is limited. "Twisted pair" telephone cable is an example used for voice communication	Bandwidth varies from low to high. Can handle video as well as other types of data	Bandwidth high to very high, when directed in a medium (fiber) can travel with minimum attenuation	Bandwidth is not a major consideration, but detection and analysis are important
Can pass through near line of sight using reflection. "Sound which is not an EM wave", it attenuates exponentially with distance and travels slower than EM wave. A classic example is "lightning followed by thunder"	Lower part of radio spectrum (MF, HF, VHF, UHF bands) can provide near line of sight coverage. Most urban coverage for cell phone occurs uses reflections from nearby buildings and objects with no "line of sight" to the base station tower	Can travel great distances through space and air but requires direct "line of sight". Light waves do not bend around corners and typically do not reflect off natural objects such as mountains, trees	Requires proximity and gets severely attenuated over distance. For example, X-ray used by a dentist operates at 1×10^{18} Hz but travels only a few Centimeters
Low-frequency signals can be mounted/modulated over radio and sent over great distances. Telegraphy and IoT (Internet of Things) are good examples	At the upper end of millimeter-wave band, radio waves behave like light. However, it is possible to radiate them with precision design of antennas	Light travels through space but in terms of communication, it is not easy to use, unless it is contained in a medium. Only short hops using light beams are practical	Mostly useful for near field and diagnostic use, but not commonly used in long-distance communication systems

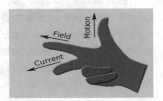

Fig. 2.2 Fleming's right-hand rule (a quick check)

2.2.4 Reflection of Radio Waves—"Ionosphere" Reflector in the Sky

Let us now consider the Ionosphere mentioned in (iii) above—which refers to different layers above the earth, that reflect different frequency ranges of radio waves. This property of the Ionosphere in the sky has been used for almost one hundred years.

Figure 2.3 depicts an imaginary view of the earth involving two separate continents. It allows radio broadcast and other communication services over long distances. This service is commonly referred to as "short wave radio" and allows communication beyond the horizon (there is no line-of-sight) [3].

This can be used in a various applications that need communication over very long distances (hundreds of km).

In Fig. 2.3, if hills on either side of large ocean were assumed to be on different sub-continents, then the curves going up towards the sky, show increasing frequency bands as they reach out farther. There is a point when finally, in the upper UHF band these waves penetrate the ionosphere and travel up towards outer space. Note that all these bands are within the radio frequency range indicated in Table 2.2. Medium Frequency MF (300 kHz–3 MHz) is used by AM radio, amateur radio and avalanche beacons. The High Frequency HF (3 MHz–30 MHz) is used for short wave broadcast, amateur radio, over-the-horizon aviation, RFID, over-the-horizon radar, Automatic Link Establishment (ALE). Near Vertical Incidence Sky wave (NVIS) is used for marine and mobile radio telephony. Very High Frequency or VHF (30 MHz–300 MHz) is used by FM radio and TV broadcast, Line-of-Sight (LoS) for aircraft to aircraft, aircraft to ground communications, land mobile and maritime mobile communications, amateur radio and weather radio. Beyond

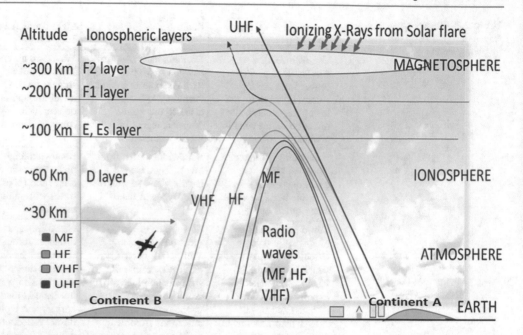

Fig. 2.3 Ionospheric layers and radio waves

this frequency band, Ultra High Frequency (UHF) and higher bands penetrate through Ionosphere and move towards outer space and used for space/satellite communication.

In Fig. 2.3, the different frequency bands are represented by different colors. For example, the lowest frequency band of MF is represented by dark blue, followed by HF in green, followed by VHF in bluish-green (Cyan) and finally UHF is represented by Red. Note that the frequency bands do not have a sharp cut off and a gap. They are gradual and continuous; therefore, an overlap is inevitable. Also, not indicated in the figure are differences in height of ionospheric layer during day and night. Warming of the earth surface plays major role in moving of these layers "up during the day" and "down during night". These are measured and well recorded; broadcasters and communication service providers adjust their frequencies such that these changes are accounted for. By inference, all frequency bands above UHF also penetrate through the ionosphere and move towards outer space. As one moves up in altitude, the earth's atmosphere consists of troposphere that has air (from earth surface and up but below 30 km where aircrafts can fly), followed by different layers of the Ionosphere. Note the aircraft shown in Fig. 2.3 is shown quite low, since aircraft needs "air" to fly and cannot sustain flight beyond 60,000 ft (less than 20 km).

At the upper end of the Ionosphere (around 300 km above earth) is the "magnetosphere" that helps to shield most of X-rays emitted during solar flares. In addition, there is radiation from space towards the earth, in other frequency bands. Well-known among them is "background noise", according to the big bang theory. NASA (National Aviation and Space Administration) has provided a window towards/from the sky, as shown in Fig. 2.4. It is an overview of the entire EM band in the sky showing parts open to outer space and parts that are

blocked [3]. Starting from the left, observe that the Ionosphere is opaque to lower frequency radio waves; this provides an opportunity to use it as a reflector (short wave radio).

In Fig. 2.4, note that part of infrared band can also pass through the atmospheric window and is used by remote sensing satellites to get images of the earth using infrared camera. This is an important feature used by remote sensing satellites to avoid problem due to clouds covering the earth. Higher frequencies of light spectrum including part of ultraviolet band, get through the atmosphere. However, X-rays and gamma rays are blocked, making it safer for residents on earth. The windows open for radio are used for satellite communication and deep space missions.

For communication using atmospheric layers indicated earlier in Fig. 2.3, an illustration of earth (somewhat exaggerated) is represented in Fig. 2.5. It illustrates part of North American continent with radio waves launched from the southwest corner traveling across the continent to the northeastern corner. Two separate sets of radio waves operating in different bands are considered. One launched at a high slant angle, gets reflected by the D or E layer. The second wave launched with a lower slant angle gets reflected by upper layers such E or the F layer. These two waves are reflected and return to the earth at different locations. When the wave launch angle is too low, radio waves travel towards the ground and these are known as "ground waves". This ground wave is the normal mode of communication for land mobile, cellular systems, and others near earth. It includes the AM and FM broadcast service as well. The higher slant angle wave in this example is used for short wave broadcast travels from Los Angeles and returns to earth at Denver. Such long-distance broadcast is useful to cover both national and international broadcasts.

Atmospheric Windows to Electromagnetic Radiation

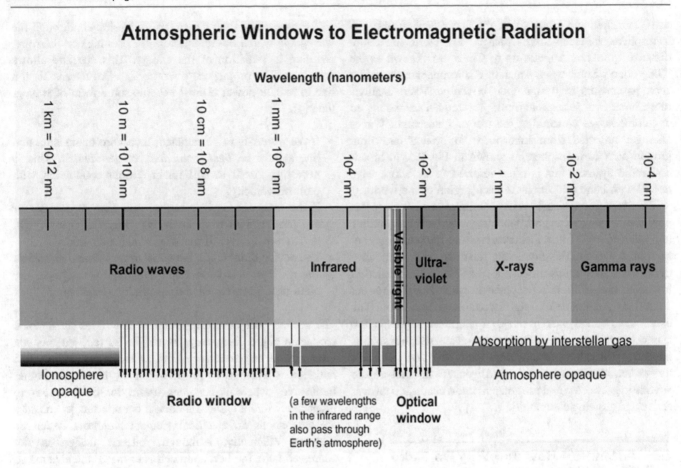

Fig. 2.4 Overview of atmospheric window to EM radiation (NASA)

Fig. 2.5 Radio wave reflections from the Ionosphere—concepts and usage

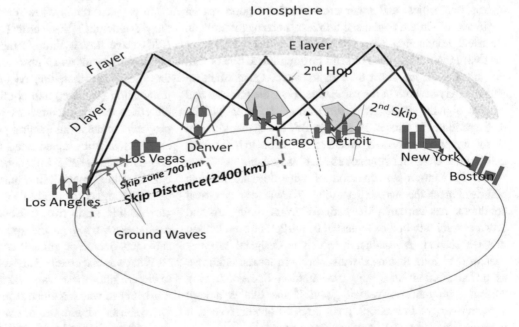

Ground waves travel from Los Angeles up to the nearby horizon; beyond horizon up to Denver, there will be no coverage. This gap is known as "skip distance". In this example, Los Vegas would be within that region of "no coverage" where neither the ground wave nor the reflected wave reaches. It is known as "the skip zone". For the other wave launched at a lower angle, the first reflection from the Ionosphere, is received at Chicago. Therefore, the entire distance from Los Angeles up to Chicago is known as the "Skip zone". Short wave Amateur Radio operators consider such parameters to design their coverage. Radio stations offer broadcast service in multiple frequencies bands to overcome issues concerning the use of ionosphere. Waves also get reflected from Ionosphere to ground and from ground back to Ionosphere as shown in Fig. 2.5. In broadcast radio systems this is often referred to as "short wave radio" with band allocations to each region of the world.

Such repetition of radio wave continues and in theory it is possible that waves can go around the globe. But after multiple hops, the radio wave becomes quite weak and therefore may not be usable. Despite this limitation, amateur radio users communicate across continents using some properties related to Ionospheric conditions. Ionosphere descends at night, making conditions conducive to long-distance communication. The usefulness of such amateur mode of wireless communication cannot be underestimated. During natural catastrophes such as hurricane or flooding, amateur radio users have saved precious lives. There are training sessions and license exams that prepare anyone to get license and use amateur radio bands. It is an ever active and growing community.

2.3 Antennas—How and Why Do Radio Waves Propagate

The word antenna refers to a sense object—antennas are on insects, butterflies, and other creatures for a good reason —"to sense". In the context of wireless, antenna refers to an electrical conducting element that can either transmit or receive signals (to receive or send information to others, and for sensing) in the form of a radio wave. Quite often an antenna may do both at the same time. How the transmit and receive signals at the same time using duplexer will be discussed later in Chaps. 14 and 19. But the element known as antenna is often made of metal. It is possible to build them using other electrical conductors such as carbon.

Let us address the question of why does an antenna radiate? Part of the answer lies in Fig. 2.1 where an electrical conductor has current (Alternating Current) going up and down several hundred or thousand or million or even trillion times a second. A conductor, which is designed to carry current over long distances with minimum losses, is known as a transmission line. Any transmission line conducting current with uniform velocity, and if the line is a kept straight over great distances, it **radiates no power**. Hence, it is termed guided wave (conducted not radiated).

Let us consider Fig. 2.6 that represents a waveguide, which acts as an antenna [4].

RF power from the transmission line travels through the waveguide which has an aperture (or opening) on the right, resulting in radiation of the energy. This aperture allows radio waves to propagate "over the air". For a transmission line to radiate power it must become some form of waveguide [4].

- If the power is to be radiated, then wire or transmission line should be *bent, truncated, or terminated*. This is necessary for it to still retain current conduction with uniform velocity.
- If the transmission line has *current that accelerates or decelerates with a time-varying constant*, then it will radiate power even if the wire remains straight.
- The device or tube, *if it gets bent or terminated to radiate energy*, then it is known as waveguide. Waveguides are often used in microwave transmission or reception.

The second part of the question is how does an antenna propagate higher frequency waves? To be fair, antenna can propagate at any frequency. But as indicated earlier, such radiation is avoided in lower frequency bands. At higher frequencies where it becomes useful for radio waves to propagate without a wired media, it is important to consider the antenna as an electrical element and how it can be designed efficiently. Although, antenna design is not addressed here, the best approach is to match the antenna so that it seems like "it is an integral part of the transmission line" and terminate it with characteristic impedance of the line. What does this mean?

In physical terms where the transmission line terminates "is the antenna". In electrical terms, the Electro-Magnetic (EM) wave travels through the line does not feel this as a terminating point as an obstruction but as a continuation of the line. When matching occurs, waves travel through the termination (antenna) and continue to propagate into external space (air or vacuum) as if the line did not end. This property "to act as an integral part of the network" is known as characteristic impedance. Most radio systems use a known characteristic impedance of a specific value. The coaxial cable (transmission line) that feeds the EM wave, is usually designed as a 50 Ω line. Waves feel the antenna as part of this same 50 Ω. Beyond the antenna, the wave continues to travel at the speed of light, whether there is cable (conducted or guided) or not (air or space).

What was discussed till now is a concept somewhat difficult to discern since these waves cannot be seen or felt—therefore it is better to watch a video from historic Bell labs where Dr. Shive shows "similarities of wave behavior" to conceptualize this with mechanical wave motion. This video is an excellent example from 1959 showing how waves of different types behave alike [5]. The video shows mechanical wave motion

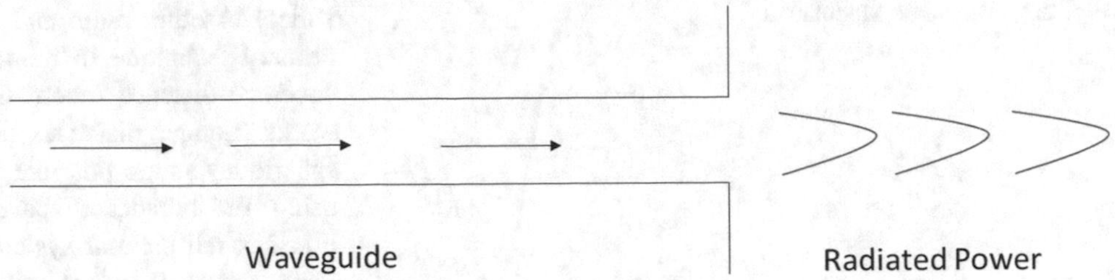

Fig. 2.6 Waveguide and power radiation

Waveguide   Radiated Power

in one plane (similar to electrical wave). It is a practical way to see wave motion and its reflection. That reflection depends on the type of termination and its characteristic impedance (matching). We will revisit this aspect in Chap. 17 on antennas.

How EM wave propagates beyond the antenna is to a large extent dictated by the type of antenna. An omnidirectional antenna is specified in Physics as a point source. Such a point source cannot be implemented in practice. Hence, the closest to this concept is a dipole or rod which looks like a short pole with the cable feeding into its mid-point. Since the wave cannot radiate along the pole axis (rod being a conductor it continues to guide waves). Only beyond its edges wave radiates and will form a doughnut shape, with the hole in middle where the dipole antenna is located. We will revisit different types of antennas in Chap. 11, Sect. 11.5, Chap. 13, Sect. 13.3, Chap. 14, Sect. 14.3 and the entire Chap. 17, offering more details. These chapters describe multiple variety of antennas that we see and use in our daily life.

Finally, a brief review of using power lines that utilities around the world use to carry electrical power to homes and offices—can these lines be used to carry electromagnetic waves? An antenna acts as a transducer between a guided wave in a transmission line and an electromagnetic wave in free space. Due to the nature power transmission lines carrying guided waves, they are useful for low throughput communication. By using carrier systems that allow communication over power lines, power engineers do send low-frequency radio waves over the same open wires. PLCC (Power Line Carrier Communication) is a method used for telecommunication using power lines. It helps communication between electrical substations. Modulated by radio waves that can operate in slightly higher bands, but sent over same power lines, the data transfer rate is quite reasonable (about 10 kbps). Such data rates are sufficient for monitoring the health of substation and other power line equipment. It is also used for basic voice communication over extended distances in rural areas. Any means of wired communication, including PLCC limits it to specific points and cannot be assumed to offer service to the public, as a conventional telecommunication network.

2.4 Brief Overview of Bygone Wireless Technologies (AM, FM, PM)

AM Radio

Almost everyone who has traveled in a car or listened to a radio at home is familiar with the AM and FM broadcast radio bands. There are knobs on every radio receiver that allow users to select between these bands and listen to programs on a radio channel of choice on any of these bands.

What is AM (Amplitude Modulation)? Recall from high school Physics lab that an audio wave when mounted (modulated) on a radio wave, can be transmitted over great distances. This wave, when received at the other end can be demodulated to recover the original audio. This concept has been in practice since early 1900s and at least since 1910 regular broadcasts using AM were made (since 1905 in France and USA). This was known as the Golden Age of Radio [6].

The concept of AM is indicated pictorially in Fig. 2.7 where the Audio wave A when is modulated with the radio wave B, will result in an Amplitude Modulated Wave C. In terms of change due to DAB, most broadcast started with a retrofit kit to move their programs away from AM/FM towards Digital Audio Broadcast (DAB). In addition to the ability to send program information (artist, program, and channel information) over digital signaling track, improved audio quality (due to DAB) makes broadcasters competitive.

Over the years, AM radio remained a regular "local radio" for many reasons. Primarily, its range was limited, and quality of sound deteriorates with distance (<150 kms). Since amplitude is where the "content" is to be found, "noise" joins the "content" and AM broadcast is therefore susceptible to noise. In the early 1990s, deployment of DAB (Digital Audio Broadcast), brought a major change to the radio broadcasting community. Quality of AM broadcast, when modified by DAB, improves considerably, making it comparable to FM radio (low noise and better fidelity). Similarly, when FM uses DAB, it improves considerably to sound more like CD player. Listeners typically look forward

Fig. 2.7 Amplitude Modulation

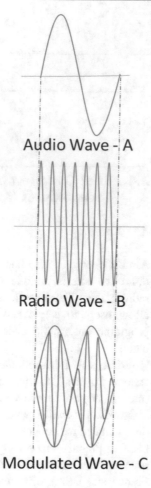

Audio Wave - A

Radio Wave - B

Modulated Wave - C

to other features, since not only program content, but other information such as text version of the news (for those with hearing impairments) can be easily incorporated. Starting off initially around 1995 this technique of using DAB to enhance AM and FM became popular throughout the world and transition was systematic.

In terms of migration of this technology—old receivers (either in a car or at home), continue to receive traditional AM quality broadcasts. New receivers support all the digital features, making them attractive for all listeners.

It is fair to state that AM as a technology has seen its best days through most of the twentieth century and is now on its exit path with a gradual phase out by DAB. The digital streams of audio also offer other advantages that we will review after a brief overview of FM radio. These are very important reasons for the rise of DAB and decline of both AM and FM.

FM Radio

Frequency modulation was one of the major inventions in the early twentieth century, developed by Edwin Armstrong in the 1930s, resulting in not only FM radio broadcast, but a variety of other radio and TV systems adopting it as their primary technique to modulate signals. But it has shorter coverage distance (about 60 kms) for most broadcast networks. Longer distances using reflections from the atmospheric layers are possible and typically "short wave radio" using FM broadcast spans great distances as shown in Fig. 2.5. All cellular systems, land mobile systems, marine radio, and aeronautical radio systems were based on the FM technique, for most of twentieth century [6]. The superior quality of FM compared to AM, made it the primary modulation method until digital modulations started to become popular in the early 1990s. There are some terms commonly defined for FM; the most common among these are:

- Modulation index m = frequency deviation/modulation frequency.
- The other one is FM deviation ratio D = Maximum frequency deviation/Maximum modulation frequency.

Figure 2.8 shows the concept of Frequency modulation. Unlike the AM, here frequency changes with the modulating Audio wave. As the Audio peaks (with positive amplitude the frequency decreases, which appears like expansion—indicating wave occurs less frequently. When the audio takes a negative peak the frequency increases that is seen as compression or occurs more frequently. This concept is indicated technically described by "frequency deviation". Therefore, if the Radio wave has a base frequency F then with modulation it changes resulting in $F \pm Fd$ where Fd is the extent by which the frequency deviates. Figure 2.8 shows the concept of FM. Note that positive part of audio wave expands or lowers the frequency and negative part compresses or increases the frequency of the modulated wave.

2.4.1 Digital Audio Broadcast—Challenges FM

In the digital broadcasting world, DAB improved FM in terms of audio quality. With FM modified by DAB, the audio stream improves to match quality of playing a CD (Compact Disk) in a quiet room. Orchestra/very high-quality music programs can be heard at studio quality, whether listener is on the road or at home. This change made FM vulnerable even in its home base of FM broadcast. Also, DAB allows recording programs—a listener can save and play them later. There is no need to miss a program if listener wasn't in the area. DAB directly interacts with the Internet—any favorite radio program can be "tuned into" from anywhere in the world; no traditional radio broadcast can match this [7].

Therefore, study of AM and FM is not very different from studying the gramophone, the turn table, tape recorder and

Fig. 2.8 The concept of Frequency modulation

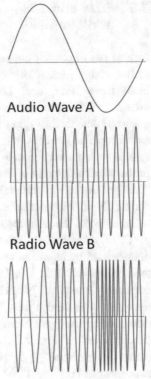

Audio Wave A

Radio Wave B

Frequency Modulated Wave C

cassette player in an audio systems course. It is useful from a historic perspective but has little practical value since AM and FM are no longer being deployed anywhere now.

The advantage of digital signal whether for audio or video is so big that analog systems such as AM or FM had to be phased out in most locations. If there are any in use, it would only be a question of time as to when they will get modified. The question to address is—is the audio/music industry alive—yes. Is the tape recorder, the turn table or cassette tape in the market today—no. Same is true for AM and FM, which indicates that they are technologies of a bygone era.

DVB standard

Digital Video Broadcast (DVB) followed as a major standard and users could choose satellite-based version, terrestrial version, or the cable version. This was a direct result of the change brought over by DAB.

In terms of impact to FM in other areas, the biggest change occurred in the cellular industry where traditionally FM was used in 1G systems. In 2G systems, FM was replaced with digital modulation techniques at baseband, using modulations such as $\pi/4$ DQPSK ($\pi/4$ Differential Quadrature Phase Shift Keying) in the North American

cellular standards, and GMSK (Gaussian Minimum Shift Keying) in the global GSM system. With this major shift, FM saw its departure from the cellular market. This was soon followed by the globally used public safety system TETRA (Terrestrial Trunk Radio) which also moved over to $\pi/4$ DQPSK modulation.

Usually, microwave links act as cellular backhaul network. All microwave links moved from FM to digital modulations using techniques such as BPSK (Binary Phase Shift Keying), QPSK (Quadrature Phase Shift Keying), and QAM (Quadrature Amplitude Modulation). Each of these modulation techniques are in widespread use today, not only in the microwave links but also in many other wireless systems. They offer throughput rates several orders higher than traditional analog FM. All of them offer digital features such as those enjoyed by DAB [8].

Satellite networks quickly followed the trend by using QPSK, QAM, and efficient digital modulation techniques to improve throughput and quality of service. Satellite services offered today are based on digital modulation techniques, not only do they connect commercial aircraft but all maritime communications. It is also the vital link providing support in times of disaster relief, since a satellite earth terminal can be air dropped, set up, and operated, within hours.

2.5   Modern Wireless Systems—Advantages of Digital

Since the age of computer and networks from early 1970s, use of bit stream (bits and bytes) has become common and is widely used globally as a means of communication. A major advantage of all these digital modulation schemes is their standard interface to IP (Internet Protocol) networks at Layer-2, commonly known as "Ethernet". Every cellular system, microwave system or other wireless system with digital modulation scheme, provides a direct Ethernet interface allowing connection to any standard layer-2 switch, making it simple to interface IP networks.

This Ethernet interface simplifies all wireless connections to existing networks that use Internet Protocols or IP. In addition, these digital modulations provide much higher throughputs depending on the received radio wave power. Such advantages far outweigh minor limitations such as "true voice" or "analog voice". In general, "analog voice" or "trunk quality voice" was a phenomenon associated with Plain Old Telephone Service (POTS), no longer supported by any major vendor. All phones (desk phones) in offices today, use voice codecs that with "Voice over IP" or VoIP and are widely supported by digital modulation schemes. Also, recording and storage of voice, music and critical conversations related to public safety, become simplified since both access and storage using computer networks and memory becomes easier.

Some of these are important reasons why both AM, and FM are regarded as analog technologies of the bygone era that are replaced wherever they exist today. Note that PM or phase modulation had only remained as concept that was not implemented in analog systems. But it is now widely adopted in its digital modulation format such as PSK (Phase Shift Keying) or QAM (Quadrature Amplitude Modulation). Digital modulation heavily relies on phase shift for many reasons. The natural use of digital modulation uses signal (a) in phase and (b) quadrature, known as I and Q.

There are other reasons why digital modulation became popular. Some of them are:

1. Better noise immunity—improved schemes to detect noise and correct signal
2. Common frame structure for all data (no distinctions of voice, video, and data)
3. Data throughput increases as signal strength improves
4. Ability of single receiver to use multiple digital modulation formats—flexibility
5. Security—use of coding, signal spreading techniques to avoid jamming
6. Based on experience, better error recovery mechanisms in noisy medium
7. Since digital processors perform the task, it is easy to miniaturize circuits.

There are some limitations, it is useful to note them as well:

1. All signals in the world are inherently analog. They must be converted to digital.
2. Due to this Analog to Digital Conversion (ADC), there is a limitation—known as "quantization error". Any quantity large or small cannot be very accurately converted into digital bit stream. There will be an error, however small that error may be.
3. Depending on pulse size, the bandwidth needed could be quite high.
4. Circuitry will be more complicated than Analog. But it is implemented at chip level. Due to this, efficiency improves, power consumption is considerably reduced.
5. The voice will sound metallic—since it is not the natural voice (constructed by vocoder). Earlier generation users will take time to get used to this, as well as a very quiet background, often they check the caller by asking "are you there"? Some systems introduce "comfort noise" to make the conversation seem more natural—as if there is an actual trunk existing.

2.6 FDM and TDM—Concepts of Duplex and Simplex

One of the important concepts that has continued from the earlier analog generation is the concept of FDM and TDM. FDM means Frequency Division Multiplex—this concept uses a specific frequency for transmit and a different frequency considerably separated from transmit is used for receiving. The concept of FDM stems from early telephone system which used separate channels for "talk AND listen". In the wireless system it translates to separate frequency for "talk and listen" at the same time. This is also known as "duplex"—in telecommunication networks.

In contrast the concept of TDM also stems from "simplex" of telecom—which stems from the "push to talk" phones used by public safety personnel. Simplex is based on the concept of "Talk OR Listen", not both simultaneously, commonly known as "Walkie-talkie". In wireless, TDM uses the same frequency for "Transmit OR Receive", which means using the Push button one "holds" dedicated channel (also known as Trunk in telephone networks). Until the button is "released" it is NOT possible to receive. These concepts of FDM and TDM play a major role in all wireless systems, particularly for voice. It plays a lesser role in others such as Internet access, as explained in later chapters. Due to the legacy systems carrying telephone traffic on to cellular, it is normal for cellular systems to use FDM and so are satellite systems, based on FDM.

In recent decades, with the advent of Internet and data streams, the concept of TDM took hold and grew into a system of its own. Digital modulation techniques provided considerable advantage in terms of packing and moving high-speed data streams. It seems appropriate at this stage to briefly visit one such modulation technique OFDM, based on work done several decades ago but took center stage, due to powerful processing techniques in chip sets.

2.6.1 Orthogonal Frequency Division Multiplex (OFDM)—Important Digital Modulation

While digital pulses have their advantages, one of the limitations indicated was "bandwidth needed could be quite high". This was addressed using an important technique called OFDM. This important digital modulation technique OFDM—Orthogonal Frequency Division Multiplex, had its origins in the late 1960s, but waited in the wings for about three decades before implementation, mainly because it involved complex computations [8]. Thanks to very complex chipsets available today, it is used today in digital television, in DSL (Digital Subscriber Line), Internet access to homes,

WiFi systems and 4G cellular as well as for 5G. What is so important about this digital modulation that endeared itself to a variety of wireless and network designers?

From Table 2.2, we observe that the radio spectrum used for wireless is somewhat limited. Although higher frequencies offer wider spectrum bandwidth, each wireless channel must be kept separate to avoid interference. This was usually achieved by FDM or Frequency Division Multiplex where each channel is kept distinct by maintaining a band gap (intentionally kept separated) to avoid interference. While this is a design necessity, band gaps are a waste of precious spectrum resource. This was not considered a problem until the need for supporting applications such as video became essential (that demand considerable bandwidth).

High-speed data rate is desired in many other applications as well. But the symbol duration reduces with the increase in data rate. Systems that use single carrier modulation suffer from severe "Inter Symbol Interference (ISI)" due to dispersive fading of wireless channels. This needs complex equalization—a method used by receiver for choosing the right signal stream, among many reflected streams that have the same information. OFDM divides the frequency selective fading channel into many narrow band flat fading sub channels in which high bit rate data is sent in parallel. This provides the advantage of avoiding ISI due to long symbol duration. Therefore, OFDM was chosen by multiple standards such as DAB and terrestrial TV in Europe, as well as global standards such as IEEE 802.11 standard WiFi, 4G cellular, and now 5G.

OFDM is expected to keep fidelity of the signal consistent irrespective of pulse shape even in tough RF environments. The original work towards OFDM was done by Chang [9], who in a fundamental contribution to OFDM, developed general conditions for the shapes of pulses. It was defined as the combination of transmitter filter and channel characteristics, with bandlimited but still overlapping spectra. Such pulses, including the full cosine-roll off pulse are shown in Fig. 2.9, making a viable OFDM system possible "without inter-channel and inter-symbol interferences." The technique used by OFDM to reduce the band gap, involves adjacent channel overlap by cleverly making them "orthogonal". Ideally, orthogonal waves do not interfere with each other, which provides the basic advantage.

In Fig. 2.9, rectangular pulse shown on the left, has a pulse spectrum indicated on its right by $\sin \frac{(f)}{f}$. This spectrum (multiple pulses and their spectrum) would normally overlap its neighbors, as shown further to its right as an OFDM spectrum. If we consider an OFDM signal block, with time interval T, it can be visualized as a sum of these sub-channel signals. We could now consider a continuous series of such blocks. If the channel has no distortion, then there will be no interference between the sub channels. If we consider the pulse to carry digital symbols, then there will be no inter symbol interference either.

In Fig. 2.9 on the right side, this is indicated by Sine wave and Cosine wave as sub channel waves. First explored by Chang in 1966, this concept had one limitation. The incoming wave (sine) required an almost instantaneous orthogonal product so that they can be processed and sent together over the air. Also, sine wave is just a tone—real-world waves are not pure sine waves. Luckily, through mathematics we learn that any wave can be represented by components, using Fourier Series, which we will briefly touch upon now (now implemented with high-speed chip sets).

French mathematician Fourier explored and found that any complex signal (or waveform) can be represented as a sum of Sine wave and Cosine wave components. His method made it easier to represent any arbitrary waveform. Depending on complexity of the waveform, frequency of Sine (or Cosine) components would also increase. In theory, an infinite number of these components can represent any complex signal digital pulse (shown in Fig. 2.10 where T = 2π cycles per second), considered as a bit stream. However, in most cases a few of these frequency components (series) can represent signals quite well. Let us consider the pulses in Fig. 2.10 that represents a signal.

This wave shown in Fig. 2.11, can be broken down into components and expressed as:

$$F(t) = a_0 + a_1 \cos(t) + a_2 \cos(2t) + a_3 \cos(4t) + \cdots + b_1 \sin(t) + b_2 \sin(2t) + b_3 \sin(4t) + \cdots$$

$$(2.8)$$

This Fourier series Eq. 2.8, provides us clues about nature of the complex wave. For example, the constant a_1 indicates

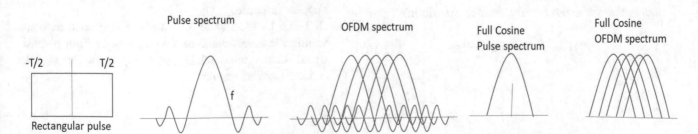

Fig. 2.9 Rectangular pulse, its spectrum, full Cosine roll off, results in OFDM spectrum

Fig. 2.10 Fourier series—representation of rectangular pulses using Sine waves

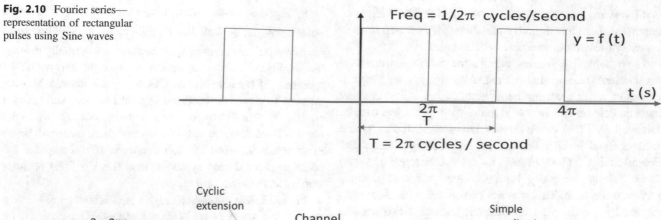

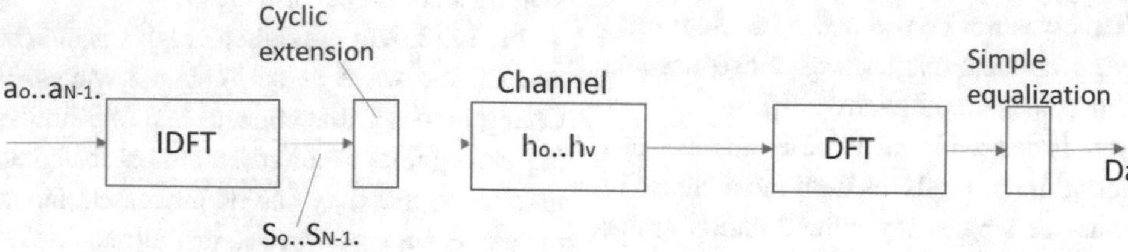

Fig. 2.11 Simplified block diagram of OFDM with DFT block

amplitude of the first orthogonal frequency fundamental, constant a_2 indicates amplitude of the second orthogonal with frequency twice the frequency. Similarly, the constant b_1 shows amplitude of the in phase fundamental frequency, constant b_2 indicates amplitude of the second in phase component that has twice the frequency and so on. Therefore, Fourier Series is a useful tool for signal processing.

We shall now consider Fourier Transform—this allows signal "transformation" from one domain to another. That means, signal represented in the time domain can be "transformed" into frequency domain and vice versa [9]. Such a transformation can be thought of a location described by "address" in one domain and by "Latitude/Longitude" in the other domain. Since both indicate the same location, they can be "transformed" or mapped from one to the other, depending on convenience. This flexibility of moving from one domain to the other is of great importance in wireless communications since both domains are used depending on the situation.

(this is known as the forward Fourier Transform)

$$F(u) = \int_{-\infty}^{\infty} F(t)e^{-2\pi rt}dt \qquad (2.9)$$

(this is known as the Inverse Fourier Transform)

$$F(t) = \int_{-\infty}^{\infty} F(u)e^{-2\pi ut}du \qquad (2.10)$$

where F (u) represents frequency and the sign "u" is $u = \frac{\omega}{2\pi}$ Hz.

The forward Fourier transform converts a function time into a function of frequency. Inverse transform does the exact opposite—it turns a function of frequency into a function of time. These two equations (2.9 and 2.10) are therefore called Fourier Transforms. Now let us consider another version DFT.

Discrete Fourier Transform (DFT) allows us to construct components using normal numbers or complex numbers (with real and imaginary parts). Since DFT is based on Fast Fourier Transform it is often indicated as an FFT (Fast Fourier Transform) block. Figure 2.11 shows a simplified illustration of OFDM system including cyclic prefix operation, where DFT block is used [10].

DFT is mainly used in wireless telecommunication for three important tasks:

1. DFT can calculate the signal's frequency spectrum. This is done by evaluating information in the sinusoids in Fourier series such as frequency, phase, and amplitude—the three quantities that specify details of the signal.
2. DFT can evaluate the frequency response based on the impulse response.
3. DFT can be used as an intermediate step such as convolution that can combine two signals to form a third signal. Convolution relates input signal, output signal, and the impulse response.

Reverting to the earlier discussion on OFDM in Fig. 2.11, major advantage of using OFDM is its ability to place cosine channel components as sub channel components. However, it does involve considerable computation. But that became possible using physically small signal processors due to progress in microprocessor technology. Such processors implement OFDM as a block within a major communication system. OFDM developed into a major technology and there are entire books/dedicated courses to provide details of OFDM theory and implementation, for example, please refer to [10]. It is beyond the scope of this book to explain details of OFDM schemes. However, there is one major technique that we will stress—related to Fast Fourier Transform. The topic, although studied as part of a course in mathematics, has great implications in communication engineering, led to a new field called "Digital Signal Processing".

2.7 Birth of Digital Signal Processing—Its Exponential Rise

Digital Signal Processing (DSP) involves extensive computation, with a basic pre-requisite—knowledge of the signal. The signal in most cases originates from the real world as sensory data. Since real world is analog, quite often an ADC (Analog to Digital Converter) precedes the DSP to convert analog signal to digital stream.

In order to lay foundation that led to exponential rise of DSP, let us start backwards—what it has done. A brief list of topics with areas of application for DSP is indicated in Figs. 2.12 and 2.13 QAM Modulator and demodulator—conceptual diagram in Fig. 2.14 shows Simplified block diagram of holographic generation using DSP. This will always remain a partial list since application of DSP keeps growing. It brings into its fold many applications that were in analog domain and those that are possible only because of DSP. The sensory signals could be analog in the form of seismic vibrations, pictures/videos, or audio, etc., which need some form of processing to bring them into a "bit stream".

Therefore, it is essential that DSP designer understands nature of the signal and the actual systems (circuits and transducers in this case). In the 1980s and, due to revolution of personal computers, marketplace demanded a variety of uses from DSP.

During the "electronics revolution" of the 1950s and 1960s, the general theme was "if any quantity can be converted to an electrical signal then it can be transported, analyzed or computed with ease—using electronics".

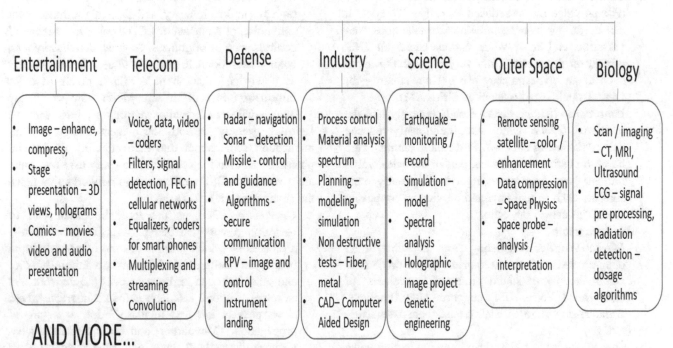

Fig. 2.12 Applications and areas where DSP is used

Therefore, transducers (components that convert one form of energy to another) viz., temperature to signal, pressure to signal, audio to signal, video to signal, picture to signal, etc., became key input elements to deliver "signal" to "electronics". The theme is now modified with DSP to state "if anything can be digitized, then it can be enhanced, improved upon, stored or presented in a variety of ways anywhere in the world" [11]. Currently, A to D Converters (ADC) or Digital to Analog Converters (DAC), provide the conversion from an analog signal to a digital stream or vice versa. Better the sampling, more accurate will be input or output. Therefore, accuracy, resolution, etc., become important depending on the type of application. There are no limits to number of applications DSP has touched in recent decades. Almost everything is touched and enhanced by DSP, as applications in Fig. 2.12 testify [11].

Specific DSP applications related to wireless systems in widespread use today are listed here.

- **Telecommunication**: Functions that are fundamental to all telecommunications:
  - (a) *Multiplexing*: Based on the number of telephones globally, multiplexing was widely based on the T1 (1.544 Mbps) or the E1 (2.048 Mbps) standards. Multiplexing in the current digital age is accomplished entirely using DSP. Conventional switching with relays or electronic switchgear is no longer needed. Ethernet is now the digital interface.
  - (b) *Compression*: Voice quality is not degraded when a 64kbps voice rate is reduced to half or 32 kpbs. It can be further reduced to 8Kbps for telephone conversation and most voice coders based on DSP operate in this range. Only very high-quality audio such as an orchestra may use 64Kbps stream—for Digital satellite radio (such as Sirius/XM).
  - (c) *Echo control*: Long-distance calls typically had the problem of echo, which was traditionally reduced using "balancing circuits". Since these circuits were based on estimated distance, part of the echo would remain. New DSP-based algorithms actually measure the echo in real time and provide an "opposing signal" to cancel out echo.
- **Audio processing**
  - (a) *Music*: Whether originating from a studio or an open-air theatre, music requires two major components—fidelity of sound produced, cancelation of background noise. DSP supports both functions. Broadcasting radio/TV stations use them extensively today.
  - (b) *Speech generation*: Based compressing human voice and replaying for a given situation, speech generation has evolved to a point where the written word

can be spoken out—widely used in all computers today. In addition, language learning packages, children's learning packages are all based on DSP.
  - (c) *Speech recognition*: This is considerably harder, since it can vary with speakers. However, learning algorithms used by DSP, make it possible to recognize speech. It is in widespread use in automated systems, order generation systems, and others.
- **Image processing**
  a. *Medical*: Perhaps the largest user area today—medical images are widely used in real time surgery. Diagnosis based on X-rays, CAT scan (Computer-Aided Tomography), and MRI (Magnetic Resonance Imaging) are perhaps the most widely known use of DSP applications.
  b. *Aerospace*: Whether coming from outer space as a black and white image, or data repainted from the infrared camera output of a remote sensing satellite (using false color), DSP has played important role in aerospace. The "head up display" in all commercial aircraft today shows images where the aircraft is relative to runway, even on a foggy night. Such images help the pilot make a safe landing. In addition, images of weather based on Doppler radar and other images generated using data from different systems are clearly displayed.
  c. *Commercial*: Systems today use excellent quality LCD screens where images are often so clear that even facial details of a news reader can be vivid. Thanks to high resolution imagery and DSP algorithms, color calibration of a plasma or LCD display has a range of controls such as brightness, contrast, detail/sharpness, color saturation/Chroma, tint/hue. It is widely acknowledged that about a trillion pixels of color information is sent to screens on how to display images with vivid details even on very large screens.
- **Cellular systems**: Perhaps the biggest change in communication came about due to availability of the cell phone. Other than the actual device every user has, there are a variety of DSP applications in networks that make the smart phone a unique device.
  a. *Convolution*: Perhaps the most important part of background work that DSP performs is convolution—a process by which it compares the input and output making an intelligent decision on how faithfully voice and other data can be brought over the network. An extremely difficult task here is the air interface where data "packets get lost regularly" due to nature of propagation. Convolution and other related mechanisms discover how many and where packets were lost and tries to compensate them, making the conversation intelligible, the pictures bring clarity, other

data used as input of digits/letters—are accurately sent over to an interactive system (such as a financial transaction over cell phone).

b. *Voice coders*: Perhaps a useful contributor that helped make the cell phone popular, is voice coders that not only carry voice but also get translated to the type of interface at the receiving phone (whether it is a VoIP phone, or another cell phone or even POTS from a different generation, etc.). DSP provides a whole process undertaken at the switch to interface capabilities of device at both ends.

c. *Picture/video coders*: Quite often these coders mimic their wired IP counterparts, but images are compressed to fit the exact screen size of the cell phone where it must be displayed yet retaining the clarity of image and accompanying voice are well synchronized (video stream segments).

d. *Base Band Unit (BBU)*: This unit marks a fundamental change from traditional heterodyne receiver system where an Intermediate Frequency (IF) was used. The baseband uses a Digital Signal Processor (DSP) to process and forward/receive audio and data streams to/from a mobile. The BBU therefore acts as a "digital streaming unit" sending/receiving digital stream which is forwarded through a radio. The BBU supports many functions typically associated with the physical channel such as: Mapping of physical and transmission channels, multiplexing/de-multiplexing, channel coding and decoding, channel spreading and dispreading, modulation and demodulation, physical layer procedures, and measurements. Many of these functions are recommended by 3GPP standards and are implemented in all cell phones and base stations deployed today [12].

2.8   Rise of Digital Modulation Towards Broadband Wireless

Traditional communication has the notion of "sensitivity of the receiver is a limiting factor for successful operation of communication system". DSP has challenged this concept with new notion of "higher received power means higher throughput using higher modulation schemes". The lowest signal levels near or close to sensitivity of the receiver will only result in the bare minimum class of modulation such as BPSK. But as the signal strength increases better modulation schemes such as QPSK, 16 QAM, 64QAM, 128 QAM, and 256 QAM are supported by the same receiver. Note that 256QAM modulation provides data rates that are 30 times the BPSK rate, for about 18 dB increase (about 60 times increase in power). We will review this aspect again in

Chap. 5 Sect. 5.4, Chap. 12 Sect. 12.2, Chap. 14, Sect. 14.2.7, and Chap. 15, Sect. 15.3.

Digital modulation gave rise to entirely new standards and performance in terms of data throughput, unforeseen even during early 1990s. The move from 2 to 3G and later standards of cellular provided dramatic increase in throughput rates, typically averaging about ten times increase per cellular generation. In parallel, the user got a second option which is use of much higher throughput data rates when in stationary environment (such as office, home).

With the introduction of WLAN standards by the IEEE (generally known as the 802.11 WiFi standards), broadband wireless system became very popular. Work on this standard began in 1997 and the first version was published in 1999 [13]. Early versions of this standard competed with DSL which at that time was considered broadband (2Mbps upwards). During the next few years release of later versions of this standard adopted OFDM, improved its performance to 54Mbps. A major change here was that it operated on an "unlicensed" band. Many were initially skeptical, but the DSP and the access protocol it used (called CTS or Clear-To-Send) made it possible for multiple users to operate smoothly within a given area, without interference, even though the same channel is used to transmit and receive. This unlicensed band of 2.4 GHz gained popularity and is the default WiFi band operated globally as an "unlicensed spectrum".

In the next few years IEEE also tried another standard known as 802.16 Metro LAN, generally known as WiMax. Its original intent was to provide fixed wireless access to homes/offices that uses a microwave link from a nearby tower to homes. Its original intent (fixed wireless access) was met by design and is still used in many places. The real wireless LAN with mobility design was IEEE 802.20 never saw light of the day. Therefore, wireless LAN is recognized as a fixed wireless technology, with mobility only at pedestrian speeds. However, the word "WiFi" has also become synonymous with "Free" since it uses "unlicensed band". Service providers, hotel chains, etc., provide it as basic infrastructure that allows customers connect to the Internet.

Since 4G cellular was deployed around 2015, cellular access also has been termed broadband access throughputs of 50 Mbps on the downlink being typical. The major difference is cellular supports both fixed and mobile access and is generally ubiquitous. Therefore, throughput differences between WiFi broadband and cellular is discernable as of now. With 5G deployments this difference is further reduced. However, cellular is known for very high security and well supported by service providers. Its use is quite widespread whenever the user is not at the destination (on the road or other areas). Its growth is likely to continue, towards 6G. An excellent overview of cellular technology and its growth is reported in a recent article [14].

All the three broadband access schemes described so far, use digital modulation schemes with FFT and can be simulated [15]. The major difference is that in the cellular system modulation schemes usually improve each generation of technology. For example, in 4G the LTE standard prescribes uses OFDM (described earlier). Details of access schemes in the up-link and down link will be described in later Chaps. 10, 11 and 13, but digital modulation is certainly very well established.

2.9 Digital Modulation Example—QAM

Till now several "digital modulation" techniques were introduced. How are these different and why are they so widely adapted? To get a better understanding, let us consider an example—the QAM or Quadrature Amplitude Modulation.

In Sect. 2.3 the concept of Amplitude modulation was introduced. Analog wave modulated in its amplitude is called AM, which was explained using Fig. 2.7. To extend this concept and incorporate digital signals of 0 and 1, the concept of Quadrature in introduced. QAM is a modulation technique which is used to combine two amplitude modulated waves into a single channel to increase the channel bandwidth [16].

Depending upon the type of input signal form (analog or digital) either analog or digital modulation schemes can be used.

To accomplish this, Fig. 2.13 shows the concept of a QAM modulator and demodulator where two individual signals are modulated and transmitted to the receiver. However, by using the two input signals, the channel bandwidth also will increase. But the advantage is that QAM can transmit two message signals over the same channel. This QAM technique is known as "quadrature carrier multiplexing".

Upper part of Fig. 2.13 shows the QAM modulator, where QAM modulator1 and local oscillator make up the "in-phase channel" (with $A * Cos\omega_c * t$). The QAM modulator2 and local oscillator (with $A * Sin\omega_c * t$) is known a quadrature channel. Both output signals of the in-phase channel and quadrature channel are summed up (shown by +) so the resultant output will be the output of QAM."

At the receiver, this QAM signal is forwarded to both the upper channel and lower channel of receiver, which separate

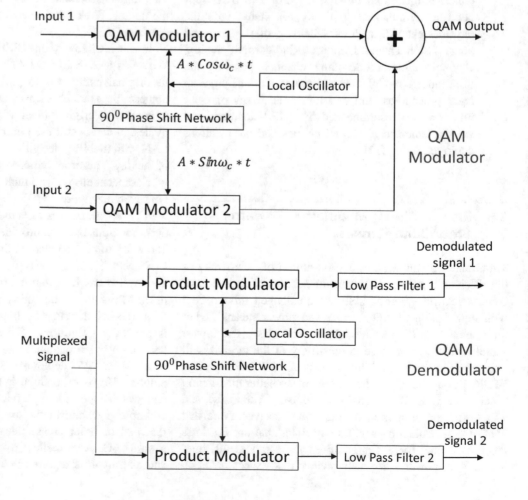

Fig. 2.13 QAM Modulator and demodulator—conceptual diagram

Fig. 2.14 Simplified block diagram of holographic generation using DSP

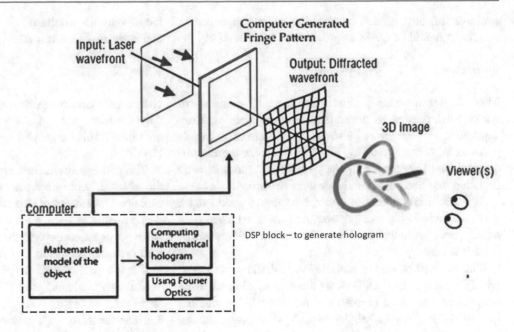

the resultant signals of product modulators in terms of in phase and quadrature. These signals and further passed through LPF1 and LPF2. These cut off frequencies of the Low Pass Filters are fixed based on the cut off frequencies of input 1 and input 2 signals. Finally, the filtered outputs of the receiver are indicated as recovered signals of the original.

2.9.1 Advantages of QAM

The digital modulation technique of QAM results in several advantages.

One of the major advantages of QAM is that it supports higher data rate. Therefore, number of bits supported by the carrier signal increases, which is useful for broadband system. It has great advantage in wireless communication networks and this will be illustrated in different applications discussed in later chapters. Due to the in-phase and quadrature split, QAM's has higher noise immunity, resulting in lower noise interference. Indirectly this results in lower probability of error or lower BER (Bit Error Rate). An example of QAM in application related to holography is shown in Fig. 2.14 where an image is broken and reassembled to appear as a 3D object to the viewer using fringe patterns. This application has been widely used in recent times, not only in broadcast, but in other picture processing applications as well.

Due to better utilization of bandwidth QAM is used in radio communications (microwave links, cellular) as well as cable television. Although QAM was shown as an example, other techniques such as BPSK, QPSK, etc., also provide similar advantages. Depending on the application, digital modulations are used in all wireless systems today.

2.10 Conclusion

Starting with basics of radio propagation, this chapter proceeded to lead the pathway to digital wireless system that prevails today. Nature of electromagnetic wave, its propagation in wired and wireless medium were introduced. The concept of how an antenna radiates, how is radio wave perceived and measured using power as the basis, were introduced. It also emphasized the need to focus on digital modulation techniques that operate in all working systems, irrespective of wireless technology. With a clear de-emphasis on old world modulation techniques of AM and FM it elaborated the reasons of why digital wireless systems are more efficient, effectiveness of DSP/and systematic replacement of analog radio taking place in all spheres of wireless communication. The complex techniques of OFDM, its advantages in terms of modulation and high throughput were introduced to underscore its use in a variety of wireless technologies. Finally, QAM was shown as an example of how digital modulation is accomplished. A simulated model of QAM is provided in reference [16].

Summary

This chapter introduced certain fundamental concepts or background needed to understand wireless systems. Concepts such as what are radio waves, how radio wave travels —on earth, towards the sky, and towards space were introduced. Use of different frequency bands of electromagnetic radiation and their characteristics were described.

We also introduced the concept of digital modulation, why and how it became popular, and its advantages over analog modulation that used to be the mainstay till the end of twentieth century.

The concept of digital modulation underlining the most widely acknowledged OFDM, its principles and some of its unique features. As a corollary of this modulation was the rise of DSP—why is digital signal processing such an important feature defining a variety of modern technologies —not limited to communications alone? Rise of DSP resulted in important changes to wireless communication network, making the device smaller, smarter, helping an entire generation of people globally recognize cell phone as their primary device for communication.

Finally, the chapter concludes with the movement of wireless towards broadband access—as a means to communicate and bringing new methods of interactions on the network. Efforts to avoid bringing communication wires/cable to the home is under way, expanding on the work that began with WiMax and continuing in 5G has brought this service directly to homes.

QAM was shown as an example of digital modulation— where shifting the signal phase by a quarter (or 90°) considerable improvement is achieved in terms of higher throughput and better noise immunity.

Emphasis on participation

Why are radio waves needed to carry on conversations when there is global telecommunication with IP network? Why was it difficult for analog radio to carry video streams?

Is short wave radio a thing of the past? How many of you have plan on getting an amateur radio license? Why many people try to get FT8 Ham radio license and compare digital and analog?

DSP Kits for the lab—if not tried already, it may be time for you to indulge in basic DSP kits from Texas Instruments or other vendors. Get a feel of how writing programs on DSP is different from conventional processors and coding.

When the trans-Atlantic cable was laid and then upgraded to include fiber, it was considered a competitor to satellite radio. Why are satellites still very popular for international and carry maritime traffic?

Knowledge gained

Nature of electromagnetic waves and their propagation— why are some parts of electromagnetic spectrum used and not others? This was explained based on lessons/exercises in Physics.

The concept of antenna and how the radio waves leave the antenna and propagate through air or vacuum was explained. The concept of waves guided through two wire or coaxial cable is itself well understood, but the concept of wave propagation beyond the antenna is essential for wireless. Important concepts that make antenna practically seem as if it is a part of the coaxial cable, were introduced.

Fundamental concept of Digital Signal Processing based on Fourier series and Fourier Transforms was explained. Reasons for using DFT (Discrete Fourier Transform) block was described and their use in enhancing performance of network was noted.

Concept of digital modulation, the impact of OFDM on modernizing wireless communication was emphasized. Reasons for OFDM development and the considerable computational power needed for its deployment were explained in the context of chip set and semiconductor growth.

Finally, how some acceleration was brought about by digital modulation resulted in schemes that allow wireless systems take the lead in access to communication networks (such as IP), was explained. Its continued growth and expected impact were noted.

Homework problems

1. Find at least ten short wave radio stations in your area and establish their frequency and coverage patterns. Are these stations serving local, regional, or international community bases? What are their strategies compared to local AM, FM stations? Do these broadcasters already have digital broadcasts (DAB or Satellite radio)?
2. Simulate a complex wave pattern and write—Fourier series to describe it (Sine and Cosine). Use reference [14] or other MATLAB/JavaScript to develop an RF wave. Develop a digital stream of duration of about 10 s. Find a DFT to describe this digital pattern.
3. What is the highest frequency band that would bounce back from the F layer?
4. Since Power Lines support Carrier communications, why don't your local electric power provider offer Internet/Communication service? What are the limitations?

5. Remote sensing satellites carry an infrared camera instead of a high-resolution normal camera. Why is this important?
6. One of the important uses of DSP in stage entertainment is making up "holograms" and presenting a "virtual person" on stage. What does this involve in terms of DSP effort? Write a block diagram to describe how a producer would develop/use this system.

Project ideas

1. Plot typical I and Q diagrams for BPSK, QPSK, 16 QAM. Find the maximum number of symbols each of these modulation schemes represent. How are I and Q diagrams inherently digital in representation?
2. Can an A to D converter be used to digitize voice? What other blocks are needed to digitize voice, so it be replayed at the other end using a normal loudspeaker?
3. When a cell phone calls your typical desk phone using VoIP, are there different coders at both ends? What are the differences? Which of these, sound better?
4. Contact your local radio station to find out if they use DAB enhanced FM or AM channels. Are these available locally? Can you try to record and replay their radio broadcasts? Is the quality comparable to CD? How would you measure their audio quality?
5. Concepts of space communication—how long did it take to communicate with voyager when it was near Jupiter? When you see the Sun, assume that there could be a switch to turn off the Sun. When would you actually see the Sun turned off?

Out-of-box thinking

1. The traditional power line pole with wires running supported by them. Do you believe it is possible to use the humble pole and lines to carry Millimeter waves? Find out which of your cellular providers is trying this concept.
2. Section 2.5 shows many DSP applications. Which one interests you? Can you take an existing application and offer an upgrade on it to make it work better? What should the DSP do to implement your enhancement idea?

References

1. Maxwell's equations. www.maxwells-equations.com/
2. NIST Technical note 1379, Direct Comparison Transfer of microwave power sensor calibrations. http://nvlpubs.nist.gov/nistpubs/Legacy/TN/nbstechnicalnote1379.pdf
3. Effects of earth's upper atmosphere on radio waves. https://radiojove.gsfc.nasa.gov/education/educ/radio/tran-rec/exerc/iono.htm
4. Different types of antennas and characteristics of antennas. Electronics Hub. https://www.electronicshub.org/types-of-antennas/?unapproved=415527&moderation-hash=766fe66feef5c292409f96cc5b1056ae#comment-415527
5. AT&T archives: Similarities of wave behavior (Bonus Edition), April 3, 2012. https://www.youtube.com/watch?v=DovunOxlY1k
6. Analog Modulation, Digital Modulation (AM, FM, PM, ASK, FSK, PSK). http://www.equestionanswers.com/notes/modulation-analog-digital.php
7. Digital Audio Broadcasting. https://searchmobilecomputing.techtarget.com/definition/digital-audio-broadcasting
8. The history of Orthogonal Frequency division Multiplexing, Stephen B Weinstein. http://cttcservices.com/HistoryofOFDM11.09.pdf
9. Chang RW, High-speed multichannel data transmission with bandlimited orthogonal Signals, Bell Sys. Tech. J., vol. 45, Dec.1966, pp. 1775–96; see also U.S. Patent 3,488,445, Jan. 6, 1970, https://ieeexplore.ieee.org/document/6769442
10. Orthogonal Frequency Division Multiplexing for wireless communications, ISBN 978–0–387–30235–5, Editors Li, Ye Geoffrey, Stuber, Gordon L. (Eds.)
11. The Scientist and Engineer's Guide to Digital Signal Processing, Steven W Smith PhD, Soft Cover, 2002, ISBN 0–7506–7444-X, Hard Cover, 1997, ISBN 0–9660176–3–3
12. System Architecture for 3GPP LTE Modem using a Programmable Baseband Processor, Di Wu, Johan Eilert and Dake Liu, Andres Nilson, Erik Tell, Eric Alfredsson, Linkoping University, Sweden. https://pdfs.semanticscholar.org/4d41/4a8cfe759965a49b29e689a7899
13. WaveLAN®-II: a high-performance wireless LAN for the unlicensed band, Ad Kamerman, Leo Monteban, Published in: Bell Labs Technical Journal (Volume: 2, Issue: 3, Summer 1997) https://doi.org/10.1002/bltj.2069
14. 6G Vision and Requirements, Klaus David and Hendrik Berndt, IEEE Vehicular Technology magazine, September 2018, pp 72–80. https://ieeexplore.ieee.org/document/8412482
15. Electromagnetic wave visualization in MATLAB, by Simulation Master, Oct 20, 2019. https://www.youtube.com/watch?v=LYVTIWKYQ9I
16. Quadrature Amplitude Modulation, Working principle and its applications, EL-PRO-FOCUS, Electronics, Project, Focus. https://www.elprocus.com/quadrature-amplitude-modulation/

Wireless Systems—Technologies

Abbreviations

WLAN	Wireless Local Area Network
MSC	Mobile Switching Center
LTE	Long-Term Evolution Cellular standard for global communication for 4G and beyond
1000 base T	Gigabit interface (copper based with RJ 45 connector)
10GBPS interface	Fiber interface only
T1/E1 Interface	Telecommunication standard interfaces of North America (T1) and global (E1)
Telstar	First communication satellite to connect North America and Europe
Geostationary orbit	Orbit in which a satellite appears stationary to an observer standing on earth
PAN	Personal Area Network
TV	Television
Hybrid-cast	Combination of broadcast and live feeds from the Internet or Local networks
IETF	Internet Engineering Task Force, an independent body that governs and regulates the Internet
WiFi	Wireless Fidelity, a WLAN system that operates as a 100 m wireless extension of LAN
AP	Access Point, a central unit of WiFi that connects to the Internet and communicates with all devices (Smart phone, WiFi gadgets such as TV, home security systems, etc.)
Multipath	Multiple paths by which radio waves travel between transmitter and receiver
Doppler shift/spread	Effect of frequency shift due to combination of mobility and atmospheric conditions, first investigated by Austrian Physicist and Mathematician C.A. Doppler
Multipath fading	Phenomenon that occurs due to signals reaching a receiver via many paths and their relative strengths and phases change
Mitigation technique	Design efforts so that frequent loss of signal or data is compensated by techniques that use creative means to recover or compensate, to allow effective function of receiver
Equalization	A statistical technique aims to remove ISI (Inter Symbol Interference) by tracking the signal regularly and compensate such that the best signal is always used by the receiver
Convolution code	A type of error correcting code that generates parity symbols by applying Boolean polynomials
Raleigh scattering	The scattering of light (or electromagnetic waves) without changing

© Springer Nature Switzerland AG 2022
K. Raghunandan, *Introduction to Wireless Communications and Networks*,
Textbooks in Telecommunication Engineering, https://doi.org/10.1007/978-3-030-92188-0_3

	wavelength. First investigated by British Physicist Lord Raleigh		theory. It addresses the problem of errors that can occur during transmission and storage of information/data
Rician model	Channel model to simulate signals that gets canceled by itself due to atmospheric propagation	Shannon-Hartley theorem	It provides the maximum rate at which data can be sent over a communication channel of given bandwidth
Log normal	Channel model to simulate random obstruction of radio signals due to large buildings and hills, in the propagation path		
Longley rice	Model to simulate signals over a very wide range of frequencies (20 MHz–40 GHz). It was developed by American scientists Anita Longley and Phil Rice almost 50 years ago and widely used		
Okamura hata	Channel model to simulate signal propagation in urban areas, developed by Japanese Scientists Okamura (in 1968) and Hata. Uses direct, diffracted and reflected signals for estimation		
Training sequence	Learning program for equalizer to establish signal pattern on different travel routes, so it could be improved upon with repetition and provide the most optimum performance		
AWGN	Additive White Gaussian Noise. Statistical noise with a probability density function equal to normal distribution, named after Carl Frederick Gauss		
ECC	Error Correction Codes. A branch of applied mathematics that is used in telecommunication, information theory, coding		

In this chapter, we will review engineering design concepts of various wireless technologies in use. Nature of such applications, their characteristics, and user base vary. Each of these systems cater to different user groups and are designed to meet their needs. Sometimes a combination of these may be in use; these will be pointed out as we progress through this chapter. Underlying commonality will be stressed throughout this chapter. The objective is to bring some logical sense of how wireless systems are designed with a menu of technology choices. It also provides the reader, the thought process behind standards and why they are useful.

3.1   Wireless System Types—Cellular, Microwave, Satellite, WLAN, Broadcast

Perhaps the most widely deployed wireless technologies today are cellular, microwave, satellite, WLAN, broadcast, Near Field Communication (NFC) and short distance technologies of RFID, UWB and Zigbee etc. This chapter addresses technology and design aspects of each. Mobility and issues related to how this is addressed in wireless systems is expanded as "FEC—Forwarded Error Correction". It involved concepts of error correction codes and related topics.

3.1.1   Cellular System

Cellular system gained popularity due to a variety of reasons. Primary among them is the ability of cell phone to be used anywhere including moving at considerable speeds. This is not the only reason. There are others, such as ability to maintain a consistent set of services that users have

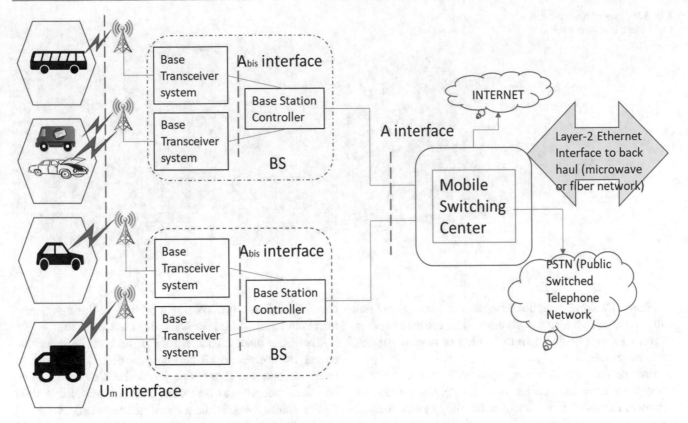

Fig. 3.1 General overview of cellular network architecture

subscribed to. Some may subscribe to a bare minimum service to contain costs; business users may prefer features that cost more but provide higher levels of service.

Figure 3.1 provides overview of cellular system with standards-based interfaces are defined at each point. For example, the air interface Um indicates how the mobile will communicate with the base station's transceiver (transmitter + receiver) system. The transceiver in turn would use the "Abis" interface to communicate with the Base Station Controller. The A interface communicates with the switch (Mobile Switching Center). Part of this work involves commonality for different vendor equipment so that any vendor who makes the mobile, can make sure that it can communicate with base station produced by any other vendor, in any part of the world. This universal method of communication and the frequency bands came about gradually. Currently, LTE (Long-term Evolution) is the fourth-generation standard universal to all countries of the world and now moving to 5G. There are other aspects that make it universal (beyond standard interfaces within the cellular network) which will be discussed in Chap. 11.

Consider the backhaul network interface shown on the right. This backhaul network could be used either as the A interface (connecting to PSTN) or beyond the switch to another network. At this level, the interface is a universal standard. It can be described as the backhaul connection with a standard Ethernet Interface (broad arrow). This interface is typically a Gigabit (1000 base T) interface but is also available as 10 Gbps interface. Both interfaces readily offer connection to either a fiber network or a microwave network (which ever the service provider chooses). Both support the 1000 Base T and 10 Gbps interfaces using the Ethernet standard. Backhaul network is very important and will be elaborated in Chap. 4, where its origin and growth will be discussed in detail.

This feature makes the cellular network easy to interface to any existing network, in any country. We will review further details of the cellular network in Chap. 11, but a brief description on foundation of mobility and security will be briefly reviewed in Sect. 3.3. Note that the architecture indicated with the interfaces, illustrates the importance of standards and universality of the cellular network. Each of the elements and the protocols continue to change with each generation of cellular. But the commonality of its elements and interfaces is retained including connection to the Internet.

3.1.2 Microwave System

When a communication link is needed between two fixed points (point-to-point) microwave link offers shortest distance (line-of-sight is required) and simplest to deploy (cost and effort).

Fig. 3.2 Typical illustration of a microwave link with conical dishes

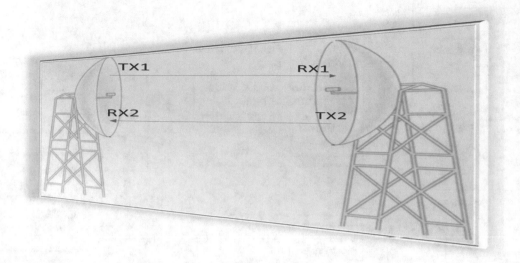

Figure 3.2 shows simplified diagram of a microwave link with two antennas facing each other. The distance between these two dishes could be up to 100 km or more depending on local terrain.

Perhaps the longest serving systems in wireless, microwave links have remained part of the telecommunication networks for about a century. It is the only system that has no direct access to the typical end user. Therefore, wireless users may see them, yet may not realize their conversations are ultimately carried through such links.

It is the service provider, or a public safety agency, who operates these links to carry bulk of the user traffic as backhaul. Most of these links operate on licensed bands of 6, 11, 23 GHz and are bidirectional. Spanning over long distances, these links typically carry traffic across very difficult terrains (mountains, large bodies of water, etc.). The interface to the microwave link follows the telecommunication network standards. It was T1/E1 interface earlier in telecom domain, now replaced by Ethernet links in the modern IP (Internet Protocol) network domain.

It is important to visualize microwave link as an extension of the A-link shown by broad bidirectional arrow in Fig. 3.1, as Layer-2 link. Basic function of this microwave link is to act as data highway for the entire data leaving/arriving at cellular/mobile switch. Therefore, it is the highway that connects one network to the next network (may be several towns away). In Fig. 3.2, TX1 transmits from antenna on the left which is received as RX1 by antenna on the right. Similarly, TX2 transmitted by the antenna on right, is received as RX2 by the antenna on the left. The frequencies of TX1 and RX1 are known as a matched pair and these are different from TX2 and RX2 which is a different pair. Both are allocated within a large spectrum at 6 GHz, or 11 GHz, etc. This bidirectional feature makes them ideal for all communication networks with separate frequencies for

transmit and receive. We will revisit it during review microwave and Millimeter wave links in Chap. 13.

There are other characteristics of microwave links that endear themselves to all service providers. It is easier to build and operate. There is no physical cable or fiber to be laid from one end to the other. The antennas focus their energy solely towards the receive antennas and avoids all obstructions including humans. Therefore, the radiated energy is always in the sky, well above the population on ground. Life expectancy and reliability of these links are generally very good. Therefore, other than periodic checks, not much effort is expended to maintain these links.

3.1.3 Satellite System

The concept of satellite communication began in the 1960s with Telstar 1 satellite. It was built as an international collaboration between AT&T, Bell Labs, NASA, the British General Post Office, and the French National Post, Telegraph, and Telecom Office [1]. A pre-cursor to the concept of space communication was a seminal paper and book by Arthur C Clarke that laid down concepts of satellite communication, space flight, and strategies of how to go into outer space, beyond our galaxy [2]. One of the important concepts that Arthur C Clarke proposed was "Geostationary orbit" which he calculated and showed would be in the Equatorial plane, placed about 36,000 km from the earth [2]. Using calculations, he indicated that when a satellite is placed in that orbit, it will appear "stationary" to a person on the earth, since it takes 24 h to go around the earth once as shown in Figs. 3.3a, b. Note that Fig. 3.3a is a side view showing the equatorial plane where the circle has radius of about 36,000 km. Fig. 3.3b is top view looking at the earth from high above the pole. Earlier during study of this orbit, it

Fig. 3.3 **a** Geostationary Orbit—
viewed from equatorial plane.
b Geostationary orbit viewed
from above (North Pole)

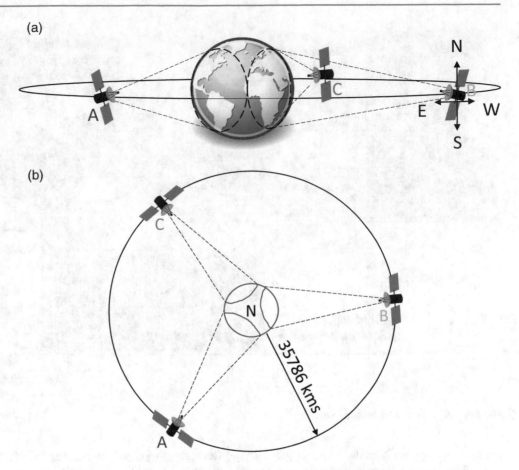

was shown that three satellites (A, B, C) placed equally apart on the geostationary orbit, could cover the globe. However, the limitation of this orbit is that extremities of the globe (near the North and South Poles) will not be covered due to the skip effect (see Fig. 3.3b).

In Fig. 3.3b, the view is from the top of earth or from below (directly over North Pole or South pole) shows spots of coverage of the earth. Observe that both in the Arctic Circle as well as in the Antarctic Circle, there is no coverage due to the gaps in the beam coverage, as indicated.

Therefore, countries that have considerable part of geographic land mass in the Arctic Circle (such as Canada, Russia, Scandinavia) will have limited coverage that will not reach their northern territories. Despite this limitation, the geostationary orbit at 36000 km on the equatorial plane, became an important resource in space and the UN (United Nations) allocates space on this orbit to all member nations. This concept was expanded in later years, and today the UN assigns slots in this orbit to most member countries, making it an international resource in space. However, the geostationary orbit is only one of the orbits that communication satellites use. There are many others, that we will review in Chap. 13 with more details.

3.1.4 Wireless LAN (WLAN)

Wireless LAN (Local Area Network) is typically a 100 m extension of the wired LAN—almost always within a building or campus. However, its widespread use as WiFi has made it very popular as a destination wireless link at home, in office, at rail stations, airports, etc. It is perhaps one of the most popular of wireless technologies in urban areas where Local Area Network (LAN) and fiber network normally exists. WLAN is a systematic extension of this network to provide flexibility to the end user. This topic includes a variety of Personal Area Networks (PAN) most are based on IEEE standards. Many such as Bluetooth, ZigBee, WiMax, and others, are in this category. They will be elaborated in Chap. 12.

3.1.5 Broadcast Systems

Broadcast by design, is a one-way communication with one transmitter and a large number of receivers. This method of communication from a broadcast station to users spread over a large area provides a service, such as radio, TV, and others. To support users who may be mobile, who may be below ground (basement for example), broadcast systems typically

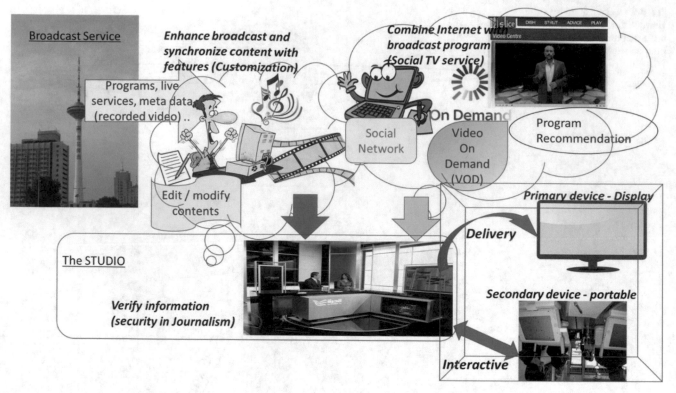

Fig. 3.4 Major parts of hybrid-cast service

transmit high power signals. There is generally an indirect method to estimate how many users are "tuned in". This estimate is based on the number of active receivers in a specific area. The topic of broadcast communication evolved and continues as a major communication method for about a century. The growth of Internet and several applications that ride over it, enhanced, and enriched the broadcast industry.

A hybrid version that can be termed Hybrid-cast is deployed globally. Figure 3.4 shows the major parts of this new Hybrid-Cast service. The traditional Broadcast service now constitutes both the programs from archives as well as live services. It does include a host of metadata that are available from achieves. The second part consists of adding/modifying content "on the fly" by synchronizing several functions as they stream through the broadcast service into the studio before going "over the air". The concept of news readers and typical TV shows, Radio shows, and others get such enhancements in terms of either a conference, or a live feed with text running below (edits), etc. The third part is the ability to provide other services such as "on demand" which may include broadcast from previous days or weeks. It could also be retrieved from archives. The viewer could also get recommendations based on search queries, for example. The Studio and its staff verify to make sure that what is broadcast is based on standards demanded

in journalism such as "verify the news is accurate" (before being broadcast). The fourth and final segment is the viewers —which in the earlier generation was considered "receive only", but now provide feedback through social network or other Internet feeds.

This has changed in recent times, due to mobile devices and dynamic updates provided by users. However, the enhancement is mainly due to "interactive sessions" not only in terms of voting during popular shows but also content and viewer ratings, inputs and many more. This should not be construed as a bidirectional service from the broadcaster. Feedback from viewers almost always follows a different path, typically a wireless service fed through a cellular provider, or a WiFi following another Internet provider's path back to the broadcast service provider.

It is important to note that the broadcast industry is closely associated with satellite service to extend their reach. The "Direct TV" concept is a specific case of receiving the broadcast, re-routed through a satellite. Similarly, cable TV is a specific case of a local provider receiving a strong signal and then re-routing contents to your home through a cable. In most nations, still TV and Radio are regulated and made available to all viewers directly "over the air". Therefore, the service providers typically make sure that at least in the local area of broadcast, their signal is quite strong and supports mobility. In terms of signal strength, what does "strong" mean?

To provide a comparative view, typical base station in a cellular system can transmit at a maximum of 100 W and no more. In contrast, any TV or radio broadcast station transmits tens of kilowatts of power. The public must realize that any TV transmitter, will transmit at least 10 times the power of a cell tower; often it is 30–100 times the power from a cell tower. This will be reviewed in greater detail in Chap. 7, in terms of public concerns and safety. However, it is of paramount importance to note that due to digital technology with high efficiency—whether it is TV, cell tower or others, "radiated power has reduced considerably in all wireless systems".

3.2 Wireless and Mobility—Complexity Due to Mobility

What is "mobility" and how is this defined? The question is paramount, since it relates to the very basis of how wireless systems are designed. The Internet standards body IETF (Internet Engineering Task Force), for example, defines it using the concept of "mobile IP". This means a mobile device that was connected an IP address (Ethernet port) is moved to another IP address. If these changes are accounted for, mobility is possible. This includes a WLAN (such as 802.11 based WiFi) where a device has mobility within the area serviced by an access point (typically within a circle of 100 m). On the other hand, the cellular industry and land mobile systems used by public safety agencies take a very different perspective. To them mobility means user is moving at high speeds and over extended distances. Therefore, the approach followed by this industry involves deeper understanding of Physics and principles behind mobility, its impact of propagation and mitigation techniques. This aspect is addressed now.

Mobility can be addressed at two levels. First, how is mobility perceived in Physics? If we consider a commuter rail or a vehicle traveling on a highway, typical speed assumed is 120kmph. This means the vehicle will travel 2 km every minute, or 1 km every 30 s. If we inspect this with respect to an access point defined by "mobile IP" (a standard defined by the IETF—Internet Engineering Task Force) and used by WiFi, then typically 10 s of association time is required by a WiFi device (smart phone for example) to associate with an access point. During this time, the user would have traveled 1/3 km which is more than the distance covered by an access point (100 m is the normal coverage range). Therefore, the user's device would continuously try to associate with the next Access Point (AP) but it will move away beyond coverage even before that association is completed successfully. Therefore, placing access point along the highway or rail track is not an option.

Another way to address this problem could be to increase the coverage from each transmitter—from 100 m to several kilometers. While base stations have enough power to do so, access points operate in higher frequency bands with limited output power; they cannot do so due to severe loss in propagation—their output is typically kept at less than ½ W and always that limits the coverage distance. Another important difference is hand-off time in cellular systems, which takes typically <50 ms. There is considerable effort spent during design to achieve consistent hand off when mobile moves from one base station to the next. This continues across large regions (such as a continent). This concept is generally termed "ubiquitous coverage" meaning radio coverage without a break during mobility.

Second part of the problem relates to mobility concerned with Doppler shift and fading. These relate to speed of the mobile unit. Mechanism to mitigate Doppler shift must be designed into the physical and MAC layer. Since WiFi 802.11 layers were not designed to include such mitigation techniques, it is unfair to expect WiFi to support mobility. However, traditional mobile networks (such as cellular, land mobile, broadcast, and satellite networks) mitigate and support mobile environments. The following discussion is limited to systems that recognize and address these phenomena related to Doppler shift and fading.

3.2.1 Doppler Shift

Let us consider the sub part related to Doppler shift. Doppler shift is defined as a signal frequency shift (or change) when an object approaches the transmitter or moves away from the transmitter. In the audio range this is heard when a siren moves towards the listener, its frequency (shrill) seems to increase. When the vehicle moves away from the listener its frequency seems to decrease. Similarly, in generic terms of an electromagnetic wave:

This phenomenon is described by the equation: $\frac{\Delta \lambda}{\lambda_0} = \frac{v}{c}$

$$(3.1)$$

where $\Delta \lambda$ = wavelength shift, λ_0 = Wavelength of the source not moving (in this case the base station), v = velocity of the source (the mobile), and c = the speed of light. It is essential that the receiver accounts for Doppler shift so that the received signal is compensated accordingly. In the digital radio domain, Doppler shift results in spread of the signal over a period or symbol frame.

$$Tc \approx \frac{1}{Ds} \qquad (3.2)$$

Fig. 3.5 Multipath propagation
—urban environment

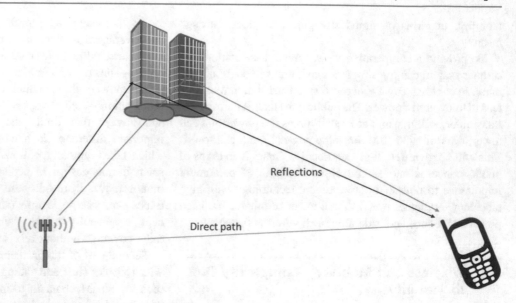

where Tc = Coherence time and *Ds*is the Doppler spread.

Observe that in Eq. 3.1, velocity of the mobile is an important factor. Hence, faster the mobile travels, higher will be Doppler shift. Perhaps the highest velocity we encounter for wireless happens to be speed of the satellite which is multiple times the speed of a jet aircraft. Therefore, satellite systems consider Doppler spread carefully during design. Land-based mobility has increased over the years. Recent high-speed rail systems in Japan, France, and other countries brought major challenges for deployment and effective operation of cellular system since passenger's travel at this speed ($\sim$ 550 kmph). Satellite links are often used to back-haul traffic from trans-continental rail systems traveling through wilderness at great speeds.

Let us now review the topic of fading. It is inherent to every wireless unit that operates in an open environment and fundamental to the technology behind widespread use of the cell phone.

3.2.2 Channel Fading

What constitutes fading? When signal propagates through the air (or vacuum), it is reflected by objects in that environment. Therefore, receiver is likely to get multiple signals, one direct ray from the transmitter but many others reflected from nearby objects. This is known as multipath, indicated in Fig. 3.5 with three rays—one direct, two rays reflected from buildings.

It is important to note that even in an open flat land area, there are always two signals—one directly from the transmitter and a second one reflected from ground. In general, multipath always exists.

Fading is made up of two main types—Slow fading and Fast fading. These terms refer to changes in magnitude of phase of the channel. Slow fading can be imposed by a large object such as hill or some large building blocking the signal between transmitter and receiver. Time difference between two consecutive measurements where there is considerable difference between measurements (as observed by the receiver)—is known as "Coherence time". Such difference could be in terms of time as well as amplitude. Usually, multipath components have both differences—difference in propagation delays and difference in attenuations. When they are summed up at the receiver, the result (due to filtering) will affect received signal. Note that different frequencies of modulated waveform experience different attenuations and/or phase changes. This overall effect is known as "frequency-selective fading". In summary, frequency selective fading occurs independent of mobility (even if user is standing still). In an open environment, whether there is movement or not, radio signals received vary and this is attributed to slow fading. By design, OFDM (Sect. 2.4) offers good mitigation to this type of fading and is used in many systems such as IEEE 802.11 WiFi, 4G LTE, satellite, and broadcast systems.

In contrast, fast fading is directly related to speed of the mobile as well as the frequency of operation. In effect, it relates to relative mobility between the transmitter and receiver. This can be described as a time-varying behavior in the propagation environment. This results in time varying signals that also are affected by different attenuations at different times. This overall effect is known as "time-selective fading". Figure 3.6 shows fast and slow fading combined—as the user's mobile travels, moving away from the transmitter.

Fig. 3.6 Fading and mobility— fast fade and slow fade

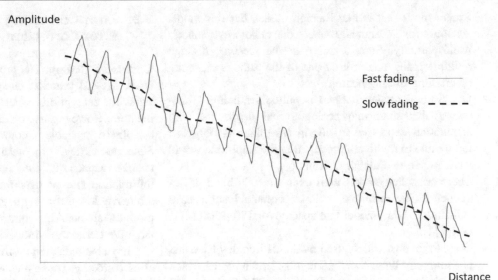

Time selective fading is due to mobility of the user—fast fading and slow fading do occur due to mobility. In fact, fast fading is related to frequency. Higher the operating frequency more will be the number of shifts occurring in Fig. 3.6, seen as sharp peaks and valleys—they are due to fast fade. For example, the commonly used 1800 MHz band of cellular (known as PCS or DCS band), will have double the number of fast fades (or nulls) that occur at 900 MHz. With higher frequency or at higher speeds, deep nulls (lows of amplitude) occur very often resulting in frame loss or parts of symbol frame could be lost. This must be accounted for during design and mitigation techniques must be developed, to account for such loss of frame/frames. This is one of the important reasons why wireless (air interface) is considered a harsh environment that needs careful design consideration, generally known as mitigation.

3.2.3 Mitigation Techniques

To support mobility there are various digital signal processing techniques developed by the cellular/satellite designers to mitigate effects of frequency selective and time selective fading. Consequently, many mitigation techniques exist to counter these effects on wireless signals. Let us consider three important techniques:

- Channel Equalization,
- Coding, (Convolution coding, Error Correction Codes),
- Diversity transmission schemes (OFDM, frequency, and space diversity schemes).

Channel Equalization: The word "equalization" was briefly mentioned in Sect. 2.4. Let us reconsider this term based on Fig. 3.5. Note that in this simplified diagram there are three distinct paths for the signal from the transmitting base station to receiver (handset). The question is which of the three signals should the receiver use? If we take the dynamic view where this cell phone is being used by someone on a train or a bus (moving vehicle) the paths regularly change. Sometimes the direct path may be cut off and only reflected paths are available. Due to more buildings, hills etc., and more than three paths may become available.

It is a monumental task for the receiver (in smart phone) to decide at any given instance which of these signals is the best. A separate, special dedicated digital signal processor is placed within the receiver (this is one of the "smarts" in the phone), that uses algorithms to predict just that—and decide which is the best signal at that instant. The algorithm regularly reviews the incoming streams that are within the symbol frame (timing) and within a signal level (amplitude) and are part of a frame sequence. Once the phone travels on a given route, this review forms a "training sequence" for the DSP. Next time, it will remember this training sequence and essentially recalls and reuses many such sequences helping the phone use the best possible signal each time user travels that route. The equalizer is used in all GSM phones today by billions of users worldwide.

Regarding the environment, there are several models based on statistical analysis of signal propagation through the atmosphere. Raleigh Scattering describes the scattering of light and other electromagnetic waves (proposed by the British Physicist Lord Raleigh). Raleigh scattering of sunlight in earth's atmosphere, causes diffused sky radiation resulting in "blue sky" during daytime and "yellowish red sky" during sun rise and sun set.

Regarding radio waves and their propagation through the atmosphere, there are several models in use. All of them do consider Raleigh scattering, but further enhance it using other parameters [3].

1. Rician model—it is like Raleigh model, but this model accounts for an anomaly where the radio wave cancels itself partially. Signals arrive at the receiver through multiple paths, and at least one of the paths is changing (shortening or lengthening).
2. Log normal model: If there is only slow fading (shadowing) then there may be log-normal distribution with attenuation expressed in dB; in case there is only fast fading due to multipath propagation the amplitude would have Rician or Raleigh distribution.
3. There are other models used over a certain band of frequencies. For example, cellular communication used "Okamura-Hata" model that spans over 150–1500 MHz range [4].
4. One of the most widely used models is Longley Rice that is commonly known as the irregular terrain model which covers a wide range from 20 MHz to 20 GHz.

Longley Rice (LR) model is popularly known as the irregular terrain model (ITM). One of its creators Anita Longley, was a pioneer who took extensive measurements while working at central radio propagation lab in Boulder, Colorado. She published them in a paper during early 1968 co-authored with a colleague—Phil Rice. Since then, it is known as LR model and used extensively by public safety agencies, aircraft communication (by pilots), and Television stations, who needed a reliable propagation model. Cellular systems too widely adopted this model during deployment of 1G and 2G.

The equalizer bases its computation on one of these models, taking into consideration how many rays are being received at given time. The models indicated generally classify geographic areas into rural (up to 3 rays including reflected), semi-rural (from 4 to 6 rays), urban (7–9 rays), and dense urban (10–12 rays). Although individual models may vary, this provides a common guideline on how much computation is needed by the equalizer.

"Training sequence" is established by each cell phone/base station radio manufacturer, using such models and routes that are known to be difficult in terms of coverage and operation. Combined effort over many years has resulted in equalizers that perform very well.

3.3 Error Correcting Codes and Forward Error Correction

Codes and Coding: The practice of using codes for security is an age-old practice that has been extended in modern times for encryption as data is sent over wired or wireless medium. However, use of Error Correcting Codes (ECC) to establish "reliable communication" grew due to Shannon-Hartley's theorem which states that states the channel capacity C (the theoretical upper bound for any information rate of data that can be communicated at an arbitrarily low error rate using an average received power S through an analog communication channel subject to Additive Gaussian White Noise (AWGN) of power N [5].

They also indicate that the channel capacity excludes the error correction codes—which means ECC will reduce the amount of "useful data". That is an important part to note, because ECC is not the useful part of data, but is important to recover data in a noisy (AWGN) environment.

Error Correcting Codes (ECC) is a branch of mathematics that addresses the problem of error occurring during transmission and storage of data. It assumes that in any environment, noise, electromagnetic radiation, and other forms of disturbances affect communication. In turn, this leads to errors. Such errors could be minor such as wrong word is received, to the extreme where nothing is received at the other end. These assumptions are true in wireless communication where both extremes are possible, due to harsh environmental conditions. Let us consider an example where a communication system is trying to send messages over a medium such as air. Let us assume, four names of different fruits are coded using two bits as shown in Table 3.1.

When the encoded message for "Apple" is transmitted over a noisy channel (Wireless), the bits for U_1 are sent. Due to noise, if one of the bits is changed during transmission then the received code is (0,1) instead of (0,0). Then the receiver does not realize that the message is corrupted and will decode it as "Mango". This process is shown in in Fig. 3.7. Then this problem could be resolved using additional codes or redundant codes.

With channel coding, this error can be detected and possibly corrected by introducing a redundancy bit making up a group of four (V_1 V_2 V_3 V_4). Therefore, the modified new codes are.

Table 3.1 Example of coding scheme

Fruit	Symbol	Code	Source encoder format
Apple	U_1	(0,0)	U_1 U_2 U_3 U_4
Mango	U_2	(0,1)	U_1 U_2 U_3 U_4
Orange	U_3	(1,0)	U_1 U_2 U_3 U_4
Pear	U_4	(1,1)	U_1 U_2 U_3 U_4

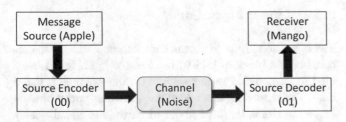

Fig. 3.7 Model of data transmission system and errors due to channel noise

$00 \rightarrow V_1 = (000)$; $01 \rightarrow V_2 = (011)$; $10 \rightarrow V_3 = (101)$; and $11 \rightarrow V_4 = (110)$. Using three bits instead of two, the probability of reducing the error is considerably improved, based on certain assumptions. Let us start with the assumption that only one bit is corrupted in the noisy channel.

The newly encoded message "Apple" happens to be (000). When this message is transmitted and an error of one bit only occurred, then receiver could get any one of the following codes: (100), (010), or (001). Observe that none of these (100), (010), or (001) is among the newly encoded messages (V_1 V_2 V_3 V_4). The receiver now knows that there was some error during transmission. Therefore, there is considerable improvement in detecting error in the received message.

Note that the above channel encoding scheme only detected but did not correct errors. It only improved the detection of error to some extent. Despite this redundant bit, if (100) is received, then it is not possible to assess whether (100) comes from (000), (101), or (110). However, if "three" redundancy bits were introduced (instead of one bit), then it would be possible to correct errors. Therefore, channel coding scheme could be modified as

$$(00) \rightarrow (00000); (01) \rightarrow (01111); (01)$$
$$\rightarrow (10110); \text{ and } (11) \rightarrow (110011)$$

Now, if the first message (00,000) were transmitted over a noisy channel and that there is only bit in error due to noisy channel, then the received word must be one of the following five: (10,000), (01,000), (00,100), (00,010), or (00,001). Since only one error occurred, each of these five codes differs from (00,000) by only one bit, and from the other three correct codes (01,111), (10,110), and (11,001) by at least two bits, then the receiver will "decode and correct the received message" into (00,000). This scheme would then allow received message to be correctly decoded into "Apple".

Depending on the assumptions and accuracy required at the receiving end, this process becomes complex and this resulted in a separate topic of mathematic known as Algebraic coding theory. To investigate efficient coding schemes, the basis of such effort lies in Shannon-Hartley equation

which indicates the capacity of a communication channel, indicated by Eq. 3.3

$$C = B log_2 \left(1 + \frac{s}{N}\right) \qquad (3.3)$$

where C—capacity of the channel (this excludes EEC), B is the bandwidth in of the channel in Hz, S is power of the Signal, N is the power of noise. S/N is known as the Signal to noise ratio. C is also known as the Shannon's capacity limit—this is the maximum capacity of the channel under ideal conditions.

Based on this limit, there were a number of developments including the Nyquist rate, which indicates that the number of pulses that can be passed through a channel is limited to twice the bandwidth of the channel.

$$f_p < 2B \qquad (3.3)$$

The two Eqs. 3.3 and 3.4 are important contributions that indicate how to transfer information and what are the theoretical limits. It led to two major fields of study "information theory" and "coding theory" [6].

Following this fundamental equation by Shannon from Bell Labs in 1948, there was considerable work on codes and first of those concerned cyclic codes used shift registers. One of the famous cyclic codes is the Hamming code developed in 1950 and is still in use today.

During the 1960s, development of multiple error correcting codes was developed by three famous mathematicians **B**ose, Roy-**C**haudhuri, and **H**ocquenghem, now popularly known as the BCH codes. About the same time, Reed and Solomon developed codes for non-binary channels, which is also widely used in wireless communication, known commonly as the RS code [7].

During the 1980s, many products, particularly memory storage and digital communication systems implemented the Hamming code. In the early 1990s, there were major inquiries into the area of coding that brought several codes, leading to efforts that tried to reach the Shannon's limit C. Several encoding and decoding algorithms were developed. Most universities started to offer Algebraic Coding as a subject in graduate study classes. Early 2003 saw a redefinition of cryptography in providing secure transmission of messages so that two or more persons could communicate in a way that "guarantees confidentiality, data integrity and authentication" [8].

Importance of cryptography was emphasized and enhanced by cellular standards body 3GPP. An open invitation was offered to researchers. Based on their response, Kasumi algorithm was chosen among the codes evaluated. This algorithm was later implemented as Confidentiality function f8, and Integrity function f9. The authentication and security algorithm in 4G communications standard known as LTE [9] uses f8 and f9. The 3GPP introduced a set of algorithms known as the MILENAGE series in its 4G standard LTE, consisting of multiple functions that guarantee confidentiality, data integrity, and authentication, that are more extensive than security schemes implemented in traditional IP networks or the 802.11 WiFi [9]. The power of coding and its ability to provide secure method of communication using the cell phone was established based on decades of work. Each generation of cellular systems now repeats this process and a new family of codes is established. This topic will be discussed in Chap. 6, as the basis of security in cellular systems.

3.3.1 Forward Error Correction (FEC)

The concept of forward error correction is an extension of the concept of error checking. In wireless communication, any channel is expected to be noisy and therefore the received information is expected to contain errors. To address this problem, FEC tends to focus on study of the channel and its noise characteristics [10].

In the example discussed earlier, Table 3.1 showed a scheme that was quite laborious and yet, not efficient. The concept of inventing codes that are both efficient in terms of number of how many redundant bits are used, yet how good they are in correcting errors, became an important objective of algebraic coding theory. Based on these objectives, algebraic coding theory is basically divided into two major types of codes: Linear block codes and Convolutional codes.

3.3.2 Linear Block Codes

These are made up of linear sum of code words. If data consists of a block of k bits (the example in Table 3.1 has 2 bits), it is modified to have n bits (in the example it had 5 bits). The redundant bits added are known as parity bits (n–k). In the example, the number of parity bits added was 3. Therefore, the Linear Block Code is now made up of 5 bits (linear sum of 2 + 3). The ratio k/n (in example this was 2/5) is known as the code rate (this should be <1). This process of liner block codes is quite tedious in terms of adding and subtracting parity bits. It also has the limitation of buffering required at the encoder and decoder. Block coding is used when a continuous stream of communication is not needed. For example, transfer of a file—it can be done sometime and if needed (in case of errors) it is possible to repeat this action i.e., reload the file. It is not used in a continuous communication stream.

3.3.3 Convolution Codes

Convolution codes on the other hand, work on a continuous data stream (serial data) by generating parity symbols using a sliding application of a Boolean function [11].

Typical convolution coder is specified as (n, k, L), where n—number of output bits of the encoder, k is the number of input bits to the encoder and L is the constraint length of the encoder. To elaborate, if we consider a convolution code (2, 1, 3), then n = 2, k = 1, and L = 3, then Fig. 3.8 shows the encoder structure for this convolution coder. The generator of polynomials uses the memory elements that are linked to achieve coding as shown in Fig. 3.8. Input is indicated by Io which is shown by input two sequences prior to the output (as I−1 and I−2). This sequence allows the data to move from input to the output with generator of polynomials taking inputs and moving the stream to output. It takes inputs and moves the stream to output based on the current state and next state.

For the (2, 1, 3) encoder two generator polynomials are used, one for each output.

$$g_0 = [101].$$

$$g_1 = [111].$$

Transition between states depends on the present input I0. The solid line in state transition diagram of Fig. 3.9 shows transitions due to the input I0 = 0 and dotted lines represent transitions due to input I0 = 1. Output bits generated during each state transition are shown along the transition lines. When number of memory elements is m then the number of states indicated by state diagram will be 2^m. For the (2, 1, 3)

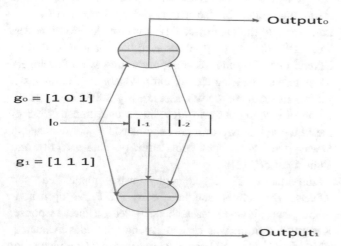

Fig. 3.8 Convolution Encoder Structure

Table 3.2 The (2, 1, 3) convolution encoder states

Input	Current state	Next state	Output
0	00	00	00
0	01	00	11
0	10	01	10
0	11	01	10
1	00	10	11
1	01	10	00
1	10	11	10
1	11	11	01

convolutional encoder the number of states will be $2^m = 4$, therefore last two memory elements are used to store past inputs I–1 and I–2.

Being a convolution coder, data "flows" through the coder and output of coder becomes the input for final transmitter of communication system. Figure 3.9 shows the convolution encoder. Different states and its function are indicated in Table 3.2.

Use of convolution coder, becomes a natural choice in most real-time communication systems such as the cell phone [12]. Wireless systems use FEC schemes based on the following assumptions:

1. The frame size has "n" bits, and bit error probability is "p" (independent of error probability of other bits).
2. Probability of correct transmission of frame $= (1 - p)^n$.
3. If frames are large, larger will be the probability of error.
4. If the probability of error is large, then it is more effective to use FEC code (Forward Error Correction). This is true for cellular systems.

5. By default, therefore, use of block codes and the possibility of receiver asking the transmitter to "retransmit" is not considered practical in cellular systems.

In a wireless channel, probability of error is always high even under the best conditions. This has been established over the years by simulations as well as field measurements. Therefore, use of FEC becomes a logical choice. Thanks to Digital Signal Processing (DSP), a variety of these functions (algorithms) were developed in the last two decades, making it very useful to implement them on a signal processor embedded within small devices such as cell phones. Perhaps the most popular of the algorithms was developed by Viterbi [13]. Although Viterbi algorithm consumes considerable resources, it does the maximum likelihood decoding, making it the prime choice in cellular systems, used in over a billion cell phones, see [14].

Channel models and use of error correcting codes discussed so far are essential in design. In addition, many improvements were made due to active feedback from the receiver, as shown in Fig. 3.10. Note that the modulation scheme adapts (changes to accommodate the channel condition) based on how good or how poor the channel is, at any given time.

The feedback in terms of specific parameters such as BER/PER (Bit Error Rate/Packet Error Rate), RSSI (Received Signal Strength Indicator), as well as power control that were used in earlier 2G/3G systems, was quite impressive. Further enhancements of these feedback parameters in 4G system resulted in better parameters such as RSRP (Received Signal Received Power), RSRQ (Received Signal Reference Quality), SINR (Signal to Interference & Noise Ratio). They provide wealth of information to the base station in terms what the mobile (cell phone/4G modem) is experiencing at any given instance. Such parameters are evaluated after every few main frames are received and then averaged using known algorithms that help in providing feedback.

It is feedback such as this (sent from mobile transmitter to the base station receiver) that not only enhances throughput

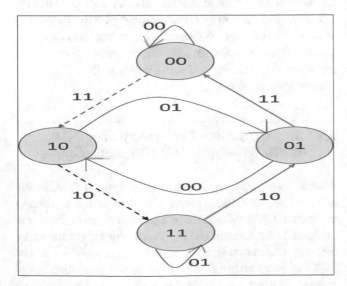

Fig. 3.9 Convolution coder states

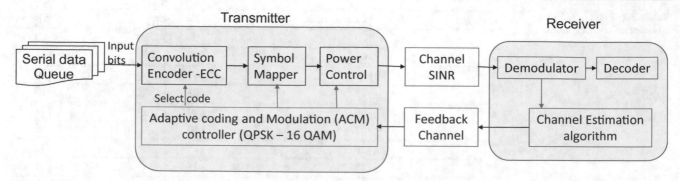

Fig. 3.10 Typical layout of encoder and FEC in modern communication systems

rate but also considerably improves overall system performance. For cellular service providers, this feature is so important that every cell phone either offers an "App" or some built-in options that test/maintenance personnel regularly use to review some of these parameters on their cell phones in real time.

3.3.4 FEC and Coding for Satellite and Intergalactic Systems

The concept of FEC and coding take on new meaning in space missions [15]. It starts with an important question—whether mission center will receive useful signals and how reliable are they? In missions with space probe is so far away from earth, signal received is so weak that despite antennas with considerable gain, there must be coding schemes to provide verifiable and accurate data. Without excellent FEC schemes, the mission will not even be successful to get signals from the space probe. Considerable effort is expended on FEC and coding in such missions.

3.4 Fixed Wireless—Advantages of Broadband at Destination

Use of mobility related features are not always deemed necessary. Once the cell phone user reaches destination (home, office, or another place), there is no need to continue with features related to mobility and the sacrifices needed such as FEC. Why? The prime reason is mobility requires compromise, particularly in terms of reduced throughput to support user traveling at high speeds. If mobility is not required, then broadband technologies such as IEEE 802.11 WLAN (WiFi as popularly known), can provide a simpler option, enhancing throughput considerably. In addition, other options such as MIMO (Multi–Input–Multi–Output) can be used to enhance the throughput rate. When the user is at destination or is moves only at pedestrian speeds, the

number of streams and delay on each stream can be calculated (by the algorithm within the mobile) and throughput can be enhanced, taking advantage of multiple streams.

Why broadband at destination? Majority of video conferences, picture-related interactive sessions, take place either at an office, or at home (if the person has home office, for example). Data rates of several hundred Mbps is practical in such broadband wireless systems. Since this uses a fixed service, several topics that were discussed earlier in this chapter (particularly related to mobility), are not needed. For example, WiFi provides a 100 m extension to the cable at home or in the office. It eliminates the need to sit at a desk with a desktop PC—provides the possibility of using either a laptop, a tablet or a smart phone instead. Despite this flexibility, user does not need the support of mobility related algorithms discussed earlier. This topic will be further elaborated in Chap. 12.

Another question could be—is it possible to use cellular service for fixed wireless? This could be the case when there is no WiFi in the area or user may not have the password to WiFi. In any case, the answer is a definitive "yes". We will review this possibility in greater detail in Sect. 3.7. In fact, this is an area that many service providers have pursued to address IoT (Internet of Things) with connections using 4G modem. But a separate set of schemes are being developed to address this situation and it is likely that this will become a major growth area.

3.5 Short Distance Technologies—RFID, UWB, Bluetooth, WiFi (WLAN), Zigbee

What is short distance? In wireless systems, it is indirectly indicated by low radiated power from the transmitter, usually lower than ½ Watt of radiated power. The coverage distance is assumed to be around 100 m. This is important for many reasons—one reason is due to use of "unlicensed" band. The 2.4 GHz spectrum is a classic example, where anyone in any country can use this band. Therefore, a variety of devices, such as toys, microwave ovens, cordless phones, door openers, and many others operate in this band. Low power is very important to minimize interference between devices operating in the same band. What else is common among them?

They are mostly consumer devices that do not need coverage over an extended area, nor do they need hand off. For example, Bluetooth is typically used for distances up to 30 m or less; it is often used as a wireless replacement to connect printers, mouse, speakers, etc., to a PC. Blue tooth offers ultra-low power devices that operate at less than 5 mW. Other Ultra-low-power technologies include UWB (Ultra-Wide Band) operating using just a few micro-Watts by using pulses over short distances. RFID also offers similar advantages using micro-watts. Both of these technologies operate over distances of just a few meters. Consequently, many of these devices have small batteries with life extending over several years. There are other short distance technologies that transmit under 400 mW, such as ZigBee (802.15.4), are used to monitor devices. WiFi (802.11) is a popular technology that can be locally used by smart phones or other devices by connecting multiple devices (TV, home security devices, etc.) to an access point. For WiFi, distance is typically limited to 100 m. These will be explained in detail in Chap. 12.

Overall, impact of the short distance technologies is considerable. They support units that are quite visible and popular as consumer devices. Even though they don't support mobility, their use in different premises for various applications has brought them to the point where their variety, type of applications they support, and customer usage continues to grow and evolve.

3.6 Near-Field Communication (NFC)—Tap and Go

What is near-field communication and what are its impacts? The concept of near-field communication was briefly mentioned in Chap. 2 as part of radio waves in Sect. 2.1.

"Near field" as defined earlier, refers to wireless systems operating within "one wavelength" of the operating frequency. What could be accomplished within such short distances? Quite a lot—using reactive mode of the magnetic field that enhances security. Since they operate over a very short distance (few centimeters), they are relatively secure and effective.

Let us consider the details shown in Fig. 3.11. Why is the distance of one wavelength considered only as a general guideline? Firstly, magnetic field is quite weak, even within one wavelength; only a small portion is in the "reactive" zone of magnetic field. This means a very small portion (0.159 times wavelength) is where the magnetic field (shown in red) is considered predominant. Rest of it which is radiative (0.841 of wavelength) is shown in light red. Therefore, the electric field does play an important role even within one wavelength and beyond (shown in light blue).

The overall dimension D of the antenna plays an important role in determining the far field. Therefore, dimensions of a near-field antenna are generally kept small compared to wavelength. In theory, beyond two wavelengths is assumed to be the far field region. To be consistent, from one wavelength to two wavelengths is the "transition region"—changing from near field to far field. Most practical systems based on "near field" applications focus on the

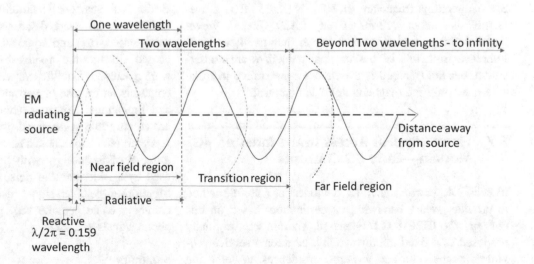

Fig. 3.11 Details of near-field and far-field regions

dimension of antenna to reduce D such that distance where magnetic field is dominant is reduced [16].

Near-field devices and their communication has been grouped under the NFC or Near-Field Communication standard Ecma-340, ISO/IEC 18,092. The standard defines an operating frequency of 13.56 MHz based on the concept of "inductive coupling". The distance is specified as a guideline upper limit—"maximum distance of 4 Cm". In Fig. 3.11, this distance is shown as the "Reactive" region where the device reacts directly with magnetic field. Examples of NFC are tapping a credit card for financial transaction (unlike a swipe or chip insert) or an identity card tapping for entry into a building. To the end user, NFC reader may seem like a bar code reader, but these two operate on entirely different technologies. While the bar code reader relies on the pictorial image, NFC reader communicates with the device (an RFID, for example) and gets a response from that specific device. Therefore, NFC uses confirmation-based approach [16].

Observe that the radiating wave is shown with dominant magnetic field within the reactive region, gradually transitioning into the dominant electric field beyond two wavelengths. We should therefore classify NFC as extremely short-distance technology (4Cm or less), much shorter than Blue tooth or UWB that operate over somewhat longer distances where electric field is dominant. Therefore, they belong to the far field region and rely on "radiative" interaction between devices. Although we discuss the "dominant" field, both the electric and magnetic fields exist along the wave and are perpendicular to each other and to direction of wave travel, as shown earlier in Fig. 2.1.1 where one wave is dominant, depending on distance from the source. Since operating frequency of NFC is 13.56 MHz, wavelength is about 22 m. Even 0.159 times Wavelength = 3.5 m, which is much longer than the 4 cm limit. Inference from this is to show that guidelines are conservative, but the standard is even more conservative to make sure that "only the magnetic field" is dominant.

3.7 Fixed Wireless Access to the Internet, 4G Modem—Early 5G Concepts

In Sect. 3.4, we considered fixed wireless as a useful method to provide greater bandwidth at destination. Based on this concept, the IEEE 802.16 standard, WiMax was originally proposed as a fixed microwave link between a service provider's central site and customer residences, to serve the rural population. It has performed this function very well and became the equivalent of cable/fiber access for rural residents. In terms of data rate, it is comparable to cable, almost ten times the speed of DSL (Digital Subscriber Line).

But as the cellular technology advanced, its ubiquitous nature helped to provide similar services to urban areas as well. Instead of residential customers, this service is focused on business customers. With a 4G modem, for example, it is possible to monitor cameras, support video or stream other broadband services. Fixed wireless was further extended to narrowband IoT which provides connections to billions of devices in offices, power plants, yards, or industrial sites where occasional monitoring of temperature, voltage, pressure, or some other parameter is vital. The major advantage of narrowband IoT is reduced cost of service to periodically access devices located in far flung facilities even across continents. Such devices could be monitored from a central/regional location. This is expected to become a major growth area in 5G as well.

Finally, 5G service plans are thought to follow the WiMax model, with much higher throughput rates using the millimeter-wave band. At these frequencies (such as 28 GHz, 60 GHz, or 80 GHz) the distance from service provider's location gets limited to one kilometer or less, but the throughput rate increases significantly reaching at least 1Gbps or more for each residential user.

3.8 Conclusion

Various aspects of systems that serve customers who are mobile and stationary, long distance and short distance were described in this chapter. For those designed to support mobility, efforts to compensate effects of mobility, were also discussed. Most systems in widespread use support a fair distance extending from few meters to several kilometers. Errors that occur due to mobility and due to the RF propagation over air, were discussed. Need for ECC and FEC were emphasized and some of its use in cellular was discussed. Changes the broadcast industry (both radio and TV) were discussed. Satellite system was briefly introduced with emphasis on the use of geostationary orbit—its advantages and limitations. However, short-distance technologies that are coming into widespread use, support distances that are less than ten centimeters. These were discussed providing a glimpse of technology with products that are visible and used by a wide range of personnel, professionals, families. Monitoring cellular signal using devices, which is not common, could become very popular, for monitoring cell phone signals see [17].

Summary

This chapter reviewed wireless technologies deployed in various regions for many user groups. It considered some important aspects such as mobility and the complications it introduces in the design of a wireless system. Methods to mitigate the effects of Doppler, slow and fast fading were considered. The effect radio waves have due to an open environment and how it worsens with mobility was further explained. Some of the popular terrain models were discussed, providing details of how these are designed. Use of codes to overcome some of these impediments and improve reliability and security of wireless was reviewed.

Changes in the broadcast industry and impact of digital radio on their programs was reviewed. Fundamental differences between broadcast and cellular communication was explained. Use of mobility related algorithms in the broadcast industry was briefly mentioned. It then explored the possibility of avoiding mobility and how performance of wireless systems could be improved when sacrifices needed to support mobility were not needed. The possibility that it creates in terms of broadband services was explored. In later chapters such broadband systems will be described in detail —these are popular and in widespread use at home and in office spaces.

Both the short-distance wireless systems and the NFC systems were reviewed. Concept of NFC and magnetic field as the predominant means to communicate was explained with examples.

Finally, the chapter concludes with fixed access as a means of communication to user homes and a primary means of connecting customers in premises was explained. It was also pointed out that this would become a major access method in 5G systems to be deployed soon.

Emphasis on participation

Why are cellular systems the primary means of communication worldwide? Why not use WiFi instead of cellular? Use of codes for security and reliable communication—is this implemented only in certain systems and not others? What are the advantages of using codes and its limitations?

Are short distance technologies popular, if so where? How many of you have set up an access point at home/check the frequency bands it operates on? Have you tried ZigBee or other narrow band IoT technologies? First, estimate the distances for a few of them and then make measurement (distance where it stops working). Provide reasons for shorter or longer than expected distance.

Get a feel of how Viterbi Convolution coding is different from conventional block coding. Do a few hands-on exercises with different numbers for symbol, code format, etc. Using Python develop and test the algorithm. See Ref. [18].

In the lab, try to set up a short link using 2.4 GHz band and see how much throughput these fixed links can provide? How much is the throughput increased when two links are used (at least two antennas). Compare this performance with using panel antennas at both ends and see how the distance between the link can be increased.

Enable your phone in field test mode and measure cell parameters? What can you check and verify with these numbers? Use Ref. [17] or other online guides to program the phone.

Knowledge gained

This chapter reviewed some of the important wireless technologies in use today. It also provided an overview of the technology and design aspects of each, where and how these are applied.

One of the important aspects of all wireless technologies is how well the data is received, in terms of voice data and streams and reliability in different applications. The technology behind making wireless systems reliable is based on coding theory a branch of applied mathematics. This was described and examples provided on how coding is developed for each application. Dramatic changes taking place in broadcast, microwave and other technologies was also discussed both in terms of its design and features that user can observe.

Mobility is perhaps the most important aspect in terms of knowledge gained. It is a key term that is not well appreciated and therefore needed much explanation in terms of why and how mobility affects wireless signals. Terms such as slow fade and fast fade were defined and explained. Equally important is the Doppler shift that distorts signals. Several coding schemes were reviewed to address errors resulting from these fades and Doppler shift. Although based on statistics, why and how these schemes have improved reliable communication in wireless systems was described. Coding was underlined as branch of mathematics that developed to the point where algorithms deliver consistent performance and the smart phone has become a reliable and essential device for mankind.

Several wireless technologies in widespread use were discussed—short-distance technologies that use the unlicensed band to offer broad band services were reviewed. These are very popular and useful as consumer devices, at home and in the office. Very short-distance technologies of NFC using both near field and how they are different from far field were discussed in terms of merits—not only in terms of higher throughput but also ease of deployment as consumer products. The far-reaching effects of these technologies and how these have become household names for an entire generation of users was underlined. Fixed wireless that could evolve into a major wireless access to homes and businesses, was briefly reviewed as part of the WiMax and its possible enhancement in 5G services.

Homework problems

1. Use your cell phone in field mode to identify base station sites and their frequencies (see Ref. [17]). How many service providers can you identify in your area? How do their operating frequencies and features compare?
2. Using a simple scheme develop convolution table for (3, 1, 4) encoder. Make a table using the Table 3.2 format and describe your scheme.
3. Does your university have a radio station and what types of hybrid-broadcast does it use? what is the feedback mechanism from listeners to let the station know their reactions?
4. Why is near-field communication not extended all the way to the full length of one wave? what is the major difference between reactive and radiative field?
5. Using different types of obstructions (such as being in an elevator, or behind a partially metallic frame, can you observe the throughput rates and BER using a mobile signal tracker?
6. You are riding a bus/train that has WIFI service. since you can move at high speeds, doesn't WIFI support mobility? if yes, why? if not, explain, why not.
7. It is Known that 802.11 WiFi Uses OFDM as a modulation scheme. Based on this information, can we assume that it supports mobility? explain whether this is true or not.

Project ideas

1. Plot typical I and Q diagrams for BPSK, QPSK, 16 QAM. Find the maximum number of symbols each of these modulation schemes represent. How I and Q diagrams are indicated using binary representation?
2. Can an A to D converter be used to digitize voice? What other blocks are needed to digitize voice and it be replayed at the other end using a normal loudspeaker?
3. When a GSM cell phone calls your typical desk phone that uses VoIP, do they have conversion coders at both ends? How about cell phone to cell phone calls? What are the differences? Which of them sound better?
4. Contact your local radio station to find out if they use DAB enhanced FM or AM channels. Are these available locally? Can you try to record and reply radio broadcasts? Is the quality comparable to CD? How would you measure them?
5. Near-field communication is being extended for security checks in a variety of locations; what are the advantages? Are there any limitations?

6. Can you set up a fixed wireless link from one building to another using WiMax or WiFi? Compare the data rates between these two technologies in the 5.xx GHz band.

Out of box thinking

The traditional power line uses pole with cables running between them. Do you believe it is possible to use the humble pole and lines to carry Millimeter waves? Find out which of your cellular provider may be try this concept.

Why is there considerable investment to move towards wireless, when wired technologies provide similar services? Is it due to lower cost, flexibility, or both? Can you create a matrix to compare them? Provide a verdict on whether the industry is doing the right thing or is it just a fad or jump on the wireless band wagon on which everyone seems to be getting on.

Do you think Nicholas Tesla's vision of providing electrical power using wireless links is practical? What are the limitations? Read about his vision of providing power free to all. Comment of whether it is feasible to convert sun rays or wind power at a central location and distribute wirelessly, within the area.

References

1. TELSTAR—the little satellite that created a modern world 50 years ago, Adam Mann, Science, 07.10.2012. https://www.wired.com/2012/07/50th-anniversary-telstar-1/
2. Ascent to Orbit: A Scientific Autobiography: The Technical Writings of Author C. Clarke 1st Edition, ISBN-13: 978–0471879107, ISBN-10: 047187910X
3. Fading channels. http://www.cs.tut.fi/kurssit/TLT-5806/Invocom/p3-7/fading_channel/p3-7_5_2.htm
4. Path loss models, S-72.333 Physical layer methods in wireless communication systems, Sylvain Ranvier/Radio Laboratory/TKK, 23 November 2004, Helsinki University of Technology, Finland
5. Shannon C (1948) A mathematical theory of communications. Bell Syst Techn J 27:379–423, 623–656
6. Fundamentals of Error Correcting Codes, W. Cary Huffman and Vera Press, Cambridge University Press, 2003, ISBN 0 521 78280 5
7. John IN, Waweru Kamaku P, Macharia DK, Mutua NM (2016) Error detection and correction using Hamming and Cyclic codes in a communication channel. Error detection and correction using hamming and cyclic codes in a communication channel. Pure Appl Math J 5(6): 220–231. https://doi.org/10.11648/j.pamj.20160506.17
8. Confidentiality algorithms. http://www.3gpp.org/specifications/60-confidentiality-algorithms
9. LTE security, how good is it—NIST evaluation report. http://cdn.preterhuman.net/texts/underground/telephony/cellular/KASUMI_Eval_rep_v20.pdf
10. Tutorial: Forward Error Correction. http://liquidsdr.org/doc/tutorial-fec/
11. "Convolution Coding", Articles on channel coding by Mathuranathan, Gaussian waves. https://www.gaussianwaves.com/2010/06/convolutional-coding-2/

12. Cheng S, Yang Z, Zhang HJ (2008) adaptive modulation and power control for throughput enhancement in cognitive radios. J Electron (China), 25(65):65–69. https://doi.org/10.1007/s11767-007-0013-4. Citation. Electronics. (China)

13. Viterbi AJ (1967) Error bounds for convolutional coding and an asymptotically optimum decoding algorithm. IEEE Tran Inform Theory 2:260–269

14. Implementation of convolutional encoder and Viterbi decoder using Verilog HDL, V. Kavinilavu and others, 978–1–4244–8679–3/11/$26.00 ©2011 IEEE

15. DSN Telecommunication Link Design handbook, NASA Deep Space Network, TMOD No. 810–005, Rev. E, Issue Date: November 30, 2000, Jet Propulsion Laboratory JPL D-19379, Rev. E. http://deepspace.jpl.nasa.gov/dsndocs/810-005/

16. NFC Forum, what is NFC?. https://nfc-forum.org/what-is-nfc/about-the-technology/

17. Cellular phone field modes. http://www.wpsantennas.com/pdf/testmode/fieldtestmodes.pdf

18. Python implementation of Viterbi algorithm. https://stackoverflow.com/questions/9729968/python-implementation-of-viterbi-algorithm

Backhaul Network

Abbreviations

Backhaul	Network that carries data to/from a major network
Central switch	Main location (usually a secure building) within a region from where calls get routed
POTS	Post Office Telephone System, but popularly called Plain Old Telephone Service
ISDN	Integrated Services Digital Network, a standard that allows telephones and computers to be interconnected and allowed conference call and a variety of services based on Signaling System 7 (SS 7)
SS7 protocol	Signaling System No. 7 is a set of telephony signaling protocols developed in 1975, which is used to set up and tear down telephone calls in most parts of the worldwide public switched telephone network
PSN	Packet Switching Network, is a type of computer communications network that groups and sends data in the form of small packets. It enables the sending of data or network packets between a source and destination node over a network channel that is shared between multiple users and/or applications
DTE	Data Terminal Equipment is an end instrument or device that converts user information into signals or reconverts received signals
DCE	Data communications equipment, it refers to computer hardware devices that are used to establish, maintain, and terminate communication network sessions between a data source and its destination. DCE is connected to the data terminal equipment (DTE) and data transmission circuit (DTC) to convert transmission signals
WAN	A Wide Area Network is a telecommunications network that extends over a large geographic area for the primary purpose of computer networking. WAN may be established with leased telecommunication circuits
PAD	Packet Assembler Disassembler, is a communications device which provides multiple asynchronous terminal connectivity to an X.25 network. It also accepts packets from the network and translates them into a data stream
NT	Network Terminator (also NTE for network termination equipment) is a device that connects the customer's data or telephone equipment to a carrier's line that comes into a building or an office. The network termination used in the specific case of an ISDN Basic Rate Interface is called an NT1
ATM	Asynchronous Transfer Mode. A method of transferring data using fixed frames of 53 bytes, using an established path
Frame relay	Frame Relay is a standardized wide area network technology that specifies the physical and data link layers of digital telecommunications channels using a packet switching methodology
OSI	Open Systems Interface, a seven-layer model established by Internet Engineering Task Force (IETF) standards body
DWDM	Dense Wave Division Multiplexer, transport method to bring commonality to

© Springer Nature Switzerland AG 2022
K. Raghunandan, *Introduction to Wireless Communications and Networks*,
Textbooks in Telecommunication Engineering, https://doi.org/10.1007/978-3-030-92188-0_4

variety of optical technologies used by fiber-optic network

MPLS
Multiprotocol Label Switching is a routing technique in telecommunications networks that directs data from one node to the next based on short path labels rather than long network addresses, thus avoiding complex lookups in a routing table and speeding traffic flows

SONET
Synchronous Optical Network (SONET) is a standard for synchronous data transmission on optical fibers. SONET is a communication protocol, that is used to transmit a large amount of data over relatively large distances using optical fiber. With SONET, multiple digital data streams are transferred at the same time over the optical fiber

SDH
SDH is a standard digital signal hierarchy that provides synchronous digital transmission, multiplexing, and cross-connection. SONET is used in the U.S. and Canada, and SDH is used in rest of the globe. SONET and SDH have relatively small differences; for example, the frame sizes are not identical, and some of the terminology is different. Otherwise, SONET and SDH are very similar technologies

GSM/GPRS
GSM is the second-generation cellular standard. GPRS - General Packet Radio Services (GPRS) is a best-effort packet-switching protocol for wireless and cellular network communication services. It is considered best effort because all packets are given the same priority and the delivery of packets isn't guaranteed. It was originally promoted as a 2.5G effort but without supporting voice, only data

Category cable
Popularly known as Ethernet cable or Category 6 cable, is a standardized twisted pair cable for Ethernet and other network physical layers that is backward compatible with the Category 5/5e and Category 3 cable standards. Cat 6 must meet more stringent specifications for crosstalk and system noise than Cat 5 and Cat 5e

OADM
Optical Add Drop Multiplexer: in wavelength division multiplexing (WDM) OADM is a device that is able to add or drop individual wavelengths, multiplexing and routing different channels of light into or out of a single-mode fiber. An OADM is an optical version of an ADM

TCP
Transport Control Protocol: It is a connection-oriented protocol, using which the data is transmitted between systems over the network. In performing this task, the data is transmitted in the form of packets. It includes error-checking, guarantees the delivery, and preserves the order of the data packets

UDP
User Datagram Protocol: It is a connectionless protocol. It is similar to TCP protocol except that it doesn't guarantee the error-checking and data recovery. Using UDP, the data will be sent continuously, irrespective of the issues in the receiving end. It is the preferred protocol in wireless where there are other methods used for error-checking and delivery

Telnet
Telecommunication Network protocol, developed in 1969, is a protocol that provides a command line interface for communication with a remote device or server, sometimes employed for remote management but also for initial device setup like network hardware. Telnet stands for "Teletype Network", but it can also be used as a verb; "to telnet" is to establish a connection using the Telnet protocol

FTP
File Transfer Protocol is a network protocol for transmitting files between computers over Transmission Control Protocol/Internet Protocol (TCP/IP). FTP is built on a client-server model architecture using separate control and data connections between the client and the server

SMTP
Simple Mail Transfer Protocol is an internet standard communication protocol for electronic mail transmission. Mail servers, other message transfer agents use SMTP to send and receive mail messages

SNMP
Simple Network Management Protocol is an application-layer protocol defined by the Internet Architecture Board (IAB) in RFC1157 for exchanging management information between network devices. It is a part of Transmission Control

	Protocol/Internet Protocol (TCP/IP) protocol suite
Front haul	A term that refers to front or access part of radio network, where data is hauled. Fronthaul, also known as mobile fronthaul, is a term that refers to the fiber-based connection of the cloud radio access network (C-RAN). C-RAN telecommunications architecture comprises the intermediate links between the centralized radio controllers and the remote radio heads (RRH) at the "edge" of a cellular network or the radio tower
RRH	Remote Radio Head, is the radio unit with multiple antennas, mounted on the radio tower. A remote radio head (RRH), also called a remote radio unit (RRU) in wireless networks. 4G and beyond infrastructure deployments will include the implementation of Fiber to the Antenna (FTTA) architecture
BBU	Baseband Unit BBU (Baseband Unit) in the cellular systems is a unit that processes baseband signals and deployed in a distributed (D-RAN) or centralized/cloud (C-RAN) network. It is one of the major components of the 3GPP system defining all equipment within the core segment of the mobile network

In the previous three chapters, the term backhaul was used a few times. In Chap. 1, and Chap. 3, there was a formal reference to what is backhaul, how it supplements services, both wired and wireless. In this chapter, backhaul is revisited in greater detail to provide a perspective of its role and importance in all communication systems including wireless networks. Backhaul is the method by which all traffic from major wireless network is carried. Should backhaul be exclusively wired or can that also be wireless? What is the capacity needed by backhaul? Does it use a different type of technology? Questions such as these will be answered in this chapter.

4.1 Function of the Backhaul Network

What is backhaul network? It is that which carries traffic from access part, whether the access is wired or wireless, to the core of the network. Since backhaul carries traffic from a large number of access units, its capacity must be inherently large. In cellular networks, backhaul is defined as the network that connects all base stations to core of the network (the central mobility switch or the Mobile Switching Center

MSC). Backhaul can be compared to a major postal truck that carries mail, large or small, while access network can be compared to the individual mail boxes in homes and offices. Backhaul network usually ends up in a switch which further streams traffic in and out to the national network. This is similar to a central postal sorting facility where mail to that area arrives and mail from that area leaves. The sorting facility puts mail going to specific regions in specific boxes and ships them out. This represents the function of a switch.

4.2 Changes to Backhaul with Each Generation of Cellular

If handling data in large quantities is all that the backhaul does, why dedicate an entire chapter to describe its function? While basic function remains, the technology of backhaul and how it handles traffic changed along with each generation of cellular network is a major part of the network function that is described here [1]. That technology has changed over a hundred years.

Figure 4.1 shows backhaul supported by older POTS (Post Office Telephone Service) network, often called Plain Old Telephone Service.

During most of twentieth century, it used only twisted pair of copper wires and carried nothing other than voice traffic, often called "Plain Old Tel Service". Recent changes to POTS were not small, they were considerable, evolving with technological change.

This change in backhaul since 1970s resulted in methods of access and changes to security schemes that will be described in later chapters. To understand such changes, consider the time before 1G network, when telecommunication system was entirely wired analog, but backhaul was somewhat moving towards digital (packet switching). What does this mean? Prior to WWII, callers relied on post office operators who connected telephone calls [2]. But later, this connecting aspect got automated which is indicated in Fig. 4.1. By late 1970s, changes had begun in the wired network when computers entered the market and had their own network. Combining the two networks resulted in use of voice coders on telephones to convert voice to data and send them as packets, similar to data packets sent between computers.

ISDN network

In wired network, PCM (Pulse Code Modulation) was introduced in voice circuits and ISDN (Integrated Services Digital Network) was an important change to voice networks. ISDN not only carried voice compressed into digital stream in the form of PCM stream but telephone also had a "data port" that allowed the telephone to be connected to a

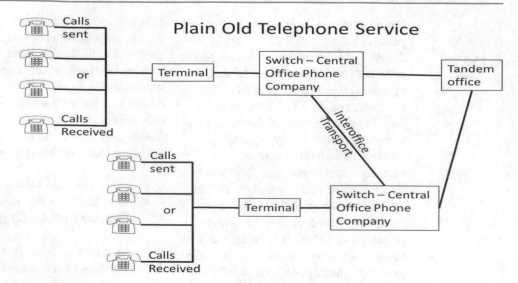

Fig. 4.1 Plain old telephone service—existed during 1G cellular network

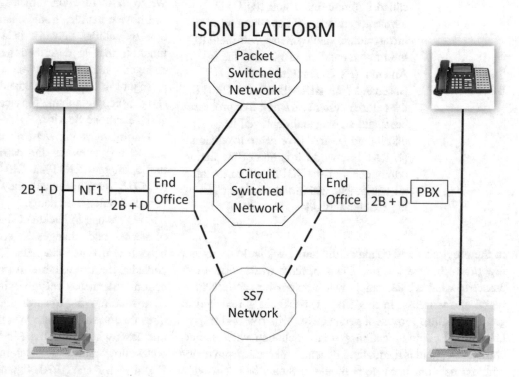

Fig. 4.2 ISDN (Integrated Services Digital Network) platform supports various networks

PC (Personal Computer). This resulted in ISDN standard shown in Fig. 4.2 that carried voice and data over the same line [3].

Such transition resulted in changes to backhaul where bulk of the traffic was voice but could connect computers using data ports. Signal System 7 (SS7) was a standard that supported services at the switch such as call hold, caller number indicator, conference call, etc., along with data link. This concept known as 2B + D is indicated in Fig. 4.2 which shows 2B = Two bearer channels or basic rate channels of 64Kbps PCM voice circuit (also known as DS0 —Digital Signal Level Zero) plus D or Data channel of 16kbps making a total of 144kbps throughput. This 2B + D

is indicated in all major links shown in Fig. 4.2. In addition, ISDN uses a Network Terminator or NT1 which is needed to take two wires from the telephone company and transpose into four wires needed to support two bearer channels. ISDN must be supported at the switch or in a PBX (Personal Business Exchange) used in commercial complexes. A bundle of 24 DS0 circuits made up DS1 or T1 of 1.544 Mbps became an important interface known as T1/E1 (2.054 Mbps) standard, where E1 was the European equivalent with slightly different rate of 2.054 Mbps [4]. This and its hierarchical upper bundles called T3 (44.736 Mbps), E3 (34.368 Mbps), and OC-3 (155 Mbps) using optical fiber, came up gradually and are still in use.

The first generation of cellular reacted to such backhaul network, by first designing a cell phone that did not digitize voice and carried voice over the air using FM modulation (much like FM broadcast). It connected to an analog or POTS line with Public Switched Telephone Network (PSTN) shown in Fig. 4.1. This was automated using switch as the center of the network. Analog cell phone used such backhaul that carried essentially voice, but towards the later part of the 1980s cellular providers introduced the concept of dial up modem that worked with POTS providing some data. During early second generation of cellular, analog modem allowed small amounts of data (1200 or 2400 bps were the data rates) to be transferred over cell phone. However, all backhaul network continued to be "circuit switched" trunk networks. Second-generation cellular widely used the T1 or E1 connections from base stations to carry voice and limited amount of data.

Trunk meant a pair of telephone cables that connected a telephone on the access network, using set of "dedicated wires" exclusively allocated to that telephone pair during the period of conversation. It represents a "connection-oriented" method.

Figure 4.3 shows the "connection-oriented" approach on the left side and "connectionless" approach on the right side. Trunk serv which is a connection-oriented approach, needs a dedicated service where the caller and the called party must both use set of wires to carry on voice conversation. Its equivalence in cellular was transmit and receive frequency pair dedicated to each cell phone to begin conversation. It was the responsibility of the switch (switch in Fig. 4.1 and SS7 circuit switched network) to assign frequency pair for that call.

Connection-oriented service therefore refers to telephone circuit which is a "circuit switched" or "trunk service" as shown on the left in Fig. 4.3, while the connectionless service is indicated on the right with data streams where packet

move from node to node without a dedicated connection between the ends (devices A, B, C, and D). This important distinction instituted a major change in backhaul network. ATM (Asynchronous Transfer Mode) which is "connection oriented" allowed control over transition path and delay, whereas IP (Internet Protocol) that uses "connectionless" service routes packets through any path that is least congested at that time. It therefore uses an MPLS—Multi Protocol Label Switching, which implements a "connectionless service" with the help of a routing technique that directs data from one node to the next based on short path labels rather than long network addresses. MPLS avoids complex lookups in a routing table hence speeds up traffic flows.

The concept of Packet Switched Network (PSN) shown in Fig. 4.3 was not implemented for some years after ISDN came into existence. The first exclusive packet switched network was the X.25 network that supported many ATM and data terminals. An overview of X.25 architecture is shown in Fig. 4.4, where user DTE (Data Terminal Equipment) consists of a PC (Personal Computer) or CPU (Central Processing Unit) [5].

X.25 is made up of network of interconnected nodes to which user equipment can connect as shown in Fig. 4.4. The user end of the network is known as Data Terminal Equipment (DTE) and the service provider's equipment is Data Circuit-terminating Equipment (DCE). X.25 routes packets across the network from one DTE to multiple DTE spread over a wide area using the service provider's switching equipment PSE (Packet Switching Equipment). Therefore, it was known as WAN (Wide Area Network). In certain cases, PAD (Packet Assembler Disassembler) which as the name suggests, assembles, and disassembles data packets was also connected to the user's DTE. This X.25 standard was originally proposed by the ITU-T as a connectionless service and consisted of three layers:

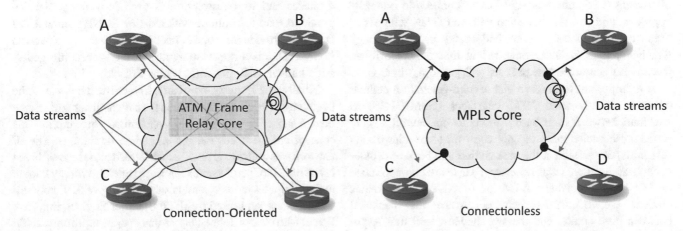

Fig. 4.3 Connection oriented and connectionless services

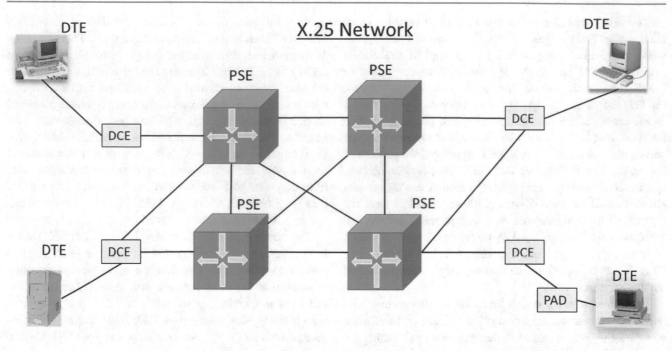

Fig. 4.4 Architecture of X.25 network—ITU-T-based model

(i). The physical layer,
(ii). The data link layer, and
(iii). The network layer.

The physical layer that consisted of twisted pair of telephone cable could handle throughput up to 100 kbps. The data link layer carried the packets that were typically either 128 or 256 bytes of data. The network layer was concerned with set up and tear down of data calls. It is important to note that this network, closely followed the model of telephone call set up and tear down even through it connected exclusively data terminals. The X.25 network became popular in the 1980s and other protocols such as Frame Relay and ATM, took advantage of low latency and low error rates offered by X.25 networks. ATM uses "connection oriented" approach, and controls transition path and delay, whereas IP uses "connectionless" service and routes packets through whichever path is least congested at that time. However, there is no guarantee that packets will get to destination.

It is important to observe that second-generation cellular networks mostly used ATM interfaces (GSM/GPRS) as backhaul between their base stations to the central switch. Since these packet sizes did not conform to cellular packet standards, conversion at the base station and back to cellular packets at the switch was necessary. This situation continued in 3G networks where ATM or SONET (Synchronous Optical Network)/ATM networks were the backhaul between base station and the switch. How is it that Asynchronous and the Synchronous, work together?

ATM has the ability to optimize incoming data and send them over a physical media whether it is a wire or fiber, it does not matter. Therefore, it can work with ISDN which predominantly uses copper wires, and SONET which uses fiber optic cable. SONET and SDH (Synchronous Digital Hierarchy) are standards for connecting telephone voice systems and computer networks that utilize the enormous bandwidth of optical fiber. They operate on precise clock sequence without bothering about whether there is data or not. That part of optimizing data in each frame is done by ATM. Together SONET with ATM optimizes throughput needed by base stations even today in many networks [6].

There are other reasons to use this combination. Fiber-optic networks are quite expensive to lay over great distances and its construction is time consuming. Service providers tend to continue with existing SONET since ATM is able to provide the control needed to deliver voice packets in time. This is a major strength that the succeeding generation known as IP network, seems to lack.

Finally, the IP network—came into existence towards the beginning of 2000 and dominated the backhaul network. Since then, it carries mostly data and very limited voice traffic. This concept has great advantages and some limitations. The IP network was built up on the X.25 and added more layers. Seven layers of OSI (Open Systems Interface) model is widely known in the computer industry and is shown in Fig. 4.5. IP network standards are supported by IETF (Internet Engineering Task Force) with clearly defined layers that operate uniformly. IETF has helped a large number of vendors to bring products that

Fig. 4.5 The Open Systems Interface (OSI) model of IP network

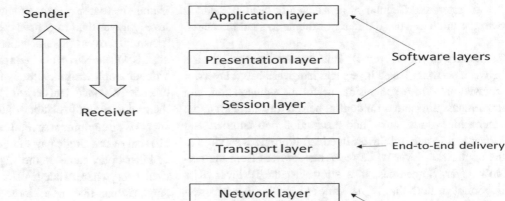

work with other products very well in a systematic manner (popularly known as plug and play).

A computer connected to the data port must have a "dedicated connection" to the other computer/computers in order to communicate. In some ways, it redefined the function of a backhaul network, as indicated at the beginning of this chapter. Unlike all other traditional backhaul network described earlier, the IP network consists of routers and switches, where each incoming data packet (or mail) has a header, which the router looks at and passes on to the next router. The path taken by each data packet, need not be the same, but so long as it reaches the destination (where they can be reassembled) it is fine. This is similar to multiple sorting centers on the way that will send the mail through different routes depending on whichever route is least congested.

The conventional model of the IP network shown in Fig. 4.5 assumes numbering starting from the bottom and counting up. Layer 1 is the Physical layer and Layer 7 is the application layer [7]. All of these layers are described for wired networks. Wireless networks do not follow the OSI layers but do interface to wired networks only at specific layers, as will be explained further in Chap. 11. They tend to address specific shortfalls of IP network such as transition delay, IP address header is not sent over the air, etc.

The lower three layers are based on hardware and in the middle is the transport layer which is responsible for end-to-end transport. It has an equivalence in wireless that is usually referred to as microwave link. The upper three layers are software layers for which the protocol, frames, and interactions are well defined. A brief description of each layer is in order at this point.

Physical Layer (Layer 1): This layer is responsible for physical connection to devices. It provides synchronization

to the network using clock signals. There are four functions of the physical layer: (a) Bit synchronization, (b) Bit control, (c) Physical topology, and (d) Transmission mode. In wireless, the IEEE 802.11 WiFi and non-mobility systems follow this model—cellular and other mobility systems do not, since they have to address mobility at the physical layer.

Data Link Layer (Layer 2): This layer is responsible for delivery of message from one node to the next node. This task is accomplished using two sub layers—Logical Link Layer (LLC) and the Media Access Control (MAC). There are five functions of the Data link layer: [a] Framing, [b] Physical addressing, [c] Error control, [d] flow control, and [e] Access Control. Cellular network uses these functions, but they address these functions very differently given the nature of "multiple access" used in wireless as opposed to "multiplex" in wired networks. Ethernet has become the common interface to connect wireless access points and base stations to the wired IP network. This commonly agreed interface provides the freedom to choose separate paths that are taken by wired and wireless technologies [8]. This may change in future as IP network imposes undue limitations on cellular systems, in terms of transition delay.

Network Layer (Layer 3): Network layer deals with transmission of messages from one host to the next. It is responsible for finding an optimum route to its destination. This involves addressing scheme where sender's and receiver's IP addresses are placed in a header in the message. It works well with wired networks and helps find an optimal route each time. The functions of this layer are (a) Routing and (b) Logical addressing. Cellular systems find this concept of IP address in the message wasteful for spectrum over the air interface. Cellular systems do not send IP address over the air. They use a different scheme to send messages.

Transport Layer (Layer 4): This layer acts as the interface between software-based layers and hardware-based layers. It receives data from either side, performs segmentation, and implements flow and error control before forwarding it to the other end. It adds source and destination port numbers and forwards it to network layer. In effect, it provides services to the application layer while receiving services from the network layer. Sometimes, it is known as the IP layer. It is important to note that IP network is a connection-less network. It makes a sincere attempt to deliver data but does not guarantee delivery.

Session Layer (Layer 5): Sessions layer coordinate to set up a session between two devices. Functions of this layer involve setting up, coordination (how long), and termination of the session between applications at the end of session.

Presentation Layer (Layer 6): This layer is responsible for presenting data translating it from application format to the network format. This layer "presents" data to the application format. For example, it is responsible for encryption and decryption of data in a secure session.

Applications Layer (Layer 7): This is the layer that users see most often. It represents data as seen by the end user in applications such as web browsers, email such as outlook or other Microsoft office products, for example. It is independent of technology (wired or wireless), but uses a software platform such as Android, iPhone, Windows, etc., that support applications.

The following section briefly reviews the progress of technology used by backhaul networks.

4.3 Progress of Technology in Backhaul

IP network started with data cables which are more efficient than conventional twisted pair of copper wires. Commonly used data cable is "Ethernet cable" or generally referred to in the industry as "Category cable". Currently, Category 6 (CAT 6) is a standard widely used. These cables are carefully designed and have been progressing in each generation with higher throughput and better noise immunity. Yet, copper used in such cables is susceptible to electrical noise and signal attenuation with distance thereby imposing a limitation on distance. CAT is limited to a distance of 100 m. Although slightly longer distances are supported, the throughput reduces, and error rates increase due to interference.

Fiber-optic cable on the other hand, has enormous advantage over copper. Light traveling through the glass tube or fiber, is immune to noise. Also, attenuation in fiber cables is significantly lower than in copper cable. Therefore, signals can be transmitted over longer distances without the need to amplify signals. Distances of hundreds of kilometers are possible, even without amplification, although the cost of laying fiber over extended distances can become prohibitive. Data throughput is the major strength of fiber-optic cable. Current technology supports throughput of several tera bits (1×10^{12} bits = 1 Tera bit) sent over a single pair of fiber-optic cable.

Fiber-optic cable is the highway of networking world, used everywhere, independent of the type of application. Applications that need large throughput such as video, demand higher data rates. In recent years, CCTV camera monitoring by public safety teams, Cable TV, video streams from a studio or outdoor/indoor video captured by a cell phone, need to be sent over great distances. This prompted wide-spread use of fiber-optic cable. Conversion of electrical signal to optical signal is necessary, using either an LED or Laser. The electrical signal is sent to such an optical device and the logic is straight forward. In E-O (Electrical to Optical) conversion, absence of light = 0, and flash of light = 1. This is achieved by optical transmitters that can trigger flash of light at very high rates of 25 Gb/s (1×10^9 = 1 Gb/s).

Optical signals may occupy different bands of light. In glass optical fibers, the signal attenuation is lowest in the infrared wavelength band. This necessitates use of light in the infrared region for telecommunication applications. Attenuation of signal within glass optical fiber is caused by two factors: scattering and absorption, both of these are carefully addressed during fiber glass design.

How did the fiber-optic cable and associated technology come into the forefront of backhaul? Note that fiber itself does not have many limitations, hence the end devices can support higher throughputs. DWDM (Dense Wave Division Multiplexer) is a technology that is bit rate independent. Also, it can carry data in IP, ATM, SONET, SDH, and Ethernet formats [9]. Optical transmission became the integrating link that provided a gradual transition from earlier sunset technologies such as ATM, SONET, and SDH into modern IP and newer gigabit and multi-gigabit links (irrespective of whether such links are made up of fiber or wireless links).

A DWDM system consists of five components (shown below). They work together to form an all optical technology and form the basic structure of DWDM technology, which serves as transport (main backhaul) an efficient optical system including "switching in the optical domain":

(a) Optical Transmitters/Receivers,
(b) DWDM Multiplex/De-Multiplex Filters,
(c) Optical Add/Drop Multiplexers (OADMs),
(d) Optical Amplifiers, and
(e) Transponders (Wavelength Converters).

The concept of using DWDM as a unifying optical link is shown in Fig. 4.6, where DWDM is shown as a protocol and bit rate-independent optical communication link. It includes all five functional components indicated above [8]. In Fig. 4.6, on the left is the transponder system. It is considered an open system since it accepts input in the form of either a standard single-mode or multimode laser pulse. The input can be from different types of physical media using different protocols and traffic types. Using the optical-electrical-optical (O-E-O) converter, wavelength of the transponder input signal is mapped to a DWDM wavelength [9]. This standardizes the signal to DWDM format. These DWDM wavelengths from the transponder are then multiplexed with signals from the direct interface, thus forming a composite optical signal which is then launched into the fiber, shown as a pipe on the right of the multiplexer.

This composite optical signal is then sent to a booster amplifier (amplifier made up of Erbium as dopant for the semiconductor) which amplifies or boosts optical signal as it leaves the multiplexer. Since the amplified signal can travel over long distances, it may be required to add or drop some of the signals on the way. An OADM (Optical Add/Drop Multiplexer) is used to drop and add bit-streams of a specific wavelength at a location on the way. Depending on the distance, additional optical amplifiers may be used along the fiber span (in-line amplifier) as shown. A pre-amplifier is used to amplify the DWDM signals before it enters the de-multiplexer [9].

The signal, having traveled over long distance (hundreds or even thousands of kilometers) is now de-multiplexed into individual DWDM wavelengths, as shown (right). Individual DWDM lambdas or wavelengths are either mapped to the required output type through the transponder or they are passed directly to client-side equipment. Such DWDM links

could be thought of as a highway with multiple lanes (or wavelengths) with exits that are add/drop where traffic either leaves or joins the highway. Amplifiers could be thought of as fuel stations that allow spent fuel to be replenished.

Table 4.1 shows the optical bands used by the fiber-optic community and standards such as the ITU (International Telecommunication Union) provide details in terms of both wavelength and bandwidth. For example, ITU-T G.694.1 provides the following guidelines:

- C-Band λ: 1530–1565 nm
- max. 360 channels (12,5 GHz Grid)
- L-Band λ: 1565–1625 nm
- max. 560 channels (12,5 GHz Grid).

Many optical bands indicated in the table are in use today. Equipment to handle these bands are widely deployed in networks. Other than the ITU recommended bands of C and L, other bands can be used. Any non-standard wavelength used, gets translated into the DWDM standard signal using O-E-O shown in Fig. 4.6. This follows the concept of open system where any optical signal can be converted to the DWDM standard signal.

4.4   Network Interfaces and Protocols

Figure 4.7 provides a practical view of how an IP network is deployed, tested, and used every day.

Protocol Layers

TCP/IP family is the practical equivalent of traditional OSI indicated in Fig. 4.5. It has equivalence in terms of protocol

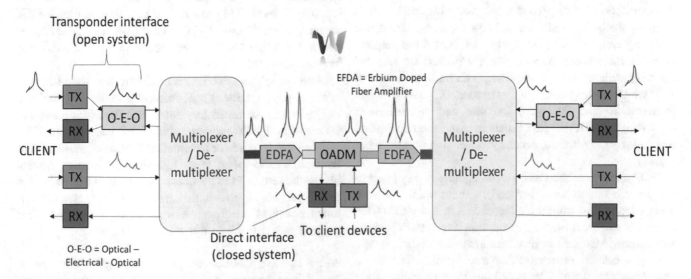

Fig. 4.6 DWDM components work together as an all optical technology

Table 4.1 DWDM optical bands
—spectrum

Optical band	Wavelengths (nm)
O (Original) band	1260–1360
E (Extended) band	1360–1460
S (Short) band	1460–1530
C (Conventional) band	1530–1565
L (Long) band	1565–1625
U (Ultra long) band	1625–1675

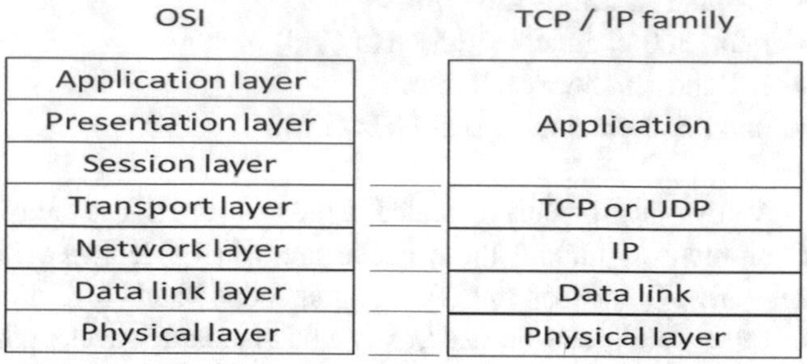

OSI	TCP / IP family
Application layer	
Presentation layer	Application
Session layer	
Transport layer	TCP or UDP
Network layer	IP
Data link layer	Data link
Physical layer	Physical layer

Fig. 4.7 TDP and UDP protocols—comparison with OSI model

—for example the transport layer where either TCP (Transport Control Protocol) or UDP (User Datagram Protocol) are used. The network layer translates to IP (Internet Protocol). All the software layers (session, presentation, and application) are grouped together as application layer protocols which are handled in conjunction with the websites using "http" which is a "stateless" protocol for session layer. Both TCP and UDP protocols are used to send packets over the Internet based on IP network protocol, but there are differences [10].

TCP is a connection-oriented protocol—it expects a physical connection is established between the two points before transmitting packets. By the same token, it expects the connection to be closed after transmitting data. It is reliable since it needs an acknowledgement from the receiving end for each packet received. It also has an error checking mechanism to make sure that packets are received correctly without error. Sequencing is an inherent feature of TCP which means packets using headers of variable length (between 20 and 80 bytes) are sent and received in a sequential order. It will resend data in case the receiving end reports errors. Therefore, accuracy of sending packets takes priority.

UDP on the one hand relies on datagrams as they come in and transfers it to the other end, without waiting for an acknowledgement from the other end. It is faster than TCP and saves resources on the transmission since error checking and acknowledgement mechanisms are minimal. It uses only a basic checksum error checking mechanism. UDP uses a fixed length header of 8 bytes and there is no retransmission

of bytes. Due to its dynamic nature, UDP supports broadcasting and is simpler than TCP. It has the ability to send more data due to its simplified mechanism. Due to its simplicity, UDP is commonly preferred in wireless networks.

Software protocol layers

Sessions layer protocol

Earlier, the session management using "http" was briefly touched upon as a "stateless protocol". Stateless means there is no built-in standard to keep record of inter-related requests. It uses client/server architecture with TCP as the transmission protocol and has two primary methods to achieve tracking across request during sessions.

(a) Request parameters
(b) Use Cookies

Request parameters: The http process uses tokens to request parameters. A token represents current state in a multiple step process, or it can identify a user. The token can be stored in server on the web page in a form field; this field can be submitted automatically each time user performs an action. In order to perform such action, the GET command of TCP is used. This token can be submitted as a GET request parameter or a POST request parameter. The request parameters using GET, are embedded in the URL and they are recorded in the browser history.

Alternatively, POST request submits the parameters as request body so it doesn't get embedded in the URL and will not be in the browser history either. For submitting sensitive information POST requests should be used, while GET requests can be used for information that is not sensitive. For example, if the token is an identifier for a user, it must always be sent in POST request. Non-sensitive examples of tokens include state identifiers, referrer values, etc. There is another way of sending identifiers in GET requests. That method uses path name instead of parameters.

Using cookies: Cookies are name-value pairs that can be stored in a browser and resubmitted whenever that user makes subsequent requests. A cookie is submitted using cookie header. This process may be familiar to most users that use the web, particularly those who use a PC or laptop or note pad, since the website would typically ask the user if cookie can be used. The cookie header sends name-value pairs separated by semicolons. The set-cookie header also contains additional directives and parameters related to the cookies. Such information helps the browser to understand how and when to submit such details.

The most common parameters are domain, path, and when it expires. The directives are "secure" and "http only". Domain parameter specifies the domain for which the cookie is valid and can be used; the cookie will also be valid for all its subdomains. The path parameter specifies the URL path. Expires is self-explanatory in the sense it is valid for that session. The "secure" directive instructs the browser to send the cookie over HTTPS only, while "http only" instructs browser to not let the website JavaScript to access the cookie. This is done to prevent exploitation of XSS (Cross-Site Scripting) that steals cookies in case the website is vulnerable to XSS. Industry provides "best practice" as security guidelines to designers who develop session related software.

Presentation layer protocol

Before sending any message, the characters, numbers should be changed into bit streams. Presentation layer addresses this task with three functions [11]. Because of these functions, the presentation layer is also called the syntax layer or semantics layer.

1. Translation,
2. Encryption,
3. Compression.

The translation function refers to the translation from the computer format to the format required by the network. This will mean conversion of an EBCDIC-coded text computer file to an ASCII-coded file. In order to carry this format in a secure manner, it uses encryption at the transmitting end and decryption at the receiving end. For example, when a user logs on to a bank account websites, the presentation layer will decrypt the data as it is received at the bank site. In order to use the network link efficiently, it compresses the data at the transmitting end and decompresses at the receiving end. Such compression plays an important role in multi-media message that may involve audio, video, pictures along with text. The presentation layer is divided into two sub-layers: (a) CASE (Common Application Service Element) and (b) SASE (Specific Application Service Element).

Quite often those who develop codes—freely work between this layer and the application layer, since both are close to how and where applications are developed.

4.5   Application Layer Protocol

Although this may seem like a misnomer, applications themselves are not handled in this layer. However, all inputs and outputs directly from the end user are handled in this layer. Commonly known protocols at this layer are Telnet (Telecommunication Network protocol), FTP (File Transfer Protocol), SMTP (Simple Mail Transfer Protocol), X window which is a graphical user interface. SNMP (Simple Network Management Protocol) widely used for monitoring equipment and devices, DHCP (Dynamic Host Configuration Protocol) which give IP addresses to hosts. At this layer the number of applications seem to grow quite regularly.

Suffice to say that in the expanding universe of global connectivity, protocols at all levels are very effective and important. There are multiple standards to support them across interfaces.

4.6   Ethernet Packets and Protocols

Ethernet is perhaps the most popular interface to LAN (Local Area Network) that allows high-performance networking among computer and other communication equipment. This has now spread to a variety of wired and wireless devices allowing all of them to interact with one another [12]. Originally proposed and adopted as an IEEE 802.3 standard, its use in industry and in residences is widespread. Every device that has WiFi or works with WiFi typically has an Ethernet connector. In households today, the Ethernet connector has been typically replaced the telephone jack. Since Ethernet is all about LAN, we will review Ethernet in greater detail in Chap. 12 that exclusively deals with wireless LAN. But the reason why it is mentioned here is because it has become the main interface that brings wired and wireless networks together.

Ethernet quickly replaced traditional backhaul interfaces such as T1/E1, OC-3, OC-48, and other telephony interfaces. The IETF standard (Internet Engineering Task Force) provides test RFC 2544 to measure throughput, latency, and other parameters between connected Ethernet ports [13]. All the telephony interfaces were based on the connection-oriented network, while Ethernet provides equivalent in connectionless packet network. Terms such as 10baseT, 100baseT, and 1000baseT and now 10000baseT have all become common to both wired and wireless networks, but are related to "connectionless packet network".

4.7 Backhaul and Fronthaul—Differences

Till now there was a fair level of discussion on backhaul. But what is "fronthaul"? It is a data network (not RF) that is within the access part of the wireless system. This may seem strange, but the reason for its introduction is because RF cable is very lossy. Anything to minimize its use, results in better design. The large increase in data rates even within the access network resulted in thinking about using fiber right at the cell tower. The logic is that once the RF communication of transmitter and receiver is completed it results in data but it is in baseband. This Baseband Unit (BBU) sits right on the tower and from the BBU "fronthaul" begins. In essence, fronthaul is the connection between a new network architecture of centralized baseband controllers and remote standalone radio heads or Remote Radio Head (RRH) at cell sites [14]. Why is this necessary and what does it do for cellular networks?

Recently there is a surge of wireless network connections leading to large number of devices getting connected by cellular systems. Throughput streams from many devices need connections with considerable data throughput. The concept is that just the radio head which serves such devices over the air, would stay in the RF domain keeping coaxial cable to the minimum to reduce signal loss. Once the signal is brought into baseband, it is all data. Recall that baseband and BBU (Baseband Unit) was discussed in Chap. 2, Sect. 2.5 d as part of cellular system. Since fiber optic network handles data efficiently (with minimum loss, capable of handling large data throughput), it is better to connect the BBU directly to a fiber optic cables, even though it does not conform to Ethernet format.

Once this data is received at the central office of the service provider, large data highway can ride on DWDM as described in the previous section. Such a concept is shown in Fig. 4.8 where only the radio head (remote radio head) resides with the antenna. Rest of the connections directly come from the radio tower but are fiber optic links. This architecture is generally known as C-RAN or Cloud based Radio Access Network, where only the RRH remains in the radio domain, on the tower. The BBU with its distribution units connects to the RRH. The other side of BBU towards central office is known as "fronthaul" which is a fiber optic network but does not conform to Ethernet protocol. Data it carries is mainly in the cellular protocol domain. It is an important feature of FTTA "Fiber-To-The-Antenna" Architecture now followed since 4G, and beyond now in 5G also.

If both the fronthaul and back haul use fiber, what is the difference? The difference is in the data format carried within the fiber. Data in the fronthaul is in the baseband format that does not conform to the OSI layers. It is closely linked to the

radio and mobility—therefore baseband is defined for each generation by the cellular standard 3GPP and its entire architecture and protocol may change with each generation of cellular technology. For example, the 4th generation of cellular 4G LTE, uses baseband processor with FFT (Fast Fourier Transforms). The powerful processors of 4G baseband are an elaborate implementation of the FFT process described in Chap. 2 and shown in Fig. 2.4 where digital modulation OFDM was briefly described. Currently 4G LTE uses OFDM as its main modulation scheme.

In simpler terms, the BBU is the equivalent of IF (Intermediate Frequency) function used in the earlier analog wireless systems. Baseband unit represents the equivalent of the analog IF amplifier stage but far more complex. Due to digital modulation, digital signal processors (DSP) perform such functions as modulation and demodulation sending/receiving a data stream, indicated as the chip rate of the system. The chip rate represents the combined output/input of all the radio networks it is connected to, as well as error correction and symbols added to support mobility. An example of baseband function is 4G LTE is indicated in [15]. Such processing units represent the BBU function shown in Fig. 4.9 and executed as part of the SIMT (Single Instruction Multi-Tasking) architecture.

An important feature of this BBU processor architecture is to issue only one instruction for each clock cycle but allow several operations to occur in parallel during that clock cycle that may extend beyond that clock cycle. Complex functions such as MIMO (Multi Input Multi Output) detection is an important task accomplished by such BBU as orchestrated by the host processor shown as connections on the right side of Fig. 4.9. It also functions as a Turbo Coder offering higher throughput within the same channel using efficient coding techniques. Such coders incorporate FEC (Forward Error Correction) and Convolution Codes for error correction needed for wireless channels discussed earlier in Chap. 3, Sects. 3.3.1 to 3.3.4. Various modulation techniques used in 4G LTE are implemented by vector load/store units that perform complex mathematical operations on the chip network [14]. The digital front end takes care of analog inputs (such as voice or other analog signals), converting them to standard digital formats using analog to digital converter and vice versa, as shown on the left side of Fig. 4.9. Note that traffic from the BBU, known as front-haul, can be carried using 60 or 80 GHz wireless millimetric wave links which have sufficient capacity to handle data up to 10Gbps, but they have distance limitation (typically in the 3–5 km range). Such wireless links are discussed in detail in Chap. 14. For longer distances fronthaul would be carried using fiber optic cable. Backhaul with DWDM could handle 100 Gbps or even 1000 Gbps (1 Tbps in the near future).

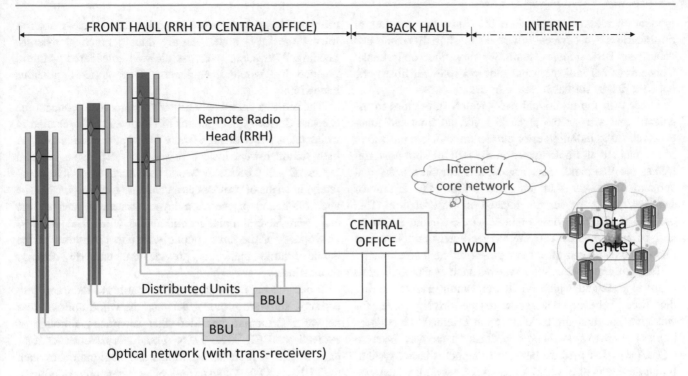

Fig. 4.8 Fronthaul and back haul for modern wireless network (IoT and 5G)

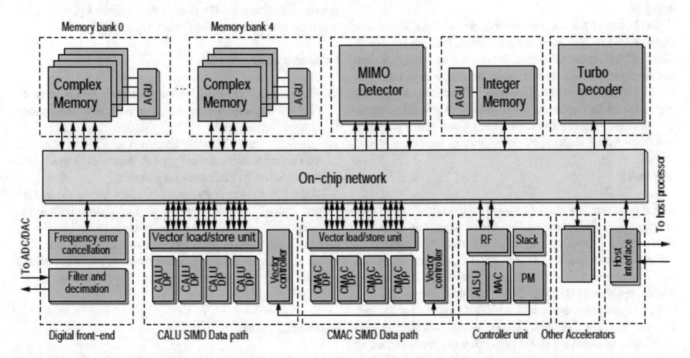

Fig. 4.9 Baseband processor architecture overview

4.8 Conclusion

In this chapter, the backhaul network that evolved with the rest of the telephony and cellular network was reviewed. Changes in the backhaul network and its format has influenced wireless network. Often wireless networks demand functions far beyond wired networks, both in terms of how data is handled, and support needed for mobility. Mobility and such complex function implementation led to the birth of fronthaul whose function is very different from back haul. Fronthaul and the role it will play a major role in modern wireless networks like 5G since introduction of millimetric wave devices will result in short RF cable and data from BBU moving entirely using fiber optic cable. Concepts of 6G indicate it has multiple roles for fiber optic network due to further increase in data.

Other than the traditional radio which stays close to the antenna and acts on the physical layer, all front-end functions related to mobility, error correction, efficient utilization of channel are all implemented in the BBU. Using powerful DSPs, the baseband units accomplish complex tasks that demand functions that are not within the realm of switches/routers or servers located in base station. This provides a different view of how wireless radio functions will be implemented and the very efficient and low loss fiber-optic cable can offer new, power-efficient solutions.

Earlier during 3G cellular systems such as CDMA had to adapt to backhaul which could not handle packets during that time. Wireless systems were the driving force, for changing backhaul from "connection oriented" to connectionless networks to handle packets. This was because CDMA was designed exclusively as a packet-based system. Initial deployment of CDMA used X.25 network which was the only packet network at that time. That trend has now changed.

Evolution of the connectionless network into IP network and the OSI model suggest alternatives, that are more efficient and the option of fiber network carrying data that does not conform to the Ethernet protocol, opens new possibilities. Now other complex functions of wireless are getting implemented in fronthaul, which was briefly touched upon. This indicates changes that are driving the entire backhaul and fronthaul networks towards supporting wireless systems.

Summary

While backhaul is a part not visible to the end user, it plays an important role in moving data across the globe. Economy of the entire human society is now linked with communication networks touching every aspect of society. Evolution of the backhaul started with analog telephony, followed by digital telephony (ISDN). With CDMA during 3G and computer networks move towards IP, it further evolved into a connectionless network. Some of the advantages and limitations of such packetized network were discussed along with the arrival of IP network (Internet Protocol), which is the mainstay of connectionless, packetized network.

Internet which initially consisted of only wired network, systematically evolved towards wireless and mobility, leading the way. Current network operators are well aware of the IP network and its evolution. But changes occurring in backhaul due to wireless was the focus of later part of this chapter. These changes are in part due to wireless not following the OSI model, mainly due to practical reasons. Sending IP headers over the air was considered wasteful practice by wireless engineers since it uses precious bandwidth.

Therefore, a completely different approach to connect the wireless device (smart phone) has evolved. This resulted in fronthaul, a major change where fiber optic cable is used on base station towers instead of the lossy RF cable. Several layers of the wireless network continued to evolve separately, in terms of protocol and hardware functions. Only the application and presentation layers, generally seen by the end user, have remained common to wired and wireless networks—for the smart phone with either Android or the Apple iPhone platforms provide the base to develop applications.

It became important to provide a hardware interface with common software protocol between the wired and wireless networks. Ethernet provided that interface—although it evolved as IEEE 802.3 LAN (Local Area Network) standard, it quickly replaced traditional backhaul interfaces such as T1/E1, OC-3, OC-48, and other telephony interfaces, allowing both copper and fiber interfaces, based on that standard. But now fronthaul uses very different protocol, but it can be carried using the fiber-optic network.

Emphasis on participation

1. Where is the backhaul for your college campus located? Find out what type of backhaul is used, is it fiber, cable, or some other type?
2. Where are the cell towers located nearby? Find out from the service provider, what type of backhaul is used from that location to their switching center.
3. TCP/IP is normally available in many laboratories. Can you "ping" known website or file server within the campus? What data rates do you see?
4. Does your campus have microwave links? Find out from the telecom supervisor whether there are plans to connect buildings with microwave links. If yes, where, If not why?

Knowledge gained

1. Important changes to backhaul network—starting with copper twisted pair, to ISDN and now on to IP networks.
2. OSI model and its importance in IP networks—common interfaces and layers in that model.
3. Importance of various layers. In particular, the application and presentation layers that are seen in every email message, social website interfaces, and messages

4. Significance of Ethernet interfaces—its widespread use in WiFi and base stations
5. Concept of fronthaul—differences between fronthaul and backhaul.
6. Baseband and significance in becoming the backbone of digital wireless systems

Homework problems

1. Research online or in the lab—find out differences between CAT 5e and CAT6 cables. What improvements were made in CAT 6 to support higher throughput rates?
2. Are there ISDN telephones in your home/college? Where are they located and what makes them useful though IP network is deployed?
3. Is DWDM deployed in your city? Find out from service providers where and what are the services it carries?
4. In cell towers that you can see, where is the RF front end and where is the baseband unit located? Is it split with RF unit on the tower or are both in a cabinet below? Draw a basic layout of units serving the cell.
5. POTS phones are still in use in many locations (home, office). Find out if it is truly served by twisted copper pair and POTS telephone line. [Hint: Is there a network adapter Terminal or NT before telephone wire connects to your phone? If so, what does it do?].
6. Every ISDN phone has a data port in the back. Is this jack being used to connect a PC/laptop as it was originally designed? Why are PC/laptops are wired directly to a separate Ethernet connector?

Project ideas

1. In your house/room, is there an Ethernet connection? Is it possible to re-wire an Ethernet 1000baseT connector to fiber media converters? With that conversion, does the Internet speed increase? Use RFC 2544 tester to measure speeds and delay between two Ethernet ports?
2. Using TCP/IP connection from the lab is it possible to make voice calls to any phone within campus? If not, what are the additional resources needed to make calls to VoIP phone?

Out-of-box thinking

1. Select fiber-optic link and between two known points. Measure the link using RFC2544 and record throughput, delay, and other parameters. Set up Millimeter wave link (80 GHz point-to-point link) and make the same measurements. What are the key differences?

2. Are there any OC-3 links in your area? How would you convert them into ethernet circuit?

References

1. Backhaul basics, a definition, network experts define backhaul networks. RCR Wireless News. https://www.rcrwireless.com/20140513/network-infrastructure/backhaul-network-definitions-cellular-backhaul-definition. Last Accessed 13 May 2014
2. Johnson C (2018) What is POTS Plain Old Telephone Service Line and Network explained. Nextiva blog. https://www.nextiva.com/blog/what-is-pots.html. Last Accessed 15 Oct 2018
3. What is ISDN, integrated services digital network? Electronic notes. https://www.electronics-notes.com/articles/connectivity/isdn/what-is-isdn.php
4. Shepler J, Why use DS3? Comparing T1, T3, OC3 & DS3 bandwidth, plus carrier ethernet. https://www.t1rex.com/ds3.html
5. Mitchell B (2019) A guide to X.25 in computer networking. Lifewire. https://www.lifewire.com/x-25-816286. Last Accessed 09 July 2019
6. Dombrowski G, Grise D (2000) ATM and SONET basics. APDG (ACM Digital Library). https://dl.acm.org/citation.cfm?id=556134
7. Shaw K (2018) Networking world, the OSI model explained, how to understand (and remember) the 7 layer network model. https://www.networkworld.com/article/3239677/the-osi-model-explained-how-to-understand-and-remember-the-7-layer-network-model.html. Last Accessed 22 Oct 2018
8. Byoung-Jo KIM, Henry J, Paul S (2012) Directions for future cellular mobile network architecture. https://journals.uic.edu/ojs/index.php/fm/article/view/4204/3372. Last Accessed 21 Sept 2019. ISSN 13960466. https://doi.org/10.5210/fm.v17i12.4204. (First Monday, [S.l Nov. 2012])
9. An overview of DWDM technology and DWDM system components, community home, by FS.com. https://community.fs.com/blog/an-overview-of-dwdm-technology-and-dwdm-system-components.html. Last Accessed 9 Apr 2015
10. Differences between TCP and UDP, geeks for geeks, computer science portal for geeks. https://www.geeksforgeeks.org/differences-between-tcp-and-udp/
11. Chapter 4, OSI model and network protocols, layer—6, the presentation layer. https://access.itxlearning.com/data/cmdata/NETPLUSN10004/Books/ec2_netplus004c04.pdf
12. Live action, ethernet packets and protocols. https://www.liveaction.com/docs/glossary/ethernet-ieee-802-3/packets-and-protocols/
13. RFC 2544 based benchmarking tests overview—tech library. https://www.juniper.net/documentation/en_US/junos/topics/concept/services-rpm-rfc2544-benchmarking-test-overview.html
14. Jonestown Jones S (2014) What the [bleep] is front haul? Light reading magazine. https://www.lightreading.com/what-the-bleep-is-fronthaul/a/d-id/707868. Last Accessed 14 Feb 2014
15. Wu D, Eilert J, Asghar R, Liu D, Nilsson A, Tell E, Alfredsson E (2009) System architecture for 3GPP LTE modem using a programmable baseband processor. Department of Electrical Engineering, Linköping University, Sweden, pp 132–137. https://www.da.isy.liu.se-pubs-diwu. https://doi.org/10.1109/SOCC.2009.5335662

Bluetooth — Personal area standard originally proposed by Ericsson of Sweden, but now part of a global standard, with ever increasing number of devices using this technology. Generally limited to about 20 m distance, but with MESH technology can extend its range considerably

3GPP — Third-Generation Partnership Project started as an international body during the third generation of cellular service. It governs all global standards of cellular service, such as 4G LTE and now working on releases of 5G LTE standard. Their standards are backward compatible

3GPP2 — Third-Generation Partnership Project was started to address the North American IS-41-based standard with definitions from IETF (Internet Engineering Task Force). This body completed its work and closed towards the end of third-generation cellular. No longer in active use

RAN — Radio Access Network, a subgroup of 3GPP standard that addresses the physical layer (air link) including spectrum aspects, and the MAC layer (Base Band) of cellular standards

SA — Services and System Aspects, a subgroup of 3GPP standard that addresses aspects such as SIM card, Codec, interfaces to different wired backhaul networks, etc.

CT — Core network and Terminals, a subgroup of 3GPP standard that addresses core network including its gateways and switch. It also addresses all aspects of terminals and user equipment that connect to the cellular network, their features and support aspects including operation and maintenance

5G NR — Fifth-generation cellular, New Radio initiative. Addresses different aspects such as multiple beam, MIMO and millimetric-wave band design and deployments

DSL — Digital Subscriber Loop. A twisted pair used to carry voice and data traffic from homes/offices in less populated areas. It is a modified version of POTS to accommodate data and voice using filters

POTS — Plain Old Telephone Service, that uses a twisted pair typically carrying voice traffic

AP — Access Point that serves as the main unit to provide WiFi in homes/offices. All WiFi devices connect to this unit and in turn AP connects to the Internet or to a LAN (Local Area Network)

Cable TV — Coaxial cable that primarily carries TV signal to homes. But it can carry traffic to/from the Internet using splitters to separate the traffic from TV channels

ISP — Internet Service Provider, typically a corporation that allows customers to connect to the Internet. ISP may use a variety of methods such as fiber-optic cable, Cable TV, DSL, or Satellite link. In recent times, it can be a cellular carrier providing 4G or 5G modem to connect customer premises to the Internet

SSID — Subscriber Set Identifier, this identifier is typically broadcast from an AP and identifies the network to which the Internet will be able to deliver or receiver traffic from. Smart phones/laptops use this SSID to identify WiFi network in the surrounding area

UWB — Ultra Wide band uses at least 500 MHz bandwidth and uses an impulse instead of continuous transmission. This allows it to support low-throughput applications such as passenger ID, location of sensors, and other applications, using lower energy but able to reach farther, by impulse transmission

K. Raghunandan, *Introduction to Wireless Communications and Networks*,
Textbooks in Telecommunication Engineering, https://doi.org/10.1007/978-3-030-92188-0_5

EMS	Emergency Management Service, typically ambulance and services related to emergency
ZigBee	A technology used for monitoring devices and equipment and also use controls such as on/off
OFDMA	Orthogonal Frequency Division Multiple Access, a method that uses both frequency and time domain signals to allocate resources to customer, using Resource Blocks (RB)
MIMO	Multiple Input Multiple Output, a scheme that uses multiple signal paths between transmit and receive antennas. Multiple paths are created by using two or more antennas in transmit and receive
Gbps	Giga Bits Per Second, very high data rates of 10^9 bits per second. Currently used in backhaul and gateways, these are data rates for bulk traffic, from a tower, it may become available to individual home users with the implementation of 5G in the millimeter-wave band
MCS	Modulation Control Scheme, a method by which receiver use higher modulation based on higher signal received. MCS also has a scheme to use multiple data streams using MIMO for higher throughput
HDTV	High Definition Television, a method of using very large number of pixels with video processing to bring vivid details of face and scenes that were not possible earlier
PTT	Push To Talk, is a method by which the phone user pushes a button and talks. It is usually simplex—user can either talk or listen, but not both at the same time. Commonly used by public safety bodies
TETRA	Terrestrial Trunked Radio, is a standard used by most public safety agencies globally. It supports voice and limited data on PTT. Based on TDMA technology, it is closely linked to GSM in terms of design

5.1 Introduction

Most wireless systems are based on standards that are globally applicable. Why is this different from wired telecommunications? An important reason is that a smart phone user is likely to move from one country to another but uses the same phone. This was neither possible nor thought of during the wired network era. Even telephone connectors were different in different countries and adapters were often used to connect laptop computers to the Internet. For international calls, only the main national gateway had to provide interfaces. Early during 2G standards development, wireless providers and product vendors realized that, to support users, it is essential to offer a common global standard. In addition, wireless also has the unique situation where the national licenses of frequency spectrum must be based on local geography.

Such licenses can only be issued and controlled by local regulators, since they pertain to use of frequency spectrum within their own regional air space. The frequency spectrum is regulated by the maximum power that can be emitted and therefore controls the distance up to which the service extends. They gain further importance in terms of how and where they are used (indoor or outdoor, for example). This is important in all major cities where cell sites serve users who are outdoor, while in-building services are different from those outside.

In this chapter, we will learn about standards that govern a wide range of wireless systems.

Allocation of frequency bands by regulators, depends on other factors—user base, terrain, and how the nation operates their network. We will use specific examples, followed by general guidelines, since it is not practical to cover each nation. An important concern for wireless is interference from nearby systems and how they are controlled. Control is achieved partly by regulating power control, but other techniques can be applied using algorithms.

5.2 Wireless Standards—3GPP, IEEE, Bluetooth, and Others

The process of "development to deployment" shown in Fig. 5.1 is common to all standards [1].

Wireless standards are in a unique situation since this technology involves changes at rapid pace (one generation for each decade). Although the general procedure for development of wireless standard follows the methodology shown in Fig. 5.1 a scheme to follow rapid pace is needed. Different approaches are needed to accommodate such rapid changes. In addition to the cellular service available to public, there are wireless technologies used by public safety and governmental agencies. These are important and will be discussed in this chapter. Finally, the next major leap of wireless is into IoT and its impact on devices, society, and standards, which will be discussed towards the end of this chapter.

This process of development and deployment has major bearing on wireless since user devices are developed in different parts of the globe, but they are deployed globally [1]. Therefore, a common body that operates independent of governments and regions is essential. Some of those bodies such as 3GPP, the IEEE, and Bluetooth forum etc., will be discussed in this chapter.

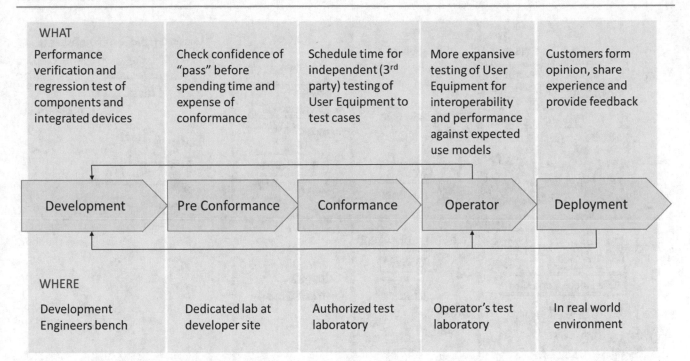

Fig. 5.1 Typical flow from development to deployment

5.2.1   3GPP (Third-Generation Partnership Project)—The Mobile Broadband Standard

What is 3GPP? Different standards address varying aspects of wireless systems. Among them an important standard is the 3GPP. It evolved from the concept of IMT2000 which projected throughputs of over 1Mpbs over the air with initiatives to make cellular the primary access to people of the world (around the year 2000). 3GPP found that traditional approach of standards was too slow and would not keep up with the pace of changes in wireless. Therefore, it formed a partnership of various institutions seeking help to write specifications to all cellular systems globally. This body started during third generation of cellular system, hence the name 3GPP. Its primary mission is to bring cellular technology across the globe.

Consequently, it continued work through fourth generation and is currently working on the fifth generation (5G) of cellular systems, which was initially released in 2018, but later versions continue to get released in the next few years. 3GPP is made up of volunteers from vendors of base stations, cell phones, and application providers. They are known as partners, since 3GPP is a partnership project. It is based on the concept of commonality of interfaces across products from a range of companies that not only build products but also service providers, applications developers of various systems, etc. Its standards are available to all partners. Earlier versions of their standard are freely available on their website. It is possible to get not only 4G

specifications but also those for 3G and 2G. Moving forward, early 6G concepts of what they could be, are also freely available on their website (www.3gpp.org). 3GPP has three major technical areas:

1. Radio Access Networks (RAN),
2. Services & Systems Aspects (SA),
3. Core Network & Terminals (CT).

5.2.1.1   RAN (Radio Access Networks)

This group focuses on layer 1 (physical—air interface) and layer 2 (known as MAC—Medium Access Control layer, relates mainly to baseband and radio protocols). RAN specifications dictate how base stations must be designed and built. "Radio Access" relates to how the smart phone communicates "over the air" with base stations and in turn how the base station communicates with the public network (including the Internet). It is the key group that decides and specifies how much throughput is possible over the air, which radio frequency bands can be used, what radio access techniques are used, etc. Relating to earlier description in Chaps. 2 and 3, mitigation techniques brought about by DSP, digital modulation methods, as well as access protocols based on the types of access (both modulation types and multiple access techniques) are part of this group's work. Details of mobility, how hand off takes place, methods, and what measurements decide hand-off, are all part of the RAN specification. Figure 5.2 shows evolution of cellular technology with focus on RAN. The upper part refers to 3GPP standard, while the lower (below dotted line) refer to the

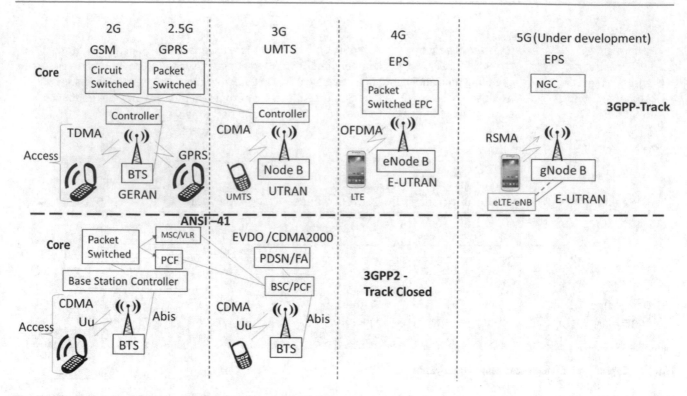

Fig. 5.2 Evolution of cellular network from 3GPP perspective

North American 3GPP2 standard. Since wireless technology brought major changes in each generation, the RAN specification from one generation to the next looks quite different.

Since 3GPP must provide backward compatibility, RAN provides methods to fall back. For example, in an area if there is no 4G service, the fall back would be to 3G service. See Fig. 5.2 for details. What type of network does 3G provide? Depending on service in that region, it could be North American EVDO (Evolution Data Optimized) which is part of CDMA technology, is shown in the bottom portion of figure, or it could be the 3GPP defined as UMTS network.

The fallback mechanism is initiated so that the smart phone seeks 4G cellular service first and if not available changes over to the appropriate 3G network. Currently, the LTE network known as 4G is operational in many parts of the world. There are some areas where the 3G network and in some cases even the 2G (GSM) network, operates. In Fig. 5.2, note that not only the access technology changed from 2G TDMA to 3G CDMA to 4G OFDMA but also other network features changed as well. In 4G, the concept of Radio Network Controller (RNC) was completely eliminated, and the RAN was evolved as a simple network made up of multiple base stations. In 5G, the access itself will become resource block-related (conceived in 3D view as frequency domain Vs Time domain Vs resource blocks) so that RAN addresses user needs in terms of resource block, by adjusting the frequency and time domains dynamically. This shows the pace of evolution that cellular industry

accommodates. Each generation of cellular is developed and deployed every decade and this trend has continued since early 1990s. In many ways, it makes consumers happy since they see an explosion of services with applications that were never known earlier. At the same time, service providers work very hard to accommodate the elderly—for them a simple cell phone with basic features are still essential.

It is important to observe definition of mobility reviewed in Sect. 3.2. To revisit that thought using Fig. 5.2, the bottom part indicates evolution of North American ANSI-41 which conformed to the wired Internet and borrowed definitions from IETF (Internet Engineering Task Force). Some important units such as Foreign Agent (FA) for roaming are examples of such definitions. That project based on ANSI-41 was known as 3GPP2, but it closed at the end of 3G. Why did a standard that started out with packet network and CDMA technology stop? Part of the reason is, to unify world standards it was essential to choose only one global standard. The 3GPP believed in "*true Internet must be based on Mobility*" and therefore laid out its own definitions, instead of borrowing IETF definitions (such as Mobile IP or FA). It ultimately won the mobility argument and now in 4G only the 3GPP standards exist as global standard. They do interface to the current Internet but have become a strong advocates to modify the current Internet towards mobility.

Figure 5.3 shows the access model that allows a variety of standards that can access the 3GPP cellular network indicated as Evolved Packet Core (EPC). Many of the other

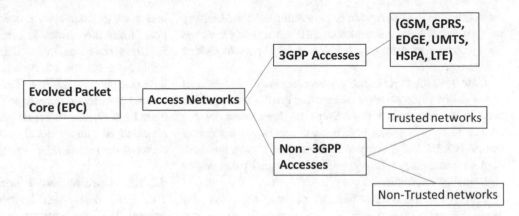

Fig. 5.3 Access model of 3GPP to show access for devices from different standards

standards need access to 3GPP (such as EVDO from 3GPP2, or IEEE 802.11 WiFi, etc.), that are based on the current Internet standard (IETF). These networks are classified as "non-3GPP" in which there are two subcategories—the "trusted" usually refers well-known standards (such as 3GPP2 which is cellular), while others that may be evolving are classified under "non-trusted". The "nontrusted" networks (such as 802.11 which is not cellular) would require more security cross-checks or fire walls before they can access the 3GPP network. The main reason for such differentiation is due to security algorithms in cellular being significantly better than 802.11, for example.

5.2.1.2 Services and System Aspects (SA)

The second group within 3GPP is SA that is responsible for definition, specification, and evolution of system architecture allowing "networks" that are trusted and non-trusted to join the 3GPP networks. It also assigns functions to subsystems throughout the network. Examples of such sub-systems include UTRAN (Universal Terrestrial RAN), SIM (Subscriber Identity Module), and CODEC (Coder and Decoders for speech, audio, video, and multimedia). The terminal commonly known as cell phone or smart phone and how it interacts with the network is also of interest to this group. A very important function of this group is security and developing cellular security which is quite different from WiFi security or IP security. This groups also provides a framework for services, for cellular, fixed, and cordless applications.

Part of this function includes responsibility to ensure an open architecture to allow IP multimedia Subsystem (IMS). Ancillary functions such as billing, network management are also part of their responsibility. The group acts as overall coordinator of all other working groups since systems and services encompass all aspects of the work done by 3GPP. To gain insight into working of this group, some examples would be helpful. The concept of SIM (Subscribe Information Module) or now the USIM (Universal SIM) is unique to the cell phone. It contains information related to

what the user expects from the service provider, but equally important is that it shall not contain any personal information. Any SIM must work anywhere in the world, irrespective of who the service provider is. Also, several security aspects such as what if it is stolen, what happens if someone tries to duplicate and many other scenarios are addressed by this group.

CODEC—in telephony, speech had remained central to the entire network until the late 1980s. The concept of data networks to interwork with computers existed, but combining the two, particularly using telephone network, evolved gradually. Consequently, in the cellular network data appeared in a short form during second generation. Later in 3G systems, data was considered the main stay with speech taking on a secondary role. In 4G and later systems, speech has taken a low priority. Yet it is important for callers in different parts of the world to call a cell phone. It is essential to support such calls no matter where it originated from—whether from a POTS phone or ISDN telephone, a modern VoIP phone or even a microphone attached to a PC.

Similarly, audio systems have evolved on their own with emphasis on music and orchestra and higher quality of sound. To support high-quality audio, it is important to provide higher fidelity coders that may need considerable bandwidth. In Sect. 3.3, additional spectrum allocation towards FEC (Forward Error Correction) was discussed. Based on that concept, conversational audio may need only 25% additional bandwidth towards FEC, whereas high-quality music may require almost 40% additional bandwidth. This group in 3GPP is tasked with the concept of where and how such resources must be allocated to support audio of varying quality.

Video and multimedia services have evolved using MPEG and other standards. For example, in Sect. 1.4, we reviewed the MPEG-4 standard which is used in most cameras (including CCTV) that allows streams of high-resolution video. It is important to consider them in the context of viewing them on a cell phone, which has both limited screen space and battery life limitations. Note that the cell phone is also capable of generating such video using its own camera and microphone. Efficient use of air waves (or spectrum) to stream such video is an important task of this group.

How does the humble cell phone compare to professional high-quality video cameras or superior quality microphones used in concert halls today? Some of these questions are addressed by this group before some compromise solution can evolve. Globally, everyone carries a cell phone, but most do not own or carry a heavy, professional quality camera, or high-quality microphone.

Security of the cellular network plays a major role, far more than any other network. The same cell phone can operate in any country, any region, irrespective of social, economic, and other factors. In many countries, cell phone happens to be the only method of financial transactions. Its security aspects take on additional importance due to such factors. In Sects. 1.1 and in 3.3, we briefly reviewed some security algorithms. Functions f8 and f9 are based on the security aspects discussed earlier in Sect. 3.3 and now relate to work performed by this group. Figure 5.4 shows the security process used by 3GPP to pick such cypher code functions, for each generation of cellular standard. What percentage of the total project time was spent on each major activity is shown in red blocks.

During planning phase for each generation, 3GPP conducts competition inviting mathematicians familiar with cypher codes to compete. The process indicated in Fig. 5.4 was used for 3G where researchers from three different universities competed. The winning code was then picked by 3GPP and further work, based on this winning Kasumi code was done by this subgroup to design the f8 and f9 functions. Note that such process repeats for each generation and just because basic details are made available in the standards website, it does not mean the one can hack into this system. There are several layers of security in addition to f8 and f9

that every cellular system uses based on detailed specifications from this group [2]. Encryption and authentication functions are so extensive that the same cell phone user is unlikely to get the same code (key) a second time during their cell phone's lifetime. Thanks to unique features of this layered security, it is perhaps the only network that hackers have been unable to crack into, so far. This topic will be reviewed in greater detail in Chap. 6 where security and personal devices will be reviewed.

5.2.1.3 Core Network and Terminals (CT)
This group deals with the two extremities of the cellular system. The core network that only the service provider touches and deploys, and the terminals which every end user uses/recognizes such as cell phone/smart phone. How these two units work together to connect everyone around the world—is the underlying task of this group. Its specifications relate to areas such as Call Control, Session Management, and Mobility Management. Traditional telecommunication network functions, such as signaling between the core network nodes, interconnection with external networks, O&M (Operation and Maintenance) requirements are also part of their specifications work.

User equipment (Terminal) includes cell phones, modems, and a variety of devices that directly access cellular network. Other than radio access, features of these devices pertaining to layer 3 and other services are the responsibility of this group. Similarly, SIM/USIM and how it interfaces to other parts of the device are specified by this group. For example, part of the new USIM card is the UICC platform (Universal IC card) that provides interfaces to a variety of services on a cell phone. Figure 5.5 shows details where applications that the user could download from a rail network operator, a commuter bus service, a bank or an employer ID, etc., would all be connected at the smart phone using the UICC. The advantage of this platform is that in

Fig. 5.4 Cellular security process for each generation of cellular

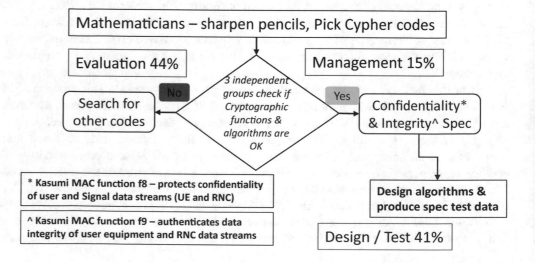

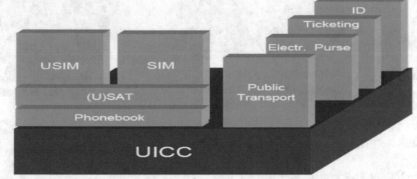

Fig. 5.5 Typical application platform on a smart phone and its architecture

terms of security, each is independent. For example, if your commuter bus ticket application were hacked at that provider's website, it would not affect any of the other applications on the cell phone. The USIM supplier provides firewalls between these applications.

Such arrangements are important in modern world where attacks on individual provider websites are quite likely. The cell phone specifications provide a safety net to protect the user community, in terms of safeguarding their identity and other critical applications.

This platform shown in Fig. 5.5 is part of the work by ETSI (European Telecom Standards Institute) whose Technical Committee proposed the Smart Card Platform (SCP) for all smart phones. How this platform interfaces with the cell phone is the responsibility of 3GPP CT subgroup. Such interfaces are quite complex and extensive in the modern cell phone.

In practical terms, the Universal SIM card (USIM) is now offered in three different sizes: Standard, Micro, and Nano—how these are interfaced within a cell phone or smaller device such as smart watch or miniature wearables used by athletes or other devices and how they in turn work with the cellular network, are specified. The number and variety of devices that use cellular network is growing by leaps and bounds and keep this subgroup of 3GPP very busy.

Also, the core network interfaces to the Internet via firewalls. How these are controlled, session protocols, translations, and many functions such as text messaging,

content sharing, how the applications are executed on a terminal/user device are thought out in detail and specified. Recently multi-mode terminals have come into greater use [3]. For example, the same smart phone today can work on cellular 3G network in rural areas would work on 4G in urban environment. Depending on availability of service, it can normally work in 4G but can switch back to 3G if that service is not available. Such multimode terminals with different technologies (CDMA of 3G or OFDM of 4G) need signaling protocols that are quite different between the two networks. Such details of how the terminal (smart phone) must work in such situations are specified by this group. Note that the same phone is capable of also working on WiFi which is based on IEEE 802.11 standards [4]. But that takes an independent path to the Internet, it may not pass through a base station, details such as these will be discussed in Chaps. 11 and 12.

5.2.1.4 The 5G—NR Initiative

Current activity for 3GPP is the NR (New Radio) for fifth generation of cellular. It started service with Release 16 (in 2019). There are several changes—the first change is inclusion of higher frequency bands, including the millimeter-wave bands of up to 40 GHz and even higher bands going up to 86 GHz. Why make a drastic change when "Laws of Physics" indicate that path loss over the air increases logarithmically with distance and frequency as shown in Eq. 5.1:

Fig. 5.6 Innovative use of
Millimeter waves and
Massive MIMO for 5G cellular

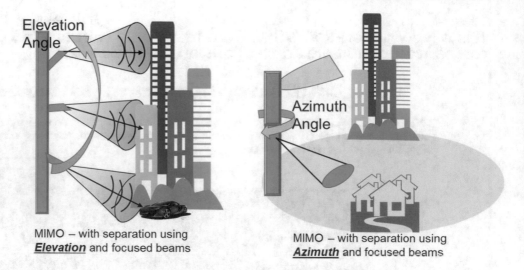

MIMO – with separation using
Elevation and focused beams

MIMO – with separation using
Azimuth and focused beams

$$Pathloss = 20\log_{10}\frac{4\pi fR}{c} \qquad (5.1)$$

where f—frequency and R is the range. Increase in frequency increases loss logarithmically. For a given path loss, the only way to limit loss over the air would be to reduce R —the range. On the other hand, higher frequency offers considerably more bandwidth and higher throughput. The compromise in 5G is going to be offer dramatically higher throughputs from a high post/power line pole directly beaming into the building or an apartment complex, reducing the range R, as shown in Fig. 5.6. It means, most neighborhoods will see a 5G tower in their vicinity, since it will have very limited range and limited power too.

The second change is extending the concept of MIMO (Multiple Input Multiple Output) to provide many streams of data in three dimensions as shown in Fig. 5.6.

The third change is use of beam forming. With multiple beams from the same radio tower in different directions, it would be possible to serve different floors of an apartment (shown on the right in Fig. 5.6). This makes it attractive—since many frequency bands can be re-used in different beams in different azimuth or elevation allowing the service provider to aggregate traffic from many clients around the radio tower. Yet each user gets very high throughput (different beams mean—no interference within the same frequency channel). It is important to stress that beam formation and steering is accomplished by DSP, there is no physical movement of antenna. Figure 5.6 illustrates combination of all these concepts to provide "massive MIMO (Multi Input Multi Output)", for details see [5].

Massive MIMO—separation of signals using focused beams allow better use of concepts such as "massive MIMO". Three axis separation involves different elevation and azimuth angles at each radio site. Another dimension is the distance between sites (typically less than ¼ of mile or

less than a km in urban canyons). Since millimeter waves, cannot travel far due to severe attenuation in the atmosphere, this provides a natural isolation, reducing interference. Does this mean increase in radiation? No, that is not true, this aspect is discussed further in detail, in Chap. 7. Power transmitted from base station is also quite low (few hundred mW). This is discussed again in Chap. 19, showing its complexity in terms of measuring and deployment.

5.2.2 IEEE (Institute of Electrical and Electronic Engineers) Standards

This is world's largest professional body of electrical, electronic, and computer engineers (with over 450,000 members globally). Most universities have student bodies affiliated to this body. One portion of this body works on "standards" making contributions in a variety of related fields. Some examples of the standards include electrical, computers, signals, biomedical, and many fields. Originally formed by merger of Institute of radio engineers (IRE started in 1912) with several related associate bodies, its Founding members from IRE were pioneers of wireless communications. The IEEE has many related societies such as Antennas and Propagation, EMC (Electro-Magnetic Compatibility), Vehicular Technology (VT), Computer Society, Communication Society and many more. Some of the standards of IEEE relate to measurement methods for RF radiation (C95.1 RF safety levels), what is considered safe for human and other living beings. The 802.11 family of standards popularly known as WiFi, and the 802.16 standard known as WiMax, as well as many of the wired standards such as 802.3 Ethernet and 802.3af Power over Ethernet are widely deployed. These are listed here but reviewed in greater detail in Chap. 12, where LAN and PAN are discussed.

The 802.11 is part of family of 22 standards called 802 Local Area Network (LAN)/Metro Area Network (MAN). This implies its use over short distances up to 100 m for wireless applications. Other widely known standards in this 802 family are:

802.1 Bridging—covers LAN/MAN bridging and management. Covers management and the lower sub-layers of OSI Layer 2.
802.2 Logic Link Control commonly referred to as LLC or Logical Link Control specification.
802.3 Ethernet—the most widely used standard across the entire IP network. It is today synonymous with Layer 2 interfaces, irrespective of wired or wireless.

802.4 Token bus—abandoned.
802.5 Token Ring—originally meant for the telephone twisted pair, it also serves fiber up to data rates of 100 Mbps.
802.6 Through 802.10—these were either withdrawn or superseded by other standards.
802.11 Wireless LAN (WiFi) Media Access Control and Physical layer. This standard has multiple versions over the years. Starting with a, b, g, n, and ac progressively moving towards higher data rates. The most popular band is 2.4 GHz that offers a total of 83 MHz bandwidth, that supports all PCs, smart phones, gadgets, wearables, and a variety of devices under the generic term "ISM—Industrial, Scientific and Medical" band. There are quite a few other bands in the 5 GHz spectrum, the most common among them is the 5.8 GHz band with 165 MHz bandwidth, available in many smart phones and devices.
802.12 Demand Priority—to increase Ethernet to 100Mbps.
802.13 Not used.
802.14 Cable modems—Withdrawn.
802.15 Personal Area Networks.
802.15.1 Bluetooth.
802.15.3 UWB or Ultra-Wide Band.
802.15.4 ZigBee—short-range sensor networks.
802.15.5 Mesh Network.
802.16 Wireless Metro area networks—fixed backhaul links known as WiMax.
802.17 Resilient Packet Ring.
802.18 Radio Regulatory TAG.
802.20 Mobile broadband wireless access—LAN standard to support mobility, never deployed.
802.21 Media Independent handoff—working group to develop handoff between 802 heterogeneous and cellular networks.
802.22 Wireless Regional Area Network.

5.2.3 The IEEE 802.11 Standard

Originally started in 1997, this standard specifically provides services like wireless LAN which extends wired LAN by about 100 m over the air. The standard deals with only the first two layers, the Physical and MAC layer, which are implemented within the AP (Access Point) [6].

All the higher layers from layer 3 to layer 7 are identical to wired LAN and not covered by this standard (including security), they are covered by IETF. One exception is 802.11i for security which has additional security features such as encryption of the air link. Figure 5.7 shows the basic concept of the IEEE 802.11 standard implemented in an AP which regularly sends out a beacon that consists of parameters listed in the figure. The STA (Subscriber Terminal/Appliance) is the wireless device such as a smart phone or a WiFi device which first sends a request to the AP to associate itself with that AP. In turn, the AP responds allowing the device (STA) to associate with the Access Point. This transaction typically needs a password and once installed in the device, is used for authorization.

The back panel of AP typically has an Ethernet connector. Using an Ethernet cable (Cat 6), the Access Point (AP) connects to some service (cable or fiber) that carries its traffic from home or office to the Internet. That service could be via fiber distribution panel (FDP), cable modem, or cellular modem or DSL (digital Subscriber Loop) which the Internet service provider or LAN provider offers to its customers. This is shown by a yellow bidirectional arrow. This is how smart phone, a laptop with WiFi capability, or any WiFi device connects to AP, and in turn connects to the Internet. The coverage of AP is shown by the circle. It allows movement of user devices within that area. The AP can transmit about 200mW (the same power as cell phone). Since it operates at higher frequencies than cellular, with limited power, but unlicensed band of common channels in 2.4 or 5.8 GHz, its coverage distance is typically limited to about 100 m.

Solid bidirectional arrow is what the user can choose from different service providers often known as Internet Service Provider (ISP). Options are fiber, DSL, Cable TV, wireless link (such as 4G LTE), or IEEE 802.16-WiMax link or in remote areas, a satellite link.

Whichever connection to Internet the customer chooses, appropriate modem is provided by the ISP. This link sets limits of throughput to the Internet—WiFi speeds are always limited to within the circle in the figure. It is this ISP link that 5G standards (from 3GPP) are trying to address since it impacts customer experience in terms of throughput (Internet speed). It also affects capacity for business users. Higher

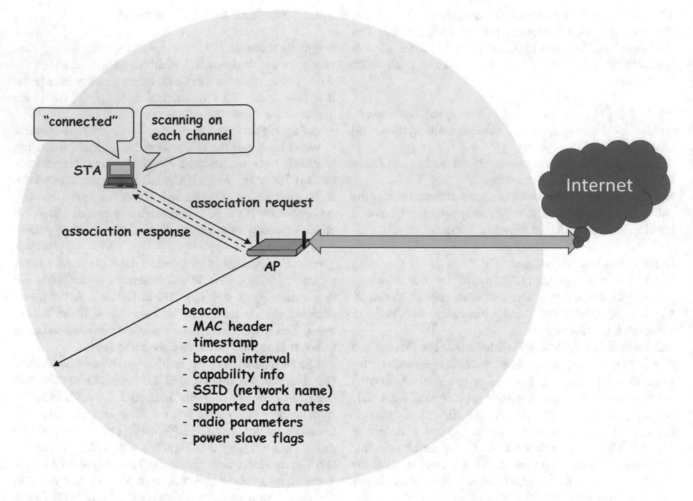

Fig. 5.7 Basic features of WiFi (IEEE 802.11 Standard)

the throughput, better are transaction speeds for all users in office/apartment. But cellular links will offer better security and lower latency.

5.2.4 Personal Area Networks IEEE 802.15 Standards

IEEE uses the term "personal area" relates to proximity of the network devices to the user. There are multiple of these personal area standards. Most common among then are Bluetooth that any cell phone user may be familiar with. It is in the ear bud and microphone with a mini battery and takes the cell phone conversation to ear and mouth—the distance in most cases is no more than 30 m. Its importance lies is use of small batteries with long life. There are many other standards for personal area such as ZigBee, UWB, Z-Wave —some of these are described now.

5.2.5 Bluetooth Standard

One of the IEEE standards 802.15.1 known as Bluetooth, makes wireless gadgets much easier and safer to use. During the last decade, a separate standard organization known as Bluetooth forum was formed, to promote and enhance the number of applications for Bluetooth. That forum exclusively handles all recent standards of Bluetooth technology, and discussed further in Chap. 12.

In simple terms, Bluetooth eliminates wires over short distances of 30 m or less, by providing a wireless connection. This could be a simple connection from audio amplifier to loudspeaker or connecting a PC to a printer. Bluetooth devices are battery powered and use short pulses to make it last several years. By carefully avoiding interference using algorithms. Bluetooth devices use the 2.4 GHz band globally unlicensed to allow wireless link instead of wires over short distances. Since blue tooth uses short pulses, even the

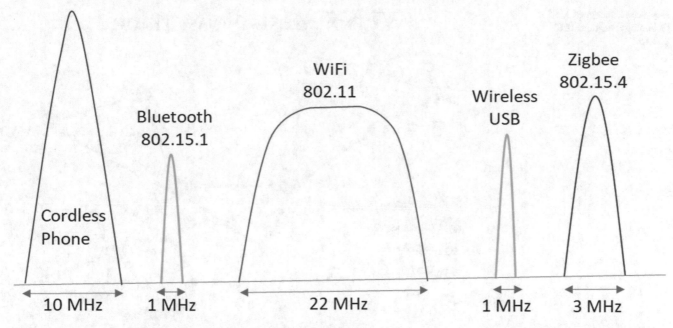

Fig. 5.8 Some of the technologies using the 2.4 GHz unlicensed band

small battery lasts many years. Power transmitted is no more than 4 or 5 mW (less than what a flashlight bulb uses). Personal area standards utilize the 2.4 GHz band for a variety of applications. This band is popular since it is unlicensed globally. Some of the technologies in 2.4 GHz unlicensed band are shown in Fig. 5.8, which are quite popular at home and in offices—observe that each standard uses different spectral bandwidth and protocol. They do not interfere with each other using clever access methods. How and why these access methods are used - is discussed further in Chap. 12.

5.2.6 UWB or Ultra-Wide Band

Designated as the IEEE 802.15.4a Standard, this technology holds promise for sensors—particularly indoor sensors providing accuracy in the range of centimeters [6]. Major difference between conventional radio and UWB is that conventional systems transmit information by varying the power level, frequency, and/or phase of a sinusoidal wave. UWB on the other hand transmits information by generating radio energy only at specific time intervals (short pulse). This allows UWB signal to occupy a large bandwidth, thus enabling pulse-position or time modulation. It supports low-throughput applications such as automatic processing of passengers who hold monthly pass or valid ID in corporate buildings, etc. Unlike WiFi or Bluetooth, UWB uses the 3.1–10.6 GHz band with a very wide bandwidth of 500 MHz, that allows it to use an impulse signal and accurately determine the location.

5.2.7 ZigBee or Low-Throughput Data Standard

The IEEE 802.15.4 ZigBee standard, on the other hand looks for a low-throughput application for sensors that need to provide very little information (such as On/Off position, basic ID in terms of text, or device health checks such as temperature, relative humidity, or battery status, etc). This information may be needed only one or two times a day, but from multiple devices, spread over an area. This is where ZigBee brings simple devices that may be spread over a wide area. Applications could also include, turning off some of the lights in public areas, late at night when traffic is less, etc. Although it is part of the low power family of standards, using Mesh network ZigBee devices can send/collect information over several hundred meters. Due to lower throughput rate (30 Kbps typical), devices can operate in low-frequency bands like unlicensed 900 MHz to reach farther. However, ZigBee normally operates in the popular 2.4 GHz band [7].

5.2.8 MESH Network Standard

Although MESH network is synonymous with WiFi, it is not restricted to any specific technology. MESH is an independent standard that allows the concept of using a device to extend its range.

It uses the physical layer (layer 1) and the MAC layer (layer 2) that are typically unique to each type of technology. Therefore, this standard provides details how to use MESH for different technologies such as WiFi, ZigBee, wireless optical

Fig. 5.9 Concept of MESH
technology standard IEEE
802.15.5

WPAN Mesh Networking

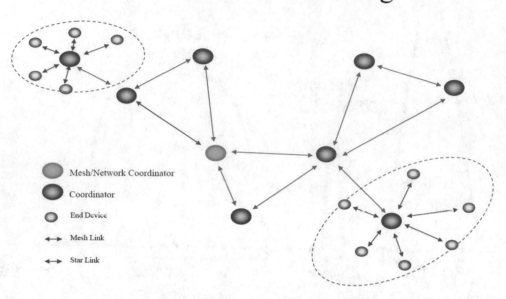

short range, etc. The concept of MESH is shown in Fig. 5.9 where the end devices can be further away from the original coordinator (such as Access Point). One of the nodes (such as an Access Point) becomes a coordinator and allows the end devices to connect to another network, where a second device acts as a coordinator. The task of a coordinator is to allow "hops" so that an end device can communicate with another end device indirectly through "hops" using coordinator devices. There are some limitations—as the number of "hops" increase the transition delay increases as well. But this concept of "ad-hoc" network in the field such as setting up a temporary network at any given location and automatically allow the devices to communicate, is the major feature of any MESH network. Since MESH is not technology specific, interconnection of peripherals such as Bluetooth is equally possible. In the IEEE 802.15.5 MESH Standard, different subgroups address applications such as low rate, high rate, low power short range, etc. [8]. Although MESH discussed here applies to WLAN such as WiFi, it is not limited to any specific technology—cellular system uses them too.

It is important to note that all IEEE standards are readily available to the public via the "IEEE get program", allowing institutions, or practicing individuals. Everyone can download these standards for reference and use. These technologies will be discussed further in Chap. 12.

5.2.9 RF Radiation—Protection Against Adverse Health Effect in Humans

Perhaps an IEEE standard that has far reaching consequence to the health of human beings in terms of exposure to RF radiation, comes from IEEE C95.1–2010. "It was established to protect against adverse health effects in human beings". It is important that every student becomes aware of this standard which contains 1143 literature references, and a 10-page long bibliography referencing 144 documents of work performed by physicists and engineers worldwide, spanning well over three decades. Although the first version of this standard was published in 1999, it has compiled and studied scientific work going all the way back to 1960s [9]. We will elaborate on this standard with more details in Chap. 7 on "public policy, health and safety" where studies by the medical community (WHO—World Health Organization) will also be covered.

This standard from IEEE has recommendations to protect against established adverse health effects in human beings associated with exposure to electric, magnetic, and electromagnetic fields in the frequency range of 3 kHz – 300 GHz. It is based on studies with results verified globally by engineers and doctors. There are many freelance writers who proclaim adverse effects and make comments about electromagnetic radiation and its effects—with no scientific basis or measurements! Effect of many such articles is felt at township meetings, resulting in denial of wireless services to citizens. It is important to note that every service provider adheres to well-established scientific standards such as this C95.1. In addition, regulatory bodies use discretion to provide and verify recommendations to protect human health.

Several national standards such as the FCC, ITU, and others have followed this standard for many years to ensure human society in their regions are limited to RF exposure that are known to be safe and well established based on scientific knowledge.

5.3 Allocation of Frequency Bands, Spectrum Licenses

The national bodies of each region/country have the important task of making sure that their citizens use frequency bands with minimum/no interference. It is also important for them to ensure the safety of all beings (human, animals, vegetation, and the environment). With these objectives, national bodies allocate, regulate, and monitor frequency bands.

Allocation of frequency bands is mostly based on the individual service and its needs. Television and radio, for example, need propagation over large distances. The user does not communicate to the transmitter. It is a broadcast service as reviewed in Sect. 3.1 (Fig. 3.4 showing Hybrid-cast service). Such services serve large areas with bands such as MF of 300 kHz–3 MHz (Medium Frequency), HF 3–30 MHz (High Frequency), VHF 30–300 MHz (Very High Frequency), and the lower UHF band 300–3000 MHz (Ultra High Frequency).

The most popular among frequency allocations is the cellular band auctioned at considerable cost to the service providers. These were limited to frequencies in the upper UHF (such as 400, 700, 800, and 1800, 1900 MHz) staying well below 3 GHz. There were obvious reasons—propagation distances are reasonable, better coverage through dwellings, sparely populated tree line, etc.

However, with the introduction of 5G service, this concept is changing, mainly due to considerable demand on the cellular bands. The increased demand is not only due to increase in human population but also penchant for higher throughput for services such as video streams. The result is a move towards SHF (Super High Frequency) bands moving up to 30 GHz and even into the millimeter-wave band (starting from 30 GHz all the way to 300 GHz) where the wavelength is all in millimeters. These bands provide considerable bandwidth, upwards of 1000 MHz, but signals can only travel short distances (less than 1 km). Due to severe propagation loss, use of focused beam, becomes a practical method of serving customers in such bands.

Other than these services, there are government agencies that get allocation of channels for important emergency (public safety such as police, fire, and EMS) and national defense. These channels must remain clear always, due to the nature of their services. These are separately allocated by the national agency (such as FCC) but maintained and monitored by the respective government agencies.

5.3.1 Need for License

Need for license arises due to multiple agencies or many individuals competing for spectrum. The legally authorized license therefore allows an individual or an agency to assert their right and file complaints of any interference. Such complaints are addressed by the national agency which allocates spectrum and typically this process works well and in favor of whoever holds the license. To keep this well regulated, the license has an expiry date by which the service must continue to use the band, renew its license, and maintain records.

5.4 Methods to Control Interference— Regulation

Why is interference a concern?

Interference makes communication susceptible to noise (in the normal case) and can even lead to complete blockage of communication. Therefore, communication channels must be kept clear of interference. In dense urban areas if this is not entirely possible, then there are other ways to mitigate and minimize interference.

The most common to control interference is based on distance. Since RF signal attenuates with distance, it is possible to keep space between transmitters thereby reduce interference. This must be carefully modeled and reviewed, since distance is only one of the factors. The second factor is height of the tower—higher the transmitter larger must be the spacing. The third factor is terrain—worst could be a flat terrain which includes arid land, large bodies of water. Quite often there could be mountains which help reduce the distance between transmitters. In general, the terrain in each area in specific and the model must account for the actual geographic properties.

A different method of interference control is through clever methods based on technology. Most common among them are "access methods" such as TDMA—Time Division Multiple Access, CDMA—Code Division Multiple Access and now OFDMA—Orthogonal Frequency Division Multiple Access (briefly referred in Chaps. 1 and 2). While these methods primarily aim to increase capacity (many users sharing a frequency channel), they also control interference. They will be discussed in detail in Chap. 11.

Methods using algorithms to test a channel and then use (when no other user is using it) have become very widely deployed in recent technologies such as WiFi, ZigBee, Bluetooth, and others. Why is this necessary? Part of the reason is that these technologies use "unlicensed" band. Therefore, every user can use this band for free. Methods to control interference must be "built-in" to the standard that gets implemented in products in the form of algorithms. It is also true of a cordless phone, and sensors based on similar technologies operating in the "unlicensed" band. Methods of interference control can vary but all of them follow the regulatory method of DFS or Dynamic Frequency Selection

—in cordless phones, blue tooth and ZigBee it could be frequency hopping; in WiFi it is "Listen before talk mechanism".

5.5 Standards Evolution to Accommodate New Technologies

Primary challenge to all standards bodies is the "galloping pace" of wireless technologies. The process of arriving at a common standard needs consensus and takes time. But the rapid pace of cellular evolution resulted in the concept of Partnership Project. This worked successfully and the current 3GPP (Third-Generation Partnership Project) has continued on to fifth generation maintaining close ties with industry partners.

The IEEE standards related to Local Area Network (LAN) created the 802.1x standards for wireless LAN, now popular as WiFi. There are other versions of this standard that will be reviewed in Chap. 12. The common base established by this group is the Modulation Coding Scheme (MCS), which provides basis for a variety of digital wireless technologies. Table 5.1 shows the current version of this scheme.

This scheme is followed in all digital wireless standards. Therefore, the concept of sensitivity, which used to be a measure of analog receiver performance, is now replaced by minimum power required to sustain a specific type of modulation. The conventional SNR or Signal to Noise Ratio also changes with signal levels. Earlier in analog system SNR = 12, was the normally required specification for voice communication, but in digital much lower SNR is acceptable as seen in the table. This is because lower data throughput (using BPSK, QPSK) at low signal levels is acceptable in many applications. While the scheme in Table 5.1 was developed for wireless LAN, it serves as the basis to operate other wireless systems such as microwave links. Notice that as the bandwidth increases, the received signal level required to support that modulation also increases. Therefore, to operate in higher modulation such as 256 QAM sustaining multiple spatial streams, very strong signal in the order of -48dBm is needed at the receiver. In cellular system, this would be possible only in the user is only within few hundred meters from cell site—quite often not practical. However, in microwave links, this may be possible using very narrow beams over short distances and using horizontal and vertical polarizations (may be 1–2 km links and not 10 or 50 km links which is normal for microwave links).

Other standard bodies such as the ATIS (USA), ETSI (Europe), ARIB (Japan), CCSA (China), TSDSI (India), and ITA (Korea) are partners in the 3GPP and continue the process of evolving standards. This partnership concept allows 3GPP to move forward rapidly with technology to provide specifications, while standard bodies of different countries take responsibility for implement the standards [10]. There are other standard bodies that 3GPP associates with—an example in Chap. 1 Sect. 1.5 alluded to JPEG, MPEG. While these standard bodies relate to photos and motion pictures where medium (wired or wireless) may not seem relevant; yet it makes sense for 3GPP to work with them, since smart phone has become the primary means for users to see photographs and motion pictures, as indicated in Table 5.1.

The smart phone and other wireless devices have become de-facto standards of generating pictures and videos from scene of incidence. They are also thrifty about usage of bandwidth, making these bodies (JPEG, MPEG) to think seriously about how their standards can be implemented in a better way to carry pictures and videos over wireless medium. Even a decade ago, JPEG and MPEG developed standards assuming only wired media, but it is no longer the primary use.

5.6 Separation of Commercial and Public Safety Users

Separation of commercial licenses (for profit) and public safety (government based, for public benefit, non-profit) are needed for many reasons. Among them are competition based on capacity and quality of service drives commercial wireless providers. Their technology evolution must necessarily keep pace with the market and often in the cellular industry this means replacement of entire networks every 10 years on an average.

Public safety and defense users are not keen on capacity, but on reliability of service, to deal with emergencies. Changing network and services need not keep pace with technology—interworking with existing network is essential to provide a seamless transition of users, retaining the same level of reliability, to handle emergency situations. Such aspects of public safety are discussed further in Chap. 15 in considerable detail.

Based on the above needs, each regional regulator provides auctions for the commercial service providers. While maintaining equal opportunity for all, the licensing process brings considerable revenue to the government. There is a

Table 5.1 Modulation control scheme used by wireless networks

802.11n + 802.11ac data rates and SNR requirements

HT MCS Index	Modulation	Coding	20MHz				40MHz			
			Data Rate (Mbps)		Min. SNR (dBm)	Receive Sensitivity (RSSI)	Data Rate (Mbps)		Min. SNR (dBm)	Receive Sensitivity (RSSI)
			GI = 800ns	GI = 400ns			GI = 800ns	GI = 400ns		
1 Spatial Stream										
0	BPSK	1/2	6.5	7.2	2	-82	13.5	15	5	-79
1	QPSK	1/2	13	14.4	5	-79	27	30	8	-76
2	QPSK	3/4	19.5	21.7	9	-77	40.5	45	12	-74
3	16-QAM	1/2	26	28.9	11	-74	54	60	14	-71
4	16-QAM	3/4	39	43.3	15	-70	81	90	18	-67
5	64-QAM	2/3	52	57.8	18	-66	108	120	21	-63
6	64-QAM	3/4	58.5	65	20	-65	121.5	135	23	-62
7	64-QAM	5/6	65	72.2	25	-64	135	150	28	-61
2 Spatial Streams										
8	BPSK	1/2	13	14.4	2	-82	27	30	5	-79
9	QPSK	1/2	26	28.9	5	-79	54	60	8	-76
10	QPSK	3/4	39	43.3	9	-77	81	90	12	-74
11	16-QAM	1/2	52	57.8	11	-74	108	120	14	-71
12	16-QAM	3/4	78	86.7	15	-70	162	180	18	-67
13	64-QAM	2/3	104	115.6	18	-66	216	240	21	-63
14	64-QAM	3/4	117	130.3	20	-65	243	270	23	-62
15	64-QAM	5/6	130	144.4	25	-64	270	300	28	-61
3 Spatial Streams										
16	BPSK	1/2	19.5	21.7	2	-82	40.5	45	5	-79
17	QPSK	1/2	39	43.3	5	-79	81	90	8	-76
18	QPSK	3/4	58.5	65	9	-77	121.5	135	12	-74
19	16-QAM	1/2	78	86.7	11	-74	162	180	14	-71
20	16-QAM	3/4	117	130	15	-70	243	270	18	-67
21	64-QAM	2/3	156	173.3	18	-66	324	360	21	-63
22	64-QAM	3/4	175.5	195	20	-65	364.5	405	23	-62
23	64-QAM	5/6	195	216.7	25	-64	405	450	28	-61
4 Spatial Streams										
24	BPSK	1/2	26	28.9	2	-82	54	60	5	-79
25	QPSK	1/2	52	57.8	5	-79	108	120	8	-76
26	QPSK	3/4	78	86.7	9	-77	162	180	12	-74
27	16-QAM	1/2	104	115.6	11	-74	216	240	14	-71
28	16-QAM	3/4	156	173.3	15	-70	324	360	18	-67
29	64-QAM	2/3	208	231.1	18	-66	432	480	21	-63
30	64-QAM	3/4	234	260	20	-65	486	540	23	-62
31	64-QAM	5/6	260	288.9	25	-64	540	600	28	-61

natural pressure to allocate more spectrum useful for commercial service. This everlasting thirst for capacity resulted in new thinking—moving towards higher frequency bands that offer considerable bandwidth resulting in multiple GBPS (Giga Bits Per Second) data rates. While that is attractive, there are many challenges to work in this frequency band. Commercial service providers are trying to use it over short distances, for applications such as provide the primary communication access to homes. It is expected that other applications such as smart driverless vehicles could benefit from these higher frequency bands—they are being actively pursued by the cellular service providers.

Related applications such as TV has also evolved. Since users in the younger generation don't use TV signals over the air, other package providers would like to offer services directly over the Internet links (could be 5G or other wired service). This move by video program industry for movies, sports and other channels makes a major impact of High-Definition TV (HDTV). It is important to observe that many of these are directly linking the smart phone and TV set at home, bypassing the conventional route of wired links to laptop or a desktop PC as the primary access to the Internet. Such changes provide both opportunity and challenge to the wireless networks.

Another major move is towards merger of cellular service bands into government agencies. Models such as these provide an opportunity to serve public service users, as if they are a separate corporate identity. These networks would be served much like private networks. PTT or Push-To-Talk phones used by public safety personnel now work on LTE 4G service. TETRA (Terrestrial Trunk Radio) the public safety standard, has laid out a clear path towards LTE. The advantage of this approach is that public safety users can retain other functions related to their own private network that is reliable yet keep pace with the latest versions of wireless standards.

5.7 Internet-of-Things (IoT) and Proliferation of Wireless

Small devices, sensors, and remote monitoring of temperature, pressure, voltage, air movement, and other parameters are important and do not take up much bandwidth. The very nature and proliferation of their use suggests that it is not practical to wire them into a network. At the same time, it is simpler to connect them through a wireless network. Examples could be wireless (WiFi-based) cameras at home that can be monitored from any part of the world where Internet access exists.

Other applications could be to turn on/off streetlights, household, or office lights, turn on equipment such as microwave oven or heaters from a remote location. This may not be new, since Lunar mission or Mars mission did accomplish such tasks on board a spacecraft several millions of kms (Kilometers) away. Yet on earth, there are remote areas that are touched by wireless signals and all that is needed is access them a few times a day.

Standards for such low-throughput, reliable service have evolved. For example, Z-Wave is widely used for home and small business automation [11]. It has over 300 product partner companies making them interoperate to support home security, health care, hospitality, and other markets. It is important to support IoT that uses environmentally conscious products such as solar panels, windmills, and others to provide green energy and to reduce carbon print. These devices

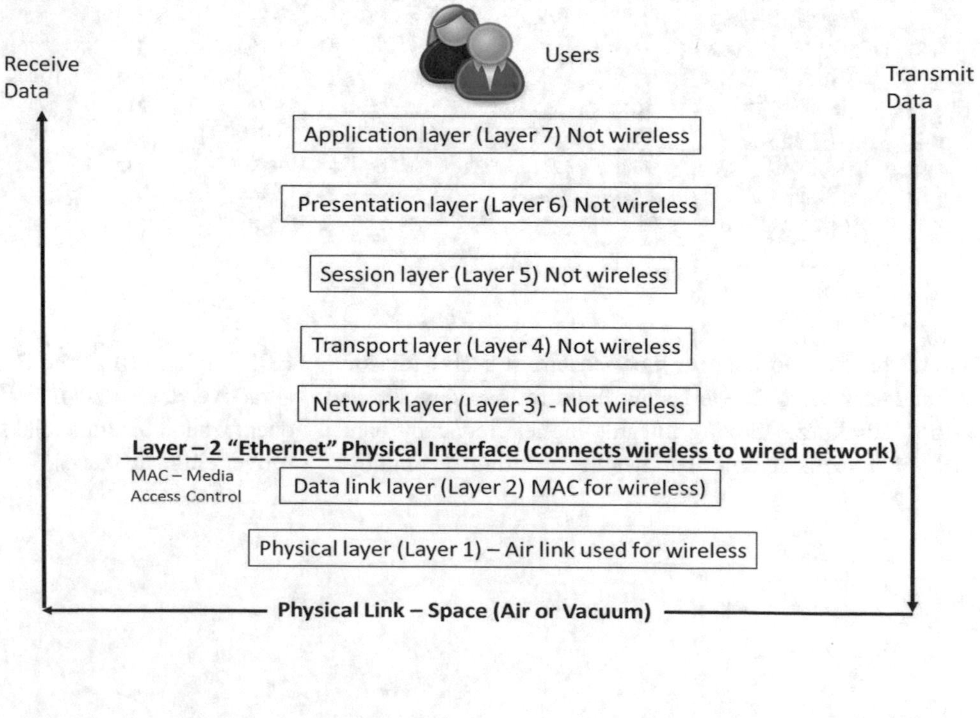

Fig. 5.10 The OSI model and its interpretation by wireless industry

require only periodic monitoring and control. Alternative power sources are ideal for supporting IoT devices that need very low power and often operate in wilderness.

Internet is used extensively by all wireless networks. It is better to reference generally used network model. Figure 5.10 shows the OSI (Open System Interconnection) model that wireless industry adapted to bring commonality between wired and wireless networks and where they are connected. All wireless standards discussed so far use this interpretation.

Layer 1—physical layer messages are different for each wireless standard, discussing power levels, handoff messages and aspects related to radio. Data link layer messages are interpreted by MAC (Media Access Control) layer protocols that relate to type of access method for radio, type of frame used and radio protocols. These protocols follow the design methodology implemented in that standard. An important interconnecting link between wired and wireless networks is the "Ethernet—IEEE 802.3 standard". Even between cellular and WiFi or any of the wireless standards this is the interconnection. Implementation of other layers could only vary slightly. For example, in cell phones iOS (Apple) uses a Human Interface implementation while Android follows material design. Such differences are minimal since users can get used to both. The OSI model continues to serve as a reference for networks, but implementation of the first two layers is considerably different in wireless. These two layers emphasize the Physics and Mathematical aspects of carrying data over the air interface (even if it were to travel through vacuum/outer space). Each wireless system uses air interface that is quite distinct, and they differ from each other. These will be described in Chaps. 11, 12, and 13.

5.8 Conclusion

This chapter focused on the need for standards and national licenses in the context of wireless communications. Various standard bodies and the work they do to make wireless systems and devices work, was elaborated. The importance of standards, their global participation to make wireless systems work anywhere is a point that cannot be overstated.

In their approach towards networks, how different parts of the system should operate, public concerns about safety measures towards RF radiation were all part of the discussion in this chapter. The dramatic growth pattern of wireless technology, measures taken by the standard bodies to not stifle but support that growth were shown using examples.

This approach by wireless standards and how it contrasts with wired networks, for example, were points made to assert the view that wireless has a bigger role to play in view

of devices and their proximity to people around the globe. The reader is encouraged to visit some of the 3GPP web videos indicated in the references to get a current understanding of technological growth [12].

Summary

Wireless standards have evolved from two major organizations—the 3GPP which is a partnership group that provides specifications allowing individual national standards to evolve their standards based on these specifications.

Second major body is part of globe's largest professional body known as the IEEE (Institute of Electrical and Electronic Engineers). It has variety of standards related to wireless, and the health and safety concerning RF radiation. IEEE has other standards—but they are not related to wireless. Standards that are related to LAN (Local Area Network) consists of 802.11 WiFi, 802.15 Personal Area Network including ZigBee, MESH, and a family of Personal area standards. The Metro Area Network standard 802.16 WiMax currently specifies access to rural areas. The IEEE 95.1C relates to health and safety of the human population, with focus on RF radiation. This is an area that is not understood by the public. Yet the IEEE standard is based on scientific study by both the engineering and medical community spanning over a period of several decades. Many national bodies base their regulatory requirements based on this important standard from IEEE.

The topic of wireless technology's progress and how it has been a challenge to bodies that make up standards was reviewed. Based on the growth of one generation of cellular each decade (started from 2G in 1990 and it is now evolving the fifth generation), the need to retain a separate partnership body was emphasized. A similar approach by Z-Wave was indicated.

Need for regulation and issue of license by national bodies was reviewed. It is based on technical aspects related to the nature of radio waves. In addition, constitutional requirements of every nation indicate the need to separate commercial and government (non-profit) organizations. Advantage and limitations of unlicensed band such as 2.4 GHz, was discussed. The word unlicensed was carefully reviewed to show that regulatory bodies provide guidelines and do have regulations such as limits on power and algorithms to suggest solutions to operate in unlicensed bands, with minimum/no interference.

Emphasis on participation

1. Can you reprogram a WiFi access point to operate in the 5.x GHz band and see how it performs in terms of throughput?

2. What is the maximum power level your WiFi can operate at? What about other devices operating in an unlicensed band? Can you place these devices near the front door of a microwave oven and measure its operation while the microwave oven is working?
3. Make throughput measurements on a MESH system and record end-to-end delay for 1 hop, 2 hop, and 3 hop MESH networks.
4. Visit 3GPP website and use of their specifications document to verify with the phone vendor some features on your smart phone. For example, look up your cell phone model and verify whose UICC card (or USIM card) is being used.
5. In 802.11n, there is a provision for MESH network using multiple antennas. Remove one antenna and measure the throughput. Conduct trial in different indoor/outdoor locations.
6. Use commercially available ultra-low-power Bluetooth module and tabulate distances. How far can the units be placed and still retain connection reliably.

Knowledge gained

1. Historic perspective of wireless standards and their growth in recent times. Conventional voice communication leading to all digital networks.
2. Standards related to wireless—specifications from 3GPP were described in detail.
3. Some of the unique security features in wireless—based on the nature of mobility of devices across the globe.
4. Cell towers in the neighborhood leading to public concerns. Standards of RF radiation health and mitigation—addressed over the past three/four decades.
5. Need for license—for radio, TV, cellular commercial Vs public safety.
6. Wide range of standards activity in standards and national bodies—methods to monitor and control level of RF radiation in unlicensed and licensed bands to control interference.

Homework problems

1. A typical microwave oven operates in the unlicensed band of 2.4 GHz. How much is the RF radiation coming out of its door? Based on the allowed level, compute distances where its power is insignificant—in the kitchen, in the house?
2. Use a typical wireless USB mouse with a pad—within 10Cm of an Access Point antenna. Does it affect its performance? If it does not—write a paragraph on the algorithm used to avoid interference—at the Access Point, at the USB mouse.

3. Who is the service provider for your cell phone? On what frequency band do they operate in your area of residence and your office area?
4. Which cryptographic function does 4G LTE follow? Does this provide sufficient privacy for your voice call? Why do you think so?
5. Why does evaluation of cryptographic function take as much time as design and test?
6. A typical SIM card today comes in three sizes—Nano, Micro, and standard. Identify wireless devices that take up each size. Why should there be three sizes?

Project ideas

1. Discuss with your supervisor and choose one of the three prime areas of 3GPP to research further—Radio Access, Systems and Service, Core Network. From their website, download a few submissions and discuss its impact on smart cell phone.
2. There are devices that the elderly are advised to wear that sends vital body-related information to the doctor periodically. How safe is this information?
3. Contact your local Police, Fire squad and find out what radio system technology they use. Who chose this system for them and what features do they like?
4. Using IEEE C95.1 can you formulate a set of guidelines to your local community board?

Out of box thinking

1. Standards should evolve towards embedding the cell phone under the skin—do you support or oppose this idea? Set up a debate to discuss this issue.
2. Several towns assert that they should not have cell phones in their town. Consequently, all cell phones in that town must operate at full power to scream and reach cell site that is farther away. Debate and discuss if citizens of towns are being misguided based on power from the tower yet forgetting about increased power from cell phone next to your head.
3. Review the IEEE C95.1 standard and check the list of references it provides. How many studies were conducted by the medical community? What are their conclusions? What guidelines would you provide to your state representative in terms of cell phone policy?
4. A team of experts hold the view that community/township that opposes erecting a cell tower in their neighborhood, is harming their community. The reasons they provide is every cell phone must now operate at peak power (literally shout out) in order to reach the far away tower. Would you support their argument and talk to your high school/college teachers to form community action groups?

References

1. The ABC's of 5G new radio standards, Keysight technologies, e Books 2000–2020. https://www.keysight.com-assets-ebooks
2. Design reuse, Kasumi f8 and f9 cores. https://www.design-reuse.com/sip/3gpp-kasumi-f8-and-f9-cores-ip-17808/
3. Widjaja I, Nuzman C (2011) Mitigating signaling overhead from multi-mode terminals, and others, published in, 2011. In: 23rd international teletraffic congress (ITC), San Francisco, 6–9 Sept 2011. INSPEC accession number: 12290053, published by IEEE
4. IEEE 802 wireless standards—fast reference. https://searchmobilecomputing.techtarget.com/definition/IEEE-802-Wireless-Standards-Fast-Reference
5. Raghunandan K (2018) 5G RAN millimeter wave initiatives. In: IEEE 5G world forum, 9–11 July 2018. Santa Clara, CA
6. Ultra-wide band for dummies, special edition by Quorvo.com. https://www.quorvo.com
7. IEEE 802.15.4 standard (Zigbee): a tutorial/primer. https://www.electronics-notes.com/articles/connectivity/ieee-802-15-4-wireless/basics-tutorial-primer.php
8. Mesh topology capability in wireless personal area networks (WPANs). https://standards.ieee.org/standard/802_15_5-2009.html
9. IEEE standard for safety levels with respect to human exposure to radio frequency electromagnetic fields, 3 kHz to 300 GHz. https://standards.ieee.org/standard/C95_1-2005.html
10. Understanding 3GPP—starting with the basics, Qualcomm. https://www.qualcomm.com/news/onq/2017/08/02/understanding-3gpp-starting-basics
11. The Z-wave alliance standard. https://z-wavealliance.org/TheIoTispoweredbyZ-wave
12. 3GPP LIVE—an overview of videos related to 3GPP and the technologies it supports. https://vimeo.com/3gpplive

Abbreviations

Block Chain	Block chain is a specific type of database, Block chains store data in blocks that are then chained together It has become popular in online financial transactions, as a ledger that records
Hacker	A person who tries to attack and break a network, website, or other online resource, with a malicious intent
CAVE	Cellular Authentication and Voice Encryption —an algorithm originally developed by Bell Labs to thwart attempts by those who would steal cell phone details and make a clone and subvert growth of cellular technology. It is the original, successful algorithm of 2G that laid foundation to future algorithms in cellular systems. In GSM, this is called authentication algorithm A3
AuC	Authentication Center—is part of the core network which is responsible for authentication process used to confirm identity of mobile and other devices on the cellular network
SIM	Subscriber Identification Module—is an integrated circuit placed in a card operating system (COS) that is intended to securely store the international mobile subscriber identity (IMSI) number and its related key, which are used to identify and its related key. Introduced originally by GSM but now universally in all cell phones
Rand SSD	Random Shared Secret Data. First step of the CAVE algorithm to authenticate a mobile. This is a number generated at the authentication center and sent by each base station to the mobiles it is serving at that time. Using a "Pseudo random number generator (PRNG)".
	PRNG is an algorithm for generating a sequence of numbers whose properties approximate the properties of sequences of random numbers. In GSM, this is a 128-bit random number known as RAND
SSD update	The number calculated by the mobile using Rand SSD, MIN, and A-key. Calculation follows the CAVE or any other algorithm specified by the standard. All IS-41 standard-based systems followed the CAVE algorithm for this calculation. In GSM, this is called Signed Response (SRES) and calculation also include the SIM of the cell phone
A-key	Authentication key, in GSM this is referred to as Ki and is placed in SIM, HLR, and VLR but never transmitted over the air
UC	Unique Challenge. Second step of CAVE algorithm. The unique challenge used the Rand SSD calculated by the mobile as the basis to generate another number unique to that specific mobile. With this step the AuC uniquely identifies the mobile confirming its identity and authenticates it
HLR	Home Location Register, a record is kept for each mobile registered with the service provider and it is placed in the home territory of where the mobile was originally registered.
VLR	Visitor Location Register, a record kept in a region visited by the mobile. It is a location different from the territory of HLR. Since users move around it is quite likely they will be in VLR region each day or even multiple times a day
MIN	Mobile Identification Number, the phone number assigned to the mobile by the service provider

© Springer Nature Switzerland AG 2022
K. Raghunandan, *Introduction to Wireless Communications and Networks*,
Textbooks in Telecommunication Engineering, https://doi.org/10.1007/978-3-030-92188-0_6

IMSI	International Mobile Subscriber Identity, a number allocated uniformly to cell phone across the globe, which includes the country code. The MIN is never transmitted over the air, but IMSI is used by internal network security elements of the cellular system to identify the mobile
TMSI	Temporary Mobil Subscriber Identity, a number transmitted over the air, which in turn is translated to IMSI by the cellular system. TMSI can vary in format depending on the service provider
GUTI	Globally Unique Temporary Identity is used to identify different versions of TMSI, thereby helping protect mobile's confidentiality and identity of user equipment
GPRS	General Packet Radio Service, a 2.5G technology that aimed at increasing the data throughput of GSM (second generation) cellular network. It was a method to connect cell phone to the Internet
SGSN	Serving GPRS Support Node provides switching functionality, security, and authentication via the HLR for GPRS users
GGSN	Gateway GPRS Support Node, which acts like a router and is part of the core network that connects GSM-based 3G networks to the Internet
EPC	Evolved Packet Core network is a framework that provides convergence of voice and data in 4G LTE
EDGE	Enhanced Data rates for Global Evolution allows increased data rates with backward compatibility to GSM
UMTS	Universal Mobile Telecommunication Service is a 3G CDMA standard, developed by 3GPP
UEA	UMTS Encryption Algorithm (used for cybersecurity, computing, computer security) in 3GPP standards of 3G UMTS and 4G LTE
UIA	UMTS Integrity Algorithm with an Integrity Key (IK) is used for computing a message authentication code
eNode B	Evolved Node B, a nomenclature used in 4G and later to denote base station, where E-UTRAN Node B is derived from the previous 3G specification
HeNB	Home eNode B is used in small cell or LTE femto cell, but with the same security functionality as the eNode B
AKA	Authentication Key Agreement, the fundamental process used by security to authenticate a user, by running same security algorithm at the user device and in the network
MME	Mobility Management Entity, the key control node for LTE access. In LTE network, it also controls earlier generation such as 3G and 2G devices
NAS	Non-Access Stratum is a set of protocols in the evolved packet system. The NAS is used to convey non-radio signaling between the User Equipment (UE) and the Mobility Management Entity (MME) for an LTE/E-UTRAN access
HSS	Home Subscriber Server is the concatenation of the HLR (Home Location Register) and the AuC (Authentication Center), which were functions existing in 2G and 3G
ASME	Access Security Management Entity is assumed by MME, which receives the top-level keys from HSS or HLR. After authentication is completed, UE and MME shared the same K_ASME key
RRC	Radio Resource Controller, K_RRc, is a key which is used for the protection of RRC traffic. It has two parts one for integrity and another for encryption
RNC	Radio Network Controller is used during 2G and 3G and it is an important network element closely focused on radio-related functions such as mobility, hand-off, and power regulation, among others
NIST	National Institute of Science and Technology, the standard body of USA responsible for maintaining and developing technologies for better standards and methods in science and technology
WPA	WiFi Protected Access, there have been three versions of this security algorithm. WPA3 launched in 2018 is the most recent. It has two versions, one for personal and another for enterprise users
Zigbee	A short distance wireless technology known as the network leader in smart home and buildings
OTAP	Over the Air Provisioning, a method extensively used by cellular providers use to update user devices (including cell phones). It uses a secure link with higher security exclusively used by the provider
ECC	Error-Correcting codes is used for controlling errors in data over unreliable or noisy communication channels. The coding will vary depending on the application requirement
AES	Advanced Encryption Standard or Rijndael (Dutch) is a specification for the encryption of electronic data (symmetric block cipher) established by the U.S. National Institute of Standards and Technology (NIST) in 2001.

Earlier version with both 128-bit encryption and the current 256-bit encryption is in use

COPD   Chronic obstructive pulmonary disease refers to a group of diseases that cause airflow blockage and breathing-related problems. Regular monitoring becomes possible with wireless sensors

BMI   Body mass index is a value derived from the mass and height of a person. It is a person's weight in kilograms divided by the square of height in meters

URLLC   Ultra-Reliable Low-Latency Communication is part of 5G specification developed for driverless vehicle and related technologies that need less than 1 ms latency and 99.9999% reliability

CCTV   Closed Circuit Television is used for monitoring certain areas in pursuit of security. Usually CCTV has multiple cameras located in different parts of premises/street with a central monitoring unit

Security is perhaps the most distinguishing hallmark of cellular technology. Contrary to general perception, cellular systems are safer than conventional computer networks. This chapter will elaborate on various aspects of security and how the cellular standards have taken it to another level, using a well thought out and mathematically intense methodology.

6.1 A Closer Look at Cybersecurity

The cyberspace is increasingly used by global population. Three major platforms dominate traffic in cyberspace today. First is the WWW or World Wide Web, which is used for connecting people with corporate bodies. Even before this IP network was widely deployed, there was an older application known to all—the email. Both have existed for over three decades. The third but more recent platform is touted to be "block chain" linked to the controversial "Bit Coin". These three are chosen as samples to consider the impact of wireless on these platforms. Each of these platforms pose a challenge to administrators since there is a large community of "hackers" who try to break into each of these platforms. They do so for many reasons such as

1. Just for pranks—to show that they are smart.
2. For bothering users—disrupt their operations and take control.
3. Hold corporate bodies/individuals to ransom—demanding some form of gratuity.

4. Nation states—trying to gain regional/global dominance.

Whatever the reason, the concept of cybersecurity and how to administer networks has become the primary concern of not only corporate bodies, but also governments, individual users, etc. The concept of hackers being criminals has partly changed. Several investigative agencies work with them to understand where the weaknesses are in each network—this is the concept of "use a thief to catch a thief". There are non-profit websites where hackers share information of what they found [1]. There is also an annual conference of these "hackers" in Los Angeles and other places. The question is—do these hackers target wireless networks or are they quite focused on applications that hit the user the hardest? The answer is not straight forward, but as we navigate through this chapter, the reader will observe that wireless network is not the obvious choice for hackers—it is harder to crack.

Primary goal of many "hackers" is to disrupt the network—most common method is to develop applications and provide links that open up the end user's resources. Other methods attempt to decrypt messages or find locations where messages appear as plain text. In the context of wireless communications, perhaps the first major onslaught occurred during the late 1980s when the 1G phones were analog and it was easy to read the messages that were sent "over the air". It was possible to clone the cell phone and make calls as if the cloned unit belonged to legitimate user. This simple concept was extended even in 2G phones when it was still easy to clone phones [2]. But 3G and later generations used Internet to surf WWW, use email extensively for personal and corporate use, and also blockchain transactions. Let us review them in terms of smart phone usage for these three examples and their security issues.

Some weaknesses continue even to this day, but often it requires the hacker to have physical access to the phone. Physical security is a major area of concern, which will be addressed towards the end of this chapter. This is the main reason why every service provider warns users to call them immediately if owners believe they lost their cell phone. All topics up to Sect. 6.6 of this chapter will focus on electronic security and physical security will be addressed in Sect. 6.7, towards the end of this chapter. With widespread use of cell phones as primary means to access the Internet, rouge nation states began sophisticated operations to hack mainly government resources such as critical infrastructure and agencies that contain sensitive information [3]. It may involve intellectual property, national defense, or other sensitive areas like nuclear or space initiatives hacked by the most organized teams. Even these teams target websites (www) or use emails and have even tried blockchain scams.

6.1.1 Cellular Security

During 2G the cellular industry caught up to cloning and introduced powerful algorithms to counteract. Perhaps the most powerful of them was the CAVE (cellular authentication and voice encryption) algorithm introduced in 2G and with the CDMA network [4]. This was briefly mentioned in Chap. 1 but let us delve into it now. What is this algorithm, how does it work, and why is it still in use after almost two decades?

An important factor is how the public key is distributed and whether someone can crack into this scheme. Details of how a device and its user are authenticated and how the network can retain such information securely is an important criterion. During early 1990s (early 2G) cellular industry was in its infancy—cloning was at its peak. Cellular industry faced an important milestone decision. If cloning could be stopped, then cellular would survive, else it would face extinction. With this dire warning, researchers designed a method which provided authentication and secure access that were unique. Figure 6.1 shows the overall architecture.

Why was CAVE algorithm different? Authentication process used by CAVE algorithm involved three distinct numbers. (1) MIN—mobile identification number or mobile phone number. (2) The mobile ESN—electronic serial number—that is written only once by the manufacturer when the mobile is built. (3) A-key (authentication key) which is a write-only number written by cellular service provider (like password it can be written but not read). The same A-key is written into HLR (home location register)—a data record located at the Authentication Center (AuC). Also HLR/AC is where the RandSSD gets generated.

CAVE algorithm consists of two steps. Step 1 is SSD (shared secret data) and step 2 is Unique Challenge (UC). With CAVE step 1, "Rand SSD, ESN, and MIN" are used as inputs to compute a new number called "SSD update". The

Rand SSD is a randomly generated number sent over the air through the base station to the mobile to compute "SSD update" using CAVE algorithm. At the network core, the same three numbers are used to compute "SSD update". When the mobile completes computation and sends calculated SSD update, it is compared with the SSD update computed by the network. If both SSD updates match, then the mobile is authenticated by the AuC for step 1. This is only first step of authentication which must be followed by step 2, which will be described in the next sub-section (continuous process).

Note that important factor here is the A-key—which is never sent over the air but is used for calculations at both ends—in the network and at the mobile. By observing random numbers over the air, it is not easy to deduce what the A-key could be. Therefore, anyone monitoring the air interface cannot use these random numbers to duplicate authentication in real time. Use of random number generation reduces chance of repeated trials to break in. Each time the air interface is observed a different set of random numbers come up. The Rand SSD and the SSD update computed by the mobile cannot be deduced by some other logic. This is one of the important foundations on which the CAVE algorithm defeated the practice of cloning mobiles. Success of the CAVE algorithm over two decades in public use proves this. Let us consider the complete authentication procedure in practice that uses two distinct but connected steps.

6.1.2 Cellular Authentication: Continuous Process

First step of authentication process begins as soon as mobile is powered up. Mobile scans to find base station. Since base station sends out broadcast messages regularly, mobile

Fig. 6.1 CAVE algorithm—cellular security architecture

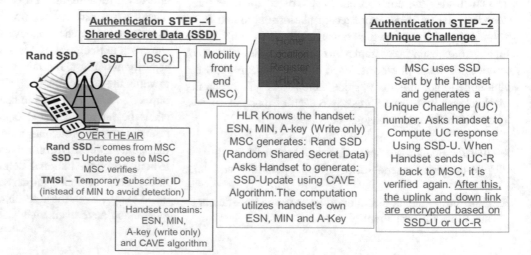

responds to this message requesting service. The network (through the base station) sends a message known as "Rand SSD"—Randomly generated number—asking the mobile to authenticate, using CAVE algorithm. At the Authentication Center (AuC), each mobile is allocated unique data record known as Home Location Register (HLR) as shown in Fig. 6.1. This HLR is always located at the home base where the mobile was primarily registered. Whenever mobile moves away to another area a VLR or visitor location register is a record created in that visited area. VLR is a temporary copy of HLR. The serving base station in visited area contacts the home base to get HLR copy and updates it as the mobile moves through different zones that are not in its home territory. If SSD is shared with the visited network, then VLR locally authenticates the roamer. Otherwise, the VLR will proxy authentication responses from roamers to their home HLR/AuC for authentication. Details of this process to track mobile will be further reviewed in Chap. 11.

The second step of CAVE authentication process begins with "Unique challenge" (shown on right side of Fig. 6.1). The starting point for unique challenge is "SSD update"— the number mobile just calculated and sent in step 1. Using SSD update, a "Unique Challenge" number is now generated by the network and sent to the same mobile to prove its unique identity. Using this "Unique Challenge" number, mobile calculates again (using SSD update generated earlier, MIN, and A-key) and sends its response or "Unique Challenge Response". When this response is received it is again verified by the network, and the mobile is finally authenticated. Either SSD update or unique challenge response is used to generate specific encryption key. That specific encryption key is used to encrypt air links (uplink and downlink) for that specific call or session.

Mobile goes through this process each time there is authentication which starts with power up. In addition, "periodic registration" is set by service provider typically every 5 min or 10 min depending on traffic in that area. This verifies whether mobile is still in the area and is it active or dormant. If user moves away, without any activity on the mobile, re-selection is used to change to control channel of another base station. This ensures service from another base station's control channel whether mobile is on call or not. Therefore, over-the-air encryption is unique to each mobile and key used each time changes with each call and also due to each of these actions. This process provides an insight into the complexity of why it is not easy to break into a cellular network. Authentication involves the physical layer (air interface) and the MAC layer where protocols related to that cellular standard operate. For a potential hacker, knowledge of these two layers become important, as expected hacker prefers weakest link—applications.

In addition, at the cryptographic level, mathematical resiliency of the code is tested to show how long it would take to break this code [5]. Such mathematical analysis shown in Fig. 5.4 earlier doesn't mean that a practical system implemented by a cellular provider can be broken into. It is essentially a pen-and-paper exercise by mathematicians to test resiliency of the code. There are many variables, including time, the SSD update received each time, unique challenge update received, etc. that make it impractical to implement crypto methods to even analyze and break service. These methods cannot be implemented or verified during a call. Therefore, every potential hacker always looks for other means to enter/break the network. These involve hacker's familiarity with IP network nodes, firewall, or application layer attacks.

6.1.3   Understanding Unique ID (Identifiers) in LTE

Before we begin with the entire set of keys in different generations of cellular, let us first focus on what are these keys, how and where are they generated and why?

In cellular systems, all transmissions between user and the network are sent over the air interface—user key must be carefully protected to protect "physical security of the user" which can be in danger. Compared to wired infrastructure this is a major difference. Therefore, service providers try to secure higher network layers and support privacy and confidentiality of users. This procedure begins with an agreement transaction known as Authentication Key Agreement (AKA). This agreement is between the user and the network. Two keys are used—the Cipher Key (CK) and the Integrity Key (IK). Both keys are driven from user's secret key (K) and revealing them may disclose some information about user's secret. Therefore, it is very important to make sure vulnerability of AKA must remain in focus all the time.

The first step is to "not send user identity over the air". Identification of mobile subscriber is provided by International Mobile Subscriber Identity (IMSI), which is a unique number linked to the mobile user but is not the actual mobile number. Even this IMSI is not actually used during a cellular call/session; instead, a Temporary Mobile Subscriber Identity (TMSI) is used. The same mobile can be allocated three "temporary identities" by different parts of cellular network that provide service to the mobile. TMSI only has local significance—only in VLR, SGSN (serving GPRS support node), and MME (mobility management entity).

The reason for this separation even within TMSI is to protect the mobile's identity and confidentiality, since the user often moves beyond home network, resulting in roaming into or visiting another network. Such movement of every cell phone user is common and may occur multiple times a day. It is important to uniquely identify the mobile and user, without revealing identity of mobile. Figure 6.2

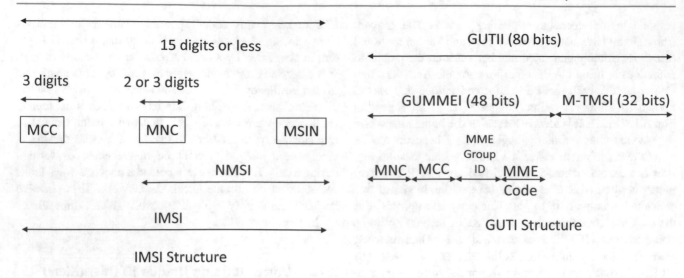

Fig. 6.2 Structure of IMSI and GUTI—used for identification of mobile

shows structure of IMSI as well as related identity in terms of GUTI (globally unique temporary identity) of UE (user equipment) and different versions of TMSI that are all used in the network [6]. A brief description of each of the terms listed in the figure and their functions to identify mobile is in order—these are explained briefly now.

Mobile Country Code (MCC) identifies where the user is domicile. MNC or mobile network code identifies the Public Land Mobile Network (PLMN), whose length can vary between 2 and 3 digits depending on MCC. Mobile Subscriber Identification Number (MSIN) identifies the mobile subscriber within MNC.

TMSI is set up based on agreements between mobile operators and mobile equipment manufacturers. TMSI consists of four octets; hexadecimal representation can be used. TMSI held by VLR will not be passed on to HLR. TMSI is used when paging (calling the mobile), and therefore VLR which is the part connecting user, uses it.

Globally Unique Temporary Identity (GUTI) is used for unambiguous identification of user's permanent identity. It is allocated by Mobility Management Entity (MME). It can be used by the network and UE to establish identity in Evolved Packet Core Network (EPC).

GUTI consists of two main parts: Globally Unique MME Identifier (GUMMEI) to identify which MME has allocated the GUTI. M-TMSI identifies the UE within MME that allocated GUTI. In LTE networks, GUMMEI portion of the GUTI consists of several parts. These are the Mobile Network Code (MNC), the Mobile Country Code (MCC), and the Mobility Management Entity ID (MME ID). The Mobility Management Entity is the primary control node for the LTE access network. MMEs are usually clustered in pools, and the MME ID identifies both the MME pool and the node within that pool.

GUTI can be shortened for radio communication. The shortened form is called the S-TMSI and which has the full M-TMSI and only a portion of the GUMMEI. The part of the GUMMEI that identifies the MME node is called the MME Code, which is included in the S-TMSI. The part of the GUMMEI that identifies the MME pool is called the MME Group ID. This part is not included in the abbreviated GUTI. Therefore, it is necessary to avoid using the same MMEC number to identify two mobility management entities that are in adjacent or overlapping pools. However, in all cases, the M-TMSI portion of the GUTI must be kept unique for each device.

6.1.4 Security Algorithms and Keys in 3G and 4G

Following CAVE algorithm, cryptographic analysis procedures were developed for 3G and continued in 4G. While process of authentication is similar, certain aspects were enhanced to streamline security. Important among them was SIM (subscriber identity module) card, which needs its own security checks. The USIM or universal subscriber identity module is a card that contains information about the user, configuration of device it is used in, text messages, and billing information, modem or other device, and each need separate authentication, as described here.

Since the same USIM card can be inserted in different devices, it works independent of the device. If a USIM card is stolen/misused by anyone other than legitimate user, there are ways to check and stop its abuse. Card's security scheme is independent of where it is inserted. During 3GPP progress to next generation, most of the RNC (radio network controller) functionality of 2G/3G was incorporated into eNode

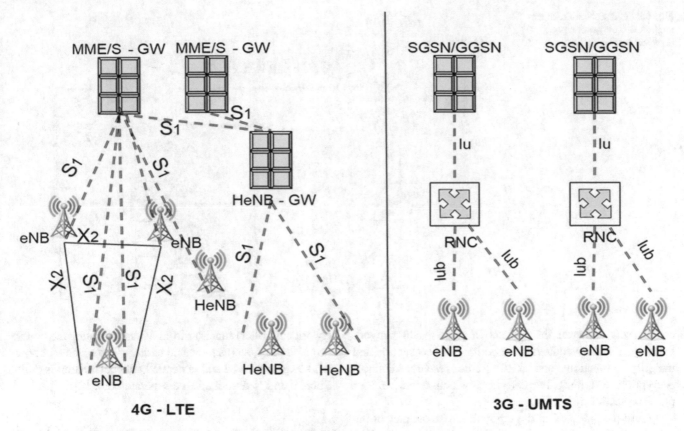

Fig. 6.3 Comparison of 3G and 4G architecture—changes in links and interfaces

B in 3G/4G. Security for 3G and 4G was developed based on a selection of family of cryptocodes. The family of code selected is known as Kasumi. KASUMI is a 64-bit block cipher which uses a 128-bit key. It forms the basis of confidentiality (f8 or UEA1) and integrity (f9 or UIA1) algorithms to provide data security for signaling and user data within GSM, GPRS, EDGE, and UMTS standards [7]. Figure 6.3 provides a comparison of changes from 3 to 4G. The flat IP structure of 4G accommodated Home eNodeB (HeNB) radio units such as WiMax/WiFi. These networks provide access to cellular network through other devices (such as access point or base station units) but introduce vulnerability due to IP [6]. Any unit that does not belong to 3GPP standard is treated separately with additional interfaces such as the Hybrid "eNodeB" gateway shown on the left, in Fig. 6.3. This feature and key allocations will be further elaborated in Chap. 11, Sect. 11.2.6.2, comparing changes in different generations of cellular.

6.1.5   LTE Security Schemes

LTE security scheme is based on separation of sharing keys at multiple layers. Figure 6.4 provides comparison of same keys used at the user side (on the left) and network side (on the

right) in 4G LTE. The authentication process follows CAVE algorithm scheme where deriving valid response is based on running the same computation at the user end and at the network end and then comparing the result. Multiple layers of deriving keys and separating them in layers is deliberate, with the understanding that a potential hacker needs several results as inputs that are not available over the air, etc.

Figure 6.4 shows how they are derived at both sides using different layers of network elements up to the Authentication Center (AuC) at the network end and USIM at user end. At the first layer, secret key K is shared between USIM (universal SIM card) on the user equipment side and on the network side it is shared by AuC/HSS (authentication center/home subscriber server). The sharing occurs using the AKA process (authentication and key agreement). In the next layer, on user equipment side, ME (mobile equipment), and on the network-side MME (mobility management entity) are responsible for paging and authentication of the mobile; they derive keys CK (cipher key) for encryption and IK (integrity key) integrity protection from K. The Non-Access Stratum (NAS) is highest stratum of the control plane between UE and MME at the "radio interface". Main functions of protocols that are part of NAS are to support mobility of the user equipment (UE) and the support of session management procedures to establish and maintain IP

Fig. 6.4 Hierarchical keys and method of key generation between entities in LTE

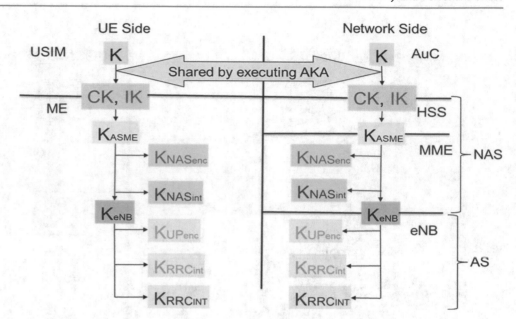

connectivity between the UE and a packet data network gateway. The derived keys are used for NAS encryption and integrity protection, and RRC (radio resource control) encryption and integrity protection, which refers to user plane protection [8].

Figure 6.4 starts off at the top with innermost part of the network and progresses downwards ending in base station and user who form the "Access Segment". Entities of each are shown on the left and right with the keys shown along the derived pattern. The Authentication Center (AuC)/Home Subscriber Server (HSS or the old HLR of 2G) form part of the core much like the earlier CAVE algorithm that used shared secret key (A-key stored in HLR). The confidentiality key derived from the f8 function of Kasumi and Integrity Key derived from f9 of Kasumi for the second layer. This is shared by UE (user equipment such as cell phone, modem, etc.) and HSS (home subscriber server). The next lower layer on both sides is access security and management which is a server but also connected to the user equipment. This key is now used by the MME (multi-media entity) to derive two separate keys: network access key's encryption part, network access key's integrity part.

The UE (user equipment) uses the ASME (access security management entity) key to derive key used by base station. Base station eNodeB then calculates three keys: the Kup encryption key, the KRRC encryption key, and KRRC integrity key. RRC (radio resource control) is deliberately kept close to the mobile by moving it to the base station. This change was needed due to removing RNC (radio

network controller) function from 3G and merging most of it into eNodeB of 4G (as indicated earlier). The scheme of how each key gets derived and generated after calculation is quite elaborate and is described in a separate book [8].

6.1.6 Why Such an Elaborate Security Scheme?

Over two-thirds of human population use cellular technology as the primary method of communication, and it continues to absorb many network applications and functions originally developed for wired networks (including the Internet). Security scheme takes an important role based on the number of threats and possible attacks on the cellular network [8]. While moving to 4G LTE, standards body 3GPP decided to make cellular network a "flat–IP network" to connect base stations using wired Ethernet layer 2. This decision had its great advantages for all IP-based applications running on a smart phone. By careful network design it supports real-time communication needed between base stations. But its limitation is now the cellular network also became vulnerable to security attacks by hackers who know IP networks well.

A closer look at possible threats and countermeasures is elaborated in Sect. 6.5 where changes from one generation to the next are described. An excellent paper by NIST shows changes in types of security threats observed by industry over the years from 2 to 5G is in [9]. Elaborate work and survey of vulnerability database shows a growing trend of defense mechanisms built to make it harder for potential attacker to crack into conversations.

Internet sessions and encrypted messages are sent as text (SMS). This area of research is expected to continue its work further.

6.2   Differences Between WiFi Security and Cellular Security

Table 6.1 provides comparison of security schemes cellular and WiFi networks.

Major difference is that WiFi (IEEE 802.11 and other standards such as 802.15, 802.16) describes radio access part (physical) and MAC layer, where "security is NOT handled". This is elaborated in the table. Let us consider the differences in security schemes used by WiFi, as shown in Table 6.1. All 802.1x standards refer security function to

IETF standards body. IEEE 802.1x is primarily concerned with the air interface. This is consistent with the fact that WiFi/WiMax are local area or personal area network extensions. These standards represent the last one hundred meters of wireless extension to wired IP network. One exception to this approach was security features incorporated in 2004 as 802.11i to provide encryption of the air link and authentication. Table 6.2 shows the basic approach of 802.11i and beyond used by the 802.11 standard.

Recent security measures such as WPA3 (WiFi-protected access) are expected to be incorporated by WiFi, during 2018. Although WPA3 enhancement is likely to appeal to defense, federal, and government agencies due to better security features, it is not backwards compatible with WPA2 or earlier versions. Millions of WiFi systems that are currently deployed may not be able to use WPA3. But WPA3 offers both personal and enterprise versions. The enterprise version can use 192-bit encrypted service for better protection.

Table 6.1 Differences between WiFi and cellular security

Cellular security	WiFi/WiMax security	Comments
Licensed bands and careful RF design are highlights of cellular communication. These two combined offer the highest level of security that typical hackers are unaware of	WiFi is typically set up by the end user or network engineer who are not trained in RF design methods. Use of unlicensed band allows a potential hacker to readily try different methods to access	By careful RF design it is possible to improve security of WiFi network. But this is rarely true in both public and private environments. Only cellular network is designed by trained RF engineers
Two aspects of data: confidentiality of user and integrity of data are handled extensively. Cryptocodes are carefully evaluated, verified before they are chosen. This is to make sure that they are mathematically resilient	Data packets sent over the air link are protected using encryption, but the real defense mechanism is password. Encryption and use of firewalls both have their limitations since potential hacker has methods to break	Cellular industry has major role of data access from a handheld device located in any environment (not a secure building). Therefore, handling of data is evaluated carefully, and defense mechanisms are built in
Modems and other devices can share USIM card making security mechanism complex. Separate functions to identify can be incorporated in the smart phone using UICC card located within cell phone	Secure corporate communication using WiFi requires verifications on users and firewall mechanisms are required to address security threats. WAP3 is expected to address this limitation	Corporate WiFi networks are in widespread use. WPA3 is a new security scheme to address limitations of 802.11i and earlier limitations seen in WPA2 schemes were hacked
Based on layers 1 and 2 (physical and MAC layers of cellular standard) using elaborate set of keys. The mechanisms of deriving keys are a defense mechanism	In 2004, the 802.11i was introduced as guideline to encrypt the air link. Security measures are in layers 3 and beyond, handled by wired IP network standards of IETF	In WiFi/WiMax security came as a late after thought. The concept of encryption emerged 15 years later; key exchange mechanisms remain vulnerable
Method of RF access is FDD or frequency division duplex for all commercial cellular systems, providing separately encrypted links for transmit and receive	WiFi uses TDD or Time Division Duplex—therefore the same link is used for both transmit and receive. Unlicensed band requires checks before transmitting	While TDD has been tried in cellular it isn't popular. TDD is similar to simplex (talk or listen not both). In addition to unlicensed band, it is another major limitation in RF link
Use of licensed bands for backhaul microwave links is popular in cellular since it provides secure links with lower latency needed by base stations for real-time communication	Licensed band WiFi for public safety in some countries had limited success. Use of wired backhaul using fiber/Ethernet cable is standard method that uses 1000baseT or 10000baseT	Microwave links use FDD and encryption. They provide low latency (<1 ms) making them favored links for real-time communication between base stations
Use of WiFi to network home or office but use cellular backhaul provides improved security	Use of "hotspot" where WiFi is locally embedded and only cellular connection is available	This scheme becomes as secure as cellular itself since hacker has no access to private WiFi

Table 6.2 Summary of wifi security over a decade (2004–2018)

Security schemes	WEP	WPA	WPA2	WPA3	Comment
Objective	Bring wired like privacy in wireless	Based on 802.11i but no need for new hardware	Implement mandatory features of 802.11i with new hardware	Announced by WiFi alliance in 2018 to go beyond WPA2	Limitation of previous version is overcome by next one
Encryption	RC 4	TKIP + RC4	CCMP/AES	GCMP 256	WPA3 Only supported by 802.11ac
Authentication	WEP–Open WEP-Shared	WPA-PSK, WPA-Enterprise	WPA2 Personal, WPA2 Enterprise	WPA3 Personal, WPA3 Enterprise	Simultaneous authentication of equals (SAE) used in WPA3, instead of pre shared key
Data integrity	CRC–32	MIC algorithm	Cipher Block Chaining, Message Authentication Code (based on AES)	256-bit broadcast/multicast Integrity Protocol Galois Message Authentication Code (BIP-GMAC-256)	192-bit cryptographic scheme in WPA3 to enhance security
Key management	None	Four-way handshake	4 Way handshake	Elliptic Curve Diffie-Hellman (ECDH) exchange and Elliptic Curve Digital Signature Algorithm (ECDSA)	Protected Management Frame (PMF) mandatory in WPA3; stops dictionary method to guess password

6.2.1 Security of Bluetooth and Other Proximity Wireless

Security schemes of wireless technologies such as Bluetooth, Zigbee, and others were briefly mentioned earlier in Sect. 3.5 but will be reviewed here. Five basic security services are specified in the Bluetooth standard IEEE 802.15.1–2002:

- Authentication: Verifying the identity of communicating devices based on their Bluetooth address. Bluetooth does not provide native user authentication.
- Confidentiality: Preventing information compromise caused by eavesdropping by ensuring that only authorized devices can access and view transmitted data.
- Authorization: Allowing the control of resources by ensuring that a device is authorized to use a service before permitting it to do so.
- Message Integrity: Verifying that a message sent between two Bluetooth devices has not been altered in transit.
- Pairing/bonding: Creating one or more shared secret keys and the storing of these keys for use in subsequent connections in order to form a trusted device pair.

Details of Bluetooth technology security features are described in reference [10]. Since Bluetooth is widely used with cell phones in various applications, features such as ultra-low power ULP (4mW) can support devices that operate at distances of no more than 10 m. This provides

security to an extent and often such links are paired and broken based on user needs. Majority of the security operations are initiated during device setup, some need user intervention. In office and home systems, these work well since devices are moved within the premises.

Bluetooth standards developed in parallel with the 802.11 WiFi standards. There are several versions of Bluetooth dating back to two decades. Backward compatibility is supported by later versions, but power levels continuously reduced in later versions (Class 3) as seen in Table 6.3.

In Table 6.3, Class 1 devices are not widely used. Devices of all other classes (1.5–3) operate over shorter distances and with much lower power. Popularity of Bluetooth is mainly due to its small, low-power, and battery-operated devices and their versatility. Table 6.4 shows BR/EDR/HS (basic rate/enhanced data rate/high speed) security mode level summary for Bluetooth devices.

From Table 6.4, it is obvious that later versions focus on security rather than throughput. In addition, later version 4.2 onwards also focusses on using lower power using wider spacing of channels. Later versions also support "dual mode" devices that support BR/EDR/HS (basic rate/enhanced data rate/high speed) as well as low-energy versions.

Ideally suited for ad hoc (as required) network setup, Bluetooth offers flexibility of using devices in various network configurations, as and when needed.

Devices like the garage door opener are considered proximity devices. They too have security schemes that are

Table 6.3 Bluetooth devices—power levels and coverage distance

Type	Power	Max power level	Designed operating range	Sample devices
Class 1	High	100 mW (20 dBm)	Up to 100 m	USB adapters, access points
Class 1.5 (low energy)	Medium/high	10 mW (10 dBm)	Up to 30 m but typically 5 m	Beacons, wearable sensors
Class 2	Medium	2.5 mW (4 dBm)	Up to 10 m	Mobile devices, Bluetooth adaptors, smart card readers
Class 3	Low	1 mW (0 dBm)	Up to 1 m	Bluetooth adapters

Table 6.4 Security features for different versions of Bluetooth technology

Characteristic	Bluetooth BR/EDR		Bluetooth low energy	
	Prior to 4.1	4.1 onwards	Prior to 4.2	4.2 onwards
RF physical channels	79 channels with 1 MHz spacing		40 channels with 2 MHz spacing	
Discovery/connect	Inquiry/paging		Advertising	
Number of Piconet Slaves	7 active/255 total		Unlimited	
Device address privacy	None		Private device addressing available	
Maximum data rate	1–3 Mbps		1 Mbps via GFSK modulation	
Pairing algorithm	Prior to 2.1: 21/E22/SAFER+ 2.1–4.0: P-192 Elliptic Curve9, HMAC-SHA-256	256 elliptical curve, HMAC SHA-256	AES 128	256 elliptical curve, AES-CMAC
Device authentication algorithm	E1/SAFER	HMAC SHA-256	AES-CCM	
Encryption algorithm	E0/SAFER	AES-CCM	AES-CCM	
Typical range	30 m		50 m	
Output power	100 mW (20 dBm)		10 mW (10 dBm)	

not sophisticated. One of the key points to remember is that a potential hacker could get to the garage door opening system, if it is connected to a WiFi network. While it may seem attractive to connect the garage door opening system to the home WiFi, it is important to consider the risk of a potential hacker gaining access to the house. Based on that risk, it may be prudent not to connect it to the home WiFi. Some may even work on Bluetooth, but they all rely on the handheld remote being the primary device to open the door. In terms of physical security which is discussed in Sect. 6.7, it is safer to use the remote and not allow physical access to the house, by connecting the garage door to home WiFi network.

6.2.2 Zigbee Security Schemes

Zigbee technology is widely used in medical and transport sectors due to its short distance access (about 20–50 m) with low-power devices. ZigBee allows many more devices to communicate within vicinity. Low energy consumption and scalability are its main features. Most of them are battery operated, thereby allowing incorporation of security within the chip (no elaborate external scheme). The IEEE 802.15.4 standard that describes Zigbee includes security protocols. Primarily symmetric key is used in its cryptography scheme, with an elaborate key protocol [11]. ZigBee uses three different types of keys for association to a network, or within a group of devices or a link between two elements.

- Master key: This is the most important key from which link keys are established. Given its importance, the initial master key must be obtained by secure means (preinstallation or key transport, both of which are possible).
- Link key: The link key encrypts point-to-point communication at the application level and is only known by elements taking part in that link. This key is only shared between two network elements and is different "link key" which is used for each pair of elements. The link key is used to minimize security risks related to distributing the master key.
- Network key: This key is used at the network level and known by all elements that belong to it. The network key is used in groups when more than two elements within a network communicate as a group.

Key management in Zigbee

Security characteristic of Zigbee is that it has a variety of key establishment mechanisms. Some of these are very attractive to locations such as hospitals where data security is paramount:

- Preinstallation: This method applies only to master keys. The manufacturer incorporates a master key into the device itself. During installation user can select one of the preinstalled keys by using a series of jumpers in the device (in devices with optional keys preinstalled).
- Key transport: The device makes a request to a trust center for a key to be sent. This method is valid for requesting any of the three types of key. The trust center can work in two ways:
  o Commercial mode: The trust center itself keeps a list of devices, master keys, link keys, and network keys. In this mode, the memory required by the trust center rises as a function of the number of devices associated to the network. Given that typically hundreds of devices are used, this method is useful in corporate buildings.
  o Residential mode: The trust center holds only the network key and controls network access; the other information is stored in each node. The memory required by the trust center does not depend on the size of the network. In this case, there is no monitoring to verify whether sequence numbers have been modified by intruders. For a city-wide street lighting, this may be better. While providing security it is not necessary to monitor thousands of locations from trust center.

6.3 Why is Cellular is Less Prone to Attacks

This may not seem obvious but major security violations occur at the upper layers of OSI (in Chap. 5, see Fig. 5.10), such as application and session layers. Security features of cellular focus mainly in the lower two layers of physical (air interface of cellular standards) and MAC (protocols used by each cellular generation). These are not commonly known to the public and even cellular providers only have a limited understanding of security implementations recommended by 3GPP. Reasons for this limited understanding are due to the following:

The physical layer messages are implemented within base stations and often embedded in chips. Only part of the control is available to the network provider since many physical layer messages are controlled directly by the base station manufacturer. The MAC layer is even more receding into chip set since embedded algorithms are implemented by protocol stacks within VLSI chip. There are only two aspects of security that typical cellular provider deals with: 1. the SIM card and 2. entering A-key at the HLR and in phone (could use OTAP—Over-The-Air Provisioning).

It is not a surprise if typical cellular provider is unable to elaborate on other security methods described earlier in this chapter. Although the provider is concerned about network security, the controls they have on their network are available mostly at the IP network layer that the hacker is already well versed. Therefore, where does hacking occur? It is not easy to get between base stations since these are essentially private networks. Attack can occur beyond the cellular switch where it connects to the Internet. Weakness in terms of IP network security gets exposed in the Internet. This interface to the Internet is where all service providers install a firewall—in terms of control, this provides the border of control. Beyond this and into the Internet, cellular provider faces the same challenges as any wired network provider.

In contrast, when some cellular providers offer all financial and sensitive communication links within their network, then it is likely to be more secure than the wired Internet network of today. This is what 3GPP and cellular network designers are hoping for and keep it as their goal. In countries where major bank transactions can only take place on cellular network, it is likely that transactions occur more securely [12]. Current attacks on cellular networks try to force the user to downgrade to earlier generation networks such as 2G or 3G. This offers a clue that current 4G and forthcoming 5G are more resilient in terms of layers 1 and 2. This process of improving security and hacker trying to challenge them will always be ongoing.

Unfortunately, major countries where the wired Internet is widely prevalent prevailed on 3GPP to adapt flat IP network in 4G. Now there is a move to a newer model where wired network becomes less prevalent than it is today. The first indication of this change is seen in devices such as "hot spot" being offered by many cellular providers. It is a user device that communicates with cellular base station (like a cell phone) but does not provide any connection to Ethernet (embedded inside the device). Instead, it directly offers WiFi as an interface to connect devices. This WiFi and its connection to devices is typical of a wireless to wireless interface. If WiFi is made secure (with WPA3) then this could be a powerful message on avoiding wired network.

The message is: Ethernet may be used within the device but may not be available to end user.

Does this offer a model where the current Ethernet will possibly move deeper allowing a "wireless Internet"? It is difficult to predict, but with some of the newer models proposed for 5G, it is quite possible [13]. It would be worth exploring, particularly for those interested in network security and its future. This topic is often mentioned as the next frontier where countries could pit against each other in terms of economic and technological models for business.

6.4 Basis of Codes and Mathematics (Cryptography)

Earlier, in Sect. 3.3, Error-Correcting Codes (ECC) were reviewed in detail using transition state in Table 3.1 and in Fig. 3.7 through Fig. 3.10. A brief exploration of mathematics was made showing use of codes by the wireless industry. While this process is common to all networks, cellular industry invested considerable time and effort in promoting this branch of applied mathematics. Textbooks that explain this topic in considerable detail are available, see [14]. We will address mathematical concepts in Sect. 6.4.3 for both error correction and encryption. In 1994, Mathematician Peter Shor (from Bell labs) came to instant fame by discovering that hypothetical devices could quickly factor large numbers. Subsequent work by physicists showed that the fabric of space–time is a quantum error-correcting code. This led to a major topic of research called "quantum computing", which is now expanding, leading to supercomputers.

Mathematics and security seem related due to advances in technology. Yet from time immemorial, messages sent through pigeons, messengers, and later through telegraphy (twentieth century)—all of them had some form of coding, to avoid someone else intercepting and reading it. This concept extends to network security and considerable effort is made to ensure any email, text message, photo, or other private communication is not intercepted or read by those to whom it was not intended. This process takes a major leap in the world of wireless for two reasons:

(1) Corruption of data as its messages move over the air and (2) confidentiality of user since the mobile device essentially belongs to an individual, unlike wired devices such as PC, tablet, or laptop or even a desk phone, since all of them can be thought of as shared devices.

When data is sent using wireless—general perception is that if sent over the air it can be intercepted—efforts were made in terms of encryption to ensure this link is kept secure. Elaborate schemes addressing possible threats and preventive mechanisms exist and are always being improved. Any number of devices on the network, such as a WiFi access point, a cellular base station or other equipment may be intercepted to retrieve data/message—but cryptographic resources include security at each of these devices as well as their interactions throughout the network. Efforts involve introducing mathematical codes in the data stream.

Engineering methods towards network reliability were developed over several decades. Although their primary goal was not towards improving security, indirectly they brought certain level of security. Secured interleaving of digital streams of CDMA or OFDMA (4G LTE) results in such clever efforts bringing considerable robustness in the network [15]. These procedures increase the security of data and reduce errors and burst. They also improve security of data stream due to elaborate design schemes that any potential hacker would not be able to step into. There is no direct access to these coding schemes, which are deeply embedded in chip design.

6.4.1 Cryptography: Key to Network Security

To obtain deeper understanding of this area, basics of applied mathematics known as "coding theory" is necessary. It is important to have background in three essential areas: mathematics, electrical engineering, and computer science [16]. In addition, physics is needed to understand cryptography particularly in the context of wireless as was observed in Sects. 6.1 and 6.2 including their sub-sections.

These areas of knowledge support cryptography as related to security. Cryptography works on basic concepts such as confidentiality, data integrity, authentication, and non-repudiation. It involves both encryption and error-correcting codes. These aspects were reviewed in one form or another in Sects. 6.1 and its sub-sections. The topic of cryptography can be abstracted as indicated in Fig. 6.5.

Cryptography is always an overhead. Therefore, network providers indicate data with two aspects—"raw data rate", "user data rate". Raw data rate = user data rate + overhead due to cryptography. That overhead can vary depending on type of medium, usually overhead is more for wireless networks since it must address corruption of data as it is sent

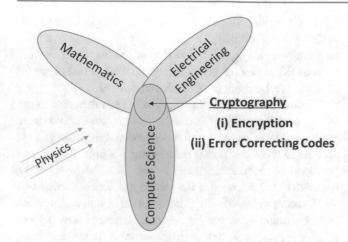

Fig. 6.5 Cryptography as subject—its component areas

over the air. How much is the overhead? It depends—if we consider broadcast network such as television, for talk shows if raw data rate = 100, then user data rate = 75, and overhead due to cryptography = 25. For high-definition TV with fine orchestra being sent over the air to an automobile, overhead can be as high as 40, reducing user data rate = 60. This is also true for space missions as well. Despite this limitation, efficient algorithms continue to evolve and have accomplished good results.

Let us first consider the encryption part of cryptography, for which definition of key terms is essential as indicated below. Later in this section, we consider systems that combine both encryption and error correction together, which is currently a topic of great interest to researchers. It is likely that future systems may adopt this combined approach to allow more efficient systems. But the approach for all wireless systems will be to use hardware-based solutions, to reduce delay, make better use of battery and other resources. Let us start with definition of key terms.

Confidentiality: User's identity, location, and other details are not disclosed, although used by the network. Elaborate schemes such as IMSI and TMSI that were discussed earlier in Sect. 6.1 addressed this aspect. Cipher Key (CK) deals with topic.

Data Integrity: Keeping the data secure, not losing or modifying it any form such that the recipient gets exactly what the sender had sent is important. This was addressed in detail using IK, whose primary goal is to retain data integrity.

Authentication: To verify the validity of user, whether the user is authorized to use, what feature of the network is the user eligible to use are all part of authentication. Starting with CAVE algorithm authentication was addressed in detail. Authentication process is jointly accomplished as an interaction as it is shared by AuC/HSS (authentication

center/home subscriber server). Exchange of dialog using the AKA process (authentication and key agreement), which was described earlier in Fig. 6.4.

Non-repudiation: If the sender were to take the position that the message was not sent by him/her, then such repudiation (sender claiming "I never sent it") must be addressed by the network provider. This is accomplished by a variety of time stamps and other evidence that links the message directly to the device that sent the message, time it was sent, application that was used, etc. In the case of wireless devices such as cell phone, it is straight forward since it has become a personal device. The only situation when repudiation is taken seriously is when it is lost and someone else is sending messages. In terms of security—such actions help law enforcement to trace the location of the device.

6.4.2 Cipher Algorithms for Encryption

Cipher algorithms are used for data encryption and can be classified into two main types of algorithms based on whether encryption key and decryption key are the same or different keys.

Symmetric algorithm where Ke = Kd meaning encryption and decryption use the same key.

Asymmetric algorithm where $k_e \neq k_d$ meaning encryption and decryption use different keys.

- Symmetric algorithms can be further classified into:
  (i) Block cipher (examples: DES, AES, PRESENT, etc.),
  (ii) Stream cipher (examples: A5, Grain, etc.)
- Asymmetric algorithms (examples: RSA, ECC, etc.).

Let us consider some algorithm-related wireless systems. Figure 6.6 shows symmetric stream cipher which uses the same key k and can operate on 8 bit (byte) at a time and operate as stream. In LTE, confidentiality algorithm EEA is a symmetric synchronous stream cipher. This type of ciphering has the advantage to generate the mask of data before even receiving the data to encrypt which helps to save time. In addition, it is based on bitwise operations carried out quickly as expected in cellular systems. The stream M entering the encryption block gets keystream as the other input making a cipher stream C. At the decryption block, the same key is used to decrypt the message (at the receiver) and users get the original stream M on their device. There are other mechanisms such as "Nonce" in authentication protocols that prevent replay. Nonce is an arbitrary number that is used just once in cryptographic communication. The pseudorandom sequence is carefully chosen to provide randomness and unpredictability to session keys k. Every key

Fig. 6.6 Diagram showing basic function of stream cipher

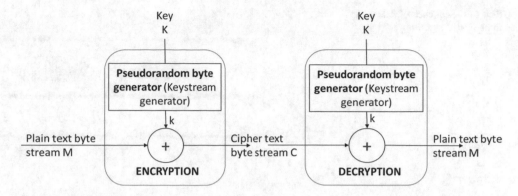

and other derived key for LTE were shown earlier in Fig. 6.4 where key generation at the mobile and at the base station were described.

Among block ciphers AES (advanced encryption standard) used in WiFi as part of 802.11i has gone through multiple iterations currently at WAP3 version. 802.1x standard mandates support for blocks in sizes of 64, 128, and 256 bytes. There are other block cipher algorithms developed for wireless sensor networks, example is shown in Fig. 6.7 (generic block cipher).

Sensor nodes are key points in sensor networks that are low-end devices with limited resources.

Limitation are memory, computation power, battery, and network bandwidth. Security protocols designed for such low-end devices consist of two blocks shown in Fig. 6.7. First is cipher encryption block and second is authentication block which is a micro-version of the Timed, Efficient, Streaming, and Loss-tolerant Authentication Protocol (TESLAP). Security protocol for sensor network uses symmetric encryption to provide data confidentiality, two-party authentication, and fresh data [17].

To send message D to the recipient, SNEP (sensor network encryption protocol) sends out E and M. This means SNEP encrypts D to E using shared key K_{encr} between the sender and the receiver to prevent unauthorized disclosure of data. It uses shared key K_{mac}, known only to the sender and the receiver, to provide message authentication. Therefore, both data confidentiality and message authentication can be implemented. Message D is encrypted with the counter C, which will be different in each message. The same message D will be encrypted differently even it is sent multiple times. Therefore, semantic security is implemented in SNEP. The MAC is also produced using the counter C; hence, it enables SNEP to prevent replying to old messages.

Figure 6.8 shows the sequence of data flow with dashed lines indicating timed sequence of key disclosure. The solid line shows timed sequence of key usage. Based on this flow it is important to note that data may be sent multiple times, without being concerned about interception by an unintended party [17]. Since sensor networks have slow data rate, such iterative block cipher process is quite useful,

Fig. 6.7 Block cipher Sensor Network Encryption Protocol (SNEP)

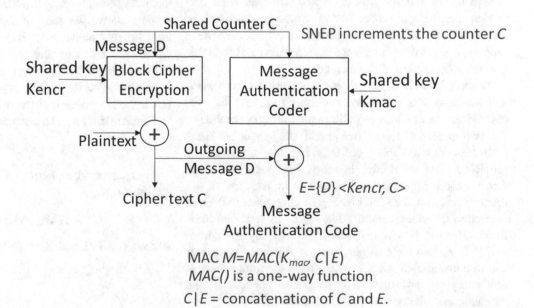

Fig. 6.8 Sequence of data flow in an SNEP algorithm

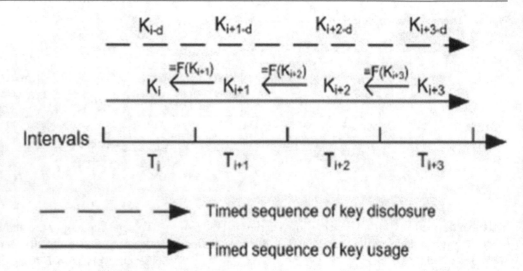

where bulk data is handled and symmetric encryption method is preferred.

Asymmetric encryption is considered more secure and uses software-based implementation. Two keys are used, the public key is available to everyone, but the private key which is not disclosed is known only to the authenticated receiver. It takes more time to implement because of complex logic involved. In contrast, a stream cipher is a key symmetric cipher where plain text digits are combined with a pseudo-random text stream as shown earlier in Fig. 6.6, where each plaintext digit is encrypted one at time with the corresponding digit of keystream [18].

6.4.3 Combining Encryption and Error Correction

In conventional secure communication system model, always the message is first encrypted and then error correction mechanism (coded using methods described in Sect. 3.3) is applied. For retrieval of the message, the reverse process is applied with message first decoded and then followed by decryption of the message.

While this has been the process in all systems (both wired and wireless), it is not always necessary to follow this process. More recent advances indicate possibility of combining the two procedures into one could be beneficial. Error Correction-Based Cipher (ECBC) has the advantage of providing data reliability, integrity, and security. These features were important considerations in selection of 4G security schemes. The combination of encryption with error correction provides certain advantages, in terms of resiliency. Even if the channel were tapped by any eavesdropper, it is not possible to get both. In Fig. 6.9, upper part shows conventional scheme where message from sender is first encrypted and then error correction applied. At the receiving end, the reverse process is used [19].

In both cases, the same key (k) indicates use of symmetric encryption. Lower part of figure shows a combined ECBC method to accomplish the same task in shorter cycle. Sender wants to send a message "u" to receiver—they share a secret key while making sure that others (Eaves dropper) have no knowledge of u. Sender does this by passing the message through the Error Correction-Based Cipher (ECBC) to obtain encoded ciphertext x with the aid of secret key k. On the channel when y is received, legitimate receiver decodes and decrypts in a single step using ECBC with the aid of key "k" to obtain the message u'. Eavesdropper does not have a knowledge of the key to ECBC, and hence the ciphertext received is not error free as shown in Fig. 6.9.

Shannon's entropy of u for this ECBC model is therefore larger than that of conventional model expressed as $H(\frac{u}{c})ecbc) > H(u)$. The encryption and decryption of ECBC are shown in Fig. 6.10. The block diagram on the left indicates encryption process and the one on right indicates the decryption process [19]. The blockchaining effect of this scheme allows the same plaintext block to be enciphered into different ciphertexts. Block chaining is a mechanism where each block of plaintext is XORed with the previous ciphertext block being encrypted.

Similarly, the decryption of a block of ciphertext depends on all the preceding ciphertext block.

Ciphertext Ci is mathematically expressed as

$$Ci = (XiGP + Zi) \quad \ldots \qquad (6.1)$$

For encryption, ciphertexts Ci for i = 1, 2, 3, and so forth, are shown as

$$C1 = f(M1 + Q0)\,GP + Z1, \quad \ldots \qquad (6.2)$$

where Q0 = IV1 and Z1 = g(IV2),

$$C2 = f\_M2 + Q*1\_GP + Z2, \quad \ldots \qquad (6.3)$$

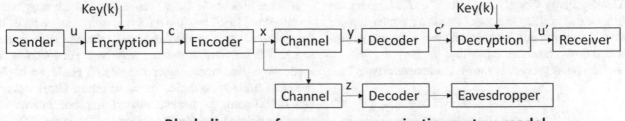

Block diagram of a secure communication system model

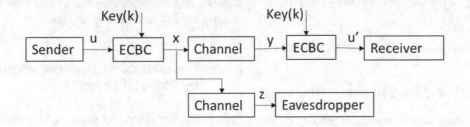

Block diagram of a secure communication system using ECBC model

Fig. 6.9 Conventional combined schemes—error correction and encryption

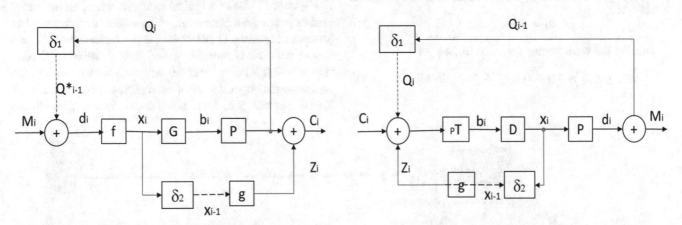

Block diagram of the ECBC encryption scheme   **Block diagram of the ECBC decryption scheme**

Fig. 6.10 Encryption and decryptions schemes used by ECBC model

where $Q1 = f (M1 + Q0) GP$ and $Z2 = g(X1)$, $X1 = f (M1 + Q0)$, and

$$C3 = f\_M3 + Q*2\_GP + Z3, \quad \dots \quad (6.4)$$

$$Ci = f\_Mi + Q*i - 1\_GP + Zi, \quad \dots \quad (6.5)$$

where $Q*i - 1 = f (Mi - 1 + Qi - 2) GP$ and $Zi = g(Xi - 1)$, $Xi - 1 = f (Mi - 1 + Q*i - 2)$.

The cipher also employs double randomization since the plaintext is XORed with $Q*i - 1$ and the permuted codeword is XORed with Zi. This also prevents construction of the generator matrix from the ciphertext.

ECBC scheme is based on the blockchaining technique. In ECBC, a k-bit plaintext block M is enciphered into n-bit ciphertext block C. A separate nonlinear element known as S-BOX is used. The S-BOX represents the nonlinear function f. This function computes the multiplicative inverse of each input byte of the state in GF (2^8) followed by affine transformation. This is a nonlinear byte substitution and consists of two transformations:

(i) multiplicative inverse in GF (2^8): the mapping of x
    $x - 1$, where $x - 1$ is the multiplicative inverse;
(ii) affine transformation over GF (2): $x \rightarrow Ax + b$, where.

A and b are constants. The concatenated output from the S-BOX is encoded using the generator matrix (G) of

Low-Density Parity Check Code (LDPC). LDPC codes are linear block codes. They are codes that have received major attention in recent years because of their excellent performance and error correction capability.

The decryption process is shown mathematically as

$$Ci = f(Mi + Q*_{i-1})\,GP + Zi. \qquad (6.6)$$

Applying the decryption process to Eq. (6.4) results in

$$Qi = [f(Mi + Q*_{i-1})\,GP + Zi] + Zi = f(Mi + Q*_{i-1})\,GP. \qquad (6.7)$$

Multiplying with the transpose of the permutation matrix, we get

$$bi = [f\_Mi + Q*_{i-1})\,GP]P^{T} = f(Mi + Q*_{i-1}))\,G. \qquad (6.8)$$

Applying the decoding algorithm to bi depending on the code employed, then

$$Xi = f(Mi + Q*_{i-1}) \qquad (6.9)$$

Applying the inverse of the nonlinear function f^{-1}, then

$$di = Mi + Q*_{i-1} \qquad (6.10)$$

Adding the error vector $Q*_{i-1}$ to di, we get

$$di + Q*_{i-1} = Mi + Q*_{i-1} + Q*_{i-1} = Mi \qquad (6.11)$$

More details of the ECBC scheme and its advantages are described in [19]. There are a few notes that can be made on this scheme. Using a blockchaining scheme for ciphertext and LDPC for error correction may be a good choice for applications that need higher security. It could be implemented iteratively using software schemes. Other options that utilize entirely hardware-based implementations are being considered making it a strong candidate for 802.15.4 Zigbee wireless application. Due to improved security schemes implementations such as this may offer improved security for IoT and related devices.

6.4.4 Geometry and Trigonometry: RF Security by Physical Layout

Perhaps another important branch of mathematics, normally not discussed in security schemes, is "**Geometry and trigonometry**" of how RF units can be used towards security. Advantage of antenna systems is that they can direct the beam towards specific spots thereby limiting access.

Figure 6.11 shows a typical example, where an industrial yard or parking lot is covered with cameras or ticket-issuing booths and sensors. Data from these can be brought back to a central node or command center. It is possible to deploy "Point-To-MultiPoint (PTMP) system" shown within the enclosure that operates even in unlicensed bands, such as 2.4 GHz or 5.2, 5.3, 5.47, and 5.8 GHz bands (these bands

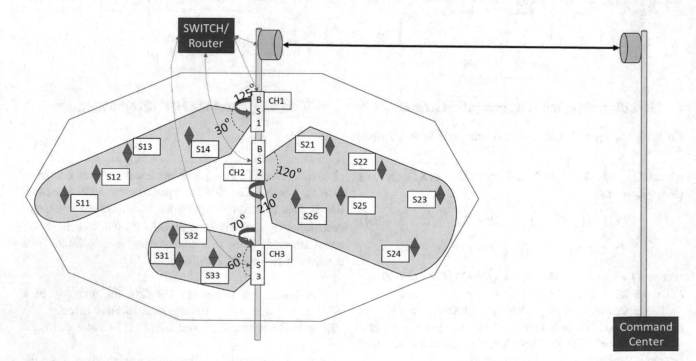

Fig. 6.11 Directional antennas on base stations provide coverage and security

vary depending on the country) using IEEE 802.11 standard security.

This looks like WiFi, yet it is different. The first difference is instead of an access point, there is a base station unit. Base station (BS) acts as central control point with subscriber units acting as multiple subscriber points. Unlike access point, base station initiates, and controls communication. The scheme is termed as PTMP (point to multipoint) network. The BS controlling the subscriber units provides the first level of security, since a rouge subscriber unit cannot talk to the BS (we will see this in the fifth level of security).

The second level is that Base Station BS1 sends commands in a sequence using TDMA protocol. BS1 would send commands on Channel 1 contacting units in the sequence S11, then S12, then S13 and S14. The sequence will repeat at a time interval set by the command center, which can be at specific times of day and perhaps each unit is allowed 3 s, for example, to communicate, etc. Subscriber unit can only respond, not initiate conversation.

The third level of security is due to elevation angles. Each base station could have different elevation angles—BS1 has $30°$, BS2 has $120°$, and BS3 has $60°$. Therefore, subscriber units must be located within those elevation angles (can be at different heights from ground but within the elevation beam).

The fourth level of security is the Azimuth (measured from true North on the pole's own axis). BS1 has $125°$, BS2 has $210°$, and BS3 has $60°$. Therefore the set of sensors are located on the same pole but looking in different directions based on Azimuth.

Finally, the fifth and important one is each of the units must be pre-registered with that base station (before deployment). This means, BS1 already has S11, S12, S13, S14 pre-registered (by the network provider) into its database. Therefore, it will not communicate with others, even if a new one is introduced within that beam it will not help, until the service provider or super user connects to BS 1 and registers the new subscriber unit into BS 1 database.

For a potential hacker, this offers series of impediments in terms of not only network knowledge, but also ladders and other equipment needed to climb into that beam. Even after that effort, the unit cannot communicate with the base station since electronic serial numbers of the subscriber unit must have been pre-registered at the base station. The microwave link presents another layer of security since it is encrypted over the air. Microwave links too have details of each other's serial number, acting as a pair and will not communicate with some other unit. Details of this link security will be discussed in Chap. 14, on microwave/millimeter-wave links.

For hackers contemplating introduction of a stealth base station, the command center can introduce a similar pre-registration process for switch where only designated, pre-registered base stations can communicate with the switch. Overall challenge to anyone trying to break into such a system is considerable, since it involves a combination of antenna pattern recognition, elevation, and azimuth angles of each antenna, base station unit's encryption mechanism including type, make, and model of that unit. Such Point-to-multipoint (PTMP) systems are commercially available with separate encryption schemes beyond layer 2. It is up to designers to deploy such secure systems; it is quite inexpensive and does not involve any special units. By choosing the right unit (readily available in the market) and connecting them carefully, considerable security can be achieved. However, this scheme needs knowledge of RF engineering design that many organizations may not have in their IT departments.

6.5 Process of Evolving Security Algorithms in 3GPP: Each Generation

The process of evaluating security threats and making improvements for each generation of cellular system is an ongoing effort by 3GPP. The procedure for evaluation is well laid out starting with selecting the ciphercode and then working details of implementation through the standards project, as shown earlier in Fig. 5.4. The 3GPP delegated the task to ETSI standards body which in turn set up the Security Algorithm Group of Experts (SAGE) to create a task force for the design and evaluation work for the 3G cryptographic algorithm.

This procedure was initiated by SAGE, during 3G and some security functions continued into 4G. During 5G, key requirements in terms of inter-vehicular communication for automobiles in driverless cars will bring new focus on security, automation, and others. There are important requirements such as its ability to provide integrity and data security at different layers. To support such provisions, Sect. 6.2.1 showed use of hierarchical keys and method of key generation between entities in LTE. Guidelines of NIST (National Institute of Science and Technology) for network security are used [20]. The following were carefully evaluated by 3GPP for 4G:

Use 128-bit encryption—this could be split. To follow 3GPP implementation choices,

Two sets were considered: 128-EEA1/EIA1 and 128-EEA2/EIA2.

AES and SNOW 3G were chosen as basis.

The intent was to keep them as different from each other as possible.

Cracking one would not affect the other.

An alternative third set EEA3/EIA3 was recommended by ETSI (European Telecom Standards Institute).

The following authentication and key generation algorithms are offered as options by ETSI:

- MILENAGE.
- MILENAGE is an algorithm set for the UMTS authentication and key generation functions f1, f1, f2, f3, f4, f5, and f5.
- TUAK.

A more recently developed second algorithm set for UMTS authentication and key generation based on the public K_{ECCAK} hash function family (which will also serve as the SHA-3 hash function standard) is also offered.

How and why are such options provided? Often, evaluation by different bodies is offered as choices for product manufacturers to differentiate. These are based on careful evaluation, an example of which is indicated in [20]. The designer first assumes the role of a hacker and tries different ways and means to crack the encryption security and data integrity. The second step is to identify the process by which authentication is conducted—to see if there are weaknesses.

Based on "privacy requirements" a threat model is developed. The threat could be active such as a fake base station introducing malicious messages, or it could be passive such as just sniffing the data, without attempting to attack. In both categories, known attacks reported by network providers are considered. In addition, new threats that have not occurred are created. The process of attack, improvement, and re-attack are repeated, until the designer is satisfied. Practical methods of attack, ability to isolate the attack, where is the attack coming from, etc. are identified. An experimental setup is offered to other teams to test and verify. Once this process is complete, recommendations are made to standard bodies who ratify and accept them. Some of the successful ones get implemented in the next release or next generation of the standard.

6.6 Applications in Medical, Financial, and Other Sensitive Areas

Wireless devices offer considerable advantage in the medical, financial, and other sensitive areas such as corporate security, national security, and espionage. Given its primary role, threats and counterthreat scenarios from each of these applications are important in evaluating security.

Medical security emphasizes patient-related information in terms of data security. It is also very sensitive to data integrity since doctors make life-altering decisions based on actual data received. At the same time, patient information must be very secure and accessed only by those providing care or only persons authorized by patient. Wireless devices are used with its own built-in security. In addition, the entire network path is evaluated to make sure it remains secure at every step of the way. Medical community, in general, has embraced 4G for a variety of reasons and 17 wireless

devices are listed as game-changing devices to improve patient care [21].

Most of them use near-field wireless technologies such as Bluetooth or Zigbee. Some use proprietary applications setup on cell phone that use these two or similar short distance technologies to report measurements over wireless network, in a secure and reliable manner. Convenience of ad hoc setup that varies every day is the hallmark feature of these technologies. There are 17 wireless devices listed as priority for medical use [21]. They are (i) ECG in your pocket, (ii) vital signs remotely monitored, (iii) wireless glucometer for diabetic patients, (iv) ingestible sensor, (v) symptoms tracker for asthma or COPD patients, (vi) insulin administrator, (vii) cardiac monitor, (viii) bio-electrical impedance monitor, (ix) grand care for the elderly, (x) early warning patient monitor, (xi) patient-generated eyeglass prescription, (xii) sole of new machine—for foot injury, (xiii) wound round-ulcer monitor, (xiv) reducing pressure ulcer for bedridden, (xv) real-life tricorder-patient monitor (xvi) checking all diabetic patients at once, and (xvii) scale that measure BMI. In addition to such device, administrative functions such as asset tracking, patient monitoring, and hospital emergency that cannot tolerate delays in conventional telecommunication network are turning to cellular network as the faster solution.

Financial information is time sensitive, and therefore evaluation of network delay is an important parameter in addition to security. In particular, superfast trading can use dedicated microwave links since they are always faster than IP networks that use cable/fiber and pass information through series of routers and switches, resulting in considerable delay. In general, the financial industry believes in 4G LTE and looks forward to 5G. There are some who believe every financial transaction should be conducted only using cellular network [22].

In 5G, new sensitive area is vehicle-to-vehicle communication for driverless cars. The "latency" parameter in URLLC (ultra-reliable, low-latency communication) of 5G has specified latency of no more than 1 ms. This may be difficult to achieve for many IP networks (based on fiber with routers/switches or traditional Ethernet Category 6 cable). In current IP networks, latency of 30 ms is typical. Alternate methods to reduce latency are being explored in IP networks, which must be evaluated for security. Using fixed wireless technology such as microwave or millimeter-wave links may offer a better option. Such links have latency of no more than a few hundred microseconds (for long distance) and even tens of microseconds (for short distance). Real-time processing of vehicle radar and automotive control gear require "low latency" as an important factor for safe travel that 5G will address. This will be addressed in later chapters such as microwave links (Chap. 14) and vehicular technology (Chap. 15).

6.7 Physical Security and Access

Till now we dealt with electronic security that concerned data when it passes through either wired or wireless network. What about the situation where someone is able to copy your credit card or corporate ID to gain access? This can lead to serious consequences discussed at the beginning of this chapter. Physical security can take different forms and some are listed here:

1. Computer theft: It is easier to steal a laptop or a cell phone that may contain not only passwords, but lot of personal data that the cybercriminal can use to gain access to bank account or corporate access to gain sensitive information.
2. Cameras and biometric identity: Commonly CCTV cameras are installed as a means to identify criminals trying to gain unauthorized access. Are the viewing angles, ambient light conditions checked? Where are these monitored—locally or remotely? These are some of the considerations that can lead to effective countermeasures.
3. Card skimming: Using powerful magnetic devices, it may be possible to skim your credit card or other magnetic strip information (including corporate ID) that might provide criminals access to these accounts or access to corporate offices.
4. Near-field communication: Since smart phones allow "tap and go" technology of near field used often in financial transaction, it is important to verify if the business uses secure channels. A potential hacker could listen in and tap personal information if the channels are not secure [23].
5. Another important risk is access to your vehicle. With COVID-19, several owners leave their cars unattended for extended periods of time. Access to many of them is keyless entry and this technology is not very secure. Thieves/burglars can steal or vandalize your vehicles [24].

In terms of access, let us revisit the three examples discussed earlier "www, email and block chain". How has wireless started to change these three cornerstones of network data drivers?

6.7.1 The World Wide Web

Having realized cell phones became the primary means of access to websites that provide news, entertainment, etc., they are being modified to accommodate mobile users. Modifications include the following:

(a) Main page must fit the small screen of a mobile.
(b) Tabs allow easy navigation through the website.

(c) Instantaneous feedback channels through social network such as Twitter.
(d) Variety of technology agnostic development tools for web developers.

In order to help WWW developers and others, BBC (British Broadcasting Corporation), for example, has put together a list of recommendations [25]. They offer guidelines on how the mobile user would perceive their website. What steps are needed to make it easier for mobile users. In terms of security, there is some difference. Tracking users and pushing advertisements etc. is less prevalent on mobiles, since those agencies also need to push their contents following the "mobile user" guidelines. Currently differences between a desktop user and mobile user are noticeable but minimal. But it is important to watch how WWW evolves in the next few years. An indication of how BBC has provided guidelines and that many websites already follow such guidelines is a good example of changes that will occur in website designs.

6.7.2 Email: Will It Remain Viable?

Unlike the WWW, email now encountered major competition from mobiles—every text message is encrypted and cannot be read until decrypted by the intended receiver. Unlike the text message, the email is largely based on free webmail programs that are easy to track, either by subpoena or by the companies offering the free tools, "E-mail is the easiest to spy on" is the verdict of interceptors [26]. Other chat-apps such as "WhatsApp" provide alternatives in terms of private conversations that are secure. The result is email may no longer remain one of the pillars of network data stream. It may continue in the corporate world for some more time. It is important to note that many of the chat applications pass through cellular network, not PSTN network of IP routers and switches. Cellular network provides an alternative route that users feel comfortable and secure due to less hacking, and encryption which they feel is more secure.

6.7.3 Block Chain

Perhaps the biggest advantage for block chain is that it was introduced after mobile had already become stable method of financial transaction. By eliminating the need to use an intermediary organization, block chain increases the level of security for mobile banking. In effect it has become enabler to promote mobile usage. Using blockchain transactions takes place directly between two parties on a fully transparent ledger, so it is immune from human manipulation or error [27]. In the next few years well over half the banks worldwide will be using block chain as a standard method in all the financial transactions.

To summarize—of the three pillars of network data traffic, the WWW is adapting itself and will perhaps survive in its future form supporting mostly mobile customers. Email, on the other hand, is on its way out and may remain in use only for specific corporate purposes—its use for private communication will diminish and the younger generation rarely correspond using email. Block chain, on the other hand, is rising with the mobile user and promoting its usage. It will become a natural partner of mobile users by increasing their confidence to transact online.

6.8 Conclusion

Network security when applied to wireless takes a major turn, providing interesting, unexpected, and insights. This chapter covered some of the initial challenges the cellular industry faced, followed by a strong come back. Some of the key algorithms and systematic evaluation of mathematical codes resulted in sustained growth of cellular. It may not be an exaggeration to state that the survival and growth of cellular industry has been the hallmark of cellular security, to the point many IT professionals in corporations as well as cellular providers are often oblivious of it. But general perception across the spectrum of society that "cellular is secure" is now accepted.

In addition to the traditional methods, other schemes not originally aimed at security brought improved security of wireless systems. This was discussed to show that there are additional ways and means to pursue. Comparison between cellular and WiFi security showed the strengths and weaknesses of each. With WAP3 WiFi seems to come of age pointing it as "the way to go" in defense, corporate, and sectors that traditionally viewed WiFi with suspicion in terms of security.

Finally, RF engineering with its ability to offer antennas with beams can bring an entirely new way of cost-efficient, secure systems that are now being deployed, but not known widely. The saga of security as a challenge and response between hackers and administrators continues to be an interesting field, for times to come. The concept of moving away from the IP layer has not gone away either.

It is worth recalling the well-known security patent US397412A by Hedy Lamar (Marky Hedy Kiesler). Hedy was a Hollywood actress but more importantly she was a gifted inventor, particularly in security and encryption. Her work described secret communication system "encryption being effected by mechanical apparatus". (for example, rotating cams, switches, key tape punches). Her Patent was published on August 11, 1942 [28]. Although her innovation was not widely deployed in her own lifetime, many looked forward to the anticipated date of her patent's expiration on August 11, 1959.

Subsequent to her patent expiration, her innovation was widely deployed and praised—often indicated by network security experts as the forerunner to modern-day network communication security of spread spectrum—widely used in CDMA technology [29]. Perhaps she would have looked at many of the present-day innovations and inventions in network security and beamed happily at her own work, which acted as an important basis and springboard for future work.

Summary

This chapter provided not only an overview of cellular security but also compared and contrasted it with other security schemes known to IT security professionals in corporations. With wireless being the largest network globally, the challenges it faces from cybercriminals continues to grow. At the same time, cellular has certain in-built robustness never enjoyed by wired network. Whether the unmatched speed of microwave links or the flexibility in deployment offered by cellular, security schemes continue to sustain and enhance growth.

This chapter highlighted some of the early challenges, followed by robust response that surprised many skeptics who always followed WiFi security but had not paid much attention to the cellular security and how it evolved. Contrast between these two was brought in Tables 6.2.1 and 6.2.2 to provide a quick overview of their path showing success/weakness of each.

Finally, some security schemes are unique to wireless (both cellular and WiFi) because of the control over the "air interface". These were explained to show innovative schemes using geometry and trigonometry of antenna beam patterns with deployments in industry, as a method to remain secure based on good simple design. It shows that deeper knowledge of RF (radio frequency) engineering can be brought to bear resulting in low-cost, yet high-security schemes. This aspect was highlighted in "Geometry and Trigonometry" aspects of deployment. Some of the air interface methods and ability to control messages sent over the air offer perspectives of not only current systems, but others that are in the pipeline. 5G may see a major deployment of such schemes for signal separation in three-dimensional space. Also use of 5G as the main method of communication for driverless cars was confirmed by all major automakers in 2019.

The chapter elaborated on coding theory, cipher, and encryption in an effort to make students understanding the level of complexity and implementation in wireless security. It strongly argues against public notion that wireless is insecure. This chapter even provided examples where professionals in the financial industry strongly feel cellular is right choice for all confidential transactions.

Emphasis on participation

During late 1980s, a student from Cambridge University had invited professionals to BBC studio where on a live program was broadcasted. That student showed to live audience, how he could successfully log into world's most important defense network. This was viewed with awe and suspicion then but has now become an accepted method to learn about weakness in any network.

1. What are your views on the three primary pillars of cyberspace: WWW, email, and block chain in terms of security and fraud? Are they the main drivers of data in networks?
2. Where are the weakness in each of these? List at least two—using security as a concern.
3. Given one corporate view that cellular is the only way to operate financial industry, offer reasons why this could be correct?
4. Telephone and voice offer their own security in terms of language and phrases known only to the speaker and listener. Why is this not popular any more, from a security perspective? Should people reduce texting and increase talking using cell phones?
5. The Cell Phone Has Provided Growth to Small Business Users, Farmers, and Laborers. Will This Continue to Sustain and How is Text Message Expected to Grow Securely?
6. Will the conventional IP network of today disappear? The cell phone does not have an IP address but desktop PC, or the WiFi access point do have an IP address? Does it make them inherently more vulnerable than the cell phone?

Knowledge gained

New knowledge gained in this chapter is surprisingly linked to old world knowledge of secure messages through traditional messengers. While the fundamentals of security do not seem to change, some methods of implementation and use of technology seem to provide a different insight into an area deemed essential to human society.

Considerable support from the world of applied mathematics, brings excellent schemes and methods to provide two terms "integrity" and "confidentiality" that are both modern and old at the same time. Various methods of generating and separating keys, how they are managed so that potential hacker is denied access, were briefly touched upon.

Various schemes and methods discussed in this chapter are meant to provide an overview of security schemes "not seen" by public, and often not known even to those who operate the network. The knowledge gained in this chapter provides an overview of effort by academics, designers in industry, and professional users as well as contribution by hackers whose singular achievement was to show weakness in the network/codes.

Homework problems

1. Bluetooth, WiFi, and garage door opener all operate on the 2.4 GHz band. Would it be possible to intercept WiFi using garage door opener? What security scheme is used to stop this?
2. A SIM card is being used in a cell phone that is being discarded. The service provider offers a new SIM card and claims the old one cannot be used. How would you verify if that is correct?
3. A new access point bought recently must be connected to 4G network. Do you need a SIM/USIM card? If yes why, if not why? Instead of access point if a cellular modem is used what are the changes expected? Does the 4G modem need a USIM card?
4. Confidentiality and Integrity Algorithms of Kasumi Are Readily Described Online. Using These Documents, is It not Possible to Crack 4G Network? Why not?
5. Vernam cipher was based on simple XOR truth table. Using Vernam code make a ciphertext of one of your own paragraphs (with ASCII code). Ask your partner to decode and tally answers.

A	B	A $\oplus$ B
0	0	0
0	1	1
1	0	1
1	1	0

Plain text $\oplus$ key = ciphertext.
Ciphertext $\oplus$ key = plaintext.

6. Make your own simple code based on Boolean logic, different from Vernam code, and without telling your partner the coding scheme, ciphertext one sentence and challenge your partner to decode it.

Project ideas

A campus football stadium has ticket counters at two of the five gates. All the gates have cameras (Camera 1–10) with motion sensor lights. All of this must be centrally monitored from sports department building 1 km away. Design a low-cost WiFi scheme and propose a network. What frequency bands would you use? Note: The campus WiFi network (on 2.4 GHz) also serves the auditorium.

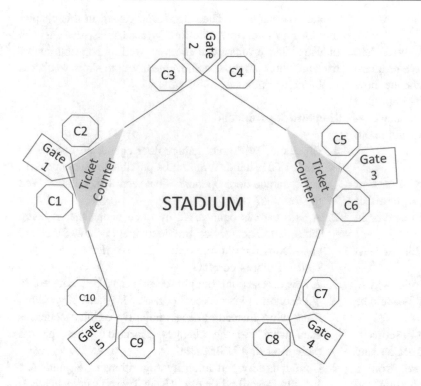

Out-of-box thinking

CDMA or code-division multiple access is one of the most robust, in terms of security. Yet a different access technology in the form of OFDMA came up in 4G. Is change in technology essential to security? Will CDMA make a comeback given its track record?

Propose a scheme not based on mathematical code that still offers better network security. Could it be based on new sense mechanism such as smell or sight, etc. in terms of perception?

References

1. Defcon DCG group—for Information technology alternatives. https://www.defcon.org/
2. How to clone a phone without touching it. https://cocospy.com/blog/how-to-clone-a-phone.html
3. Nation states—why they hack, edge wave. http://www.edgewave.com/wp-content/uploads/2015/04/EdgeWave_NationStates_WhyTheyHack.pdf
4. Gauravam P, Millan WL (2004) Improved attack on the cellular authentication and voice encryption algorithm (CAVE). Research Gate, Article. https://www.researchgate.net/publication/27466264_Improved_Attack_on_the_Cellular_Authentication_and_Voice_Encryption_Algorithm_CAVE
5. KASUMI 3GPP f8 and f9 cores. https://www.design-reuse.com/sip/3gpp-kasumi-f8-and-f9-cores-ip-17808/?login=1
6. LTE and beyond, IMSI, TMSI and GUTI—how they are created. http://www.lteandbeyond.com/2012/02/imsi-tmsi-and-guti-how-they-are-created.html. Last Accessed 15 Feb 2012
7. Cao J, Ma M, Li H, Zhang Y, Luo Z (2014) A survey on security aspects for LTE and LTE-A networks. IEEE Commun Surv Tutor 16(1 First quarter):283–302
8. Forsberg D, Horn G, Moeller WD, Niemi V (2013) LTE security (NSN/Nokia series) book. John Wiley publication. ISBN 9781118355589
9. The evolution of security in 5G, overview of 5G security architecture in 3GPP. https://www.5GAmericas.org/files/8515/.../5G_Americas_5G_Security_Whitepaper_final.pdf
10. Guide to bluetooth security, NIST special publication 800–121, revision 2. https://nvlpubs.nist.gov/nistpubs/SpecialPublications/NIST.SP.800-121r2.pdf
11. Incibe (2016) Security in Zigbee communications. https://www.incibe-cert.es/en/blog/security-zigbee-communications. Last Accessed 26 Apr 2016
12. Siciliano R (2011) Is mobile banking more secure than traditional online banking? ATM Marketplace. https://www.atmmarketplace.com/blogs/is-mobile-banking-more-secure-than-traditional-online-banking/. Last Accessed 2 Aug 2011
13. Cichonski J, Franklin J, Bartock M, LTE Security how good is it. National Institute of Standards and Technology (NIST). https://csrc.nist.gov/CSRC/media/Presentations/LTE-Security-How-Good-is-it/images-media/day2_research_200-250.pdf
14. Stallings W (2005) Cryptography and network security. Prentice Hall, Fourth Edition. ISBN-13: 978-0-13-187316-2
15. Parhi K (2012) Verifying equivalence of digital signal processing circuits. In: 46th ASILOMAR conference of IEEE in Pacific Grove. CA, USA, pp 4–7
16. Robeiro C, Cryptography and network security. IIT, Madras. http://www.cse.iitm.ac.in/~chester/courses/16e_cns/slides/01_Introduction.pdf
17. Borgaonkar R, Hirschi L, Park S, Shaik (2019) A new privacy threat on 3G, 4G, and upcoming 5G AKA protocols

18. Rong C, Cheng H, Vacca J (2014) Wireless network security. Elsivier, pp 291–317. ISBN 978-0-12-416689-9. https://www.sciencedirect.com/topics/computer-science/stream-ciphers

19. Adamo O, Varanasi MR, Joint scheme for physical layer error correction and security. International scholarly research network. ISRN Communications and Networking, Pomalaza-Ráez C (ed.), Volume 2011, article ID 502987, 9 pages. https://www.hindawi.com/journals/isrn/2011/502987/. https://doi.org/10.5402/2011/502987

20. ETSI work program for 2020–2021, the future for tomorrow. https://www.etsi.org/images/files/WorkProgramme/etsi-work-programme-2020-2021.pdf

21. Scher DL, Cheasanow N (2014) 15 Game changing wireless devices to improve patient care. Medscape News & Perspective. https://www.medscape.com/features/slideshow/wireless-devices#16. Last Accessed 23 Oct 2014

22. How secure is 4G? Opengear. https://opengear.com/articles/just-how-secure-4g

23. Security concerns with NFC technology. http://nearfieldcomm-unication.org/nfc-security.html

24. Sfeir J (2020) What is the difference between bluetooth low energy, UWB and NFC for keyless entry? Microwave and RF magazine. https://www.mwrf.com/technologies/systems/article/21140621/whats-the-difference-between-bluetooth-low-energy-uwb-and-nfc-for-keyless-entry. Last Accessed 31 Aug 2020

25. BBC future media standards and guidelines, mobile accessibility standards and guidelines v1.0. https://www.bbc.co.uk/guidelines/futuremedia/accessibility/mobile_access.shtml

26. Graham J Texting or e-mail: which gives you more secure communication? USA Today (2019). https://www.usatoday.com/story/tech/2019/10/13/how-keep-your-chats-private-whatsapp-signal-viber/3909981002/. Last Accessed 13 Oct 2019

27. Mobile banking a breakthrough: block chain and fintech. https://blockchainlabs.asia/news/mobile-banking-a-breakthrough-blockchain-and-fintech-development/

28. Secret communication system—Hedy. https://patents.google.com/patent/US2292387A/en

29. Chapman G (2000) Hedy Lamarr's invention finally comes of age. gary.chapman@mail.utexas.edu. Los Angeles TIMES. https://www.latimes.com/archives/la-xpm-2000-jan-31-fi-59503-story.html. Last Accessed 31 Jan 2000

Abbreviations

Ionizing radiation	Electromagnetic radiation that has a secondary effect on knocking out the outer valency electron in a molecule. This results in ionization of the molecule
Non-ionizing radiation	EM radiation that may only result in temporary warming of matter due to torsion or rotation of the molecule, with affecting the actual property of the molecule
Photo ionization	Ionization produced in a medium by the action of electromagnetic radiation
Compton Scattering	Discovered by Arthur Holly Compton, it is the scattering of a photon after an interaction with a charged particle, usually an electron
Ultraviolet	Ultraviolet (UV) light has shorter wavelengths than visible light. Although UV waves are invisible to the human eye, some insects, such as bumblebees, can see them
Infrared Radiation (IR)	It is a region of the electromagnetic radiation spectrum where wavelengths range from about 700 nm (nm) to 1 mm (mm). Infrared waves are longer than those of visible light, but shorter than those of radio waves
Phantom skull	Model of human head filled with liquids and gel composition similar to human brain. Used for measuring RF radiation absorption by the head
SAR	Specific Absorption Ratio: SAR is defined as the power absorbed per mass of tissue and has units of watts per kilogram (W/kg).
Maximum Permissible Exposure	Limit of RF power that a human/living being, which can absorb without causing ill health. It is an SAR-based method that indicates exposure limits
MRI	Magnetic Resonance Imaging is a commonly used technique in hospitals. MRI does not use ionizing radiation (high-energy radiation that can potentially cause damage to DNA, like the X-rays used CT scans). There are no known harmful side effects associated with temporary exposure to the strong magnetic field used by MRI scanners
CT scan	A CT scan of the abdomen (belly) and pelvis exposes a person to about 10 mSv. A PET/CT exposes you to about 25 mSv of radiation. This is equal to about 8 years of average background radiation exposure
RFA, Radio Frequency Ablation	A medical procedure commonly used for back pain treatment that is using RF radiation (non-ionizing) to destroy the nerve fibers carrying pain signals to the brain. This can result in permanent relief to the patient
WHO	The World Health Organization is a health standards body within the United Nations that has a clear commitment to achieve better health for all being around the globe
Unintentional radiators	Electronic or other units designed for some other function but emits RF radiation as a secondary effect.

K. Raghunandan, *Introduction to Wireless Communications and Networks*,
Textbooks in Telecommunication Engineering, https://doi.org/10.1007/978-3-030-92188-0_7

RFID

Switching power supplies, chokes, and PC may have such radiation Radio Frequency Identifier units are smart labels/tags that has integrated circuit with an antenna. They are pre-coded with digital content. RFID reader extracts this digital information, using a range of technologies known as Automatic Identification and Data Capture (AIDC). RFID tags can be passive (no battery) or active (with battery-powered unit)

7.1 Health Concerns: Public Perception

Currently almost in every town, city, and even rural areas, cell towers are visible. This has raised health concerns and there is a strong public perception that RF radiation can cause serious health issues. Without taking a position, it is important to first review what has been done so far. Did those who designed the cell phones and base stations think about public safety? Was it an afterthought that was acted upon in response to public complaints?

In this case, the answer is quite clear. Those designers not only thought of public health but did quite detailed studies over the past four decades. In fact, those studies started way before cellular was even thought of. Such research spans all major countries of the world and includes engineers, physicists, and medical doctors. There are well over 1400 such research studies all of which point to the same conclusion. That RF radiation in levels being used at all these networks poses no serious health concerns to humans, flora, and fauna. Does this mean any transmitter with any amount of power is within these limits?

To answer these questions, it is necessary to delve into the fundamentals of Physics that defines electromagnetic radiation. In Chap. 2, basic review and electromagnetic spectrum and nature of radio waves were described. Section 2.1 delved into the Physics aspect of these waves that range all the way from our power lines to beyond the light spectrum. We shall now discuss more on the interaction of these waves with biological matter. By that we should consider its interaction with humans such as muscular mass, blood, and tissues. We should also consider other reactions such as plants and vegetation, animals and birds that are exposed to this radiation.

For our review, let us consider Fig. 2.1 of Chap. 2 where a pictorial representation of Electromagnetic (EM) waves and Table 2.1 that described various frequency bands and

power levels being used. We shall now revisit the research work that physicists and engineers conducted and how medical doctors helped in defining the interaction with biological matter.

7.2 IEEE Standards of Radiation Safety

Biological effects of EM waves can be considered based on their fundamental interaction with matter. That not only includes biological matter, but all other materials as well (solid, liquid, gas). EM interactions are based on a fundamental notion—is the EM wave ionizing or "non-ionizing". These two terms are fundamental to public awareness, and therefore let us consider them in detail. Figure 7.1 shows different wavelengths of radiation and how it affects matter at the molecular level. Microwave and RF radiation are at the bottom of the chart in terms of radiation energy levels. Its effect in the radio band is limited to molecular rotation and torsion. Earlier in Table 2.1 these were represented in the first two columns with frequency ranging from 1 Hz to 300 GHz. The third column represented visible light spectrum and the very last column represented waves beyond visible light. Why is ionization so important? Because ionization can fundamentally alter the molecular structure leading to long-term effects.

Let us physically consider what happens. With non-ionizing radiation, the only effect observed is warming. In Fig. 7.1, on the right-hand side are "non-ionizing" radiation (marked in green) that includes not only radio waves but also visible light and infrared waves. Hence, as soon as the source of RF energy is turned off, the heating stops. There is no further or secondary effect. In Fig. 7.1, longer the vertical arrow, higher is the ionization energy. Radio waves and microwaves used in the wireless technology have the lowest energy levels (on the right bottom corner) and can cause only molecular rotation and torsion of the cells. Since the energy is very low, even warming of the body is not felt. The next higher level (to its left) is infrared rays coming from the sun that can warm up matter. But as soon as the source is removed or the sun goes down, matter starts to cool down. This is witnessed by all of us daily. Compared to the wavelengths used in wireless, infrared radiation wavelengths are shorter, resulting in molecular vibration; more warming of skin can result in temporary redness of the skin. But it heals, in course of time. Infrared radiation is used as room heater in homes, and in beauty salons.

Ultraviolet rays, on the other hand, have much higher energy level resulting in electron level changes, to an extent modifying the molecule. Doctors warn us to use UV lotions to protect the skin, since UV rays are part of sunlight to which we are exposed during very warm days outdoor.

Fig. 7.1 Background to EM radiation—its effects on matter

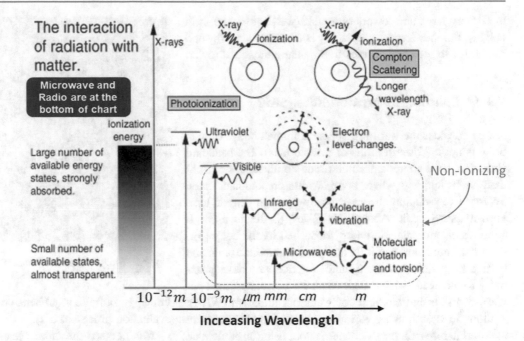

That brings us to the waves that cause ionization (in Chap. 1, last column in Table 2.2) which happens to come from X-ray, gamma ray, and these naturally occur in outer space. Luckily, they are blocked by the ozone layers of the Earth. But man-made X-ray or gamma ray is used in medical equipment as part of radiation therapy. Effect of ionization is felt "Even after the source is turned off". They are represented in the left top (small wavelength) in Fig. 7.1 showing ionization energy is highest with X-rays; gamma rays resulting in secondary effect such as "Compton Scattering". Although not shown in Fig. 7.1, alpha particles, beta particles, neutrons, gamma rays, and cosmic rays are all known as "ionizing radiation", meaning that when these rays interact with an atom (or molecule) they can knock off an orbital electron. The loss of an electron can cause problems, including everything from cell death to genetic mutations (leading to cancer), in any living being. This is the reason why "ionizing radiation" is always taken seriously. But "non-ionizing" radiation does not have these issues as already explained and shown in the bottom green box.

Observe that RF and microwave have low energy that can only cause molecular rotation which causes warming muscles—wireless technology wavelengths are in mm, cm, or m in wavelength. In recent times, our society uses the word "radiation" in the negative sense, forgetting the fact that electromagnetic waves coming from the Sun and observed as visible light is radiation essential to the very survival of all species living on Earth. But it is also readily seen that such EM waves from the Sun mostly contain "non-ionizing" radiation (except UV as indicated earlier). Thanks to the ozone layer that blocks most of the "ionizing radiation" from outer space. Now that the differences between "ionizing"

and "non-ionizing" radiation are clarified, let us focus on the "non-ionizing" radiation which covers the entire wireless spectrum as well as most of the visible light spectrum.

7.3 National Bodies (FCC and Others) and Regulations

Since wireless consists of EM waves that are "non-ionizing" radiation, why do national and governmental bodies regulate its power levels? Is it not true that regulatory bodies limit the amount of RF energy sent over the air and don't they specify how much RF energy can be radiated in populated areas?

To answer these two questions, we should begin with the basis of how these regulations were set up and why? The origin of these regulations' dates to 1970s when majority of the citizenry were neither aware nor concerned with wireless or radio waves. Broadcast radio and TV do transmit power levels well over few thousand watts, which begs the question, what does this do to humans who live nearby? During 1950–1974 there was an extensive study of Naval Korean War veterans, and effect of radar exposure among them [1]. It is well known that the Second World War used radars that were high powered (kW or MW but in the form of short pulses), and the defense industry was concerned about the safety of soldiers who were in the vicinity [2]. Some of these studies particularly by doctors and scientific research community suggested use of regulations by national bodies to limit human exposure. Broadcast industry, on the other hand, has their towers set up in open areas that provide them better access to microwave or satellite links which connect their sites to the overall network. Size of broadcast antennas tend to be very large and towers quite tall (well above 100 m in height). For these reasons, citizenry is usually not nearby and rarely do we see communities living near a broadcast tower.

7.3.1 Specific Absorption Rate (SAR)

Let us consider the warming effect on human body. Human body is not uniform in terms of muscle mass. The head and abdomen have more muscular tissue (brain and stomach) than hands and legs where bone has the predominant mass. To measure radiation absorbed by the body, a generic term known as "Specific Absorption Rate" (SAR) is used. It signifies electromagnetic energy absorbed by the body [3].

But the area most carefully analyzed organ is the brain and human head since handheld radios operate very close proximity to the head. SAR is defined as the power absorbed per mass of tissue and has units of watts per kilogram (W/kg). Although research using various models of the human head are used in research, major concern about using an adult male head model resulted in further questions. The head of a child and the dielectric properties of its smaller skull are known and established to be more vulnerable. In fact, research confirms that radiation absorption into a child's head can be over two times greater, and absorption into child skull's bone marrow can be ten times greater than adults [4]. SAR and how laboratory tests comply with regulations depends on three things: (1) the model, (2) the method, and (3) the results. With these guidelines, standards were developed based on modified SAR and head model (not male head model).

IEEE 1528 standard provides guidelines on how to develop a human head model. Typical model shown in Fig. 7.2 is an example from "environmental health trust". This trust represents an independent team of citizens that work with research groups. A sample model in Fig. 7.2 was developed using material whose chemical composition closely matches the human head. Table 7.1 shows one such model [4] with composition of "phantom head" and what its contents are.

With increased use of wireless devices and their proximity to humans, the concept of SAR and providing guidelines by regulatory bodies will continue. How are these limits arrived at? To answer this question, we must consider primary effect of radio frequency on humans. Its only effect is heating of muscle tissue as illustrated in Fig. 7.1 classified as non-ionizing RF energy (microwaves). Researchers consider specific power level at which the increase in temperature (heat) is observed in muscular tissue. This is determined for a band of RF frequencies and tabulated. Then the limit is set at 1/100th this power level—just as a matter of precaution. A comprehensive version of such standards is the IEEE 95.1c that originally started in 1966 and revised in 1971 and again in 1982. The latest version was approved in 1991 and evaluation process of arriving at recommended

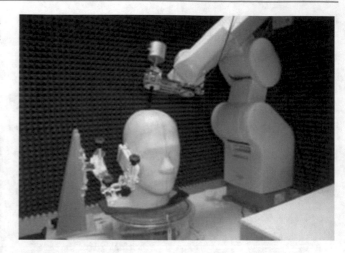

Fig. 7.2 Model of human "Phantom skull" used in SAR test

limits is quite comprehensive (see Fig. 7.3 of how the evaluation process work).

It is perhaps the largest team with the most comprehensive evaluation of all literature of published work by qualified scientists, physicians, and engineers. This work is therefore beyond dispute and used as basis by most national/international regulatory bodies such as FCC and ITU. This work represents professional findings in the frequency range 3 kHz to 300 GHz, encompassing all electromagnetic waves designed to prevent harmful effects in humans. The SCC28 indicated in Fig. 7.3 is open to anyone who can work and provide inputs to the group. Detailed tables and values are provided with measurements over the different frequency ranges, at various power levels [5]. For general understanding, it is useful to note some generic values (basic guideline). This is indicated as Maximum Permissible Exposure (MPE) in Table 7.2 but separated for occupational and public.

Why is there a big difference between the two groups? The question can be answered in a logical manner. Those who are occupational users such as Police, Fire, EMS (emergency medical service), and others, use it for one shift of 8 hours a day. They use PTT (push-to-talk) which transmits RF only when they are speaking. Also, they go through professional training on how to use and care for their handheld device. In contract, all others (general population) use cell phones during most of day and night. Therefore, it is safe to assume that they could be exposed to RF radiation 24 h a day. Also, they are not trained in its usage, and it is important that they are set to much lower exposure limits. Note that all regulators don't strictly follow IEEE 95c but do use it as a reference, and modifications are made as needed.

A cursory glance at the table indicates that safe limit for a newborn baby (about 2 kg) is 3.2 W of RF energy. To put this in perspective, the normal cell phone today can only

Table 7.1 Composition of phantom materials (percentage by weight)

Ingredient	Equivalent tissue		
	Brain	Muscle	Eye
Water	68.40	75.44	88.84
NaCl (sodium chloride)	0.61	0.91	1.21
Polyethylene powder	23.33	15.2	–
Super "Stuff" (gelling agent)	7.66	8.45	9.95

Fig. 7.3 Evaluation process used by radiation health safety standard of IEEE

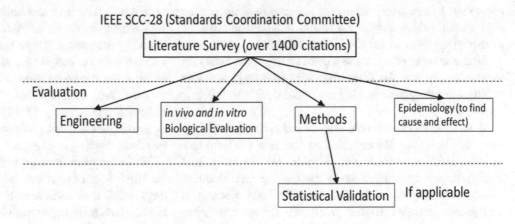

Table 7.2 Maximum permissible limits of SAR (W/Kg)

Occupational/controlled exposure limits		
Whole body	Partial body	Hands, wrists, feet, and ankles
0.4	8.0	20.0
General population uncontrolled exposure limits		
Whole body	Partial body	Hands, wrists, feet, and ankles
0.08	1.6	4.0

transmit (maximum) 200 mW or 1/5th of Watt. This limit would require 16 of those phones operating at peak energy transmitting only towards the newborn (not possible since cell phones transmit equally in all directions). Similarly, the WiFi access point at home also can only transmit 200 mw and the same is true. The key here is that MPE limit is set so conservatively that even in the worst situations, it is not possible to be exposed to this level of radiation.

Federal regulations by agencies such as FCC specify the radiation limit (MPE) based on distance as well as ERP (effective radiated power). The following is an excerpt taken from a recent FCC regulation [6]. It states:

SAR-Based Exemption. For fixed, mobile, and portable RF sources near a human body, where the separation distance is normally between 0.5 and 40 cm and may be less than l/2p, the commission adopted the new RF exposure evaluation exemption formula shown here for time-averaged power thresholds (specified in mW) for exemption of single portable, mobile, and fixed RF sources at 0.3–6 GHz.

A source is exempt if each of the maximum time-averaged available (matched conducted) power and effective radiated power (ERP) is no more than

$$P_{TH}(mW) = \left\{ ERP_{20\,cm}\left(\frac{d}{20\,cm}\right)^x d \le 20\,cm \right.$$

$$P_{TH}(mW) = \left\{ ERP_{20\,cm}\, 20\,cm \le 40\,cm \right. \tag{7.1}$$

where $x = -\log_{10}\left(\frac{60}{ERP_{20\,cm}\sqrt{f}}\right)$ *and* f *is in GHz*

And $\{ERP_{20\,cm}(mW) = \{2040f\ 0.3\,GHz \le f < 1.5\,GHz$

$\{ERP_{20\,cm}(mW) = \{3060f\ 1.5\,GHz \le f < 3\,GHz,$

where d *is the separation distance.*

This type of analysis though based on IEEE 95.1c accounts for proximity devices that are very close to human beings. This considers the distance as well as ERP (Effective Radiated Power) to define what is admissible.

7.3.2 Beneficial Uses of RF Energy

Perhaps the most directly seen example of useful RF energy device is the microwave oven used in the kitchen around the world today. Its value in precisely targeting and heating food is well understood. Telecom regulations clearly restrict indicating no more than 5 mW of energy should leak through the door (this is less than what is normally transmitted by a cell phone). This leakage is tested, verified, and approved, before every model of microwave oven can be sold to the public. Figure 7.4 shows a typical microwave oven that operates in the 2.4 GHz band today.

Manufacturers of microwave oven also give the safety tip —"don't peep into the door when it is operating" and also confirm that the power leaking out of the door is well below 5 mW.

With advances in medical science and accurate modeling using Multiphysics, RF radiation has become a helpful tool in diagnosis and treatment. Signals in 915 MHz and 2.45 GHz are considered ideal for heating and ablation, many such devices prefer the 2.4 GHz band since it is unlicensed globally. Higher frequency signals are better suited for treatment of skin and liver cancer, as well as treatment of heart to treat arrhythmia [7]. Use of multiphysics requires considerable accuracy and skill so that treatment such as MRI (magnetic resonance imaging) can target the area precisely by understanding interaction of electromagnetics in different layers of human body. Unlike MRI, the CT (computerized tomography) scan uses ionizing radiation like X-ray producing accurate images of the body, locating precise area of tumor or its growth [8]. Later, doctors may proceed with specific treatment using other medical procedures. Therefore, medical hospitals use both ionizing and non-ionizing radiations for diagnosis.

Such beneficial use of RF energy resulted in a standard by the IEEE known as SCC34 whose objective is to understand precise use of RF energy exposure. IEEE Standard Coordination Committees continue to define and enhance the IEEE 95.1 standard (IEEE 95.2 and 95.3 are being worked) widely recognized and used by regulatory authorities throughout the world.

One of the important features of this standard SCC34 is an open public comment method by which anyone concerned with RF health and safety can provide input and work with experts in the field. Figure 7.5 shows the process followed by the standard, note that on the right-hand side each national standard can use the same format and provide inputs to the standards board as well as adapt the standard to their national use. For example, the SCC28 had about 100 members representing 13 different nations. The IEEE is indeed an international professional body with majority of its members residing in countries outside of USA.

While the work done by SCC28 and the results published via IEEE 95.x have stood the test of time (about half century) not all their results are based on modeling only. In the initial years, there were multiple tests conducted on animals, in particular, different species of monkeys [9]. This included studies related to effects on the body, on organs, and tissues such as.

1. Nervous system.
2. Visual system.
3. Endocrine system.
4. Immune system.
5. Hematologic and cardiovascular system.
6. Animal carcinogenesis.

Such studies conducted over a wide range of frequencies were pursued by researchers in various countries, with similar conclusions. This study was quite elaborate in terms of building a GWEN (ground wave emergency network) consisting of many sites [9]: (a) input/output stations capable of transmit and receive, (b) Receive-Only stations (RO), and (c) Relay Nodes (RN). Each of these could operate over a wide range of frequencies and power levels. Studies were conducted using detailed calculations of pulse width, modulation, and signal formats. The power levels were set to the limits indicated in Table 7.2 and effects on primates were evaluated to cross check cause and effect. It also references works of 80 other publications. Reinforcing work was done in this area which were carefully evaluated and re-affirmed by work done by others as well. The body of knowledge from researchers, physicians, and engineers from various countries offers a consistent view. The basis of radiation limits set in Table 7.2 is based on multiple studies that are together deemed quite conservative.

Fig. 7.4 Typical Microwave Oven—energy-efficient cooking device in kitchens

Fig. 7.5 RF safety standard
setting process

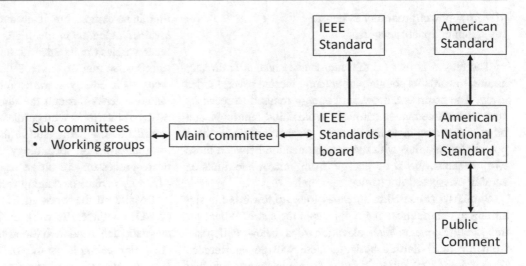

7.3.3   Did Wireless Proximity to the Public Increase Health Concerns?

Arguably this topic may seem altruistic—the fact is that work on RF safety started 50 years before public concern was raised in greater social circles. This is not an odd co-incidence but when studies began originally in the 1950s and 1960s, transmitted signal levels were quite high, yet studies concluded that other than the predominant effect of warming the body, no harmful effects were noticed. Initial deployment of all RF systems was either in the military or in the broadcast industry where antennas and dishes were located away from public view. This was not a deliberate attempt, but a practical one, given the large size of antenna towers and dishes needed. Over a period, the need to improve reliability of communication dictated smaller and more efficient devices. The reduction in size and power was not a result of public health concern, but to develop efficient wireless systems to accommodate more users (capacity).

With the growth in wireless technology, cell phone towers and WiFi and proximity devices have raised concern from citizens all over the world. Their concern has focused on what they see in their homes and neighborhood. There is a general lack of understanding that this concern has been addressed with extensive ongoing work for half a century. In some townships, questions are raised on whether manufacturers/designers are concerned about RF safety. The questions are relevant, but answers have been available for long, as briefly reviewed in the previous section. If this is of any consolation, by the time citizenry became aware of and were concerned, the power levels of transmitters have dropped dramatically, quite often 1/100th of what it used to be in the early 1950s and 1960s. This substantial drop in power is due to growth of technology and rise of digital systems.

While the physics of propagation at a given frequency cannot and will not change, transmit levels have reduced since many devices are often battery powered. At the receiver, improvement in digital technology resulted in reliably decoding signals that are about 1/100th to 1/1000th of earlier values (compared to 1950s and 1960s). Modern systems operate very well since digital design is focused on use of minimum power needed to communicate reliably and such designs avoid use of excessive power since that potentially interferes with its own network. Perhaps in the very first chapter, Fig. 1.1a), (b) would serve as reminders to readers, how dramatic the reduction has been, where cell phone transmit power reduced from 5 W to $\frac{1}{5}$ W.

The RF safety standards are conservative and have not changed since they are based on the physics of interaction of RF with biological matter. The result is that invariably, every survey confirms that limits set by RF safety standard C 95.1 since they are conservative and even these levels are not observed in a practical system. In addition, most countries insist on field measured results to confirm that they comply with their national standard. Such certification is used as a pre-requisite to permit construction of a radio tower, or a wireless unit to be sold in the market. In addition, there is another standard—International Commission on Non-Ionizing Radiation Protection (ICNIRP, 1998)—which has made similar recommendations. With another 10 years of study and findings until 2016, the World Health Organization (WHO) has taken the view that although no health effects on human or living beings were found due to base stations or mobile, they will continue further studies to see if there would be additional findings.

In terms of RF radiation safety, there are certain systems that the public need not be concerned about. Some of the following may not be obvious, but noteworthy:

(a) Microwave systems.
(b) Satellite systems.
(c) Receiver systems.

(d) Low power/proximity systems.

(e) Unintentional radiators.

The reason is that (a) the microwave antenna is directly aiming towards its counterpart antenna located several miles away. The beam is narrow, and energy focused to reach its counterpart antenna—it carefully avoids hills, buildings, and other structures including human population since these potentially interfere with the communication between these two antennas. Bulk of energy from microwave links is carefully designed not to reach ground.

Similarly, (b) satellite antennas look up towards the sky aiming to communicate with objects in the space. Whenever the satellite antenna/dish elevation dips below 6 degrees from the local Earth's horizon, there will be interference from ground, resulting in poor communication. Satellite dishes carefully avoid all interfering objects such as hills, buildings, and tall structures, so that they can communicate with the satellite or space vehicle with a clear line-of-sight. Other dishes looking towards the sky could be large antennas used by Astrophysicists trying to receive signals from far away objects in galaxy/outer space such as a star or asteroid or another body. They do not transmit energy towards Earth.

(c) All receiver systems are based on a variety of technologies. The most common receiver radios are used in cars and at home, television sets used at home and offices, GPS units used in cars, in cell phones and integrated with other applications, RFID tags used at toll booths, super markets, and other locations and a host of consumer tags/badges based on RFID. All receiver systems only tap into energy already available in the vicinity. They do not produce any new RF energy. Perhaps the largest of these in size are radio telescopes which could be very large, perhaps 500 m in diameter. Despite their large size, their primary focus is towards the outer universe and study of objects in the galaxy to receive electromagnetic waves emitted from outer space [10].

(d) Typical low-power/proximity systems characterized by battery-powered, miniature units whose range is about 30 m or less. These include Bluetooth, ZigBee, and personal area communication technologies such as UWB and others. Most of them emit very low power in the range of 5 mW or less, often comparable in energy to small bulbs used in toys. These devices can range from commodity products such as wireless microphones, alarm signals from household appliances, to road telematics installed on buses that can communication with traffic lights. The ITU (Int'l Telecom Union) lists a wide range of such devices that readers can refer to [11].

(e) Unintentional radiators are electronic or electrical units that emit radio waves, but their primary use is something else. Typical examples are fluorescent lamps with chokes that emit radiation. The primary purpose is to illuminate lamps, but it produces RF radiation. While the level of radiation is not high, it can cause interference to other wireless units such as handheld units used by public safety personnel or home TV. Similarly, microwave ovens focus their energy towards food inside. Small amount of radiation leaks through the door, which is no more than 5 mW, but interferes with cordless phone. All such devices are certified before being sold in the market [12]. This includes most information technology devices that use high-speed micro-processors, resulting in radiation that must be contained by careful packaging techniques.

Despite all the work so far, World Health Organization (WHO) continues its work on long-term effects of electromagnetic fields and their "long-term" impact on public health [13]. For example, some of the non-thermal effects by low-power signals could produce calcium ion mobility, which are responsible for transmitting information in tissue cells.

7.3.4 Digital Unplug: Away from RF

While it may not be easy to find places that do not have satellite coverage, it is possible to find locations that have no cell phone coverage. The intent may not be for reducing exposure to RF/cellular coverage, but this trend in the cellular world is catching up. Until the cell phone became popular, it was easy to relax and keep away from telecommunication or wired infrastructure. There are still many locations without a telephone/cable connection. But cellular coverage is widespread and so is the satellite signal. Although signal from satellite is considerably weaker in power it covers most of the globe, including the oceans.

Like those who use sunshade and lotion on their skin to avoid "sunburn" due to ultraviolet rays, there are those who prefer to take precautions. There is nothing wrong in turning off your WiFi access point at night, or even cell phone can be turned off at night to sleep peacefully. RF like all other electromagnetic waves reduces in power as square of the distance, at any given frequency. To illustrate this point, consider the absolute worst case with the cell phone or access point next to your body (head, for example). The maximum energy towards the body can only be 100 mW since out of 200 mW (max) only half can come towards the body and the other half radiates away from you. If user keeps the same cell phone 10 m away, the radiation reduces to 1 mW (not 1/10th but 1/100th). At 100 m, it reduces to 0.01 mW or 10 microwatts. This is how quickly RF exposure decreases. The same is true about cell towers. If the cell towers transmit 10 watts and you are only 100 m away (quite close), the power reduces to 10/10000 or 1 mW. This example is given as an illustration, but signal attenuation (reduction) also increases with increase in frequency. Free

Space Path Loss (FSPL) described earlier in Chap. 2, provided Eq. 2.5 showing power loss with distance as well as frequency. It is provided here again as Eq. 7.2 for quick recollection.

$$FSPL(dB) = 20\log_{10}d + 20\log_{10}(f) - 147.55. \quad (7.2)$$

Why is this reduction so dramatic? The clue is in simple geometry.

Radiation expands as a sphere in all directions and therefore can be thought of as an expanding sphere. The human head would be just a dot on the surface of sphere, and the RF source is at the center. When the sphere expands with distance, the signal spreads in all directions equally and what is received at a distance "d" can be thought of as larger sphere on the surface of which the human head sits like a small spec. This is illustrated in Fig. 7.6, where the human head is initially close to the source (on the left) such as cell phone's transmitter. As the person moves away from the source (cell phone), the radiation expands through "free space" and at 10 m RF field is reduced to 1/100th compared to the one of left. At 100 m, RF field reduces to 1/10000th. This is also illustrated by the red balloon expanding to seem like a pink balloon (color diluted) and further to very light pink (color further diluted). The color dilution is akin to illustration of the RF field dilution. The same amount of power from the source expands through space/air equally in all directions. The thumb rule is signal reduces as square of the distance. This was indicated in Eq. 7.2 as $20\log_{10}(d)$. It also reduces the same way with frequency, as square of the frequency as well, but user has no control on the frequency at which a device operates.

For the service provider, the cell phone operating at 800 MHz could send signal to a tower much farther than the typical WiFi and 2400 MHz which can only send signal to a WiFi hotspot (access point) no more than 100 m away. In general terms, higher the operating frequency more will be the attenuation through free space. Keeping that as the thumb rule for now, it helps to note that keeping cell phone or wireless device away from you at night helps in your exposure to RF, when you are asleep. This concept of attenuation with distance will be elaborated again in later Chaps. 11 and 12 when we discuss cellular and WLAN systems in detail.

Hence "Digital unplug" is a conscious and deliberate attempt to keep the device away from you whenever possible. Remember—the cell phone is just device, and there can be life without it. Try it for a day, several days, or even parts of the day when you sleep, when wireless device is kept away. This conscious effort is likened by many to "peace", "away from network", or silence that allows the users to be by themselves and not "dictated by the device". Many corporate bodies actively encourage this which improves quality of lifestyle and productivity in the long term.

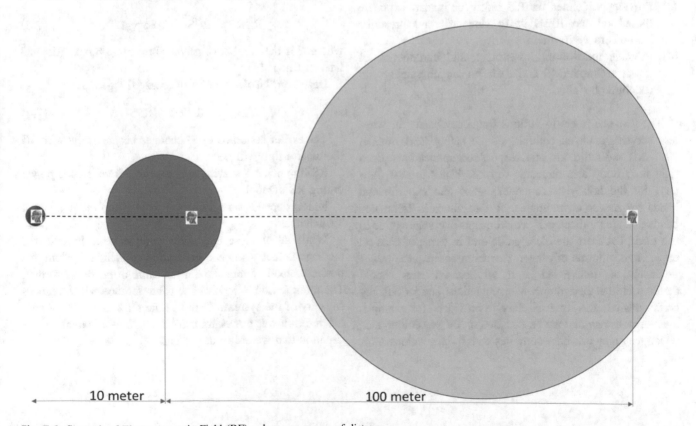

10 meter   100 meter

Fig. 7.6 Strength of Electromagnetic Field (RF) reduces as square of distance

7.4 Verification and Imposition of Safety Regulations

This is an important topic that carries much weightage for manufacturers, service providers, and users. Public concern and debate on this topic have been extensive—some of their concerns are genuine and others arise due to lack of understanding. Regulatory authorities pay considerable attention towards how products comply with well-established standards. For wireless products sold worldwide, there are documented measurement methods to evaluate and screen devices before they enter the market. We will discuss each of them now.

7.4.1 Regulatory Bodies

Regulatory bodies such as ITU-R, the FCC, and others have a multi-fold function towards wireless services involving broadband cellular, satellite, and remote sensing technologies:

(a) Manage the frequency spectrum efficiently, its best use within national boundaries, and monitor their legal operations.
(b) Standardization—seek inputs and publish standards based on inputs from manufacturers, users, and groups with social responsibility towards use of technology.
(c) Provide guidelines on RF safety or human exposure (based on the IEEE 95.1c standard) and prescribe methods to verify them.
(d) Promote innovation, investment, and introduction of newer technologies, and their impact on society and economic benefits.

The first one refers to various frequency bands for wireless services such as cellular, public safety land mobile, television and radio, air, and seaport communications. Each regulatory body also allocates separate bands for the military, for the hobbyists (amateur radio) and experimental bands for newer technologies. It also allocates frequency bands that are "unlicensed" which means everyone is free to use them, but there are clear guidelines in terms of caps on upper limit in terms of power, frequency usage, etc.

Except the unlicensed band, all licensed users should demonstrate that they comply with regulations, and periodically certify devices and services. Service providers, for example, must demonstrate that the base station or TV transmitter, etc., which they may purchase complies with the regulations. This proof usually takes the form of a certificate issued by a neutral test laboratory that tests the base station or transmitter to its national regulatory standard and certify how they were measured and to what extent it was tested. Therefore, the vendor chosen by service provider must provide documents that their equipment (base station) meets the standard and submit certification of approval. It is often necessary to specifically state whether newer version may require recertification. Many neutral test laboratories are used as qualified entities that offer competitive measurement and evaluation services—the service provider therefore gets a choice of vendors who get certificates from such qualified laboratories.

Why is this procedure followed? Often the reasons are embedded in technology and details. With new technology measuring instruments and methods could change. The service providers and equipment vendor often write letters to the educate the regulatory authority indicating changes. For example, power density in a channel was measured in old base stations, purely as an RF carrier wave measurement. In 4G, for example, power contained within the useful bytes is measured. Therefore, what used to be RSSI (received signal strength indicator) shown as bars on a cell phone, now gets measured in three distinct quantities as RSRP (reference signal received power) and RSRQ (referenced signal received quality) and RSSI. But the cell phone will not indicate this to the end user, but bars are adjusted accordingly. Relation between these quantities are as follows:

$$RSRP = RSSI - 10\log_{10}12^N, \tag{7.3}$$

where N is the number of physical resource blocks (defined later in Chap. 11).

Based on this the quality of received signal is

$$RSRQ = RSRP / (RSSI/N). \tag{7.4}$$

The earlier notion of RSSI changes because now with 4G the relationship will be

RSSI = noise + serving cell power + interference power during RS symbol.

RSRQ depends on serving cell power and the number of Tx antennas.

While all of these may seem complicated, the essential message is that with each generation of cellular system, the power seems to either remain the same or go down further. The main reason is to avoid self-interference which reduces capacity of the system. Health of the citizenry is also better off due to lower power transmitted. This is the dual benefit situation that we are in at present.

7.4.2 Human Exposure to RF EM Fields: Regulation

To address public concern, regulatory bodies make it mandatory to submit an RF emission analysis report as pre-requisite to getting license. What this means is that for each radio site, a certified engineer must analyze (using regulatory guidelines of how the calculate/measure) and submit a report. After verifying the report to ensure the emission levels are within limits, the regulatory body provides the license. Therefore, the public perception that radio sites can be constructed without much oversight from the government is not true.

A classic example of such bulletin issued by government is OET 65, issued by the Federal Communication Commission with jurisdiction on all United Sates of America. It expects details such as location (latitude and longitude), height of the tower, power level, details of RF including modulation, channel assignment, type of antenna and its characteristics, amount of power radiated into the air, and closest point where humans could have exposure and radiated emission there. Indeed, such details help the regulatory authority assess impact of radio towers in densely populated areas.

7.4.3 Verification and Imposition of Regulatory Limits

How can the regulator exercise its authority? Monitoring all installations within its jurisdiction, although possible, is not practical. But the authority has a clear rules and review process to register all radio sites and provide details in public domain [16]. The usual means of enforcement comes through a competitor or someone else filing a complaint. Since all radio sites should be on the national database with details, if anyone notices a tower coming up that is not registered, a formal complaint can be launched. Also, citizen's action group can stop and even order dismantling a tower that did not go through environmental and public review process. That provides the authority valid reason to verify and take legal action. Such violations and appropriate action (such as imposing fine) are publicly posted after it has gone through the legal process [17].

FCC has recognized safety guidelines and evaluated RF environmental exposure, other agencies such as Environmental Health Trust (EHT) are moving towards methods to verify and enforce compliance by national law [18]. Many countries in Europe have focused on limiting use of cell phones among younger generation (in the age group of 14 years or less).

7.4.4 Measurement Methods and Technology Growth

Measurement techniques continuously evolve with technology. They are based on inputs from product development houses, instrumentation companies that entirely concern themselves with accuracy and verification with nationally/internationally established standards, as well as regulatory bodies who evaluate and assign permission to independent laboratories. Let us consider use of proximity devices that fire-fighting professionals wear on their body. It is very useful since it alerts the system if fire fighter is in distress and lying without moving for 30 s.

This is now also extensively used by the chemical industry to protect workers who work in areas that could have toxic chemical fumes. What are the challenges? The challenge in design to fit unit within the headgear or close to the body. But it leads to lowering the transmitted output, antenna no longer seen outside, and output cannot be measured accurately using conventional methods. While no one questions the efficacy of such device, it poses major challenges to accuracy with which the RF power can be measured and approved [19]. Fortunately, national standards such as NIST work on the physics, engineering, and measurement aspects to be a reliable test method. Then such methods are standardized and used by all independent laboratories to certify wireless products.

In the ever-changing world of cell phones, this challenge grows further since vendors would like to provide approved products within a short time. Challenge of 5G is millimeter wave, for which established measurement methods don't exist. Then how would the independent laboratory certify. NIST or other national laboratories establish new measurement method by first identifying the problem areas followed by solutions [20].

For simple confirmatory checks in terms of whether the radiation levels are within regulatory limits, small portable instruments do exist. These are general-purpose instruments covering a wide frequency range, with software to conform to national regulatory levels, for example, see [21]. Such instruments are used to satisfy safety departments, unions, and others whose primary concern will be to make sure that workers are not exposed to high levels of RF radiation.

7.4.5 EMI/EMC Compliance

This chapter reviewed public policy, health, and safety. There is need to establish safety of electronic products as well. Questions from product vendors "do my products get be affected if they happen to be near wireless units" must be addressed. Why is this essential? For example, a pace maker placed near heart, a hearing aid, or some other assistance device may be affected by wireless units nearby. Electronic devices could be any appliance whose unintended radiation affects other units. Radiation from cell phone or power supplies, or ballast used in a fluorescent lamp or other device could affect them. There is a whole topic of compliance known as EMI/EMC (electromagnetic interference/ electromagnetic conductance) that is evaluated. The International Telecommunication Union (ITU) lists many countries and their regulatory schemes to validate appliances, see [22]. Some of the measurements described are as follows.

1. **Emission Testing (deals with measurements to evaluate)**

- **Radiated emissions (EMI testing)**

Purpose of the test is to measure the electromagnetic field strength of the emissions that are unintentionally generated by the product. This is typically followed by comparison to see if it is within regulated limit, otherwise the device vendor will have to modify device to comply.

- **Conducted emissions**

Any device that creates electromagnetic energy would have a portion of it will be conducted onto the power supply cord. Test purpose—to restrict the amount of interference the device can couple back onto a power supply.

2. **Immunity Testing (deals with ability of device to withstand)**

- **Radiated immunity**

Purpose of the test is to see how well the device performs when it's subjected to an electric field of a specified amplitude (measured in volts/meter) across a range of frequencies.

- **Conducted immunity**

When cables are bundled together, they have capacitive and inductive coupling. Purpose of this test is to check if a given cable/wire can withstand frequencies of current carried in an adjacent cable.

- **ESD—Electrostatic discharge**

Most common reason for accumulation of electrostatic charge is walking on a carpet . The purpose of this test is to apply electrostatic discharges to any area that is normally accessible to a human touch and check if there is a clear discharge mechanism so that the operator can safely touch the equipment/device.

- **EFT—Electrical Fast Transient**

Electrical fast transient could be a sudden change in current or discharge due to high-current appliance such as motor, fluorescent lamp ballast, or a toggling switch nearby that can affect your TV or computer peripheral. Purpose of the test is to ensure such appliances are immune to fast transients.

Since electronic/wireless devices are manufactured and sold globally, such compliance tests have become mandatory and certification is needed in order to sell equipment anywhere in the world. This insures not only safety, but uniform level of expectation from the public about how an equipment operates.

7.5   Conclusions

The topic of public policy and safety involves cross section of society beyond engineering and physics. This chapter reviewed important areas such as government regulations, public awareness groups, and other social impacts. Fortunately, the amount of research work so far is so extensive and varied in nature that evidence it provides is irrefutable. The IEEE 95.1 is an important standard that summarizes work performed by medical doctors, engineers, and physicists over a period of several decades. It is therefore used by most nations as the reference standard from which local variants are derived.

There are two important EM sources to observe. One is the largest and most visible source of radiation, "the Sun". All sources on Earth including RF equipment are minuscule in terms of power compared to the Sun. There are many celestial bodies that also emit RF, which is grouped under "background noise", another important factor considered by every communication system. This chapter also delved into positive aspects of RF radiation. How it is helpful and useful for medical treatment. Sun's radiation, for example, is a natural source of vitamin D, essential to humans and living beings. Similarly, there are devices using RF radiation to diagnose (such as MRI or CT scan) and help in treatment. Another example is the humble microwave oven in the kitchen. Over the past few decades, it is the most visible source of RF in kitchens around the world, for a good reason. It is clean, the most energy-efficient electrical device with RF power completely focused on food inside, with very little RF leaking out of its door. Manufacturers take pride to show how well they have designed the door to shield such powerful RF energy and not even 1/10000th leaks of the door.

Instead of addressing all communication systems under one category, five distinct systems were considered to show how some of the systems do not radiate towards humans at all. This is important since public perception of combining all wireless systems together raises concerns that are not logical and therefore invalid. Antennas used for receiving only will have no effect at all, since they are just tapping into existing signals available in the air, which are typically 1/100th or even lower than suggested radiation limits (Table 7.2). All microwave and satellite systems are under this category since they are not directed towards public, and humans/living beings are not affected. Those antennas directly focus RF power towards the receiving antenna.

For human body (as well as other animals) the amount of radiation that is considered safe was established based on extensive scientific trials. These were reviewed and Table 7.2 provided a bird's-eye view on safe levels acceptable to humans. Knowing such limits helps everyone; humans can consider factors to reduce exposure by just moving a short distance away from the source. This was illustrated in Fig. 7.6, showing how the RF power quickly reduces by square of distance. It also helps government-controlled regulatory bodies to monitor and even penalize those who violate such guidelines [17]. Several societies and governments use such examples to initiate guidelines to user community.

Summary

The chapter began with a review of ionizing and non-ionizing radiation, to show that RF energy is in fact non-ionizing. Its impact is that as soon as the source is turned off, the warming effect stops. There are no secondary effects such as those found in ionizing radiation such as ultraviolet ray, X-ray, or gamma ray. Figure 7.1 showed that ionization energy in RF is the lowest and therefore has the least effect compared to others considered in that figure. It is important to note that many of the ionizing radiation comes from the Sun. This understanding was used by early research on how to model the human head and measure effect of cell phone next to the head.

This chapter further considered how the standard was developed, how the safe limits were set about four decades ago. There are further improved versions incorporated into later releases of the standard, but Table 7.2 shows the guideline numbers. The open nature of this safety standard allows everyone to provide inputs, how public can offer comments and express their concerns were also shown in Fig. 7.5 and briefly discussed.

Despite research work and well-established standard, guidelines to further reduce exposure to RF signal were indicated. Given the human nature to be careful, turning RF sources at home or in the office, when it is not used, is good practice. The global community understands and appreciates Sun and the energy it provides, at the same time they do take care in terms of extended exposure by finding shade, or using cream to cover their skin, etc. Similarly, it is a good practice not to have RF power when it is not needed.

National regulators (such as FCC or ITU) provide limits based on the IEEE 95.1 standard, which in turn is based on global research by over 1400 physicists, medical doctors, and engineers, spanning several decades. Their regulations set very conservative limits addressing concerns of human health and social aspects. Reaction towards violators can be quite rigorous. FCC has imposed hefty fines and some regulators have established methods to limit RF exposure to children. There are good practices with uniform global guidelines established for unlicensed frequency bands such as 2.4 GHz (common band for WiFi).

Finally, efforts made to establish test methods for each new technology and product were reviewed. Often these originate from research units that are either autonomous or government funded. They are typically not from any product vendor and therefore provide neutrality and credibility. Independent test laboratories adopt these test methods that are accepted by regulators and their certification further establishes credibility of products sold in the market.

Emphasis on participation

Does your university have a WiFi system? Where are the access points located? Where is the nearest cellular tower in your neighborhood? Do you know the RF power from these?

How much is the received power on your cell phone? Use the field test mode on your phone to measure it. Use instructions from websites such as: www.ubersignal.com to set up your phone.

Communicate with your local cellular provider or TV broadcaster. Find out what is the power level and where in the national database is it registered?

Such national registration database not only offers the exact location of the transmitter, but it also provides the height, amount of power, frequency, and other details. Have you talked to your community leaders about these and do they understand the importance of Fig. 6.6 in terms of where is the highest received power from the transmitter?

There are many conservation efforts for animals that involve use of a collar—on animals such as lions, tigers, or bears. What is the power in these transmitters, since they are worn on the neck of animals and regular transmission from the collar is used to track them? Have you talked to wildlife preservation groups and how does it compare to Table 7.1?

Knowledge gained

This chapter focused on scientific research on RF radiation and public health. It elaborated on the fundamental distinction between ionizing and non-ionizing radiations. It stressed the difference between the two and every student is expected to realize that nuclear reactors, ultraviolet radiation from the Sun, X-ray, and gamma rays are all "ionizing radiation". This means after these sources are "turned off", their secondary effect "continues". Unlike these, the RF radiation from radio waves is "non-ionizing" which means there is no "secondary effect". Warming effect on human body stops as soon as the RF transmitter is turned off—whether it is a cell phone, WiFi, microwave oven, or a broadcasting station, it does not matter. All of them have only "non-ionizing" RF radiation.

Despite this distinction, all government bodies regulate it in one way or another, to reduce human exposure to RF radiation. Unlike the Sun or other celestial bodies on which we have no control, all RF radiation sources created by humans can be controlled. If not practical to turn it off, there is always the alternative to move further away from such sources, thereby decreasing exposure.

The primary intent of this chapter is to educate all members of society. Irrespective of their background and interest, everyone uses cell phone and is entitled to know what RF radiation is, how to reduce exposure to it, and know what measures all vendors of equipment and device take to reduce exposure. It is also important for them to know the extensive research and methods already in existence for the past several decades, it is best to follow them.

Homework problems

1. Using your cell phone download a software application that can measure the received signal. Check your weight in Kg. Using these two inputs compute the SAR in watts/kg and compare it against Table 7.2. Are you within safe limits in the location you reside and sleep?

2. In your area where is the nearest broadcast tower? Find out from database or local sources. Find out from the broadcaster the power transmitted. Knowing your weight generate the SAR watts/kg, compare it against Table 7.2. Are your service providers operating within safe limits?

3. Where is the nearest cell phone tower in your neighborhood? Locate the number of antennas on that tower and identify them using national register? How much is the power from each? Add all of them to get the total RF power from that tower and check whether the closest

residence is within limits set out in Table 7.2. This is what residents in your area should be concerned about.

4. Locate four different types of electronic equipment that may radiate RF unintentionally (such as florescent light) and verify whether they are certified and comply with regulations (for example, FCC 47 CFR part 2, Sect. 15.3 (z) is mandatory for equipment sold in USA). Does it interfere with your TV or computer peripheral? How close can you place this unit before you can see any effect on TV?

5. Why were adult male phantom models not considered to be true representation of children? What are the changes you can suggest based on your literature search involving such studies?

6. Many people sleep with their cell phone under their pillow. Can you perform proximity studies and connect them to radiation health? Would you recommend turning cell phone off before sleep or even placing them away from you (several meters away)? What are your answers to those who find it perfectly fine since cell phone operates under nationally accepted limits?

Project ideas

1. Are there military establishments in your area that fought during World War II? Did they use radars, and do you find them or pictures in local museums? Trace their history and transmit levels.

2. Table 7.1 shows materials used in one phantom model. What other phantom models can you find and how are the compositions different?

Out-of-box thinking

1. Follow a major broadcaster who has operated studios for many decades. How have their transmitters changed in terms of power? Do they find the new digital broadcast better? Can you suggest alternatives to completely get rid of radio transmitters, since Internet can stream programs directly from studios?

2. Some vendors have suggested the use of cellular chip embedded in human body. This can eliminate the need to carry a cell phone. Can you make an argument to support them, yet silence skeptics who would be concerned about health aspects?

3. Should your home WiFi get turned off at night/when you are not home? What are your arguments for and against making it a regular practice?

References

1. Cherry N (8 June 2000) Probable health effects associated with mobile base stations in communities: the need for health studies. Lincoln University, Canterbury, New Zealand. https://researcharchive.lincoln.ac.nz/bitstream/handle/10182/4090/Health_effects_base_stations_in_communities.pdf?sequence=1&isAllowed=y

2. DeFrank J, Bryan PK, Hicksw CW, Sliney DH (2021) Non-ionizing radiation. Occupational health: the soldier and the industrial base. Chapter 15. U.S. Army Environmental Hygiene Agency, pp 539–580. https://ckapfwstor001.blob.core.usgovcloudapi.net/pfw-images/borden/occ-health/OHch15.pdf

3. Cleveland RF, Athey TW (1989) Specific Absorption Rate (SAR) in models of the human head exposed to hand-held UHF portable radios. Bioelectromagnetics 10:173–186. https://doi.org/10.1002/bem.2250100206

4. The SAR test is inadequate–why do scientists state that the SAR is inadequate to protect cell phone users? Environmental health Trust. https://ehtrust.org/sar-test-inadequate/

5. Mason P, Murphy M, Petersen R (2001) IEEE EMF Health and Safety Standard. IEEE-EMF-HEALTH-SCC28. https://www.who.int/peh-emf/meetings/southkorea/IEEE%20EMF%20HEALTH%20-%20Mason.pdf

6. Human exposure to radio frequency–federal register. https://www.federalregister.gov/documents/2020/04/01/2020-02745/human-exposure-to-radiofrequency-electromagnetic-fields-and-reassessment-of-fcc-radiofrequency

7. Browne J (Aug 29, 2013) Medical aid draws from RF and microwave energy. Special report. Microwave & RF magazine. https://www.mwrf.com/technologies/systems/article/21845172/medical-aid-draws-from-rfmicrowave-energy

8. Jin JM, Yan S (April 2019) Multiphysics modeling in electromagnetics–technical challenges and potential solutions. IEEE Antennas Propag Mag, 14–26. https://www.researchgate.net/publication/331460477_Multiphysics_Modeling_in_Electromagnetics_Technical_Challenges_and_Potential_Solutions

9. Assessment of the Possible Health Effects of Ground Wave Emergency Network, National Research Council (US) Committee on Assessment of the Possible Health Effects of Ground Wave Emergency Network (GWEN),Washington (DC), 1993. https://www.ncbi.nlm.nih.gov/pubmed/24967486

10. Frenzel L (March 5, 2019) What you need to know about radio telescopes. Microwave and RF today magazine

11. Recommendation ITU-R SM 1538, Technical and operating parameters and spectrum requirements for short-range radiocommunication devices

12. Making EMI Compliance measurements, Agilent technologies, Application note, March 9, 2011. https://www.ccontrols.ch/cms/upload/applikationen/EMC-Messplatz/5990-7420EN.pdf

13. Electromagnetic fields and public health: Radars and human health, fact sheet No.226, https://www.who.int/peh-emf/publications/facts/fs226/en/

14. Why you really need to unplug while on vacation https://www.entrepreneur.com/article/247799

15. Evaluating Compliance with FCC Guidelines for Human Exposure to Radio frequency Electromagnetic Fields, OET 65

16. Antenna and siting, Federal Communications Commission. https://www.fcc.gov/wireless/bureau-divisions/competition-infrastructure-policy-division/tower-and-antenna-siting

17. FCC fines A-O Broadcasting $25000, for violation of RF radiation limits and other commission rules. https://www.fcc.gov/document/fcc-fines-o-broadcasting-25000-violations-rf-radiation-limits

18. Environmental Health Trust: International policy actions on wireless. https://ehtrust.org/wp-content/uploads/International-Policy-Precautionary-Actions-on-Wireless-Radiation.pdf

19. Development of Laboratory Test Methods for RF-Based Electronic Safety Equipment: Guide to the National Fire Protection Association 1982 Standard. https://nvlpubs.nist.gov/nistpubs/TechnicalNotes/NIST.TN.1937.pdf

20. Bringing precision to measurements for millimeter wave 5G wireless, Kate Remley. https://ws680.nist.gov/publication/get_pdf.cfm?pub_id=924708

21. Non ionizing radiation monitor, Personal monitor, Nardalert 23. https://www.atecorp.com/atecorp/media/pdfs/data-sheets/narda-nardalert-s3-personal-rf-monitor.pdf?ext=.pdf

22. International Telecommunication Union, Mandatory Conformity Assessment Schemes. https://www.itu.int/en/ITU-T/C-I/conformity/Pages/Aschemes.aspx

Engineering Economics

PC	Personal Computer, general terminology to describe a desktop or a laptop computer used by an individual
Tbps	Terra Bit per second (10^{12} bits per second), a very high-throughput rate supported by optical fiber
DSL	Digital Subscriber Line, a method of using existing twisted pair (telephone cable) to carry data, by separating the voice and data circuits at the end with filters but using twisted pair to carry both
Cable TV	Distribution network to carry television signals to individual homes or offices. Also used for communication of data from homes connecting to the Internet. Channels to Internet are shared between users in each neighborhood
WiMax	Wireless Metro Area network IEEE 802.16 is used in rural or semi-urban neighborhoods, with one central distribution point with RF coverage to serve an area of residential neighborhood
URLLC	Ultra-reliable Low-Latency Communication. Part of 5G standard to address automate car and automated machine on shop floor with highly reliable network of 99.9999% with latency of <10 ms.
Wireless LAN	Wireless Local Area Network. Typically adhering to the wired IETF network standards, it creates a local area network interfacing to the Internet or Intranet
ITU	International Telecommunication Union. An international body that addresses telecommunication-related standards. It provides guidelines for regulators of all member countries to follow
WLL	Wireless Local Loop. A network enclosed within a cable loop antenna. All users within the loop will be served by the system—typically it uses cellular bands to provide service within that area
VHF/UHF	Very High Frequency/Ultra-high Frequency. VHF bands typically use the 30–300 MHz range and UHF operates in the 300–3000 MHz range
RRL	Radio Relay Link. Microwave links that use "point-to-point" communication between towers laid out in the form of a relay line, to carry signals over very long distances (usually limited by horizon). Each relay hop can be as far as 100 km if the land is flat with no obstruction, but typically 50 km or more
MIMO	Multiple Input Multiple Output. A method using multiple antennas and multiple beams to split and combine traffic thereby increasing the throughput
Traffic (telecom)	Generally categorized with two terms, the signaling traffic and user traffic. But most models consider the user traffic that indicates how many users are being served. The most common term used is Erlang which indicates capacity of the system
IoT	Internet of Things. A category of service originally known as machine-to-machine network, serving devices that provide measurements such as temperature, voltage, pressure, etc., a few times daily
WEEE directive	The purpose of WEEE directive is to prevent waste electrical and electronic equipment (WEEE) and to facilitate the reuse, recycle, or recover of such wastes
Space debris	Objects such as final stage of rockets, satellites that stay in orbit beyond end of their life, and objects created due to collision

© Springer Nature Switzerland AG 2022
K. Raghunandan, *Introduction to Wireless Communications and Networks*,
Textbooks in Telecommunication Engineering, https://doi.org/10.1007/978-3-030-92188-0_8

with other orbiting objects are together termed space debris (or waste). All such objects that are beyond Earth's immediate gravitational pull (~ 300 km) will not fall back to Earth

LEO   Low Earth Orbit (between 300 and 400 km around the Earth) is where the space station is parked. Other satellites use it as a "transfer orbit" a transition point from where they are moved to other orbits

GEO   Geostationary orbit (at $\sim 36{,}000$ km on the equatorial plane) an orbit where satellite is placed. For an observer on Earth, the satellite will be perceived as stationary since it takes 24 h for the satellite to orbit the Earth once

This chapter considers the cost of not only cell phone or other wireless devices, but also network and other equipment essential to provide wireless service. It considers some of the other factors, such as labor cost, maintenance, and cost to recycle—an important part since cell phones tend to recycle faster than their wired counterparts like the telephone, the PC, or notepad. It provides a glimpse to the service provider's view of factors to consider when deploying wireless service.

8.1 Cost of Design and Deployment: Wired Versus Wireless

There is intense debate between the wired and wireless engineering communities regarding cost of deployment, design, and long-term use of communication systems. The reason why these are contested is partly historic, yet it is straight forward. Until recently, the wired network with its deployment and use was considered "the only way". Therefore, cost of its design and deployment was considered something "essential to have telecommunication infrastructure". This was challenged by the introduction of wireless which took major market share from these providers. In current times, when "voice" is no longer a revenue earner whether in wired or wireless industry, the real difference seems to be the humble device—the cell phone. Literally over 80% of global population has cell phones at individual levels, while many of them may have phone only in the office and not at home, may have a PC in the office not at home, etc.

8.1.1 Fixed Links Between Two Locations

With the introduction of wireless links—this thought process is questionable and often the argument is wireless costs so much less, so why is wired infrastructure necessary?

Numbers quoted by the wireless community vary but cost of millimeter-wave links could be between 1/5th and 1/10th of the cost of laying fiber [1]. Why is this cost disparity so large?

There are a few reasons:

- Although cost of fiber-optic link or the cable itself is not significant, the process of laying it in ground is labor intensive and adds to cost in terms of time and expenditure.
- In urban areas time taken for digging streets can be considerable, due to infrastructure existing below streets (such as sewer, electrical power, and others).
- Any change in location calls for repetition of the entire process.
- A microwave link between the same two points can be erected in a matter of hours and labor cost for installation is therefore low (1–2 days is normal).
- In areas where fiber network does not exist microwave wireless link becomes the obvious choice.
- If the distance between the two points is large, then microwave link becomes a compelling choice. Curvature of the Earth is the only limit in desert areas, for example.
- Finally, most customers need between 20 and 500 Mbps that microwave links can deliver, with far less transition delay compared to fiber-optic cable.

In that case why are fiber-optic networks preferred and what is are its benefits? The fiber-optic network is capable of handling enormous amounts of data, with throughput ranging from 10 Gbps all the way to 10 Tbps. Given its fundamental nature of carrying information using a light source, it is immune to electrical noise. Another advantage of light as opposed to RF is attenuation is minimal. Signals can travel great distances without need for amplification. Also, there is merit to the argument that if electrical signal can be transported using light, very large amount of data can be transferred. In short, fiber link is like the multi-lane highway of telecommunication networks that can carry considerably more data flowing between major cities, for example.

Microwave links have existed since late 1940s with links between cities such as London and Frankfurt, New York, and Chicago [2]. Weather plays a role in these links, but design of any microwave link readily takes this into account and the required performance is based on factors such as how much signal is lost during snow, rain, fog, and absorption by foliage. Early microwave links carried analog signal with traditional modulation methods. In the past 20 years, microwave links are purely digital and adjust modulation types and power levels to match atmospheric conditions. With that the throughput may reduce during severe weather but will get back to full rate when normalcy is restored. Various frequency bands are allocated by regulatory bodies for microwave links [3]. This is discussed in greater detail in Chap. 14 on microwave links.

8.1.2   Services to Premises: Residential/Business

Until recently, telecommunication service, whether residential or for business, was always via fiber-optic link or cable TV service or DSL (digital subscriber line using twisted pair). They were often regarded as the only method to provide service to any specific location, whether in rural or urban areas [4]. With the advent of wireless links there are other options.

This has changed recently due to two major developments in the wireless industry.

(a) The IEEE 802.16 WiMax standard offers a digital link to serve customers in an area which could be rural town or an urban housing community. This fixed link with a beam serving the area offers excellent digital telecommunication service.
(b) Fixed wireless products commercially available in the millimeter-wave bands of 60 GHz and 80 GHz offer similar service with throughputs of 1 Gbps and 10 Gbps, respectively.

Both of these were able to carve out a market by providing excellent residential and commercial service in specific locations [5]. Now the 5G standard of cellular has set primary focus on eliminating wired links to homes and replace them with beams from antennas directly beaming to homes/business premises with a minimum of 1 Gbps guaranteed service to every home/office. Conceptually this is like WiMax but implemented using millimeter wave that offers huge bandwidth. In effect this could be viewed as a combination of (a) and (b) indicated above.

Again, the cost of laying cable, etc. follows similar cost argument making fixed microwave service a better choice. Currently well over 50% of the global cellular network providers use microwave as the backhaul link, due to its simplicity and reliable, long-term use with minimum maintenance. In effect 5G is focused on making efforts to address customer premises access and to limited extent backhaul network, moving away from the fixed wires and towards wireless. Another feature of 5G is support of driverless cars using URLLC (ultra-reliable, low-latency communication) that is looming large on changing the Ethernet standard of wired industry.

8.2   Economics of Wireless Growth: One Generation/Decade

Wireless industry seems to have shaken the very roots of how business was conducted in the telecom industry. From the time Graham Bell telephone was installed in the twentieth century up until late 1980s, there were incremental improvements. Improvements involved moving away from copper wire towards coaxial cable (cable TV). Later, fiber-optic links brought in a major revolution offering almost unlimited bandwidth on thin glass tube. With the first generation of cellular, analog cell phones came into existence. No one anticipated its rapid growth until late 1990s when second generation began. After that each generation of cellular took an obvious jump each decade. With cellular services growing rapidly, major focus was removal of major parts of the cellular infrastructure network and retaining parts of it became the rule. At the same time, replacing individual handset or the cell phone, every 3–4 years also became the norm and not an exception.

Obviously, businesses should have revolted against such rapid changes, but in sharp contrast they did not. Observing its growth, they fully supported removal and replacement of major parts of network. This was justified since cost of installation and removal led to benefits for the customer and profits to the operator. These two factors weighed heavily in helping such moves, resulting in billions of dollars/euros or other currencies transacted towards cellular service. These moves were rewarded by rapid economic growth in terms of expansion of telecom network. In many countries, telecommunication network barely existed, but cellular offered the first opportunity towards rapid economic growth, with financial, critical infrastructure access such a road, food distribution making remarkable turnaround, thanks to the cell phone. In most countries, cellular network has become the primary access to Internet.

Important enablers were regulatory bodies of each country—it was in their interest to bring revenue to the government by auctioning frequency spectrum that helped country's economic growth and businesses establishing themselves as leaders in technology. There were other factors for the national regulatory body (part of the national government) in terms of creating spectrum. It became possible to auction spectrum as a resource that could be allocated fairly to competing cellular providers in different bands. This was a major effort in terms of not only technical but offering great business aspects. This means all the investment is recovered and still profits are made—according to market research reports. Wired service providers in most countries today are losing market share due to better offers from cellular service providers. Their revenue from wired service continues to diminish, as revenue earned by wireless service providers increase.

8.2.1 Growth of Unlicensed Spectrum

An important change emerged in terms of providing "unlicensed" spectrum—most notably 2.4 GHz band that became a globally accepted "unlicensed band". While not earning revenue in terms of spectrum auction, this motivated the concept of devices and services that are universal in nature. Its effect on services such as WiFi, Bluetooth, and similar technologies were readily appreciated by humanity. To a large extent, business leaders recognized utility such as these that promote economic growth by sharing information readily across the globe [6]. Value of just the WiFi service alone is estimated to be $1.6 trillion in 2018. Such high economic values prompted regulatory bodies to identify other bands that could be unlicensed to provide economic growth. An example of this is the 5.9 GHz band, expected to play a major role in road infrastructure allowing communication for vehicle-to-vehicle safety [7]. There are several other bands in the 5 GHz range that are "unlicensed" and in use today. This will be elaborated in Chap. 12 where wireless LAN services are discussed.

Equally important is the unlicensed band of 60 GHz that has seen much growth. There are products in the microwave/millimeter-wave link, which is a major contribution for backhaul, it has also seen considerable interest and product growth in home networking [8]. Products in this unlicensed band can become the backhaul network within the house, routing every device connected to WiFi, including traditional PC, TV, and cameras connecting to the Internet.

8.2.2 Cost of License and Fee Structure

Both the licensed and the unlicensed bands of spectrum have contributed to economic growth. The advantages and limitations of auction with revenue compared against unlicensed with major economic impact on society are discussed in the report by ITU [9]. The legislative and legal aspects as well as how countries could arrive at decisions on how spectrum should be managed are explained in detail in that report. The report includes formulas for calculating spectrum fee, for different types of wireless services such as fixed point-to-point service, mobile satellite service. It provides examples of how fee was calculated for 2G and 3G cellular services. Chapter 4 of this report shows examples of how to calculate the fee and formulas used for various services.

It would be helpful to reiterate some calculations in order to focus on effort needed to develop the formula for fee calculation. Let us consider the following and briefly review them.

1. Administrative fees—Administrative fees are intended to cover all the costs, related to spectrum planning, management, and monitoring.
2. Establishing spectrum fees—there are five general steps to calculate spectrum fees. First, in order to arrive at basis of calculation, simple formula with multiplication using currency "k" used by that country (national currency). The next four steps involve demand assessment, cost assessment, choosing the fee, and determination of fees.
3. Different services and how to assess their fee: important wireless services considered for fee assessment include
   - Point-to-point fixed service assignment.
   - Point-to-point fixed service allotment (unlike the above, here allotment of total area served by national territory is considered).
   - Wireless local loop (WLL) for fixed services, where WLL consists of an area encompassed by the antenna in the form of cable deployed as a loop inside which service is available.
   - Fee applied to an Earth station within the fixed or mobile satellite service (annual fee).
   - Fee applied to fixed or mobile satellite service (for an area).
   - Fee applied for private networks in the mobile service.

Typical examples of how to develop the formula and calculate fee for current day 3G mobile service is provided in the report. In order to understand and apply this to wireless service, let us consider details in the report [9].

The following equation could be used for determining the annual amount of revenue from the service, R_s of the spectrum fee:

$$R_s = t\% * CA, \qquad (8.1)$$

where

CA: represents the operator's turnover for the corresponding year in respect of the 3G mobile service frequencies.
t%: represents the percentage to be levied (perhaps as tax or duty) on the operator's turnover.

To this, annual fee is added as "entry ticket", payable upon allocation of the license. The amount of the entry ticket, which may be proportional to the allocated bandwidth, should be set with particular reference to the number of users and amount of area serviced in order, as the case may be, not to hamper the deployment of new entrant's networks.

An important part of cost model involves efficient use of spectrum. What this means is that given the same size (say 1 MHz) an analog system may provide only two channels while an efficient digital system may accommodate 20 or even 200 channels in the same spectrum size utilizing latest techniques of access and modulation. They can therefore serve either large number of users or provide a variety of data services using the same spectrum.

This efficiency factor is applied to real-world services such as VHF/UHF sound or a TV radio broadcast; it also applies to mobile services such as maritime radio service, land mobile service, aeronautical mobile, radio navigation, and radio locations (primary radar) services.

Fixed services use radio relay links and calculations are based on the number of RRL. Fee calculations performed by different countries and their experiences in applying principles outlined in this report are helpful. To the student, this provides an understanding of some economic aspects of different wireless services, whether for private, corporate, or government agency services.

8.3 Cost of Using Green Energy Devices to Deploy Cellular

The concept of using green energy devices such as solar panels or other devices based on geothermal, wind energy, etc. was discussed in cellular standards to address needs of countries that lacked major electrical grid infrastructure. While this initial interest addressed challenges associated with deploying cellular/wireless systems in remote areas, the study steadily moved towards energy efficiency of wireless systems.

The question of how much power is needed to operate a base station is answered at multiple layers [10]. A framework of energy efficiency evaluation is based on the 3GPP standard model. Energy efficiency linearly relates to bandwidth. Therefore, green energy refers to both the power consumption by base stations (in the network) as well as battery efficiency in the handset/smart phone. It is imperative that these two are directly related—if the serving base station is efficient, it provides the best service to consumer based on efficient methods of wireless transfer.

Models for efficiency begin with power efficacy. Current LTE networks utilize MIMO (multiple input multiple output) making them efficient in terms of both bandwidth and power. This is shown in Fig. 8.1 as a model where electrical power is input to the base station. Field installations need air conditioning system to keep systems within specified temperature. Irrespective of the type of AC source (single phase 110 V/220 V), it is important to convert this into clean DC power, with battery back-up. Since cellular service is now deemed an "essential service" in many countries, battery back-up of 8 h or more is routinely expected by every regulatory authority.

In addition to the throughput (in terms of data rate), the quality of service parameters require optimal performance under various conditions. This includes ability to tolerate interference from other RF systems (including competing cellular providers) and perform well under mobility conditions. For example, base station serving along highway routes need to regularly hand off users to the next base station, without loss of service to any customer. Similarly, base stations serving high-speed rail must support mobility of speeds exceeding 300 km/h. Those operating in hilly terrain must effectively deal with reflections and possible shadow regions. Factors such as area served, number of customers, terrain, and other factors are also used in arriving at efficiency.

Overall reduction in transmitted power occurred due to improvement in technology. In terms of semiconductors used as well as circuits to transfer power resulted in reduced outputs (right side of Fig. 8.1). LTE base stations typically range in RF power from 50 W (for major cell site with power radiating over three sectors) to less than half of that in a femto cell (or small cell that has only one sector).

However, when all different systems shown in Fig. 8.1 (all the boxes) are considered with AC mains operating power used, macro-cell site consumes about 1300 W, but a femto cell consumes about 10 W. The biggest change in recent technologies is the "Sleep mode", during which it is not serving any customers (early mornings, for example), when it needs only 0.1% of the total power used. For the operator, this can result in considerably reduced utility bills, and much smaller battery power back-up. It also provides an indication of alternate energy sources such as solar or wind power usage to supply power to base stations. Such systems directly produce DC power, eliminating the need to covert power from AC to DC, further improving efficiency of power transfer (laying cables from utility, use of transformer, converter, etc.).

8.3.1 Traffic Models Based on Area

Traffic depends on number of people served/unit area. Generally accepted models (for Europe) are segregated as shown in Table 8.1 [10].

Obviously, dense urban areas such as major cities form bulk of network traffic, but cover very small area, as compared to the total geographic area of a region. Sparsely populated areas form over half of the entire land mass that produces very little traffic. In Nordic countries, such as Russia, Canada, Australia, and Alaska, these numbers get even more exaggerated, since major areas of those regions on Earth belong to the "sparsely populated" category. Consequently, the last category dictates deployment of large

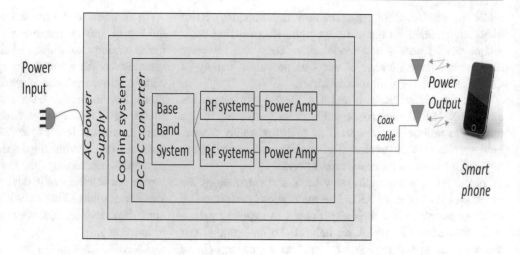

Fig. 8.1 Typical base station transmit/receive system (AC power input, RF power output)

Table 8.1 Deployment areas in Europe (excluding Nordic and Russia)

Deployment d	Population density P_d (Citizens/Sq. km)	Area covered a_d (as percent of total area) (%)
Dense urban	3000	1
Urban	1000	2
Suburban	500	4
Rural	100	36
Sparsely populated or wilderness	25	57

cells, since largest possible area covered is key to efficiency, yet very little traffic gets generated. There is no need to have three sectors, only one is sufficient. Physics demands the use of lower frequency bands, since higher the frequency, more will be the propagation loss; therefore, sparsely populated areas must be served by these bands.

8.3.2 Traffic Demand Based on Terminal Types

General perception of cellular system serving a cell phone is quickly vanishing. Whether user has a PC or any other type of device, all users are being served by a cellular system. The perception in current/future years should be telecommunication needs of population worldwide will be served by cellular network. Hence, the following table serves as a good reference [10].

Table 8.2 provides insights on how data consumption per user has increased over the years, with PC taking the highest spot. Will this trend continue? Perhaps it will and expect this traffic to be supported more by cellular network over time. Data consumption by users has another category that has started to come up of late. It is the "hot spot" device that connects cellular network to WiFi allowing home networking of all the terminals at home. In some way, this device can be thought of as a gateway into the premises. It also supports

terminals not listed in the table, such as smart TV, home security devices, and others that operate on a WiFi network but will use cellular as the main backhaul to connect to the Internet.

In Table 8.1, the category of sparsely populated areas was covered. A different interpretation of this service is emerging, which is known as "Internet of Things (IoT-narrowband)". These are devices spread over a large area, such as a mine or devices operating in ocean or other areas. They do not produce much data—perhaps less than the reference PC mentioned in Table 7.2. Yet they are of important since periodic inputs such as temperature, wind, or some other parameter need monitoring once/twice a day. The importance of wireless coverage to such units is vital and considered a major growth area due to millions of devices expected under this category. Are the data rates changing with 5G deployments? Yes, smart phone is quickly taking the place of Tablet moving to the 2nd position and PC is already getting replaced by laptop.

In that case how are the fee rates modeled? Using the ITU report mentioned earlier, formula can be developed to accommodate newer services [9]. Maintenance and security personnel are increasingly relying on wireless services and it may become a premium revenue earner although the data rate could be small [10]. Frequency spectrum that could become handy for such services may use "TV white space radios" [11]. These are channels normally allocated for TV

Table 8.2 Estimated traffic demand ranges of terminal types (Europe—2015)

Terminal type k	Average rate r_k (in Mbps) range	Daily consumption (MB)	Monthly consumption (GB)
PC	$0.5 \rightarrow 2$	$512 \rightarrow 2048$	$16 \rightarrow 64$
Tablet	$0.25 \rightarrow 1$	$256 \rightarrow 1024$	$8 \rightarrow 32$
Smart phone	$0.0625 \rightarrow 0.25$	$64 \rightarrow 256$	$2 \rightarrow 8$
Reference PC (2010)	$0.03125 \rightarrow 0.125$	$32 \rightarrow 128$	$1 \rightarrow 4$

The ranges are rounded to the power of 2 for notational convenience

service, but unused channels in such service bands are used for other wireless services. When applying for such licenses the regulatory authority typically expects applicant to prove that their device will not affect TV transmission in that channel. Quite often this is true—that channel may be broadcast by another station located far away and not meant for this area. Such gaps in TV spectrum are useful since they offer services in lower frequencies with inherent advantages to cover sparsely populated area, as well as narrowband IoT (Internet of Things).

How do these base stations get power? Off the grid power usually needs use of diesel generators or propane gas-based systems. Of late there is better likelihood of using solar panels for one prime reason—cost and reliability [12]. Wind power or geothermal power are other options.

8.4 Cellular Handsets: Cost of Recycle and Reuse

With the increase in number of cell phone comes the economic aspects of what it costs to manufacture them, their use, and recycling process. Unlike many other electronic products, the life cycle of cell phone is typically no more than 3–5 years. That brings the fore, the economic aspects of how to keep our planet clean and safe by recycle of these devices.

There are several studies that document the process well. A typical reference paper is given in [13]. Materials used in cell phones recent models consist of roughly 25% metals, 30–50% plastics with the remainder being glass, ceramics, and epoxy. Most of the metal mass is made up of copper, steel, and aluminum. Depending on the region, it is even possible to recycle most cell phones, so that they can directly get back into the manufacturing process.

Compared to many other electronic products, cell phones have a thriving recycling market. Some studies indicate almost 65% reuse, providing employment to many. The European Commission on Environment has directives Waste Electrical and Electronic Equipment (WEEE) for recycling and reuse. Cell phones are part of category 3 of this directive. Major part of the recovery is copper, but some units also try and recover some heavy metals. In general, the European directive is considered very strict and difficult to meet due to economics of recovering ceramic and other parts. It is also recognized that cell phones may have to be designed differently in order to meet European directive. This is an ongoing effort.

There are well-established procedures on how to turn off your device, its tracking features, etc., before sending it finally for recycling or drop off bins with incentives to return retired phone for recycle [14]. In many regions, there are multiple methods. Either the service provider or the Original Equipment Manufacturer (OEM) may set up methods to receive retired devices. Almost all recycling includes removal of batteries that are sent out separately to dumps that are part of non-hazardous waste. For most recyclers, cell phones form a small part of their business and profits earned seems to be around $0.5/handset, indicating profits are earned mainly based on volume.

Another growth area similar to handset but somewhat different is the use of security cameras at home, in office space, and public areas. Although many cameras are currently wired, newer installations increasingly rely on cellular or WiFi infrastructure based on cost. If the camera is close to a communication switch supported by backhaul fiber network, then there could be a cable installed from camera to the switch. The cable length has a limit of 100 m. Any distance longer than that needs either a fiber installation or wireless installation. At that point, often use of cellular or WiFi may be preferred, due to cost considerations, which was already discussed in Sect. 8.1 comparing cost of wired versus wireless.

Due to extensive use and change of handsets every few years the question of packaging and related material such as handbooks, gadgets such as chargers, headphones, and others become part of the recycling program. This part is not well defined—the outlets being varied it is not easy to hold just the service providers responsible for recycling them. More than other consumer products, cell phone manufacturers have responsibility towards recycling. The key slogan "by 2050 there could be more plastic in the sea than fish" must strike a chord to all users [15].

8.5 Satellite Launch, Lifespan, and Usage Patterns

Unlike electronic devices used and recycled on the planet, satellites and their launch follow a different pattern. The process of building satellites is well established and practiced by many commercial enterprises. However, the business of launching is still limited to and controlled by governments or regional bodies. Therefore, profits are made in launching based on probability of successful launch. There are several vendors who provide risk insurance, given the fact that a failure during launch could mean loss of satellite as well as likelihood of missing that spot in the orbit (allocated by the United Nations) [16].

In terms of waste and possible hazards, there are some that are quite serious. The first few stages of the launch vehicle do fall into the ocean along launch trajectory and could be recovered. In fact, SpaceX is a company that focused on recovering lower stages of the launch vehicle [17]. However, the final stage that launches satellite in orbit does not fall to the Earth. It becomes part of the space debris. The satellites themselves also become part of the space debris and do not return to Earth. This is a known fact, ever since the first telecommunication satellites were sent in to orbit. Therefore, it is important to understand the entire cycle and its long-term implications.

The process of building and launching a communication satellite usually ranges from 5 to 10 years in design and planning. It begins with planning by applying for a space resource in the geostationary orbit (about 36,000 km from Earth, in the equatorial plane). This gets processed by the ITU or Int'l Telecom Union [18]. Approval method is based on constitution of the ITU, article 12 of which states "To ensure rational, equitable, efficient and economical use of the radio frequency spectrum by all radio communication services – including those using the geostationary satellite orbit or other satellite orbits – and to carry out studies on radio communication matters".

The approval process follows article 44 of ITU, which recognizes "Radio frequencies & satellite orbits are limited natural resources". Therefore, "Rational, Efficient, Economical Use of this resource is important" in order to provide "equitable access" to all ITU members. ITU carefully evaluates the application using its own Radio Regulations (RR) whose main purpose is to insure "Interference-free operation of Radio Communications". Once approval from ITU is obtained, the process of building satellite begins adhering to the spectrum and location allocated; this is followed by launching it into orbit within the time slot allocated by ITU.

8.5.1 Cost of Space Components and Building Space Craft

Electronic components are rated from consumer (lowest) to commercial (higher) to industrial (higher and ruggedized) to military (very high reliability and very ruggedized) to space qualified (to extreme temperatures, vibration, shock, and ability to withstand radiation). Cost of space qualified components are multiple times the cost of military hardware due to harsh environment in space [19]. Compared to the consumer product the cost for space products is typically about 10–100 times more, due to extra cost incurred in design as well as rigorous test procedures.

Other than hardware, cost of the satellite itself, there are incidental costs such as insurance against launch failure, in orbit failure of components due to hit by space debris or higher than expected radiation levels (gamma or X-rays emitted due to solar flares, for example).

Most communication satellites have a life span of 15–20 years, which is limited by amount of control fuel on-board satellite. Once the satellite runs out of control fuel (control thrusters use gas or liquid fuel), it gradually drifts away from orbit and will not support sustained communication. At this time, the ITU is informed and the next satellite to replace it gets that slot. In almost all cases, the electrical power from solar panel is still adequate to sustain satellite operations, but lack of control fuel and subsequent drift indicates "end of life" for satellites.

There are quite a few satellites that are lost in geostationary and other popular orbit. Important among them those in other orbits are GPS satellites, remote sensing satellites, and those in other orbits used by countries with land area closer to the North Pole, for example (such as Russia, Canada). The ITU uses RR (radio regulation) as a process to regulate and control interference from/to these satellites in other orbits as well. These will be addressed in Chap. 13 where the topic of satellite communication is reviewed. Irrespective of the orbit, all satellites and their final launch stage components become space debris orbiting around Earth for long periods of time. Some trials under way to reuse rocket stages but note that this does not contribute to space debris. Used rocket stages generally fall into the sea and are not normally recovered.

This is a serious problem since space debris that remain in orbit can cause more fragments after each collision. This is depicted in Fig. 8.2 to show how the Earth's atmosphere was clean and clear before space exploration when satellite launch (Sputnik) began in 1957 and subsequent spread of space debris over the years. There is an effort by academics, and industry consortium funded by NASA, to clean up space debris [20]. Skeptics offer a doomsday scenario about space debris that are serious enough to stop all future space launches [21].

Space debris continues to increase due to two factors: (a) abandoned satellites and (b) units crashing into one another causing splintered parts, meaning when two objects crash into each other additional pieces are created due to collision. Size of space debris is counted in mm or cm. Debris larger than 1 cm has the potential to crash into another object and cause splinters [22].

Table 8.3 shows the impact of how small pieces of debris in the Low Earth Orbit (LEO—about 300 km from Earth surface) can cause serious damage. It is helpful to recollect that objects in space move at enormous velocities since injection velocity needed to sustain an orbit is high. In LEO, impact speed of 10 km per second is common (about 36000 kmph). At those speeds, table shows that small object of 10 cm can cause damage similar to large bomb [22].

Whether cleanup can happen is not only dependent on the space industry but also on individual countries and their policy to adhere to common guidelines to help all living beings on Earth. It is likely that cleanup efforts will eventually begin, resulting in better space traffic, regulating debris. Currently, there are even concerns about human travel in space due to the sheer number of debris and possibility of some of them colliding with a space station/satellite, in orbit. Table 8.4 shows the quantity and number of debris in space which provides an overview of how much of cleanup is needed and what is practically possible (what can be tracked).

In future, this can add recovery costs to every satellite launch since there could be possible regulations on how to recover the disused satellite at the end of its life and bring it back to Earth. While it is relatively less expensive to gather

and bring pieces from LEO (300–500 km from Earth), to go all the way up to 36,000 km in the geostationary orbit, collect the debris left there, and bring it back can cost quite a lot. In future, this additional cost could dissuade corporations to send satellites into geostationary orbit (GEO—at 36000 km in the equatorial plane) which is already crowded, and the debris left there today is enormous. Table 8.4 shows another problem that objects smaller than 5 cm cannot even be tracked, which millions of those fragments will remain in orbit even after a cleanup [22].

On a positive note, the reliability of launch as well as services from satellites is now very well established. Life span of satellites continues to increase. Considerable telecommunication traffic around the globe is due to satellite systems, and its market size was estimated at about $65 billion in 2020.

8.6   Conclusion

In this chapter, we reviewed the economic strategies not only of vendors, but governments, standard bodies, and others as they evolved with wireless as a technology, as a business, and as an economic engine for national/international growth.

There is little doubt that many nations and regions are grateful that this technology allowed communication far beyond their national, regional, and international territories. It has indeed brought what was the norm (wired network) to a sideline and occupied the prime spot that drives the economy. In many parts of the world, it has brought about a financial network that did not exist earlier. The fact that ITU brought a major detailed report just to address the fee structure, options with detailed examples of how specific countries encountered the economic aspects, needs special

Fig. 8.2 A dramatic view of how the space around Earth is inundated with space debris

Table 8.3 Hazards of space debris

Debris size	Mass (g) aluminum sphere	Kinetic energy (J)	Equivalent TNT (kg)	Energy similar to
1 mm	0.0014	71	0.0003	Baseball
3 mm	0.038	1910	0.008	Bullets
1 cm	0.41	70,700	0.3	Falling anvil
5 cm	176.7	8,840,000	37	Hit by bus
10 cm	1413.7	70,700,000	300	Large bomb

Table 8.4 Size and quantity of space debris

Debris size	Quantity	Impact
1–3 mm	Millions	• Cannot be tracked • Localized damage
3 mm to 1 cm	Millions	• Cannot be tracked • Localized damage • Upper limit of shielding
1–5 cm	500,000 (estimated)	• Most cannot be tracked • Major damage
5–10 cm	Thousands	• Lower limit of tracking • Catastrophic damage
10 cm or larger	Hundreds to low thousands	• Tracked and cataloged by space surveillance network • Catastrophic damage

mention. It was elaborated in this chapter, with possible methods for other models.

In effect the ITU report provides a window to students who want to learn and practice the business and financial aspects of how wireless network and applications can be conducted as a viable business. What is also interesting is that while wireless jumps from one generation to the next in each decade, there is a prosperous business that buys, reuses, and sells wireless equipment and systems. This major recycling business has reduced environmental concerns and an industry that is willing to reinvent itself.

With the move from wired to wireless comes the interesting aspect of how network operators now sell their services very differently. The focus has shifted to how much data is consumed each day and month. While demand and supply provide financial stability, there is an entirely new market hitherto unheard of the narrowband IoT. Something that was earlier relegated to "if possible" kind of monitoring, to "essential now" type of approach. It has attracted market attention of all cellular providers worldwide.

Finally, the business of launching, operating, and retiring satellites poses questions that are very different from rest of the wireless industry on Earth. The growing mass of space debris, how to deal with and clean it has caught the attention of not only physicists and aerospace engineers, but environmentalists who worry about carrying humans into space in later decades.

Summary

This chapter starts with cost comparison between wired and wireless systems. Despite the higher initial cost, wired infrastructure has lower maintenance costs. Yet the initial cost of deployment becomes prohibitive depending on distance. Gradually wireless network has taken over rural parts of countries and with the fifth-generation plans to bring direct access to residences and business premises also wireless.

This was followed by detailed cost comparison based on area, population served, and the type of service. It has now reached a point where consumption of data is fairly stable with each type of end device, dictating the amount of data consumed by the user. This pointed to an interesting perspective that large parts of the world are sparsely populated and can only be served by wireless. This concept of low throughput but large area coverage using wireless is incidentally taking off as narrowband IoT. Devices spread over large area but need communication of no more than a few hundred bytes a few times a day is now projected as the growth market.

The next section touched on how regulators in different countries prepare cost estimates on how service providers, institutions, and individuals would be charged. The ITU provides a detailed report that is useful for such an exercise. It can be used by students to better understand the entire process and economics of wireless industry as it moves forward. A few examples were discussed in order to emphasize this point.

Cost of manufacturing cell phone, its short life span, and usage lead to financial questions about this consumer product, and the society that views it as a key commodity. This is part of the reason why many countries continue to adapt a subsidized model in order to enhance their national economic growth. Inference of this approach is that cellular communication has become the essential means for communication and commerce interweaved into social structure of every nation. Recycling of cell phone has become well established with stable methods; it doesn't seem to be a major concern to environment.

Finally, the topic of building and launching satellites was considered. Since global community views satellites as essential infrastructure, its cost and particularly the debris it leaves in outer space continue to be a matter of concern. Although there are some trials to tackle this question, there is no easy answer as to how this would be resolved. The global community needs consensus in terms of space traffic, cleanup, and related issues. This must be considered a matter of concern to the entire world population of not only humans, but all living beings on Earth.

Emphasis on participation

Make a list of current cell phones in your class. Inquire from vendors and service providers, what methods do they use to recycle phones? Find out how a used phone is valued and bought back.

The GPS unit in your phone or other device uses satellites. Where are they located and how much is the cost of building and launching them? Who pays for this and find out how often these satellites need replacement. In Europe, the Galileo satellite system is used as an equivalent to the original GPS. Find out the cost of this system for nations and how it is funded.

Knowledge gained

This chapter introduced the topic of economics and cost of different wireless systems. It showed why wireless is gradually replacing wired infrastructure. While use of cell phone has become universal, use of PC and other devices that connect to the Internet were reviewed in terms of how much data is consumed. Wired infrastructure is currently geared towards consumption of large amounts of data such as cameras used in security services.

Major growth area of narrowband IoT to cover vast areas shows a clear preference to wireless, based on cost. In terms of other wireless services, this will continue to be an important area to serve wilderness and sparsely populated areas.

Satellite services are in a stable market, with ITU providing both allocation of slots in space and controlling interference between all spacecraft. Yet the cost of launching, removal of space debris will continue to be a challenge to all of us. With future plans towards space missions to Mars, cleanup effort may receive public attention and efforts may begin soon.

Homework problems

(1) A university football stadium has decided to install live streaming system of the match with cameras at eight locations evenly spread out near front row seats—connecting to control room at one end of the stadium. Sports action video needs 2 Mbps/camera. What type of system would you propose—wired or wireless? Justify with explanations of your proposal, with cost structure for each.

(2) A wildlife reserve study project needs communication links to university department (100 km away), wildlife reserve head office (5 km away), and field outpost (100 m away). Cellular provider offers the following:
   a. $100/month for full video capacity of 20 Mbps.
   b. $50/month for mixed service of 2 Mbps (20% video, rest data, and voice).
   c. $15/month for narrowband IoT service of 100 Kbps. What types of services would you propose for each of three locations and explain with reason.

(3) A wireless operator of 3G service has an annual turnover of $250,000. The government expects a revenue of $50,000. How much is the tax levied on this operator. [Hint: Use Eq. 8.1.]

(4) A satellite service provider plans to lease international corporate communication with $500/month per satellite location. Market study indicates 5000 such locations of which the provider wants to take at least 30% of the market. Cost of satellite is $1.5 million and cost to launch is $2.5 million. Launch insurance costs $5000/month spread over 5 years. Satellite builder offers 10 years in orbit guarantee that increase satellite cost by 10%. Operational expenses are 0.5 million/year that includes spectrum revenue charges. How many years will it take the service provider to recover investment and begin make profit?

Project ideas

(a) Inmarsat offers International maritime satellite services to ships and commercial aircraft that are listed in "InmarSat Financials". Their cost model is based on owning and operating the satellite. Viasat is currently leasing channels from Inmarsat and claims that it is cheaper and more cost-effective to lease satellite service rather than owning satellites and operating them.

(1) Perform a market study (online research) to check whether Viasat claim is true. Consider a 10-year strategy since each generation of GX (Global Express) satellite would operate for a minimum of 10 years.

(2) Compare the Inmarsat model to the Intelsat model that serves residential and business customers in remote areas. Their financials and fact sheet are also available online.

(b) Using Report ITU-R SM.2012–5 (06/2016), perform a study to show how a new cellular service provider could enter the market in a rural area by leasing channels at bulk rate and offer 3G service. Assume the government in that country offers revenue (tax levied) further reduced by 30% to offer essential cellular services in rural areas.

Provide details of assumptions including formula used and justifications. How long will it take for the provider to become profitable?

Out-of-box thinking

Since 5G is thinking of offering direct connections to homes from a light pole in front of the house—develop two/three different cost models and monthly rate for Internet access.

1. Antenna beam focused on house/apartment. From pole to next pole continue with fiber-optic backhaul. Cost of running fiber bundled along ground wire of power transmission. What are the precautions? What would be the cost? Base it on locally advertised rates.

2. Pole uses multiple antenna beams directed towards different houses in the neighborhood. If serving a high-rise apartment, each beam can focus on and serve two/three floors. Backhaul from the pole continues with fiber-optic cable strung along ground cable.

3. Same arrangement as 1 and 2 but backhaul using line-of-sight millimeter-wave links at 60 GHz or 80 GHz to the next pole and after 3–5 poles install a switch with gateway connection to Internet. 60 GHz offers 1 Gbps throughput, 80 GHz offers 10 Gbps throughput.

For all of these scenarios provide cost comparisons—assume uniform distance of 150 m between poles. Assume

pole height of no more than 20 m. It is important to recall that fiber-optic network introduces considerable delay (up to 30 ms) while millimeter-wave links typically provides <50 microsecond delay for such distances.

References

1. Frontier Business: Fiber Optic Vs Wireless. https://business.frontier.com/blog/fiber-optic-vs-wireless/

2. Secret world of microwave networks, Sebastian Anthony, 11/3/2016, ARS Technica. https://arstechnica.com/information-technology/2016/11/private-microwave-networks-financial-hft/

3. TCO the forgotten factor: How to ensure microwave costs do not sink your network. https://www.vizocom.com/ict/5-key-factors-in-designing-a-point-to-point-microwave-link/

4. Pro's and Con's of Microwave Internet service. https://www.techwalla.com/articles/pros-cons-of-microwave-internet-service

5. Fifth generation Telecommunication technologies: Issues for the Congress. Jan 20, 2019. https://fas.org/sgp/crs/misc/R45485.pdf

6. What is the value of WiFi. https://www.wi-fi.org/value-of-wi-fi

7. The economic importance of unlicensed spectrum in the case of 5.9GHz, Diana Carew, August 14, 2018, Carew, Diana, TPRC 46: The 46th Research Conference on Communication, Information and Internet Policy 2018. Available at SSRN. https://ssrn.com/abstract=3140726

8. What's next: Opportunities in WiFi with 60GHz, Carol Ansley, Charles Cheevers, Arris. https://www.arris.com/globalassets/.../white.../opportunities-with-wifi-with-60-ghz.pdf

9. Economic aspects of spectrum management, Report ITU-R SM.2012–5 (06/2016), International Telecommunication union. https://www.itu.int/dms_pub/itu-r/opb/rep/R-REP-SM.2012-5-2016-PDF-E.pdf

10. Auer G, Giannini V, Desset C, Godor I, Skillermark P, Olsson M, Imran MA, Sabella D, Gonzalez MJ, Blume O, Fehske A (2011) How much energy is needed to run a wireless network. http://citeseerx.ist.psu.edu/viewdoc/download?doi=10.1.1.468.2886&rep=rep1...

11. White space–Federal Communications Commission, 2014. https://www.fcc.gov/general/white-space

12. The rise of green mobile telecom towers, Jan 2013, https://www.thisisxy.com/sites/default/files/articles/xypartners_green_telecom_towers_0.pdf

13. Geyer R, Doctori Blass V (2010) The economics of cell phone reuse & recycling. Int J Adv Manuf Technol 47:515. https://doi.org/10.1007/s00170-009-2228-z

14. European Commission on Waste management, Categories 3, IT and Telecomm. Equipment. https://eur-lex.europa.eu/legal-content/EN/TXT/PDF/?uri=CELEX:32011L0065&from=EN

15. By 2050 the oceans could have more plastic than fish, by Rebecca Harrington, Jan 26, 2017, Business Insider. https://www.businessinsider.com/plastic-in-ocean-outweighs-fish-evidence-report-2017-1

16. Space and Satellite Insurance. https://www.marsh.com/uk/industries/aviation-aerospace/space-and-satellite-insurance.html

17. SpaceX Falcon Heavy Rocket Lofts 24 satellites in 1st night launch, By Meghan Bartels, Spaceflight, June 2019. https://www.space.com/spacex-falcon-heavy-stp2-launch-success.html

18. Orbit spectrum allocation procedures – ITU Registration mechanism, Yvon Henri, 12 June 2008. https://www.itu.int/en/ITU-R/space/symposiumWroclaw2008/Wroclaw_YH.pdf
19. Making the grade: from COTS to full space grade, Robert Baumann, Sept 27, 2018. Radiosity Solutions. https://radiositysolutions.com/making-the-grade-from-cots-to-full-space-grade/
20. Cleaning up the cosmic neighborhood: NASA grant to advance technology, help remove space debris from orbit, July 9, 2019. https://www.purdue.edu/newsroom/releases/2019/Q3/cleaning-up-the-cosmic-neighborhood-nasa-grant-to-advance-technology,-help-remove-space-debris-from-orbit.html
21. Why We May Not Be Able to Visit Space in the Future, Brilliant publications. https://www.youtube.com/watch?v=X-QSZhh_YWg
22. A space debris primer, by Roger Thompson, Cross Link magazine, Fall 2015. https://aerospace.org/sites/default/files/2019-04/Crosslink%20Fall%202015%20V16N1%20.pdf

Abbreviations

GSM	Global Society Mobile, Second-generation European cellular standard
NFC	Near-Field Communication, usually used within 4 cm from the source, is useful for transactions that could be in magnetic field, traditionally used for payment, patient monitoring, and related applications
VoIP	Voice over Internet Protocol, a standard based on IETF and voice coder used is ACELP. A VoIP phone uses digital voice encoded and decoded at the sender and receiver ends
PBX	Private Business Exchange, a telephone exchange meant to support telephone and related services such as voice, fax, and conference services within a business complex or a campus
SARSAT	Search-And-Rescue SATellites, a family of satellites operating in low Earth, medium Earth, and geostationary orbits that support emergency search and rescue operations
NOAA	National Oceanic and Atmospheric Administration, an agency with mission to understand and predict changes in climate, weather, oceans, and coasts, to share that knowledge and information with others, and to conserve and manage coastal and marine ecosystems and resources
GLONASS	Global Navigation Satellite System—is a space-based satellite navigation system operated by Russia as part of a radio navigation-satellite service. It provides an alternative to GPS and is the second navigational system in operation with global coverage and of comparable precision
LBS	Location-Based Services—is a general term to denote software services utilizing geographic data and information to provide services or information to users
BLE	Bluetooth Low Emission: Bluetooth devices that have a particularly low transmit power of about 4 mW. BLE devices use GATT (generic attribute profile) to conserve energy
VR wireless	Virtual reality usually experienced with a headset that provides audio visual experience allowing full freedom of movement and fast response due to low latency of the wireless network (such as WiGig at 60 GHz)
FAST	Five-hundred-meter Aperture Space Telescope. Currently, the world's largest space telescope that receives signals from space

This chapter deals with the social and lifestyle impact of wireless and how user demand shapes devices, types of batteries and their functionality impacting human society. The convenience of wireless prompted customers not to get tagged down by wire even when they are at home. That aspect allows user to freely move around within and outside the house, replacing even the cordless phone.

It is important to recognize and acknowledge behavioral changes wireless devices brought in the younger generation. For example, voice communication has now become secondary, which used to be the primary mode of communication during telephone era. Service providers no longer count number of voice calls or the duration. It is no longer a major revenue earning stream, rather a baseline service. Handwriting is now replaced by typing or texting (use of short words, not sentences). Data is the primary service that counts towards revenue for the provider. In technical terms, the baseband network which moves information within smart phone, no longer differentiates between voice and data. Everything in baseband stream is data, including voice.

K. Raghunandan, *Introduction to Wireless Communications and Networks*,
Textbooks in Telecommunication Engineering, https://doi.org/10.1007/978-3-030-92188-0_9

Therefore, voice is sent and received by the phone/device as data. It is only converted to voice at each user's end device.

9.1 The Wireless Generation and Wearables

This concept had an impact on the size of cell phones. Early twenty-first century saw some of the smallest phones. Yet in later years, cell phone size increased considerably! To get a perspective, Fig. 9.1 shows a Tiny T1 cell phone next to an Android phone [1].

The Tiny T1 measuring a minuscule 46.7 mm × 21 mm × 12 mm, weighed only 13 gms working on 2G network such as GSM. It is important to realize this no longer appeals to majority of the cell phone users, because its focus is on voice, not data. The reason industry moved on from 2G to 3G and 4G and now 5G is due to preponderance of data and its applications, such as pictures, video, and text messages which must be displayed on a larger screen for user comprehension.

Small size may seem fashionable, but in terms of human interface and psyche of younger generation is not inclined to voice. Although engineers who design small phones will be quick to point out that its microphone can pick up the speaker's voice very well, a human speaker will not feel comfortable with their voice being picked up over their cheek bone. The second and essential feature is reading the tiny display and use of very tiny keypad with our fingers that are larger—both are practical challenges. This shows us why the cell phone moved back to a larger size, with usable soft key pad with large display that works well for voice calls but is much more data centric, with video calls, pictures in websites, and applications that can do a lot more than what a

Fig. 9.1 Tiny T1 in foreground with a modern Android phone in the background

PC is capable of these days. Large display is indeed the central part of smart phone design.

Not all applications are stacked up against small size. Small is useful in devices that need to be worn on the person. Something that customers may need infrequently which are the family of devices known as wearables. Key features of such wearables are very small data rate and very small battery to drive the device. Being light in size and weight such tiny devices have their place in many applications. Due to very low transmit power, they are quite safe even when worn on the body 24 hours a day.

In terms of functional use, wearables have a unique, well-established feature set. They are often related to health, with basic checks such as pulse rate, heart rate, blood pressure, and other body parameters sent to health professional offering a periodic report about the patient. Those who perform physical exercises such as running, walking, swimming, etc. use it to monitor themselves. They often transfer such data to cell phone or other Bluetooth/WiFi devices capable of moving such data over personal network where devices are placed within a few meters.

Wearables come in the form of a ring, or a wristwatch or a mouse worn along finger, head band, arm band even for babies, and so on. The list of wearables is endless but there are six important wireless technologies that play important role in building such devices [2]. They are

- Near-Field Communication (NFC).
- Bluetooth Low Energy (BLE).
- ANT.
- Bluetooth classic.
- WiFi.
- Cellular.

Typically, NFC or near-field devices operate over very short distances, no more than a few centimeters from the source. They are useful in transferring data over a short range and consume very little power (battery operated). NFC typically operates within magnetic field and requires limited power, allowing small batteries to operate them over several years. Close proximity is the key. This was discussed earlier in Chap. 3, Sect. 3.6 in detail.

Bluetooth Low Energy (BLE) became popular recently since they operate within 30 m (typical) and transmit less power (about 4 mW). Since this is a widely deployed technology it is possible to pair it to a smart phone or other Bluetooth-enabled device to transfer data. Battery in a BLE could typically last several years and is convenient for not only sports, but also other applications such as finding an athlete location within a crowded area. This technology was also discussed in Chap. 3, Sects. 3.5 and 3.6 where its applications over short distance were indicated. It will be

Fig. 9.2 Wearables and bio sensors—in sports/athletic activity

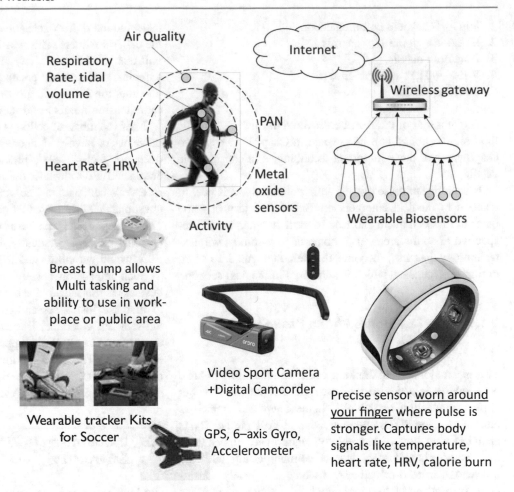

Air Quality

Respiratory Rate, tidal volume

Internet

Wireless gateway

PAN

Heart Rate, HRV

Metal oxide sensors

Activity

Wearable Biosensors

Breast pump allows Multi tasking and ability to use in work-place or public area

Wearable tracker Kits for Soccer

Video Sport Camera +Digital Camcorder

GPS, 6–axis Gyro, Accelerometer

Precise sensor worn around your finger where pulse is stronger. Captures body signals like temperature, heart rate, HRV, calorie burn

discussed again in Chap. 12 as part of wireless LAN—yes, it can form a wireless LAN.

ANT—this is a proprietary technology gaining popularity particularly for sports wearable sensors such as heart rate, speed, cycling power, distance, etc.

Figure 9.2 illustrates typical parameters monitored in sports and athletic and other home activities.

Some of these are widely used in races, events, and everyday exercises. Others such as the breast pump have brought considerable relief to working women (mothers of young toddlers). Video sport camera and camcorders brought a change in journalism. Unlike heavy duty cameras these can be easily carried and used at the scene of action. Precise sensor worn around the finger helps monitor the elderly, since their health parameters may have to be monitored daily based on the status of their health.

Bluetooth classic: This technology has been deployed for many years. Compared to BLE, this provides more throughput since it consumes more power. Bluetooth-enabled devices are used in cars, audio systems, printers, and many other fields, making it useful to pair wearables with many of these devices.

WiFi deployment at home and offices is well known—sometimes WiFi is deployed temporarily for major events. It is possible to restrict such networks to participating personnel only, making it fairly secure. Range of coverage is longer, typically about 100 m. It is possible to deploy WiFi access points right along the running/bicycle track making it a useful method to communicate not only data but also voice.

Cellular—the obvious choice for water rafting, skiing, and similar events where other networks are not practical. The greatest benefit of cellular is there is no particular limit on distance. In remote regions, this could be satellite link as well. Pairing of devices including the cell phone is straight forward.

The wireless gateway shown on the right top in Fig. 9.2 typically uses a cellular modem to connect to the Internet. All cellular modems provide WiFi for networking devices within its vicinity. However, the event planners could use multiple locations for such gateway devices making use of WiFi for local coverage and wired access to the Internet. The popular wireless technologies used in all these wearables and sports are as follows:

1. NFC—Near-field communication.
2. BLE—Bluetooth low energy.
3. Bluetooth classic.
4. WiFi—802.11 and variants.
5. Cellular—LTE.

The first four of these will be covered in Chap. 12 and the final one "cellular" will be covered in Chap. 11. These two chapters cover these important technologies in considerable detail.

In majority of these events, an important aspect is how to respond to medical emergencies. Either an injury or some other unforeseen event can lead to such an emergency. Irrespective of sports arena or at home or in an office, wireless technology has now become the default method to alert, during emergencies, which is addressed in the next section.

9.1.1 Wireless Home Phone Base Station and PBX

This is a major shift in thinking due to cellular. Home phone base station is a device that uses the cellular network but allows connection to the old "home telephone" which is either an analog POTS phone or a digital VoIP phone. The is an interesting turn of technology, since many homes used the "home phone" as a way of communicating with the household, instead of cell phone which is typically considered "private" and for an individual. Why "home phone" has turned to "cellular network" as a solution to an old existing technology? The answer is somewhat interesting. The "old technology" provided by wired network providers such as cable TV, DSL, or fiber-optic network, needs regular maintenance. This is due to the fact that most homes earlier had a twisted pair, which can go bad due to storm, rain, or even simple aging. They find this cost of maintenance somewhat high since voice no longer earns revenue. The old concept also implied the faith people had in "home phone" that it would stay on even if electrical power went down. That is no longer true since the "central office" of the service provider no longer maintains batteries to power the ringing current of old phones.

With such changes over the years, the risk of losing "home phone" is disturbing to the earlier generation, who still want to believe that cellular is not as reliable as "home phone". The irony now is that home phone is now served by cellular network, which is designed to support cell phones. In times of major weather change or emergencies "cell phone" has become the device people rely on.

Similar is the impact on businesses. The PBX or personal business exchange is an equipment that serves customers within an office building or university campus, for example. In early days, there used to be operators who would connect callers to the right "extension" within the complex, based on the caller's request. This was later replaced by an automatic unit that would have a few telephone lines coming into the complex and it would connect to the right extension based on either the last four digits or the machine would ask caller for extension number and then connect.

Now the mode of action is completely different. Since all office users have cell phones and the younger generation does not feel a "desk phone" as necessity, all that the automated PBX has to do is to forward the call to the receiver's cell phone. The very role of PBX then becomes questionable since the caller can directly call the cell phone. The concept of desk phone as office phone and cell phone as personal phone is gradually fading away. Desk phone and PBX would get eliminated in the not too distant future.

Such impact on society involves a change in the way business is transacted. Businesses large and small can afford to first use just a server and almost no staff to offer the old-style PBX service. But as callers realize this, they will directly start to call the cell phone and help in eliminating the PBX itself. The other option corporations consider is to move the PBX service to the cloud. This is a different direction, not involving wireless, but is a possibility.

9.2 Introduction to Wireless Tracking During Emergency

Not long-ago emergency care—whether medical, police, or fire always relied on the phone number from which an emergency call was initiated. That phone had a physical address to which they could dispatch personnel to help. It was not common for the caller to describe where he/she was located. If asked, they could always provide the address or the telephone number.

All of this changed completely as the cell phone became the primary means of communication during an emergency. There is no particular reason to believe the caller is at home or in an office. Since the caller could be anywhere, often in places they may not know the local address, it became important for the system to know where the caller was located.

During the early 1990s when 2G cellular was deployed worldwide, this became a major issue that every region and local authority had to resolve. Cellular network already knows where the caller is—why? Because when someone calls a cell phone number, the system must know where to deliver that call. Cellular standards had a clearly identified method on how to route the signal to get to the cell tower that serves the receiving person's cell phone at that time. This will be described in detail in Chap. 11.

This was indeed a good starting point. Therefore, regulatory agencies such as FCC, ITU decreed that the cellular provider must provide the location to emergency unit.

However, there was another practical difficulty. A typical cell could cover areas that are anywhere between 12 and 50 km^2. It was not practical for emergency personnel to scan such a large area, looking for the caller. Precious time would be lost since the emergency could be medical, fire, or someone being threatened or under duress. Fortunately, in 2G CDMA system (TIA-95) mandated the use of GPS signal at each cell tower. This factor considerably improved the location of the caller.

Accurate location of the caller is identified by the base station using a process known as differential GPS. Since the cell tower serving the user already knows its own GPS location, it instructs the handset to report timing difference between signals from neighboring towers as well as its own. Based on report received from the cell phone, cell tower uses triangulation of signals from self and neighbors, as seen by the mobile. With this, the user can be traced within about 50 m, making it easier for emergency providers to locate user.

Readers may wonder why the cell phone does not directly use GPS. There could be limitations such as in an urban area cell phone cannot see three GPS satellites (minimum needed). This can also happen in mountainous terrain. The differential method considerably improves accuracy in any case, which is very helpful during an emergency. In particular, if user is in an unfamiliar territory or is unable to report the location to the authorities, this becomes a major point of contact during emergency. In recent years, incidents such as this have become common when caller may be stranded on a highway or remote region and may not even know about the specific location within the region.

This leads us to the next type of emergency involving the elderly, who may be living at home by themselves. It is now common to observe agencies, such as the local police or providers, require the elderly to have a tag with a button on their wrist. In case of a fall or urgent need they could just press the button, which in turn sends a wireless signal to the home WiFi (access point) or cellular system to alert authorities. Before the advent of wireless technology, medical emergencies such as this were not effectively addressed unless the elderly person could actually walk up to home phone and call.

Another major emergency involves natural disasters such as floods or forest fire or snow that can isolate typical cell user. Those living in vulnerable zones such as low-lying areas near water are better served by keeping an emergency satellite phone. During natural disaster, the local cell tower also could be damaged or without power, but satellite services remain unaffected. There are dedicated services known as SARSAT ("Search-And-Rescue-SATellites") that offer emergency service. They look for distress signals at 406 MHz worldwide to locate those stranded on Earth. Almost all sea-going vessels, aircraft, and other vehicles are fitted with this distress alarm beacon. National Oceanic and Atmospheric Administration (NOAA) works with all member nations and their local rescue agents to recover those stranded using multiple satellites [3]. It also provides an international platform where all distress beacons are registered. Emergency Position Indicating Radio Beacon (EPIRB) is used by all ocean-going vessels. The 406 MHz EPIRBs are divided into two categories. Category I EPIRBs are activated either manually or automatically. The automatic activation is triggered when the EPIRB is released from its bracket. Category I EPIRBs are housed in a special bracket equipped with a hydrostatic release. This mechanism releases the EPIRB at a water depth of 3–10 feet. The buoyant EPIRB then floats to the surface and begins transmitting. If it is Category I EPIRB, it's very important that it should be mounted outside the vessel's cabin where it will be able to `float free' of the sinking vessel.

Category II EPIRBs are manual activation only units. If one of these are used, it should be stored in the most accessible location on board where it can be quickly accessed in an emergency.

406 MHz beacons are digitally coded and begin to transmit distress signals without delay. This means that even a brief inadvertent signal can generate a false alert. To avoid getting a call from the Coast Guard make sure that the EPIRB is tested regularly following the manufacturer's recommendations carefully. In general, all maritime vessels follow these guidelines for general beacon testing and inspecting procedures. A 406 MHz distress frequency signal is sent via satellite and Earth stations to the nearest rescue coordination center.

These satellites are launched and maintained in orbit by many major space agencies listed below.

Low-Earth Orbiting Search-And-Rescue (LEOSAR) Satellites

- NOAA Polar Orbiting Environmental Satellites (POES) —known as "SARSAT".
- ESA/EUMETSAT Polar Orbiting Meteorological Satellites (MetOp)—known as "SARSAT".

Geostationary Orbiting Search-And-Rescue (GEOSAR) Satellites

- NOAA Geostationary Orbiting Environmental Satellites (GOES).
- ISRO Indian National Satellite (INSAT).
- ESA Metosat Second Generation (MSG).

Medium-altitude Earth Orbiting Search-And-Rescue (MEOSAR) Satellites

- United States' Global Positioning System Satellites (GPS).
- Russia's Globalnaya Navigazionnaya Sputnikovaya Sistema (GLONASS) Satellites.
- European Space Agency GALILEO Global Positioning Satellites.

Each category has its own function and benefit in terms of search and rescue. The LEOSAR goes around the Earth in about 2 h and provides location and movement of distress. MEOSAR provides about seven times the area of coverage offering a broader view of location. GEOSAR satellites are stationary with respect to Earth. Therefore, they provide a continuous and uniform view of the same location. All of these satellites work closely with GPS to provide accurate location. It is therefore important that those who obtain a distress beacon for their use must register with the agency, irrespective of whether it is for a sea-going vessel, an aircraft, a vehicle in remote region, or even for someone who ventures into wilderness as a hobby [4]. Such registration allows the authorities to quickly identify the individual/organization and alert their respective contact persons about the situation, even as they handle the emergency.

Figure 9.3 shows typical details of what is involved in the rescue of an individual, an aircraft, or ship in distress. The three types of satellite systems shown (LEOSAR, GEOSAR, and MEOSAR) are on the top as the major location systems. Depending on the region, one of the satellites (USA or Europe or Russia) will operate in the rescue operation.

Mission control center from SARSAT coordinates with the rescue coordination center and the local terminal to not only identity the vessel or individual but rescue them. This effort has significant impact during natural disasters and is a systematic setup for those venturing out into the wilderness, either by foot, by aircraft, or by boat. Disaster beacon is common to all of them.

Figure 9.3 shows scenario where either a ship or an aircraft or even a stranded individual could alert the authority with a distress call. Arrows show multiple cases of wireless links, each of which operate at different frequency bands using different technologies but are well coordinated to communicate. This has significantly improved emergency response to disaster and recovery, since timely help is of essence.

9.3 Location-Based Services and Public Perception

In contrast to the emergency situation where location of the user is critical and must be accurately known, there are other services that only utilize marketing and other information for those willing to receive them.

This should not be confused with any type of emergency. The general public opinion on why the cellular system must know the user location is well founded. It is not true that the service provider shares this information to other commercial agencies. The location is used by the system to deliver calls. It is also used by the system to share the location detail with applications that the user wants, for example, a navigation service that helps the user go to a certain destination.

Fig. 9.3 Rescue operation using satellite and local resources (Curtesy NOAA SARSAT)

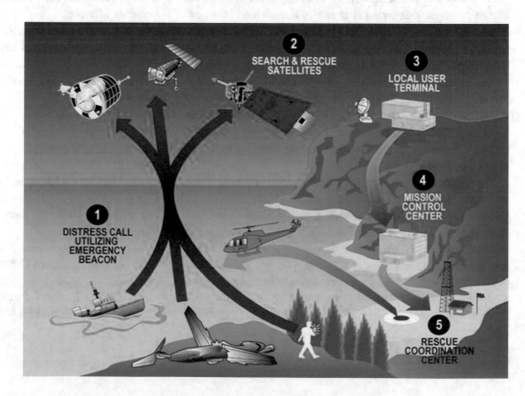

Perhaps most users have no problem using it for navigation, since it helps them. But many may also sign up for applications such as local points of tourist interest, or local restaurants or similar services. These agencies would have access to location of the user. They use this to "push services" where they may offer discount on products or tickets or similar commercial service.

Public opinion was initially against this concept and was seen as an intrusion into their privacy, but gradually that view has started to change. Every cell phone user must realize that cellular system definitely "has to" know the location in order to provide the basic cell phone service, as well as support emergency service providers during an emergency call. This is not an option, it is mandatory for the service provider. However, all other services including navigation are optional and whether to share their location or not towards specific services is completely at user's discretion. Details of how Location-Based Services (LBS) work, its components and working are well documented in [5].

It is useful to note that location information is not limited to outdoor or cellular. It also encompasses user presence in large buildings, airports, or campuses where other wireless services such as WiFi, RFID, or Bluetooth may operate. There are several vendors who operate within such buildings and can use these products towards "push services" to the cell phone. At the same time, Bluetooth technology, in particular, BLE (Bluetooth low emission) can be used to either find friends in a crowded area or even navigate the user through a complex building such as an airport with multiple terminals, Railway station complex, or commercial complex.

Such services may use mobile maps of shopping complex, providing users a clear view of where different services are located, with reference to user's current position. Some navigation services have evolved into sophisticated systems, providing voice instructions to user or may support navigation with virtual reality by superimposing text or pictures on user's cell phone.

9.4   Virtual Reality and Its Use in Media

Virtual reality is a term normally used in the visual context, where a screen or some form of interface makes the user feel that they are in that frame. The concept originally started in training simulators for pilots, drivers, and other operational staff. The screen in front of them was video that would change with inputs from user that involved steering wheel, up/down sensors, and others that included audio. The driver would be trained to drive the virtual aircraft/bus/train safely and later feel comfortable in real-life system, when on the job.

There are many games that followed a similar concept. The biggest change in recent times is the use of wireless systems where transponders allow actual movement of the

person within a given scenario. This allows for dynamics in situations where the person being training is not sitting but moving in a given area [6]. Why is this dynamic situation important and is it just another video game? There are quite a few real-life training scenarios involving assembler training who work on large bundles of cable in aircraft assembly, involving hundreds of feet. It could also be used to train workers in storage depots where operators move lift trucks involving heavy loads to be carried around safely, when there may be other workers or trucks in the aisles. Use of wireless sensors offer not only flexibility, but range of motions that were not possible with fixed wire. Such situations may involve tracking cameras that are not stationary, sensors worn on the body such as those described earlier in Sect. 9.1 including those used by workers with disabilities.

In the media, TV programs, and news cast, virtual reality takes a different hue. The focus shifts to entertainment, with creation of scenarios based on facts gathered while an incident is still happening [7]. This need moving cameras that are not wired to anything. Wireless links from a moving camera can be dynamic video streams of 2 Mbps providing 30 frames/second with high resolution (see Sect. 1.5 for details). Moving within an arena such as a football field, creating yellow line where one doesn't exist but seen by the TV viewers, is now an accepted reality.

A major portion of the audience may be watching the sport/event on their cell phone. How do the dynamics of the game get carried on to a mobile device? This is tested by broadcasters in order to deliver quality programs. The path delay from the simulating device/senor to the user is typically due to network. Use of IP networks which is common, results in path delays of anywhere from 20 to 80 ms with throughput of no more than 2 Mbps [8]. Cellular network today is quite capable of supporting about 10 such streams, in 5G this is likely to increase by about 50 times, supporting 500 such sensors simultaneously.

Why would designers choose wireless instead of wired? Other than flexibility, it is due to considerable reduction in path delay. Unlike IP network where signal flows from node to the next passing through servers or switches, wireless link takes it directly over the air resulting in a maximum of 1 ms delay. This dramatically improves performance of virtual reality and experience of end users. Therefore, industry involved in virtual reality looks forward to use of wireless as its ultimate solution that not only solves problems of connectivity but is dynamic in nature. It offers considerable improvement in quality due to sharp reduction in path delay.

What does this mean to the end user? Immediate effect is seen in headgear—covered glasses within which video is displayed. This change resulted in VR (virtual reality) headsets that allow user to become completely mobile and have only the cell phone with headset [9]. Wireless headsets eliminate the need for gloves and can track head movement with great accuracy. This leads to many training applications that can directly move into wireless headsets and make training lessons closer to real-life situations.

A good example of very high-throughput connection would be use of millimeter-wave links for wireless rather than WiFi or Bluetooth. Millimeter wave links offer 1–10 Gbps data rates over short distances and path delay is typically around 100 μs. Such short latency is not achieved on any IP network today. There is active research in this area that could bring new products that truly stand out in wireless VR a major thrust area [10].

9.5 Communication with Outer Space: Possible Only with Wireless

Since the early days of space flight, the only means of communication with space modules has used wireless. This had considerable impact on all beings on Earth. Early space craft had a few systems that were essential to keep in touch with them. This did not involve audio, but wireless signals that were needed for tracking, commands to maneuver, links that monitored health of on-board systems and periodic commands to move the spacecraft to required spots in space. Voice communication with astronauts/cosmonauts began with manned missions, initially to in orbit flights, followed by missions to the moon.

Equally important and dramatic were missions that took spacecraft not only to Mars but planets well beyond. Great distances that these flights traveled, resulted in considerable path delay for the signal, often measured in hours, not minutes. Perhaps most dramatic of all space missions is the deep space mission of Voyager that is currently so far away beyond solar system that it takes signal (traveling at the speed of light) over 20 h to just reach the Earth. The current position and view is possible only using mission control pictures of the spacecraft, see [11]. The time taken to hear from Voyager will only increase as it continues to move away from the Earth.

More practical with daily usage involves satellites around the Earth, which are no more than 36,000 km away from Earth. The round-trip delay is about 320 ms and is in general use for international traffic. They had the most impact on humanity. The low Earth orbit provides pictures of the Earth and surveys mineral, forests to such great detail that many nations have come to rely on it for their long-term planning.

GPS (and its European counterpart Galileo) is perhaps the ones with most impact on every cell user around the globe. Dependence on this navigation satellite system is so intense that people plan their travel solely on the hope of getting navigation support from GPS. So do aircraft and ships; all of them have practically shifted from traditional methods such as

maps and compass. Gone are the days of the "Bermuda triangle" where ships and aircraft used to get lost due to malfunctioning of navigation instruments of the earlier generation [12].

Equally important are services from geostationary orbit satellites beam TV and computer traffic across continents. Each nation has its own set of satellites for their region. Impact on business and individuals is huge. Their livelihood, entertainment, and family connections across continents depend on reliable satellite service. In Chap. 13 on satellite communications, many of the details will be discussed at length.

Outer space looks very different when RF receiver is tuned to different RF bands on a given day. The galaxy provides clues on how many of the celestial bodies send RF signals. Figure 9.4 provides an overview of what is likely on a given day [13]. This not only includes radiation from Earth as a whole, but the Sun, Jupiter, and some local transmissions as well.

Beyond our solar galaxy are some celestial things of great interest to astronomers who want to view much farther than where Voyager has gone till now. Space exploration using large radio telescope has an abiding interest in receiving signals. Shown in Fig. 9.5 is a massive space exploration antenna dish that was assembled together in China during 2015 and it has 500-m aperture.

It is so large that this humongous dish was built into a natural depression in Earth, surrounded by several hills. Known as FAST (five hundred meter aperture space telescope), its surface is made of 4450 triangular panels, 11 m (36 ft) on each side, assembled together in the form of a geodesic dome. There are 2225 winches located underneath making it an active surface [14].

Why such radio telescopes are usually located in remote regions? The answer is partly found in Fig. 9.4 which shows Earth's atmosphere that has many radio signals including man made (cell phone included). The telescope operating in the range 70 MHz–3.0 GHz has to be located far away from radio signals in this range to only receive the signals coming from outer space.

There are a number of outer space missions from Europe, China, India, and Israel—all of them involving astronomy or telescope or some other experiment in outer space [15].

Impact of such space programs is felt both in science and society—quite often information from such scientific missions are shared between many nations. It is used to design better satellite systems, which improved our understanding of the universe. In the long term, they continue to track new asteroids and light that is radiated from the sun and other stars.

Modern view of the universe is not based on religious beliefs, but facts based on measurements. They have provided evidence that there are billions of galaxies. It also continues to explore the question—are we alone in the universe? Perhaps not—but the exploration continues.

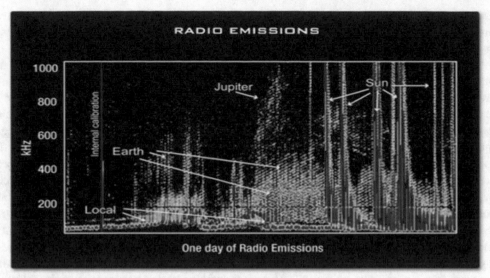

Credit: NASA/GSFC Wind Waves Michael L. Kaiser

Fig. 9.4 RF radiation from the sky

Fig. 9.5 Aerial view of World's largest telescope in Pingtang Astronomy Town, China

9.6 Conclusion

This chapter focused on the social impact of wireless technology and how it continues to transform human society
across the globe. Different sections focused on social behavior
and changes wireless brought about in the new generation,
their communication methods. Most visible impact is the
reduced use of telephone and its replacement; voice conversation supplemented with a variety of methods such as text,
social media, and conversations in an outdoor environment.

Satellites had a major impact on all Earthly beings
bringing the global community closer. It has also helped
emergency services, providing search and rescue

communication services on an unprecedented level irrespective of nature of disaster. Satellite services are essentially wireless in nature, with local wired links possible in some cases.

Other than emergency, several Location-Based Services (LBS) brought awareness to the public not only about cell phone, but other wireless devices as well. Some of these services are actively sought after by travelers and socially active urban population. They find these services not only useful, but quite exciting, in terms of knowing the local business, educational, and recreation spots.

Virtual reality, with its initial impact through games, opened a new world of imagination. Gradually it has moved into other service areas such as training drivers, entertainment including broadcast and media service. It is expected to grow into a completely new area on its own due to use of wireless technology.

Outer space has always fascinated human beings. With the advent of space technology wireless communication has given a new meaning to space exploration. It not only supports voice and pictures, but has brought back images of galaxies so distant that were not discovered until radio signals from distinct pictures sent by Voyager spacecraft. It also provides a better image of RF noise environment, mapping both natural and man-made noise in the atmosphere. Radio telescopes, in general, brought a new level of awareness regarding universe, beyond our galaxy.

Summary

This chapter brings an entirely different perspective—the social and cultural aspects of wireless technology, the devices and how it transforms society. It opens into an ongoing debate, with opposition from conservative groups, yet they too admit the positive contribution of wireless.

Perhaps different services that wireless offers provide an overview of its useful impact on society as a whole. Nations that barely had telecommunication infrastructure are thriving today both economically and socially, thanks to wireless technology. Businesses dealing with wireless services are forced to adapt continuously, moving away from wired services to wireless, in order to sustain as a business. Impact of wireless is felt on how people travel (navigation service, online reservation while traveling, etc.), how emergency services are rendered, and how wearable devices have impacted sports are all visible to everyone today. Changes continue and society, in general, seems to welcome wireless devices. It is important to note that wearables include devices and antenna within the human body. For the elderly and those with chronic conditions, this has come as a boon since medical personnel can keep a tab and provide help when needed.

Virtual reality emerged in game shows were offered as entertainment, but its impact in training is being recognized. There are clear indications of its growth due to wireless technology and its ability to make the trainee mobile, bringing them closer to the real environment.

Outer space and satellite services not only fascinate physicists but offer extensive support to aircraft and maritime with extended dependency on navigation satellites. The importance of GPS was so strongly felt that a new generation of services using Galileo series of satellites are now ushering an era of interweaving this service into every mobile vehicle and cell phone. There are other related services such as weather, indicating to pilots of an impending storm ahead, or warning ships and small vessels of an oncoming weather condition, while they are at sea.

This chapter helps the student appreciate why wireless technology is not just about telecommunication, but a variety of services that it supports society both directly and indirectly. It can be viewed as a transformation tool, essential for business, entertainment, and social life. It also helps everyone understand opposing points of view, since wireless has brought about a permanent change in the things which are planned and executed.

Emphasis on participation

In your township how are the emergency services set up? Do they know how many of their citizens have the disaster beacon registration?

What virtual reality games have you played? What is unique about them and where are wireless devices used?

Develop two location-based services for tourists planning to visit your university. Can you develop a virtual map of the campus showing visitors how to move around the campus?

What are the latest findings from astrophysicists? Is it possible to predict the path of oncoming asteroids using the large radio telescopes? What are the basic studies for which they use them?

Knowledge gained

- The chapter provided a good understanding of wearable devices and their importance in health and sports. It was a way to introduce social impact due to wireless.
- How search and rescue operations are conducted in different parts of the world for which set of satellites are used—these were briefly reviewed.
- Use of virtual reality as a training tool—the concept moved from games and entertainment to something directly useful to business, which was explained.

- An overview of location-based services indicated applications beyond the conventional navigation aid. Services that are useful to many urban residents were explored.

Homework problems

1. Why are satellites for search and rescue in different orbits? What are advantages of each orbit?
2. Why is 406 MHz chosen as a disaster beacon signal? How does the system recognize whose beacon is sending the signal?
3. Using virtual reality, what lines are drawn in a typical football game? Why is this line not on ground but only seen on your TV screen?
4. Traveling at the speed of light, a spacecraft sends signal that it is near Jupiter. Ground control sends order to take pictures. How long does it take to complete this entire transaction? [Hint: Distance between Earth and Jupiter, and speed of light can be checked online].

Project ideas

1. Without using GPS navigation, plan out a trip to the city about 250 km away. Write instructions to a driver in simple sentences of how to reach the destination. Using online resources translate directions into audio. Would you use this method as a back-up for navigation? What are its advantages?
2. Use a systematic method to transfer medical data from an elderly patient with known heart condition and diabetes. What parameters do doctors need and how often? Indicate a list of commercially available devices used in the project.

Out-of-box thinking

1. There is a stream ahead of you and there are no bridges. The navigation software on your cell phone indicates the street that crosses the stream is about 20 km further south. Since you are on an adventure trip, you want to accurately estimate the width of the stream. How would you do this?
2. Fire fighters indicate that during forest fire the cell phone is not helpful. What alternatives would you suggest for them to communicate among themselves? What infrastructure should the fire department build as part of future plans?

References

1. The world's smallest phone, introducing the Zanco Tiny T1, by Zini Mobiles Ltd. https://www.kickstarter.com/projects/1500916193/the-worlds-smallest-phone-introducing-the-zanco-ti
2. Soderholm T (May 24, 2016) 6 wireless technologies for wearables. Get [Connected] Blog. https://blog.nordicsemi.com/getconnected/wireless-technologies-for-wearables
3. Search and Rescue satellites–NOAA. https://www.sarsat.noaa.gov/satellites1.html
4. Register your beacon–the most important step NOAA. https://www.sarsat.noaa.gov/beacon.html
5. Huang H, Gao S (2018) Location-Based Services. The geographic information science & technology body of knowledge (1st Quarter 2018 Edition). Wilson JP (ed) https://doi.org/10.22224/gistbok/2018.1.14. University consortium for Geographic Information Science. https://gistbok.ucgis.org/bok-topics/location-based-services
6. Foxlin E, Harrington M, Pfeifer G (1998) Constellation: A wide range wireless motion–Tracking system for augmented reality and virtual set applications. http://citeseerx.ist.psu.edu/viewdoc/download?doi=10.1.1.333.2895&rep=rep1&type=pdf
7. Augmented reality transforms broadcast industry, by Bill Dow, CIO review 2019. https://ar-vr.cioreview.com/cxoinsight/augmented-reality-transforms-broadcast-industry-nid-13684-cid-135.html
8. Sundararaj A, Dindia P (2004) Towards Virtual networks for Virtual machine computing. Northwestern University. https://pdfs.semanticscholar.org/b1a4/dabf247fdcc11ded6fdd5e719b4b01a6131f.pdf
9. Levsky Y Virtual Reality finally goes wireless. App Real, Ha Melaka St, Netanya, 24505. Israel. https://appreal-vr.com/blog/virtual-reality-goes-wireless/
10. Circuit breaker–MIT's new method of radio transmission could one day make wireless VR a reality; By Chaim Gartenberg@cgartenberg Nov 28, 2016. https://www.theverge.com/circuitbreaker/2016/11/28/13763912/mit-radio-transmission-millimeter-wave-wireless-vr
11. Voyager mission status, where they are now, Jet Propulsion Laboratory. https://voyager.jpl.nasa.gov/mission/status/#where_are_they_now
12. Bermuda Triangle, area North Atlantic region. https://www.britannica.com/place/Bermuda-Triangle
13. Tour of the electromagnetic spectrum, NASA. https://science.nasa.gov/ems/05_radiowaves
14. China built world's largest telescope, then came tourists. https://www.wired.com/story/china-fast-worlds-largest-telescope-tourists/
15. Space missions to watch in 2019, Sky and Telescope magazine, Dec 2018. https://www.skyandtelescope.com/astronomy-news/space-missions-to-watch-in-2019/

Abbreviations

LED	Light Emitting Diode. Miniature semiconductor diode whose main function is a light source. Originally used in measuring instruments as indicator, now widely used as efficient light source to replace the fluorescent and incandescent lamps
DIP switch	Dual In-line Package switch, miniature manual switch usually used on a printed circuit board alongside other circuitry. Available in rotary, rocker, and slide type of configurations
OpenHab	Vendor agnostic cloud-based open source automation software platform www.openhab.org
Radio Collar	A neckband equipped with a small radio transmitter and attached to an animal for tracking its movement in the wild
Anti-snare	Snares are anchored cable or wire nooses set to catch wild animals. Collars are specially designed to avoid the animal getting caught in such nooses
GPS GSM collar	Introduced in 2002, this collar sends out text messages on the location of animal, since a GSM modem is integrated into the collar
PTT	Platform Transmit Terminal. A unit that contains a satellite transmitter, usually designed for specific satellite type. They are either attached or inserted into the body of animal or bird, helping direct telemetry link to satellites
Lat/Long	Abbreviation for Latitude and Longitude. All GPS system provides these two as standard parameters based on a globally accepted standard
GPS	Global Positioning System: Based on the original satellite system from USA, the term is generally used also to represent data from other satellites Galileo (Europe), and GLONASS (Russia)
EMS	Emergency Medical System usually consists of ambulance and related systems personnel. Along with Fire and Police, they form the primary public safety system
EAM	Electro Active Materials used in medical applications. A range of materials like smart polymers, bio dopants, internally charged polymer foams for biosensing, memory effect of material for stints, and other sensor applications
CNT	Carbon Nano Tubes, since carbon is an electrical conductor, they can be used as

K. Raghunandan, *Introduction to Wireless Communications and Networks*,
Textbooks in Telecommunication Engineering, https://doi.org/10.1007/978-3-030-92188-0_10

Loop antenna implant	yarn to support the growth of damaged nerves, or power devices or sensors in vivo Loop antenna as embroidery inserted into the body, for power transfer to implanted wireless devices within the body	PA-DSS	security throughout the transaction process Payment Application Data Security Standards apply to those who are developing payment applications (such as Google Pay)
Medical Device Radio Communication service	Generally known as MedRadio, different bands in the 402–405 MHz and additional spectrum at 401–402 MHz and 405–406 MHz for a total of five megahertz of spectrum for implanted devices as well as devices worn on the actual body are allocated by the FCC	PCI-PTS	Payment Card Industry PIN Transaction Security applies particularly to how Personal Identity Number (PIN) is transacted securely
HDMI	High Definition Multimedia Interface, it is a consumer electronic interface standard for simultaneously transmitting digital video and audio from a source, such as a computer or TV cable box, to a computer monitor, TV, or projector	PPSE	Proximity Payment Service Environment applies to near field and contactless payment methods and it ensures that they follow the EMVCo standard methodology
MHL	Mobile High definition Link is an industry standard for a mobile audio/video interface that allows the connection of smartphones, tablets, and other devices	EMVCo standard	Originally meant for Eurocard, Master Card, and Visa, it merged into one uniform EMVCo standard and applies to all mobile transactions
ETA	Electronic Trade Association, an organization that provides standards for mobile payments	VoIP	Voice over Internet Protocol is a standard set by IETF (Internet Engineering Task Force). It uses voice coders such G 711 and more recently G 729 that allow toll-quality voice carried over IP network
PCI DSS	Payment Card Industry Data Security Standard, it provides methods of how to securely send payment-related transactions over the network	WSN	Wireless Sensor Network refers to networks that are spatially dispersed but dedicated sensors that monitor and record the physical conditions of an industry product or human or other entities
PCI-SSC	Payment Card Industry Data Security Standards Council, established on September 7, 2006 to manage the ongoing evolution of the Payment Card Industry (PCI) security standards with a focus on improving payment account	Z-Wave	A proprietary standard that is widely used in home security systems
		SOC	System on a Chip is a concept extensively used in the semiconductor industry. Vendors who have developed chips for a specific task or application may offer them to larger, integrating chip vendor who collects many such useful

chips and integrates them together while also adding other useful system operating software or other features. Such a concept is extensively used in a smartphone, that resulted in multiple applications working on the cell phone (examples—direction finder, air temperature, etc.

The previous chapter focused on the social and cultural impact of wireless devices. This chapter explores other aspects of how wireless has become visible due to proximity devices. We will explore a variety of applications, all of them have one thing in common. The device is never far away from the living being (humans, animals, birds). Along with proximity comes responsibility for the designer to reduce RF power, since it applies to all devices addressed in this chapter. The devices shall have lower power to ensure operation even in unlicensed band based on guidelines by the regulatory authority such as FCC. Professionals and armatures refer to this as "FCC Part 15 compliant devices". We will review the details of different devices under Part 15 in this chapter.

10.1 Remote Monitoring and Control: Home, Garage Door, and Others

RF link extends the range of control for every user. Such links can be perceived as a short invisible cable over the air. The TV remote control has now gone wireless, so is the computer mouse. Why are the red LED lights no longer used? The answer lies in the ability of RF to go through objects that visible light (red LED) cannot penetrate. The other reason is that unlike red LED light that must be pointed towards the receiver, RF can use an omnidirectional antenna. Such antenna transmits in all directions, and it is no longer necessary to point the remote control towards the TV set. Wireless remote works the same way irrespective of which direction it points to, thanks to the omnidirectional antenna inside.

This principle holds good for a variety of other devices—the garage door opener, for example. FCC regulations focus on how to limit the radiated power over the air, which is a combination of power from the transmitter multiplied by gain of the antenna [1]. Since these devices at home are intended to radiate, FCC classifies them as "intentional radiators" under Part 15 subparts C through F and H. It

regulates even the antenna gain to make sure power and range are limited. It regulates those hobbyists who may want to build their own device. They too should comply with this regulation. ITU provides recommendation Rec. ITU-R SM.1538-1, which is similar to the FCC, but allows member nations to regulate such devices using their own national standards [2].

The garage door opener should also have built in security codes so that only authorized persons of the household, with a matching device inside the garage can operate the door. The garage door happens to be the entry door to many houses in urban and semi-urban areas. Therefore, security lock using code becomes important. There are many design methods, but an example would be to use a keypad with 1024 possible combination of 4-digit codes. The same code must be entered into the receiver inside the garage door controller. Cracking the code can be made harder for the hacker, using programmable DIP switches. It could also be further enhanced using a home-based WLAN (Wireless Local Area Network) that the home owner can control and monitor from anywhere.

There are many other devices used within the house such as wireless microphones, baby monitors, extended Bluetooth loudspeakers, and remote controls. FCC also includes low-power broadcast transmitters such as those used within a small arena. Many of them don't need access to home WiFi and are considered less risk prone. Units such as garden lights, sprinklers for the lawn, and driveway lights, can all be turned on and off from a remote location, if they are on WiFi. Artificial intelligence has allowed such devices to sense your arrival, turn on lights, adjust air conditioners, start the coffee machine, and turn on TV/music of choice. It is possible to accomplish such actions with voice-commanded assistants such as Alexa (voice service from Amazon) and Siri (by Apple) are popular in many households.

Other than WiFi, there are many wireless technologies such as Zigbee, Bluetooth Low Energy (BLE), Z-Wave etc. that support residential devices such as window sensors, night security vision camera, and thermostat for house temperature control. To connect multiple technologies, there is also integrating platform known as OpenHab [3].

Smart home network was just a concept one decade ago but has taken off well with active support from home users. Most of them are networked using home WiFi that often includes Smart security network—connected to township police departments in many cases. Smart home networks use technologies that support distances of no more than 100 m. The intent is to limit its spread to shorter distances, helpful from both security and practical points of view.

In arenas that support events, it is possible to set up entire audio and video systems with wireless microphones, movement sensing cameras with light control, all integrated into a central console.

10.2 Monitoring Animals in the Wild, Their Care

Natural wildlife and its study have been a source of education to nations around the globe. In recent years, sophisticated methods of tagging animals and birds have led to understanding their habitat and helping them survive and recoup their population in game reserves. The most common tagging used is known as the radio collar. There are different varieties depending on the type of animal, but all of them work on battery kept inside the collar [4].

10.2.1 Standard VHF Collar

This is in widespread use, and perhaps has the lowest cost in terms of its installation, long-term use, and retrieval.

Operating in the VHF bands (30–300 MHz), these contain transmitter that sends out unique pulses that can both identify the location as well as the animal wearing that collar. Due to nature of open forest reserve, it is necessary that the transmitted pulses reach several kilometers away to wildlife conservationists who monitor them either by air, or on land. Figure 10.1 shows a Cheetah wearing a VHF radio collar that has an internal antenna. The antenna could be internal (hidden within the collar package) or external. Depending on the type of animal, collars are chosen by professionals to have antenna inside or outside the collar.

Fig. 10.1 Cheetah wearing a radio collar with internal antenna

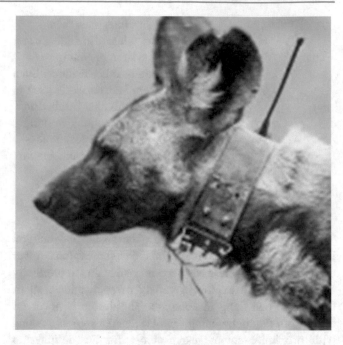

Fig. 10.2 Endangered Wild dog wearing a collar with anti-snare (antenna external)

10.2.2 VHF Collar with Snare Protection (Antenna External)

Some animals may need a collar with an external antenna. Although these collars function similar to the one with internal antenna, collars with external antenna offer slightly higher gain and are therefore monitored over greater distance. Usually, these are used on animals that move over a wider territory, spanning hundreds of square kilometers (such as the wild dog). Designers may use anti-snare to avoid animal collar getting caught in poacher wires that can choke the animal (see Fig. 10.2).

Collars used on animals need to be of simple design, with battery included in it. How long the battery lasts is a major factor that makes design of wireless collar—a challenge. Smaller the collar, smaller will be the battery and therefore shorter battery life. Since it is necessary to temporarily immobilize the animal using anesthesia to change battery, it is important to manage battery time. Such immobilization often involves taking blood or saliva samples to ensure animal is in good health.

Collars are therefore carefully selected not only based on the wireless aspects such as power, range, and tracking but also other factors. Table 10.1 shows the range of collars usage in preservation of wildlife [4]. Anti-snare is a feature that defends the animal from poachers who may try to trap the animal. GPS/GSM collar makes it easier to track since it sends text messages. Such collars have been developed for many animals such as predators, Lynx, elephants, and others making it easy for animal conservation groups and researchers to use them.

Table 10.1 Types of collars and usage—wildlife monitoring

	VHF collars	GPS/GSM collars	GPS satellite collars
Method of locating the collar	Monitor searches for the radio signal from the collar, then records a location on a handheld GPS. More time-consuming in field work	Collar send GPS locations via GSM mobile phone coverage, or remotely downloaded with a handheld UHF device. Visual mapping data, available via, Google Earth	Collar sends its location data to monitor by e-mail. Visual mapping data available via Google Earth. Less field time required
Data acquired	As a visual sighting, data are collected on home range, habitat usage, animal behavior, population demographics, and diet	Position of animal is recorded to determine home range and habitat use; however, no behavioral data are recorded, unless it has a VHF attachment	Position of animal is recorded to determine home range and habitat use. Ideal to use in inaccessible areas. However, no behavioral data are recorded
Mortality sensor	Monitor discovers a faster radio pulse if animal has not moved for more than 4 h	Position of animal is recorded to determine home range and habitat use; however, no behavioral data are recorded, unless it has a VHF attachment	Position of animal is recorded to determine home range and habitat use. Ideal to use in inaccessible areas. However, no behavioral data are recorded
Lifetime of batteries	3 years	2 years	2 years
Age of technology	Over 20 years	Over 10 years	2 years
Approximate cost[a]	ZAR 5000	ZAR 10 000–ZAR 35 000	ZAR 45 000

[a] One South African Rand = 0.074 USD (rate during mid-2021)

From Table 10.1 the older technology of VHF collar seems the least expensive. But it is also a basic necessity if animal behavioral data are needed. Modern methods with GPS satellite location allow tracking for web-based applications—these are elaborated in the following subsections.

10.2.3 Satellite Telemetry Tracking

Satellite telemetry tracking has been in use for a few decades even before the advent of GPS satellite tracking. Platform Transmitter Terminal (PTT) is either attached or implanted into the animal. The PTT sends signals to satellite that provides the Latitude, Longitude (Lat/Long) of the animal being tracked. The satellites are often in polar orbit that crosses over the poles. Whenever the satellite is in the visible range, it tracks the animal using conventional telemetry receiver. The satellite then uses the received signal to calculate the Lat/Long, based on Doppler shift and ranging methods commonly used in all satellite systems [5].

Important distinction is that the accuracy of location can vary depending on the quality of signal received, and statistical methods use repetitive data received from the animal PTT. There are seven different classes of accuracies (ranging from about 150 m to several kilometers) that the satellite may provide, and local tracking must supplement data with more accurate location.

The biggest advantage of this method is that it makes it possible to follow migrating bird and animal population, traveling over great distances (hundreds or even thousand kilometers). Most common band used for satellite telemetry is the UHF band—quite common in other satellite systems too.

Currently, there is widespread use of GPS along with VHF transmitter. But satellite telemetry methods are typically used in specific applications. It uses a wide range of battery technologies that can support birds, sea-going animals in addition to mammals, rodents, and others. It can also provide physiological data over extended periods of time [6].

10.2.4 GPS Tracking Collars

GPS telemetry is the more recent trend since these navigational satellites are a complete constellation and are capable of tracking the animal over a long period time. The data available provide information on its entire territory. However, limitations of GPS include limited battery time and considerable cost. Its advantage is that the researcher or tracking organization gets the data over a computer. There is no need to travel to the territory. GPS receiver has thus become a popular research tool. Typically, conservation team members who are local to the animal's territory use other methods (such as VHF, satellite telemetry) to accurately locate and rescue/treat animal.

At the beginning of this chapter, the FCC regulation part 15 was mentioned and subsections for intentional transmitters were indicated. How are these being applied when the user of proximity device is an animal/bird? The response seems quite interesting—it may not be a concern but there are some issues—we will review them now.

10.2.5 Tracking Methods and Improvement in Transmitters

Early transmitters were based on VHF transmitter standards of 1960s. They have come a long way in terms of reducing power requirements. Also, battery technology has improved with extended life that lasts over 4 years. Newer concepts such as dropping the collar off at the end of battery life are used. But this poses some risks. If not recovered, valuable data could be lost, since historic data are preserved within the collar [7].

The transmit power of VHF is within FCC limits but similar to earlier era handheld units used by emergency squad personnel fire, EMS etc. Current versions of VHF transmitter may not be on par with sophisticated smartphone technology but has certainly reduced power compared to earlier versions. In all these cases, the driving principle seems to focus on battery life and therefore reduce battery use by making it as a beacon, rather than a smartphone or portable handheld where voice conversation is the focus.

Power cannot be considerably reduced since the receiver unit is likely to be several kilometers or in the case of satellites, quite far away. This limitation plays against GPS tracking units, which tend to last for no more than a year. Also, GPS tends to be about 10 times the price of VHF transmitter. The VHF transmitter is still the least expensive technique in widespread use.

Ability to remotely turn off the receiver also helps in extending battery life. One solution is to Turn off the transmitter during night time when the animal sleeps or during winter hibernation (bears, for example) when it may sleep for extended period of months. This can considerably extend useful life of the collar.

Adverse effect of transmitter on the animal is not due to radiated energy but more due to weight on its back (for birds) and for some mammals like deer it may offer possible detection by predator. It seems to have resulted in reproductive problems for birds. In general, the beacon with 450 ms pulse in VHF is not significant concern in terms of radiation safety. With techniques to turn off transmitters as often as practical, RF radiation is not something the conservationists would be concerned about. In any case the collars meet the FCC part 15 regulated limits.

10.3 Surgery and Medical Procedures: From a Distance

Let us shift our focus to medical procedures. Witnessing surgical procedure "prerecorded with a camera" is not new. It has been shown on TV and in training classes, for a long time. However, the use of satellite network for live coverage of surgery opened a new area in telemedicine. This not only became common to rural doctors serving local population but also across nations.

However, the major beneficiary seems to be plastic surgery that requires clear view of the patient and precise remote operation to achieve good results in conditions such as burns, traumatic injury, wounds etc. [8]. The use of modern wireless glass (such as Google glass) by the doctor allows the remote surgeon to not only see real images of the patient but also perform surgery using tools that are remotely controlled. Broadband wireless connection allows the Internet to actively support critical phases of surgery. Wireless connections may involve WiFi 802.11ac and Bluetooth technologies locally over short distances, but connected through satellite or cellular links to cover the long distance where the specialist surgeon may be located.

10.3.1 Implanting Devices in the Human Body

There are many wireless devices that could be implanted in the body allowing control of bodily functions such as heart rhythms, monitor hypertension, provide functional electrical stimulation of nerves, operate as glaucoma sensors, and monitor bladder and cranial pressure.

Research area in medicine is actively pursuing the use of wireless transponders enclosed in tablets (capsule) swallowed by mouth that can allow doctors to monitor drug usage within the body.

Figure 10.3 is an illustration of various wearable, minimally invasive, and implantable medical devices with the sensing and communication paradigms elaborated further in Ref. [20]. In addition to those, Wearable smart textiles incorporate Electro Active Materials (EAMs) that allow the conversion of mechanical energy to electrical energy and vice versa. They include carbon nanotubes (CNTs), conductive polymers, ion polymer-metal composites, ferroelectric polymers, and dielectric elastomers. There is significant material development work, devoted to developing energy sources that can efficiently harvest energy from the

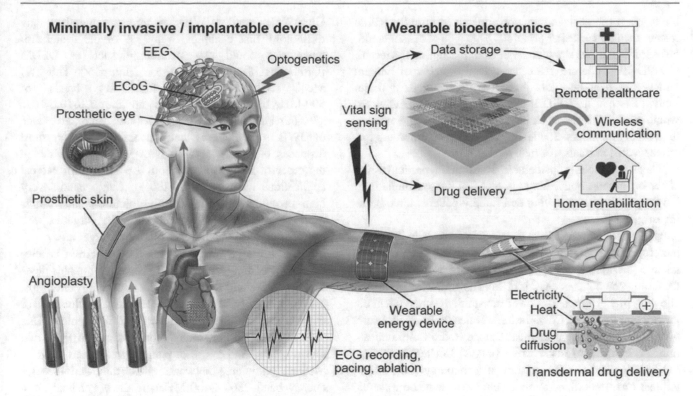

Fig. 10.3 Illustration of some minimally invasive device with wireless technology

environment and store it efficiently to power wearable electronic devices. Energy sources that can be utilized for this purpose include the sun, motion, wind, thermoelectric, electrochemical, triboelectric, etc.

The conversion of mechanical energy to electrical energy is common for micro/nanoscale transduction mechanisms such as piezoelectric and triboelectric generators and elastomers. These can be utilized to power sensors, instead of traditional batteries. In addition, RFID offers very small devices that are of considerable help to the patient and supports the welfare of the elderly.

Like cell phones, wearable and implantable RFID technologies need to conform to international and national safety

guidelines for the specific absorption rate or SAR, which was elaborated in Chap. 7 earlier [9]. In general, the concept of RFID is viewed as a method to identify—materials, animals, birds, and now human beings. There is some controversy on whether to allow RFID sensors to be inserted in humans; many groups have opposed it as an invasion of privacy and plan to bring laws to oppose it. But regulators like FCC have taken a supportive stand since it aids health and lifestyle.

Figure 10.4 shows one possible version of RFID chip that could be inserted in human body. On the right is the basic concept of how RFID works. The RFID tag (RFID chip shown on left) is interrogated by an external RFID reader. The tag itself need not be active (no power or battery needed)

Fig. 10.4 RFID chip in humans —could be inserted in hand

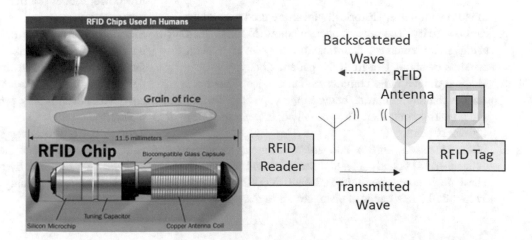

but it just responds using the power received from the RFID reader, using backscatter principle. This type of sensor could be useful in various situations. For example, during the recent COVID-19 pandemic, there are a large number of human bodies stored at hospitals that were unidentified. If these individuals had an RFID inserted when they were alive, it would have been much easier to identify them and inform their family members. Such RFID tags are in common use for domesticated animals, for many years now.

There has been considerable research in particular its effect on human body. Various types of sensors including loop type of antenna having embroidery patterns have been investigated [9].

Is this another area where wireless devices are in the body therefore regulations apply? Indeed, they do and FCC has taken a broader view of this dramatic impact of wireless technology on patients. FCC is making amendments to part 15 allowing higher limits for power transmitted. Amendment of Parts 2 and 95 of the Commission's Rules do Provide Additional Spectrum for the Medical Device Radio Communication Service in the 413–457 MHz band, ET Docket No. 09-36. FCC recognized this major turn in technology [10]. Their impression is underlined by statement recognizing how useful this technology could be, which states "The video camera captures images that are converted into instructional signals by the video processing unit and are sent back to the eyeglasses" which is wirelessly transmitted to the implant location within the body (near the eye). OET's order will permit the device to exceed the Part 15 limits for intentional radiators when the data signals are transmitted from the eyeglasses to the implanted patient who is visually impaired but now will be able to see due to this technology. The document states the importance of allocating frequencies for this great cause by stating "Helping a blind person to see. Empowering a paraplegic to stand. That's the power of wireless technology. And that is why the FCC will continue working around the clock to harness this power to improve the lives of people".

FCC regulations support various types of wireless devices used in the medical profession [10]:

1. Inductive implants: Historically these are used to monitor cardio activity. The sensors operate below 200 kHz band using small antennas and transmit only over a short distance of about 1 m from the patient's body.
2. Medical device radio communication service: These are devices that communicate between the implanted devices and a nearby controller. They usually operate in the 401–406 MHz band.
3. WiFi, Bluetooth, ZigBee, and similar technologies in the unlicensed ISM band (Industrial Scientific and Medicine). These devices operate in the 902–928, 2400–2483.5, and 5725–5850 MHz bands at distances up to about 100 m.

4. Ultra-Wide Band (UWB): A term attributed to any technology that occupies over 25% of the center frequency—it could use technology such as OFDM (Orthogonal Frequency Division Multiplex). The "WI Media Alliance" has defined 14 bands of 500-MHz bandwidth to divide up the 3.1–10.6 GHz spectrum allocated for Ultra-Wideband. Major advantage of UWB in medicine is that it operates over very short distances, but with vast throughput (very high data rate in the order of 500 Mbps to about 1.6 Gbps). In medical applications particularly those that need very high-resolution cameras with high picture quality this is useful. For example, it could be used by surgeons who can operate with much better clarity of target area.
5. Medical micro-power networks: New spectrum is allocated to accommodate implanting micro simulator devices that could help restore movement of paralyzed limbs. The frequency band 413–457 MHz band is allocated for this purpose that can transmit up to a few meters. This was indicated earlier, under modifications to FCC rule allowing the blind to see or paraplegic to stand.
6. Medical body area networks: FCC allocated the frequency band 2360–2400 MHz, to allow Personal Area Network (PAN) at distances up to a few meters. This could be used in intensive care units to get data from patient carrying it to medical personnel.

Why are so many bands and wireless technologies needed for medical profession? The answer perhaps lies in several factors:

(a) Devices need to operate no more than a meter away from the body and are readily available today—thanks to VLSI and battery technology.
(b) The task of implanting device within or on the body communicating to nearby receivers is ideally accomplished by wireless.
(c) There should be no interference from the device or from outside to the device. Since the device operates on very low power it does not travel far enough to interfere with other devices.
(d) Since specific licensed bands are allocated by the regulatory body—this makes it easier for device manufacturers to accomplish the task reliably operating the device without concern of interference between different devices even if they were all present inside the same patient.
(e) Perhaps the biggest advantage is that wireless allows patients to move out of the bed and out of the hospital and still be monitored by medical personnel, collecting accurate data, and reliably maintaining privacy [11].

10.4   Use of Smartphone as the Only Communication Device

The cell phone replacing conventional telephone is widely accepted by the society and has become an established fact. But its use as the only communication device, takes a new meaning that builds and extends across all human interactions. What does this mean and how does it affect various professional areas including information technology?

To warm up to this fact and changes in the not-too-distant future, let us consider a few realities, in terms of what the cell phone has effectively replaced.

10.4.1   TV Screen Getting Gradually Replaced by the Cell Phone

This has become a fact although TV is not entirely replaced until now. It is possible to watch news, see favorite TV program on smart cell phone, but size of its screen makes it acceptable for individual viewer but not a family or community. Consider the fact that what is available on a cell phone can be projected to larger screen using WiFi. This is already true with streaming applications such as Chromecast and many others. Stand-alone smartphone projectors are available and allow cell phone screen to be projected. There are many smartphones with built-in projector software—it allows powerpoint presentations to screen up to 50 inches wide with project video resolutions of 640-by-360 pixels.

There are other products that collect over the air broadcast signals and stream them through the Internet. The smartphone connected to the Internet can view them on its screen. To extend this to a larger screen, a wide range of wireless screen mirroring technologies are available. These include "Miracast", which can be thought of as HDMI (High-Definition Multimedia Interface) over WiFi—which is part of standards released by WiFi alliance since 2012. Mobile High-definition Link (MHL) uses the micro USB port of cell phone. These technologies now offer a cable free method of watching what is available on the smartphone, to be projected on a wide screen in the living room [12]. In effect, the smartphone is gradually replacing the TV, which is becoming just a screen that only projects everything from a smartphone. That screen is no longer a communication device, it is just a display.

10.4.2   Cell Phone as Identity

Many nations have adopted smartphone for identity of individuals. It is already used in airports and in other locations to provide important government-issued identifications such as digital driver license yet maintain privacy and provide police

and law enforcement agencies real-time information on the person [12]. Personal identity raises questions on how citizens are protected. This needs the digital keys to be independent of service provider. Transactions between smartphone and the verifying device (Government or other entity) provide secure messaging, authentication, and verification. These steps are spelt out in standards such as one from "Global platform" consisting of industry partners involved in making chips, devices, and secure interface platforms [13]. In effect, the future of driver license, passport for travel could all be integrated and replaced by the smartphone, and it communicates securely to identify the individual.

10.4.3   Cell Phone as Your Car Key

Several auto vendors are actively considering applications that allow the smartphone to be used as car key [14]. Identity of the owner, the car make, and model would be specific. It is important that only authorized person with the correct digital key can open and operate the car. Several cars in the market already use digital key residing in a chip, and to drive. Proximity sensors (based on Bluetooth Low Power or similar technology) would detect when the phone is near the car or inside it. This allows the doors to open and the engine to start. The smartphone is now a device that communicates with the car and allows valid driver to use the car.

10.4.4   Cell Phone for Keyless Entry into Buildings

Widely deployed using several platforms, smartphone is actually very secure for transactions that need extensive verification. It is also used in many locations for keyless entry into public and private buildings such as offices [15]. It can eliminate a great deal of formalities such as issuing cards to employees, identity verification by building management, security guards, etc. Near Field Communication (NFC) described earlier in Chap. 3, Sect. 3.6 provides an excellent choice since it requires the individual to be physically present at entry of the building. In addition to consuming very little power (less drain on the cell phone battery), it operates only in very close proximity of no more than a few centimeters. Cell phone is already very well trusted by many security and financial agencies as a communication device that provides verifiable information.

10.4.5   Cell Phone Versus Laptop and Desktop

It has been a progressive transition from desktop PC to laptop (quite popular) to the cell phone.

Advantage for laptop or PC is faster processor due to larger size and drawing power from mains.

But the limitations are that they always need a landline link such as WiFi or a direct Ethernet cable to connect to the Internet. If they have to work anywhere else, then they need the help of a cell phone, which accesses Internet with cellular network over the air. Smartphone can work on WiFi too, and as a hot spot it can support laptop or PC. It has other major advantages. It is a portable camera; it is capable of several control tasks described so far and works everywhere —much of the world population is supported by cellular signals. In areas that are not, smartphones with satellites link capability do exist. In course of time will the PC and laptop go away? The answer is "perhaps yes". The PC or laptop may remain for specific video detailing and other heavy-duty computing functions that are power intensive. The cell phone processor though powerful must continue to operate on battery power but is capable of replacing PC and laptop in course of time. Future of information technology may include "not using IP network", which is the mainstay of computer communication today. That network protocol is extensively based on "wired network" with inherent assumptions of IP address and Ethernet cable connections. It does not represent the reality of cellular network where cell phone does not have an IP address.

10.4.6   Cell Phone: Secure Mobile Wallet

Cell phone as the primary method of payment is accepted by the younger generation. More than convenience, it offers a method that is secure, retains privacy, makes it easier to share costs using applications that reside on the phone, easier to pay for services (such as taxi/cab) anywhere. It neither needs a desk or desk counter nor does it need other gadgets that are wired into a building. The Electronic Transaction Association (ETA) provides standards for mobile payments [16]. For example, Payment Card Industry Data Security Standard Council (PCI-DSS or PCI-SSC) is followed by companies like Visa and Master card that issue credit cards. Payment Application Industry standard (PA-DSS) applies to those who develop payment applications, transactions, etc. PIN Transaction Security (PCI-PTS) is a set of standards that describe how the PIN (Personal Identification Number) is securely transacted. Similarly, Point-to-Point Encryption (P2PE) is a standard that ensures data are securely encrypted and transacted. Chips embedded in the cell phone follow the EMVCo standard—originally meant for Eurocard, Master Card and Visa, it merged into EMVCo standard and applies to all mobile transactions. This includes NFC (Near Field Communication) or contactless published in 2010, as well as PPSE (Proximity Payment Service Environment) and application management for secure environment, published

in 2017. The cell phone has effectively taken the place of credit cards and key features of financial transactions [17].

10.4.7   Cell Phone: Replacing Digital Telephone

Perhaps obvious and true that the wired telephone that evolved into VoIP phone (Voice over IP) has limited capability. It cannot be compared to the cell phone or smartphone. It still has applications in offices or residences where wired IP networks are available. Currently, it is estimated that only half of the population still uses landline phone at home. Their number is decreasing as younger generation prefers not to have a wired phone at all. However, VoIP is considered less expensive and reliable. The old telephone that used to be powered by reliable battery banks located in central telephone exchange (in power outage conditions) is no longer available in many parts of the globe, this was elaborated in Sect. 9.1.1 of Chap. 9.

The advantages of cell phone and their application as a communication device seem to grow with time. It is important to recall that most of what was discussed in this section existed in some form earlier but was replaced by cell phone. Even text messages (or SMS) did exist in the wired networking world in the form of instant messaging, which was offered by a variety of application providers. Its usage dwindled as cell phone handles the bulk of such traffic during social media interactions. Prior to that, it existed as telegraph that is mostly forgotten now. It is important to emphasize that cell phone although used as a phone resulted in so many other applications and benefits that its voice usage is now only a small part of the feature set.

10.5   Sensors and Their Integration into Wireless: Bluetooth Extensions

The world of sensors is quite vast—Wireless Sensor Network (WSN) is based on tiny devices that are only a few cubic millimeters in size operating on about 10 milliwatts of power [18]. The basic sensor node, which is miniature in size, is built on a chip area of few square millimeters and has only few components as indicated in Fig. 10.3 where layers of chip are shown next to the human shoulder. Basics of the sensor are shown in Upper part of the diagram in Fig. 10.5, shows three essential blocks—the sensor S, the processor P, and the Communication device C. Lower part of the figure shows their details. The sensor resides within a tiny part on the chip, connected to a Central Processing Unit that has its own tiny operating system, with a small memory. Communicating device essentially is a radio capable of transmitting using very small power in the range of 1–10 mW. Therefore, small battery (button type) would be sufficient to power such

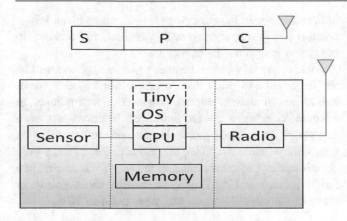

Fig. 10.5 Representation of miniature sensor network

a sensor network. Though small, such sensors are useful in a wide variety of applications with physical parameters such as: acceleration, vibration, gyroscope, tilt, magnetic, heat, motion, pressure, temp, light, moisture, humidity, barometric

- Chemical parameters such as: CO, CO_2, radon,
- Biological (not limited to humans),
- Micro-radar etc.
- Dynamic motions such as: actuators too (mirrors, motors, smart surfaces, micro-robots).

Application of such sensor networks has now become widespread in all of the above-listed applications that continue to evolve [18]. Although the radio could operate using technology such as Bluetooth Low Energy (BLE), there is no particular reason to limit it to that technology (see Fig. 10.5).

Many of these sensor networks can integrate into existing networks that could be based on other technologies such as WiFi 802.11 or ZigBee, Z-Wave, or others. The physical interface to such networks is usually Ethernet. Biological sensors could be inside the body in which case communication would be wireless as described in Sect. 10.3 in this chapter. The options are seemingly endless, but sensors have truly made their impact due to wireless. What does this mean? Sensors are located where the action is—the point that needs monitoring. It may not be convenient or even possible to run wires to such locations particularly if they are moving. Wireless resolves those problems making sensors free to be placed where they should be.

An interesting application of Bluetooth Low Energy is to sense friends in a crowded arena. If there is a friendly phone nearby it can alert the owner. This is done more by communicating with Bluetooth devices that friends have on their phone and if are within vicinity. Although this application does not fit the conventional sensor model, with a smartphone it becomes an excellent feature.

Another example for application of wireless sensors is in monitoring agricultural land—using wireless sensors for air and soil temperature, humidity, luminosity, solar visible radiation, wind speed and direction, rainfall, atmospheric pressure, leaf wetness, and fruit diameter [19]. These sensors convey information to monitoring stations locally using technologies such as ZigBee, WiFi, Bluetooth, etc. Such data from the field are distributed to farmers, technicians, specialists, manufacturers, and retailers. Even devices implantable in human body are becoming available through manufacturers—they conform to international and national safety guidelines [20].

When such useful applications using sensors are on tiny chips would it be possible to incorporate them in a smartphone. The answer is a resounding "yes" —that is already being done. Semiconductor vendors use the concept of SOC (System-On-a-Chip) to incorporate whatever is needed, on a platform. Multiple small chips sit on that platform. Finally, the platform itself is made of a semiconductor and integrating all these small chips together makes it a system, hence the name "System On a Chip". Some functions already available as applications on the smartphone are compass or direction finder, local air temperature, etc. It is possible to incorporate not only sensor but also many of the other functions—which is why it is termed "smartphone".

10.6 Conclusion

This chapter brings to fore the concept of proximity devices that wireless technology brought forth to the society as a whole. It is perhaps the most widely witnessed part of wireless since the device is visible and quite often its applications are apparent to those using them.

However, there were a few exceptions where entirely new applications were reviewed. The cell phone and how far it has gone beyond being just a phone, reinforces the concept of this device becoming more powerful and useful as time goes by. The intent was to touch upon some aspects to illustrate how different applications on the smartphone work together to make it highly sought-after device in society today. While capable of communicating over great distances, the cell phone can also operate as a proximity device.

Use of wireless sensors and changes they brought about in wildlife conservation is often seen in documentary shows but how its manufacturers and maintainers view these transmitters was the focus of our review. While low energy is a necessity due to battery life, the limitation it poses and alternatives were evaluated. It showed that the VHF transmitter using a time-tested technology, has retained its place over several decades. GPS receiver while providing better accuracy seems to have limitations in terms of lifetime for the battery and higher cost.

Whether surgery and medicine were benefitting largely using satellite services for telemedicine, which was a matter of interest in some societies. However, with modern wireless devices implanted in the body, their use and focus of the medical community have shifted considerably. The importance of this is felt in the FCC focusing on providing specific bands just for wireless devices in the medical profession. The number of devices implanted or staying close to human body has grown. At the same time their importance in life transforming applications has also grown (for blind to see and others to stand up—are good examples).

Sensors and their use in process control have considerably history dating back to 1960 or even earlier. However, the use of wireless technology focuses on combining the sensor with the second-generation processor such as Intel 8086 that came into market as a 16 bit processor during 1980s and provided enough computing power using a very small area within VLSI chip. Combining it with a low-power transmitter such as proximity technologies indicated in this chapter brought about a major change in the use of sensors. It has indeed brought applications that were traditionally considered impractical such as large areas of agricultural land.

The chapter concludes by showing the breadth of low power, low throughput technology in wireless, which contrasts with the mainstream focus on large gateways and an ever-increasing demand for throughput. In industry and professional circles, this area of proximity devices and low power wireless is forecasted as about 10 times the size of smartphone market—showing the potential it holds.

Summary

The chapter provides a wide-ranging view of devices that are within the proximity of where the action is. Whether it is wildlife conservation, surgery and medical applications, power of the smartphone for wireless sensors, the impact is evident. This has been the focus of this chapter, which takes the reader through a journey of various fields on endeavor and how each of them benefits from wireless.

While the previous chapter focused on wireless devices their social and cultural aspects, this chapter focused on its proximity and low power. Reasons why each professional embraces wireless is one aspect that runs as a common thread—the freedom it provides and ease of deployment compared to wired system with similar capabilities. The use of battery that can last several years makes a considerable impact in terms of user perception, for planning and long-term maintenance. This was indicated both in wildlife preservation and the medical profession. Low power technologies such as BLE (Bluetooth Low Energy), ZigBee, and others can be compared to small spec of light seen in an indicator lamp, in terms of power. Such comparison brings comfort to those concerned with radiation, and others asking, "how long does the battery last"?

Making use of wireless features to serve applications that are important and don't need much power makes a major impact to an entire product range. The smartphone, in addition to being a phone, embraced functions not even thought of as possible just a decade ago. In spreading its wings, the wireless simplified the way a variety of tasks were done in society, whether replacing the way TV is viewed in a family room, checking identity at security check points, or many of the other functions that were listed in 10.4.

Emphasis on participation

Make a list of devices that you can remember from elementary school days that have either changed or totally replaced by wireless technology.

Contact your local Zoo or wildlife conservation group and find out how many tracking devices are used on animals of various types. How much is the battery life and do they seek further improvements in terms of performance—which parameters do they want to improve?

Which new applications on your smartphone do you find most useful? Are there other features you want to see on your phone to make daily life easier?

Knowledge gained

- Proximity devices are essentially low power wireless devices that operate well but only over short distances of no more than 100 m.
- Most proximity devices use battery as power sources—efforts are made in design to increase battery life, by turning on the device only when needed.
- Wireless devices have made a lasting impact on a variety of professionals and their applications—doctors, wildlife conservationists, and hobbyists, among others.
- Wireless sensors have made an impact on processing instrumentation by monitoring and gathering data in remote locations—with ease of deployment and maintenance.

Homework problems

1. A collar used by Cheetah lasts 3 years and the conservationist plans to improve upon it, by turning off the transmitter when possible. How many hours should it be shut off to improve battery life to 4 years?
2. Why is FCC changing rules to 47 CFR with amendments to part 2 and 95 allowing higher power for devices? Which technologies could benefit from these changes?

3. List at least two standards that address security aspects of mobile wallet transaction using cell phone?
4. Using wireless sensors for agriculture—list at least four types of sensors that could help an agriculturist in getting better crops/harvest.

Project ideas

1. FCC has designated new spectrum, allowing new types of MedRadio devices to access 24 megahertz of total spectrum in the 413–419 MHz, 426–432 MHz, 438–444 MHz, and 451–457 MHz bands on a secondary basis. Develop concepts of an implantable device that could help patients who have paralyzed limbs or organs. Coordinate with medical services (Doctors) to develop this concept. Use FCC rules 47 CFR parts 2 and 95 (link enclosed) to get details. https://www.govinfo.gov/content/pkg/FR-2012-01-27/pdf/2012-1540.pdf.
2. Based on the concept in Fig. 10.3, develop a conceptual model of sensor with computational network and transmitter for a new application in a field of your choice. Develop details of what the sensor would do, what is its input to the CPU, what type of tiny operating system is proposed, and the wireless technology used for transmitting.
3. Develop (in the lab) the same application for transmitting short bursts of data (no more than 1 kilobyte at a time) three times a day—one using Bluetooth Low Energy device, another using ZigBee device. Compare and contrast their performance in terms of battery power consumed, distance of reliable transmission, and estimated market price.

Out of box thinking

1. Why should a wild animal wear a collar? Why not use only implantable device that could be inserted under its skin quickly. If this is allowed in humans (FCC does allow based on view of medical community), then why not animals? Can this not be done to groomed animals such as horses or sheep that freely roam around large areas?
2. During para-jumping would it be possible to measure parameters such as ambient temperature, wind speed around the body, pulse rate, skin temperature—for a novice (who needs to jump with an instructor)? Using sensors how would you store these data so that it can be retrieved (a) in real time (b) from a storage device, after they have reached the ground?

References

1. Understanding the FCC regulations for low power, non-licensed transmitters. Office of Engineering and Technology (OET), October 1993 (Edited and reprinted in February 1996). https://transition.fcc.gov/Bureaus/Engineering_Technology/Documents/bulletins/oet63/oet63rev.pdf
2. Technical and operating parameters and spectrum requirements for short-range radio communication devices, Recommendation ITU-R SM.1538-1, 2001–2003
3. 17 Smart home products that don't need home WiFi. https://futurism.com/17-smart-home-products-dont-need-wi-fi
4. GPS and VHF tracking collars used for wildlife monitoring, 17 Apr 2014, Wildlife ACT, News, conservation and research. https://wildlifeact.com/blog/gps-and-vhf-tracking-collars-used-for-wildlife-monitoring/
5. Perras M, Nebel S (2012) Satellite telemetry and its impact on the study of animal migration. Nat Edu Knowl 3(12):4. https://www.nature.com/scitable/knowledge/library/satellite-telemetry-and-its-impact-on-the-94842487
6. Wired Life, John Welsh, Bob Welsh, Minnesota State conservation magazine Nov–Dec 2003. https://files.dnr.state.mn.us/mcvmagazine/young_naturalists/young-naturalists-article/telemetry/wiredlife.pdf
7. A Critique of wildlife radio-tracking and its use in national parks, by David Mech and Shannon, Barber Meyer, US geological survey. https://www.researchgate.net/publication/237291937_A_CRITIQUE_OF_WILDLIFE_RADIO-TRACKING_AND_ITS_USE_IN_NATIONAL_PARKS
8. Telemedicine and plastic surgery–a pilot study: Hindawi Publishing Corporation Plastic Surgery International Volume 2015, Article ID 187505, 4 pages. https://doi.org/10.1155/2015/187505
9. Kiorti A (Oct 2018) RFID Antennas for Body-Area applications–from wearables to implants. IEEE Antennas Propag Mag, 14–25. https://ieeexplore.ieee.org/abstract/document/8453266
10. Wireless networks of implanted medical devices, FCC Takes Steps to Introduce New Advanced Medical Technologies to Treat Neuromuscular Disorders and Traumatic Injuries. November 30, 2011. https://www.fcc.gov/document/wireless-networks-implanted-medical-devices
11. Mahn T (Sept 2014) Wireless medical technologies. Fish and Richardson, Navigating Government in new medical age. https://www.fr.com/files/Uploads/attachments/regulatory-wireless-medical-technologies.pdf
12. Digital Trends, Home Theater - Here's how to mirror your smartphone or Tablet on to your TV, Quentyn Kennemerand Ryan Waniata, posted on July 26, 2019. https://www.digitaltrends.com/home-theater/how-to-mirror-a-phone-on-your-tv/
13. Mobile ID solutions for Government-to-Citizen applications, Security Document World, Whitepaper. http://www.securitydocumentworld.com/creo_files/upload/article-files/hid-mobile-id-sols-wp-en.pdf
14. Global Platform–the standard for secure digital services and devices. https://globalplatform.org/
15. Valdes-Dapena P (24 Feb 2016) Your phone could soon be your car key. CNN Business. https://money.cnn.com/2016/02/22/luxury/volvo-phone-app-start-drive/
16. Penny J (19 Jul 2018) Simplify access control with smartphones. Build Mag Smarter Build Manag. https://www.buildings.com/news/industry-news/articleid/21265/title/simplify-access-control-with-smartphones

17. Mobile Payments Security Standards Whitepaper: Executive Summary, Electronic Transactions Association (ETA). https://www.electran.org/mobile-payments-security-whitepaper/

18. Wireless Sensor Networks-Prof. Jan Madsen, Informatics and Mathematical Modelling Technical University of Denmark, Richard Petersens Plads, Building 322 DK2800 Lyngby, Denmark. https://www.ida.liu.se/~petel71/SN/lecture-notes/sn.pdf

19. Stamenković Z, Randjić S, Santamaria I, Pešović U, Panić G, Tanasković S (2016) Advanced wireless sensor nodes and networks for agricultural applications. In: 24th telecommunications forum TELFOR 2016, Serbia, Belgrade, November 22–23, 2016. https://gtas.unican.es/files/pub/wsn_agriculture_telfor16.pdf

20. Fiber type solar cells, nanogenerators, batteries, and supercapacitors for wearable applications. Advanced Science open access 1800340, June 2018. Published by WILEY-VCH Verlag GmbH & Co. KGaA, Weinheim. Sreekanth J. Varma, Kowsik Sambath Kumar, Sudipta Seal, Swaminathan Rajaraman, and Jayan Thomas*. https://doi.org/10.1002/advs.201800340

Abbreviations

Smartphone	Term indicates the cell phone that uses multiple applications to make it sense and provide services such as GPS location, Bluetooth tethering supported by Android or iPhone operating system
Mobility	Used in the context of cellular system to mean supporting high-speed trains of 550 kmph and support low flying aircraft
AMPS	American Mobile Public System, the first-generation cellular network of North America
ECC	Error Correction Codes are used for controlling errors in data when they are transferred over noisy channels such as wireless (over the air)
FEC	Forward Error Correction, is a method of obtaining error control in data transmission in which the source (transmitter) sends redundant data and the destination (receiver) recognizes only the portion of the data that contains no apparent errors
Ubiquitous coverage	It is the distribution of communications infrastructure and wireless technologies throughout the environment to enable continuous connectivity
Frequency reuse	Using an allocated frequency band within the area, causing minimal or acceptable interference. The goal of reuse is to maximize the use of frequency bands repeatedly to serve a large number of customers with good quality service
FDMA	Frequency Division Multiple Access. A Scheme to use a pair of separated frequency bands one for uplink and another for downlink. The frequency pair maintains a constant difference (say 45 MHz) between them so that there is no interference between uplink and downlink
PBX	Personal Business Exchange. A small telephone exchange that typically caters to about 1000 users located in a corporate building or a campus, so that its users can set up private conversations or have a private LAN (Local Area Network), which does connect to outside network only at specific points
MSC	Mobile Switching Center. A telephone exchange with front end hardware/software to cater to mobile users. MSC connects to other cellular networks and the Internet at specific, carefully selected points
SS-7	Signaling System 7, a generation of signaling protocols used by wired telephone systems that connected to non-IP, exclusively to wired, public service telephone network (PSTN)
BSC	Base Station Controller: A cellular network unit that controls a number of base stations in an area, including mobility and signaling, connecting them in turn to a Mobile Switching Center
RNC	Radio Network Controller, is a unit that supports many base stations controllers in an area in terms of cellular mobility-related activity such as handoff, signal strength on various channels in base stations, directing traffic based on the number of users at a given base station etc. RNC in turn reports to MSC which acts as the central switch that controls cellular traffic flow for an entire metro area (typically million or more users)
SONET-ATM	Synchronous Optical Network is a standard for synchronous data

© Springer Nature Switzerland AG 2022
K. Raghunandan, *Introduction to Wireless Communications and Networks*,
Textbooks in Telecommunication Engineering, https://doi.org/10.1007/978-3-030-92188-0_11

transmission on optical fibers. ATM: Asynchronous Transfer Mode, uses data cells of specific size, which is carried over SONET network. In European system SONET has its equivalent as SDH (Synchronous Digital Hierarchy) with slightly different data rates. The combination uses the concept of controlled transfer of data with clearly established transition delays

T3, E3 and OC-3
A T3 is a dedicated phone connection supporting data rates of about 43 Mbps, E3 is its European equivalent with 34 Mbps. OC-3 is a network line with transmission data rate of up to 155.52 Mbps. Depending on the system OC-3 is also known as STS-3 (electrical level) and STM-1 (SDH)

Registration
Process used by cellular system to authenticate the user, and know where the cell phone is located, follow movements of the user, and provide cellular service using the base station in that area

Call origination
Procedure used to start calling someone. The caller "originates" a call using a cell phone where the procedure to establish a call involves the allocation of frequencies for the caller and the called party (if called party also is a cell phone). If called party is a landline phone, then it is passed on to a landline telephone exchange or switch from where the called party is signaled using SS 7

Call Termination
Procedure by which a cell phone receives a call, for which the network knows where the cell user is and sends signaling to that particular base station to contact the user. Due to mobility of the user, call termination is a more involved process to accurately establish where the cell user is at a given moment (if that user is on a high-speed train for example)

HLR
Home Location Register. A database record of cell user whose cell phone and subscription details are kept in the area where the user is registered. HLR is used to establish the validity of the user, as a basic prerequisite to allow use cellular system

Call timer
Usually set to 5 s, it is a timer to account for gaps during a cellular call—if the timer exceeds 5 s of inactivity (Base

station does not hear mobile, or mobile does not hear base station) then call is terminated. This mechanism allows mobile to reacquire service rather than "assume" it has service, which was cut off due to environment in which mobile traveled (hills, tunnels, or in under long canopy)

TDM
Time Division Multiplex is a method of sharing time on the same wire/cable. This method is exclusively used in wired networks. It is not used in wireless systems, since timing is strictly based on the assumption that signal from different users arrive at the prescribed time and does not vary. Such strict timing is not possible in wireless system due to user mobility

TDMA
Time Division Multiple Access, is a method by which multiple users share the same channel, but at different times

CDMA
Code Division Multiple Access, is a method by which multiple users share the same channel at the same time but using different codes

UMTS
Universal Mobile Telecommunication Standard, is a third-generation cellular standard specified by 3GPP, that also specifies the telecommunication interfaces. UMTS supports HSDPA (High Speed Digital Public Access), which is a broadband, high speed standard for 3G cellular. It is also known as UTRA-FDD or Universal Terrestrial Radio Access with Frequency Division Duplex

WCDMA
Wide Band Code Division Multiple Access, is a standard used globally based on the 3GPP specification supported by ETSI standards body. It is based on CDMA technology but does not provide the high speeds of HSDPA offered by UMTS

CDMA 2000
CDMA system during 3G, used in North America with specifications from 3GPP2. It used the concept of 1xRTT which evolved into 1xEVDO (Evolution-Data Optimized), comparable to UMTS of the 3GPP specification

MAHO
Mobile Assisted Hand Off, is a method by which base station collects information about the neighboring cells where mobile is located. This is done by base station

	asking mobile to measure signal strength on channels of neighboring cells and report back on a periodic basis (every 5 s)
GSM	Global System for Mobile, is the second-generation TDMA system used throughout the world for cellular communication. It is a standard developed by the ETSI—European Telecommunication Standards Institution
DSP	Digital Signal Processor, is an integrated circuit used in base stations, cell phones, and other telecommunication systems. It is a special microprocessor capable of handling complex mathematical operations such as Fast Fourier Transform that allows signal processing used in audio, video and picture processing, radio detection, channel tuning, and other functions
PN sequence	Pseudo Noise sequence used in CDMA, as part of spread spectrum code. Minimum PN sequence offset used is 64 chips, that is, 512 PN offsets are available to identify the CDMA sectors (215/64 = 512)
Equalizer	Equalizer is a dedicated DSP whose main function is to choose the best signal available in a multipath environment. Two classes of non-linear adaptive equalization structures possible. Maximum Likelihood Sequence Estimation (MLSE) and the Decision Feedback Equalizer (DFE)
SOVA	Soft Output Viterbi Algorithm, is often implemented in GSM handsets since it allows flexibility in terms of improving performance of the receiver
RAKE receiver	It is a radio receiver designed to counter the effects of multipath fading. It is implemented in decoding units known as fingers. Each finger independently decodes a single multipath component. Later these are combined together to recover the modified (improved) signal
RAT	Radio Access Technology, indicates the type of technology such as TDMA, CDMA or OFDMA depending on the network design
EPS	Evolved Packet System represents the evolution of the 3G/UMTS standard 3GPP standard committee, to move towards LTE
EEA	Represents the EPS encryption algorithm specified by 3GPP
EIA	Represents the EPS Integrity algorithm specified by 3GPP
Chinese Cypher ZUC	Zu Chongzhi is a stream cypher algorithm that uses a 16-stage linear feedback shift register. The ZUC algorithm is the core of the standardized 3GPP Confidentiality and Integrity algorithms 128-EEA3 and 128-EIA3
CMEA	Cellular Message Encryption Algorithm is used to encrypt messages (voice channel data) when the cell phone is used for financial transactions. This task is used when the Mobile is on a Voice Channel, to encrypt and decrypt some portions of digital messages transmitted to the BS. Specified in ETSI TS 151 011—V4.1.0, H 3.6
S-NAPTR	Straight forward Named Authority Pointer, is a stripped-down version of NAPTR (IETF) procedures that simplifies and reduces complexity associated with the regular NAPTR procedures
IMS	IP Multimedia System, is an architectural framework to deliver multimedia services. It does not define any application thereby providing freedom to any application present or future to be served by this system. Initially introduced in GSM (release 5) it has functioned as a flexible framework
PGN-GW	Packet Data Network Gateway, used in IP networks and also used in cellular IMS core
SGW	Serving Gateway used in IP networks and also used in cellular IMS core
MME	Mobility Management Entity which is s responsible for the tracking and the paging of User Equipment (cell phone) in idle mode. It is the termination point of the Non-Access Stratum (NAS)
HSS	Home Subscriber Server: It is a database that contains user-related and subscriber-related information. It also provides support functions in mobility management, call and session setup, user authentication, and access authorization. It is based on the pre-3GPP Release 4—Home Location Register (HLR) and Authentication Centre (AuC)

VoLTE	Voice over LTE or Long-Term Evolution, is the 4G and 5G specification to carry voice over the cellular network. Enhanced Voice Services (EVS) is a super-wideband speech audio coding standard that was developed for VoLTE that offers up to 20 kHz bandwidth. The application areas of EVS consist of improved telephony and teleconferencing, audiovisual conferencing services, and streaming audio
TDD	Time Division Duplex, is a concept where the transmit and receive using the same frequency band. Therefore, the user can either talk or listen not both at the same time, making it a "simplex" network
FDD	Frequency Division Duplex, is a concept where a frequency pair is used on at transmit and other at receive. The frequency pair maintain a constant separation between them (usually 45 MHz) so that there will be no interference between them. This allows talk and listen at the same time and has been used in cellular systems extensively
Control Plane	A control plane comprises a set of protocols responsible for dynamic provisioning of connections and other specific features, such as traffic grooming, QoS, and survivability strategies
User Plane	It is defined in 3GPP technical specification 23.501, the UPF (User Plane Function) provides: The interconnect point between the mobile infrastructure and the Data Network (DN)
CUPS	Control/User Plane Separation (CUPS), in mobile networks refer to the complete separation between control plane functions (which take care of the user connection management, as well as defining QoS policies, performing user authentication, etc.) and user plane functions (which deal with data traffic forwarding). It was first introduced by 3GPP's Release 14 at packet data gateway level, concerning the 4G Evolved Packet Core (mobile core for 4G networks) and has then been introduced for the 5G System (5GS) in the latest 3GPP's Release 15
RB	Resource Block, is the smallest unit that can be allocated to any user in LTE. It is also the minimum unit that can be transmitted on an OFDM system used by 4G and 5G. An RB is one segment of the OFDM spectrum that is 12 subcarriers wide for a total of 180 kHz
MIMO	Multiple Input Multiple Output, refers to antennas at the transmit and receive ends. In general, each antenna must be separated by a distance of at least half wavelength. The incoming data stream is separately sent over each beam and recombined at the other end. The objective of MIMO is to increase throughput using multiple beams (or rays) of signal
eNodeB	Evolved Node B or Enhanced Node B, refers to a base station in 4G and beyond. It is the base station component of the LTE network providing coverage for mobile broadband network
RSRP	Reference Signal Received Power, is the energy measured over the band of the subcarrier
RSSI	Received Signal Strength Indicator, is the energy measured over the entire band and is therefore always higher than the RSRP
RSRQ	Reference Signal Received Quality, is a carrier to Interference (C/I) measurement of the signal considering the RSSI and the number of resource blocks. RSRQ = 10 log (N) + RSRP (dBm) − RSSI (dBm)
eMBB	Enhanced Mobile Broadband, part of the 5G release it incorporates millimeter wave band, Massive MIMO (up to 64 rays) providing throughput rates from 1 to 10G bps to individual homes
URLLC	Ultra Reliable, Low Latency Communication, provides one-way latency of 1 ms and end to end latency of no more than 10 ms. It also provides much higher reliability of 99.9999% needed for driverless cars, precision automation needed in medical surgery etc. It is an important part of the new 5G feature set
mMTC	Massive Machine Type Communication to incorporate millions of machines that can provide status and control signals such as temperature, pressure, motors,

relays, and status of sensors that part of sensor networks. Generally termed IoT or Internet of Things, this will be the third important feature of 5G

UWB   Ultra-Wide Band, is a technology related to location of persons or machines with a very high accuracy. There are other applications expected out of UWB, but is likely to become part of 5G in later releases

QAM   Quadrature Amplitude Modulation, is a digital modulation technique that supports very high throughput communication, often in the range of 100 Mbps–10 Gbps. When combined with MIMO, this type of modulation allows higher throughput rates

SIP   Session Initiation Protocol, is a signaling protocol used for initiating, maintaining, and terminating real-time sessions that include voice, video and used by IP network to initiate a voice call or session

Throughout this book, various wireless technologies were discussed in the first ten chapters. All of them helped the global community to become a better place to live. The highlight of wireless becoming a globally accepted technology is certainly due to "Cellular Systems", which is the topic of this chapter. It is truly remarkable that over a period of about four decades, engineers and professionals relentlessly brought out five generations of cellular systems, to cover most of our inhabited world with cellular signals. Therefore, this chapter deserves a more detailed review.

The focus of this chapter will be cellular systems—how this change was brought about and how each generation of cellular has taken the human society to much higher levels of communication using the cell phone (now popular as "smartphone"). Major feature of this system is it supports mobility at enormous speeds including the current high-speed rail operating at 550 kmph. It is appropriate to begin this chapter by focusing on mobility and how this is handled.

11.1 Fundamentals of Mobility: Major Considerations

What makes the cellular system stand out from all other wireless systems? In simple terms, its seamless support of mobility. Users moving on land, air, and water bodies can continue to get cellular service, since this system is designed to support such mobility, mostly without break.

In Sect. 3.2, mobility was reviewed, and some of its impacts on wireless were considered with Physics-related mechanisms using analysis. In short, these principles, such as Doppler shift, channel fading, and mitigation techniques described there should all be incorporated into the system. The practical aspect of how-to layout the system on land including its terrains, water bodies (such as lakes, rivers etc.), and air (flying aircraft on continental routes or hot air balloons) since these should be planned carefully.

Cellular systems address this in their physical and MAC layers (layer 1 and layer 2) by incorporating capabilities of Error Correction Codes (ECC), doppler correction, and Forward Error Correction (FEC) techniques, on how the data frames are laid out. In essence, messages going over cellular systems must be visualized and forecast during design to address such effects due to mobility. Any information, whether voice, data, text, symbols, pictures, video, or anything else that passes through the cellular "channel" over the air must incorporate such techniques by design. The cellular channel starts from the user handset, extends (over the air) into the base station, though the user may or may not see base stations most of the time. In turn, the same technique should be incorporated at the base station. Since mobility inherently does not specify at what speed the user is moving, it becomes an important task of the network to find out and compensate accordingly. In general, 5G supports high-speed trains up to 550 kmph, but earlier generations supported speeds of up to 250 kmph.

The first inference of such mobility results in adjoining base stations that provide radio coverage to the entire area. Such coverage today extends to most parts of the populated global land mass. Second is the concept of cellular coverage or radio coverage in the form of honeycomb structure. It has been observed by physicists and engineers that this honeycomb structure can effectively cover the undulating landmass, the flat-water surface as well as lower part of the terrestrial air space (up to a few hundred meters). What is "ubiquitous coverage"—it means without having gaps, therefore offering continuous coverage and reduce interference at the same time. To plan such terrestrial coverage—two key parameters are considered: Capacity (how many users) and Coverage (how much area covered). Usually, the first term means system is for urban areas with heavy population, the second term is for rural expanse where population is low but area is large.

11.1.1 Coverage Versus Capacity Planning

Before cellular systems, there were public safety systems for police, fire, and emergency services. These wireless services started with coverage over the widest area as major consideration. With tall towers that transmitted high RF power (typically around 500–1000 W), these systems aimed at "coverage" of largest possible area—between sites, each spanning 20–30 km were common.

Soon it became apparent to designers that the channel allocated to such transmitter could not be used "again" in nearby areas. The next adjacent transmitters could only use a different channel, which was farther away in frequency spectrum. This separation was essential to reduce interference—and needed careful planning. Initially, this was not an issue since regulators provided the frequency needed as the demand was low. Over a period of time as number of sites increased, frequency spectrum became crowded particularly around major cities. That led to a different type of design commonly known as "capacity planning".

Capacity planning takes a bird's eye view of a vast coverage area—with a basic concept. How can a certain frequency band be "used again" to serve maximum number of customers within that area. It considers schemes like "sectors" within a cell coverage area. With such techniques to increase capacity, it accommodates the use of the same frequency again after two adjacent cells. This concept known as "frequency reuse" with sectorized cells is elaborated later in this chapter using Fig. 11.6. The end goal is to serve the largest number of users. Such capacity planning became the fundamental principle that most cellular service providers practice today. Earlier generations used the concept of frequency re-use but modern systems use multiple, clever techniques, with an equivalent effect.

11.1.2   Inception of Cellular Service

The first generation of cellular is worth mentioning. It laid down some concepts of cellular as a method to achieve ubiquitous coverage, although it has some limitations.

In Fig. 11.1, the concept called FDMA—Frequency Division Multiple Access is illustrated with FM modulation. What does that mean? It means, the frequency of the uplink channel (which is the channel mobile uses to look "up" to the base station) and the downlink (the channel base station uses to communicate "down" to the mobile). These two channels are kept separate in the frequency spectrum and are considerably apart in frequency to avoid self-interference. In simpler terms uplink is the channel on which the user "talks", while downlink is the channel on which the user "listens". Left side of Fig. 11.1 shows downlinks indicated by Blue arrows coming down from tower to each vehicle (f1d, f2d, f3d). Similarly, uplink going from each vehicle up to the tower (F1u, F2u, F3u) is shown by upward red arrows. Notice on the right, that the f1d and F1u form a channel pair (talk/listen pair) to and from the car. The downlink frequency f1d is 45 MHz higher than uplink

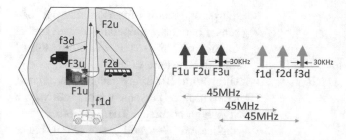

Fig. 11.1 Concept of FDMA (Frequency Division Multiple Access)

frequency F1u (in the North American system). Similarly, f2d and F2u to the bus form a pair; f3d and F3u to the van form a pair.

Such pairs of frequencies (indicated by blue and red arrows) are used at each cell site, thereby keeping each conversation separate in terms of the mobile user's talk and listen paths. This convention of FDMA is universally used throughout the communication industry. Separation between channels may not be as wide as 45 MHz, it may be different (smaller or larger). But frequency separation does exist, and licenses are always issued in pairs for FDMA. The concept of cell and coverage and functions of cellular network were discussed earlier in Chaps. 1, 2, 3, 4, and 5. The first generation of cellular system was analog, both at the base station, over the air and in the handset.

With this basic scheme defined, how does the cellular system operate? The base station shown in Fig. 11.1 has a small shelter next to tower, which initiates communication. How does it start? It regularly sends out broadcast messages over the air, known as "pilot" or "control signal". During 1G, this signal was a low-frequency audio signal mounted on a RF carrier using FM, and the procedure was called "tone signaling". When any mobile entered the area, its cell phone received this broadcast message or "tone" and responded. This process was known as registration—after which the mobile user sees a cellular signal available on the user display. That registration process was explained in Sect. 6.1 where the authentication process and interaction between the network and mobile were explained using Fig. 6.1. After successful registration and the mobile gets service, it can make or receive calls. In order for that to happen, there is an extensive infrastructure consisting of network elements.

11.1.3   First-Generation Cellular

In order to understand the very first generation of public service using cellular some systems from the early 1980s can be reviewed. Since European nations and Japan also had a 1G network, it is important to choose one, for the sake of understanding 1G network. For that purpose, let us consider

Fig. 11.2 First-generation
cellular network infrastructure:
basic architecture

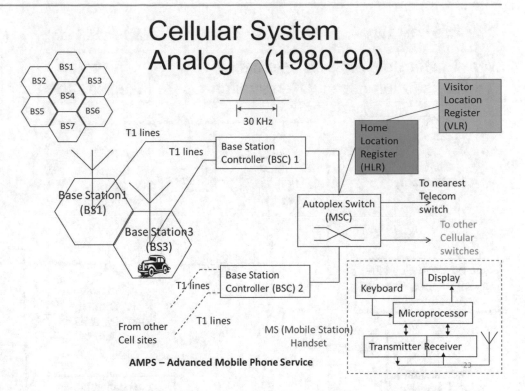

the North American network that was deployed by AT&T with handsets by Motorola set up during early 1980s. These are indicated as a hierarchical structure in Fig. 11.2, used in North American AMPS (American Mobile Public System) system. At the center of such network was the "Autoplex", which was a modified version of PBX (Personal Business Exchange) used by wired network, capable of handling at least 1000 phones. This switch had limited capability and was replaced with the more powerful switch 5ESS switch during 2G. But for the limited traffic handled during 1G, this simpler PBX switch served well but had some limitations.

In Fig. 11.2, the mobile station MS (Mobile Subscriber or Handset) would be in communication with a given base station. As user moves, mobile gets "handed off" to the neighboring cell (Base Station). There are any number of such neighboring base stations (based on cellular structure shown in the top left). This process of handoff involves not only the BSC (Base Station Controller) but also the Mobile Switching Center (MSC). In early systems of 1G, observing mobility of user (MS) and how network could serve the user was decided by the MSC (Mobile Switching Center). In this hierarchical structure, MSC remained master of the network for that entire region, similar to the wired telephone network of those times.

The uplink frequency received from the MS and the downlink frequency sent from BS were also allocated by the MSC. This architecture was directly derived from the wired telecommunication network infrastructure where the "Switch" was always the master of the network that controlled

all telephones and their signaling (see SS7 signaling in Chap. 4, Sect. 4.2). During 1G, the switch was not fully aware of field conditions where mobile traveled but made an estimate based on received signal strength and used its knowledge of nearby (neighboring) cell sites already deployed. It was not uncommon to see handoff failure. There were other limitations, in terms of excessive signaling since MSC made all decisions.

1G being an analog network, backhaul between each Base Station and BSC used traditional T1 or E1 that was part of the telephone network standard used at that time (for details of backhaul see Chap. 4). Even from BSC to MSC, it would typically use a T3 or E3 backhaul and for longer distances an OC-3 used optical fiber. SONET-ATM networks came into existence towards the end of 1G time frame—which was around late 1980s. There were 12 different standards in Europe alone, and North America and Japan had their own analog standards (as indicated earlier). There is no need to go into those standards since none of them exist now. All of them were generally based on similar technology. Let us focus on several concepts that 1G laid down as foundation, which helped future generations of cellular systems.

The most basic contribution of the first generation is the "registration" indicated in Fig. 11.3. It is the very basis by which any cellular network "knows" where the mobile is located at any given time and where to direct incoming calls to the mobile. This entire registration sequence assumes the mobile is not on call (switch uses a different method if it is "on call").

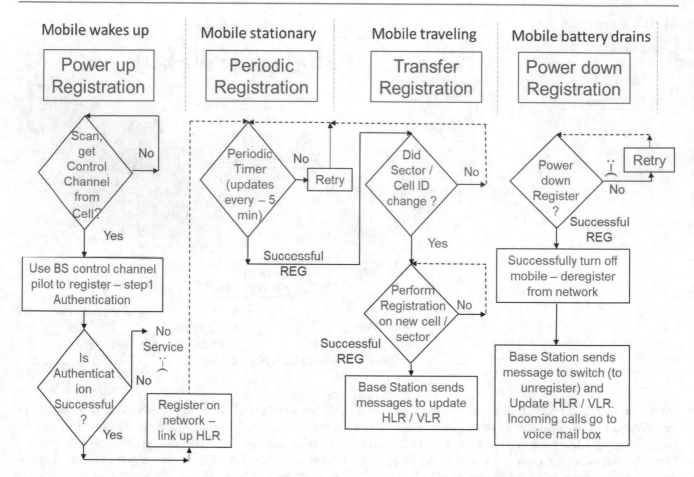

Fig. 11.3 Registration of the mobile and how network stays informed

The process begins with Power up registration (on the left) when the cell phone is "turned on" after batteries are charged. It goes through the registration process by first scanning for signals over the air and obtains the "pilot" or "beacon" sent by the BS from the cell tower. This beacon stream from the tower is always present as broadcast and informs the mobile its own BS ID (Base Station Identification) and describes the process by which mobile can communicate with that BS. The process can be slightly different depending on the type of mobile (also known as mobile class).

Then the mobile sends its response requesting registration —the MSC asks mobile to send its ID and capabilities. During second generation of cellular, authentication process described in Sect. 6.1 and Fig. 6.1 was added to provide additional security. Using this registration process, MSC initiates the authentication procedure. Such an authentication process not only verifies whether it is a valid user but also uses "unique challenge" to rule out any attempt to use a duplicate mobile set. In general, registration process informs the network on features and capabilities of the mobile, the services user is subscribed to. During 1G, service was limited to voice traffic, but in 2G and later generations, it included various forms of data services.

The next step in Fig. 11.3 is periodic registration—it may not seem obvious, but this is an essential step. If the user stays in one location (such as office or home), it is important to know how long the user continues to stay there. This is verified by periodic registration, typically every 5 min. During peak hours, service provider may change this to, may be, 10 min, in order to limit signaling traffic. It provides a means for the network to know when the user eventually moves (travels) from that area. With any such movement, the transfer registration process begins.

Transfer registration is the step by which any cellular network supports user mobility. It does use a set of inter-active functions. When mobile moves towards a new base station, it has to register with that base station again. But it can be achieved by different methods. In the simplest method, it is done by a process known as neighbor cell reselection, which we will review shortly.

Reselection to the new cell allows the system to update location of the mobile and current service available from new base station. Transfer registration (or reselection) happens at a very fast pace when user travels on a highway or train. Starting with 2G, cells along that path continuously track and register the mobile in a methodic order to keep up with its progress at high speeds.

Right side of Fig. 11.3 describes another essential function related to mobile being turned off either intentionally or due to natural process of battery draining away. In any case, it is important for the mobile to say "goodbye" to the network before turning off. The network must know that, so it need not waste resources trying to find the mobile. All incoming calls can be routed to the user mailbox, and message sent to the caller stating the called user "is not available". When the mobile is charged again or turned on, the process of power-up registration begins, bringing the mobile back into the network.

Cellphone call origination and termination (incoming call) were systematically developed during 1G. Note that registration of the mobile is a pre-requisite for any of these to occur. HLR (Home Location Register), which is a database record of the mobile, is the fundamental element that always resides at a specific switch where mobile was originally registered by service provider. HLR is needed to authenticate mobile, for call origination to take place as indicated in Fig. 11.4.

Call origination sequence begins by mobile scanning for service provider's base station signal and then sending request to the BS to originate a call. During 1G, the "authentication" sequence described in Sect. 6.1 did not exist, only a basic verification using HLR was being conducted as part of the registration process. Why is registration a pre-requisite to originate a call? The logical reason is that it provides the current location of the mobile to the network. In addition, registration allows the base station to know capabilities of the mobile (type of voice coder, type of service the user has subscribed to etc.). These are needed for providing the appropriate service to the mobile and also the billing process indicated later in the sequence.

It is important to note that once the user is "on call", the traffic channel is active and signaling (handles by control channel) works in the background to support the call (including handoff).

One of the important differences between wired and wireless is use of a "call timer" in wireless, which notes whether the user and base station are in touch with each other. The "call timer" starts whenever there is a loss of signal indicated by the absence of "ACK" or acknowledgment to signals sent by the BS. In cellular systems, control signal is the most reliable part of the signal with additional resiliency (in the digital frame). If this control signal is not heard by the mobile (due to terrain or other conditions) then it may be necessary to begin a sequence to get call has to be terminated. The "call timer" starts that process at the BS.

Mobile call origination

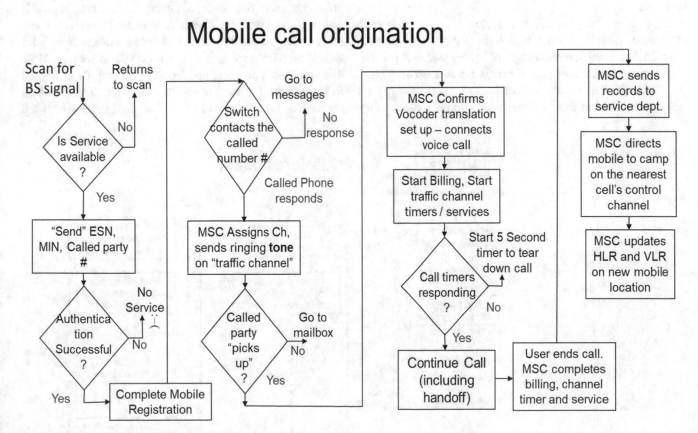

Fig. 11.4 Mobile call original sequence

Similarly, when the MS does not hear back from the BS, it starts its own "call timer". The timer is another way for both as a reminder to attempt registration process as soon as the control signal is available and send ACK to the BS.

The intent is to make sure that the user does not waste time thinking the unit is "on call" while the radio link may have broken due to any reason. Often, the "call timer" recovers, when the user gets the signal back—perhaps after passing under a bridge or dense canopy of tree leaves. Once the acknowledgment is received by BS or MS, they inform each other to stop the "call timer" and the call continues. Sometimes, if the user is in a hilly area or goes through a long tunnel with no service, such "call timer" help in abandoning the call, so that those frequencies (allocated to the user mobile) can be reused elsewhere.

The last part of Fig. 11.4 deals with administrative duties performed by the network. There is also a need for the mobile to camp or register with the nearest control channel, which could be different from the one where call ended. Network typically directs the mobile to go to the control channel that is appropriate to serve the mobile. If mobile roams beyond home territory then VLR (Visitor Location Register) is updated to show the latest serving BS in the roaming area. Then periodic registration process shown earlier in Fig. 11.3 begins and continues with rest of that registration sequence continuing, such as transfer registration.

Let us review the call termination (or incoming call) sequence. This is illustrated by the logical sequence in Fig. 11.5. Calling a mobile starts off with a search for where the user is located right now. If user is in home territory HLR is contacted. The importance of looking elsewhere can be gauged by the fact that user may not be in the same country

where the mobile is registered. Therefore, VLR or Visitor Location Register provides the latest update on where the user is and details of the current BS serving that mobile. This information greatly reduces the search problem. The process is more involved: It depends on whether the called user is from landline phone or a mobile. For landline, call control will check with the HLR and since there is no record, it will be sent to the nearest landline switch (not MSC). If the called number is mobile then HLR of the area helps direct where to route the call, to the specific BS where the terminating MS is located.

If the intended MS is roaming then the HLR will redirect the call to the VLR where the MS is currently located. Thus, MS's HLR is always kept informed about the VLR area where the MS happens to be at any time. For MS originated call, it follows a similar process. In VLR, there are two steps, VLR1 that is last updated VLR (periodic like every 5 min) and VLR2 the most recent known location (latest sync with mobile using registration). The system is fairly complex in terms of protocols and timing, and keeps up with the mobile location regularly updated when MS is roaming.

The paging sequence is directly sent to base station currently serving the mobile. Once the mobile acknowledges the paging from BS, it undergoes authentication procedure to validate user. Following these are tasks such as which vocoder is used. This is needed for the switch to translate voice streams at both ends. Knowing that the landline caller can be using a desk VoIP phone, or a cell phone with some other coder, translation of coded voice streams from both ends must take place at the switch (MSC). After this, MSC assigns channel and ringing signal is sent over that assigned traffic channel. If the caller continues to wait till call is set up (usually up to two rings in terms of time, after which the

Fig. 11.5 Call termination sequence

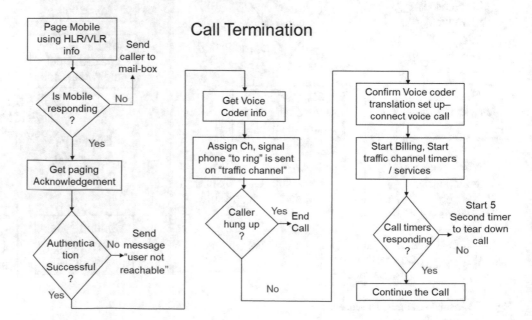

Call Termination

called party's phone begins to actually start ringing), voice coder translation is confirmed and then voice call is established. The call would continue normally and call timers of 5 s duration are established when users start any cellular call. This is followed by the administrative duties such as billing at end of the call.

11.2   Second and Third Generations: TDMA and CDMA (2G, 3G)

So far, we established the basic functionalities of mobile and base station interactions. With the second generation of cellular, major changes occurred—establishment of digital cellular and widespread use of cell phones both in the North American market and GSM expanding its reach globally. This necessitated innovative techniques to substantially enhance capacity leading to growth of MSC into a full-fledged switch catering not 1000 but almost a million users in a region. It further enhanced the process of registration using "reselection" where mobile was able to choose the control channel. Further enhancements in radio meant active involvement of the mobile in the handoff process—known as MAHO (Mobile Assisted Hand Off). We will now review these enhancements. Figure 11.6 shows the features and enhancements made in 2G network. All major enhancements are indicated in red—starting with the cell structure as the **first change** resulting in splitting cells into

three sectors—generally termed alpha, beta, and gamma sectors of 120° each.

This concept of sectoring further enhanced capacity and allowed reuse of same frequency in every third cell sector. That meant cell sizes were reduced (lower transmit power), so that two adjacent cells later, the third adjacent cell could reuse the frequency of the first one. In simpler terms, lateral cells could operate on frequencies as f1, f2, f3, and f1 used again after two adjacent cells.

The **second change** was introduction of two competing digital access technologies—TDMA and CDMA shown at the top of Fig. 11.6, in terms of bandwidth. Both of these access techniques were derived from satellite systems where they had proven track record. CDMA was introduced in North America as the TIA-95 standard, while Europe introduced TDMA using the GSM standard. In 1990–95, there was also an alternative TDMA introduction in North America known as TIA-54 (digital traffic channel, but analog control channel) followed by another TDMA standard during 1995–2000 known as TIA-136 (digital traffic channel and digital control channel). The North American TDMA (30 kHz bandwidth) was meant to interwork with AMPS of 1G, providing a means to transition from analog to digital. However, this system did not stay for too long—after about 5 years of operation, North American TDMA (TIA-136) operators decided to abandon it and join GSM where mature products and services had become available and were in widespread use. Figure 11.6 shows slot sequence, where

Fig. 11.6 Second-generation cellular network (digital: TDMA/GSM and CDMA)

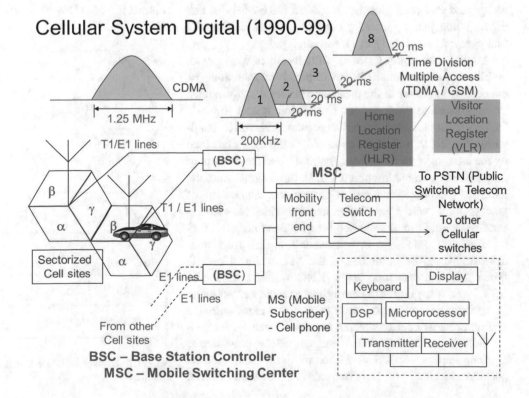

TDMA carrier of 200 kHz is for GSM which supported 8 time slots. North American TDMA (TIA-136) was channelized in 30 kHz (compatible with AMPS) which supported 3 TDMA slots (not shown in Fig. 11.6).

The **third change** was "mobility front end" at the MSC. The switch's signaling functions became extensive, requiring a dedicated mobility front end to work on "mobility". This was later shared with another unit called RNC (Radio Network Controller), which acted as a buffer working unit between BSC and MSC but focused on mobility aspects of the radio network. This meant several BSC units reported to the RNC, and a few of those RNC units reported to the MSC.

The **fourth change** was a new interface that directly worked with other cellular MSC—since there were many cellular networks now. The existing interfaces to a wired network were specified in terms of PSTN interface, indicating it was meant for existing old-world wired networks. The reason for this split was mainly due to CDMA that used a **packetized radio**, but the backhaul network was still a **trunk switched network**. The cellular switch, on the other hand, directly handled packets using an X.25 network and later using ATM/SONET, followed by IP network.

The **fifth change** was the introduction of DSP in the handset to enhance capabilities far beyond what users and their service providers had even imagined. It boldly introduced pictures, video, and other services in the handset that were thought to be the exclusive realm of wired networks using computers. These two applications transformed the cell phone into a camera and a video unit, now widely recognized and used globally. It altered the concept of cell phone being just a phone, it became valuable commodity that everyone used daily as an essential gadget.

A few points about CDMA—initially, there was considerable skepticism about its widespread introduction. Its concept was counter to the established norm that noise must be suppressed in order for communication system to work. CDMA recommended spread spectrum and wider bandwidth, leading to a different concept. Multiple rays received by the handset were considered "helpful", and the concept of rake receiver became important to rake those signals into the handset. The bigger and visible change was the concept of "soft handoff" where the handset decided which base station stream it would use. It was used for both handoffs as well as for diversity combining. In soft handoff, signals from multiple BS would be combined at the MS, while in uplink, frame selection was used at the RNC to select whichever BS's frame was good. This meant there would be "no break in data stream", an important feature to download software and other items of data streams from websites.

North American region encouraged cellular providers to abandon analog AMPS and directly jump into "digital packetized technology" fully in pace with the future of wired

Internet. The soft handoff meant that the website could "bi-cast" its stream into the cell phone via two adjacent base stations, and the "handoff" did not result in any "break" in data stream. In contrast, the TDMA technology (both TIA-136 and GSM) had this limitation that there would be a "break in data stream" before handset could contact the new base station. Every handoff meant a break in data stream. This feature led CDMA to become the common technology for both North American and European standards in 3G. North America marketed it as CDMA 2000 and in European marketed as UMTS/WCDMA. There were differences between these two (not compatible with each other) but the operating technology became CDMA in all of 3G.

Since handoff is so important to all cellular systems, let us consider the differences between these two technologies. The next two subsections will lay out the concepts of TDMA and CDMA from a handoff perspective, to highlight differences between hard handoff and soft handoff. Since the mobile moves through adjacent cells, it results in an essential handoff process. Cellular networks use the seven-cell structure shown earlier in Fig. 11.2 (top left) to support contiguous coverage.

11.2.1 TDMA Concept

TDMA (Time Division Multiple Access) is a technique by which multiple mobiles use the same channel that is "separated in time". This separation is known as "time slot", which is usually of 20 ms duration, shown in Fig. 11.7 and contrast it with TDM of wired networks.

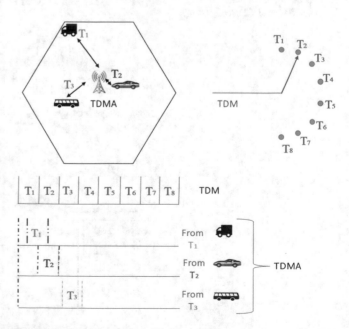

Fig. 11.7 Concept of Time Division Multiple Access versus TDM Multiplexer

A globally known standard of TDMA is the GSM system used by millions of subscribers. It prescribes the use of 200 kHz bandwidth for each channel. Such a channel can have up to eight active mobiles. Therefore, mobile 1 uses slot 1, followed by mobile 2 on slot 2 and so on, with mobile 8 on slot 8. Then the cycle repeats with the first mobile coming up on slot 1 again. Although this process seems similar to multiplexing in wired systems, it is actually more complicated. It is important to understand this concept from a wireless perspective, as indicated in Fig. 11.7.

On the right side of Fig. 11.7 is the traditional TDM (Time Division Multiplexer) used in wired networks, where the switch uniformly rotates collecting inputs from each point for time duration T1 through T8 at equal intervals. The result is indicated below as uniform time slots of T1 through T8 as expected. Note that all multiplexer inputs are on the same panel and signals arrive precisely at the expected time. When the same method is attempted at a cell site, the arrival of signals begins to vary. The reason is—signal from T1 (from the van) starts from the edge of cell coverage area. It actually starts from the van precisely at the beginning of time sequence T1, but it takes time for the signal to travel over the air, up to the cell site and when it arrives at the cell tower it is delayed (as shown). It starts to arrive almost at the middle of time interval T1 and completes arrival towards the middle of T2. In contrast, signal from T2, (the car) starts at the beginning of its allocated time slot T2 and arrives almost instantly at the cell site since this car is very close to the cell tower. The third signal (from the bus) has some delay since it is in the middle to cell coverage area—starting at the beginning of time interval T3, its signal arrives later in T3, and continues into slot T4 (as shown).

Such variations in time of signal arrival at the cell site are due to "access delay" or signal traveling varying distances causing non-uniform delay. This is characteristic of all cellular systems when signal path is considerable (several kilometers) and could vary continuously as vehicles travel. The cell sites (Base station), therefore, need to continuously monitor and adjust the timing of each cellular device as it moves around cellular coverage area. Farther the mobile is from cell site, earlier should be their beginning of transmission, to compensate for signal arrival at the cell site (Base Station) such that they maintain time and don't extend into the next slot. Such time adjustment happens all the time in cellular systems, where distance traveled by signals can vary from few meters to several kilometers. The GSM system being a TDMA system does such adjustment to each mobile all the time, and it works very well.

The concept of "access delay" exists in satellite and cellular systems where signal travels over the great distance before arriving at the receiver (at cell site/satellite terminal). Therefore, the term "multiple access" refers to such wireless systems. Both TDMA and CDMA were concepts that originated in satellite systems in 1980s and were carried over to cellular systems during 1990s.

11.2.2   CDMA Concept

In contrast to TDMA, CDMA uses a much broader bandwidth and expects all mobiles to operate on "the same channel at the same time". Until this concept was introduced, communication engineers would liken this to "co-channel interference", which means no matter where the transmission originated from (even if 100 km away), the two signals can interfere with each other. CDMA on the other hand treats this as an advantage by using code that is orthogonal to frequency and time, as shown in Fig. 11.8. It uses Direct Sequence Spread Spectrum (DSSS) technology, which assumes both transmitter and receiver have knowledge of spreading code, and both ends use the same code [2]. With multiple users on the same channel, each cell phone is allocated a specific code by the system. It is not possible to intercept that conversation, without knowing that code. Due to complex coding scheme, the CDMA signal appears as "Gaussian noise", and it is not possible to decipher how many users are there on the channel and also difficult to jam. With these advantages, military used it for secure communication over many years. An important concept used in CDMA is "spread spectrum", which was indicated earlier in Chap. 1, as an invention by Hedy Lamar, who had patented it in the 1940s.

CDMA system spreads low throughput data from each user, by scrambling it over a wide spectrum. User's data to be transmitted are multiplied by with a high data rate bit sequence and then modulated over an RF carrier, to produce

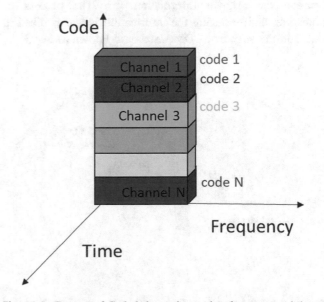

Fig. 11.8 Concept of Code being orthogonal to frequency and time

a signal that has a much wider bandwidth than user data itself. It offers two-fold advantage—(a) makes it more secure and (b) it increases capacity for a given channel. These ideas were very well enhanced by Irwin Jacobs, during 1980s, who invented CDMA. Irwin claimed that this would become a forerunner to highly secure and instant data communications globally, increasing capacity of a channel multiple times that of an analog channel [3]. It was not only shown mathematically but also demonstrated by a practical system using chipsets that worked well in the lab and in field trials.

He was ably supported by Andrew Viterbi who is the inventor of the now famous "Viterbi Algorithm" used in every cell phone and satellite TV receiver [4]. Viterbi joined Irwin Jacobs as co-founder to form the company Qualcomm. Viterbi made major contributions to CDMA making it practical to deploy CDMA technology in commercial cellular systems. Later their company Qualcomm became synonymous with CDMA, leading this technology globally to become the common baseline for all 3G systems worldwide.

Basic concept of how CDMA works is shown in Fig. 11.9, where the base station transmits "composite signal on the downlink" that contains coded information for all mobiles (Code 1 through Code N). Since each mobile is provided with a unique code, only that mobile will decode and receive information pertaining to it. Each mobile then transmits using their own assigned code, and the base station receives such information from different mobiles and sorts the information stream based on code. Though, in concept, this looks simple, it involves complex computations in terms of how information gets coded and the entire system is kept secure. While the information to be sent by each user has a relatively slow data rate, CDMA systems spread it with a fast chip sequence, before transmitting it over the air and to retrieve the original information [5]. The process of spreading and retrieving the information is standardized in detail in the very first CDMA standard known as IS-95.

In general, CDMA standard indicates that three main codes are used for this process. They are:

- Walsh codes, 64-chip orthogonal sequences are used in each base station of IS-95. This enables the base station to have 64 separate channels for communication with mobiles. Similarly, in CDMA 2000 (3G standard), 256 codes were used allowing up to 256 users for each base station.
- A short code: $2^{15} - 1 = 32{,}767$ chips long, which has the property of "being orthogonal to any nonzero offset of itself". It is called PN (Pseudo Noise) offset in CDMA systems. PN Offset is a characteristic of the signal from a cell on a tower. Phones select a PN Offset to select the cell that they want to listen to. This is used to synchronize the forward as well as the reverse links. These links are shown in Fig. 11.9 as downlink and uplink. In IS-95, "Short code" is also used to identify base stations and this sequence is repeated 74 times every 2 s.
- A long code: This has a throughput rate of 1.2288 Mb/s with chips that are 42 bits long. It is used for 'encryption and spreading". Encryption is achieved by using Long code mask created by 64-bit A-key assigned by CAVE algorithm, which was described earlier in Sect. 6.1 as well as in Fig. 6.1. The long code takes about 41.2 days to repeat its sequence.

During 2G when CDMA was introduced as IS-95 standard, it used a Walsh function matrix of 64. The first line of this matrix contains all zeros (used as pilot), while the subsequent lines that are orthogonal to each other contained different combinations of 0 and 1. When implemented in the system, it allowed each mobile user to use one line that had equal representation of binary bits, providing zero cross-correlation among users. There are detailed descriptions of how each of these codes are used in CDMA technology—please refer to [6].

Let us compare the two technologies—TDMA and CDMA since both of them became well established as 2G standards serving global population. They are likely to continue serving users in the form of GSM network or the CDMA network for the near future.

11.2.3 Comparison of TDMA and CDMA

Since the differences need considerable description, instead of using the A and B tabular form, the following paragraphs are used based on topics of differences—about five of them are:

(a) Channel usage
(b) Handoff

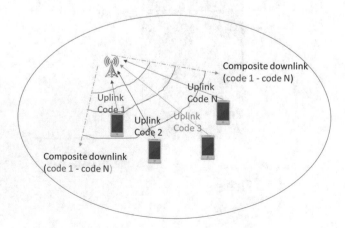

Fig. 11.9 CDMA—concept of communication over the air link

(c) Handling interference

(d) Reselection—of control channel

(e) Security schemes—by each generation.

(a) **Channel Usage**

At the outset, this seems fine for TDMA, since its "time division" offers well-defined method of how many users can use a given channel. For example, in GSM, each channel of 200 kHz is time shared by 8 slots. The North American IS-136 has 30 kHz bandwidth in which it has six time slots. The GSM standard effectively uses 25 kHz for each user, and its vocoder known as ACELP (Algebraic Code Excited Linear Prediction coder) sends 13.2 Kbps stream of data for duration of 577 ms for GSM [7]. The IS-136 standard utilizes a different codec known as VSELP (Vector Sum Excited Linear Predictive coder) of 8.1 Kbps data stream. The IS-136 standard is capable of a total stream of 48.6 kb/s stream divided into six slots (but combined 1 & 4, 2 & 5, and 3 & 6 to make it a three-slot system). Its total frame size of 40 ms is divided into these six slots, each of which was allocated 6.67 ms, and each slot accommodates 324 bits [8]. VSELP, whose voice quality was not much liked by users, is no longer in use. The ACELP coder although considered less efficient (design efficiency of coder is to use lowest bit stream rate possible), it became widely popular and used by a variety of standards including desktop telephones (commonly known as VoIP phones). In general, the number of users is based on slots and does not change. Designer only has the option to send voice or data efficiently but has no control on the number slots and therefore number of users' capacity for a given channel is already fixed by the TDMA standard.

In contrast, CDMA technology uses the "systems approach" to control interference and thereby accommodate users. What is this approach—it can be described as "power control" of mobiles on the cell and adjacent cells. By control of reverse link or the power that comes back from each mobile, CDMA base station accommodates more or less users, depending on network traffic conditions. The reverse link power control has three layers: (1) the open loop power control to estimate the received power and measurement by Base Station Transmitter and Receiver or BTS before closed loop Power Control takes over, (2) closed loop power control that keeps all MS's received power to be equal, and (3) outer loop power control keeps power level to achieve the desired frame error rate. This is a continuous process that results in variable capacity that can compromise on interference noise. Although this noise is not noticeable to users, it can be clearly measured and is used by the system to accommodate more users during peak hours of traffic [9]. Since CDMA uses spread spectrum, it is economical only if number of users are more (metro or urban areas).

There is another aspect—the ability of receiver to harness multiple reflected signals, both at the base station and the mobile, which provides considerable advantage. This will be further discussed under the topic of handling interference.

(b) **Handoff (or handover)**

Handoff is defined as moving the mobile from the current cell site/sector to another cell/sector when user is "on call". That call can be a voice or data call, but user is actively utilizing the air interface to communicate and needs a clear communication channel. TDMA uses the hard handoff method for handoff, CDMA uses the soft handoff method. These are described below.

11.2.4 Hard Handoff (TDMA)

Key features of hard handoff are based on actions of BS that currently serves mobile (let us assume Cell 4 in Fig. 11.2 at the top). Although BS4 in the middle of seven cell configuration, directs the entire operation, it uses a process that is based on inputs from its own neighboring cells and the mobile as well. Let us assume that all cells are sectored (alpha, beta, and gamma as shown in Fig. 11.6) Handover uses a well-defined process (per GSM standard) that is carried out uniformly for all mobiles in its current service:

1. Base station 4 (Fig. 11.2) is always in communication with its neighboring cells BS (1–7) to obtain potential handoff channels, it can handover a mobile to, when needed.

2. It asks each neighboring cell to provide a list of possible channels that the mobile can be handed off to and updates such channel list periodically.

3. After getting the list, BS4 compiles a total list of about 24 potential candidate channels (a combination of channels from all the neighboring base station channels) and asks mobile to measure RSSI on all those channels. This is known as the "neighbor list".

4. Based on measurements received from a specific mobile, BS updates making a short list for that mobile. Let us assume its short list consists of some "handoff channels" of Cell 3, since measurements from the mobile showed strong signals from Cell 3 (indicating mobile is moving towards cell 3). Such handoff process is called MAHO (Mobile Assisted Hand Off).

5. Once handoff channel is chosen, Cell 4 negotiates with Cell 3 to get the final choice of one specific channel with its channel # and a specific time slot # within that channel (hard handoff means "break before make")—there will be break in voice data stream that can last up to 100 ms.

6. The current serving base station 4 issues a "Handoff message" to the mobile that contains the new channel # and time slot # it obtained from Cell 3.

7. Handoff takes place only if mobile is still "on call" (on traffic channel). Mobile confirms this status by sending an acknowledgment message to BS 4 and gets ready for handoff. It knows that a trial transmission is needed on this new channel and sector and opens its receiver to that channel.

8. Cell 4 (let us assume, sector beta) issues command to mobile to tune into "handoff channel # and slot #" at Cell 3 (let us assume—sector gamma) and send (transmit) an acknowledgment to confirm that mobile can follow this specific handoff message.

9. Mobile acknowledges "goodbye" to BS4 and then tunes to the new channel # slot # and sector, starts with a short trial transmission to BS3. The base station acknowledges, sending messages to mobile adjusting its timing and other essential parameters as needed.

10. After such an initial period of test "burst" (called pre-amble) that contains no user voice/data, the new cell confirms and asks mobile to resume normal voice/data traffic. Mobile acknowledges to BS 3 and resumes its voice/data stream on new channel # and slot #.

Note that all these ten steps are accomplished with little or no disturbance to user conversation, because steps 8–10 are implemented in no more than 100 ms (usually 50 ms). Some users may hear a "click" if the transition took 100 ms.

11.2.5 Soft/Softer Handoff (CDMA)

Soft handoff occurs in CDMA, while the mobile is "on call". Unlike TDMA, where mobile has to obey commands from the base station, in CDMA, the mobile is allowed to communicate with two base stations simultaneously. Based on the quality of communication (including power control), the

mobile decides and chooses the base station and sector that it will go to [10]. There are many possible scenarios but the most common are two that are shown in Fig. 11.10 where the mobile finally decides to move (Scenario one), or it decides to remain with the current base station—(Scenario two), depending on the quality of signal [11].

The concept of soft handoff and softer handoff are both related to CDMA. Soft handoff is a situation where the mobile is able to communicate with two adjacent cell sites both of which are on the same channel but have different code sequences. This technique is used in CDMA to provide options to the mobile so that it could move to the site where quality of service is better. During peak hour traffic, this technique could also be used to off load mobiles to an adjacent site that may be relatively less loaded. In that case, it is perceived as part of power control scheme.

Two common scenarios are considered in Fig. 11.10, the first one shows the normal case where the mobile actually moves from Base Station 1 (or BS1) to BS2. The process followed is: first mobile is communicating only with BS1, followed by communication with both and then communicating with BS2. The second scenario is similar except that the mobile decides to stay with BS1, its original base station itself since it is better than BS2 (decides to stay back on BS1). The concept of moving from one sector to the next within the same cell is known as "softer handoff" (indicated in Fig. 11.10 on the right). This may be done by the cell to offload traffic from one sector to the next during peak hours, for example. The three sectors α, β, and γ are shown on the right to indicate each sector. Hand off between them is known as "softer handoff". Even in CDMA, the mobile assists the base station in the handoff process by measuring and reporting the strengths of received pilots. It looks for pilots on the current CDMA frequency assigned to it to detect the presence of CDMA channels and measure their strengths.

When the mobile detects a pilot of sufficient strength that is "not associated with any of the Forward Traffic Channels assigned to it", it sends a Pilot Strength Measurement Message or an Extended Pilot Strength Measurement Message to that base station. That base station then assigns a Forward Traffic Channel associated with that pilot to the mobile and directs the mobile station to perform a handoff. Whether to "perform a handoff or not" is decided by mobile.

This unique sequence leading to "soft handoff" brings a major plus point to CDMA since data stream is "not broken" during handoff. For example, a download from a website does not get cut. Instead, both base stations send bi-cast stream of same data from that website. Due to this inherent advantage, CDMA technology became the default in 3G, and two separate standard bodies formed CDMA2000 and W-CDMA, both implemented it globally. In practice, the CDMA2000 from 3GPP2 uses 1xEVDO (Evolution-Data

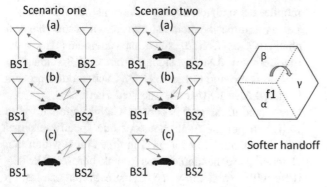

Fig. 11.10 Soft handoff and softer handoff in CDMA

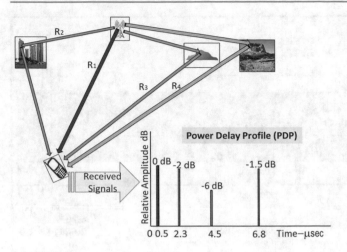

Fig. 11.11 Multipath with signal levels and delay spread

Optimized) specification. It contrast the UMTS from the 3GPP has HSDPA (High Speed Digital Public Access) specification. Both are based on CDMA but do not inter-work with each other.

11.2.5.1 Handling Interference

Interference in communication systems is not a new concept. Any unwanted signal whether it is due to another signal operating on the same channel, an adjacent channel, or even two channels away, could interfere. For example, multipath was described earlier in Chap. 3, Sect. 3.2 indicating multiple rays of signal arriving at the receiver. Let us consider multipath in the context of GSM, which is a TDMA system, and how it handles interference.

In Fig. 11.11, there is a snapshot view of typical suburban area where the cell phone user receives a direct signal Ray R1 (red), from the base station tower, a reflected signal Ray R2 (blue) from an apartment building, another reflected signal Ray R3 (green) from a hill nearby, and a reflected signal R4 (gray) from a rocky mountain that is somewhat far. Overall, the receiver in the handset receives four separate signals at this instance and given the symbol frame structure of GSM (on the right), it must choose one of these four.

The difficulty is that all of them are relatively strong, the weakest being R3 coming from the hill (−6 dB), which may have been scattered since the hill may contain trees. At the very next instance (as user travels), the scenario could change dramatically. The direct ray R1 may not be visible since the vehicle took a turn and lost sight of the tower, reflected ray from the rocky mountain may become the strongest. Unfortunately, without an "equalizer" (a special DSP processor), these multipath profiles cause inter-symbol and inter-channel interferences, which cannot be corrected.

Since such changes occur regularly and in an unpredictable manner, a separate signal processor dedicated to

choosing which should be the strongest signal at each instant becomes important. Such a signal processor dedicated to the complex task is termed "equalizer", and it uses extensive and complex mathematical algorithm to carry out the task. Equalization was a topic addressed while introducing TDMA in Chap. 3. It was mentioned that using a separate signal processor known as "equalizer" was essential. Function of equalizer was indicated in Chap. 3, Sect. 3.2, and the explanation used Fig. 3.5 as reference.

The GSM standard data frame is shown in Fig. 11.12, where each transmitted slot includes data bits along with a training sequence, $str(t)$, that is already known to the receiver. Such training sequence enables the receiver to perform channel estimation and equalization. This method is pre-established in the standard so that all network elements (at the base and at the handset) are aware and follow this method.

In GSM, the function of the equalizer takes center stage since handling interference is an integral part of convolution and coding scheme as shown in Fig. 11.12, where the entire GSM frame is illustrated from a coding perspective [12]. It shows 26 bits used for "training", in the middle of the multiframe. Two classes of non-linear adaptive equalization structures are possible: the Maximum Likelihood Sequence Estimation (MLSE) and the Decision Feedback Equalizer (DFE).

Based on field test and studies, GSM makes use of "training sequence" for channel estimation and ISI (inter symbol interference) cancellation by use of "Viterbi equalizer". Figure 11.13 shows the coding sequence that includes the Viterbi equalizer (last unit). At the receiver, the received training sequence $r_{tr}(t)$, is a convolution of the signal transmitted training sequence, $S_{tr}(t)$, or in other words, it is a modified version of the transmitted signal.

The modification occurs due to the changes that occur to the signal while it passes through the air. Such changes are termed convolution that is modified by the channel's impulse response, $hc(t)$:

$$r_{tr}(t) = S_{tr}(t) \otimes hc(t) \qquad (11.1)$$

The received training sequence in the digital domain, $r_{tr}[k]$, is fed into a digital matched filter with an impulse response, $hmf[k]$, that is matched to Str [k]. The matched filter output, $he[k]$, can be written as:

$$he[k] = r_{tr}[k] \otimes h_{mf}[k] = S_{tr}[k] \otimes hc[k] \otimes h_{mf}[k] = Rs[k] \otimes hc[k] \qquad (11.2)$$

where $Rs[k]$ is the autocorrelation function of $Str[k]$. In order to make it efficient, the GSM standard uses training sequences that are engineered so that $Rs[k]$ is a highly peaked (impulse-like) real function. Therefore, $he[k]$ is a

others, reflected. In urban areas, often the direct ray is not visible but multiple reflected rays of continuously variable delay and power levels serve the mobile. Models for rural areas, hilly terrains, and other environments have been used to simulate different situations in which GSM mobile typically operates [13]. Such models show the use of delay spreads up to 20 μs and relative power varying up to −14 dB; it is common to find 12 taps or up to 12 different rays. In terms of spread, the general thumb rule for speed of EM wave is about 3 μs/km, which means a signal may travel an additional 6 km (delay) before arriving at the receiver but becomes part of the received frame that must be considered.

11.2.6   Rake Receivers: CDMA

In CDMA, the task of equalization is achieved with a different approach—its wide bandwidth offers the possibility of "adjusting delay of multiple rays and grading them". The concept of "rake receiver" is applied to different rays that are graded to bring their energy into the receiver. This concept was first proposed and patented by Price and Green in 1956. The "Rake Receiver" is designed to counter the effects of multipath, using several sub-receivers known as "fingers". The term is derived from the garden rake, which uses its fingers to rake leaves.

Figure 11.14 shows such a rake receiver with three fingers. It is possible to use more than three fingers. But if too many fingers are used, the advantage of gathering multiple rays is nullified by combiner losses. This block diagram represents rake receiver used in the first commercial CDMA system as part of 2G standard (1.25 MHz bandwidth) known as the IS-95. Refer to the multipath diagram shown in Fig. 11.11, where three reflected rays were considered [14]. For this discussion, let us consider the instance when the direct ray R1 was not available—only the reflected rays R2, R3, and R4 are available. Among them, R3 is "diffracted" due to trees on the green mountain. Each independent ray has a different delay with varying power levels or complex time variant gain G. The entire block diagram with three fingers in Fig. 11.14 can be summarized as "**timings and finger allocation, with matched filter**" as indicated by one block at the bottom of Fig. 11.14.

Referring to the block diagram in Fig. 11.14, functions of each rake receiver finger are as follows:

- **Matched filter**
  - Impulse Response Measurement
  - Largest peaks to RAKE fingers
  - Timing to delay equalizer
  - Tracks and monitors peaks with a measurement rate depending on speeds of mobile station and on propagation environment

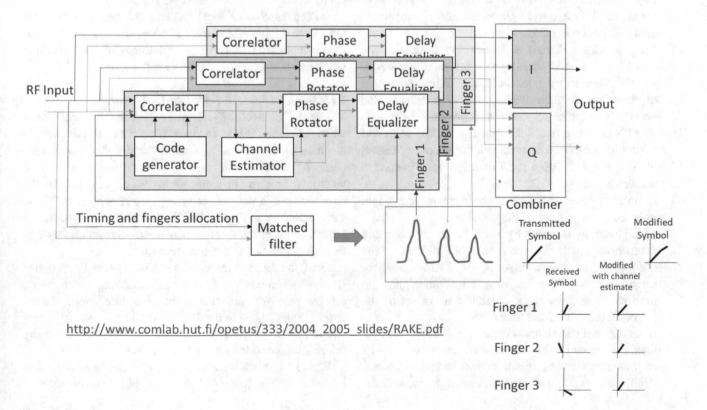

http://www.comlab.hut.fi/opetus/333/2004_2005_slides/RAKE.pdf

Fig. 11.14   Block diagram of Rake Receiver of IS-95 standard

- **Code Generator**
  - PN codes for the user or channel
- **Correlator**
  - De-spreading and integration of user data symbols
- **Channel Estimator**
  - Channel state estimate—channel effect corrections
- **Phase Rotator**
  - Phase correction
- **Delay Equalizer**
  - Compensates delay for the difference in the arrival times of the symbols in each finger
- **Combiner**
  - Adding of the channel compensated symbol—multipath diversity against fading

The rake receiver (in phone) is implemented on the downlink using the following procedure [15]:

1. On the downlink, the cell phone receiver in Fig. 11.14 uses a search engine that scans in the time domain and finds the expected time of a multipath pilot signal.
2. The CDMA base station sends such pilot regularly, which allows coherent detection of signals.
3. The search engine informs the mobile phone's processor, where to find the strongest replicas of the signal and their signal strengths.
4. The mobile's processor has a correlator located in each finger that separately detects strong multipath components. Each base station signal is uniquely identified using its PN code offset.
5. Each correlator detects a time-shifted version of the CDMA signal and correlates to a portion of the signal, which is delayed by at least one chip in time compared to other fingers. Such estimation of time delay is also known the impulse response measurement.
6. The PN chip rate for IS-95 system is 1.2288 MHz and allows for resolution of multipath differences of 1.2288×10^{-6}, which is 0.814 µs. This information is used to decide chip time.
7. In order to generate weighting coefficients, each finger uses the actual output of the correlator.
8. Code tracking is done by each finger to track and compensate small deviations in multipath delay. The code generator in each finger of Fig. 11.14 is used for scrambling and time aligning multipath signals to properly align them at the combiner, as shown on the bottom right side of the figure.
9. If another cell site signal becomes significantly stronger than the current pilot signal then the mobile unit's controller processor initiates the soft handover process. With this process, both cell sites begin to transmit the same call data (bi-cast from switch) on all traffic channels for that receiver.
10. Each finger can dynamically provide symbols that are quite different from the original transmitted symbol, as shown on the right bottom of the figure. The difference can be both in phase and magnitude but are time synchronized so that they can be combined to form modified but useful symbol.

During 3G, the W-CDMA standard was implemented as part of the 3GPP specification. It used wider bandwidth of 5 MHz thereby offering improved performance with adaptive channel estimation. Rake receiver was also implemented in the uplink where base station typically uses receiver diversity implemented by two antennas. This is quite different from the 1.25 MHz bandwidth and single antenna used in the 2G CDMA system IS-95.

Figure 11.15 shows the W-CDMA version of the rake receiver used in base stations, with diversity using two antennas. Impulse response measurement helps allocation of signals to different fingers [15]. Channel estimator along with combiner allows the soft decision process as mobile moves with stronger/weaker pilot signals from two adjacent base stations. The high chip rate of 3.84 MHz produced by the A/D converter (Analog to Digital Converter) offers 260 nS for Rake finger allocation resolution, which is significantly better than 814 ns used during 2G.

Implementation of Rake receiver and several other features such as power control, PN code, and many other features established CDMA as the technology of 3G. This topic will be elaborated further in the next section.

11.2.6.1 Reselection of Control Channel

The term "reselection" relates to "control channel" used when the mobile is idle. In general, any time after the mobile is powered up, it has to remain "camped" on a control channel. Whenever the mobile user initiates a session of voice call or data, or someone makes a voice call to the mobile, this idle state of "camping" is a pre-requisite. These were reviewed in Figs. 11.3, 11.4, and 11.5 showing the logical sequence of how the mobile moves from idle state to an active state under different scenarios.

Here, the focus shifts to how mobile selects the "control channel" when it is idle (user is neither making nor receiving call but moving) and what is the procedure. Since control channel selection is indicative of mobile "not on call", reselection is a procedure executed by the mobile, using information provided by the base station.

Figure 11.16 shows a simple reselection sequence starting with mobile "camped" on a specific "channel A" but

Fig. 11.15 Rake Receiver in WCDMA system

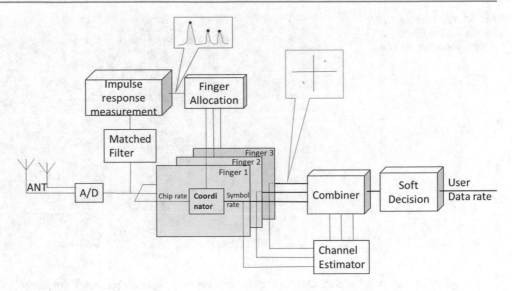

Fig. 11.16 Reselection process —mobile in idle state

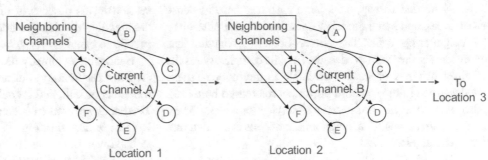

surrounded by many neighboring control channels being potential "candidates".

If we consider Fig. 11.16 using registration of mobile was described in 11.1.3, power up registration follows steps such as scanning and attempt to register (with authentication and other steps). Once the mobile registers, it camps on that channel, which is represented in Fig. 11.16 by channel A. When the mobile stays idle on channel A (Location 1), that control channel informs the mobile regarding neighboring channels described here as B, C, D, E, F, and G. Each of these "control channels" belong to a neighboring cell. Each cell site not shown in this figure but is shown in Fig. 11.2 top left as 7 cell cluster. Always imagine the current serving cell to be in the center. For example, channel B belongs to cell site directly north of the serving cell site A. Channel E is located directly south of cell site A. Channels C and G are North East and North West of channel A. Similarly, Channels D and F are South East and South West of channel A.

Now let us assume that the mobile moved directly north towards channel B but remains in idle state. Then "transfer registration" described earlier in Fig. 11.3 takes place and mobile registers with Base station B (Location 2). With this new site connected to the mobile, Channel A has now become one of the neighboring channels. Also note that channel G is no longer a neighbor to cell site B, it is now replaced by Channel H, which is a neighbor to B but was not a neighbor to A.

This process of reselection has evolved but has remained common to both TDMA and CDMA technologies. It has also remained fairly stable and is backward compatible with the different generations of 2G, 3G, and 4G services. This process of reselection allows the mobile to find out the type of service available at that base station. For example, a mobile capable of operating on either 3G or 4G today, may find out that in some rural location only 3G service is available. Therefore it uses 3G (UMTS/HSPA+) since it is a 3GPP standard mobile. Similarly, switching between 2G (GSM) and 3G (UMTS/HSPA+) is possible for a earlier generation mobile that can operate on 2G or 3G. Later in Sect. 11.3.1, we will also review the 4G standard LTE followed by 5G NR (New Radio) which allows a mobile capable of operating in either 4G or 5G. Such backward compatibility is needed in every cell phone (mobile) since urban areas use latest standard infrastructure while rural areas may have an older infrastructure, mainly due to capacity and cost limitations.

Fig. 11.17 Selection and handover for different Radio Access Technologies (RAT)

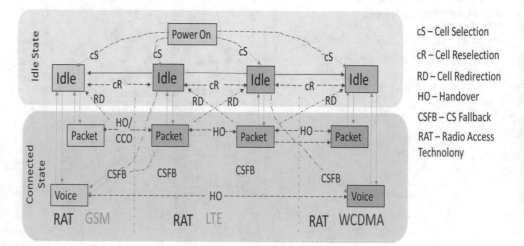

cS – Cell Selection

cR – Cell Reselection

RD – Cell Redirection

HO – Handover

CSFB – CS Fallback

RAT – Radio Access Technolony

The process can be complex due to terrain and higher capacity needed to support a large customer base in dense urban areas, and some models for reselection may offer up to 24 neighboring cells. The nature of cell reselection has evolved over the past two decades, and its complexity results in different types of service available to the mobile. These are described in the standard document referenced here [16]. Also due to cells from various service providers being available in a given area, neighbor list of cells is divided into four general categories:

(a) Acceptable cell—mobile can camp on to receive limited service

(b) Suitable cell—mobile can camp on to receive normal service

(c) Barred cell—mobile is not allowed to camp on (barred from camping on it)

(d) Reserved cell—reserved for cellular operator's use.

All the four categories refer to the idle state are shown in Fig. 11.17, which indicates options such as "selection, reselection, or redirection". Selection is something the mobile is allowed to do, based on criteria set forth by cellular provider. Reselection, on the other hand, is due to moving from cell where mobile had camped and going to another cell. This "reselection" may happen due to weightage given on the channel, and there could be a channel that is "preferred" over the current one. Reselection could also be due to mobility or slight change in position when signal on the currently camped channel is becoming weak and another cell is becoming stronger.

On the other hand, handover procedure takes place only when mobile is in "connected state" or "on call". At the end of a call, the mobile might have moved so far away from the original cell, the entire neighbor list indicated Fig. 11.16 may become irrelevant. To avoid searching by the mobile, base station helps the mobile by "directing" it to camp on a specific channel (message included at the end of call or session). This process of directing the mobile is known as "redirection", which provides a quick means of getting the mobile to camp on a "suitable cell".

Handover, as already described, is based on RAT (Radio Access Technology) during "connected state" or when mobile is on call, and it could be "hard" or "soft" handover. The handover process is completely different for each of the RAT as indicated in Fig. 11.17. However, it is also possible to handover from one technology to another using specific sequence. GSM on the left bottom is a TDMA technology that uses "hard" handover, whereas WCDMA, on the right side, uses "soft" handover. LTE has different procedures for handover, which will be described in Sect. 11.3.1.

Cell Selection Fall back (CSFB): This was a specific procedure defined for use during early stages of LTE deployment when continuity of LTE coverage was not there so a UMTS capable UE (User Equipment/cell phone) could fall back to UMTS circuit switched service. Under those circumstances, CSFB was used as a process where if mobile moves from idle to connected state and gets back to the cell in the same area, which does not need "redirection". This means mobile could use the known list of "neighbor cells", which it was aware of, prior to getting "on call".

While such sequences may seem intriguing, they are well thought out and clearly defined in the standards. Mobile uses the built-in logic to handle these messages in a pre-defined manner. It is mostly due to such defined sequences, that mobile is prevented from "dropping off the network" no matter where and how it moves. Such sequences and procedures are the foundation of what cellular providers indicate by stating "always connected". It allows voice call to come through, even when user is browsing or downloading software or in the midst of multiple data sessions.

Backward compatibility of RAT and cellular services is important and essential to provide minimal support such as voice and low data rate even in remote areas. The mobile must be able to operate at least in "minimal mode" since it has become a basic device for communication. Based on information provided by different cellular providers, 2G and 3G services may continue for many more years [17]. There is another important reason—most narrow band IoT services that require very little bandwidth are essential and can operate on 2G and 3G technology networks.

11.2.6.2 Security Schemes: By Each Generation

While Chap. 6 reviewed the security of wireless systems, it is important at this stage to compare the different generations of cellular in terms of security. In addition to growth, there are inherent differences in approach to security. This is described in detail in a publication describing the security schemes of the fourth-generation standard LTE, see [18]. A brief review of 2G, 3G, and 4G from a cryptographic viewpoint is in order now.

In 2G cellular systems, cryptographic algorithms used to secure the air interface and perform subscriber authentication functions were not publicly disclosed. However, this was partly described in Chap. 6. The GSM algorithm families pertinent to cryptographic discussion are A3, A5, and A8. A3 provides subscriber authentication, A5 provides air interface confidentiality, and A8 is related to A3, since it provides subscriber authentication functions, but only within the SIM card.

When 3G systems were developed, UMTS/WCDMA introduced the first publicly disclosed cryptographic algorithms used in commercial cellular systems. The terms UEA (UMTS Encryption Algorithm) and UIA (UMTS Integrity Algorithm) are used within UMTS as broad categories, to describe cryptographic functions. UEA1 is a 128-bit block cipher called KASUMI, which is related to the Japanese cipher MISTY. UIA1 is a Message Authentication Code

(MAC) and is also based on KASUMI. UEA2 is a stream cipher related to SNOW 3G, and UIA2 computes a MAC based on the same algorithm [19]. LTE builds upon the lessons learned from deploying cryptographic algorithms and while operating the 2G and 3G systems. The SAGE (Security Algorithm Group of Experts) is a team from ETSI (European Telecom Standards Institute) that is responsible for selection, support and enhancement of all security related algorithms.

During the fourth generation, LTE introduced a new set of cryptographic algorithms and significantly different key structures, as compared with GSM and UMTS. 4G started with three sets of cryptographic algorithms for both confidentiality and integrity. These were termed EPS, Encryption Algorithms (EEA), and EPS (Evolved Packet System) Integrity Algorithms (EIA). EEA1 and EIA1 are based on SNOW 3G, very similar to algorithms used in UMTS. However, EEA2 and EIA2 are based on the Advanced Encryption Standard (AES), which evolved in the Internet network standards. EEA2 is defined by AES in CTR mode (e.g., stream cipher) and EIA2 defined by AES-CMAC (Cipher-based MAC). EEA3 and EIA3 are both based on a Chinese cipher ZUC [20]. Although new algorithms were introduced in LTE, network implementations commonly include 2G and 3G algorithms to allow backward compatibility for legacy devices and cellular deployments. This helps in the selection and handover indicated as part of Fig. 11.17 where handover or reselection may trigger security/authentication procedures.

Many keys in LTE are 256 bits long, but in some current implementations, only the 128 least significant bits are used. The specification allows system-wide upgrade from 128-bit to 256-bit keys. In LTE, the control and user planes may use different algorithms and key sizes. Figure 11.18 shows the

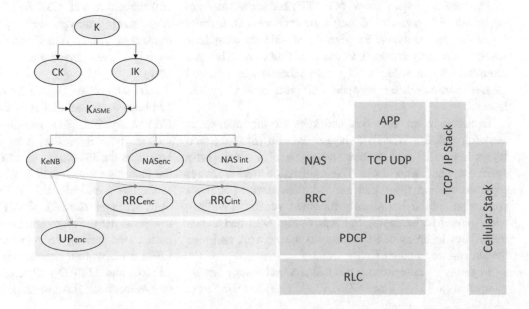

Fig. 11.18 Keys protecting the network stack

Fig. 11.19 CAVE algorithm and Key management in CDMA systems

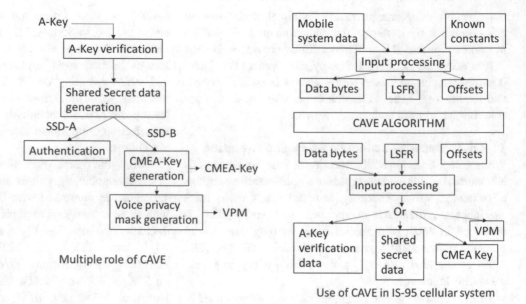

Multiple role of CAVE

Use of CAVE in IS-95 cellular system

Table 11.1 Initial loading of CAVE

CAVE component	A-Key verification	SSD generation
LFSR	32 MSBs of A-key	32 MSBs of RAND SSD
Sreg [0,1,....,7]	A-Key	A-Key
Sreg [8]	Algorithm version	Algorithm version
Sreg [9,10,11]	24 LSBs of A-key	24 MSBs of RAND SSD
Sreg [12,...,15]	ESN	ESN
Offset_1	128	128
Offset _2	128	128

various keys alongside their use for an appropriate proto-col. Note how this relates to the LTE key sequence shown in Fig. 6.4 of Chap. 6.

Figure 11.18 shows the important keys used by the 3GPP cellular system (now 5G LTE) and how they are organized and connected. Observe the traditional OSI model layers appear partly on the right-hand side showing how wireless currently connects to the wired network. This is a modified version of Fig. 4.6.1 where Ethernet was indicated as the common layer between OSI and cellular system layers.

In this diagram, protocols and keys are the main focus showing the IP, TCP/UDP layers as well as application layers in parallel with cellular stacks (on left), which partly spans these and other stacks. The Application layer is always separate and, in the wired and wireless devices, appears somewhat similar but slightly modified versions (Widows and Android for example). In Chap. 6, Fig. 6.2.1 had shown these keys in terms of separation on the network and user device (phone, modem etc.).

In terms of comparison, the CDMA technology imple-mented in IS-95 (2G), and CDMA2000 (3G) had a different approach, using CAVE algorithm discussed earlier in Chap. 6, Fig. 6.1. Basic method of key generation is shown in Fig. 11.19 where CAVE algorithm combines authentica-tion and key generation [21]. Key generation and use with authentication and CMEA (Cellular Message Encryption Algorithm) encrypt cell phone keys for financial transactions, and Voice Privacy Mask (VPM), are part of an overall system based on CAVE algorithm described in Chap. 6.

Left side of the Fig. 11.19 shows the basic process, that had been described earlier in Chap. 6. Main components of CAVE algorithm are shown on the right in Fig. 11.19, indicate how it was used in the first version of CDMA known as the IS-95 system during 2G. Main components of the algorithm are sixteen 8-bit data registers, two 8-bit off-sets offset 1 and offset 2, and one 32-bit Linear Feedback Shift Register (LFSR). CAVE operates in four or eight rounds as per the requirements of a specific application with each round having 16 register update phases. The 32-bit LFSR contains four separate register bytes LFSRA, LFSRB, LFSRC, and LFSRD with a primitive feedback polynomial whose feedback function is defined as:

$$L_{t+32} = L_t \oplus L_{t+1} \oplus L_t + 2 \oplus L_{t+22}$$

For each phase, CAVE uses bytes from the LFSR, the offsets, and two 8*4 LUTs or S Boxes (each table has 256 nibble values) to modify one of the registers. Offsets offset_1 and offset_2 act as pointers into the low and high CAVE tables parts shown in Table 11.1, indicated here.

Despite such differences, both the CAVE algorithm of CDMA2000 and the UEA or UMTS Encryption Algorithm of WCDMA have co-existed during 3G cellular systems. The next section describes the future of two separate cellular standards that existed during 2G and 3G, and a brief account of how the one cellular body has evolved during 4G and beyond.

11.3 Cellular Standards Body: 3GPP and Evolution of Core Network

Throughout this chapter and in earlier chapters, references were made to CDMA standards of North America and Europe. While the two were based on the same technology, there were marked differences. The North American standard used 1.25 MHz bandwidth while WCDMA used 5 MHz bandwidth. We will also address IP network—connections between wired network and the cellular core network, so that evolution of both network will become clear to students.

3G standards in North America and in Europe were developed by different bodies. The North American standard was developed by 3GPP2 (Third Generation Partnership Project – 2), which was based on the concept of "Mobile IP" [22]. It supported two major standards of CDMA, the 2G standard of IS-95 (later TIA-95) and CDMA2000 for 3G. "Mobile IP" was a concept it borrowed from IETF standard (Internet Engineering Task Force), which is the mainstay for Internet and wired networks that we referenced in Chaps. 4 and 5. The specification 3GPP2 was supported by five national standard bodies:

ARIB—Association of Radio Industries and Businesses (Japan)
CCSA—China Communications Standards Association (China)
TIA—Telecommunications Industry Association (North America)
TTA—Telecommunications Technology Association (Korea)
TTC—Telecommunications Technology Committee (Japan).

However, its deployment around the globe was limited and when 4G technology was ushered, 3GPP2 made unsuccessful efforts to consolidate both the organizations (3GPP and 3GPP2). Since it didn't happen, 3GPP2 was closed down. Its documents are available in [22], but further work towards 4G was exclusively by 3GPP. The next section will focus on LTE becoming global standard by 3GPP, which actively pursued efforts towards future cellular standards including 5G.

11.3.1 Moving to 4G: Evolved Packet Core (EPC)

The core network of cellular systems directly interfaces to the backhaul network, described in Chap. 4, where Sect. 4.2 explained in detail how the backhaul network evolved with each generation. Further, Sect. 4.3 discussed the progress of technology in the backhaul network. Digital Cellular systems at the core of the network have a "packet core" and its evolution started with 2G CDMA, since it was designed as a packet network. At that time x.25 was the only packet network available for backhaul (not widely implemented). Therefore, during 2G, majority of the backhaul connections to wired networks were interfaced with circuit switched.

Cellular networks are used to convert from their packet core to the switched network (T1/E1, or ATM, or SONET etc.,) for transport and again convert back to packets when they reached cellular core network at the other end. Gradually, all cellular systems started their progress adjusting to developments in backhaul for transport. The IETF (Internet Engineering Task Force) as a standard body governed all of Internet and developed during 2G time frame [23]. It promoted the use of IP or Internet Protocol as a common baseline. During 3G, cellular systems started using packet network at the core with IP interface to make them compatible with the Internet standards. Later during 4G and 5G, the IP interface evolved further with Ethernet as the common interface.

The use of Ethernet connection was described in Chap. 4, as common interface to all IP networks. Evolution of cellular packet core started with release 5 of 3GPP in 2003 and later evolved through releases 6 and 7 (2006–2007). Many changes were made in the core network, to make it possible to support all Internet services on the cellular network. Services available on wired network use the Domain Name System (DNS) as defined in IETF RFC 1034. Therefore, 3GPP decided to incorporate DNS concept of records is in the "core network" of cellular system.

Instead of creating new DNS records to map a host name to a node name, this specification defines how host names can be constructed and used in S-NAPTR (Straight forward —Named Authority Pointer) procedure within 3GPP EPC [24]. The S-NAPTR procedure or the "Straightforward-NAPTR" procedure is defined by IETF in RFC 3958. It describes a Dynamic Delegation Discovery

Fig. 11.20 Basic EPC architecture with E-UTRAN access

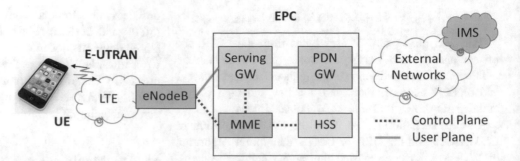

System (DDDS) application procedure on how to resolve a domain name, application service name. It describes application protocol to dynamically interact with target server and port by using both NAPTR and SRV (see IETF RFC 2782) resource records [24].

Figure 11.20 shows the basic architecture of Evolved Packet Core (EPC). Here, the EPC consists of Serving Gateway (SGW) and the Packet data network Gateway (PGW) that connect the user plane to external networks on the User plane. The MME (Mobility Management Entity) and HSS (Home Subscriber Server) operate on the control plane and are responsible for signaling functions. Due to the inherent mobility aspects, MME interacts with serving gateway on the user plane. The control plane (red dotted line) handles all signaling and the user plane handles services such as voice, video, and multimedia [24].

The User Equipment (UE) or cell phone connects to the base station (eNodeB—enhanced NodeB of UMTS) and the UMTS Terrestrial Radio Access Network (U-TRAN) supports access to the current LTE network. From the core network, Packet Data Network Gateway (PGW) uses the IMS (IP Multimedia System) to reach external networks. This is done by using SIP (Session Initiation Protocol) as the signaling method to allow voice, text, and multimedia services to traverse all connected networks. 3GPP works closely with experts in IETF to ensure maximum re-usability of Internet standards, preventing fragmentation of IMS standards. Services and System Aspects (SA) group within 3GPP is responsible for producing the Common IMS specifications.

11.3.2 Multimedia Services and Role of IMS

Till now, multimedia services were mentioned in different models—but how do they rely on the IMS (IP Multimedia System)? Each generation of cellular from 1G through 5G now, had to contend with one basic feature—to carry the human voice. For 1G, it was straight forward—it used analog and only carried voice traffic. With 2G and later generations of cellular, each standard had "Vocoder" or Voice Coder. Coding of voice is essential in order to

conserve precious bandwidth over the air (also known as spectrum). In the very first chapter, using Table 1.3 it was pointed out that 2G and WiFi both use ACELP (Algebraic Code Excited Linear Prediction), but 3G used EVRC (Enhanced Variable Rate Coder), and later 4G used VoLTE (Voice over LTE) coder as shown in Table 1.2 for Android based cell phones, which has now become the common denominator. Such changes in type of vocoder result in the use of IMS as a means to carry voice as well as video and picture (shown in Table 1.4) during each generation of cellular.

While 3GPP extended the concept to other services such as SMS/MMS (short message service/multimedia service), these services used control channel, and a method was introduced to carry them over IP (during 2007, release 7). These are in widespread use throughout the global cell phone community.

Figure 11.21 shows the initial introduction of sending SMS over IP when an IP message Gateway was used instead of MSC (Mobile Switching Center) or SGSN (Serving GPRS Serving Node) in a GSM network [25]. At that time, SMS was not meant for real-time voice/multimedia calls. Such messages can be delivered with or without the use of IP network [26]. Even to this day, in remote regions where there is only 2G network and no IP backhaul and PSTN (Public Switched Telephone Network) networks operate using SS7 signaling (explained in Sect. 4.2) on all wired networks, delivery of SMS/MMS using MSC used to be the standard method from 2G that is still used. For such sessions, even in well developed areas that has IP network, use of SIP (Session Initiation Protocol) and other features of IETF standard are used to set up voice calls with IP networks. In Fig. 11.21, the IP-MESSAGE-GW (Gate Way) provides the protocol interworking for delivery of short message between the UE and the GSM/UMTS network. The functions of this key network element are:

- To connect to the GMSC (GSM MSC) using established MAP protocols over SS7, appearing to the GMSC as an MSC or SGSN using the E or Gd reference points.
- To connect to the SMS-IWMSC (Interworking MSC) using established MAP (Mobile Application Part)

Fig. 11.21 Architecture for SMS support with an IP attached terminal

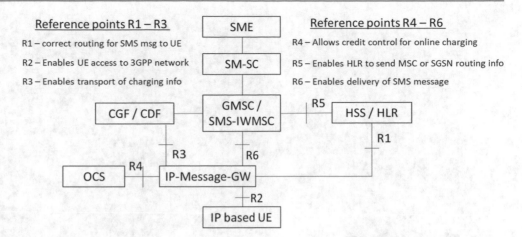

Reference points R1 – R3

R1 – correct routing for SMS msg to UE

R2 – Enables UE access to 3GPP network

R3 – Enables transport of charging info

Reference points R4 – R6

R4 – Allows credit control for online charging

R5 – Enables HLR to send MSC or SGSN routing info

R6 – Enables delivery of SMS message

protocols over SS7, appearing to the SMS-IWMSC as an MSC or SGSN using the E or Gd reference points.

- To communicate with the UE using IP-based protocols maintaining the format and functionality of the SMS message. It is intended that existing messaging protocols supported by the UE should be reused for this purpose.
- To maintain the association between the MSISDN (Mobile Subscriber ISDN) and the IP address of the terminal.
- Support registration and authentication of the UE for SMS services.
- Support of security associations between UE and IP-MESSAGE-GW.
- The proper reliability of IP connection between IP-MESSAGE-GW and UE should be assured.
- Provide interworking between SMS/MMS service in the CS/PS domain and the IP-MESSAGE-GW.
- The IPMESSAGE-GW interconnects, for example, two PLMNs, where one PLMN provides SMS/MMS over IP, and the other PLMN (Public Land Mobile Network) does not support or is not willing to transport SMS/MMS over IP.

Important among its functions are synchronization and recognition of the type of user device. User device may be a modern laptop or tablet (type of vocoder used), or the user has a smartphone or a modem carrying services to other subnet entities in a local network. Based on the type of device it is able to deliver the SMS/MMS message to the end device.

The unique feature of 3GPP is how it defines the IMS— **"the IP Multimedia system does not define applications"**. It is an architectural framework to deliver multimedia services. IMS tries to use IETF protocols wherever possible, so that cellular units (cell phone, modem, and others) can access and use multimedia and voice applications.

From its first release during GSM (release 5), IMS has retained this philosophy. Some recent examples of this

approach are "MMtel or Multi-Media telephony" that allows fixed-line telephone and mobile multimedia service resulting in VoLTE (Voice over LTE) and "WiFi calling" that uses GAN or "Generic Access Network" protocol to extend mobile voice, data, and multimedia applications over IP network leading to (VoWiFi) or voice over WiFi. The commercial name for this is Unlicensed Mobile Access (UMA) leading to the popular applications such as "WhatsApp" that allow users to make international calls by just using WiFi and IP network [26]. The impact is enormous since international telephone calls using telephone networks are becoming a feature of the past generation.

Earlier in this chapter, we reviewed several generations of cellular. The first and second generations essentially connected to telephones on the PSTN (Public Switched Telephone Network) that use SS-7 signaling as explained in Chap. 4. Right from 2G, the SIP (Session Initiation Protocol) was used by IMS to connect cell phone to the PSTN dial up phone. The IMS used a media gateway switch with SIP for all signaling functions. The SIP server has features of the MSC in terms of signaling functions—it knows the number dialed, how to route the call, initiate ringing and when receiver picks up, to complete rest of the network connections.

The SIP is a text-based protocol and does not know content of the call, it only initiates signaling.

This is similar to "http" where the browser only knows the web address to land on but is unaware of the contents at that web site. Whether the call is voice only, or multimedia call that includes video, pictures, text, and other related features (such as in applications like "WhatsApp") are decided by end user device applications [27]. They also allow the mobile to move from dense urban area, which may have LTE service (4G), to a suburban area, which may have WCDMA service (3G), or to remote area which may only offer GSM service (2G). In all cases, the reverse is true as well (calls from rural/remote to urban are supported similarly).

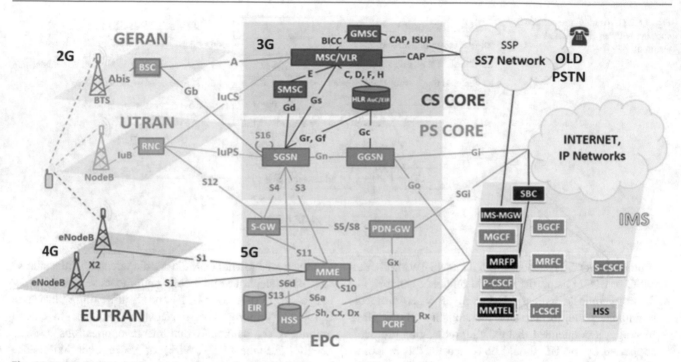

Fig. 11.22 Evolution of access and core networks—role of IMS

In order to understand the importance of SIP and how IMS brings different features of EPC and earlier generations of cellular and the PSTN network, reference to earlier networks is useful.

Figure 11.22 shows the evolution of cellular core network starting from 2G when IMS was first introduced [28]. On the left are 2G, 3G, and 4G illustrations of radio access networks. In the middle are the corresponding core network with key units of 3G, Packet core connecting IMS and other IP networks and 5G with similar connections that were described earlier. On the right is IMS working closely with IP networks as well as traditional SS7 signaling network (Old PSTN), which evolved gradually over these three generations of cellular networks. It also shows many functions it has brought together—voice features, authentication, service authorization, call control, routing, and interoperability with PSTN, billing, additional services, and VAS (Value Added Services).

None of these functions exist in the EPC (Evolved Packet Core), which is why EPC without IMS cannot process a voice call. With IMS, such operations become straight forward—the Media gateway would directly contact earlier generation phones using SS7 network signaling. Control element would be a SIP server, and UE would be a media gateway (user plane). SIP is a control protocol in the application layer; hence, there is no need for switching centers/servers.

Let us consider a live conference call based on MMS—where three parties may/may not use smartphones. To handle this call, we assume that two users (UE-1 and UE-2) are on call and then UE-1 decides to add UE-3 making this a multimedia call. The IMS elements that handle this are P-CSCF, S-CDSF, and AS/MRFC. Flow of messages is logical and is indicated in the figure.

Figure 11.23 shows these message exchanges between three users (UE1, UE2 and UE3) and the IMS elements and how multimedia call between three users is set up by IMS using SIP. There are multiple session scenarios that can be created by IMS, using SIP, that are described in [29]. The messages sent back and forth are detailed in standards, but the key concept of this Fig. 11.23 is to show the level of complexity in setting up such calls. It is important to realize such calls, sequences have become common now—thanks to IMS and the services it uses all these message sequences.

11.3.3 Service-Based Architecture of 5G

With the approach taken by 3GPP standards to make 5G evolve into a core based on service architecture, some steps in the evolution are necessary [29]. This may be a two-step process.

Figure 11.24 shows the two versions—the first version is the upper portion, which shows reference point of the architecture. Traditionally, standard body 3GPP would define the reference points in standard (in this case 5G), indicating different functions in boxes and the interfaces shown by numbers connecting the boxes. In contrast, the

Fig. 11.23 Showing message sequence during third-party multimedia call

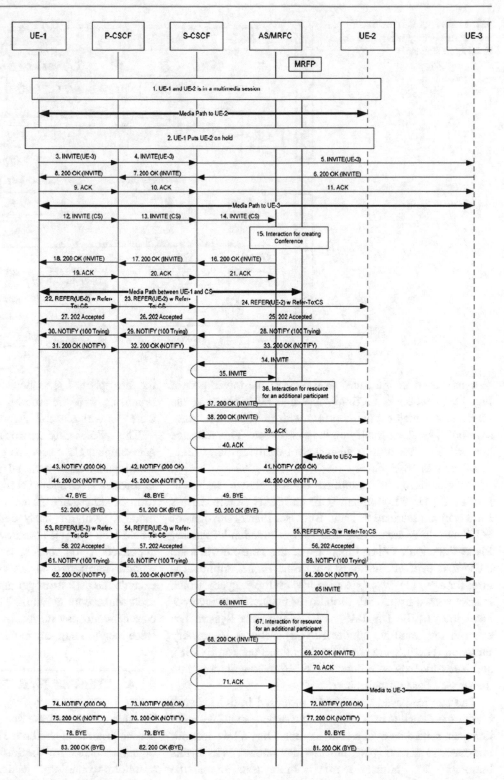

Fig. 11.24 5G Service-based architecture—may usher new era in telecom standards

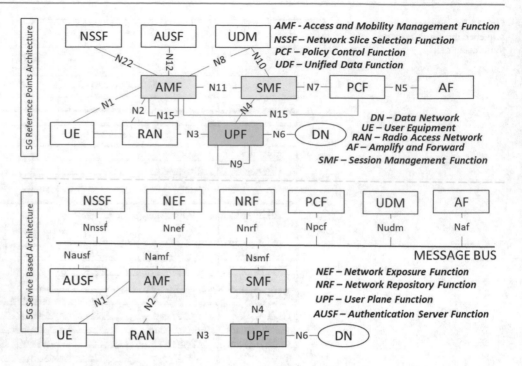

"service-based architecture" is shown in the lower part of Fig. 11.24 where a uniform "message bus" connects the services. All major units interface with the common "message bus", greatly simplifying the architecture. This message bus provides freedom to designers in evolving other functions such as NEF, NRF.

Compared to the simplified version shown earlier in Fig. 11.20, User Equipment (UE) and RAN (Radio Access Network) are retained as such. But User Plane Functions of SGW and PGW are merged into a single function UPF (User Plane Function). Other functions of the core network are those that evolved during 3G–5G time frame. Authentication, for example, was expanded into multiple layers, which is now shown by AUSF. Mobility Management is incorporated into AMF. The IMF or IP Multimedia System has evolved into multiple entities such as Network Repository Function (NRF), Network Slice Selection Function (NSSF), and Network Exposure Function (NEF) making its way into the service-based architecture.

The service-based architecture proposed in 5G network brings the concept of a message bus where multiple functions can send data over a common bus. Only a few network interfaces such as N1, N2, and N4 may remain. Why is this important? The primary reason is more users around the world access Internet using the smartphone and base stations. Less number of users have the conventional PC desktop or even the laptop both of which access the Internet using some wired network (may have a short WiFi link that connects to wired network). Therefore, Internet access methods that support mobility take the center stage to define new functions and methods for devices that are mobile (like

the smartphone). The new Service-based Architecture is a promising step towards an API-based architecture for the Core Network Control Plane.

The cellular core network is where such mobility access gets defined. User services such as voice, text, multimedia, social interactions etc., all rely on the smartphone. User Equipment (UE) shown at the bottom in Fig. 11.24 takes the form of either smartphone or a device that uses 4G or 5G modem. Without such cellular mobility devices—users would look for WiFi wherever they go. Users may get disappointed if that beach, or national park or some other outdoor location on lakes or oceans without WiFi would exclude the user from getting on the Internet if they did not use cellular core network. The functions defined in cellular core network may be readily accepted by satellite networks since they too support mobility.

11.4 LTE: The First Global 4G Standard

So far, we reviewed the first generation that was analog; the second generation had both TDMA and CDMA as competing digital standards, and the third generation settled on CDMA as a uniform technology standard. With each generation, radio engineers focused on an important requirement "how much throughput can the technology deliver over the air and what is the level of complexity that cell phone can handle". This is the fundamental requirement for cellular—the backhaul or wired part of cellular network carries what the radio can support. Backhaul has no role in defining the cellular generation. Why is this important? Let us ponder over this further.

If we consider an aircraft, for example, its speed, weight, whether it is long haul or short haul, its body shape etc., are all defined by aero engineers who design its parts that directly confront air. The comfort of seats inside, the galley and windows are details that are adjusted to fit within requirements provided by the air frame, if it is used for passenger traffic or just the luggage hold if it carries only cargo, also, size of cockpit whether it is used for military or commercial purposes. Similarly, cellular networks are defined by radio engineers to address the air interface and provide details of its data throughput from the phone and base station. Fiber optic or other types of backhaul networks are adjusted to carry data, as reviewed in Chap. 4.

With this approach, radio engineers pondered over user requirement, and design radio to increase data rate by 10 times for each generation, as basic criteria. Other criteria for 4G came from users in offices who had shifted data usage pattern from wired and connected PCs, towards cell phone/smartphone. Cell phone had started to provide functions that any windows-based PC or Apple computer could offer. If broadband data could be handled by smartphone, then extensive wired network within each building could be avoided. The use of access points to serve the smartphones using the WiFi mode within buildings but can continue with the same smartphone outside using cellular broadband mode, helped users move in and out of the building. It slowly evolved towards eliminating Private Business Exchange (PBX) that wired telephone providers used.

Such requirements brought to fore the possibility of using OFDMA (Orthogonal Frequency Division Multiple Access) as an access technology for 4G to develop a uniform global standard. We reviewed OFDM in Sect. 2.4 as a technology developed by Bell labs during 1960s which later came into public view during development of Wireless LAN 802.11 standards. However, this standard was based on supporting indoor area of no more than 100 m, and mobility is not supported. It used Time Division Duplex (TDD). Cellular providers often prefer to use FDD (Frequency Division Duplex). 4G offers both options, TDD as well as FDD, and, in addition, it has to absolutely support mobility for outdoor use [30].

OFDM was chosen as the modulation since it is less susceptible to indoor multipath; it offers improved performance within buildings. Powerful DSPs (Digital Signal Processors) are used to implement the entire scheme in smartphones. The topic of multiple access was discussed in Fig. 11.1 at the beginning of this chapter. By applying this concept, OFDMA as an access scheme was used for downlink. OFDMA is complex yet offers more flexibility to users in terms of throughput rates.

To keep the system practical, the uplink and downlink schemes were separated using different frequency allocations schemes, as indicated in Fig. 11.25 with colored channel schemes [30]. It shows individual users send their data from smartphones on the uplink frequency band using SC-FDMA (Single Carrier Frequency Division Multiple Access) architecture, which does not need much power and can be operated using simple battery in smartphones. The base station, on the other hand, provides the downlink that uses conventional OFDMA (Orthogonal Frequency Division Multiple Access). This scheme requires linear power amplifier that results in low Error Vector Magnitude (EVM) —on the separate downlink frequency band, an important parameter for transmitters. The uplink and downlink are deliberately kept different in terms of access.

This scheme needs more power at the base station but offers better overall efficiency to serve individual users with varying data needs. Different methods for uplink and downlink shown in Fig. 11.25 resulted in the basic foundation to move forward for future generations of cellular— hence the name "Long Term Evolution (LTE)". Now 5G is evolving based on this concept of 4G, with dissimilar uplink, and downlink access, but also includes millimeter wave bands [31].

Since LTE is a major network in most countries, some concepts of network engineering were incorporated in this standard. These are illustrated in Fig. 11.26 depicting the control and user planes.

This conceptual diagram in Fig. 11.26 relates to protocols and how they are handled separately in the two layers (Radio layer and Transport layer). This is formally known as CUPS or Control and User Plane Separation. Radio network layer

Fig. 11.25 Different architectures for uplink and downlink (FDD)

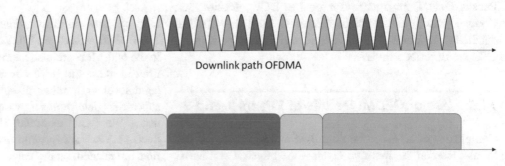

Downlink path OFDMA

Uplink path SC-FDMA

Fig. 11.26 Control plane and User plane—relation to radio network

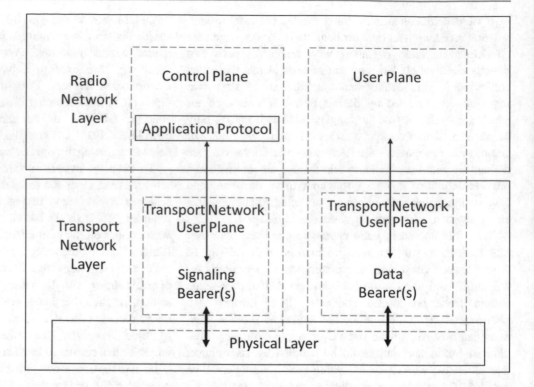

shows the control plane from which all protocols related to radio signaling are handled. Note that the application protocol also is part of this control plane since the way an application works on a cell phone relates to not only the radio protocol but also the type of handset, and whether it is currently operating on 2G, 3G, or 4G. Some applications that are data intensive may not be supported on 2G for example. In such cases, the signaling bearer from physical layer (radio air interface layer) would indicate this limitation. Radio Resource Control (RRC) would handle such information since it handles signaling messages between base station and the handset (smartphone).

The user plane handles protocols such as TCP, UDP, and IP that are protocols commonly used in all wired IP networks. This enables the phone as well as the base station to work directly with IP networks and applications that exist in wired networks and servers.

There are other protocols that process information from both, before passing them to the physical layer—these are: Packet Data Convergence Protocol (PDCP), Radio Link Control (RLC), and Medium Access Control (MAC). The detailed description of protocol stacks and their functions in LTE are described in [25].

11.5   Use of OFDM: Details of Layers 1 and 2

Complexity of LTE can be described at various levels. But let us focus on the first two layers—the physical and MAC (Media Access Control) layers that define radio and its

baseband protocols at the incoming data stream level. These two layers are parts of the radio protocol architecture. In LTE, separation of user's data and control signaling are addressed by using the terms "user plane" and "control plane" [32].

Six major topics that could be addressed in these layers are:

- Frequency and bandwidth
- Modulation techniques
- MIMO
- Data rate
- Access
- TDD-LTE.

Each of these will be briefly reviewed here.

11.5.1   Frequency and Bandwidth

LTE operates on conventional cellular bands of 700, 800, 1800, 1900 as well as 1700/2100 MHz bands. Other bands in the 600 MHz are also supported. However, the bandwidth used in 4G could have considerable variation, at the baseband level or number of subcarrier level. The number of subcarriers determines how much resource is needed by each user. We will examine this in the immediate next Sect. 11.5.2. For high end users, a large number of subcarriers are needed, as indicated earlier by colors in Fig. 11.25. The main RF carrier bandwidths are used accordingly from

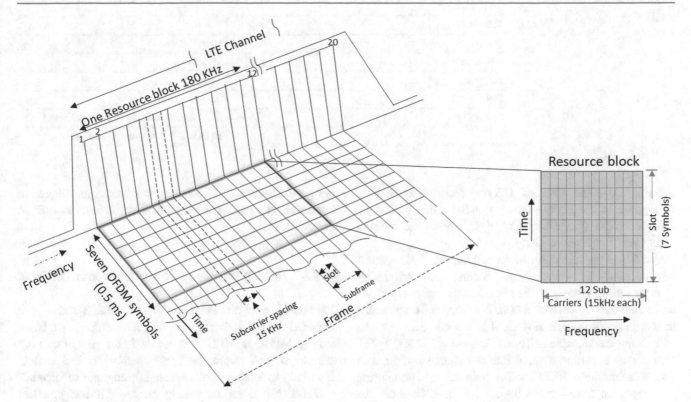

Fig. 11.27 Concept and usage of Resource Block in LTE

1.4 MHz all the way up to 20 MHz, which is beyond the widest bandwidth supported in any previous cellular systems, such as CDMA. The LTE standard allows multiple options for RF bandwidth such as 1.4, 3, 5, 10, 15, and 20 MHz and up to 100 MHz with carrier aggregation. Such wide variation in bandwidth allocation to user is accommodated using a concept introduced in LTE called Resource Block (RB). The concept of an RB is shown in Fig. 11.27 where an RB is illustrated with 12 separate streams resulting in one block of $12 \times 15 = 180$ kHz. This represents the basis of bandwidth allocation in an LTE subcarrier.

11.5.2   Modulation Techniques

Modulation with OFDMA uses the scheme of "subcarriers" with a channel bandwidth of 15 kHz, each, resulting in small subcarrier-based "Symbol frame", which is $1/15$ kHz $= 66.7$ μs. The high-speed data stream that comes from each user's handset is first divided into such multiple smaller streams. Each smaller stream is modulated by a separate subcarrier. For example, in 5 MHz (or 5000 kHz) bandwidth, there would be $5000/15 = 333$ subcarriers that can be used. In reality, it has 25 resource blocks, i.e., about 300 subcarriers (instead of 333) due to guard band requirements.

This Resource Block (RB) is the smallest unit that can be allocated to any user. For example, text-based traffic may need very few RB units, interactive web browsing may need tens of RBs, but video from users may need hundreds of RBs. But that does not mean that all the data sent by a user in mounted on one RB or one frame. User data is spread using multiple Resource Blocks over several LTE channels, thereby making sure that data is not lost due to multipath even if some frames may get corrupted, entire data stream is not. In addition, the issue of multipath due to mobility with Doppler Effect is addressed by different approaches. Instead of conventional use of equalizers at the receiver, the concept of Cyclic Prefix (CP) is used. CP is used with every transmission as part of the symbol frame to handle time dispersion due to multipath. Also, transmit diversity is provided to offset fading issues in addition to Cyclic Prefix [33].

Figure 11.27 shows the concept of Resource Block. LTE transmits data by dividing it into slower parallel paths that modulate multiple subcarriers in the assigned channel [33]. The data are transmitted in segments of one symbol per segment over each subcarrier. Data to be transmitted are allocated to one or more resource blocks (RBs). An RB is one segment of the OFDM spectrum that is 12 subcarriers wide for a total of 180 kHz. There are seven-time segments per subcarrier for a duration of 0.5 mS. Data are then transmitted in packets or frames, and a standard frame

Table 11.2 Time units used in LTE

Time unit	Value
Frame	10 ms
Half-frame	5 ms
Sub-frame	1 ms
Slot	0.5 ms
Symbol	0.5/7 for normal CP, 0.5/6 for extended CP
Ts	1/(15,000 * 2048) sec $\approx$ 32.6 ns

contains 20 time slots of 0.5 ms each. Frequency is orthogonal to the paper—hence the RB when picked up from the paper (green block on right) shows the time on Y-axis and frequency on X-axis.

An RB is the minimum basic building block of a transmission, and most transmissions require many RBs. This concept can be viewed in the time domain starting with the basic time unit Ts = 1/30720000. All other units are related to this basic time unit. How is this number derived? The radio frame has a length of 10 ms (Tframe = 307,200 $\times$ Ts). Each frame is further divided into ten equal sub-frames of 1 ms (Tsubframe = 30,720 $\times$ Ts). Scheduling is done using algorithms on a sub-frame basis. It is used for both the downlink and uplink. Each sub-frame is further divided into two equal slots of 0.5 ms (Tslot = 15,360 $\times$ Ts). Each slot in turn consists of OFDM symbols, which can be either seven (normal cyclic prefix) or six (extended cyclic prefix). Cyclic Prefix (CP) can be thought of as a duplication of a portion of the symbol end [33].

For the normal mode, the first symbol has a cyclic prefix of length TCP = 160 $\times$ Ts $\approx$ 5.2 μs. Remaining six symbols have "Time cyclic prefix" of TCP = 144 $\times$ Ts $\approx$ 4.7 μs. The reason for different CP length of the first symbol is to make the overall slot length in terms of time units divisible by 15,360. For the extended mode, the cyclic prefix is TCP-e = 512 $\times$ Ts $\approx$ 16.7 μs. The CP is kept longer than the typical delay spread of a few microseconds encountered in an extended range. It is useful to remember thumb rule that radio waves travel through air at $\sim$ km/μs. Most cells are less than 10 km hence transition delay is no more than 3.3 μs. The normal cyclic prefix of seven symbols is used in urban cells and high data rate applications while the extended cyclic prefix of six symbols is used in special cases like multi-cell broadcast or in very large cells (e.g., rural areas, low data rate applications). Basic time units used in LTE are shown in Table 11.2.

Any method of countering multipath involves extra time/symbols; it is considered "overhead". The use of CP results in overhead of about 7.5%, which is nominal. In earlier generations of cellular using equalizer, convolution etc., overhead could be from 15 to 25%, depending on algorithms. In LTE, subcarriers have very small bandwidth compared to the mobile coherent bandwidth and therefore experience "flat fading" with no need for equalizers used in 3G systems.

11.5.3 MIMO: Multi-Input Multi Output (MIMO)

This concept was discussed earlier in Chaps. 3 and 4, also explained in 5.1.4 as part of the 5G initiative in beam forming. MIMO in 4G LTE is used with the goal of sending multiple streams from the base station to the mobile (downlink) to increase downloading throughput or speed.

While this is an interesting concept, it has practical challenges. Challenges arise due to size of the handset, which must fit in the human palm. To accommodate four or eight distinct antennas and receive each of their beams effectively, some designs are being proposed, see [33]. In addition, mobility brings challenges since multiple beams travel over the air and change dynamically with movement of mobile at each given instant. The concept of using multiple antennas at the base station is shown in Fig. 11.28. The LTE standard defines 4 $\times$ 4 MIMO configuration, which is difficult to implement in the handset. If four such antennas were used in the same polarization at the base station (eNode B), by careful beam forming, it may be possible to receive at the User Equipment (smartphone) two orthogonal set of beams over the air. This configuration of 8 $\times$ 2 beam forming is shown in Fig. 11.29, where orthogonal set of beams from four antennas are conceptually shown over the air (four blue combined into one beam and four green form the other orthogonal beam). Two orthogonal beams over the air must be perceived as combination of signals from the four vertical antennas (blue) and four horizontal (green) antennas.

The concept was tried successfully, thanks to technological innovations in recent years [33]. At the handset (UE) in particular, the use of 3D-printed antennas has started to bring a new genre of antenna technology. By directly integrating such antennas with high-density chip sets in UE, innovations to improve data rates using MIMO have become possible. Similar efforts will continue to increase data rate using MIMO, and it is expected to be an active area of

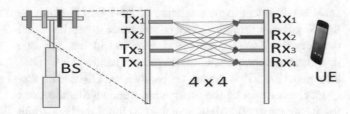

Fig. 11.28 Antenna configuration to optimize MIMO data rate

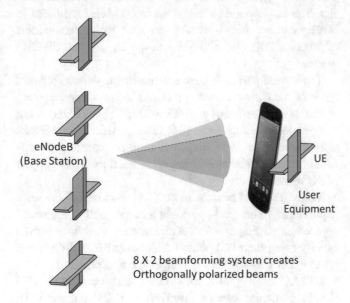

Fig. 11.29 An 8 × 2 beam forming system at BS creates orthogonal beams at handset

research for industry and academicians alike. Testing multiple signals over the air is quite difficult. The difficulty can be partly visualized from Fig. 11.28 where base station must maintain $\lambda/2$ distance between antennas. But at the handset, this is difficult. In the current cellular bands of 700 or 800 MHz full wavelength λ would be 0.428 m and half wave $\lambda/2$ will be 21.4 Cm. This is more than the length of the handset. That is part of the reason why UE (handset) attempts to receive only two orthogonal beams as depicted in Fig. 11.29, resulting in 8 × 2 beam and not an 8 × 8 beam.

We will revisit this issue again, during description of MIMO in 5G later within this chapter. The use of "Millimeter wave" could change this scenario and improve throughput data rates. Reasons why this is becoming a major research area for RF engineers will also be explained [34].

11.5.4   Data Rate/Throughput

Data rates comparable to wired access (such as DSL, cable, fiber to home, and others) have become possible in LTE. This has resulted in major changes in the telecom industry in general. With data rates of 100 Mbps becoming common in

4G, many users have started to use 4G modem as the primary device for Internet access to homes, and, in some cases, this modem is sufficient to provide limited backhaul for network application involving narrow band systems such as public safety, IoT devices. Data rates nearing the theoretical limit of 100 Mbps are deployed in 4G LTE-Advanced, providing a viable alternative to many residential users.

Recall that different data rates and throughputs were discussed in Chap. 5, and Table 5.4 showed a variety of data rates at different signal levels. A major shift in 4G is the method by which the signal level is measured. There are three separate measurements that indicate received signal strength (which was RSSI in most systems). 4G introduced two new concepts—RSRP (Reference Signal Received Power) and RSRQ (Reference Signal Received Quality).

These three measurements are concerned with data rate and throughput. They are related to each other, using concept such as Resource Block, illustrated earlier in Fig. 11.27. While RSSI is measured as energy over the entire RF bandwidth, RSRP is measured over narrower bands of the subcarrier. It is therefore related to RSSI with the concept of Resource block (which has 12 subcarriers) using the equation:

$$RSRP = RSSI - 10 \, \log\left(12^N\right) \qquad (11.3)$$

Similarly, RSRQ is also related to these two using the equation:

$$RSRQ = 10 \log(N) + RSRP\,(dBm) - RSSI\,(dBm) \qquad (11.4)$$

In Eqs. 11.3, and 11.4, N is the number of Resource Blocks. Therefore, RSRP is always lower than RSSI, and modern 4G smartphones measure signal in RSRP and not RSSI. Also, RSRQ is based on the number of Resource blocks, it will be smaller. It is possible to measure these quantities on a modern 4G or 5G cell phone using "field engineering mode".

For ready reference, it may be noted that, N—Number of Resource Blocks is based on the bandwidth (indicated in parenthesis) as indicated below:

N = 6   (1.4 MHz),   N = 15   (3 MHz),   N = 25   (5 MHz), N = 50   (10 MHz),   N = 75   (15 MHz) and the largest N = 100 (20 MHz). All of these bandwidths are used in the LTE standard.

The Resource Block is therefore central to how LTE operates, and allocation of N resource blocks described above can vary depending on what the user's current tasks are. In many ways, this makes the concept of signal measurement directly relate to the way LTE operates. Higher the throughput rate lower will be the value of RSRP indicated.

This may seem confusing to those used to RSSI, but the concept of RSRP and RSRQ is based on how LTE uses the OFDM signal.

11.5.5 TDD-LTE

Since TDD was not used in 2G or 3G why use it in 4G? This question has implications particularly for countries with large population that serves users almost exclusively with cellular technology. Due to limited availability of spectrum, this is a good alternative for many countries. According to Global mobile Suppliers Association (GSA), over 56 countries have deployed the TDD version of LTE [30]. It is worth recalling that many countries did not have a robust telecommunication infrastructure using wired network. Cellular technology brought these countries into the global economy as discussed earlier in Chaps. 7 and 8. TDD will provide an excellent enhancement to service providers of these countries. Also, certain bands are exclusive to TDD only and are not paired (uplink and downlink on the same band). Another benefit is that the TDD systems enjoy DL-UL channel reciprocity, which can be taken advantage off in MIMO systems. This is similar to WiFi, which is based on TDD.

The chipset used in any smartphone for 4G will continue to support FDD since that is the technology deployed in most countries of the world, irrespective of whether those countries have TDD or not. Those countries with TDD deployed will be able to use both technologies.

11.6 Work in Progress Towards 5G

Work towards 5G started with Release 15 of 3GPP standard during 2015–2018. It brought remarkable perspectives on how cellular technology would serve the global population in the coming years. Based on 4G LTE, a new radio system is being defined. Rel 15 5G became a standard in 2018 and was deployed. Rel 16 was standardized, and released in 2020, with some major frequency allocations and feature enhancements. The biggest change was allocation in the "Millimeter wave band". In technical terms "Millimeter

wave band" refers to wavelengths of 10–1 mm (~from 30 to 300 GHz). Currently 28 GHz and other upper bands all the way to 64–71 GHz are being considered for one important reason—they offer "huge bandwidth". Data throughput in RF is based on bandwidth. Higher the frequency, more will be the bandwidth. Bandwidth allocations for 5G are up to 400 MHz, which is 20 times the bandwidth used in 4G. 3GPP specification "TS 38.101 series 1 to 4" lists the frequency bands and bandwidths supported globally. The larger bandwidths of up to 400 MHz are listed in 3GPP radio specifications n257 to n261 listing bandwidths of 50, 100, 200, and 400 MHz in the 26, 28, and 39 GHz bands.

The second part of feature enhancements—three different types of radio systems are proposed to accommodate very different features. These are indicated in Table 11.3 showing three major radio systems [34]. The generic term 5G NR (Fifth-Generation New Radio) is used to represent all different radio architectures being considered by 3GPP as part of 5G releases.

Table 11.3 addresses many features based on user requirements. There is little doubt that over half of the world population participates on social media websites, almost entirely using their cell phones. Hence eMBB radio gets top priority and was worked as part of the first release.

URLLC radio gained importance since connected vehicles in particular cannot afford such delay (currently IP networks may have 30 mS delay, with end-to-end delay of 100 mS). URLLC takes advantage of direct links over-the-air (generally known as V2X—Vehicle to everything) reducing delays to microseconds (maximum over the air latency allowed will be 1 mS). This approach can reduce the cost of laying cable along roadways. Efforts are under way in Ethernet standard also to reduce latency in IP networks. URLLC radio expects higher reliability to 6 digit rather than 5-digit reliability of earlier networks (IP networks are designed for 99.999%). This requirement is due to connected vehicles moving at great speeds and their safety. They depend on network reliability for safety and avoid collision. URLLC supports other applications such as factory assembly line, or Virtual Reality surgery, or smart grid, where low latency and high reliability are essential needs.

Table 11.3 Three major radio systems to address different features of 5G	Enhanced mobile broadband (eMBB)	Ultra-reliable and low latency communication (URLLC)	Massive machine-type communication (mMTC)
	All data, all the time—two billion people on social media	Ultra-high reliability, ultra-responsive	30 billion connected "things", with low cost, low energy
	500 kmph mobility (high speed rail), 10–20 Gbps peak data rates	<1 ms air link latency, 5 ms end-to-end latency, 99.9999% reliable, 50 Kbps–10 Mbps	105–106 devices/km^2, with 1–100 kbps/device, 10-year battery life

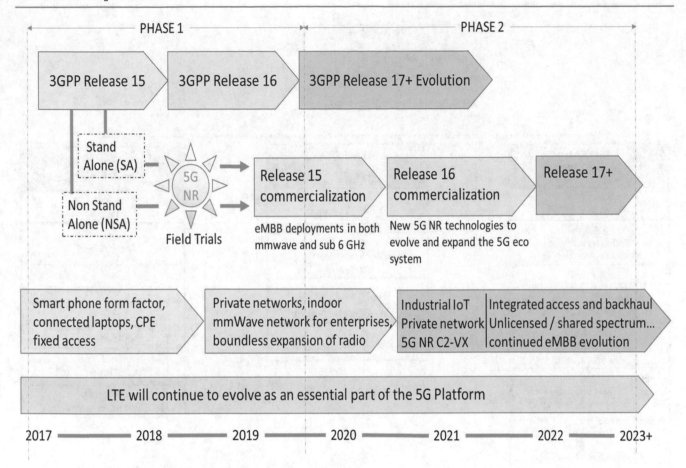

Fig. 11.30 Evolution of 5G—timeline and basic features

mMTC is an extension of IoT that was described earlier in several chapters. Massive number of devices appearing in smart cities, smart manufacturing and smart monitoring of large areas use sensors and devices that will increase to about least 10 times the human population. Yet their throughput requirement is low. They are used less often (few times a day) and due to remote locations, need reliable battery technology. All of these features are being addressed under the third category of mMTC, which is seen by many cellular providers as their biggest growth. Low data rates will not need additional capacity in cellular network, yet extensive coverage supported by cellular systems is useful. Extended battery life is needed due to constraints such as visiting remote sites, for system maintenance. Remote monitoring of mMTC includes periodic report of device health. This will be a major improvement to many such systems worldwide.

The first phase of 5G release focused on further expansions of 4G LTE followed by the first 5G NR (New Radio) specification in release 15 issued by 3GPP in June 2018. Since then, there has systematic effort to bring 5G network towards public deployment, with Release 16 in late 2019 with initial trials followed by commercial release. This can be perceived as the beginning of phase 2. Since 5G addresses a variety of features in planned during future releases, highlights of 3GPP timeline are indicated in Fig. 11.30. Note that backward compatibility (known as non-stand-alone) and forward compatibility (or Stand-alone) are important objectives of 3GPP releases. Hence, official deployment of 5G was started in 2019 by major cellular providers. But the deployment sequence shows that 5G rollout will continue through most of 2020s culminating in some futuristic concepts of 6G [34].

A useful key while referring to 3GPP is their nomenclature. Specifications that begin with TR indicate Technical Reports, while those with TS indicate Technical Specification. Since documents of 3GPP are available to public, it is important to acknowledge that Technical Specification is for the purpose of advancing and understanding technology. Such document portrays cooperative work by different member units with the concept of advancing technology. TR documents by nature tend to provide description and

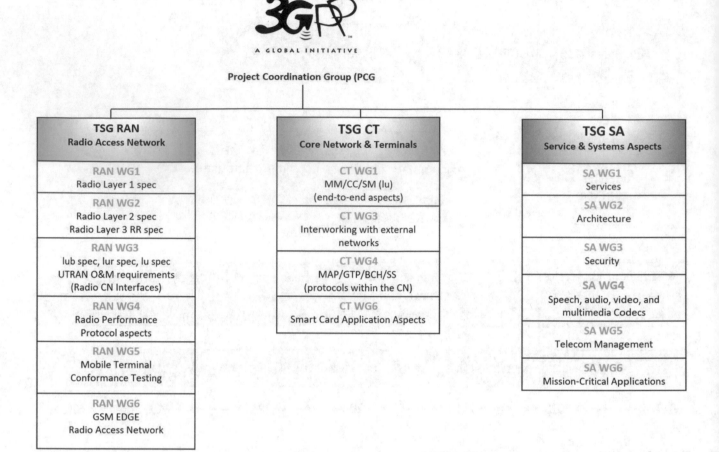

Fig. 11.31 Organization of 3GPP standard and working groups

evaluation methods, whereas TS is a specification that is binding on products and networks so that they inter-work as a system.

So far, we focused on Radio Access network, which is a major change that 5G will address. There are several other aspects that the body has addressed—such as the core network, types of terminals (smartphone is one of the terminals, but there are others), and finally the system and application part. Together, they address different types of services being planned as part of 5G releases during various phases. Figure 11.31 shows the organization and structure of 3GPP to coordinate work on its various specifications and how to release them as documents.

The Radio Access Network (RAN) has six working groups (WG) to address different layers as indicated in Fig. 11.31. Releases from each of these working groups are coordinated so that radio designers of various products can foresee how a radio for 5G can be designed, at the same, it has to be compatible with 4G LTE networks. It is worth noting that some work goes all the way back to GSM which is a 2G network. The intent is to make sure that cellular networks of earlier generations are carefully reviewed for

backward compatibility. GSM networks serve major part of IoT today and could become the foundation to move forward on mMTC.

The core network and terminal group have four working groups addressing different aspects. Some of them, for example, WG6—address existing products and enhancements. Others such WG4 will work on protocols within 5G networks. Interworking with external networks includes not only wired IP network, ISDN, and other types of backhaul deployed in various parts of the world but also public safety networks many of which are still operating in analog mode (such as AM, FM with SONET/ATM backhaul).

Finally, the System Aspects (SA) group also has six working groups that address the overarching system aspects such as security, services, and mission critical applications that will address public safety networks (Police, Fire, EMS for example). Several new multimedia devices operate on cellular networks today. They are currently part of Local Area Network that needs cellular connection to extend their reach in interworking on a global market. This could mean, for example, a robot or an automated device that operates locally (whether in a machine shop or a sophisticated

Table 11.4 Improvements in cellular by generation

Cellular generation	Radio network (RAN)	Core network (CT)	System aspects (SA)	User/provider benefits
2G	TDMA/CDMA. GSM and TIA-95. Both with vocoders and neighbor lists for handoff. Bandwidth 200 kHz	Packet core— introduce IMS to handle signaling on both SS7 and IP network	Security introduced (CAVE algorithm). Use of SIM card to separate user and network	Three cell sectors each supporting 64 users. Voice and limited data (19.2 Kbps). SMS text msg
3G	CDMA as 3G standard. Soft handoff support for data. Inherent security due to CDMA spread spectrum. Bandwidth 5 MHz	IP core network— IMS and EPC interworking with SIP for voice and data sessions	Pictures with JPEG 2000 and limited MPEG4 video. Social networks and websites specific for mobile use	Cell capacity increase to 512 users, increased data (1.04 Mbps) and reliable network
4G	OFDM with variety of bandwidth and subcarrier. Concept of Resource blocks (RB) MIMO deployment in BS and UE. Interleave user data on subcarriers. Bandwidth $n \times 20$ MHz Use of TDD/FDD	IMS and extensive IETF standard for all IP functions. Separation of user data streams only at end points Public safety users join 3GPP	Seven-layered security— separation of services. Extensive use of mobile. Use of 4G modem as back haul	Cell capacity enhancement to 1024 by multiple services within sector. 50–100 Mbps for commercial users
5G	Multilayered, multi-beam radio using mm wave radio. URLLC and 64×64 MIMO antennas at BS. Antenna with IC chips beams and streams. Direct high capacity beam serving residential users. Bandwidth 400 MHz. Latency is important	Predominant method of Internet access with mobiles. Multiple applications in Android and Apple phones. Public relying on cellular during extreme weather conditions	Extensive deployment of "mobile only" applications. URLLC exceeds land line network spec in latency and reliability Support to SCADA network	Cell capacity enhancement of up to 10,000 users with multibeam MIMO. Exclusive support to driverless vehicles. Beam with throughput up to 1 Gbps

medical/surgical unit), should be controlled remotely by a specialist [34].

Such current and futuristic ideas require collaboration of several fields in which 3GPP and its cellular teams address. For anyone with innovative ideas, 3GPP acts as a friendly, neutral platform, to interface with experts in different fields. Systems and services address such ideas from a range of industries. In an attempt to summarize the accomplishments starting from 2G through 5G, Table 11.4 captures key points of progress in cellular technology. It also provides an overview of contributions in each generation—not limited to the throughput rates. In each generation, collaboration for various industry and academic partners contributed to bring improvement in services to benefit users and service

providers. These represent highlights since the details of implementation in each generation of cellular were extensive. The last column indicates what users and service providers observe. They may not know how such exponential progression in throughput was achieved, but they can readily feel it.

Such improvements (that users perceive) are due to more bandwidth used in each generation. But it is also due to efficient utilization of bandwidth—not only in radio design but core network support in signaling and system aspects with unique applications. In brief, each generation of cellular takes about one decade to fully deploy. 2G from mid-90 s to 2003, 3G during 2004–2014, 4G from 2013 to 2018, and 5G from 2019. It is normal for public to see the claims of the

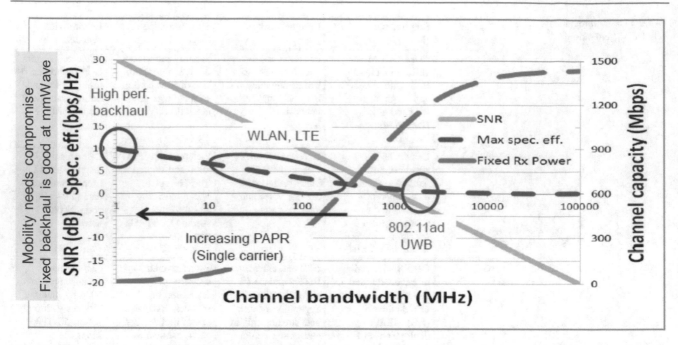

Fig. 11.32 Signal to Noise reduces as channel bandwidth is increased

next generation (for example, 6G being talked about now), well before current generation is deployed. It does not mean that the next generation is deployed in parallel. It only means ideas of what it should be included in the next generation are being proposed. For example, 6G plans to deploy augmented human being (virtual reality) to be present in a meeting somewhere as avatars [31].

11.7 Advantages and Concerns About Millimeter Wave

There are many advantages of using Millimeter wave. Primary impact on all broadband radio systems is the enormous throughput it can offer. From initial systems that operated with voice as the only objective just 25 kHz or 30 kHz was considered sufficient (in 1G and 2G time frame). 5G systems today consider bandwidth as large as 400 MHz in the Millimeter wave band. Using such large bandwidth, throughputs of several Gbps are projected in 5G. If this band offers such advantages why it was not considered earlier? There are some practical challenges that are known to RF engineers, for quite some time. With increase in frequency (such a Millimeter band), the path loss increases substantially (as square of the distance). In Chap. 7, using Fig. 7.6, this was described in the context of how RF energy dramatically attenuates with increasing distance and increase in frequency. This has a huge impact on Millimeter waves, since useful distance gets limited to no more than a few hundred meters. This is the first limitation, which can be overcome to

some extent using sophisticated directional antennas that track the user, or fixed services to homes. Both are based on the concept of providing reasonable power at the receiver. The second limitation is fundamental: There is a general assumption that more bandwidth leads to more throughput. This is true only if Signal to Noise Ratio or SNR > 0, which means signal is high enough to support data rates. However, higher the BW, higher will be the noise floor. Using the fundamental equation for noise, $Pav = kTB$, where Pav is the average power contained in the noise spectrum, k is Boltzmann's constant, T is the absolute temperature in Kelvin, B is bandwidth in Hz. When B = 1 GHz the noise floor will be − 93 dBm (quite high).

Note that this is close to normally receive signal level in most cell phones. There is a point beyond which increase in B will decrease throughput, since SNR for a given transmit power, falls to zero. The throughput to bandwidth ratio Bps/Hz is important but so is µJ/bit (energy per bit) which is a key measurement that shows how much energy is needed from battery to energize mobile devices [35]. Figure 11.32 shows that about 400 MHz of bandwidth is a reasonable compromise that is currently accepted in 5G. Perhaps, the compelling curve in Fig. 11.32 is the red dotted SNR line that drops linearly from SNR = 10 dB to almost zero around 1000 MHz bandwidth. The curve shows RF power—in order to retain maximum spectral efficiency, increase in Peak to Average Power Ratio (blue curve) is needed. Therefore, methods other than just increase in bandwidth must be considered [34].

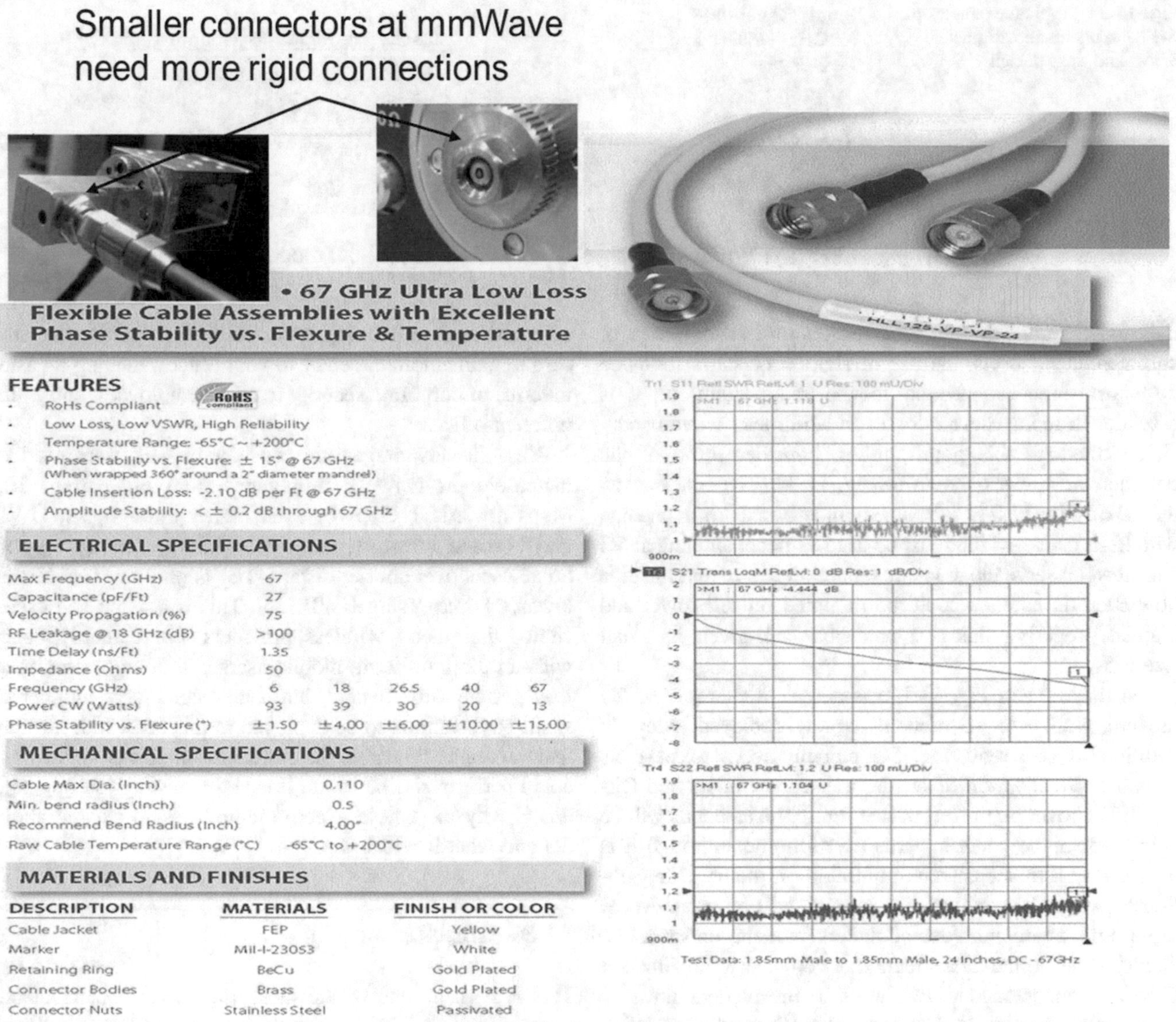

Smaller connectors at mmWave need more rigid connections

• 67 GHz Ultra Low Loss

Flexible Cable Assemblies with Excellent Phase Stability vs. Flexure & Temperature

FEATURES
RoHS Compliant
- RoHs Compliant
- Low Loss, Low VSWR, High Reliability
- Temperature Range: -65°C ~ +200°C
- Phase Stability vs. Flexure: ± 15° @ 67 GHz
  (When wrapped 360° around a 2" diameter mandrel)
- Cable Insertion Loss: -2.10 dB per Ft @ 67 GHz
- Amplitude Stability: < ± 0.2 dB through 67 GHz

ELECTRICAL SPECIFICATIONS

Max Frequency (GHz)	67				
Capacitance (pF/Ft)	27				
Velocity Propagation (%)	75				
RF Leakage @ 18 GHz (dB)	>100				
Time Delay (ns/Ft)	1.35				
Impedance (Ohms)	50				
Frequency (GHz)	6	18	26.5	40	67
Power CW (Watts)	93	39	30	20	13
Phase Stability vs. Flexure (°)	±1.50	±4.00	±6.00	±10.00	±15.00

MECHANICAL SPECIFICATIONS

Cable Max Dia. (Inch)	0.110
Min. bend radius (Inch)	0.5
Recommend Bend Radius (Inch)	4.00"
Raw Cable Temperature Range (°C)	-65°C to +200°C

MATERIALS AND FINISHES

DESCRIPTION	MATERIALS	FINISH OR COLOR
Cable Jacket	FEP	Yellow
Marker	Mil-I-23053	White
Retaining Ring	BeCu	Gold Plated
Connector Bodies	Brass	Gold Plated
Connector Nuts	Stainless Steel	Passivated

Test Data: 1.85mm Male to 1.85mm Male, 24 Inches, DC - 67GHz

Fig. 11.33 Typical cable assembly and rigid connectors for Millimeter wave band

The second option is to consider higher modulation. Higher modulation comes at a cost—in terms of higher signal power needed at the receiver. This was discussed in Chap. 5 and illustrated in Table 5.4. As a practical example, let us consider Table 5.4 at 40 MHz bandwidth with two spatial streams. In Fig. 11.29, receiver (handset) can discern the two orthogonal streams. This involves the option of MIMO or the use of separate streams over the air [35]:

- BPSK provides 15 Mbps at–79 dBm (basic modulation type when received power is low).
- 16 QAM provides 90 Mbps at−67 dBm (higher modulation needs 12 dB increase in received power). This is equivalent of about 15 times increase in power.
- 64 QAM provides 150 Mbps at−61 dBm (much better modulation when minimum received power is very

good). This is equivalent of almost 90 times increase in power.

Other limitations include the use of cable to connect the antenna—at the base station and in cell phone, this becomes an issue. Figure 11.33 shows a commercially available cable assembly [35].

Cable losses become severe at higher frequencies. To illustrate this with an example, the short 3 ft (1 m) flexible cable shown in Fig. 11.33 has 6 dB loss at 67 GHz, which means only ¼ of input power comes out at the other end. In current cellular frequency bands cable loss is small, and it is common practice to use such flexible cables to finally connect to the antenna port.

The consequence of severe cable loss at Millimeter frequencies is a compelling factor in practical deployment,

Fig. 11.34 Typical examples of 5G components in Millimeter Wave and their design

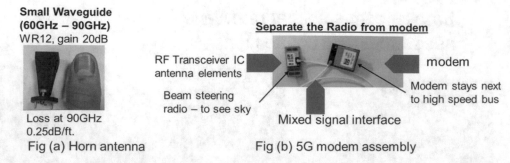

Small Waveguide (60GHz – 90GHz)
WR12, gain 20dB

Loss at 90GHz
0.25dB/ft.
Fig (a) Horn antenna

Separate the Radio from modem

RF Transceiver IC antenna elements

Beam steering radio – to see sky

Mixed signal interface

modem

Modem stays next to high speed bus

Fig (b) 5G modem assembly

measurement, and design. Whether it is for test equipment or commercial handsets, the use of rigid connectors followed by short direct connection to the radio input/output is essential to reduce severe losses in Millimeter wave band.

To illustrate this point further, consider some of the commercial devices shown in Fig. 11.34. The first one on the left, Fig. 11.34a is a waveguide-based horn antenna with high gain—which must be used as panel antenna at 5G cell sites. Observe the dramatic reduction in size and imagine how such an antenna could be mounted on cell tower and aligned properly—this is a precision task given its small size [35].

On the right in Fig. 11.34b is a typical example of 5G modem, similar to 4G modems widely deployed today. In addition to size reduction, 5G modem needs to have its antennas directly printed on the transceiver Integrated Circuit (IC) shown on the left of that figure. Its antennas will be able to steer the beam towards user's home, although it is even difficult to see antenna imprinted on the chip. Integration between RF circuitry and antenna will be tight. At the same time, modulation/demodulation (modem) unit must be placed closer to the data stream that connects to an Ethernet Giga-bit connector. The two units in the modem must be connected using mixed signals such as IF-based modulation signals (not I and Q), DC, or digital control, and a frequency reference signal. This illustrates changes that product vendors must incorporate—it is not an upgrade; it is a complete change in construction of the product itself with precision not perceived in previous generations. Also steering of antenna is not physical, beam steering is done by DSP software.

In parallel, the concept of UWB or Ultra-Wide Band has gained popularity in another sphere—accurate measurement over short distance is, for example, of its application. Using the NFC (Near Field Communication) concept, it has found its way into handsets. Is this a 5G application? Perhaps not —but it does provide highly accurate measurement to locate friends or colleagues in crowded places. These are some of the reasons why service providers are slow in introducing "Ultra-Wide Band" (UWB) in 5G market. It does not mean that UWB will not be deployed, but it should be viewed as not part of 5G standard [36]. But it may offer innovative

solutions such as locating entry points and the use the SIM card to gain automatic entry to your offices, making it easier in terms of corporate security to allow employees, entry into different offices.

Yes, the advantages are considerable—if users need to increase their network throughput from the current 100 Mbps towards 1 Gbps or even higher throughputs, UWB may become technology of choice. Initially, it may support home users over shorter distances by deploying fixed beams towards homes using UWB [36]. This is a quick and easier choice that many wireless providers will certainly begin with. In the long term, mobile users within close proximity may get the same benefit, but it may need large number of small base stations to be deployed in that area. Indeed, they may no longer resemble the current tower-like structures but could perform similar tasks. It is an interesting journey, and this is why the whole telecom industry is so excited about 5G and what it will bring.

11.8 Conclusion

This is perhaps one of the most important chapters of the book. It started with an introduction to "what is mobility" and how does it affect radio signal received by users. Coverage versus capacity, interference, and unpredictable nature of radio wave over the air and how this is controlled to bring predictable, reliable communication to global citizens, was discussed. This was followed by a brief account of cellular systems that began with 1G, mostly following the wired telephone network model. But cellular quickly got transported to digital system with clear break away from wired telephone network that began with 2G in early 1990s. It is considered important to review each generation starting with 2G since GSM, which is 2G technology is in operation in many parts of the world and may continue to serve IoT in large areas. Also, 3G network has remained the backup network to LTE in most countries and remains operational in the foreseeable future.

The competition between TDMA and CDMA as access technologies during 2G and how CDMA gradually asserted its presence as the main technology of 3G was reviewed in detail. While these transformations continued, there were two competing standard bodies 3GPP and 3GPP2 trying to control the specifications of cellular systems. Finally, 3GPP remained as the only body to define 4G cellular systems and established LTE as the single global standard for cellular. 3GPP will remain active as the global body as it brought 5G specifications starting in 2018 and continues today.

Both 4G and current 5G technologies made a significant impact on the global wireless network in terms of keeping a common platform for future, hence this is known as Long Term Evolution (LTE). The concept is to make sure that subsequent generations of cellular systems are completely backward compatible—a major concern in an industry that has seen each generation of network changing every decade. Several new concepts are being brought new technologies—major among them is MIMO that was also used in 4G, TDD which is now deployed in many countries foreseeing considerable growth. Perhaps, the big impact of 5G will be the introduction of Millimeter wave technology and Ultra-Reliable Low Latency Communication (URLLC) that tends to avoid the use of wired backhaul altogether.

The chapter concludes with some of the challenges faced by product developers and service providers in introducing Millimeter wave components at base stations and in handsets. These are short-term challenges with a path forward and have solutions to overcome these challenges.

Summary

Cellular systems remain the fundamental concept in minds of the global population as a technology synonymous with the word "wireless". Although there are many other wireless systems described not only in this book and elsewhere, but this concept is also strongly rooted in one device that global citizens encounter every day—the cell phone. Although, that device performs many other functions and interfaces to other networks such as WiFi, Bluetooth etc., that is not its primary function. Cellular and mobility remain at the heart of cell phone design, which is addressed in detail by this chapter.

A brief introduction to mobility indicated at the beginning of this chapter lays the foundation for such a design. Mobility is perceived by cellular designers as transportation speeds of trains that operate around 500 kmph today. To encounter such speeds, the effects of Doppler shift, fading, and other issues related to the concept of Forward Error Correction (FE) were reviewed—since that is the basis of cellular design.

Concepts of how a cell phone processes calls, how the network is laid out in order to avoid interference, how to track the user discreetly so incoming calls could be delivered, were all concepts that were well laid out during 1G. Since 1G had separate standard in many countries, none of that cell phone could operate outside that country, but the concepts were brought out in 2G designs—helping cellular become a digital network.

Many modifications were implemented during 2G that included improved security, and authentication procedures, and concept of MAHO (Mobile Assisted Hand Off) and also sectored cells. All of these were incorporated by common standard bodies of 3GPP and 3GPP2. While 3GPP continued the path of TDMA, 3GPP2 exclusively addressed CDMA as the technology to move forward. It succeeded in making CDMA the common technology for 3G, but 3GPP had its own version of W-CDMA, and the two systems did not interwork.

This was reviewed in detail to show how 3GPP prevailed as the only global standard beyond 3G, for all cellular systems. Reasons that drove this common standard were related to operation of networks and inconvenience to global customer base in having different cellular systems around the globe. 3GPP also laid the foundation of moving forward with an established customer base and 4G was termed "LTE or Long-Term Evolution". It laid to rest, concerns from a variety of service providers and product developers. All future cellular systems would be backward compatible with 4G LTE.

Each generation of cellular brought technology that is complex and very sophisticated. In terms of security, it is far superior to wired networks, which is why it is paving the way to future communications around the globe. In particular, the details of how 5G is addressing this were discussed, showing three separate radio future capabilities that address different requirements and market segments. All the three will work from the same platform yet address issues that are quite different.

The first one known as eMBB (enhanced Mobile Broad Band) shows an advanced version of LTE. It will incorporate newer technologies such as Millimeter wave, Massive MIMO, and others. The second radio feature, known as URLLC (Ultra Reliable Low Latency Communication) addresses some of the well-known limitations of wireless network—"end-to-end latency". In order to address the growing market for driver-less cars, this technology plans to by-pass traditional IP backhaul and use one wireless network as access and another as backhaul. The third radio feature known as "massive Machine-Type communications", or mMTC, is for IoT (Internet of Things) on a very large scale. These features will be addressed later in Chap. 15 on Vehicular technology, and in Chap. 16, where details of IoT and SCADA will be reviewed.

Since cellular technology has set the pace for fast changes and improvements, this chapter is expected to be in flux for quite some time. It may be best for readers to keep up with the changes using the website: www.3GPP.org. To help reader, some hints of its organization and documentation were offered towards the end of this chapter, in terms of how the standards body 3GPP is organized and how to understand their documents.

Emphasis on participation

- Where are 4G LTE networks deployed in your area? Find out which 3G network acts as backup. Are there 2G networks such as GSM, operating in your area? What are they used for?
- Visit the website 3GPP.org and watch their latest video—based on the video, write a short report on what 3GPP is addressing in 5G, currently.
- Using 3D models in your laboratory, create sequences of antennas using rectangular blocks. With the concepts indicated in Sect. 2.2, how many separate paths over the air can you create? If the receiver moves, how do the patterns change?

Knowledge gained

- Mobility aspects and how they were implemented over the past decade in each generation of cellular, were reviewed to address issues of increasing capacity, improve reliability.
- Access concepts of TDMA, CDMA, and OFDM were reviewed, with examples of how they work in 2G, 3G, 4G, and 5G networks.
- The use of advanced techniques such as MIMO, higher modulation methods, and larger bandwidth of 5G were reviewed in the light of increasing throughput over the air.
- Differences between TDD and FDD were reviewed to show why TDD is acceptable in places where user density demands.

Homework problems

1. Create a call sequence where both users are on cell phones. Create another sequence with desk phone. Compare this with one user with a desk phone. What are the differences?
2. Why are there three sectors in a modern cell? Is it possible to create 6 or 12 sectors using the same cell? How much would be the increase in capacity per cell?
3. Using Millimeter wave band of 30 GHz, it is necessary to create antennas spaced half wavelengths apart as shown in Fig. 11.28. How far apart should they be placed? Provide the precise distance and a procedure to make sure they are exactly aligned.
4. Communicate with measuring equipment vendors such as Keysight, Rohde & Schwarz, or Anritsu. Find out which antenna and cable testers do they use for Millimeter bands? List at least two challenges in making accurate measurements in the 28–39 GHz band (UWB).
5. Using the Eqs. 11.5.4.1 and 11.5.4.2, calculate the following: For a user currently using 5 MHz bandwidth, RSSI measured on power meter is −85 dBm. What would be the calculated RSRP and RSRQ measured on user's smartphone?

Project ideas

- In the lab investigate methods to measure transition delay (latency) while transmitting different types of frames. RFC 5544 of IETF indicates how this should be set up and measured. (Ref: www.IETF.org) Make two measurements for delay and jitter. 1. Using network connected by wires/fiber optic cable. 2. Using wireless network (cellular). If both ends of the cellular network are not accessible, log in to a website such as "Oopla" to see the "Internet speed" or throughput and delay. If possible set up a wireless link of the same length as fiber optic cable/wires. Record the difference in delay and jitter.
- In the project described above, how do you compare the two networks, under varying traffic conditions? Make a table and verify against the URLLC specifications. How much is the gap? List a few applications where URLLC is essential. (hint: If an IP network is set up avoid router/switch if possible).

Out of box thinking

Without waiting for URLLC, try to monitor a simple robot or a keypad remotely, using a campus network. Find methods to increase the network delay and include units such as a solenoid that has a turn on time of 20 mS to simulate situations—consider this as an experiment to control units remotely. The objective is to observe why URLLC is essential for driverless vehicle—why is "low latency" important without which it could become difficult to control the vehicle on the road in real traffic.

References

1. Evolution of mobile generation technology 1G to 5G, by Ms. Lopa J Vora. International journal of modern trends in engineering and research, vol 2, issue 10, Oct 2015. ISSN (online): 2349–9745, ISSN (print): 2393–8161. Semantic Scholar. https://pdfs.semanticscholar.org/4dfd/40cc3a386573ee861c5329ab4c6711210819.pdf
2. What is CDMA–Code Division Multiple Access tutorial, Electronics-notes, magazine. https://www.electronics-notes.com/articles/radio/dsss/cdma-what-is-code-division-multiple-access-tutorial.php
3. The world changing technology that almost wasn't. https://www.qualcomm.com/invention/stories/world-changing-technology

4. The Viterbi Algorithm demystified, by Andrew J Viterbi, Viterbi magazine, Spring 2017. https://magazine.viterbi.usc.edu/spring-2017/intro/the-viterbi-algorithm-demystified-a-brief-intuitive-approach/

5. Chapter 7, Walsh codes, by Thomas Schwengler, University of Colarado, 2018. https://morse.colorado.edu/~tlen5510/text/classwebch7.html

6. Codes used in CDMA, by Mathuranathan, 28 Feb 2011, Gaussian Waves, Signal processing simplified. https://www.gaussianwaves.com/2011/02/codes-used-in-cdma-2/

7. GSM time slot and frequency specifications, RF Café, Data derived from ETSI EN 300 910. http://www.rfcafe.com/references/electrical/gsm-specs.htm

8. Networking fundamentals: Wide, Local and Personal area communications, by Kevah Pehlavan and Prashanth Krishnamurthy. Wiley Publications book, pp 172–174

9. CDMA Radio with Repeaters, Shapira J, Miller S (2007) Section 2.2.3 power control, Springer publications book, pp 30–34

10. CDMA quality and capacity optimization, Adam Rosenberg and Sid Kemp, McGraHill professional book, Feb 23, 2003, p 152

11. Soft handoff in CDMA systems, Daniel Wong, T. J. Lim. IEEE Personal Communications, Dec 1997. citeseerx.ist.psu.edu

12. GSM Channel Equalization, Decoding, and SOVA on the MSC8126 Viterbi Coprocessor (VCOP) by Manoj Bapat, Dov Levenglick, and Odi Dahan. Application note AN2943 Rev 0, 5/2005, Freescale semiconductors

13. Jeruchim MC, Balaban P, Sham Shanmugam K (2002) Simulation of communication systems: modeling, methodology and techniques. Kluwer Academic Publications. , pp 614–621. http://ebooks.kluweronline.com. ISBN 0-306-46267-2. ebook: 0-306-46971-5

14. Rake Receiver, JPL's Wireless communication Reference Website. Section CDMA. http://www.wirelesscommunication.nl/reference/chaptr05/cdma/rake.htm

15. Rake Receiver, by Tommi Heikkilla, S-72.333 Postgraduate Course in Radio Communications, autumn 2004

16. Technical Specification. LTE: Evolved Universal Terrestrial Radio Access (E-UTRA), User Equipment (UE) procedures in idle mode, 3GPP TS 36.304 version 9.1.0 Release 9, ETSI TS 136 304 V9.1.0 (2010–02)

17. Upcoming 2G and 3G global cellular network sunset dates, Mike Bleakmore, Sept 17, 2019. https://www.digi.com/blog/upcoming-2g-and-3g-global-cellular-network-sunset-dates/

18. Ciconski J, Franklin JM, Bartok M, Guide to LTE Security. NIST Special Publication, pp 800–187. https://doi.org/10.6028/NIST.SP.800-187

19. ETSI/SAGE, Specification of the 3GPP Confidentiality and Integrity Algorithms UEA2 & UIA2. Document 1: UEA2 and UIA2 Specification, Version 2.1, March 16, 2009. https://www.gsma.com/aboutus/wp-content/uploads/2014/12/uea2uia2d1v21.pdf

20. 3rd Generation Partnership Project, System Architecture Evolution (SAE): Security Architecture, 3GPP TS 33.401 V12.12, 2014. http://www.3gpp.org/DynaReport/33401.htm

21. Gauravaram P, Millan W (2004) Improved attack on the cellular authentication and voice encryption algorithm (CAVE). In: Proceedings international workshop on cryptographic algorithms and their uses. Gold coast, Australia, pp. 1–13. https://eprints.qut.edu.au/secure/00004701/01/cave_final.pdf

22. 3GPP2 approved documents Library. https://www.3gpp2.org/Public_html/Specs/

23. LTE radio protocol architecture: Control plane and User plane, 2019. https://www.tutorialspoint.com/lte/lte_radio_protocol_architecture.htm

24. 3GPP TS. 29.303 Technical Specification Group Core Network and Terminals, Stage 3, Release 12. https://www.arib.or.jp/english/html/overview/doc/STD-63V12_00/5_Appendix/Rel12/29/29303-c40.pdf

25. Support of SMS and MMS over generic 3GPP IP access, Release 7. 3GPP Technical report, 3GPP TR 23.804 V2.0.0 (2005–05). https://www.3gpp.org/ftp/tsg_sa/tsg_sa/TSGS_28/Docs/pdf/SP-050349.pdf

26. Core network evolution–5G service based architecture, by Karim Rabie, Saudi Telecom Company, Netmanias, Dec 11, 2017. https://netmanias.com/en/post/blog/12967/5g/core-network-evolution-5g-service-based-arhcitecture

27. Part-3: What happens when a user performs a voice call from an LTE/4G network? – VoLTE, November 28, 2016 | By Leonardo Zanoni Pedrini, Telefonica Brasil. https://www.netmanias.com/en/post/blog/10907/lte-volte/part-3-what-happens-when-a-user-performs-a-voice-call-from-an-lte-4g-network-volte

28. Technical specification ETSI TS124 605 V12.4.0, Digital cellular telecommunications system (Phase 2+); Universal Mobile Telecommunications System (UMTS); LTE; Conference (CONF) using IP Multimedia (IM) Core Network (CN) subsystem; Protocol specification. https://www.etsi.org/deliver/etsi_ts/124600_124699/124605/12.04.00_60/ts_124605v120400p.pdf

29. LTE in a nutshell–the Physical layer, White paper TSI 100402–004, by Frank Rayal, CTO, TSI Wireless, Canada. https://frankrayal.com/wp-content/uploads/2017/02/LTE-in-a-Nutshell-System-Overview.pdf

30. LTE TDD network deployments. https://www.qualcomm.com/media/documents/files/lte-tdd-the-global-solution-for-unpaired-spectrum.pdf

31. 6G a generational shift to augment human existence, Peter Vetter, Head, Access and Devices research, Bell Labs. https://www.youtube.com/watch?v=uHABTfdOEPc

32. Meet the challenges of testing eight-antenna LTE, by Kang Chen. Microwave and RF magazine, June 02, 2012. https://www.mwrf.com/technologies/components/article/21844389/meet-the-challenges-of-testing-eightantenna-lte

33. An introduction to LTE-Advanced: The real 4G, by Lou Freznel, Jan 08, 2013, Electronic Design magazine. https://www.electronicdesign.com/technologies/4g/article/21796272/an-introduction-to-lteadvanced-the-real-4g

34. The ABC's of 5G new radio standards, Keysight Technologies, e Books 2000–2020. https://www.keysight.com›assets›ebooks

35. Raghunandan K 5G RAN Millimeter wave initiatives, IEEE 5G tutorial. IEEE 5G Forum, July 9–11, Santa Clara, CA

36. NXP Secure UWB deployed in Samsung Galaxy Note20 Ultra Bringing the First UWB-Enabled Android Device to Market. https://media.nxp.com/news-releases/news-release-details/nxp-secure-uwb-deployed-samsung-galaxy-note20-ultra-bringing/

Abbreviations

WLAN	Wireless Local Area Network, a network based on wireless protocols in some layers and wired or IP protocols in the upper layers
WiFi	Wireless Fidelity, a term commonly used to refer to the IEEE 802.11 series of standards used to create and operate a WLAN
UWB	Ultra-Wide Band refers to a technology that operates within a short distance using low-energy pulses to gauge distance and locate, using very large (>500 MHz) bandwidth
WaveLAN	Commercial name of an early WiFi product developed by AT&T—based on 802.11x standard (a, b, g versions)
PCMCIA	Personal Computer Memory Card International Association, the size of a business card, the first set of products supported WiFi 802.11x used for many years on PC and laptop
CardBus	Current implementation of WiFi 802.11xx within PC or laptop or other products that is not externally visible to users
Bluetooth	A separate standard originally promoted by Ericsson, and now used globally as a wireless technology for short distance communication. It is supported by SIG (special interest group) that supports and promotes all standards and products of Bluetooth technology
LLC	Link Layer Control, one of the sublayers which is part of the data link layer in the seven-layer OSI (open standards interface) model
DSSS	Direct Sequence Spread Spectrum is a spread spectrum modulation technique primarily used to reduce overall signal interference, and also prescribed by regulators. The direct-sequence modulation makes the transmitted signal wider in bandwidth than the information bandwidth, since it spreads the data over a wider bandwidth
FHSS	Frequency-Hopping Spread Spectrum is a method of transmitting radio signals by rapidly changing the carrier frequency among many distinct frequencies occupying a large spectral band. The changes are controlled by a code known to both transmitter and receiver. FHSS is used to avoid interference, to prevent eavesdropping, and to enable code-division multiple access (CDMA) communications
OFDM	Orthogonal Frequency Division Multiplexing is a transmission method to encoding digital data on multiple carrier frequencies (known as subcarriers). The subcarriers are orthogonal and reduce interference. It is used in a variety of applications like digital television and audio broadcast, DSL Internet access, wireless networks, power line networks, WiFi, and 4G/5G mobile communications
ISM band	Industrial, Scientific, and Medical band is part of unlicensed spectrum available to everyone for use. The regulators usually provide guidelines to reduce interference (using techniques like FHSS or DSSS) and also put a limit on RF power transmitted over the air (4 W over the air in 2.4 GHz band)
UNII	Unlicensed National Information Infrastructure with multiple allocations

© Springer Nature Switzerland AG 2022
K. Raghunandan, *Introduction to Wireless Communications and Networks*,
Textbooks in Telecommunication Engineering, https://doi.org/10.1007/978-3-030-92188-0_12

DSRC

in different bands, mainly in the 5 to 6 GHz frequency band

Direct Short-Range Communication is a frequency band specially allocated to support automation in the transport industry (5.850–5.930 GHz). It supports applications that depend upon transferring information between vehicles and roadside devices as well as between vehicles themselves

WAVE

WAVE (wireless access in vehicular environment) is a standard protocol stack which was developed for automobile technology by using a combination of two IEEE standards 1609 and 802.11p. It has specific spectrum allocated to critical safety, control and service channels, and high power for public safety

BLE

Bluetooth Low Energy is a terminology used for devices that use very low energy to transmit as well as use sleep algorithm to conserve energy. BLE devices typically retain a battery for many years using such careful energy management. Most common use is in audio systems

PAN

Personal Area communication Network is a term used for the 802.15.x standard that provide support from the lower layers (physical, MAC, SSCS, and LLC). The upper layers (network and application layers) define type of application such as ZigBee, Bluetooth, UWB, etc. and cover distances of about ~ 30 m or less

Wireless HART

Wireless version of HART (highway addressable remote transducer) is a protocol standard used in automation and industrial control applications

WiMax

Wireless Metropolitan Area Network is 802.16 standard that was popular about a decade ago, but not in widespread use. The concept is meant to support a point (tower) to multipoint (neighborhood) communication for urban and semi-urban communities

Smart Grid Technology

Part of the IEEE 802.24 standard, the Smart Grid Technology initiative by the IEEE has been working since 2012 towards newer methods to make the

power grid efficient and response to changes in the global scene

TDD

Time Division Duplex is a channel usage method that allows either transmit or receive but not both at the same time. This method used by the 802.11 WiFi and many other systems

PTMP

Point-To-Multipoint system uses the same 802.11 standard but instead of AP uses Base Station BS that controls multiple Subscriber Units (SUs) using polling algorithm and operating all its SU units on specific channels. Unlike AP, the BS initiates polling addressing each SU using TDMA protocol

WiFi alliance

It is an alliance of all vendors who make WiFi products or implement them in networks. WiFi alliance-certified devices and their interoperability with various networks

GigE Vision

The GigE Vision standard is a widely adopted camera interface standard developed using the Ethernet (IEEE 802.3) communication standard

Wireless Local Area Network (WLAN) is arguably the most popular form of wireless networks which all urban/suburban dwellers use daily. The simplest way to visualize this is to think of cordless phones. The way a cordless phone was used in the earlier telecom era, WLAN provides similar short distance wireless flexibility to a digital Local Area Network (LAN). That may be an oversimplification, since in digital era, voice is one of the small applications. There are a variety of others based mostly on data. In fact, many perceive WLAN as a method to access the Internet.

In this chapter, we will review the concept of WLAN from mid-90 s and how it has grown into an important technology most wireless users are familiar with. It is useful to review all wireless LAN networks such as WiFi, Zigbee, Bluetooth, UWB, and Z-wave. We can compare them with cellular systems which will be addressed under the topic of "technology at your destination".

12.1 The IEEE 802.11 WiFi Standards

Early work on wireless LAN began in mid-90 s in Bell Labs, Netherlands [1]. An early version of this product was known as "WaveLAN-I" and it was actively supported by AT&T

Bell Labs, with installations in its own buildings. Either desktop computers or laptops had a PCMCIA card inserted into the mother board to provide the WiFi access shown in Fig. 12.1.

Typical image of PCMCIA card on the left (size of a business card) and the Access Point (AP) on the right are shown in Fig. 12.1 and were marketed as WaveLAN-II system. This card (on the left) was issued by corporate network departments to their employees throughout late 90 s and early 2000. WaveLAN-I was an earlier model that system did not get much of commercial exposure. The PCMCIA (personal computer memory card international association) card is now defunct but exists as a chip within each laptop or PC today, many users may only know it as PC card or "CardBus".

This card-based system provided businesses a glimpse of what wireless LAN could do, for example, businesses did not have to install cable conduits to every office suite to extend Ethernet cable. It also provided Internet access in open areas within office buildings such as foyer or walkways. Due to such flexibility, the "bronze" card shown in Fig. 12.1 was soon replaced with a "Silver" card that had improved security implementation and better coverage, followed by the "Gold" card with much better security (128-bit encryption) and higher throughput. Due to such rapid progress as a consumer product, AT&T actively shared details of its design through IEEE standards committee to open up this technology of wireless LAN to the general public [1].

A separate standard under the IEEE banner known as 802.11 series was established in 1999. At that time, 802.3 wired Ethernet was dedicated to IP networks. One of the differences between "802.3 Ethernet" and "802.11 wireless" is the frame size. Size of the 802.3 frame is 1,522 bytes with a data payload of 1,504 bytes. The 802.11 frames are capable of transporting frames with an MSDU (MAC service data unit), payload of 2,304 bytes of upper layer data.

Early work of IEEE 802.11 standards and reference model as shown in Fig. 12.2, comparing it with the generic OSI model for wired networks. Assuming the medium (either wired or wireless interface) as the basis, it is seen that the lower two layers of OSI (open system interface) map to the physical, MAC (medium access control), and LLC (link layer control) layers of 802 model. It shows the scope of work within the IEEE 802 family of standards. The data received on the LLC sublayer consists of the MSDU which is a data payload that includes IP packets and LLC data. The LLC sublayer is the same for the 802.3 (Ethernet) and 802.11 (WiFi). This data frame is now called the MPDU (MAC payload data unit).

Details are available in the standard itself (open and available to the general public). All IEEE 802.xx LAN standards have a common physical sublayer as shown. This helps to build products such as access points with Ethernet connector to physically connect 802.11 wireless to the wired medium with equivalent physical interface of Ethernet 802.3. This is why users of 802.11 air interface perceive it as a short extension (~100 m) of Ethernet cable. Wireless extension assumes there is "no line of sight" that means user device may not see the access point. Now on to a brief overview of the lower layers.

Fig. 12.1 WLAN as PCMCIA card on left in and WLAN-II Access Point on the right

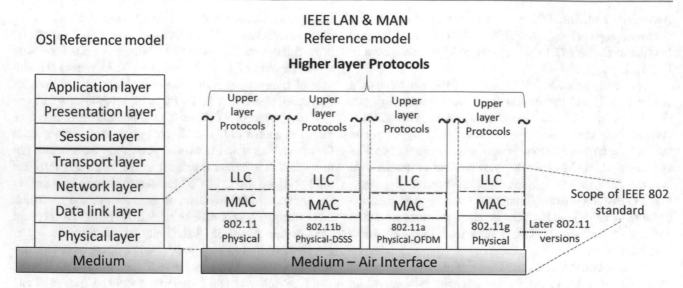

Fig. 12.2 IEEE 802 reference model

(1) LLC Sublayer: Function of the Link Layer Control (LLC) is to manage and ensure integrity of data transmissions, which can be accomplished in three ways:

– Type 1: Unacknowledged connectionless (supported in SRP or secure remote password).
– Type 2: Acknowledged connection oriented—typically wired (e.g., ATM/frame relay).
– Type 3: Acknowledged connectionless—which is typically wireless or packetized (MPLS).

(2) MAC Sublayer: Media Access Control (MAC) that provides protocols interfacing different physical layers using functions as shown below:

– Access control functions and multiple access contention resolution.
– Addressing and recognition of frames in support of LLC.
– Frame check sequence.
– LLC Protocol Data Unit (PDU) delimiting.

(3) Physical Sublayer: The physical layer evolved with the 802.11 standard. The first version of 802.11 began with different types of the frequency band access such as FHSS, DHSS, and IR (frequency-hopping spread spectrum, direct-sequence spread spectrum, and infrared). It was soon clear that different versions of the standard had to be backward compatible. There were clear guidelines by the FCC on what an unlicensed band is and how it should be accessed, to avoid interference. Some of the protocols (FHSS, DHSS, etc.) were suggested by the FCC. This was soon followed by

ITU and almost all countries readily accepted the 2.4 GHz band as "unlicensed". This was followed by allocations by FCC and others in the 5 GHz unlicensed band.

In the North American region, there are other unlicensed bands such as 902–928 MHz and the 5 GHz band under "ISM or Industrial, Scientific and Medical" services. By using the technology of OFDM (orthogonal frequency division multiplex), IEEE 802.11 standards realized that use of higher frequency "unlicensed bands" in the 5 GHz range would derive better throughput. Different versions of 802.11 standard and their key features are shown in Table 12.1.

Based on workings of the 802.11 standards committee, different versions of the standards are being released at regular intervals, with the current 802.11 ax being the fastest (work on "802.11 ay" standard is under way now). Older versions are available, but only 802.11a is not widely available. The 802.11 ax borrowed some good concepts from cellular systems such as spatial reuse, allowing identification of traffic from neighbors known as coloring scheme. This helps each Access Point (AP) to have a different color so that its traffic does not clash with its neighbors. It increases throughput up to 3.5 Gbps. This is designated as WiFi -6 or the sixth generation of 802.11. This has nothing to do with the cellular generation of 6G which is a cellular standard of 3GPP.

The unlicensed bands are classified as UNII (unlicensed national information infrastructure) with multiple allocations in different bands. Due to popular demand and its use in many other standards, these bands cover almost the entire 5 GHz to 6 GHz range. Channel allocation under UNII allows different technologies to design equipment using these channels (see Sect. 12.5).

Table 12.1 Different versions of 802.11x standard

IEEE standard	Frequency/(Bandwidth)	Throughput/Speed	Physical channel
802.11 a	5 GHz, (100 MHz)	Up to 54 Mbps	OFDM
802.11b	2.4 GHz, (83 MHz)	Up to 11 Mbps	FHSS
802.11 g	2.4 GHz, (83 MHz)	Up to 54 Mbps	OFDM
802.11 n	2.4/5 GHz, (40 MHz)	From 300 to 600 Mbps	MIMO/OFDM
802.11 ac	5 GHz, (80 MHz)	Up to 3500 Mbps	4X4 MIMO/OFDM
*802.11 ax	2.4/5 GHz (160 MHz)	Up to 9.5 Gbps	8 X 8 MIMO, OFDMA

*Note For details of channel allocation in the 5 GHz band, please see Fig. 12.15

12.1.1 Some of Other IEEE 802.1x Standards

Why are the other IEEE 802.1x standards an important part of this chapter? The answer lies in the fact that IEEE 802.1x family of standards provide many essential underlying layers found in almost every computing device made today. They are also used in the auto industry and are expected to support autonomous vehicles.

Figure 12.3 shows the DSRC/WAVE standard protocol stack.

Let us consider some examples of such computing devices. Computing device could be a desktop computer, laptop, tablet, or smartphone, or even "automobiles" where DSRC (dedicated short-range communication) standard

based on 802.11p is fitted. This standard was introduced for vehicles to communicate with each other. Let us begin with this specific one, the 802.11p.

Figure 12.3 shows the DSRC/WAVE (wireless access in vehicular environment) standard protocol stack (lower part of the figure) which was developed for automobile technology by using a combination of two IEEE standards 1609 and 802.11p. Upper part of Fig. 12.3 shows the frequency spectrum allocation for 802.11p, in the UNII-3 band of 5.850–5.920 GHz (see Fig. 12.15 for more details). It has dedicated channels for different functions elaborated in the lower part—WAVE protocol stack. These layers are divided into the management plane which consists of WME (WAVE management entity), MLME (MAC layer management entity),

Fig. 12.3 DSRC/WAVE spectrum and protocol stack for 802.11p standard

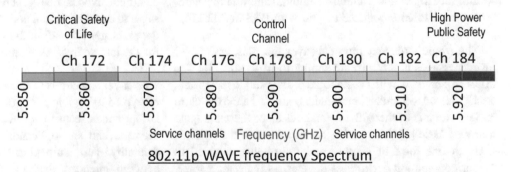

802.11p WAVE frequency Spectrum

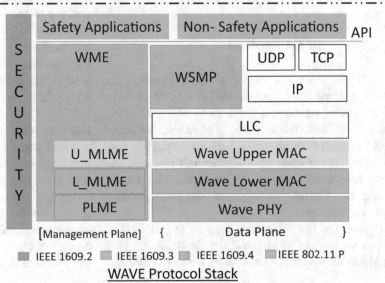

WAVE Protocol Stack

and PLME (physical layer management entity). The and data plane consists of WSMP, TCP, UDP, IPV6, LLC, MAC, and PHY layers. Security is not part of the stack and is common to all of 802. Security is common for all IP networks as laid out by IETF, and not handled in 802 standards.

Functions performed by each of the layers in IEEE 802.11p protocol stack are as follows:

- IEEE 1609–2 (in gray) defines security services for application messages and management messages in WAVE.
- IEEE 1609–3 (in green) defines connection setup and management of WAVE compliant devices.
- IEEE 1609–4 (yellowish green) is above the 802.11p layers. It enables upper layer operational aspects across multiple channels without knowledge of physical layer parameters.
- IEEE 802.11p (in blue) PHY layer handles modulation/ demodulation, error correction technique, etc. Physical layer is modified to support 10 MHz bandwidth, improved performance of WAVE compliant receiver, and improvement in power transmission mask.
- IEEE 802.11p (in blue) MAC layer handles messages to establish and maintain connection in harsh vehicular environment. It also defines signaling techniques and interface functions. Stations communicate directly without need to communicate or join with BSS in 802.11p.

The frame structure, physical layer modules, and MAC layer messages in 802.11p are similar to IEEE 802.11a. The 802.11p standard will be elaborated later in Chap. 15 and Sect. 15.3 on vehicular communications. It behooves us to review variants of the 802.xx standards since there are quite a few of these [2]. Many of them are related to the wireless LAN in one form or another and are often embedded in modern computing or wireless devices, as indicated earlier.

12.1.2   Bluetooth Technology

The concept of Bluetooth was originally conceived by Ericsson as a low-power communication link to connect devices, instead of connecting them using wires. In 1998, Ericsson, Nokia, IBM, Toshiba, and Intel formed the Blue-tooth Special Interest Group (SIG). Subsequently, 1999 was the year of the first release of the Bluetooth protocol; the next year, four other companies joined the SIG group: 3COM, Agere (Lucent Technologies Microelectronics), Microsoft, and Motorola.

In that year, the first Bluetooth headset, from Ericsson, appeared on the market. In March 2002, the IEEE 802.15.1 working group adopted the work done by Bluetooth SIG and made it an IEEE standard. But IEEE no longer maintains the standard. Bluetooth SIG standards is the main body that promotes and maintains this technology. Its current version is Bluetooth 5 which is the latest standard [3]. Bluetooth can be described as a LAN but it uses some unique techniques to extend the range of its low-power devices. Unlike other wireless LAN technologies, Bluetooth uses a layered approach to form and maintain a network.

Figure 12.4 shows the capillary gateway concept in (a) and the system architecture in terms of layers shown in (b). The overall system architecture is shown separately in Fig. 12.5 where how switches, fans, and temperature controller in a typical home are connected using Bluetooth concepts. The capillary concept helps to extend the range and form mesh area network [4]. It is based on the following important functions:

(i)   The publish/subscribe model.
(ii)   Two-layer security.
(iii)   Flooding with restricted relaying.
(iv)   Power saving with "friendship".
(v)   Bluetooth low-energy proxy.

Publish/subscribe model: Exchange of data within the mesh network is described as using a publish/subscribe paradigm. Nodes that generate messages publish the messages to an address, and nodes that are interested in receiving the messages will subscribe to such an address. This allows for flexible address assignment and group casting.

Two-layer security: Messages are authenticated and encrypted using two types of security keys. A network layer key provides security for all communication within a mesh network, and an application key is used to provide confidentiality and authentication of application data sent between intended devices. The application key makes it possible to use intermediary devices to transmit data. Messages can be authenticated for relay without enabling the intermediary devices to read or change the application data. For example, a light bulb should not be able to unlock doors, even if the unlock command must be routed through the light bulb to reach the lock.

Flooding with restricted relaying: Flooding is the simplest and straightforward method to propagate messages in a network using broadcast. When a device transmits a message, it may be received by multiple relays that in turn can forward it further. But Bluetooth mesh includes rules to restrict devices from re-relaying messages that they have recently seen and to prevent messages from being relayed through many hops. Also features like TTL (time-to-live) can be initiated to make sure there are no infinite loops.

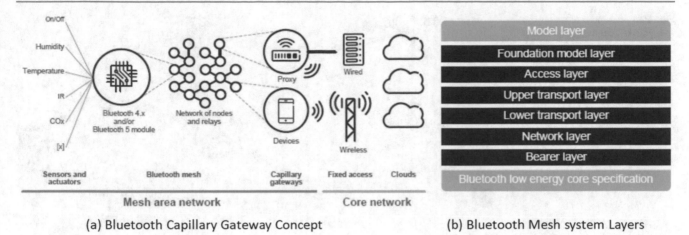

Fig. 12.4 Bluetooth capillary gateway concept and mesh architecture

Power saving with "friendship": Device that needs low-power support can associate itself with an "always-on" device that can store and relay messages on its behalf, using the concept known as friendship. Friendship is a special relationship between a low-power node and a neighboring "friend" node. Friendship is first established by the low-power node. Once established, the friend node performs actions that help reduce the power consumption on the low-power node. The friend node maintains a cache that stores all incoming messages addressed to the low-power node and delivers those messages to the low-power node when requested. In addition, the friend node delivers security updates to the low-power node.

Bluetooth Low-Energy (BLE) proxy: Some Bluetooth devices such as smartphones may not support the advertising bearer defined by Bluetooth mesh natively. To enable such devices within the mesh network, Bluetooth mesh profile specifies a proxy protocol using legacy Bluetooth connectivity, over which mesh messages can be exchanged. It leverages a robust frequency-hopping spread spectrum approach that transmits data over 40 channels. The BLE radio provides developers a considerable flexibility, including multiple PHY options that support data rates from 125 kb/s to 2 Mb/s, multiple power levels, from 1 to 100 mW, and multiple security options up to government grade. Devices that operate on battery and want to conserve battery power spend most of the time in sleep mode, allowing proxy devices to communicate.

Core specifications of Bluetooth Low Energy (BLE) are defined by the seven layers illustrated in Fig. 12.4b on the right. Each layer has functions and responsibilities to serve the layer directly above it, details of which are available in Bluetooth 5 specifications [5]. The network and transport layers are essential for network design and deployment strategies. The network layer handles aspects such as the addressing and relaying of messages, as well as network layer encryption and authentication. The lower transport layer handles segmentation and reassembly and provides acknowledged or unacknowledged transport of messages to the peer device at the receiving end. The upper transport layer encrypts and authenticates access messages and defines transport control procedures and messages. This layer is used to set up and manage the friendship feature, for example.

Figure 12.5 shows an example of Bluetooth where lamps and fans are operating using power from mains. They have bearer channel advertising and relaying information from other nodes as basic features implemented in them. They can be treated as Bluetooth 5 feature-rich devices. The low-temperature sensor is an example of a BLE device that established friendship with neighboring lamp. That lamp is now allowed to relay the message on behalf of temperature sensor.

Consider the cell phone shown in Fig. 12.5 as an older model that does not support the advertising bearer channel features of Bluetooth but establishes a proxy relationship with another lamp which supports advertising bearer communication. The mains powered devices can scan the advertising channels for incoming messages and act as full-fledged Bluetooth devices. Differences between BLE and Bluetooth classic are shown in Table 12.2.

Observe that BLE is able to implement limited features of Bluetooth and also uses lower power levels, allowing its battery to last as long as few years. It has found a niche in the audio space with billions of cell phone headset connections, wireless speakers, light bulbs, and others. Devices such as printers, or loudspeakers and handsfree operation in automobiles are other popular applications of Bluetooth. Most common use of Bluetooth technology is seen in cell phone ear-plug speakers (ear buds). Recent low-energy versions are finding many applications in the medical and sports/fitness world to monitor an individual's physical status.

Fig. 12.5 Example of topology to highlight friendship and proxy features

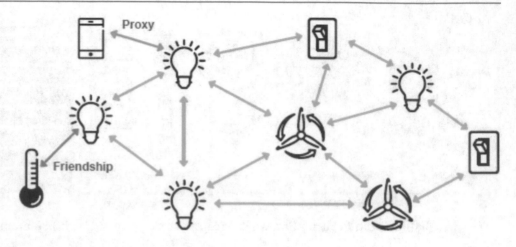

Table 12.2 Comparison of BLE and Bluetooth classic

	Bluetooth low-energy BLE	Bluetooth classic
Frequency band	2.4 GHz ISM Band (2.402–2.480 GHz Utilized)	2.4 GHz ISM Band (2.402–2.480 GHz Utilized)
Channels	40 channels with 2 MHz spacing (3 advertising channels/37 data channels)	79 channels with 1 MHz spacing
Channel usage	Frequency-Hopping Spread Spectrum (FHSS)	Frequency-Hopping Spread Spectrum (FHSS)
Modulation	GFSK	GFSK, $\pi/4$ DQPSK, 8DPSK
Power consumption	$\sim 0.01 \times$ to $0.5 \times$ of reference (depending on use case)	1 (reference value)
Data rate	LE 2 M PHY: 2 Mb/s LE 1 M PHY: 1 Mb/s LE Coded PHY (S = 2): 500 kb/s LE Coded PHY (S = 8): 125 kb/s	EDR PHY (8DPSK): 3 Mb/s EDR PHY ($\pi/4$ DQPSK): 2 Mb/s BR PHY (GFSK): 1 Mb/s
Max Tx power*	Class 1: 100 mW (+20 dBm) Class 1.5: 10 mW (+10 dBm) Class 2: 2.5 mW (+4 dBm) Class 3: 1 mW (0 dBm)	Class 1: 100 mW (+20 dBm) Class 2: 2.5 mW (+4 dBm) Class 3: 1 mW (0 dBm)
Network topologies	Point-to-point (including Piconet) Broadcast Mesh	Point-to-point (including Piconet)

Given the enormous success of Bluetooth, it is expected that SIG will continue to evolve Bluetooth with more features, better security, and other functions over the coming years.

12.1.3 IEEE 802.15 PAN (Personal Area Network) Standards

Popular among the 802 standards is the 802.15 which focuses on low-cost, low-throughput communication between devices. Yet it operates in the same frequency bands (unlicensed). Originally adapted from Bluetooth and known as 802.15.1, this standard is now termed Personal Area Communication Network (PAN). Later variant of this

standard is the 802.15.4 commonly known as Zigbee, which has found another niche area of its own.

Devices operate well within short distances of 10 m to 30 m and use a small battery. "Ultra-Low-Power" output versions operate on very low power (no more than 4mW of transmit power) to conserve battery power, allowing device to be quite small that can last many years using small batteries [6]. The 802.15.4 Zigbee is used for control of streetlamps within a small area, and other devices such as computer mouse, USB dongle, and other multiple port devices. It can also be used for applications such as electrical power meters and smart parking meters.

During major events, locating persons or vehicles in busy urban areas, PAN has become a technology of choice for monitoring and control. In recent years, IoT (Internet of

Fig. 12.6 Coordination of layers and Zigbee MESH built using 802.15.5

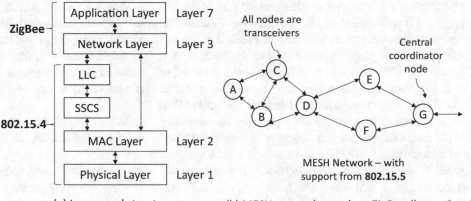

(a) Layers and structure   (b) MESH network – such as ZigBee (layers 3 to 7)

Things) also known as "M-M" (machine to machine) relies on such devices since they are quite adept at sending short bursts of text-based information as needed, using very little power. How can such low-power devices communicate? There are other protocols defined within 802.15 family for specific applications. For example, the 802.15.5 standard defines a mesh networking capability for 802.15.4. This interface standard allows devices to talk to each other forming a reliable MESH network [5]. Let us understand this using layers and network shown in Fig. 12.6.

Figure 12.6 shows that 802.15.4 standard uses only the first two layers plus the Logical Link Control (LLC) and Service Specific Convergence Sublayer (SSCS). These are additions to communicate with all upper layers as defined by additional standards such as Zigbee. The goal of the standard is to provide a base format to which other protocols and features could be added using the upper layers (layers 3 through 7). To support applications such as ZigBee, the 802.15.5 defines MESH network capability that supplements 802.15.4 standard.

Table 12.3 shows the frequency of operation for ZigBee in different parts of the world. The lower frequency bands are useful in applications that need a longer hop distance. Also, it is supported by the basic modulation of BPSK that can operate with very little power. The strength of ZigBee lies in the number of applications that are available to

support this standard. ZigBee has a wide portfolio of applications that we will now review.

While three frequency assignments are available, the 2.4 GHz band is by far the most widely used (see Table 12.3). Most available chips and modules use this popular ISM band. ZigBee supports wireless sensor networks using the MESH topology. ZigBee is also available in a version that supports energy harvesting where no battery or AC mains power is available.

One of the key benefits of ZigBee is the availability of predeveloped applications in upper layers. These upper layer software additions implement specialized uses for ZigBee. Some of these specialty applications include the following:

- Building automation for commercial monitoring and control of facilities.
- Remote control (RF4CE or RF for consumer electronics).
- Smart energy for home energy monitoring.
- Health care for medical and fitness monitoring.
- Home automation for control of smart homes.
- Input devices for keyboards, mice, touch pads, wands, etc.
- Light link for control of LED lighting.
- Retail services for shopping related uses.
- Telecom services,
- Network services related to large mesh networks.

Table 12.3 Options for frequency assignments

Graphical regions	Europe	Americas	Worldwide
Frequency assignment	868 to 868.6 MHz	902 to 928 MHz	2.4 to 2.4835 MHz
Number of channels	1	10	16
Channel bandwidth	600 kHz	2 MHz	5 MHz
Symbol rate	20 k Symbols/s	40 k Symbols/s	62.5 k Symbols/s
Date rate	20 kbps	40 kbps	250 kbps
Modulation	BPSK	BPSK	Q-QPSK

The ZigBee Alliance (zigbeealliance.org) also offers full testing and certification of ZigBee-enabled products to ensure interoperability with hundreds of certified products.

There are other applications that ride on the 802.15.4. "Wireless HART" which is a radio version of the wired HART (highway addressable remote transducer) protocol standard used in automation and industrial control applications. It defines a time multiplexed protocol for accessing multiple nodes of sensors and actuators. International Society of Automation's ISA100.11a industrial control standard is used in process control applications. It adds channel hopping, variable time-slot multiplex options, but uses MESH networking to the 802.15.4 base. Multiple applications such as these ride on the 802.15.4 standard often using 802.15.5 which defines the MESH interface. These will be discussed further in Chap. 16 and Sect. 16.5.

12.1.4   Ultra-Wide Band (UWB)

Popularity of UWB grew in the 2013–2015 time when use of Bluetooth devices to locate users within a certain proximity became common in cell phones. Now part of the 802.15.3z standard, primary focus of UWB is towards knowing position/location of a device with very high accuracy. Unlike triangulation used by GPS location service, UWB uses technology similar to RADAR which requires multiple receivers. Having calculated the time of flight, UWB knows the distance. Having calculated the angle of arrival, UWB knows the direction. By combining those two, UWB can determine the exact position with time of flight and angle of arrival. By performing this operation multiple times, UWB can then determine the speed and direction of travel. What is the distance limit for this new technology? UWB's RF range is 50 to 70 m, whereas Bluetooth's range is only 10 to 20 m for location applications and WiFi is 40 to 50 m. Also, the accuracy of WiFi and Bluetooth drops rapidly as the devices get farther away from the infrastructure, whereas UWB's accuracy is constant whatever the distance. What are the differences? First difference is the wide bandwidth and the second difference is that it uses an impulse for measurement rather long data frames.

To illustrate the advantage of UWB, particularly in indoor locations with large open areas, where user may want to know a person (sensor), or a store or an activity within the facility UWB sensor can locate with much better accuracy. Also, note in Fig. 12.7 the number of radio units (anchors) needed in the huge auditorium is far less than other technologies. UWB needs just four radio units (anchors), while WiFi needs nine and Bluetooth needs twenty five anchors to cover the same auditorium and precisely locate users and guide them.

Although UWB uses a higher frequency band compared to WiFi or Bluetooth (both on 2.4 GHz), the impulse mode allows better range estimation with higher accuracy. It is expected that due to this advantage UWB is likely to find a niche in location services [6]. It is being supported by later release of 5G as one of the important applications.

12.1.5   IEEE 802.16 WiMax

By far the most controversial among 802 standards has been WiMax (wireless metropolitan area network) or wireless MAN (metro area network), which saw considerable activity 10 years ago that gradually subsided. Original intent of 802.16 was clearly for metro area communication—particularly "point to multi point service" in some areas. One base station—typically on a tower—was to serve a community of people in metropolitan or urban area. Its several versions 802.16a, 802.16b, 802.16c, 802.16d, version e, g, k, h, and many others were all either withdrawn or superseded. The latest version of 802.16 released in 2017 is a roll up of versions p, n, q, and s.

Why were there so many versions and where was the confusion? This technology was expected to operate over a very wide frequency range of 10–66 GHz, which is too widespread. Later this was reduced to operate in the limited range of 2.3 GHz–5.8 GHz. This standard was designated to become a competitor to cellular using 802.16e that tried to incorporate mobility resulting in considerable challenge. Field trial of this technology indicated dismal performance [7]. After this failed trial in Baltimore, no more attempts were made to deploy it as a "mobile technology".

This seems to have been due to a short-sighted approach since the physical and MAC layers were not originally designed to support full mobility. Rework of these layers was needed to support mobility, resulting in lack of backward compatibility within different versions of the standard itself. Even after the rework, hand off between cell and many other features normally available in cellular networks could not be incorporated.

It is also a connection-oriented technology trying to bring several sublayers such as Ethernet, ATM, and IP which were encapsulated in its air interface. In short, it attempted to interwork with multiple wireline technologies with its air interface. To date there is insufficient information about future of WiMax. But in standards-based architecture and in WiMax forum, there is some support. Whether WiMax will survive in the long term is a question that remains to be seen. But as a fixed wireless service, whether in metro or in suburban communities, it may be used and has stable performance.

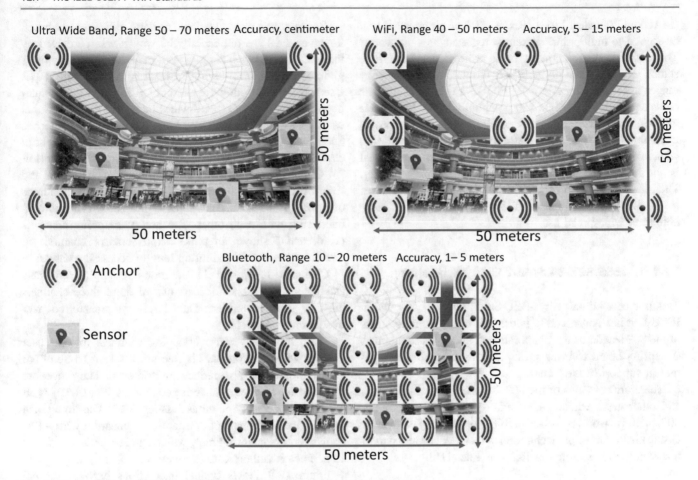

Fig. 12.7 Comparison of UWB with WiFi and Bluetooth technologies: location sensor

12.1.6 IEEE 802.20 Mobile Broadband Wireless Access

This was truly earnest effort to make the 802.xx standard join the mobility market. While the standard was well written and well intended, it never materialized in the marketplace. The last version of this standard written in 2008 remains, but there were no developments since then.

With no further versions of this standard, the concept of WiFi or any of the 802.xx standard becoming truly mobile remains speculative. It is prudent to realize WiFi as a wireless service for fixed environment. Mobility if indicated assumes no more than pedestrian mobility, over short distances of a few hundred meters.

12.1.7 IEEE 802.21 Media-Independent Handover (MIH) Service

There are others that are related to LAN such as 802.21, meant for software development to allow handover. Its implementation in any 802 standards may not increase the cost significantly.

It is possible that when MIH is implemented in network devices, it results in improved performance and possibly interworking between some of the wired and wireless technologies.

12.1.8 IEEE 802.22 Wireless Regional Area Network (White Space)

The IEEE 802.22 is a major activity that started from concepts of University of Cambridge [8] with the objective of cognitive radio as a new technology to search for "white space", which means "absence of broadcast TV channel in an area". The IEEE 802.22 is based on this concept and designated Wireless Regional Area Network (WRAN) to pursue technology of white spaces. The technology could be successful in specific markets. But with many broadcast TV services reducing (due to streaming channels over Internet), this activity is likely to pick up.

There is another potential area for growth—communication in subway or mining or even in remote areas where there is no TV signal. This standard specifies the air interface, including the cognitive Medium Access Control layer

(MAC) and Physical Layer (PHY) [8]. The air interface will be "point-to-multipoint" wireless regional area networks. The network would consist of a professional fixed base station with fixed and portable user terminals. The frequency ranges are based on current VHF/UHF TV broadcast bands between 54 and 862 MHz and potentially in the 1300 MHz to 1750 MHz and 2700 MHz to 3700 MHz [9]. Approval from the local regulatory authority is needed. Typically, what is needed is to show that a particular TV channel signal is not present and obtain approval from the local authority. Then it is possible to use this service. Commercial devices are available and efforts are under way to bring this technology to widespread use.

12.1.9 IEEE 802.24 Smart Grid Technology

This is a new initiative by IEEE Smart Grid, a body within IEEE that has considerable interest in using multiple teams of the 802 standards to collaborate and make this standard as the spring board to bring multiple LAN technologies together in support of the "smart gird".

There are activities of the 802.24 standard including IoT, and other areas where this standard may apply. Currently 802.24.1 is working as a TAG (Technology Advisory Group) listing multiple technologies that can become part of this standard. For more details, please see [10].

12.2 Widespread Use of WiFi and Its Advantages

In the 4G time frame and years since then, WiFi 802.11 has come into widespread use. One of the key reasons for its success is that it operates on unlicensed frequency bands, resulting in public perception that WiFi is "free". In reality, this is not entirely true, WiFi is only a short extension of either cable, or fiber or some other form of wired link that connects users to the Internet through one of these services. It is equally possible that WiFi may connect to a 4G modem which in turn takes the user traffic to the Internet.

But in public parks, stations, airports, offices, and homes, its advantages are felt exactly the way it was perceived by the 802.11 standards. It provides a major advantage to users giving freedom from sitting in front of a desk with a PC. Users have the option of not only using laptop or smart phone, but a host of other devices including digital TV that can directly interact with WiFi and connect to the Internet. Such widespread access is accomplished by thoughtful channel allocation, shown in Fig. 12.8 where the channels of 2.4 GHz band are carefully grouped into "overlapping" and "non-overlapping" channels. This allows literally hundreds of users to use an access point at the same time without interfering with each other. The non-overlapping channels (1, 6, and 7 shown in pink) act as control channels to broadcast basic information on how to access the band.

Channels 1, 6, and 11 are non-overlapping channels available to all users—at least one of these three channels must be used for access. Data is always transferred over overlapping channels.

This allows the non-overlapping channels to help user identify whether a channel is free with "Clear-To-Send" or CTS protocol that also reduces interference. The procedure followed by devices connecting to Access Point (AP) is an essential procedure known as "association". The time taken by a device or client unit to establish connection with AP is known as "association time", which can last between 10 and 15 s, since it authenticates the device.

Figure 12.9 shows typical interactions between an AP and the client (laptop shown is typical). There are three distinct transactions. AP typically sends out beacon regularly. After scanning and recognizing beacon, the client device sends a probe request to see if that AP is able to serve. AP sends back a response stating service is available. The client device acknowledges its receipt.

This is followed by an authentication request. Note that authentication is not part of 802.11 itself and is therefore handled by upper layers not shown in Fig. 12.9, but described earlier in Chap. 5, Fig. 5.7, where the connection to Internet was shown as the basic feature of 802.11 standard. These upper layers are typically part of the wired IP network. After authentication is received from the Internet

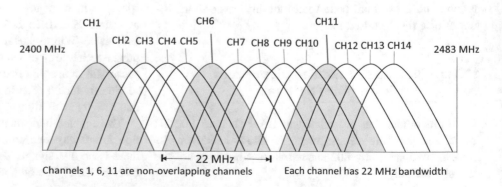

Fig. 12.8 Overlapping and non-overlapping channels of 2.4 GHz band

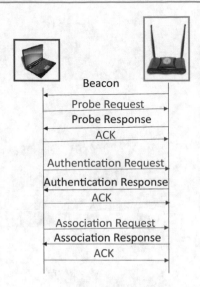

Fig. 12.9 Association procedure between AP and a client device

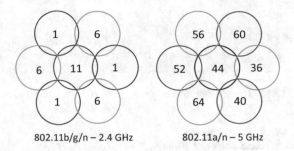

Fig. 12.10 Typical 20 MHz channel selections

switch (could be on premises or outside), AP sends authentication response received from the IP switch. Client device acknowledges its receipt.

Finally, client device sends an association request. The AP sends back an association response, which the client device again acknowledges. After this third and final step, the client device is allowed to use the AP for routing its traffic to the server on the Internet/Intranet as desired. This entire process can take up to 15 s as indicated earlier. Each network can be configured depending on user needs. A home user may have a simpler network where only devices used within that house have access. In contrast, a university network could have different layers where guests may have lower levels of authentication, while those with valid ID (faculty and students) may get higher levels of authentication.

The concept of segregating unlicensed bands into separate channels as shown in Fig. 12.10 simplifies connection. For example, in busy urban areas, multiple access points serve hundreds of customers at a time with adjacent AP units extending coverage over large areas of a city. This is the primary reason why WiFi is so popular—it is simpler to set up and operate. Using schemes of such allocating different channels, it is possible to make sure that all users, irrespective of their location within the area served by AP, will have at least one non-over-lapping channel to serve their client device. In the 2.4 GHz band three "non overlapping channels" are shown on the left side. In the 5 GHz band channels has 24 such channels yet not all of them are used. They are arranged to keep them segregated as shown on the right side in Fig. 12.10. In the 2.4 GHz band, at least one "non overlapping" channel must be used by each user to set up access.

It is pertinent to review the basic mechanism of how a WiFi user communicates with an AP and the network

beyond [9]. In order to operate as a WiFi network, the 802.11 standard prescribes three types of frames:

(a) *Data frames.*
(b) *Control frames.*
(c) *Management frames.*

User information or data is sent using *data frames* shown in Fig. 12.11. Such frames haul data from station to station. Depending on the type of network, several flavors of data frames are used for operations. For example, they can help in channel acquisition, carrier sensing (whether that channel is being used by someone), acknowledgement that data was received, etc. In conjunction with *control frames*, data frames work to deliver data reliably from one station to the next.

The main focus of *Management frames* is to perform supervisory functions. These frames are essential when a user unit joins and leaves wireless networks. If there are multiple access points adjacent to each other (in large public areas) management frame helps to move associations of the user unit from access point to the next access point. In this scenario, any large area consisting of multiple access points adjacent to one another is perceived as one contiguous cell.

Figure 12.12 shows communication between 802.11 Access Point (AP) and typical client devices. These devices include a smart phone, or a laptop or even smart light, smart speaker, or others shown as clients 4, 5, 6, 7. All of these client devices communicate with the AP. Since this standard is based on TDD (time division duplex) radio, it can either transmit or receive, but not at the same time. On the left Fig a is the part where client device transmits and AP receives the frames. On the right in Fig b part where the AP transmits and client devices receive. During transit mode each client sends its Source Address (SA) and Transmitter Address (TA). SA and TA may be the same (in management frame) or different (in data frame). Similarly, in the receiving the Receive Address (RA) and Destination Address (DA) could be different. The BSSID (Basic Service Set Identifier) is a unique identifier of the BSS (Base Station Server), for example MAC address of the AP in an infrastructure network.

2	2	6	6	6	2	6	0 - 2312	4
Frame control	Duration ID	Address 1 (Receiver)	Address 2 (Sender)	Address 1 (filtering)	Sequence control	Address 4 (Optional)	Frame body	FCS

Fig. 12.11 Generic data frame

Fig. 12.12 Communication between access point and client devices

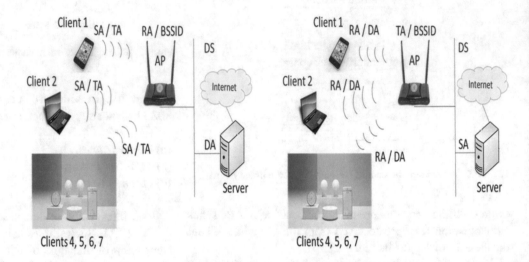

a. WiFi Access Point as Receiver   b. WiFi Access Point as Transmitter

Although the communication is only in one direction at a given time, it happens so quickly that users do not perceive this. The second and perhaps important reason why WiFi is popular is the simplicity of such communication over frequency bands that are free and unlicensed throughout the world. Every AP comes not only with a simple user manual, but also provides a basic page on any PC or laptop that someone who is a basic user can set up and start using the WiFi network at home. It is straight forward to allocate the frequency band desired using such instructions. That does not mean that networks in corporate or field environments are simple. Due to multiple options during corporate networks setup, they need more detailed knowledge. But it is certainly not that complex as setting up a cellular network.

Now let us address connections on the right side of AP shown in Fig. 12.12 (both in Fig. a and b). It shows connection to a server—what is this server? In simple terms, it is a device usually provided by the Internet Service Provider (ISP). Therefore that server could be from a cable network server, it could be from a fiber network server or even a satellite network server. In recent years, a cellular network server using 4G/5G modem (commonly known as cellular modem) could be a server with a switch in it. It is this server that truly connects the WiFi to the outside world or Internet. Without that server the WiFi network cannot communicate with any network outside that building. But it is possible that corporations may use some servers internally for offering

connections to the Intranet or private corporate network. Even those networks would eventually connect to the Internet using a firewall.

To elaborate on protocols, consider both Figs. 12.11 and 12.12 together and understand what is communicated. Security Association (SA) query by the device is a request to AP to grant permission, following which AP performs authentication procedure to verify whether the device is qualified. This was indicated in Fig. 12.9. Authentication procedure is used to avoid certain types of attacks such as spoofing or replay attack by rogue user. BSS (base station server) is a specific signature of that base station or AP. It allows client devices to distinguish this AP by its ID or identification number. This helps in avoiding confusion when multiple AP units operate in the area. This scenario is typical of any city since every residence may have an AP or multiple AP units. When there are multiple servers on different floors of a corporate building, these servers may communicate using bridge mode. These are transactions handled by *management frames*.

The Distribution System (DS) is a flag set by the server showing in which direction the data is traveling. From DS is set to "1" and To DS is set to "0" when the frame is traveling from the AP to the wireless network. From DS is set to "0" and To DS is set to "1" when the frame is traveling from a wireless client to the AP. This correlates well with our previous understanding of how TDD is accomplished. It also

802.11n + 802.11ac data rates and SNR requirements

HT MCS Index	Modulation	Coding	20MHz Data Rate (Mbps) GI = 800ns	GI = 400ns	Min. SNR (dBm)	Receive Sensitivity (RSSI)	40MHz Data Rate (Mbps) GI = 800ns	GI = 400ns	Min. SNR (dBm)	Receive Sensitivity (RSSI)
1 Spatial Stream										
0	BPSK	1/2	6.5	7.2	2	-82	13.5	15	5	-79
1	QPSK	1/2	13	14.4	5	-79	27	30	8	-76
2	QPSK	3/4	19.5	21.7	9	-77	40.5	45	12	-74
3	16-QAM	1/2	26	28.9	11	-74	54	60	14	-71
4	16-QAM	3/4	39	43.3	15	-70	81	90	18	-67
5	64-QAM	2/3	52	57.8	18	-66	108	120	21	-63
6	64-QAM	3/4	58.5	65	20	-65	121.5	135	23	-62
7	64-QAM	5/6	65	72.2	25	-64	135	150	28	-61
2 Spatial Streams										
8	BPSK	1/2	13	14.4	2	-82	27	30	5	-79
9	QPSK	1/2	26	28.9	5	-79	54	60	8	-76
10	QPSK	3/4	39	43.3	9	-77	81	90	12	-74
11	16-QAM	1/2	52	57.8	11	-74	108	120	14	-71
12	16-QAM	3/4	78	86.7	15	-70	162	180	18	-67
13	64-QAM	2/3	104	115.6	18	-66	216	240	21	-63
14	64-QAM	3/4	117	130.3	20	-65	243	270	23	-62
15	64-QAM	5/6	130	144.4	25	-64	270	300	28	-61
3 Spatial Streams										
16	BPSK	1/2	19.5	21.7	2	-82	40.5	45	5	-79
17	QPSK	1/2	39	43.3	5	-79	81	90	8	-76
18	QPSK	3/4	58.5	65	9	-77	121.5	135	12	-74
19	16-QAM	1/2	78	86.7	11	-74	162	180	14	-71
20	16-QAM	3/4	117	130	15	-70	243	270	18	-67
21	64-QAM	2/3	156	173.3	18	-66	324	360	21	-63
22	64-QAM	3/4	175.5	195	20	-65	364.5	405	23	-62
23	64-QAM	5/6	195	216.7	25	-64	405	450	28	-61
4 Spatial Streams										
24	BPSK	1/2	26	28.9	2	-82	54	60	5	-79
25	QPSK	1/2	52	57.8	5	-79	108	120	8	-76
26	QPSK	3/4	78	86.7	9	-77	162	180	12	-74
27	16-QAM	1/2	104	115.6	11	-74	216	240	14	-71
28	16-QAM	3/4	156	173.3	15	-70	324	360	18	-67
29	64-QAM	2/3	208	231.1	18	-66	432	480	21	-63
30	64-QAM	3/4	234	260	20	-65	486	540	23	-62
31	64-QAM	5/6	260	288.9	25	-64	540	600	28	-61

Fig. 12.13 MCS data rates for 802.11n and related standards

indicates how transactions shown in Fig. 12.9 are accomplished. Also, Receiver Address (RA) and Destination Address (DA) are used within the data frame shown in Fig. 12.11. It is important to understand that the DA is not likely to be the server, but the actual destination of a website or a mail server that user wants to send data to. This is typically filled by interacting website or mail server site which first sends data frames to the end user device. Subsequent transactions logically follow the interchanging of such address during the "From" and "To" showing the direction of *data frames*. The number of frames, with MIMO streams and at what signal level, which rate is supported is

elaborated in the following MCS table which was originally developed for 802.11n standard but is a useful indicator.

The usefulness of this table in Fig. 12.13 is readily observed. For example, in MCS (modulation and coding scheme) at an index of 15 (first left column), to support 64 QAM and 5/6 coding related 300 Mbps data rate, the signal strength received must be at least -61 dBm (strong signal). However, if the signal drops to -79dBm (low) then only BPSK modulation may be supported, with only 30Mpbs (one-tenth the throughput). Throughput dependency on signal level is clearly illustrated by the table—a very useful tool for any potential designer. The Guard Interval (GI) does play an important role, it changes the throughput rate. A longer GI of 800 ns provides slightly more reliable but lower throughput. Observe how SNR (signal-to-noise ratio) plays an important role in the throughput. Generally accepted SNR = 12 is considered "Good" for voice calls, but lower SNR may be acceptable when signal levels are low.

It is not practical to review every type of interaction in WiFi networks. There are textbooks, the 802.11 standard (available free to everyone), to describe the basic process and how each AP and client device can be set up [8]. Work on 802.11 is an ongoing effort in IEEE and many of the earlier standards published are freely available to the public. Using the options provided it is possible for home network or corporate network provider to set up the network. In general, almost everyone uses the authentication procedure with a basic password as a point of checking.

12.3   Why is WLAN the Technology at Your Destination?

So far, the WiFi, Bluetooth, Zigbee, UWB, and similar WLAN design and operations were discussed. But a fundamental question to be addressed is why should these be a technology at your destination, why not throughout your travel route, along a highway or a rail route or any other route? This question has come up repeatedly within the wireless LAN community and attempts were made to project it as an alternative to cellular, so that it becomes synonymous with everything in wireless technology. After all, it provides enormous throughput, easy to setup, etc. We will use WiFi as just an example—similar arguments hold for Bluetooth, Zigbee, and others.

Let us examine some of the basic concepts and understand why cellular and other wireless technologies are still in service. A cursory glance at Fig. 12.12 shows that the server and rest of the network it connects to is wired networks. The only part that is wireless is the air link between AP and the client device. How far is this air link? Normally 100 m in order to maintain the channel selection shown in Fig. 12.9, this is typical. In order to cover a city, it needs thousands of AP all connected to one controller, in order to maintain continuous movement (mobility). If the user is pedestrian, or may even use a bicycle, this may work. But using even a car, the user could travel at 120kmph, which translates to about 2000 m or the equivalent of 1000 circles of Fig. 12.9 in 1 min.

It is obvious that even a large WiFi network will find it impractical to handle that level of mobility. High-speed trains that move above 500kmph are completely out of question. There are a variety of other challenges that mobility brings (Doppler effect, losing frames, to compensate that special FEC design is needed, etc., discussed in

Chap. 3, Sect. 3.2). Overall, the conclusion would be to use the network for what it is designed for and not attempt things it was never meant to do. This was explained in 12.1.4 and what happens if we impose mobility on WiMax network can be seen in [7]. It is prudent to use WiFi or Bluetooth or ZigBee network for the purpose it is designed—high throughput in fixed situations where there is little or no mobility. There are enormous benefits in using any WLAN at destination—many home users readily observe these benefits.

Primary among them is use of smart phone as a controller for many devices shown in Fig. 12.12. This makes it easy to not only watch a camera or listen to a radio station, or dim a light, but even remotely control them. While at the destination (home or office) there is really no need to use cellular since it is primarily meant for mobility. It reduces cellular bills and to the cellular provider it frees up channels that they can use to serve other customers.

With the advent of online radio and TV streaming services, bringing programs over WiFi has become very popular [11]. Since the ISP (Internet service provider) moves the traffic to the Internet, use of WiFi allows combining multiple services over the incoming Internet connection. That includes TV, radio, telephone, and many other services—all can be programmed or monitored according to individual requirement. Although these services may come from different sources, WLAN becomes the gateway from your home or office, to multiple devices. Fig. 12.14 shows on the left multiple technologies such as cordless phone, bluetooth, 802.11 WiFi, Wireless USB and Zigbee technologies. On the right in Fig. 12.14 is WiFi and how co-exists with Bluetooth. WiFi uses the common channels of 1, 6 and 11 (wideband in blue) and the operating channels of Bluetooth (in grey) shift based on activity in WiFi. The operating channels of Bluetooth can move dynamically, to use operating channels that are free at that time. This shifting is shown by dotted lines. Similar actions are taken by other technologies operating in the unlicensed 2.4 GHz band.

Since there are so many WLAN technologies in any home/destination, won't there be interference? Yes, there will be—that is why all of them implement schemes to first scan, monitor, and then choose a channel where no signal is present. Such controls take different forms of hopping sequence algorithms, but eventually every one of the WLAN technologies do operate well due to such interference control by design. Occasionally if the microwave oven is turned on (it also operates in the same 2.4 GHz unlicensed band), some device nearby (such as cordless phone) could experience interference. But that will pass once the microwave is turned off. Cellular does not have this issue since service providers buy licensed channels at considerable cost and perform interference studies prior to installation of the network.

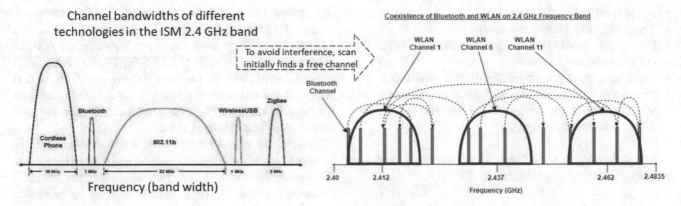

Fig. 12.14 Various WLAN technologies in the unlicensed band—control of interfere

In that case why not use WLAN within the transport such as bus and train. While this is truly the approach of many, it is important to take a step back, and think. From the bus or train how does WLAN connect to the Internet? Referring to Fig. 12.12 the server is located within the vehicle. From there, connection to the Internet has to be through either a cellular connection or a satellite connection, because there is no cable/fiber going out of the vehicle. Compared to WiFi both of these streams are slower. The situation is similar to everyone on your bus or your rail car using just one big cell phone. The result will be a slower throughput for each user. Satellite streams are generally slower than cellular since satellites are farther away compared to cell towers. Therefore, use of WiFi in transport (although tried out extensively) is no longer popular. 4G/5G data stream directly to your phone is likely to be better in terms of throughput and performance.

12.4   Security Challenges and Improvements

There is a common perception that WiFi is not secure. Is this really true and to what extent? If we strictly go by the definition of WiFi being a technology as defined by the 802.11 standard,

then the standard clearly mentions that security is not addressed in that standard. Authentication indicated in 12.2.2 really takes place in the upper layers, such as the "Sessions layer".

But that does not mean that other layers are not vulnerable. The physical layer, for example, can be secured by biometric authentication, shielding, or other locking mechanisms. These provide some resiliency to "Denial-of-Service" attacks [12]. Similarly, data link layer has its own vulnerabilities in terms of spoofing the network interface and identifying MAC address which are some of the techniques potential hackers try. Strong firewall rules try to protect the network and transport layers. Even in this case

there are several methods that potential hackers try to use. Some of the common methods of attack are: port scan, network traffic flood, malformed network packets, and IP spoofing.

WiFi devices (AP and others) usually have WPA2 or WiFi-protected access protocol, the first version of which was easily broken. After much effort WPA2 was brought out, which also had its own limitations. The more recent version WPA3 with 128-bit encryption is better but still has some weaknesses [13]. Efforts to improve security of WiFi at the higher layers of IP network continue, but so are efforts by hackers who seem to be successful in breaking almost any fix that is installed. List of such fixes were shown earlier in Chap. 6, where Table 6.2.2, Summary of WiFi security over a decade (2004–2018), is a reminder that this will be an ongoing exercise.

Cellular technology, in general, has a more systematic approach towards security. With 4G and now 5G multi-layered approach to security ensures the hacker can at best have limited ability to even attack the network. Hence, financial institutions often recommend cellular, but not WiFi.

12.5   Licensed and Unlicensed Bands Used by WiFi: A Comparison

The general perception that WiFi uses unlicensed frequency band is true most of the time, but there are some exceptions. For example, the 4.9 GHz band (4940–4990 MHz) represents 50 MHz bandwidth licensed by the FCC for public safety usage in USA [14]. This band is used by public safety agencies for a variety of uses such as.

- Wireless LANs for incident scene management (ad hoc mobile networks).
- Mesh networks.
- WiFi hotspots.

- Voice over IP (VoIP).
- Temporary fixed communications.
- Permanent fixed point-to-point/multipoint links that deliver broadband service.
- Permanent fixed point-to-point video surveillance.
- Permanent fixed point-to-point/multipoint backhaul links of broadband traffic originating from 700 MHz public safety broadband networks.

The following types of uses have secondary status to primary uses of the 4.9 GHz band:

- Permanent fixed point-to-point/multipoint links that deliver narrowband traffic.
- Permanent fixed point-to-point/multipoint backhaul of traffic originating from public safety bands not designated for broadband (i.e., public safety VHF, UHF, narrowband 700 MHz, and 800 MHz) to other networks.

Are there differences compared to the unlicensed bands? Yes, they use the same 802.11 protocol, but operate in a licensed band. WiFi devices used by general public (such as smart phone, laptop, etc.) will not work in this band. There are devices specifically developed for this band. Each of the public safety agencies must separately register every device on the network in order for it to authenticate on the network [15] following the standard licensing process. Since it is licensed, there is considerable interest among public utilities to use this band to join the "smart grid" for example. Earlier, Sect. 12.1.8 briefly reviewed the new group IEEE 802.24 that is working on smart grid.

There are other unlicensed bands used by private and public agencies. What are these and how do they differ from the popular unlicensed 2.4 GHz band? UNII bands (Unlicensed National Information Infrastructure) specifically address these. They have considerable bandwidth and are grouped as shown in Fig. 12.15 [16]. General designation indicates terms widely known by many and IEEE 802.11 standards committee uses channel numbers for devices that follow their standard [17]. Each of these frequencies have considerable bandwidth as shown.

Limitations exist—for example, weather radar could be present in some areas where it will be necessary to sense and stay out of those channels in that area. Also, the Dynamic Frequency Selection (DFS) which is used to efficiently combine channels in some areas may not be always be exercised to combine all the 80 + 80 MHz to make up 160 MHz bandwidth (aggregation). Then 802.11ac, for example, may provide maximum throughput in a very limited area [18] (see Fig. 12.15) for details of channel allocations in the 5 GHz band.

For all unlicensed bands, limitations are imposed in terms of maximum emitted power "over the air". For example, 2.4 GHz has an EIRP limit of 4 W "including antenna gain". That means the transmitter cannot send anything more than 1 W at it output. With an antenna gain of 6 dB or 4 times the input, the maximum "over-the-air" power must be limited to 4 W. This is indicated as EIRP or Equivalent Isotropic Radiated Power. For example, in Fig. 12.15, UNII-1 has power limit of 50 mW with an EIRP of 200 mW meaning power "over-the-air" is limited to 200 mW (antenna gain of 6 dB max is allowed, similar to 2.4 GHz band). UNII-2a and 2c allow 250 mW from transmitter, with an antenna gain of 6 dB. These are indicated at the top of Fig. 12.15 as Max Power with EIRP shown in parenthesis.

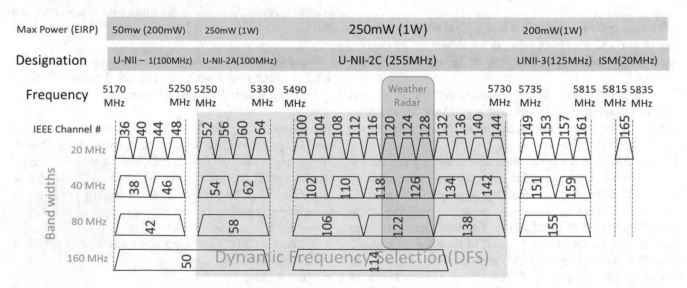

Fig. 12.15 Unlicensed frequency bands in the 5 GHz range

Similar limits to control interference are also set by ITU. Due to considerable bandwidth aggregation of channels is allowed, using which the 802.11AC has brought even higher throughput rates into marketplace [19]. Resolution 229 of ITU Regulations specifies the restrictions for use of the 5 GHz spectrum band as follows:

- 5150–5250 MHz: Sets power limit and stipulates indoor-only use.
- 5250–5350 MHz: Sets power limit, specifies use of DFS9, requires administrations to take measures to ensure the predominant number of RLANs are operated in an indoor environment (with antenna masks for those stations permitted to operate outdoors), transmitter power control or reduction of permitted power by 3 dB.
- 5470–5725 MHz: Sets power limit, DFS, transmitter power control, or reduction of permitted power by 3 dB.

The Common Regulatory Framework comprises the Framework Directive (Directive 2002/21/EC), the Authorization Directive (Directive 2002/20/EC), the Access Directive (Directive 2002/19/EC), the Universal Service Directive (Directive 2002/22/EC), and the Directive on privacy and electronic communications (Directive 2002/58/EC), as amended by the Better Regulation Directive (Directive 2009/140/EC) [20, 21].

12.6   Point-To-Multipoint Systems (PTMP): Methods to Implement Security

So far, we discussed Wireless Local Area Networks (WLAN) based on Access Points (AP). In Fig. 12.16, there is an example of different type of system that uses a Base Station (BS) instead of an Access Point (AP). Normally in WLAN, AP serves everyone in immediate neighborhood. It normally uses omni-directional antennas to spread energy uniformly in all directions up to a distance of about 100 m, which is true of an Access Point (AP). The PTMP in contrast, uses "directional antennas" so that only "subscriber units" within its beam are addressed by the Base Station (BS). There are other differences as described below.

Base Station Unit (BS) shown in Fig. 12.16 works with a TDMA access scheme and uses directional beam. Unlike typical user devices (such as smart phone, camera, etc.), it uses subscriber units (not WiFi client such as smart phone or camera) that are physically directed towards BS. Such subscriber units can only respond to commands from that base

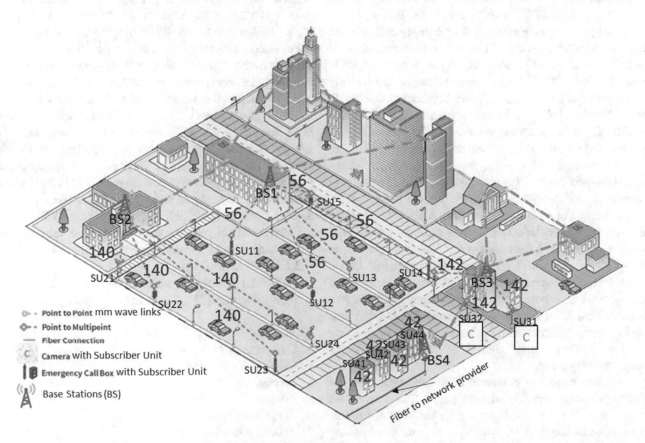

Fig. 12.16 Typical city block with point-to-multipoint communication system

station and therefore cannot "access" the base station on their own. In this figure, BS1 serves about five parking meters while BS2 control four parking meters/emergency call boxes in an adjacent lot. BS3 controls perimeter cameras for the building on right bottom. BS4 serves four apartment residents for carrying their Internet traffic. The green dotted lines indicate wireless beams from each BS unit.

How do such systems work? The base station works as the master controller in this network and "polls each of the Subscriber Units" (SU) at regular intervals. During an allocated "time slot" each BS commands its own SUs using specific channels in the 5 GHz band. In response to the command, SUs send data to their BS. Since BS1 and BS2 are on adjacent city block buildings it is essential that their channels are kept considerably apart (to avoid interference) using beams of 60 degrees or 120 degrees depending on coverage area needed. Similarly, BS3 and BS4 are close to each other and their channels must be kept apart—but they both handle considerable traffic and need channels with larger bandwidth.

Note that each SU in Fig. 12.16 is designated by the BS. For example, BS3 communicates with SU31 and SU32. Each type of client device, whether a meter, a camera, or a call box are all directly integrated with its own subscriber unit.

Channel numbers indicated earlier in Fig. 12.15 show allocation of smaller 20 MHz channels for BS1 and BS2 since they serve parking meters and emergency calling booth that need smaller bandwidth. Larger bandwidth channels of 40 MHz are allocated to BS3 and BS4 since they get traffic from cameras. Each BS uses channels that are deliberately kept apart from another BS to reduce interference among BS units. Finally, traffic from these points can be carried (backhaul) to the network provider, using point-to-point wireless links indicated by yellow dotted lines. These dotted lines can be microwave links, they may use highly directional beams in the 5 GHz range, (channel 114, for example) directional antenna of very narrowly focused beam.

The impact of such design on "smart cities" with green infrastructure is considerable. Notice minimal use of fiber only on the roof of large building and the cable provider to the apartment complex. Other than these two, wired infrastructure has been avoided in this design. Currently, such deployments have become common in many cities, for two reasons. First is, of course, marked reduction in labor cost (laying fiber), second is the simple method of carrying backhaul traffic avoiding conduits in buildings or digging streets to carry cable.

An added benefit to cities and corporate bodies that use such network is "security and efficiency". This network no longer relies on the OSI layer for authentication. Instead, manufacturers introduce an extra layer of protection by using an outdoor router protocol. Conventional WiFi router uses

CSMA/CA (carrier sense multiple access with collision avoidance) access protocol, which reduces throughput. In this case, due to TDMA and the base station initiating conversation with subscriber units, that burden is avoided, bringing much higher throughput within the same bandwidth.

Such specific protocols require a Subscriber Unit (SU) to register with the Base Station Unit (BSU), using an MD-5 secret string in the authentication process. Without the appropriate MD-5 key, rogue subscriber units cannot join the network [22]. MD-5 is not an encryption but it is a one-way transaction that is almost impossible to reverse engineer an MD-5 hash to retrieve the original string. Many manufacturers implement up to 256-bit AES encryption on all data sent over the wireless link.

Such sophistication in design results is secured communication even though units operate on unlicensed bands. The term unlicensed therefore need not result in insecure system. Earlier, in the chapter on security, Fig. 6.5.1 showed how BS units at different heights and directions can completely isolate its communication and keep it quite secure. What is needed is innovative thinking in terms of design that takes advantage of unlicensed bands and its wide bandwidth.

An AP can be configured as a "Wireless Bridge". It is a mode that AP can be set, in order to connect two separate WLAN networks. Therefore, it typically connects two AP units which act as master controller of that local WiFi network. Bridge mode allows two wireless Access Points (APs) to communicate with each for the purpose of joining multiple LANs. Some wireless bridges support only a single point-to-point connection to another AP. Others support point-to-multipoint connections to several other APs.

Finally, let us touch upon an important development of WiFi known as the WiFi alliance [20]. When the IEEE already sets standards and provides technical details why another alliance? The reason is the far-reaching popularity of WiFi where over 18 billion devices were sold in 2018 alone! There has to be a global body that verifies testing of such devices, their compatibility, and interworking. WiFi alliance is a global body of vendors who not only build the hardware, but also those who provide applications that ride over the 802.11 WiFi networks.

WiFi alliance currently certifies products in a broad range of eight categories [23]. These are.

1. Building.
2. Computer and accessories.
3. Gaming, media, music.
4. Phones.
5. Routers.
6. Smart homes.
7. Tablets, eReaders, Cameras.
8. TV and set top boxes.

There are currently 12 laboratories across the globe to test and verify devices/software that comply with their specifications [24]. In addition to making free telephone calls over WiFi ("WhatsApp" is a good example), WiFi also provides location services, and "direct connect" which means devices can connect to each other without the need of an access point.

For example, during construction of buildings—contractors, builders, corporate personnel, and field workers can talk to each other and send messages/pictures, without the need to have cellular service. Such infrastructure is easier to build and dismantle (at the end of the project) and has become popular in areas such as basements, mines, and other locations where cellular service is either unavailable or intermittent.

WiFi alliance also certifies devices that operate on both the 2.4 GHz and 5.8 GHz bands that are often used in the same package—such as routers, phones, tablets, and cameras [24].

12.7   Conclusion

This chapter reviewed one of the most popular wireless technologies—that general public knows as WiFi. Various standards under the general umbrella of 802.xx were reviewed showing how all of them interwork well and communicate well with wired system using 802.3 the well-known Ethernet standard of the same LAN family of IEEE standards.

The 802.xx family consists of not only 802.11 WLAN but others such as WiMax, white space communication, and others with good potential and their own LAN capabilities. These were reviewed in order to indicate a family of wireless standards all of whom connect to the wired network using Ethernet. Ethernet primarily connects using CAT 6—Category 6 cable or fiber-optic cable both of which interface to a variety of speeds with connectors such as 10baseT, 100baseT, and 1000baseT, 10000baseT (commonly known as GigE).

The importance of these standards can be felt in terms of their complete dominance in the wired networking industry that eliminated T1 or E1 interfaces of the telecom network. Dominance of the IP network that provided ease of connection to all of these standards that continue to evolve and expand into WLAN and other wireless networks.

Summary

This chapter focused on basic interworking of WiFi and related wireless networks. Although individual manufacturers may use specific versions either to enhance throughput or improve security, "unlicensed frequency band" remained the common theme of this chapter. Some government agencies prefer the licensed version of WiFi, which was reviewed.

Local Area Networks (LAN), being the first level of contact with the user, offer an extremely powerful view projecting the image as the only "wireless" network. While this is not true, such perceptions exist in both the networking community and among general public.

Basic approach to unlicensed band, use of non-overlapping channels, offered an approach that made WiFi a global standard. Irrespective of the region, every country uses the 2.4 GHz band uniformly, allowing the smart phone (which has both cellular and WiFi) to be readily used. The protocols, frame structure, and how it relates to the upper layers of OSI model were reviewed to provide an understanding of how these closely relate to the IP network. Why WiFi has become the technology at destination (office, home, etc.) was discussed with the intent of how different applications use WiFi in home and office environments. This is probably the most powerful reason why it has become popular among the global user base.

Variety of other LAN family, such as PAN (personal area), smart grid, WiMax, RAN (regional area), or cognitive networks, have all their own advantages and serve different aspects of wireless networks. These were reviewed as part of the 802.xx family of standards. Many of these products often interwork with WiFi and may complement some of the capabilities—Bluetooth is a typical example since it is available along with WiFi in phones, in cars, and computers, for example. Similarly, Zigbee is often available in monitoring applications.

"Point-to-Point (PTP)" and "Point-to-Multipoint (P2MP)" network offer a different perspective of wireless LAN useful to cities and corporate bodies. This was reviewed since it is technically of similar category but uses different protocols with better security. In some cases, these serve as a complement to wired networks in urban areas. This will be reviewed later in Chap. 15 where SCADA is discussed.

Emphasis on participation

1. Smart phones can operate in the 5.8 GHz band. Is it possible to operate them exclusively in this band? Work with the campus network administrator to implement a separate network where specific applications operate only on the 5 GHz network, leaving 2.4 GHz WiFi for general use.
2. White space radio is becoming popular—using spectrum analyzer identify TV channels not operating in the area. Provide a write up to the local authority to allow operation in this band. Set up a point-to-multipoint network in

the white space band and identify some potential applications.

Knowledge gained

There are several key points of understanding that emerge from this chapter.

1. The IEEE 802.xx standards of which both Ethernet and WiFi are parts. Members of the same family of standards include Zigbee, Bluetooth, cognitive white space, WiMax, and others.
2. Unlicensed band, its popularity, and how it is controlled and managed by devices without intervention from any regulatory body. Some of the guideline regulators provide and how vendors realize it is in their own interest to follow these guidelines.
3. Use of WiFi in a variety of applications and the need for WiFi alliance. How the standards continue to evolve but retain backward compatibility.
4. Identification of bands for unlicensed spectrum, use of non-overlapping channels for global usage, and recognition by devices.
5. Point-to-point and point-to-multipoint networks—how they are part of LAN but operate differently. Special efforts to make them more secure than conventional WiFi networks.
6. Use of TDD (time domain duplex) and protocols to control interference. Use of WiFi 802.11 standard only for the first two layers, while maintaining connection with other layers of OSI model-based IP networks.

Homework problems

1. Use an access point at home—measure the received signal level at different locations. Turn the AP on its base and observe variations. What do you see as the pattern in the X-plane (parallel to floor) and the Y-plane (vertical to floor). Smart phones operate on the vertical plane—how would you increase coverage area?
2. What version of 802.11 does your phone and AP use? How many antennas do they have? Find out how many MIMO streams are needed for 802.11n and 802.11ac? Is it practical to get that many streams from AP to the phone? How far would the high throughput be sustained?
3. Using Bluetooth mode on your phone compare the maximum distances it can support for a loudspeaker, for another Bluetooth device. What is the maximum distance specified for PAN. Does your phone Bluetooth meet this distance?

4. Assume the area of Fig. 12.16 is 900 m X 750 m. How does the base station operate over distances beyond 100 m? What is the maximum distance it can support if they operate on the UNII-3 band indicated in Fig. 12.15?

Project ideas

- Set up a point-to-multipoint network from your department to the farthest block possible within the campus. Using only wireless links for point-to-point (PTP)which frequency band would be use for this link? If PTP and P2MP are on the same/overlapping bands, will there be interference? Can the interference be avoided?
- Integrate a few devices that belong to different standards of 802.xx such as Zigbee, Bluetooth, and WiMax.

Out-of-box thinking

Discuss with your university/college administrator and find out how much of the network uses cable/fiber-optic network. Is it possible to reconfigure the entire campus network using P2MP and PTP network? How much would be the cost reduction? Are there any changes in network operations and maintenance?

Can this be followed up by using a WiMax link from campus to the local ISP (Internet service provider)? Would they be willing to consider such a proposal?

References

1. Kamerman A, Monteban L (1997) WaveLAN II—a high performance wireless LAN for the unlicensed band. Bell Labs Techn J Summer
2. 6LowPAN Vs Zigbee—two wireless technologies explained, Published Nov 20, 2014. LinkLabs. https://www.link-labs.com/blog/6lowpan-vs-zigbee
3. Bluetooth core specifications, Version 5 feature overview. https://www.bluetooth.com/bluetooth-resources/bluetooth-5-go-faster-go-further/
4. Bluetooth mesh networking Ericsson White paper, 284 23–3310 Uen | July 2017. https://www.ericsson.com/en/reports-and-papers/white-papers/bluetooth-mesh-networking
5. What's The Difference Between IEEE 802.15.4 And ZigBee Wireless? Lou Frenzel, Communications editor, Electronic Design magazine, Mar 22, 2013. electronicdesign_10481_differencebetweenieee80215andzigbeewireless.pdf
6. Ultra-Wide band for dummies, special edition by Quorvo.com. https://www.quorvo.com
7. "Worst of the Week: A bold look back at the Sprint WiMAX network", by Don Meyer, RCR Wireless News, April 1, 2016. https://www.rcrwireless.com/20160401/opinion/worst-week-bold-look-back-sprint-wimax-network-tag2

8. Doyle LE, Essentials of cognitive radio, published May 2009. ISBN: 9780521897709 Part of The Cambridge Wireless Essentials Series

9. 802.22 draft approved by IEEE standards association, Dec 4, 2019. https://standards.ieee.org/standard/802_22-2019.html

10. IEEE 802.24 Verticals technologies TAG scope. http://ieee802.org/24/

11. 802.11 Wireless networks, "The definitive guide", chapter 4, 2nd Edition by Matthew S. Gast, O'Reilly Library online. https://www.oreilly.com/library/view/80211-wireless-networks/0596100523/ch04.html

12. Official IEEE 802.11 working group project timelines 2020–02–04. http://www.ieee802.org/11/Reports/802.11_Timelines.htm

13. Digital Trends: Cord cutting 101: How to quit cable for on line streaming video, By Simon Cohen and Quentyn Kennemer January 29, 2020. https://www.digitaltrends.com/home-theater/how-to-quit-cable-for-online-streaming-video/

14. Application layer security and the OSI model, Finjan team, Oct 11, 2016, Blog, Cybersecurity. https://blog.finjan.com/application-layer-security-and-the-osi-model/

15. Why it is time to refresh your WiFi hardware with WPA3 Devices, Brien Posey, Jul 18, 2019. https://www.itprotoday.com/security/why-its-time-refresh-your-wi-fi-hardware-wpa3-devices

16. Public safety outlines strong use of 4.9GHz use cases, debate around frequency coordination, By Sandra Wendelken, Editor, Mission Critical Communications, Friday, August 10, 2018 https://www.rrmediagroup.com/News/NewsDetails/NewsID/17221

17. Who can get a 4.9GHz public safety license and how by Courtney Hambey, April 16, 2015. https://www.avalan.com/blog/about-a-4.9ghz-public-safety-license

18. New Rules for Unlicensed National Information Infrastructure (U-NII) bands KDB78033, KDB644545, Tho Nugyen, OET, Lab division (FCC). https://transition.fcc.gov/bureaus/oet/ea/presentations/files/oct14/51-New-Rules-for-UNII-Bands,-Oct-2014-TN.pdf

19. 802.11 AC channel planning, Revolution WiFi, March 20, 2013. http://www.revolutionwifi.net/revolutionwifi/2013/03/80211ac-channel-planning.html

20. Unlicensed National Information Infrastructure (U-NII) bands, by Dushmantha Tennakoon, OET Lab division, 31 part 15-Panel-UNII-Updates DT, Federal Comm. Commission. May 3, 2017. https://transition.fcc.gov/oet/ea/presentations/files/may17/31-Part-15-Panel-UNII-UpdatesDT.pdf

21. OFCOM—Consultation, Improving spectrum access for consumers in the 5GHz band, published in UK by OFCOM, 13 May 2016. https://www.ofcom.org.uk/__data/assets/pdf_file/0037/79777/improving-spectrum-access-consumers-5ghz.pdf

22. MD-5 Hash algorithm, Introduction to Message Digest Algorithm 5. https://www.educba.com/md5-alogrithm/

23. Introduction of WiFi standardization and Interoperability certification test, by Zhifang Feng CTTL-Terminals, CAICT, 31 Oct. 2017. ITU International. https://www.itu.int/en/ITU-D/Regional-Presence/AsiaPacific/SiteAssets/Pages/Events/2017/Oct2017CIIOT/CIIOT/7.Session3-2%20Introduction%20of%20Wi-Fi%20Interoperability%20Certification%20Test-%E5%86%AF%E5%BF%97%E8%8A%B3V3.pdf

24. WiFi Alliance, Product finder, https://www.wi-fi.org/product-finder

Satellite Communication

Orbit	In physics, an orbit is the gravitationally curved trajectory of an object, such as the trajectory of a planet around a star or a natural satellite around a planet. In the context of satellite communication, it usually refers to an artificial satellite built and launched by humans, and orbit is around the earth.
Satellite	In the context of spaceflight, a satellite is an object that has been intentionally placed into orbit, with human endeavor. The very first satellite was "Sputnik" launched in 1957 sending signals to the earth.
Kepler's laws of planetary motion	There are three laws by Johannes Kepler. (1) First one describes how planets move in elliptical orbits with the Sun as a focus, (2) The second law states—a planet covers the same area of space in the same amount of time no matter where it is in its orbit, and (3) the third law states—a planet's orbital period is proportional to the size of its orbit (its semi-major axis).
Perigee and apogee	In the context of artificial satellites around the earth, the point of the orbit closest to Earth is called perigee, while the point farthest from Earth is known as apogee.
Escape velocity	The velocity required to escape from the gravitational pull of the earth. This escape velocity is about 11.2 km per second, or approximately 33 times the speed of sound: MACH 33.
Injection velocity	The velocity at which a satellite is injected into orbit. This is always less than escape velocity. Higher the orbit (farther from earth) lower will be the injection velocity needed, since earth's gravitational pull reduces as satellites are placed farther away.
A.U Astronomical Unit	The astronomical unit (symbol: au, or AU or AU) is a unit of length representing the distance from Earth to the Sun and is physically about 150 million kilometers, or ~ 8 light minutes (since light takes 8 min to travel from the Sun to the Earth).
LEO	Low Earth Orbit, generally considered as 200–2000 km from the earth surface and apogee of up to 1000 km, where many remote sensing satellites and space stations are parked. The inclination within this orbit can vary considerably, including polar orbit which is 90° (over north and south pole), that allows "sun synchronous" orbit, meaning satellite visits the same spot on earth every few days.
MEO	Medium Earth Orbit, considered as 2000–35,000 km from the earth surface and apogee of up to 20,000 km. Many satellites including the GPS are in this orbit and highly inclined orbits that cover the globe (Molniya and Tundra) in which geosynchronous orbits (satellites take 12 h or 24 h to go around the earth, but are highly inclined and require antenna tracking) are in this orbit as well.
GEO	Geostationary orbit, is a popular orbit in the equatorial plane at 35,786 km from the earth. When satellites are parked in this orbit, they appear stationary to viewers on earth. The antenna remains

© Springer Nature Switzerland AG 2022
K. Raghunandan, *Introduction to Wireless Communications and Networks*,
Textbooks in Telecommunication Engineering, https://doi.org/10.1007/978-3-030-92188-0_13

	almost stationary with very little or no tracking. This feature helps communication satellites.	Tundra orbit	geostationary and requires tracking antennas from ground. A highly elliptical orbit but with "apogee dwelling" feature. This orbit also is geosynchronous and with multiple satellites it covers a region on earth for 24 h. Sirius satellite radio used this orbit to provide broadcast capability for the North American region.
Navigation satellites	Satellites that help in location and directions for observers on earth, three major satellite systems used for navigation are GPS, GLONASS, and Galileo.		
Remote sensing satellites	Satellites that continuously scan the earth surface and send pictures back to the earth. They can be programmed to highlight features such as minerals, forestry, ocean, and others depending on user needs. They use infra-red cameras to avoid the problems of cloud cover.	Station keeping	It is an orbital control process to maintain a satellite stationary within a given orbit by using small control thruster rockets on satellites. This also refers to a routine exercise performed regularly by all satellite agencies, to keep satellite in an allocated position in space.
PAA—phased array antenna	A phased array refers to an electronically scanned array of antennas, which creates a beam of radio waves that can be steered in a specific direction without physically moving the antennas. Such steering is usually accomplished using multiple antennas and by dynamically changing the phase of signal to each antenna element.	GEO belt	A term commonly used to refer to a library listing by the planetary society that lists all the satellites stationed in the geostationary GEO orbit. This 10-frame mosaic of Earth's geostationary satellite belt covers approximately 35° of sky around the GEO orbit.
Sentinel-2 mission	Copernicus Sentinel-2 mission comprises a constellation of two polar-orbiting satellites placed in the same sun-synchronous orbit, phased at 180° to each other. It aims at monitoring variability in land surface conditions, and its wide swath width (290 km) and high revisit time (10 days at the equator with one satellite, and 5 days with two satellites under cloud-free conditions, which results in 2–3 days at mid-latitudes) will support monitoring of Earth's surface changes.	Uplink limited	Normally refers to the noise of the uplink that limits the bandwidth available in a communication system. This term refers to both satellite communication and cellular communication.
HEO, highly elliptical orbit	Indicates an elliptic orbit with high eccentricity. This usually refers to eccentric angle around Earth. Such extremely elongated orbits have the advantage of long dwell times at a point in the sky during the approach to, and descent from, apogee.	Cryogenically cooled amplifier (LNA)	Low Noise Amplifier uses Cryogenic cooling of receivers to reduce their noise temperature, which is especially important in radio astronomy and in satellite and deep space communication. An antenna is "looking up " into the sky and, in the absence of strong celestial sources (Sun, Moon, planets, Cassiopeia, Cygnus, Taurus, Virgo, Orion, and the galactic plane) in the antenna beam "sees" a very cold sky: 2.725 K of the cosmic microwave background radiation modified by the presence of atmosphere. The antenna temperature required is, therefore, one order of magnitude less than those seen in terrestrial applications (300 K). Reduction of receiver noise by cryogenic cooling offers an effective way of improving radio astronomy system sensitivity.
Molniya orbit	A highly elliptical orbit extensively used by the erstwhile Soviet Union (Russia). This orbit is geosynchronous since each satellite offer 12 h of continuous coverage of a region on earth. It is not		

Antenna quality factor G/T	Antenna gain-to-noise-temperature (G/T) is a figure of merit in the characterization of antenna. It is the ratio of gain of the antenna divided by the antenna temperature (or system temperature if a receiver is specified).
VSAT	Very Small Aperture Terminal, is a small-sized earth station used to transmit/receive data, voice, and video signals over a satellite communication network, excluding broadcast television. VSAT antenna is 1.2 m–3 m diameter and operates in the Ku, C, and Ka bands. It offers high-bandwidth, bidirectional VSAT services for enterprise, government since many users will increasingly migrate to VSAT satellites.
Transponder	It is a device for receiving a radio signal and automatically transmitting a different signal (usually at a different frequency). In communication satellite, transponder refers to the series of interconnected units that form a communications channel between the receiving and the transmitting antennas.
EIRP	Equivalent Isotropic Radiated Power, represents the actual RF power in the air in a specific direction It considers transmitted power, effective antenna gain, and cable losses into account. EIRP is most commonly indicated in decibels over isotropic, dBi.
HPA	High Power Amplifier: In satellite communication refers to the high-power amplifier (HPA) provides the RF power for a payload downlink from the satellite. Before the signal goes into the HPA, there is a preamplifier that boosts the signal to a level proper for input to the HPA. These two together form the HPA subsystem. There are two types of HPA subsystem: the traveling-wave tube amplifier (TWTA) subsystem and the solid-state power amplifier (SSPA). Satellite downlink beams have very high power (up to 100, 000 watts of EIRP in spot beams). TWTA is more often used. Some ship-borne heavy duty systems also uses simpler HPA in the uplink (about 100 W EIRP).
DRM	Digital Radio Mondiale, is a set of digital audio broadcasting technologies designed to work over the bands (<30 MHz) currently used for analog radio broadcast including AM, shortwave, and FM bands. DRM is spectrally efficient compared to AM and FM and supports more stations with high-quality digital sound. It also provides two-way communication to users (in lower frequency bands). DRM is also the name of the international non-profit consortium that designed the platform and now promotes its introduction. Radio France Internationale, TéléDiffusion de France, BBC World Service, Deutsche Welle, Voice of America, Telefunken (now Transradio) and Thomcast (now Ampegon) are part of DRM consortium.
GLONASS	GLONASS refers to a system of navigation satellites and consists of an orbital constellation of 24 satellites covering the entire Russian territory, since 2011.
GPS	Global Positioning System consists of satellite constellation arranged into six equally spaced orbital planes surrounding the Earth. The U.S. Space Force has been flying 31 operational GPS satellites for well over a decade.
Galileo	The Galileo System is a global navigation satellite system that went live in 2016. It was created by the European Union through the European Space Agency and operated by the European Union Agency for the Space Program.
BeiDou	BDS is a global navigation system constructed and operated independently by China. The BeiDou navigation satellite system BDS has completed the constellation deployment and started to provide global services.
Telemetry	Telemetry is measurement at a distance. In the context of satellite communication, it refers to measurement made on satellites and sent back to the earth (downlink only). It is used as a health monitoring system as well as a method to confirm commands sent from earth are implemented.
Telecommand	In the context of satellite communication, Telecommand

	represents a system that sends commands to the satellite (uplink only).
Tracking	Satellite tracking consists of a system that involves multiple methods such as tone ranging, use of PN code to confirm the actual position and movement of the satellite system. In recent years, Satellite laser ranging has become a proven geodetic technique. It is the most accurate technique currently available to determine the geocentric position of an Earth satellite.
Drone of UAV	Unmanned Aerial Vehicle is an aircraft without a human pilot, it is often referred to as a drone. UAV is a part of the system that consists of ground control station for remote control of UAV. They are used in a variety of applications, reconnaissance, agricultural operations, and others.

The topic of satellite communication opens fascinating concepts about the universe and stars yet practical aspects of connecting citizens of the world, no matter where they are located on earth. In terms of setting up the system, there is considerable risk in getting satellites into any orbit. It is important to understand different orbits and how satellites are visible to different parts of the globe. These factors play a major role in the design of not only the communication system but also how they are launched, maneuvered into locations in space, and oriented in these orbits. It is appropriate to initiate this chapter with a review of such considerations in space.

13.1 Satellites and Orbits: Considerations

When "Sputnik" was launched on 4 October 1957, it broadcast basic radio signal at 20.005 MHz and 40.002 MHz, discernable to all amateur radio enthusiasts and others. The "Sputnik" satellite was in low earth orbit and transmitted signals only for 3 weeks. Only the basic concept of how to launch, monitor, and receive signals from space was established.

Later trials involved other orbits and interest in satellite communications steadily grew. How did this effort come about and what factors led to communication satellites taking center stage in just a decade after Sputnik? What are the different orbits used by satellites? To answer such questions, a quick review of celestial mechanics is useful. It directly affects the wireless aspects of each orbit and their coverage/ communication aspects. Although the economics of launching satellites was discussed in Chap. 8 and some mention was

made in Chap. 9, that this is an important wireless technology, a systematic study of orbits and usage is now in order.

Johannes Kepler—the seventeenth century mathematician and astronomer, proposed three laws of planetary motion based on his study. They have a direct impact on satellites launched from earth as well as communication system design. In order to recall, his laws are:

First Law (Law of Orbits): All planets move about the Sun in elliptical orbits.

Second Law (Law of Areas): The radius vector joining any planet to the Sun, sweeps out equal areas in equal lengths of time.

Third Law (Law of Periods): The "square of the orbital period" of a planet is directly proportional to the cube of the semi-major axis of its orbit. This captures the relationship between the distance of planets from the Sun and their orbital periods.

Now on to application of the "First Law". Orbits of planets around the Sun taught in schools is worth reviewing by applying the first law. Planets orbit around the Sun are actually in "elliptical orbits" and NOT "circular orbits". Due to elliptical orbits that all planets revolve in, SUN is at one focal point of the ellipse—"Sun is not the center of the circle". In Fig. 13.1, an ellipse is shown with the major axis AB and minor axis CD. If AB becomes equal to CD, then the orbit becomes circular. The focal points of the ellipse would come closer and merge into the center of the circle. Figure 13.1 shows the model where all celestial bodies including the earth are in elliptical orbit. It shows a somewhat exaggerated ellipse. Sun is at one focal point (on the left) and other focal point (on the right) is marked X in space, which has no celestial body (blank). In mathematical terms, every ellipse has two focal points, but the circle has only one center point, since the two focal points would merge into one (major and minor axis become equal forming diagonal).

For earth, currently, the maximum distance (from the Sun) called Aphelion is 152 097 701 km and the minimum distance (from the Sun) called Perihelion is 147 098 074 km. For Venus and Mercury that are closer to the Sun, the difference between these two is even less and their orbits appear to be "almost circular". For planets farther away from the Sun, the difference is considerable, which means their eccentricity is more, and they appear clearly elliptical [2]. This point was made by Kepler's teacher and guide Tyco Brahe, who showed with measurements that planets around the Sun do not have circular orbit, but elliptical, as stated in Kepler's first law.

Figure 13.2 shows the orbit of typical satellites as they are launched from earth (along dotted line from earth). Since satellites orbit around earth (similar to moon's orbit, but much closer to earth), terms used here are "Perigee", which

Fig. 13.1 Celestial bodies orbit —Sun at one of the focal points of ellipse

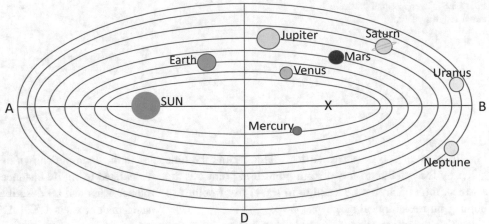

Celestial bodies elliptical Orbits – Sun at one focal point

means the position of satellite when closest to the earth, and "Apogee" which is its position when satellite is farthest from earth. Since earth revolves from west to east all satellites are launched towards east in order to initially gain earth's rotational momentum. View shown in Fig. 13.2 is from top of the earth observed directly above the North Pole. How does satellite travel in this orbit and what happens to wireless communication during this orbit (known as transfer orbit)? Due to initial velocity imparted by the rocket engines during launch, satellite moves up to "Perigee" shown on the right. At that point, satellite is "injected" into orbit—known as point of injection. At Perigee, earth's gravitational pull takes over and satellite starts to fall towards the earth but misses the earth's surface and continues to go away from earth, till Apogee, as shown in Fig. 13.2.

Its velocity comes to zero when it reaches Apogee, where it is again horizontal to other side of the earth surface. From there it starts to accelerate towards the earth (bottom of ellipse). Due to earth's gravitational pull, it accelerates towards the earth with "increasing velocity". It misses the earth again and goes to the perigee where it again becomes horizontal to earth surface. This results in an elliptical path

during which satellite's "velocity continues to decrease" (slows down) since earth is pulling it back. The cycle repeats and this motion results in an orbit.

Since satellite is continuously acted upon by earth's gravitational pull it moves with varying velocity along the elliptical path. Such variation in velocity results in varying Doppler frequency shift, which must be compensated by the satellite radio receiver on earth. It is not practical to use directional antennas since the satellite is re-orienting all the time. Only an omnidirectional antenna can be used to communicate to and from the satellite.

In summary, due to Kepler's First law—it implies that every launch from the earth results in such an "elliptical orbit of the satellite" around the earth. From this elliptical orbit (often known as transfer orbit), satellite is maneuvered later into other orbits. We will expand on this concept further and see how wireless communication design is adjusted during such orbital transfer.

Kepler's second law states "the body sweeps out equal areas in equal lengths of time", which results in continuously varying speed during elliptical orbit, which can be calculated. According to Kepler's second law, as shown in

Fig. 13.2 Satellite launched into orbit revolves in elliptical orbit around earth

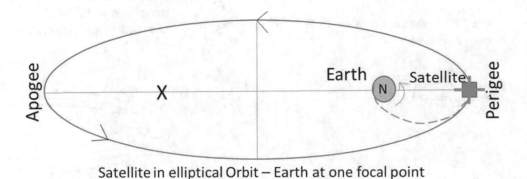

Satellite in elliptical Orbit – Earth at one focal point

Fig. 13.3 Sweeps equal areas in equal lengths of time

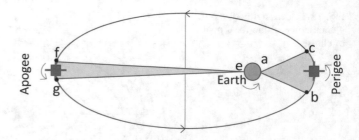

Fig. 13.3, area "a b c" = area "e f g". This means the time taken by the satellite to travel from point b to point c is the same as time taken by the satellite to travel from point f to point g on the elliptical space.

With this information, it is now possible to infer that satellite travels slowly (lower velocity) when it is far away from the earth (approaching Apogee) since f to g is a small distance yet it takes same length of time to move. Satellite travels faster (higher velocity) when it is closer to the earth (perigee) since distance from b to c is more than that of f to g, but time taken for both is the same. This provides an estimate of "Doppler shifts", which can be compensated by design, during various portions of the elliptical orbit. For the purpose of basic estimate, it is fair to assume an initial injection velocity of 11.2 km/s (Mach 33), which is the escape velocity from earth's gravitation pull [3].

Kepler's third law describes the connection between distances and orbital periods—which means how long does it take for the satellite to go around the earth. This not only affects tracking the satellite but variable angles and signal levels that become important parameters in the design of earth stations and mobile satellite terminals.

Figure 13.4 shows the relation between semi-major axis (half the distance from Sun to the orbit) and the period, which represents one revolution around the Sun. Semi-major axis can be described as shown or another common unit is AU or Astronomical Unit. The semi-major axis of the earth is taken as one A.U, which is 1.496×10^8 km, as shown in Fig. 13.4. Using Kepler's third law:

$$P^2 = a^3 \qquad (13.1)$$

where "P" is the period of earth, and "a" the semi-major axis.

In the example shown in Fig. 13.4, period of Mars can be calculated using its distance from the Sun. "Period" P of the planet Mars can be described in terms of earth's days. Mars semi-major axis is 1.524 A. U and using Eq. 13.1, period for Mars $(P_{Mars})^2 = (1.524)^3 = 3.54$ and $\sqrt{3.54} = 1.88$ earth years, which means ~ 687 earth days for Mars to revolve around the Sun once. Similarly, it is possible to calculate Period P, for any satellite that goes around the Sun (in terms of earth time of days hours etc). If it goes around the earth, like the moon, then Period of moon to go around the earth could be calculated in hours. Instead of relating to earth and AU, general terms may be used to describe such calculations.

The general equation is:

$$P^2 = \frac{4\pi^2}{G(M+m)} (a)^3 \qquad (13.2)$$

This general equation applies to all planets of our solar system. P is the period of the planet or time taken for one revolution around the Sun. M is the mass of the Sun (central body), which is typically over 1000 times the mass of any of its planets. "m" is the mass of the planet, and G is the gravitational constant, from Newton's law of gravitation and "a" the semi-major axis.

$$\left(F_{grav}\right) = \frac{GMm}{r^2} \qquad (13.3)$$

If we apply this equation to something that revolves around the earth (such as moon), then P is the period of Moon or another satellite, "a" is the semi-major axis of moon or another satellite, M is mass of the central body (earth) = 6×10^{24} kg, and "m" mass of the moon/satellite and finally the gravitational constant G = 6.67×10^{-11} m³ kg⁻¹S⁻².

Fig. 13.4 Relation between semi-major axis and orbital period

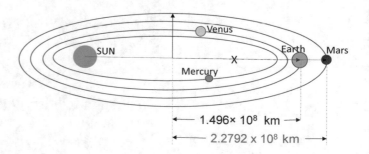

These equations have profound impact on calculating satellite orbits and how space agencies send artificial satellites into orbit. One of the most popular orbits is the geostationary orbit, which allows the satellite to revolve around the earth once in 24 h. Satellites in this orbit appear stationary to a person standing on earth. This orbit was originally proposed by Arthur C Clarke and most countries around the globe use this orbit for communication satellites [4].

13.1.1 Types of Orbits: And Global Resource

Why different orbits are considered important global resources? Why does the UN (United Nations) have a separate committee to allocate space for satellites from different nations and bodies [5]? The reasons are somewhat complex yet straightforward. Orbits can be classified into four different types based on distance and inclination. Table 13.1 shows the different types. Any orbit can use varying inclinations depending on the mission. Depending on the demand, allocation of space to satellites allow control of interference and avoid possible collision of satellites.

Although inclination is not specified in the table, HEO has a very steep inclination (discussed later).

Figure 13.5 provides a general concept of different orbits of satellites around the globe. For example, although LEO and MEO are shown with no inclination, polar orbit would be directly over the poles with a 90° inclination to earth's rotational axis. Relevant details will be indicated during the discussion of those orbits and their application, including how radio communication is affected by each orbit.Orbit:

Since all orbits typically start out with Perigee (point closest to earth) that is usually considered Low Earth Orbit (LEO), it is commonly used as "the transfer orbit" for any other orbit—it is the transition used by all satellites. Let us consider satellites that reside in Low Earth Orbit (LEO).

13.1.2 Low Earth Orbit

Low Earth Orbit is used by a variety of satellites and space stations. Staying in LEO has certain advantages. Being closer to earth the satellite revolves around the earth more

often, typically taking between 90 and 120 min per revolution (period). If pictures are taken, resolution is likely to be better. Several systems including space orbiting stations tend to stay in this orbit. It is in fact becoming the orbit where new space race may begin, see [7]. The main advantage is shorter round-trip delay for both voice and data customers. Iridium satellites (66 of them) have been in service in this orbit for several decades. But several new entrepreneurs are coming up with the concept of building a major network in this orbit to serve customers in areas and regions that are very difficult to be connected by fiber or cellular service. Perhaps the largest number of satellites in this orbit consists of microsatellites being launched by universities and scientific satellites [8]. The current estimates in the orbit are over 3000, which are expected to grow to almost 12,000 if all the planned satellites get launched.

Astro Physicists and large corporations in the communication business express concerns about the ease with which rockets and payloads can be built. Major changes in space technology came about due to 3D-printed rocket stages with no moving parts [9]. Also, microsatellites that are no bigger than a bottle of soft drinks are able to perform major experiments useful to scientific community. Multiple satellites in LEO are expected to crowd this orbit to the point where further launches may be a struggle—since there is little regulation to stop launching and parking satellites in this orbit.

At the same time, LEO is essential for all longer space missions such as going to Moon or Mars. It is important for longer missions to first arrive at the transfer orbit, which is GEO (as discussed earlier). LEO offers the right point of Apogee to take off further towards a interstellar mission like Lunar or Mars mission or even to other planets. Different types of orbits within the LEO range are always to be found, due to University-based scientific satellites [10] and military missions (often not disclosed).

With so many satellites orbiting close to the earth, the prospect of space debris continues to grow. This was discussed earlier in Chap. 8 and Fig. 8.5 and provides a stark image of how space debris has grown over the years since the days of "Sputnik". There are schemes and proposals to sweep this debris down with the hope that most of it would burn up as it enters earth's atmosphere. ESA (European Space Agency) has clear plans to start this activity within the next 5 years [11].

Table 13.1 Different types of orbits and altitudes	Orbit		Altitude (km)	
			Perigee (Period)	Apogee
	LEO: Low Earth Orbit		200–2000, normally	600–1000
	MEO: Medium Earth Orbit		2000–GEO normally	10,000–20,000
	GEO: Geostationary Orbit		Same as Apogee (24 h)	35,862 km
	HEO: Highly Elliptical Orbit	Molniya (12 h)	–500	–40,000
		Tundra (24 h)	–24,000	–48,000

Fig. 13.5 Different orbits of
satellites around the globe

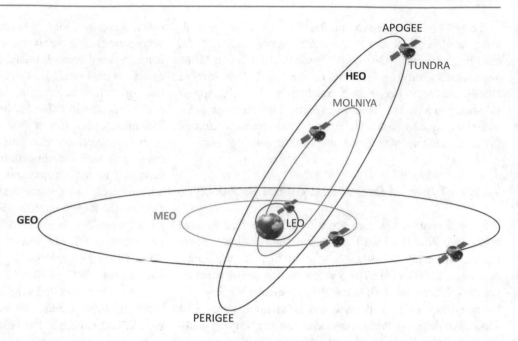

13.1.3 Medium Earth Orbit

Medium Earth Orbit in contrast to LEO is a stable region in
space that has seen important satellite systems such as Global
Navigation System GPS / GLONASS/Galileo, remote sensing satellites such as IRS, SPOT, and LANDSAT (although
some of them operate at the outer edges of LEO).

MEO offers a range of orbits, some are circular, others
highly elliptical. The most common missions used by most
countries and industry are navigational satellites and remote
sensing satellites. At the same time, the highly elliptical
"Molniya" orbit or its counterpart "Tundra" orbit are used as
well. Let us consider each of them to review specifics.

13.1.3.1 GPS/Navigation Satellite Constellation
GPS or Global Positioning System consists of at least 24
satellites at any given time. Current estimate is around 33
satellites, the balance of 9 being "standby" or "enhancement" that is not meant for "essential" service. They are
arranged in "Six equally spaced circular planes" around the
earth. The satellites are placed in approximately 20,200 km
altitude from the earth surface and circle the earth twice a
day. Each plane is occupied by four satellites in prefixed
"slots" on that planar orbit with ascending nodes of the
orbital planes separated by 60° and the planes are inclined
55°. This arrangement allows any user on earth to view at
least four satellites from virtually any place on earth. Since
2011, GPS standard allowed for an expansion of number of
satellites to be increased to 27 instead of 24. However, at any
time, 24 satellites are operational and the other three may act
as buffer to replace an aging satellite [12]. The use of three

additional satellites improves accuracy of user's position,
but it is not an objective of the "expansion".

Figure 13.6 shows the typical orbital pattern of GPS
satellites, in 6 different orbital planes around the globe,
having 4 satellites in each of those orbital planes, making a
total of 24 satellites.

GPS satellite system was originally started by the US
DoD (Department of Defense) but was made available for
civilian use in the late 1980s. Their popularity resulted in
two other similar navigation satellite systems—both of them
operate well. GLONASS operated by Russia (earlier Soviet
Union) has been in service since late 1980s and "Galileo"
operated by European Space Agency is very recent. In view
of their importance for location-based services not only for
every cell phone but also many others such as aircraft,
maritime, and station keeping (time synchronization) of
international networks, there is international cooperation
between these three agencies. All Galileo receivers are
compatible with GPS (no extra hardware required).

Figure 13.7 shows the frequency bands these three systems operate [12] along with protocol allocations. GPS,
which is the earliest navigation service, operates in three
distinct bands:

GPS L1 Band: 1575.42 MHz with a bandwidth of
15.345 MHz.

GPS L2 Band: 1227.6 MHz with a bandwidth of 11 MHz.

GPS L5 Band: 1176.45 MHz with a bandwidth of
12.5 MHz.

Fig. 13.6 GPS satellites orbit
around the earth

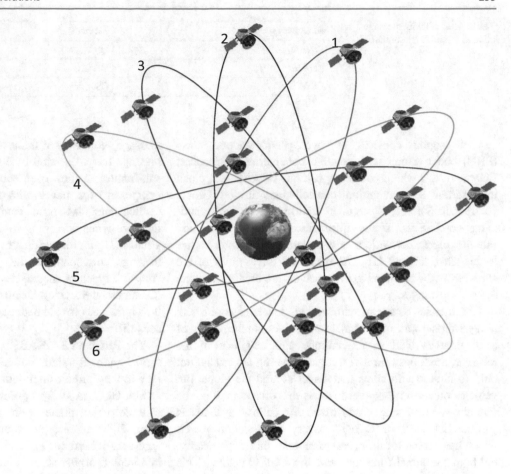

Fig. 13.7 Navigation service
providers and the frequency
bands

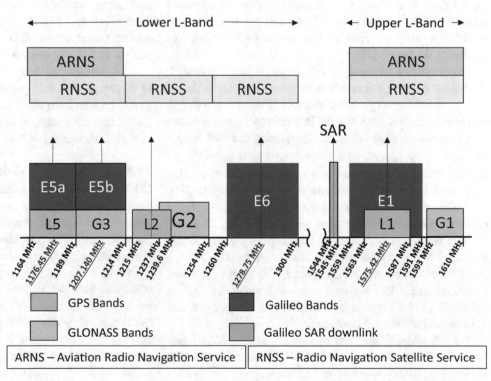

Table 13.2 GNSS message allocation and data content

Message type	Services	Content
F/NAV	OS	E5a-1
t/NAV	OS/CS	E5b-1 and E1-b
C/NAV	CS	E6b

GPS signal consists of two pseudorandom noise (PRN) codes, along with satellite ephemerides (Broadcast Ephemerides), which contains position, velocity, and clock information for navigation constellations in the Global Navigation Satellite System. In addition, ionospheric modeling coefficients, status information, system time, and satellite clock corrections are superimposed onto carrier frequencies, L1 and L2. The measured travel times of the signal from the satellites to the receivers are used to compute the pseudo ranges.

Commercial receivers with various levels of sophistication can interpret such signals and obtain information of local time as well as the latitude and longitude of the receiver. Such position and time information are crucial not only to aircraft and ships that travel around the globe, but vehicles on earth that travel in areas that drivers may not be familiar with. A major advantage for cellular systems is "differential GPS or DGPS". Every base station tower, which has a fixed location, serves mobiles that move nearby, but both see signal from the same three GPS satellites. This allows them to compute the difference to obtain much better accuracy—this is known as DGPS—that has same level of accuracy as military vehicles (for which GPS system was originally designed) [13].

Fundamentals of navigation are based on a downlink (broadcast only) signal received by users on earth, in the air and at sea. At the very core of every GPS satellite is a highly accurate frequency-time source. It is referenced and adjusted to a primary national laboratory frequency standard, which maintains time accuracy of better than 1 part in 10^{-15} per day. It translates to time accuracy of better than 1 ns per day [14]. Such highly accurate clocks are referenced to the GPS, to estimate on-board clock inaccuracy to an accepted national reference time but used around the globe. There are Cesium beam standard sources commercially available and used in different countries (known as working standards)—all of them periodically check against such master reference clocks maintained by national standard laboratories such as the NIST [14]. This will be elaborated in Chap. 19 where time measurements are discussed.

Transmissions from GPS or Galileo or GLONASS systems provide frames that arrive as RF signals at the receiver. In Galileo GNSS system, for example, such navigation message consists of frames—one of them, the F/NAV message has a duration of 600 s and it is composed of 12 sub-frames. In turn, each sub-frame has a duration of 50 s composed of 5 pages with a duration of 10 s [15]. The Galileo signal-in-space data channels transmit different message types according to the general contents shown in Table 13.2 providing different services. The F/NAV types of message correspond to the OS (Open Service) and the "t/NAV" types of message correspond to both OS and CS (Commercial Service). Details of Galileo system's protocols how it tries to provide better accuracy are indicated in detail, see [15].

The Russian GLONASS was the second system that has served entire Russian continent with claims of better accuracy in the northern high latitude areas including the Arctic Circle. GLONASS Space Segment consists of 24 satellites in three orbital planes with ascending nodes that are 120° apart. Eight satellites are equally spaced in each plane with argument of latitude displacement of 45°. These satellites are in circular orbits at an altitude of 19,100 km, with an inclination of 64.8°. Each satellite orbits the earth in approximately 11 h 15 min. Such spacing of the satellites provides continuous global coverage of earth's terrestrial surface and the near-earth space (such as aircraft) [16].

Navigation signal in general has become the primary tool to drivers and traveling public. In major cities around the globe, travelers use GPS or Galileo-based guidance even as pedestrians to walk around the city or within a building complex. Its usage is expected to increase substantially.

13.1.3.2 Remote Sensing Satellite Constellations

Orbits used by remote sensing satellites vary in altitude and period but tend to focus on synchronous orbits. Design criteria tend to revolve around the Sun's position at a given place on earth. This provides an opportunity to visit that spot around the same time of the day.

Most remote sensing satellite follows the north–south transition path, for a good reason. The earth below it traverses west to earth—therefore the satellite's beam/scan follows a swath that covers the whole globe over a given period [17]. This feature is particularly helpful in scanning the earth based on a specific theme—also known as "thematic mapping". Perhaps the most important of such orbits is

Fig. 13.8 Polar/Sun synchronous orbits of remote sensing satellites—swath on globe

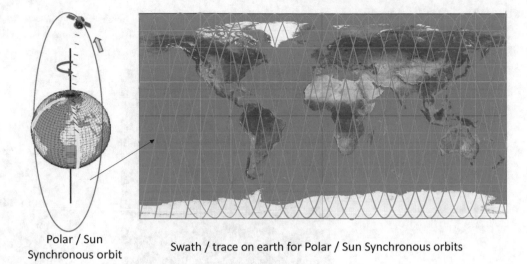

Polar / Sun
Synchronous orbit

Swath / trace on earth for Polar / Sun Synchronous orbits

the "Polar Orbit" in which the satellite moves over the North Pole and South Pole creating a "swath" on earth. Figure 13.8 shows typical polar orbit and its swath.

The scan from the satellite and earth's rotation on its own axis result in "swath" or a path that satellite's camera (usually infra-red) takes as an input. In Fig. 13.8 on the left is the polar axis in which satellite moves over North and South poles at a fixed altitude. On the right is its scan path or swath on the earth. Note that the scan has a repetitive pattern, which can be made "sun synchronous". That means the satellite can appear over the same place on earth every "x" number of days. That "x" depends on the altitude of polar orbit. This repetitive pattern illustrated on the right shows how every part of the globe gets scanned, with repetitive scans. Each scan cycle shifts east due to earth's rotation. Hence, next scan of the same area is shifted further east in latitude, ultimately after (say 21 days or x days) satellite scans the same spot at the same time.

In terms of communication signals back to earth, most of it comes down as camera output that is first compressed at the satellite and then sent on down link to the earth station below. The use of infrared camera is common, in order to avoid "cloud cover". By adapting different filters, the camera can be "tuned" to scan for minerals, for forest cover, for trace of water (in deserts) or petroleum reserves. Such information is valuable to various agencies and the downloaded camera signal is first decompressed and later further analyzed. Subsequently, images get color-coded to make it easy for identifying various resources. Such coded images offer considerable value—which is why Landsat, SPOT, CartoSat, and many other satellites became generations of satellites that continue to provide such services around the globe Fig. 13.8.

Since there is camera feed information sent to earth, data throughput rate required on such links tends to be high. For example, CartoSat-2 was a remote sensing satellite launched in 2007, into sun synchronous orbit, provided camera feed of 54.5Mbps in I and 54.5Mbps in Q (I- In phase, Q— quadrature) in the digital JPEG format. It used a compression ratio of 1:3.2 with Reed Solomon encryption [18]. Its downlink provided throughput rate of 105Mbps using Phased Array Antenna (PAA) that is directional, controlled from a dual-gimbal platform. The spectral range for its panchromatic camera is 0.45–0.85μ m. This spectral range used in 2003 by CartoSat-2, represented a limited range of sensing objects on earth. In comparison, Table 13.3 represents the Sentinel-2 satellite sensor of European Space Agency, which uses much wider range of the Spectrum. Sentinel-2B was launched in March 2017, for details see text book and web page [19].

Table 13.3 shows the resolution of images from Sentinel-2 satellite; its resolution depends not only on the orbit and altitude but also on the remote sensor on board. In every satellite senor, it is necessary to store images on board, compress, and encrypt them on board and after that send them to the receiving stations on earth. This complexity results in images sought after by multiple agencies spread across the globe, with varying interests in types of image, varying resolution, and usage. Remote sensing by satellite has grown into an industry of its own with multiple applications in different bands as indicated in the table. Polar orbit in MEO is one of the best ways to provide a wide range of services from such satellites.

13.1.3.3 Molniya and Tundra Satellite Constellation

Not all communication satellites can use the Geo-stationary orbit. The high latitudes of northern Arctic Circle and the southern Antarctic Circle cannot be covered by geostationary satellite (to be discussed in the next section). Therefore, Russia started using the Molniya orbit, where satellite orbits

Fig. 13.9 Molniya and Tundra orbits (Sirius satellites in Tundra orbit tracks)

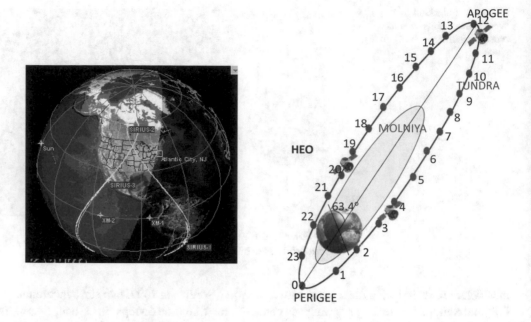

Table 13.3 Table of sentinel-2 multispectral imager

Band	Central wavelength (Micrometers)	Description (Representation of the electromagnetic spectrum)	Resolution (m)
Band 1	0.433	Coast aerosol (ultra blue)	60
Band 2	0.490	Blue	10
Band 3	0.560	Green	10
Band 4	0.665	Red	10
Band 5	0.705	Vegetation red edge	20
Band 6	0.740	Vegetation red edge	20
Band 7	0.783	Vegetation red edge	20
Band 8	0.842	Near infrared (NIR)	10
Band 8A	0.865	Narrow near infrared (NIR)	20
Band 9	0.945	Water vapor	60
Band 10	1.375	Shortwave infrared (SWIR)—Cirrus	60
Band 11	1.610	Shortwave infrared (SWIR)	20
Band 12	2.190	Shortwave infrared (SWIR)	20

high above the North Pole for considerable time. Figure 13.9 shows the Molniya and Tundra orbits (right side). Let us review these two orbits in order to understand their advantages and limitations.

Picture on the left shows three broadcast satellites—Sirius 2 and Sirius 3 are in daylight part of North America, while Sirius 1 (bottom of left picture) is behind southern side of globe in night part. Positions of Sirius satellites are shown over a period of 24 hours on the right side in the "Tundra orbit". Observe that the three Sirius satellites in Tundra orbit are positioned 8 hours apart. Figure 13.9 left side is still image from extracted from an online video (see reference [44]) that shows how the three satellites orbit with one of them behind the earth—or in night time. Both Molniya and Tundra are in HEO (Highly Elliptical Orbit) with an inclination of 63.4° to the earth's rotational axis, as shown. Since both orbits are "geosynchronous", it is important to delve into some details. How are these different from the popular "geostationary" orbit? Why are these considered more specific and not widely used? Some of these questions are relevant and may point to an important role for these orbits. Table 13.1 indicated the salient features of different satellite orbits. In highly elliptical orbits, Apogee and Perigee have considerable differences. What is meant by geosynchronous—that they take either 12 h (Molniya) or 24 h (Tundra) to go around the earth once. But to any observer on earth, they are moving continuously and therefore need tracking, which means they are not

Table 13.4 HEO (Highly Elliptical Orbits) Tundra and Molniya orbits—comparison

Orbital parameter	Tundra	Molniya	Comment
Period	24 h	12 h (1/2 day)	Tundra takes one sidereal day
Longitudes	Same longitude	Separated by 180°	One orbit over USA, next orbit over Russia
Inclination of orbit	63.4°, or $Sin^{-1}\frac{\sqrt{4}}{5}$	63.4°, or $Sin^{-1}\frac{\sqrt{4}}{5}$	Inclination chosen to cancel out perturbation caused by earth's bulge at equator
Synchronous	Geosynchronous	Geosynchronous	But not geostationary
Apogee/perigee	48,000/24,000 km	40,000/500 km	Tundra is "Apogee dwelling"
Features	Avoids passing through Van Allen belts	Needs less launch energy. Operates in two parts of globe	Molniya was used as comm. Satellite over Russia and as spy satellite over USA

"geo-stationary". By inference, their signal levels vary considerably as well, which means the receiver must be designed to operate over a wide dynamic signal range.

Both orbits cover the northern hemisphere over extended period, spending very short periods on the perigee (southern hemisphere). Hence, their use for coverage of northern hemisphere (Canada, Russia, Scandinavia, Japan) is excellent. Although it requires tracking on earth, there are advantages to using these orbits for communication satellites.

What are the differences between Molniya and Tundra orbits? Comparison between them is shown in Table 13.4 that shows some interesting commonalities and differences.

Some important facts emerge from this table. Satellites in the Tundra orbit spend considerable time at the Apogee—also known as "Apogee dwelling". Observe in Fig. 13.9 that the satellite is visible about 16 h for those living in the northern hemisphere. In theory if there are two satellites in that orbit, it should be sufficient to provide coverage.

But Sirius Satellite Radio used three satellites in that orbit for a practical reason. System concept was to switch satellites when it dipped below 45° elevation in the sky. This is because users of Sirius Satellite received radio broadcast channels while travelling in vehicles. Many of the roads in USA and Canada do have buildings or trees next to them. Therefore, a high elevation angle (above 45°) provides clear, continuous coverage from the satellite [20].

There are other advantages of Tundra orbit—like a geostationary orbit, Tundra orbit serves a specific region. Sirius satellite radio serves North America covering Canada, USA, and Mexico. Similarly, it is possible to use Tundra orbit to cover Scandinavia or Japan for example. There are some limitations—due to high velocity with which satellites travel, it is essential to provide doppler compensation. Due to varying distance from earth, it is necessary to use high gain amplifier near the receive antenna on vehicle roof. To receive Sirius radio channels, amplifiers with over 40 dB gain (10,000 times enhancement of signal) were needed and such amplifier would be embedded as part of the antenna mounting on cars or trucks [20] and also reference [44] is an archive where several satellite orbits are animated showing how they orbit the earth—it includes the three Sirius satellites orbiting the earth.

Molniya orbit, on the other hand, has a swinging apogee (12 h over Russia and the next 12 h time over North America, for example). Due to this swing, it could be shared by two regions on opposite sides of earth—one for daytime (maximum of 12 h, but in reality, about 8 h) while another during night. By placing two satellites in the same orbit, it would be possible to switch between them.

13.1.4 Geostationary Orbit

Earlier, in Fig. 13.5, different orbits were shown—the last and most interesting in them is GEO or Geostationary Orbit. This orbit was also in Fig. 3.3a and b. It was discussed in Sects. 8.5 and 9.2, for good reasons. It is the most widely used orbit for communications satellites.

Like any other orbit, it has some limitations—to show its advantages and limitations let us consider Fig. 3.3a and b together, reproduced here as Fig. 13.10 with a few more conceptual details.

From the equatorial plane, the three satellites cover different parts of the earth—satellite A covers the Western hemisphere area of American region. Satellite B covers most of Europe and African region. Satellite C covers most of Asian region (other side of globe). Both the top view and equatorial view indicate that the northern Arctic Circle and the southern Antarctic Circle are not covered. Figure 13.10a shows the coverage beams from the satellite skims over the globe (tangent to the earth's curvature) at these edges. Considerable parts of northern territory of Russia, Scandinavia and Canada are not served. Also, these satellites serve

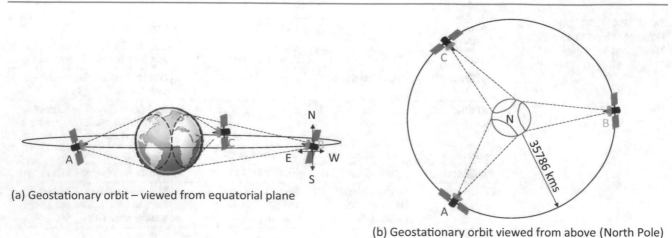

(a) Geostationary orbit – viewed from equatorial plane

(b) Geostationary orbit viewed from above (North Pole)

Fig. 13.10 Views of Geostationary orbit—side view and top view

better if population lives closer to the equator—in those countries the satellite dish would directly look up (well above 60 or 70°) offering an excellent view of the satellite, with no steering required. But those in the northern or southern extremes—the look angle is quite steep. For example, at the southern part of Norway (Oslo city for

example), the look angle is 21.7° [21]. That is quite a low angle to look up and depending on weather, good reception is not likely. This is true for most parts of North Canada, Alaska, Russia, and other land mass in the Arctic Circle. Figure 13.10a indicates two types of natural drift of the satellite. As an example on satellite B-the North–South drift

Fig. 13.11 Geostationary orbit—crowded with satellites in different frequency bands

is indicated by up and down arrows, and the East–West drift indicated by the left and right arrows. Such drifts occur on every satellite over a period of time due to variations in gravitation pull and other forces acting on the satellite in orbit. Usually, satellites carry small thrusters or gas bottles that are used to counter the effect of such drifts. These thrusters are used to compensate for such "drifts". This is known as "station keeping". Typically, satellites at the end of their life are allowed to drift, gradually reducing their availability from 24 h a day, to 23 and then 22/day etc. Eventually, the on-board systems are turned off to allow a new satellite to occupy its original spot. Unlike low earth orbits, the GEO orbit is so far away and being circular, there is no way that these satellites will fall back to earth. They will become space debris in that orbit's region.

On the positive note, major populations of the globe live further south and closer to the equator (most countries in Asia, Australia, South America, Africa etc.—they would like to use this orbit. Competition to get a spot on this circle is quite severe, therefore they have to apply for a slot on this orbit [22]. Due to such fierce competition, certain practical aspects immediately come to the fore. If satellites are placed too close to each other (on this circle), there will be interference. It has been determined that separation is required due to beam width of uplink transmitters from ground [23]. Figure 13.11 shows a view of satellites placed in this GEO orbit (517 listed at end of 2019).

There are companies such as INTELSAT (International Telecommunication Satellite) who are mainly in the business of launching and parking satellites in this orbit, which can be leased [24]. All international communications typically go through such satellites. Globally, INTELSAT operates over 50 satellites (mostly in the GEO orbit) and also has its own terrestrial network to extend its operations using fiber and extend its network to customer locations. Such companies may place multiple satellites close to each other, by placing adjacent satellites in frequency bands considerably separated. For a current list of satellites in this orbit, please see [25]. Such staggering numbers show how crowded this orbit is. At an altitude of 35,786 km, this separation translates to roughly 1 km apart.

13.2 Global Markets and Allocations

Global market for satellite communication, remote sensing, GEO, and other applications is very active. Scientific and other communities (including amateur radio) have an interest in this satellite launching as well. It is important to note that number of companies capable of building satellites is quite large and diverse. However, agencies that can launch satellites are somewhat limited—only seven countries are in the launching business—USA, Russia, France (ESA), China,

India, Kazakhstan, and Japan. The risk of launching is quite high. Therefore, launch insurance is proportionately high as well [26]. There is a serious attempt to change this situation —in particular, SpaceX is a company challenging conventional launch vehicles, by building re-usable rockets. With a large number of launches to its credit, the company has a goal of making space travel a commercial venture [27]. There are several innovative concepts including 3D printing of rocket engines that could change the concept of space launch/travel much easier.

Allocation of spots on a specific orbit is regulated by the UN with its agency ITU (International Telecommunications Union). All commercial satellites going into orbit must follow a well-regulated, dated, process that allocates parking rights to specific satellites [28]. It is not possible to regulate either military launches or others that are specific to a short science experiment. However, enforcement of these parking rights becomes an issue, since a body that has the ability to enforce is needed.

Those countries with a large market have also enacted laws to govern. The "Open-Market Reorganization for the Betterment of International Telecommunications" Act, or ORBIT Act, is an important law that governs the satellite industry in the USA. The "ORBIT" Act was signed into law in the year 2000. This law came into effect due to privatization of two historically intergovernmental satellite organizations—Intelsat and Inmarsat. Since both of these organizations control sizable allocations in the GEO orbit, ORBIT law was designed to manage privatization of major satellite constellations. The ORBIT Act is cited as a benchmark for auctioning of satellite spectrum, which is at the heart of satellite industry in terms of interference.

13.3 Uplink and Downlink Considerations

Now on to the practical aspect of using the frequency spectrum itself. In early days of satellite industry, the satellites were small, had limited battery power, and could only deploy small antennas, whether it was omnidirectional (simplest form is a metallic stick—much like those on traditional cars), or a parabolic dish directed towards the earth. The onus, therefore, fell on receiving stations, which needed very large dishes (typically 20 m or larger radius). The word "satellite earth station" still brings visions of very large dishes mounted on massive structural frame, with sophisticated systems to track the satellite.

13.3.1 Satellite Earth Stations

Such satellite earth stations are in use today although antenna size has reduced. They work as gateways through which satellites are controlled and their services extended on earth using landline networks. At the heart of these receivers is a "Low Noise Amplifier" or LNA. What is "Low Noise" and why is this necessary? When a receiver is wide open and waits to receive a very weak signal coming from a faraway satellite, it is bound to receive noise along with that signal.

$$\text{Noise Power } P_N = \int_B^B No/2\,df. \qquad (13.4)$$

Where N_0 is noise density in Watts/Hz, and P_N is equal to N_oB and B is bandwidth in Hz.

This Noise power can be expressed in Watts as.

$$P_N = 4kTB \ (\text{W}). \qquad (13.5)$$

Where,

P_N = Noise Power, in Watts

k = the Boltzmann's constant (1.38 X 10^{-23} J/K),

B = Bandwidth of the receiver.

T = Absolute temperature (Kelvin) ($17^o C or 290 K$).

Equations 13.4 and 13.5 are fundamental to all telecommunication systems. If we consider the signal arriving from a typical satellite, it has typical bandwidth of B = 500 MHz or more which needs quite a wide opening in the receiver bandwidth. Since k is a constant, the only option to reduce noise power will be to reduce T—the absolute temperature.

In early satellite systems, such temperature reduction was achieved by using receiver amplifiers kept in a cryogenic liquid such as liquid Nitrogen. Such cryogenics help reduce T from 270 to 20 K for example. While the large dish antenna manages to receive the weak signal, whose level so low, that signal must be amplified by at least 100,000 to 1000,000 times! With such large amplification, the noise also amplifies. In order to limit the noise power, it is necessary to keep

receiver in a pool of cryogenic liquid [29]. If we consider signals arriving from galactic missions such as those from Mars or beyond, the signal is extremely weak. Amplification using this method and cooling to very low temperatures become essential for all galactic communication.

Since the uplink to the satellite was from major satellite ground stations, the assumption was that transmitted power can be high. This is true and uplink power in hundreds of watts with focused beam towards satellite was considered in such links. Even if transmitted power was in kilowatts, the signal received at the satellite was only of reasonable level, since antenna on a satellite cannot be very large. Such cryogenically cooled systems on satellite are not practical either. In view of such limitations, all satellite systems are "*uplink limited*". *This means the signal level received at the satellite determines the limit of bandwidth and data throughput that can be supported.* This is a key statement applicable to all cellular systems as well [30]. Therefore, any signal sent from a cell phone (in cellular) and received at the base station, determines whether communication is possible. Note that in earlier chapter on cellular, it was mentioned that the cell phone can hardly send 200mW on to the air. What is received at the base station site is "limited" by how far the cell phone is from the base station.

The G/T or Gain/Temperature ratio is an important figure of merit for all satellite earth stations. What is G/T and how is it relevant? To answer this question, let us consider uplink/downlink.

In Fig. 13.12, the uplink (on the left) consists of HPA (High Power Amplifier) that enhances the power from satellite transmitter (Transmitter + Receiver = Transceiver) and sends that signal to the antenna. The output power P_t from that HPA is quite high, as indicated earlier. The Gain of the uplink transmit G_t of the antenna is an important specification. EIRP (Equivalent Isotropic Radiated Power) of the

Fig. 13.12 Satellite link budget with key parameters

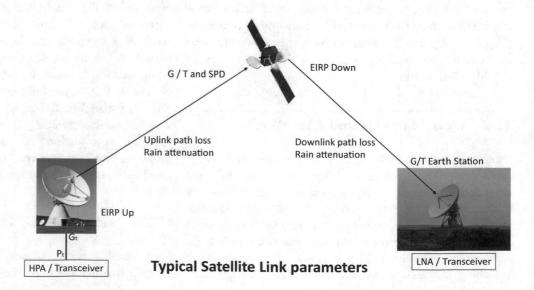

Typical Satellite Link parameters

uplink is the actual power P_t multiplied by the Gain G_t of the antenna. This represents the radiated power (relative to the ideal isotropic power). What EIRP means is Radiated Power squeezed into the narrow beam that is sent towards the satellite.

On the way to the satellite signal incurs path loss (about 36000 km if satellite is in GEO) and other attenuation such as those due to rain. Free Space Path Loss (FSPL) is common to all wireless systems and is calculated by the equation:

$$FSPL = \left(\frac{4\pi df}{c}\right)^2 \quad (13.6)$$

where, FSPL = Free Space Path Loss, c = speed of light, d = distance from transmitter to receiver (in meters), f = frequency in Hz. In addition, signal incurs rain attenuation, which can be severe at higher frequency bands. The detailed description of how to calculate the path loss, various factors involved in the uplink and down link are described in both references [30, 31]. Satellite antenna aperture is closely related to quality factor (G/T value) of any earth station. G/T value and the satellite power demand, which is equivalent of bandwidth, have logarithmic linear relationship. The value of equivalent is stated as "bandwidth increases when antenna aperture is narrower".

The figure of merit (G/T) of an Earth Station is maximum when there is no de-pointing loss, no feeder loss, no polarization mismatch, and no rain attenuation [31]. The maximum value will be:

$$\left(\frac{G}{T}\right)_{ES\,max} = (G_{R\,max})ES\left(\frac{1}{T_{D\,min}}\right)(K^{-1}) \quad (13.7)$$

OR when expressed in dB, it will be

$$\left(\frac{G}{T}\right)_{ES\,max}(dBK^{-1}) = (G_{R\,max})dBi - 10\log T_{D\,min} \quad (13.8)$$

where,

$$T_{D\,min} = T_{sky} + T_{Ground} + T_R (K) \quad (13.9)$$

The combined effect of feeder loss and rain attenuation is represented by DELTA (dB), which is given by

$$DELTA(dB) = L_{FRx} + 10\log T_D - 10\log T_D min \quad (13.10)$$

In Eq. 13.10, T_D represents the downlink system noise temperature and T_A represents the Antenna noise temperature given by the following equation.

$$T_A = \frac{T_{sky}}{T_{Rain}} + T_m\left(1 - \frac{1}{A_{Rain}}\right) + T_{Ground}(K) \quad (13.11)$$

L_{FRx} represents Receiver's Feeder Loss. The feeder loss can, and rain attenuation, vary depending on frequency of operation. In addition, rain attenuation varies considerably depending on different regions of the earth. Due to such factors, the values in Eqs. 13.7, 13.8, and 13.9 will be modified by DELTA shown in Eq. 13.10. It is beyond the scope of this book to undertake a detailed exercise in calculation of losses for specific region on earth. Usually, such calculations are performed at the time of design by consulting rainfall and DELTA tables that are available in [30, 31].

In view of intricacies involved in arriving at G/T, when selecting earth station aperture, we cannot use simple logic—"the smaller, the better". Earth station aperture is selected as careful compromise between space overhead (equivalent bandwidth) and ground overhead (antenna aperture) to achieve optimum allocation. Such a balanced system is VSAT (Very Small Aperture Terminal). To illustrate this point, let us consider the values shown in Table 13.5, [31].

Observe that the value of gain is somewhat proportional but in higher frequency bands the G/T doubles (3 dB gain is double) when diameter is increased from 6 to 10 m. Erecting such large dishes on mobile platforms like car or truck or small boats is not practical. Small terminals are therefore an

Table 13.5 Typical G/T for Hub Stations (elevation angle E = 35 deg)

	C Band		
Antenna diameter	3 ma +	6ma +	10 m
G/T clear sky	19 dB/K	25 dB/K	29 dB/K
G/T (rain (0.01%) *	18 dB/K	23 dB/K	28 dB/K
	Ku Band		
Antenna diameter	3 ma +	6 m	10 m
G/T clear sky	26 dB/K	33 dB/K	37 dB/K
G/T rain (0.01%) **	22 dB/K	29 dB/K	34 dB/K

* *Tropical or equatorial climate location,* ** *Temperate climate location + No tracking— "a" is aperture of fixed antenna.*

excellent compromise. This is why VSAT (Very Small Aperture Terminal) became popular in satellite communications. VSAT terminals vary from 1.2 m to 3 m is diameter and serve well providing reasonable throughput supporting voice and data communication.

13.3.2 Transponders on Satellites: Noise Considerations

Every communication satellite uses "transponders"—what is a transponder? Unlike typical communication systems on earth where a "transceiver" is used, there are additional constraints when they are received on satellite. It was already indicated that satellite link is "uplink limited". In addition, it was also seen that there is a considerable difference between power transmitted and received. These two factors come into their own during satellite communication system design. It becomes important to first receive the signal, "frequency shift it" and then transmit it back to earth. This concept of shifting the frequency provides two-fold advantage.

1. Terminals on earth can observe that satellite "locks" on to the signal sent from that ground station. This helps in keeping signal tracked due to considerable delay in the transmit/receive sequence resulting in a round trip delay ($\sim$320 mS).
2. The frequencies of transmit and receive are kept far apart, typically 1500 MHz or more, such that the FSPL of the two are somewhat different. This helps in balancing the uplink and downlink power levels.

The word transponder indicates the communication system onboard satellites does two functions "translation" and "response". Frequency translation (shifting it considerably) and responding to the signal sent from earth, by locking on to that signal as a frequency source to lock on to.

Later during the concept of "translation" grew into another concept known as "active satellite". What this meant was instead of being a passive "transfer and response" unit, the unit took on the "active" role of actually "decoding the signal". What this meant was that the uplink received from ground, would be demodulated, the data stream of layer-2 taken through a buffer, re-modulated using a different on-board method and sent back to earth. Why do such an exercise? The reason again was "to reduce noise". By the time signal traveled 36,000 km, it becomes weak and must be amplified by the satellite receiver unit. This results in increased noise, as was explained earlier using Eqs. 13.4 and 13.5. By demodulating the signal first and then mounting it on a different carrier, the uplink noise is eliminated [31].

This can mean several things. Uplink receiver can use TDMA while downlink could use CDMA. It also meant the method of FDM (Frequency Division Multiplex) is only the first step, followed by step 2 which is to demodulate and use a separate modulation while sending back. It fundamentally relieves the burden of carrying on with noise that came due to uplink. Such "Active Satellites" helped greatly in providing broadband services, better equipped to handle digital streams.

13.3.3 VSAT: Terminals

VSAT (Very Small Aperture Terminals) uses the concept shown in Table 13.5 and serves a wide range of users such as those in remote areas, to provide Internet access as well as TV services. They use conventional dish antennas that are not very large but with narrow aperture. They can be in a fixed location, or mounted on smaller marine (boats, yachts, and others), even mounted on vehicles such as trucks or recreation vehicles. VSAT units are a compromise between the large, heavy, commercial ship borne antenna units and the pocket size user terminal.

13.3.4 VSAT Satellite Users: Portable, Vehicle Borne, and Ship Borne

Now, let us address the practical consideration of supporting a wide range of satellite services throughout the globe, by taking examples of Intelsat and Inmarsat. Commercial satellites now are quite large, typically weighing a few tons while the ground segment is small, mobile, and agile—which means the ground-based user terminals are mounted on cars, trucks, or ships. Satellite terminals also include portable handsets, which are the same size as cell phones. Intelsat and Inmarsat are only used here as examples—there are multitude of service providers that use other technologies and not in GEO orbit, such as Iridium (constellation of 66 satellites) and also CDMA.

To support a wide range of uplink power levels coming from variety of user terminals, Inmarsat offers BGAN (Broadband Global Area Network). This supports customers such as military, explorers visiting remote areas, forest services, and other Government services including disaster recovery. The battery used in some of the smaller, portable devices set the lower limit on transmitted power on uplink which is 2 W or less [32]. Many ground-based portable user units use simple stub antenna (like 3G or earlier model cell phones). At remote locations, users may use Panel Antenna whose gain is higher than a simple stub—effective EIRP in the range of 5 W or more, an example of which is shown in Fig. 13.13 (left).

In Fig. 13.13b, on the right side shows vehicular users who may be explorers using a sophisticated dish on top of

Rugged laptop with BGAN panel antenna   Roof mounted antenna with satellite tracking

Fig. 13.13 Broadband Satellite service examples portable panel, vehicle-mounted dish

Fig. 13.14 Above Deck Equipment (ADE) and stabilized antenna platform

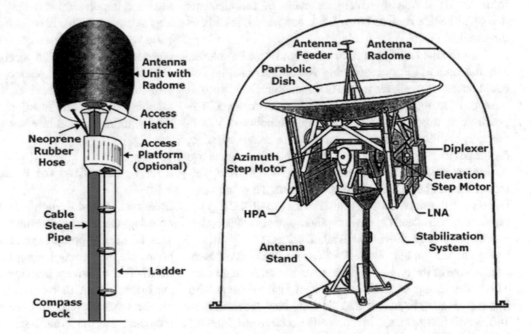

the vehicle. Such dishes though small are three-axis stabilized allowing antenna to be oriented towards satellite using smaller aperture providing much higher uplink power in the range of 10 W or more. These are medium sized in terms of power on the uplink. Much large units shown in Fig. 13.14, that are found on large ships where they are essential link to the outside world.

At the upper end of uplink power are the ship-borne units. Ships are capable of carrying very heavy antenna structures supporting a wider range of services. Some of these sophisticated antenna units incorporate "Dynamic Beam Tilting" technology, which allows satellite signal lock using intelligent real-time beam analysis to eliminate signal skips, pixilation, and loss of service [32]. Figure 13.14 ADE (Above Deck Equipment) shows on the left, how such a large unit is built on the deck of a ship. On the right is an image of what is inside the antenna unit (with radome outside). Note that the HPA (High Power Amplifier is right next

to the antenna, to maximize the signal that can be transmitted. Similarly, the LNA (Low Noise Amplifier) is right next the receiver section of the antenna to minimize noise. The diplexer which combines the transmit and receive signal also is on the same mount. All of these unit functions will be discussed further in Chap. 14 (microwave systems) and in Chap. 18 (Antenna and transmission lines). Salient features of such ship borne units are:

- Automatic Skew Control—to improve horizontal radiation reception
- Antenna Control Unit—beam and direction control, helps targeting correct satellite
- Built-in HDTV Module—allows high-quality service
- Built-in Trisat Function—search and lock for three satellites in C, Ku, and Ka bands
- Built-in GPS System—position, navigation

- GPS Interface—connection to other maritime communications
- Compatible Satellite TV Service—select from global list of satellites in the area
- Radome covering the antenna—passive cover to shield from rain, snow, wind.

Figure 13.14 shows the three-axis stabilized antenna platform commonly used on ships [33]. On the left is a conceptual view of how this platform is enclosed within a Radome and mounted on steel pipe above the deck. The steel pipe also carries cables connecting to the antenna system. Right side of Fig. 13.14 shows the stabilized antenna platform (inside Radome), which has many of the elements described earlier, such as HPA, LNA, and parabolic dish, as indicated in [33].

What are the benefits of such heavy units used in marine communications? For ocean-going vessels, the satellite link could be the only reliable means of communications, the unit must provide focus on reliability. With such features, it is possible to provide high-quality, high-definition TV services, in addition to normal voice and data communication, for all passengers on board such ships. Ship-borne units use directional antenna mounted on a three-axis stabilized platform that can be locked on to the satellite. They are large (antenna dish can be 240 cm) and heavier (up to 700 kg) but providing considerably more uplink power towards the satellite (typically above 100 Watts EIRP).

Again, the "uplink limited" statement holds good here, irrespective of the type of antenna system used from ground. This is the result of earth-based antennas from users who generally tend to be mobile and therefore may provide about 100 W of EIRP (max). In comparison, large satellites of Inmarsat [32] provide maximum sustainable downlink power with EIRP values from the satellite being 20 Watts (43 dBm) for global beams, 800 Watts (58 dBm) for regional beams, and 10000 Watts (70 dBm) for spot beams, respectively. Spot beams therefore provide considerably more power towards specific areas, than any of the uplinks. In each of these cases, uplink power from users tends to be lower than these downlink values.

13.4   Applications: Broadcast, Telecom, Navigation

There are several applications of satellites services. Four major services are reviewed here showing considerations important to user in terms of system operation. Major services are:

- Broadcast—this includes radio and TV, but both are "receive only" (downlink)
- Navigation—includes GPS, Galileo, and GLONASS but are "receive only" (downlink)
- Remote Sensing—using cameras on the uplink and video stream on downlink
- Telecommunication—majority of services use both uplink and downlink.

How do these applications affect wireless design? For the end users, it has considerable impact—since they will not have to transmit towards the satellite. Given that Broadcast and Navigation are one-way (from satellite to ground) service, it is important that the power on this downlink must be kept substantial. This is mainly due to simple receiver design (receiver chipset focuses on small battery size and simple, embedded antenna). The receiver on ground only has the ability to receive and cannot communicate back to the satellite. The service provider will take care of uplink to the satellite that has other functions. Let us consider a few more details, to understand how such services are implemented.

13.4.1   Satellites for Broadcast

Broadcast TV and Radio via satellites are excellent examples of using satellite to distribute audio and video programs. This is a fairly complex venture since studios where programs are made may belong to multitude of agencies. The distribution also may involve multiple agencies. However, the biggest factor that drives the broadcast industry to use satellites is widespread coverage (several continents). The programs may be broadcasted not only nationally; it may serve different regions in the area and also provide feeds to the Internet for global coverage.

In order to get a better understanding, let us consider the overall system diagram. This will provide an overview of communication subsystems and different agencies that may own and operate them. Uplink from such agencies is typically in the upper GHz bands in order to provide ample bandwidth for sending a variety of programs. The downlink from satellite will be usually in the lower bands, to accommodate simpler handsets and mobile users.

Active transponders discussed earlier in Sect. 13.3.2 come into effect for such large bandwidth, or broadband satellite systems. Active transponders allow onboard transponder to handle different data streams mounted on carriers at different frequency bands. The active transmitter on satellite generates a downlink carrier phase and frequency coherent with the uplink carrier. It allows measurement of

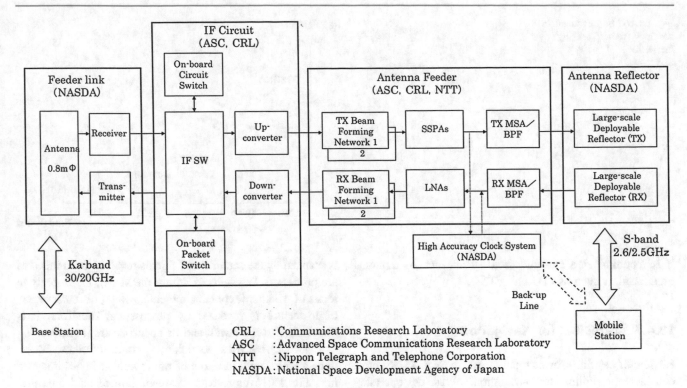

Fig. 13.15 Configuration of mobile communication and broadcast system—agencies

Doppler by the ground station, thereby helping the location of satellite more accurately. This feature indirectly helps prolong the life of the satellite.

An example from Japan is provided in Fig. 13.15 where different user agencies are listed [34]. Several radio components discussed earlier in this chapter, such as LNA, Satellite Signal Power Amplifier (SSPA), Beam forming network—are all indicated in the diagram.

1. The uplink is provided using satellite base station of the National space development agency (NASDA).
2. Since on-board systems involve new technologies such as on-board switching, the Advanced Research lab develops many of these parts (ASC).
3. Since the satellite uses a sophisticated beam forming network covering Japan with multiple beams—provided by multiple agencies such as space research lab, communication research lab, Nippon Telegraph, and Telephone Corporation, participate.
4. Antenna reflectors on the satellite that directly provide links to users—are serviced by the national space agency (NASDA).
5. The national space agency also provides a high accuracy clock system—this is because the broadcast studio sends a stream whose header must be accurately in sync with different parts of the region (from eastern islands of Japan

to western islands of Japan that are close to south china sea.

In the industry, similar efforts by different companies are common. Users of such satellite broadcast systems have unique advantages. Their favorite TV programs are available "Direct-To Home (DTH)"—this is much different from the cable TV or other methods to bring programs to home users. Over the air broadcast using FM for TV and AM/FM for audio, used to restrict services to no more than 50 or 60 km distance at best. Also, mobile users who commute to work can use satellite signals to cover large regions where their favorite radio or TV is now available.

Over the past two decades, there was a global realization that digital audio satellite coverage allows broadcasting agencies to offer programs globally. This resulted in the culmination of global standard known as DRM (Digital Radio Mondiale). DRM digital radio is the global standard that offers new broadcasting services to local, regional, national, and international audiences. It is internationally recognized system for digitizing radio in all frequency bands [35]. Broadcasting radio programs on short wave, medium wave, and long waves (AM frequencies), as well as in the VHF frequencies improves quality of broadcast. DRM is a non-proprietary, open standard (no company owns the technology), which can be implemented by any broadcaster.

Fig. 13.16 Improvement in accuracy (remote sensing) over the years

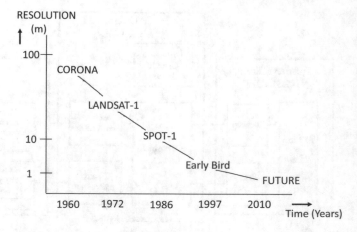

It is recommended and endorsed by relevant international organizations such as ITU and ETSI.

13.4.2 Satellites for Navigation

Review of satellites for navigation was in Sect. 13.3. Let us consider the uplink for navigation. What do operating agencies provide on the uplink to satellite that makes them so useful and accurate?

The concept of navigation is based on satellite ephemeris, which is a complex satellite range scheduling problem. Navigation data to user include satellite ephemeris, clock error, satellite almanac, and ionosphere parameter. These must be included in uplink to navigation satellite through ground stations, so as to improve the accuracy of Positioning, Velocity, and Timing (PVT). In addition, there will communication between these ground stations. Inaccuracies in connection between them using either Inter Satellite Link (ISL) or through fiber-optic links must be estimated well. Special algorithms are used to calculate such inaccuracies and add them into the uplink [36]. This complexity has to be included since most navigation satellites such as GPS, GLONASS, Galileo, or "BeiDou" are in the MEO orbit. Uplinks stations must be connected and sent from multiple locations. All agencies providing navigation services—use algorithms similar to the one cited in [36] and make continued efforts to improve accuracy. This is one of the important features (better location accuracy) readily observed by users.

13.4.3 Satellites for Remote Sensing: Accuracy

Satellites used for remote sensing were reviewed in Fig. 13.8 and it was observed that uplink essentially consisted of a swath. What is this swath and how accurate are its boundaries? The swath is a strip of land/sea on earth which

is scanned by the satellite and signals received from this area are processed. Whatever is within the swath, its property in terms of whether it contains minerals, water, or forest or any other feature is revealed by processing. Therefore, it is important that the swath and its accuracy are improved.

Figure 13.16 tries to address some of these questions about swath, in terms of improvement in accuracy. In fact, Fig. 13.16 represents the resolution of land area precisely measured during repetitive visits to the same area by the same satellite—higher accuracy is due to improvements in technology [37]. The improvement in terms of accuracy and resolution is indicated as milestones in remote sensing satellites over the years. Fundamental to such accuracy are designs of how carefully cameras mounted on the satellite frame reduce jitter due to movement of the satellite. After all, even minor vibrations on the satellite, in arc seconds translate into several meters of change over a distance of 20,000 km from where it sees the earth. Such design issues were addressed systematically over the years by satellite builders, improving accuracy of not only land observation but maintaining accuracy in orbit, which is affected by variations in gravity of earth, moon, and solar wind.

13.4.4 Satellites for Telecommunication in Various Regions

Satellite communication comes into its own in emergency situations, where there are no other options. It can be at sea or even in towns on hill tops or anywhere on the countryside with vast farmlands where cellular signal does not reach. Emergency responders, farmers, trekkers, scout camp organizers will be quick to point this out. There are many satellites such as "Iridium satellites", Globe-star, and others that provide both voice and data traffic. There are others focused on bringing DTH (Digital TV to homes) or IoT (Internet of Things). A concise table with examples of some satellite service providers and features are listed in Table 13.6.

Table 13.6 Satellite communication for various services—worldwide

Provider	Orbit	Devices	Quality/Cost	Comment
INMARSAT	13 satellites in the Geostationary orbit	Various types, handheld to major ship-mounted units	Highest quality/reasonable cost. Allows full broadband Internet access	Land mobile service + global maritime services except in polar circle
Globe-star	48 satellites in the 1400 km orbit with two earth facing antenna arrays	BYOD (Bring Your Own Device) + Sat Fi	Low latency (60 ms) for real-time Internet + voice service	Simpler devices and portability throughout the globe
Iridium	71 satellites in low earth orbit at a height 781 km inclination of 86.4°	Popular with a variety of portables and sets	Very reliable and reasonably priced. Low latency (50 ms)	Extensively used by emergency responders
SES Satellite	56 satellites in Geostationary and 16 satellites in MEO	High-definition TV/video	Direct-To-Home (DTH), Cruise, and DAB radio broadcast	Commercial provider with services at high end
Intelsat	Over 70 satellites in geostationary orbit	DTH and high throughput Internet. Wide range of devices	World's largest satellite network has support for disaster recovery	Commercial service with a variety of distribution systems
Eutelsat	35 satellites in geostationary orbit	Broadband, data, video, mobility, government	DTH and mobility as the major focus	Commercial provider with services for six global regions
Telespazio	10 satellites in LEO	Focus on data security as a service	Space intelligence service	Various space technology services
INMARSAT	13 satellites in the Geostationary orbit	Various types, handheld to major ship-mounted unit	Highest/reasonable. Allows full broadband Internet access	Land mobile service + maritime services except in polar circle

The table confirms several statements made earlier in this chapter—notice considerable operations in LEO orbit, and very crowded geostationary orbit.

13.5   Telemetry, Telecommand, and Tracking

This is perhaps an area hardly known to users, but at the core of work performed by space operators. These three provide the means to launch, maneuver, and maintain satellites in orbit.

Telemetry—as the name indicates "measurement at a distance". Telemetry is a downlink from satellite to the earth station. It provides a continuous stream of measured values of voltage, temperature, pressure, and other parameters on board the satellite (health parameters). This is essential for the ground station crew to know the health of different systems and sub-systems. It is possible to make minor adjustments to improve satellite performance, using "telecommand".

Telecommand is an uplink that helps orientation, three-axis stabilization, and other maneuvering activities necessary to keep the satellite at the correct orientation as well as at the right orbit. It helps open up the solar panel, orient it towards the Sun for maximum efficiency. Similarly, telecommand to unfurl the satellite antenna dish, its solar sail boom or other parts are very much the core of how the satellite operates efficiently.

Tracking is a complex and on-going effort to know exactly where the satellite is—in three-dimensional space. Why is this complex? Because the earth we are sitting on is itself rotating and at various times other celestial objects nearby such as the moon and celestial phenomenon such as solar wind—move the satellite from its intended position in an irregular fashion. Tracking helps the earth station operators and technologist know what to predict and how to navigate the satellite, much like how pilots operate the aeroplanes in flight on a stormy night. Satellite uses infrared sensor to orient towards the earth and during eclipse, it is important this is turned off. Similarly, operation of the Sun sensor is affected by solar eclipse. Otherwise, it can result in the satellite getting confused between earth and another celestial body (the sun or moon).

How does this fit into the wireless communication scheme? The frequency band and reliability of these links are as important as the service it provides—this may not seem obvious, but true. This is because TTC (Telemetry Tracking and Command) is essential to the very survival of the satellite—independent of when and how it provides service. Recall that there are satellites that encounter unexpected events (sun flares), as well as expected events such as solar eclipse or lunar eclipse that affects its service. To recover from such events too, TTC is essential [38].

Telemetry, for example, requires the following, in a typical flight:

Frequencies.

–S-band (2.2–2.3 GHz).

–C-band (3.7–4.2 GHz).

–Ku-band (11.7–12.2 GHz).

•BER = 10^{-5}.

These bands cannot be used for any other satellite services. Similarly, for missions that operate far away, such as intergalactic missions, distance would dictate operating at much lower frequencies, in order to limit path loss. Observe that satellite telemetry requires much higher accuracy in terms of BER. Typical BER tolerated in cellular systems is 10^{-2}. Therefore, the telemetry BER requirement of 10^{-5} would be considered "excellent quality".

Telecommand system, in comparison, is even more stringent in its BER requirement. The reason is telecommand can determine whether the satellite survives or is lost in space. For example, during solar eclipse, if the telecommunication is not used to switch over to an omnidirectional antenna, satellite dish orientation after the eclipse could be facing away from earth. Therefore, no one on earth can communicate with that satellite if that ever happens. These are not theoretical, but practical experiences in space missions, which have led to very strict methods of communicating, orienting, and operating satellites. Telecommand requirement for a similar flight as above would be:

Frequencies.

–S-band (1.6–2.2 GHz).

–C-band (5.9–6.5 GHz).

–Ku-band (14.0–14.5 GHz).

•BER = 10^{-6}.

Also note that the frequencies used within the same band are considerably away from each other so that the uplink frequency (telecommand) and much farther causing no interference to telemetry on the satellite. Telecommand systems perform the following key functions.

- Power on/off subsystems
- Change subsystem operating modes
- Control spacecraft guidance and attitude control
- Deploy booms, antennas, solar cell arrays, protective covers
- Upload computer programs.

Carrier Tracking is a two-way communication between satellite and ground. It uses coherent communication by establishing a frequency lock mechanism on-board, with the following:

- Transmitter phase locks to the received frequency
- Transmitted frequency is a specific ratio of the uplink frequency
- Easy to find and measure the frequency received on the ground
- Doppler shift provides range rate (distance to satellite and direction of movement).

Range rate of a satellite as a function of time: range rate is zero at closest approach. The Doppler shifts are almost a factor of 7 times smaller than for Earth.

Tracking is a technique commonly used by the military to determine whether an aircraft is "friend or foe" by using similar "lock in" mechanism. In the case of satellite link, the distance results in also establishing the range, since it must be accounted for during tracking. Ranging is accomplished as follows:

Uplink pseudo-random code is detected by the satellite and retransmitted on the downlink.

- Turnaround time provides range.
- Ground antenna azimuth and elevation determines satellite angular location (polar coordinates).

Interface requirements for TT&C subsystem is indicated in Table 13.7 [38].

TTC remains to be the most reliable part of every space mission—which is generally not discussed in satellite user forums, it is exclusive to satellite builders and operators.

13.6   Satellite Support to Maritime and Aeronautical Operations

Since satellites in geostationary and LEO / MEO orbits support global operations, why are these two services (aero and marine) separately noted? Its importance is in terms of satellite becoming an essential and only communication for

Table 13.7 Telemetry telecommand and tracking interfaces (On board)

Subsystem	Requirement
Attitude determination and control	Antenna pointing
Command and data handling	Command and telemetry data rates Clock, bit sync, and timing requirements Two-way communication requirements Autonomous fault detection and recovery Command and telemetry electrical interface
Electrical power subsystem	Distribution requirements
Thermal/structural	Heat sinks for power amplifiers Heat dissipation of all active boxes Location of TT&C subsystem electronic boxes Clear field of view and movement for all antennas
Payload	Storing of mission data RF and EMC interface requirements Special requirements for modulation and coding

these two types of transportation. While operators such as the pilot and ship's captain have other means of communication for long time, it is only since the late 70 s that passengers on board also communicate to the outside world, using satellite.

There are some differences between the two. Ships by their very nature are capable of handling heavy antenna on a gimbled platform (Fig. 13.14), allowing high-quality broadband services such as the BGAN from Inmarsat—which was discussed earlier. Mounting a fairly large dish on a stable platform is not an issue on a ship and routinely done.

That is not the case for an aircraft. Due to its range of motions in the air and increase in air traffic there is considerable effort being put in to have a separate satellite-based system to service aircraft [39]. This includes pilot/aircraft crew and passengers. For aircraft navigators, such satellite-based navigation services can become a good supplement particularly as aircrafts approach major airports in cities where air traffic other than long haul jets, may be considerable. Major airports already allow aircraft landing and take-off every 90 seconds—such navigation support service from satellites can complement airport-based navigation controllers and to some extent ease their tense work during peak hours.

There is another dimension to maritime and aeronautical operations—watching them using a satellite. While this may look suspicious, the intent is to scan the planet earth for its RF frequencies and help reduce congestion and interference in the long term. A new generation of satellites called the Hawkeye has started such an effort [40]. Using three small satellites in LEO (about 500 km) the Hawkeye series of satellites hopes to make it safer for aircraft and sea vessel travel around the globe.

13.7 Remote Sensing Drones and Their Growth

Drone or UAV (Unmanned Aerial Vehicle) started out as a hobby but quickly grew into a major industry of its own. Replacing reconnaissance aircraft, flying in war zones, taking pictures of difficult terrains but at close range are some of its strong points.

The first generation of remote sensing was essentially performed by pilots using cameras to take pictures (mostly by the military). This was followed by the second generation where earth-orbiting satellites (discussed earlier in this chapter—remote sensing) took extensive pictures of earth. Such satellites were typically in a Low Earth Orbit (LEO) for better resolution. The third generation is considered to belong to UAV, which offers a method of remote sensing that may provide pictures with much better resolution than the first two generations [41]. In fact, the International Journal of Remote Sensing considers it a separate branch of remote sensing with its own section to deal with research and operational paper submissions.

In terms of wireless communication, two factors must be considered— (1) type of camera and its throughput, (2) type of wireless system on board and its transmit power/frequency. In addition, there are other considerations for any designer such as keeping number of components to a minimum and reduce weight, high-quality image and

Table 13.8 Typical specifications of camera used in a drone

Ground sample distance	8 cm per pixel (per band) at 120 m (~400 ft) AGL
Sensor resolution	1280 × 960 (1.2 MP per EO band)

whether there is enough storage capacity on board. A practical method of doing so is shown in an example at [42]. Such cameras with the specifications indicated in Table 13.8 can fit into a small drone.

Most drone users prefer to use an FPV (First Person Visual) camera, which downloads directly to the user, using its on-board transmitter. There are software packages that allow the user to feel as if you as the user is sitting in the drone. Such radio units with FPV transmitter have 200 milliwatt, 5.8 GHz, 40 channel video transmitter, it produces an omnidirectional radiation pattern from its circularly polarized antenna [43]. This is typical of any Access Point at home. It is best to use circularly polarized antenna since the orientation of drone will not result in losing contact with the transmitter. This is essential due to the telecommand and telemetry, as discussed in the previous section. It is also important to choose a transmitter whose frequency is not very high—this is directly related to range from the user. It is also possible to use a combination of omnidirectional as well as high gain antenna. Once near the target area, receiving the spectral feed from the camera on directional beam provides better throughput over longer distances. This has to be an interactive, gradual exercise for every student.

Overall, drones provide a first-hand feel of flying a camera, which needs multiple skills in terms of navigation, control, and tracking. There are several volunteer organizations that teach such skills to make every student a responsible aircraft flying amateur. In addition to possibly a thrilling activity, it allows the operator to learn about basic aspects of aerospace and technical details of how satellites are able to sustain, orient themselves and serve humanity. In addition to providing a simulated view of orbiting Sirius satellites, this reference provides useful information to such enthusiasts [44].

13.8 Conclusion

What started as a fascination to explore the universe with Sputnik, quickly evolved into a major necessity by helping citizens of the world communicate. The topic of satellite communication is integral to and center piece of wireless communications. Its system design dictated by orbits and effort to launch them has resulted in something that is quite complex. This chapter tries to touch just the surface, showing the level of complexity and challenge.

The number of references at the end of this chapter varies from concepts in Physics, dynamics of flight, and orbits that often need 3D view of how it all comes together. The use of modern video simulations is all a post-effort to recall what has been achieved.

Initially approached with caution, the field quickly evolved and has become a very stable and broadband communication system to regions inaccessible to any other form of wireless system. The fact that signals from the edge of galaxy have been sent successfully and accurately, which provide testimony to design and conceptualization of flight mechanics and how accurately they are measured at great distances using telemetry.

Use of satellites whether for Navigation, Remote sensing, TV and Radio broadcast, or vital telecommunication to emergency service personnel, it is an integral part of modern-day life for everyone. Those in business, entertainment, exploration, or disaster recovery—increasing rely on the services provided by satellites.

Will there be a limit to the number of satellites in each orbit? Everyone skeptical has been proved wrong by continuous use of ever-improving technology. It is therefore difficult to conclude that there will be any limit to what satellites can do. If anything, it is moving in the positive direction, with more companies, Universities, and amateur radio enthusiasts participating in this grand technology that truly uses rocket technology to get up there into orbit.

Sufficient effort has been made to provide references and pointers to those who want to pursue this fascinating area of knowledge. There are several applications that entirely depend on satellites for navigation and guidance. Similarly, aircraft and maritime vessels depend on satellites as an essential means of communication with those onshore and in the air. VSAT terminals have become the mainstay of such satellite users. The growth in this field is expected to be substantial in the future.

Summary

This chapter captures the essence of both spaces, celestial bodies, and laws of orbits as applied to wireless communication. Ever since the Telstar communication satellites were introduced in the late 60 s, this topic of satellite communication resulted in exponential growth. It is a continuum of fascination that humanity has always had towards outer space—combined with a practical sense of what artificial satellites provide, developed and launched by humans. It was observed that the laws that govern celestial bodies, govern artificial satellites too—they too go into similar elliptical orbits, therefore communicating with them needs a similar understanding.

This followed details such as different orbits—low earth, middle earth, and geostationary orbits. It was shown that multiple satellites are in these three orbits—their characteristics and how and where they can be seen from earth and when. Such a review revealed the observation that with the exception of geostationary orbit, satellites in all other orbits rise from the horizon and move and dip on the other horizon as seen from earth. Therefore, all other orbits need tracking antennas or means to adjust by compensating for Doppler effects—while communicating.

Advantages of geostationary orbit, why satellites in this orbit seem stationary to observer on earth were reviewed using multiple diagrams, observations, and references to video simulations. It was shown that even satellites in geostationary orbit have limitations—they cannot effectively cover the Arctic and Antarctic regions of the earth. Separate orbits such as the Molniya and Tundra orbits were discussed, with their own unique features. It was shown that despite limitations, these orbits are useful and continue to have satellites.

Popular satellite applications such as GPS for navigation, Satellite feeds for TV, audio streams around the globe that broadcast industry has embraced were reviewed. Communication with maritime and aeronautical users revealed another dimension of emergency links using satellites in LEO and MEO orbits.

Other areas that benefited and have grown, such as remote sensing and telecommunication across the globe, experienced exponential grown with support from satellites in various orbits. These were extensively reviewed with examples from industry.

Finally, the often-unseen part of satellites—telemetry, telecommand, and tracking which mark the foundation of all space operators, was touched upon. The importance of this branch of space technology, why TT&C requires the highest reliability in the satellite system, was indicated with examples. They continue to form the basis of deploying, orienting, and controlling satellites throughout their mission.

The chapter concluded with the recent approach of remote sensing using drones—which touches upon the human fascination of flying something in the air and observes things that hitherto were possible only by birds or creatures could see things from the air. It has opened up an entirely new era into remote sensing, complementing methods used by satellites.

Emphasis on participation

1. Contact your local satellite TV service provider and make a few measurements on the received signal. What is the antenna gain they use and the frequency band?
2. Using your cell phone and Fig. 13.7, check how many GPS satellites are there at a time? It may be necessary to have a good radio receiver in the bands indicated in Fig. 13.7 to know how many distinct satellites are in view. Mark some subway / below ground locations where GPS signal is lost.
3. Many space agencies provide guided tours—write to any space agency in your area. Can they support some of your summer projects—with ideas?
4. Using references provided under 13.7 drones section—find out if it is possible to build and fly drone with cameras.

Knowledge gained

1. Different types of orbits, their advantages, and limitations.
2. Geostationary vs other orbits—how many communication satellites are deployed in non-GEO orbit. How service providers cope with tracking and handover.
3. Simplification of ground-based satellite antenna and receiver. Improvements in satellite technology to support mobility.
4. Satellite applications for (a) broadcast, (b) navigation, (c) remote sensing, (d) emergency communication.
5. Considerations for satellite beams in different orbits, types of antennas user can have.
6. Advantages of VSAT dish and terminals, difference between high-end broadband service for ships and aircraft.

Homework problems

1. Using Eq. 13.3.2 to calculate the noise power in a normal cell phone with 4G service that has 20 MHz bandwidth. Compare it to a satellite dish that uses 500 MHz bandwidth.
2. Using Eq. 13.3.3 to compare the FSPL for a satellite in 750 km LEO orbit with the same satellite placed in GEO orbit of 36000 km, operating in the C band at 6050 MHz on the uplink. How much is the difference in signal level received? If the satellite sends its downlink at 3800 MHz what is the space loss?
3. In Fig. 13.7, satellite in polar orbit is shown. Trace the swath (on the right side of figure) for two satellites in polar orbit, one at 900 km altitude and another at 1600 km altitude. How often does the satellite swath revisit the same place?
4. Using Molniya orbit, a satellite is placed to serve Tokyo, and rest of Japan. Beam is designed to serve that country starting at daytime 6 am till 6 pm. Where can the same satellite be used from 6 pm till next morning 6 am?
5. If the satellite indicated in problem 4 above was placed into Tundra orbit, what happens to the coverage? Can it be used elsewhere during what times?

Project ideas

1. Using Eq. 13.2 to calculate the period of satellites placed in the transfer orbit that has the following perigee and apogee (distance from earth surface):

Perigee	Apogee	Period
300 km	2000 km	?
350 km	36000 km	?

Hint: Use videos in reference 6 to understand the transfer orbit.

2. Develop a software program (MATHLAB) for a circle and ellipse. Using the program place two satellites on opposite sides of the ellipse and circle. Drop a 30° beam on earth from each satellite and observe how that beam traverses a swath when diameter of circle is 20000 km, ellipse with major axis 20000 km and minor axis 2000 km. Use Ref. [21].

Out of box thinking

1. For recent launch—rocket motors have been built using 3D printers. Use a printer program to design and build a satellite frame structure capable of supporting a 500 kg satellite remote sensing satellite.
2. Small satellites were built, and space agencies sometimes offer a free launch for putting them in low earth orbit. Propose a project to support a satellite with scientific payload and approach different launch agencies for orbiting it in LEO.

References

1. Kepler's laws – Khan Academy (partner with NASA). https://www.khanacademy.org/partner-content/nasa/measuringuniverse/orbital-mechanics/a/keplers-first-law
2. Eccentricity of Earth, Astronoo, Updated Jun 1, 2013. http://www.astronoo.com/en/articles/eccentricity-earth.html
3. Space Environment, What is escape velocity, Northwestern University. http://www.qrg.northwestern.edu/projects/vss/docs/space-environment/2-whats-escape-velocity.html
4. How to get a satellite into geostationary orbit, Jason Davis, Jan 17, 2014. The planetary society. https://www.planetary.org/blogs/jason-davis/20140116-how-to-get-a-satellite-to-gto.html
5. Finch MJ (1985–1986) Limited space: allocating the geostationary orbit, 7 Nw. J. Int'l L Bus 788 (1985–1986). http://scholarlycommons.law.northwestern.edu/njilb
6. National Space Security Institute – Orbit types: 5/7/2018 http://educationalaids.nssi.space/play.htm?Orbit_Types
7. Ritchie G (2019) Why Low Earth orbit satellites are the new space race. Bloomberg, 2019. https://www.washingtonpost.com/business/why-low-earth-orbit-satellites-are-the-new-space-race/2019/08/15/6b224bd2-bf72-11e9-a8b0-7ed8a0d5dc5d_story.html
8. USC satellite database – published Dec 8, 2005, updated Dec 16, 2019. https://www.ucsusa.org/resources/satellite-database
9. Urrutia DE (2019) Relativity space to launch satellite 'Tugs' on 3D-printed rocket, September 20, 2019. https://www.space.com/relativity-space-to-launch-momentus-tugs-2021.html
10. Surrey Space Center, A pioneer of modern "small satellites" since 1979. https://www.surrey.ac.uk/surrey-space-centre
11. ESA commissions world's first space debris removal, 9/12/2019. ESA/Safety and Security/Clean Space. https://www.esa.int/Safety_Security/Clean_Space/ESA_commissions_world_s_first_space_debris_removal
12. Everything RF – GPS frequency bands. https://www.everythingrf.com/community/gps-frequency-bands
13. GPS overview, Page created for Remote Sensing course, University of Texas at Austin. http://www.csr.utexas.edu/texas_pwv/midterm/gabor/gps.html
14. Jefferts SR (2017) Primary Frequency Standards at NIST, NIST Time and Frequency division, 19 December 2017. https://www.aps.org/units/maspg/meetings/upload/jefferts-111517.pdf
15. European GNSS (Galileo) open service, Signal in space interface control document, for public consultation. https://www.researchgate.net/publication/280696714_The_Galileo_Frequency_Structure_and_Signal_Design
16. GLONASS general introduction. European Space Agency, Navipedia. 22 June 2018. https://gssc.esa.int/navipedia/index.php/GLONASS_General_Introduction
17. Satellites and sensors - Satellite characteristics: Orbits and swaths, Natural Resources Canada, Nov 20, 2015. https://www.nrcan.gc.ca/maps-tools-publications/satellite-imagery-air-photos/remote-sensing-tutorials/satellites-sensors/satellite-characteristics-orbits-and-swaths/9283
18. Sharing Earth Observation Resources, Eo Portal Directory, European Space Agency, CartoSat-2, https://earth.esa.int/web/eoportal/satellite-missions/c-missions/cartosat-2
19. Nelson SAC, Khorram S (2019) Image processing and data analysis with ERDAS IMAGINE. CRC Press, Taylor and Francis Group. ISBN-13: 978–1–1380–3498–3. https://sentinel.esa.int/web/sentinel/missions/sentinel-2
20. DiPierro S, Akturan R, Michalski R (2010) Sirius XM satellite radio system overview and services. In: 2010 5th advanced satellite multimedia systems conference and the 11th signal processing for space communications workshop, Cagliari, pp 506–511. https://www.semanticscholar.org/paper/Sirius-XM-Satellite-Radio-system-overview-and-DiPierro-Akturan/a7dae381b27bf9fd6f4c5c70d03d7b9b5b697d7b
21. Ground control, Global Satellite communications, Satellite look angle Calculator. https://www.groundcontrol.com/Satellite_Look_Angle_Calculator.htm
22. Geostationary orbit, legal issues, United Space in Europe, European Center for Space Law. http://www.esa.int/About_Us/ECSL_European_Centre_for_Space_Law/Geostationary_Orbit._Legal_issues
23. How closely spaced are satellites at GEO? Space Exploration Beta stack exchange. https://space.stackexchange.com/questions/2515/how-closely-spaced-are-satellites-at-geo
24. The INTELSAT globalized network: overview. http://www.intelsat.com/about-us/overview/
25. Satellite Internet home page, List of satellites in geostationary orbit. http://www.satsig.net/sslist.htm
26. Foust J (2019) Space insurance rates increasing as insurers review their place in the market, Sept 14, 2019. Space News, https://spacenews.com/space-insurance-rates-increasing-as-insurers-review-their-place-in-the-market/
27. SpaceX, design, manufacture and launch of advanced rockets and spacecraft. https://www.spacex.com/about

28. Zoller J (2018) Satellites and Spectrum management – chapter IV. The Aspen Institute. https://csreports.aspeninstitute.org/Roundtable-on-Spectrum-Policy/2018/report/details/0345/Spectrum-2018

29. Romanofsky RR, Bhasin KB, Downey AN, Jackson CJ, Silver AH, Javadi HHS (1996) Integrated Cryogenic satellite communications cross-link receiver experiment, NASA technical memorandum 10710. In: 16th international communications satellite systems conference, sponsored by the american institute of Aeronautics and astronautics. Washington DC. https://ntrs.nasa.gov/archive/nasa/casi.ntrs.nasa.gov/19960012195.pdf

30. Satellite communication – link budget, Lecture 9, RF Café. https://www.rfcafe.com/references/articles/Satellite-Comm-Lectures/Satellite-Comms-Link-Budget.pdf

31. Maral G (2003) VSAT networks, 2nd edn. Wiley Publications, p 205. ISBN:0–470–86684–5. http://index-of.co.uk/Networking/John%20Wiley%20&%20Sons%20-%20VSAT%20Networks.pdf

32. Exhibit D, INMARSAT 4F1 and 4F3, Technical description. http://licensing.fcc.gov/myibfs/download.do?attachment_key=-143731

33. Ilsev SD (2012) Shipborne antenna mount and tracking systems, Durban University of technology, (DUT), Durban, South Africa. TransNav – Int J Marine Navigat Safety Sea Transport 6(2)

34. Shin-Ichi K (2003) Configuration of mobile communication satellite system and broadcasting satellite systems, Chapter 3.2. J Natl Inst Inf Commun Technol 50(3/4)

35. Digital radio mondale – what is DRM. https://www.drm.org/what-is-drm/

36. Tang Y, Wang Y, Chen J, Li X (2016) Uplink scheduling of navigation constellation, based on immune genetic algorithm. PLoS ONE 11(10):e0164730. https://doi.org/10.1371/journal.pone.0164730Li2. https://www.ncbi.nlm.nih.gov/pmc/articles/PMC5063407/pdf/pone.0164730.pdf

37. Fu W, Ma J, Chen P, Chen F (2019) Remote Sensing satellites for digital earth. Manual Digit Earth. Springer link, First online: 20 November 2019. ISBN: 978–981–32–9915–3. Print ISBM: 978–981–32–9914–6. https://doi.org/10.1007/978-981-32-9915-3_3

38. Keesee JE (2003) Satellite telemetry, tracking and control subsystems, course 120 satellitettc, MIT lecture notes. https://ocw.mit.edu/courses/aeronautics-and-astronautics/16-851-satellite-engineering-fall-2003/lecture-notes/l20_satellitettc.pdf

39. IRIS: satcom for aviation. Video by ESA (European Space Agency). https://www.youtube.com/watch?v=2aSF_WDRBPI

40. Scoles S (2018) New satellites will use radio waves to spy on ships and planes. Sci Mag. https://www.wired.com/story/new-satellites-will-use-radio-waves-to-spy-on-ships-and-planes/

41. Milas AS, Cracknell AP, Warner TA (2018) Drones – the third generation source of remote sensing data. Int J Remote Sens 39 (21), 7125–7137. Published online: 19 Nov 2018, https://doi.org/10.1080/01431161.2018.1523832

42. Robrecht Moelans (2019) How to integrate a multispectral camera under a low-cost drone, A story about drones, camera and sensor. MAPEO Multispect. https://blog.vito.be/remotesensing/multi-spectral-camera-under-a-low-cost-drone

43. Corrigan F (2019) Understanding drone FPV live video, antenna gain and range. In: Dronzon. Drones, drones technology, knowledge, news and reviews, September 11, 2019. https://www.dronezon.com/learn-about-drones-quadcopters/learn-about-uav-antenna-fpv-live-video-transmitters-receivers/

44. SiriusXM satellite radio (Info for Techies) http://www3.sympatico.ca/n.rieck/docs/sirius.html

Abbreviations

Millimeter wave	Refer to RF signals in the 30–300 GHz range, in which an unlicensed band of 60 GHz and licensed band of 80 GHz are used for millimeter wave links
Microwave link	Refers to a pair of towers with parabolic or other antennas mounted. The antennas on both of them together act as a pair. One transmits and the other one receives and vice versa. The tower in an urban area or near the control center is designated primary, and the other tower could be in the wilderness.
Fresnel zone	Named after physicist Augustin-Jean Fresnel, is one of a series of confocal prolate ellipsoidal regions of space between and around a transmitter and a receiver. The primary wave will travel in a relative straight line from the transmitter to the receiver. Other deflections can occur from any object located within that zone.
Frequency division multiplex (FDM)	A method of separating the transmit and receive frequency bands so that such simultaneous communication becomes possible. Transmit frequency F1 and receive frequency F2 are kept considerably apart to ensure that there will be no interference between them
U-NII bands	Unlicensed-National Information Infrastructure bands are defined in the USA by the Federal

	Communications Commission and the allocated bands are in the 5–6 GHz range
Microwave licensed frequency bands	The bands widely licensed by regulatory agencies globally are in the regions of 6 GHz, 11 GHz, 18 GHz, 23 GHz, and 80 GHz. Well-established microwave link products are provided in these bands, by most vendors
TE mode—transverse electric mode	This waveguide mode is dependent upon the transverse electric waves; sometimes also called H waves, characterized by the electric vector (E) being always perpendicular to the direction of propagation
TM mode	Transverse magnetic waves, also called E waves, are characterized by the magnetic vector (H vector) being always perpendicular to the direction of propagation
TEM mode	The transverse electro-magnetic wave cannot be propagated within a waveguide, but it can be propagated within a coaxial cable. It is the mode that is commonly used within coaxial and open wire feeders. The TEM wave is characterized by the fact that both the electric vector (E vector) and the magnetic vector (H vector) are perpendicular to the direction of propagation
Cross-polarization or XPD	Cross-polarization is the polarization orthogonal to the desired polarization. For example, if the fields from an antenna are meant to be horizontally polarized,

© Springer Nature Switzerland AG 2022
K. Raghunandan, *Introduction to Wireless Communications and Networks*,
Textbooks in Telecommunication Engineering, https://doi.org/10.1007/978-3-030-92188-0_14

	the cross-polarization in this case is vertical polarization. If the polarization is right hand circularly polarized (RHCP), the cross-polarization is left hand circularly polarized (LHCP)	Automatic transmit power control (ATPC)	It is a feature in microwave systems that automatically increases the transmit power during "Fade" conditions such as heavy rainfall. ATPC can be used with adaptive code modulation (ACM) to maximize link uptime, stability, and availability. When the "fade" conditions (like snow or rainfall) are over, the ATPC system reduces the transmit power again. This reduces the stress on the microwave power amplifiers, which in turn reduces power consumption, heat generation, and increases expected lifetime (MTBF or mean time between failure)
Orthogonal mode transducer (OMT)	Orthogonal mode transducer (OMT) is a passive microwave component that separates a signal into two linear orthogonal polarized signals received from a common port or, vice versa, it combines two such signals from vertical and horizontal ports into a common port. The OMT supports circularly, elliptically, and linearly polarized waveforms		
Earth's curvature and K-factor	Earth's bulge is a term used in telecommunications that refers to the circular segment of the Earth profile that blocks off long-distance communications. Since the geometric line of sight passes at varying heights over the Earth, the propagating radio wave encounters slightly different propagation conditions over the path. The usual effect of the declining pressure of the atmosphere with height is to bend radio waves down towards the surface of the Earth, effectively increasing the Earth's radius, and the distance to the radio horizon, by a factor of around 4/3. This K-factor can change from its average value depending on the weather	QAM or quadrature amplitude modulation	It refers to a family of digital modulation methods widely used in modern telecommunications to transmit information. QAM consists of two digital bit streams that change (or modulate) the amplitudes of two carrier waves using amplitude-shift keying (ASK). The two carrier waves of the same frequency are out of phase with each other by 90°, known as orthogonality or quadrature. The transmitted signal is created by adding the two carrier waves together. At the receiver, the two waves can be coherently separated (demodulated) using their orthogonality property
Fade margin	It is an allowance provided during telecommunication link design that accounts for sufficient system gain or sensitivity to accommodate expected fading. The objective is to ensure that the required quality of service is maintained	Low latency	The term applies to the time taken by the data stream to go from one point to another. Microwave links, in general, have the lowest latency (between 100 and 1 ms) that meets the 5G requirement for URLLC
Scintillation effect	Scintillations are rapid fluctuations in the phase and amplitude of an electro-magnetic wave. They occur due to local rapid variations in the refractive index of the medium through which the wave is traversing. Scintillation is mainly caused by the local variation of the ionospheric electron density	Antenna wind load	Wind load applies to antennas of all shapes and sizes that are mounted outdoors. It can be reduced by reducing the "frontal area" using mesh design or similar design methods. The generic formula for calculating wind load is $F = A \times P \times Cd$, where F is the wind load, A is the surface area of the antenna (usually given in square feet), P is the wind pressure (calculated from

another formula), and Cd is the drag coefficient

Remote tower   Among the two towers used for microwave links, usually one is in a populated where the traffic begins/ends. Other tower or towers can be in complete wilderness, either in agricultural land, forest, mountain top, etc. Such towers are described as the "remote tower" which means access and maintenance of such towers requires planning in terms of getting there, how it is operated, etc.

This chapter deals with communication links seen everywhere including the countryside. It is about tall towers with dish antennas. Such dish antennas are oriented towards their counterpart, several kilometers away (usually not in the vicinity). Quite often, such towers are located on top of hills, or tall buildings, etc., in urban areas. The primary purpose of microwave link is to carry telecommunication traffic (backhaul) such as voice and Internet data traffic. Millimeter-wave link is of similar category but uses much smaller dishes with shorter distances (link distance is no more than 5 km away). Let us begin with the basic considerations to design such links. Initial deployment of microwave links began in 1947 with links connecting New York to Chicago [1]. From that time period, till now the 6 GHz band has remained a popular choice for microwave links, although many higher frequency bands were added in later years. The majority of the global cellular network uses microwave links as backhaul—its importance is therefore paramount.

14.1 Basic Considerations: Throughput and Reliability

The concept of two towers with microwave dishes is indicated in Fig. 14.1. It captures the concept of a microwave link, using a simple transmitter/receiver pair.

The transmitter on the left sends a signal up the tower (on left) to the microwave dish antenna. From that dish antenna, the signal is sent over the air and received by the microwave dish antenna on the right. The signal received by the antenna is sent to the receiver. There are some basic questions that remain once this concept is established. What is the maximum distance allowed between the two towers? Should the

height of both towers be the same? What if there is a hill or buildings in between the two towers? Finally, why is the microwave link so popular, despite a fiber-optic network?

To answer the first question, the first thing to consider would be the frequency band. The second thing to consider would be the power from the transmitter. The third thing to consider would be the microwave antenna gain. All of them together will determine the distance. To answer the second question, the height need not be the same, but the line of light without any obstruction is necessary. It is possible to tilt one antenna slightly up or down in order to achieve a "line of sight". Yes, and details of this will be explained in later sections of this chapter.

The third question is answered by "Fresnel zone". Its fundamental basis is that intrusions may cause a reflected ray to enter the receiver. The details of this will be explained in the next section.

The answer to the final question is digital microwave links offer the same Ethernet connection that any fiber-optic network offers. Modern digital wireless networks use single-mode fiber as a common interface. Yet they provide throughput similar to fiber (10 Gbps) with lower latency, unmatched by any IP network—latency of less than 1 ms. It costs about one-tenth of the fiber network and is simpler to deploy and operate. Typically, MW links can work reliably for 40–50 years, with minimal maintenance.

14.2 Link Design: Choice of Equipment

Before we begin the concept of link design it is important to establish whether the microwave link is a frequency division multiplex (FDM) discussed earlier in Sect. 2.4.1. The majority of the microwave links are duplex and use FDM. Only in specific instances TDM or simplex is used. Therefore, Fig. 14.1 will get modified showing transmitter and receiver on both sides, indicating "send and receive"—the concept indicated in Sect. 2.4.1 as the equivalent of a duplex. They work as a matched pair (matched by the equipment manufacturer).

Now let us consider the link design, which involves transmitted and received power levels [2]. There are two power levels: one that the transmitter can handle, but the other one, more important, is what your local regulating authority allows. Since most microwave links operate on "licensed bands" it is essential to adhere to what the national regulator allows. There are several steps needed to design the link which will be described in the next few sub-sections.

Fig. 14.1 Basic concept of a
microwave link

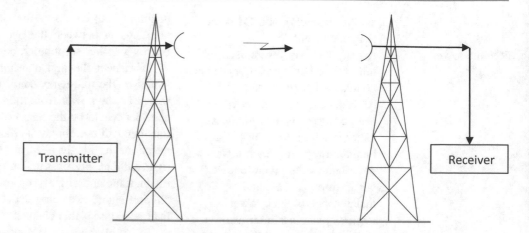

14.2.1   Link Design: Frequency and Losses

Figure 14.2 shows the popular frequency bands of microwave links and other applications. It shows some of the key
indicators for design (throughput, antenna size, spectrum,
and susceptibility to rain). From the L band on left to W on
the right, the frequency increases and their bandwidth
increases too. The first arrow (green) shows that throughput
increases with increasing bandwidth. The second arrow
(blue) shows that antenna size depends on the wavelength;
therefore lower frequencies have larger size antenna and
higher frequency uses smaller size. Third arrow indicates
(orange) that Spectrum bandwidth increases from L to W.
The last arrow (dark blue) indicates Susceptibility to fading
is more as the frequency increases. These were indicated earlier in Chaps. 2 and 3. In Chap. 11, it was pointed
out that millimetric wave spectrum bandwidth is huge (in
GHz). This fact will be reinforced again in this chapter, when milli-metric wave bands are discussed. Earlier it
was mentioned that the 6 GHz band was popular. In the

USA, FCC has been actively pursuing the task of making
this an unlicensed band to extend the growth of WiFi (UNII
services) and Internet services. Tropical countries with
considerable rainfall (South Asia and far east) have a strong
preference for the C band for both microwave and satellite
systems due to better rain susceptibility. Some of the
advantages of the C band are:

- Less interference from heavy rain fading
- Cheaper bandwidth compared to other bands, which will
  be discussed later.

Let us review the basic design using the popular C band
that is currently licensed in the range 5.926–7.125 GHz by
FCC for microwave links. This band is further divided
into lower and upper 6 GHz bands. Lower L6 is 5.925–
6.425 GHz and the Upper U6 is 6.425–7.125 GHz. Let us
work with the U6 which offers 40 MHz channel bandwidth.
But let us assume our local regulator allowed only 30 MHz
bandwidth at that site. High-performance antennas up to 6 ft

Fig. 14.2 Microwave frequency
bands defined by IEEE—key
features

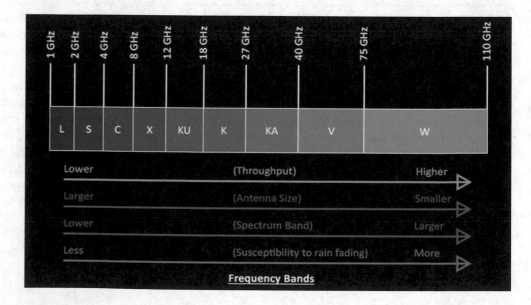

(1.8288 m) or size with a gain of 30 dB (1000) are available in this band. Let us assume that a maximum transmit power of 2 W is needed, which is + 33 dBm [2]. Transmitter's power includes an antenna gain of 30 dB (×1,000 amplification), therefore we need only 2 mW from the transmitter, thereby providing $2 \times 1000 = 2000$ mW = 2 W or +33 dBm. This may be acceptable to the regulators. However, let us review many other design considerations, in Sect. 14.3.1 we will revert further with this thought of link design.

Although the metric system is used in this book, antennas were designed originally using the British FPS system which has continued till date, therefore antenna diameter expressed is in feet.

14.2.2 Link Design: Polarization, Cable/Waveguide

With an antenna output of +33 dBm (EIRP or Equivalent Isotropic Radiated Power), the next calculation would be cable loss. There are two options for cable: coaxial cable or waveguide. Each has its own unique features. A coaxial cable is a semi-rigid type connected with a standard RF connector (N connector). It supports the TEM mode. A waveguide is a hollow metallic pipe whose side wall is calculated to support the frequency range of interest. It has a major advantage; it can support the TE and TM modes. Let us understand these modes and why they are useful. After this understanding, let us return to our thought of basic link design.

Using a rectangular waveguide as an example, Fig. 14.3 shows the TE and TM modes of a waveguide. One waveguide can support two separate channels simultaneously. In theoretical terms what this means is that the link can now carry double the throughput [3]. In practice, it will be less than double but considerably increased. In practice, only the TE mode is dominant in all far-field

communications, as shown earlier in Chap. 2, Fig. 2.1. Following the same theme of far-field waves, three modes of TE variants are shown in Fig. 14.3 on the right side. The first mode shows TE10 which indicates the minimum or cut-off frequency, it is also the most commonly used version of TE wave in all waveguide applications. Frequencies lower than this will have longer wavelengths that cannot be supported by that waveguide. In order to calculate the cut-off wavelength of a waveguide, the following equation is used:

$$\sqrt{(m/2)^2 + \left(\frac{n}{b}\right) b^2} \qquad (14.1a)$$

where "a" is the broad side of the waveguide and "b" is the short side of the waveguide. "m" is the number of half-waves on the broadside and "n" is the number of half-waves on the short side. For the dominant TE_{10} mode, m = 1 and n = 0; therefore, $\lambda \cdot c = 2a = 2$ (broad dimension) = 2a. Cut-off frequency.

$$f_c = \frac{c}{2a} \qquad (14.1b)$$

where c is the speed of light and a is the broadside dimension of the waveguide in meters.

Before proceeding further with design, let us ponder on choices between coaxial cable and waveguide and consider the advantages and limitations that waveguide has over coaxial cable:

- At higher frequencies such as those used in microwave links, the waveguide has lower loss compared to RF cable.
- Waveguide has no central or inner conductor and is usually air-filled, hence it is easy to manufacture. Since it is air-filled, no power is lost through radiation and the dielectric loss is negligible.

Fig. 14.3 TE and TM modes of a rectangular waveguide

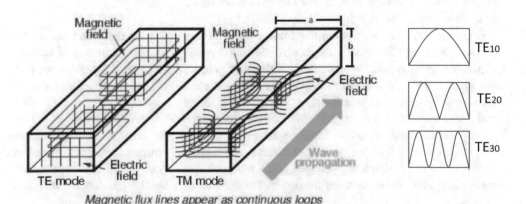

Magnetic flux lines appear as continuous loops
Electric flux lines appear with beginning and end points

- Waveguide can handle higher power compared to coaxial cable. This is because the waveguide cavity is filled with air as dielectric. Air has a breakdown voltage of 30 kV/cm. This increases the power handling capacity of the waveguide. It is useful for radar and other high-power applications.
- Since the wall of the waveguide is metallic, it is bulky, heavy, and rigid. The coaxial line is smaller in size and lighter in weight. Hence coaxial is useful for many microwave applications.
- The waveguide wall is not perfectly conducting, so there is some power loss as heat will occur in the wall of the waveguide, but this loss is less than that of coaxial cable.
- The bandwidth of the waveguide is narrower; the coaxial lines are used for broadband application.
- Waveguide must maintain a uniform cross-section. If the dimension changes then mode conversion occurs or higher-order mode may be generated in the waveguide.
- In recent years elliptical waveguide that is flexible allows techniques similar to cable to hoist it on the tower and connect the waveguide with a flange connector.
- Major limitation of a waveguide is that it needs positive air pressure (usually dry nitrogen gas) to minimize losses in the wall and avoid the formation of moisture inside.
- With millimeter-wave bands for the links and in 5G, the waveguide is making a major come back since cable losses are considerable, and waveguides offer limited loss even at high frequencies.

14.2.3 Cross-Polarization and XPD

Based on the differences, let us select the waveguide for now and consider how the vertical and horizontal polarizations can be integrated into one dual-polarized antenna to increase capacity. The convention of vertical or horizontal always refers to the position of the electric field. Hence, in Fig. 14.3 we would consider the waveguide on the left as "vertically polarized". If this is flipped such that the waveguide stands on its short edge, then it is horizontally polarized [4].

There are other things to consider from a practical perspective. Although in theory the horizontal and vertical polarizations are expected to be perfectly isolated, in practice they are not. There will be some part of the other polarization always present. The practical term used to indicate the level of isolation between the two modes is called "cross-polar discrimination" or XPD. Expressed in dB, the XPD is formally defined as "cross-polarization discrimination". It is the difference between the peak of the co-polarized main beam, and the maximum cross-polarized

signal over an angle twice the 3 dB beam width of the co-polarized main beam". We will separately discuss the 3 dB beam width while studying antenna beams later in this section.

XPD in transmit mode shows the proportion of signal that is transmitted in the orthogonal polarization to that required, while in receive mode, it is the antenna's ability to maintain the incident signal's polarization characteristics.

$$\text{XPD } (dB) = -20 log \, 10 \frac{E_{11}}{E_{12}} \qquad (14.2)$$

where E_{11} is the co-polar strength, and E_{12} is the cross-polar strength at the receiving end. An important unit which separates vertically and horizontally polarized signals for simultaneous transmission and reception is known as orthogonal mode transducer (OMT).

A dual-polarized antenna should include the OMT (Orthogonal Mode Transducer) as an integral part of the system to provide satisfactory performance [3]. The top portion of Fig. 14.4 shows OMT (yellow part) as an integrated part of antenna assembly and the bottom portion shows just the OMT (rectangular waveguide) unit. In the bottom part notice that the vertical and horizontal ports have similar flange but rectangular ports are orthogonal to each other and the third port connects to the antenna [4].

Having chosen the antenna with the dual-polarized configuration, let us consider the path loss and how far such units operate to provide reliable microwave links.

The free space path loss (FSPL) in dB is given by the following equation:

$$\text{FSPL} = 20 log_{10}(\text{d}) + 20 log_{10}(\text{f}) + 20 log_{10}\left(\frac{4\pi}{c}\right) - G_{Tx}$$
$$- G_{Rx}$$

$$(14.3)$$

where

d is the distance between the two antennas;
f is the frequency of operation in MHz;
G(Tx) is the gain of the transmitting antenna in dB;
G (Rx) is the gain of the receiving antenna in dB.

The IEEE Std 145–1993 defines FSPL as "The loss between two isotropic radiators in free space, expressed as a power ratio". FSPL is one of the constraints that limit signal flow to the other antenna. It was also discussed briefly in Sect. 2.1.1. There are other constraints as illustrated in

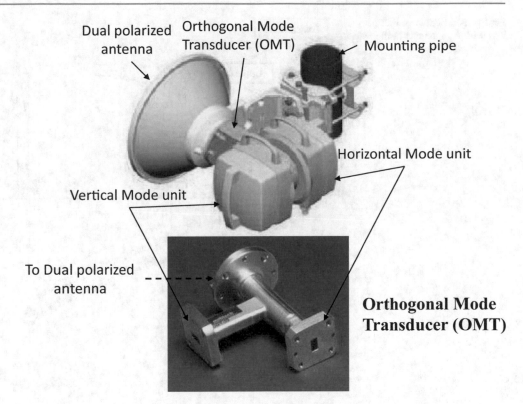

Fig. 14.4 Integrated dual-polarized antenna with OMT

Fig. 14.5 that indicate signal changes along the path as part of path loss calculation.

The bottom portion of Fig. 14.5 shows how the signal increases or decreases at each point along the link based on whether the signal gains or loses at each point. Observe that antenna gain, which was alluded to earlier at the end of Sect. 14.2.1, does help to overcome some loss. There are factors such as fading which must be included. It is the environmental factor to be discussed in the next section. So is the Fresnel zone, refractive index of atmosphere, urban buildings, etc., which are all factors that contribute to path loss. Let us consider all of them to form a realistic model of the link. It is still important to bear in mind that modern digital microwave links offer automatic power control; this means whenever there are atmospheric changes (such as rain/snow) that reduce signal, power from the transmitter is increased to compensate and maintain a reliable link. Components such as the duplexer shown in the figure will be reviewed in detail later, in Chap. 19, where RF measurements are described.

So far, the TE and TM modes using waveguides were described. Quite often, for microwave links deployment at lower frequency bands (6 and 11 GHz), the use of coaxial cable is common due to its flexibility. The coaxial cable only supports the TEM mode.

Figure 14.6 shows a typical coaxial cable and how it supports the TEM mode. Observe that the TE or transverse electric field is radial as shown on the right but is supported within the cable around the two conductors. The TM or transverse magnetic mode is supported between the inner conductor and the outer conductor (radiating out from the inner but contained by the outer conductor). Both of these are orthogonal to the direction of propagation, which in this case is into this page. This three-dimensional method of describing radio wave propagation is quite universal and commonly used. Earlier in Chap. 2, Fig. 2.3 described such TEM waves using Fig. 2.1, where the electro-magnetic waves propagated in the direction orthogonal to the electric and magnetic waves.

14.2.4 Fresnel Zone and Other Factors

The reason to form such models is that microwave links operate in the terrestrial environment where many other signals are present. Some of them are also in the same range of frequencies, one of them being satellite communication signals. The general guideline is to operate satellite antennas looking up (6° or higher towards the sky) and keep microwave dish elevation angle horizontal with no more than 5° elevation when looking up towards a dish located on a hill or

Fig. 14.5 Microwave link path loss with components in the path

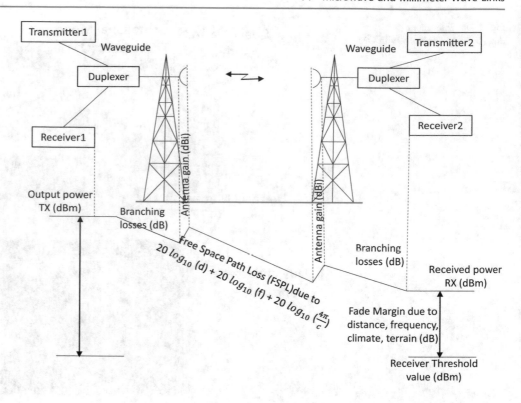

Fig. 14.6 TEM mode supported by coaxial cable

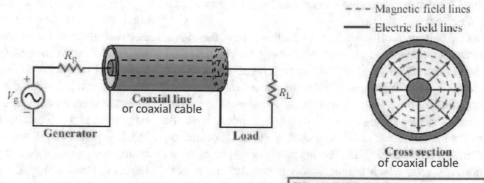

high-rise building. In urban areas other structures such as buildings that are not in line of sight but maybe reflecting the microwave signals to the receiving end impose a constraint. The path loss model evolves as a result of many such constraints included in the path. "Fresnel zone (FZ)" is a solid elliptical model within which the possibility of such reflections is considered.

In Fig. 14.7, the "radius of Fresnel zone" is shown as the ellipsoid within which the signal propagates. Examples of a high-rise building and a hill are shown lower than the antenna optical sight (blue line) but are located within the solid elliptical cone. The deflected waves due to obstacles

within the "Fresnel zone" such as high-rise building and hill are illustrated as d1 and d2, respectively. At any point within the ellipse, deflections may occur and the radius of deflection (in meters) can be evaluated by Eq. 14.4:

$$R = 17.3 \cdot \sqrt{\frac{d1 d2}{f(d1 = d2)}} \qquad (14.4)$$

where "f" is the frequency in GHz and d1 and d2 are deflections from the main beam in meters. The extreme radius of deflection R, for any microwave link is R_{max},

Fig. 14.7 Fresnel zone—the possibility of signal reflection

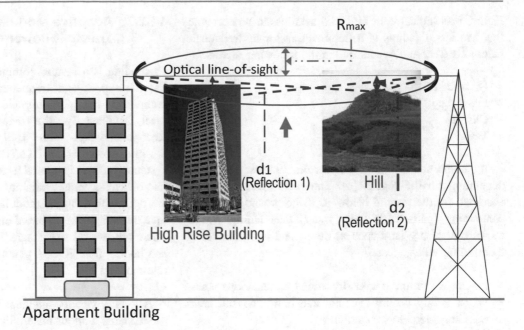

"Fresnel zone radius", which would be at the path's midpoint (where deflection is maximum and shown as *Rmax*):

$$R_{\max} = 8.66 \cdot \sqrt{\frac{D}{f}} \qquad (14.5)$$

where D is the distance between the towers in kilometers and f is the frequency in GHz. The rule of thumb for clearance is 60% of the F.Z. radius. To simplify, most designers use *Rmax* over the entire path (e.g., for map overlays) as a conservative model. The signal beam as a cylinder can be modeled using Google Earth showing whether some building rooftops on the path are within the Fresnel zone. Horizontal polarized beam is likely to suffer more in urban areas [5]. This is because any metallic object (such as a pole on a rooftop) may be within the "near field" region. The far-field (which is required) can be calculated by the following equation:

$$R_{FF} = 2\frac{D^2}{\lambda} \qquad (14.6)$$

where D is the largest diameter of the antenna in feet and λ is the wavelength in feet. Using this equation, a 4 ft dish operating at a 6 GHz band would have 195 ft as the distance, within which any metallic object in any building will affect the signal path.

14.2.5 Earth's Curvature and K-Factor

The other factor to be accounted for in F.Z. is Earth's curvature. There are many software programs readily available

to incorporate this and is commonly known as the "K-factor". K-factor shows the refractivity in the atmosphere that tends to bend the ray "towards" or "away from" the Earth. K-factor = effective Earth radius/true Earth radius. K-factor > 1 means beam is bent towards the Earth, and K = 1.333 indicates the normal refraction of the atmosphere. K < 1 means the signal is bent away from the Earth. For different regions the K-factor should be carefully chosen. While K = 4/3 is normally chosen as default, in wet coastal regions K = 0.5 is common, which means taller towers are needed in such areas. The K-factor with a tendency to refract the beam towards the Earth should be verified since it may touch a hill, which would have otherwise passed the F.Z radius test, for example.

Interpretation of the Fresnel Zones:

- Radio line of sight requires no intrusions into the first Fresnel zone within the first and last 1 km next to the receiving and transmitting antennas.
- Want no intrusions more than 40% into the first Fresnel zone at any point (optimal clearance requires 60% of the first Fresnel zone).
- At least grazing line of sight must exist during adverse refraction (when K = 1 or K = 0.667).

14.2.6 Fading and Variation with Regions

Fading has been discussed in relation to what wireless networks experience when RF energy moves over the air. Fading occurs due to environmental factors and distance.

Fading was indicated in Fig. 14.5 as a loss to be accounted for. Multipath fading is a dominant factor in frequencies below 10 GHz and occurs due to the following factors:

- Distance of the path
- Frequency
- Climate
- Terrain

A combination of such factors affects differently depending on the region [6]. The radio energy is partly absorbed by the ground below it. If the ground is wet and conducting (particularly salty water) then fading will be more. If it is dry land such as desert and less conducting, then fading will be less.

- Fading is minimum over dry, rocky mountainous areas.
- Worst fading occurs over hot and humid coastal areas, with considerable vegetation.
- Inland temperate regions are somewhere in between.
- Less fading is found in flat terrain (such as deserts and lakes) due to minimum incidence of reflections.
- Higher fading occurs in irregular, hilly terrain or forests (more reflections).

There are extensive models of fading and propagation loss including models from ITU. For practical purposes, the fade margin (Fa M) can be calculated using the following equation [6]:

$$FaM = 30 \log(d) + 10 \log(6fK) - \text{margin}(1 - R) - 70 \text{ dB} \tag{14.7}$$

where d is the link distance, f is the frequency, K is the terrain correction factor, and R is the reliability required. For example, if we need a 10 km path at 4 GHz with a link reliability of 0.999 999 over desert terrain of K = 0.2 then using Eq. 14.7 we need a Fa M = 27 dB.

Finally, the scintillation effect, which is of significance to satellite communication, has some effect on microwave link as well. Rain and cloud scintillation can be quite high during storms. This is modeled as a statistical variation normally with 1 dB variation. But for links that have lower path angle in mountainous terrains this can be significant, causing up to 5 dB variation during storm activity. During storms such links may incur disturbance and intermittent failures but recover. Currently, digital microwave links incorporate automatic power control to take care of variations in path loss. At each site, radio transmitters can automatically increase/decrease power levels.

14.2.7 Adaptive Modulation Coding: Auto Transmit Power Control

Flat fading is a feature common to all microwave systems based on path and frequency of operation. In order to compensate for flat fading and to adjust the received power level, National Spectrum Managers Association (NSMA) Recommendation WG 18.91.032 prescribes automatic transmit power control (ATPC) in all digital microwave systems [7]. This is offered as a standard feature in all digital microwave systems available today. Microwave radio units are built with adaptive modulation coding that allows multiple throughput levels based on weather and link conditions. There are several advantages in using adaptive modulation and coding that FCC is planning to indicate as part of its rule-making:

- Allowing smaller antennas in certain Part 101 antenna standards without materially increasing interference
- Exempting licensees in non-congested areas from efficiency standards to allow operators to increase link length in rural areas
- Allowing wider channels, including 60 MHz in the 6 GHz band and 80 MHz in the 11 GHz bands
- Revising waiver standard for microwave stations near the geostationary arc to align with ITU regulations
- Updating the definition of payload capacity rules in Part 101 rules to account for Internet protocol radio systems

Table 14.1 shows typical levels and modulations being used by vendors. This table indicates power level adjustable at the output of transmitter (20 dB range with 1 dB steps).

It can be adjusted in steps of + or −1 dBm depending on weather conditions. In order to better understand the modulation and coding scheme, see Fig. 14.8, where the highest throughput is provided on a clear bright day (Sun shining). With the worsening of weather conditions (fog, light rain, steady rain, and thunderstorms shown by symbolic colors) transmit power increases and the transmission continues with lower throughput. Such degradation in throughput is acceptable in all IP networks, where traffic varies throughout the day depending on the demand. Figure 14.8 shows the modulation types and how the system adapts. Quadrature amplitude modulation (QAM) is a form of modulation that uses two carriers—offset in phase by 90°—and varying symbol rates (i.e., transmitted bits per symbol) to increase throughput. As a generic example, using 1 + 0 (no redundancy) and 256 QAM, it is possible to achieve up to 215 Mbps/215 Mbps in a 28 MHz channel bandwidth.

With extreme weather such as snowstorm or heavy hailstorm, the link may reduce to basic QPSK modulation

Table 14.1 Transmit power level (dBm) at RF output port

Modulation	4–8 GHz				11 GHz	13 GHz	Tolerance
	40 MHz (Bandwidth)		30 MHz (Bandwidth)		30 MHz (Bandwidth)	40 MHz (Bandwidth)	+ or −1 dB
	STD*	HP*	STD*	HP*			
QPSK	+29	+32	+30	+33	+30	+27	
8 QAM	+29	+32	+30	+33	+30	+27	
16 QAM	+29	+32	+30	+33	+30	+27	
64 QAM	+29	+32	+30	+33	+30	+27	
128 QAM	+29	+32	+29	+32	+29	+27	
256 QAM	+28	+31	+28	+31	+28	+26	
512 QAM	+27	+30	+27	+30	+27	+25	

STD—standard, HP—high performance.

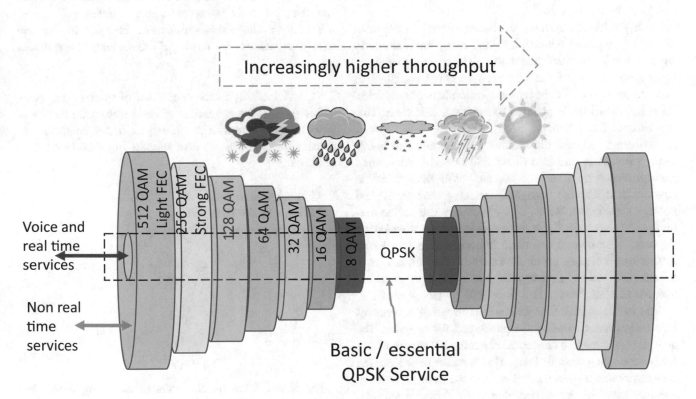

Fig. 14.8 Adaptive coding and modulation (ACM) in microwave links

that allows the lowest signal levels at which a receiver can operate. This will still provide essential voice and real-time services. But as weather conditions improve, throughput increases and gradually resumes to the full 512 QAM level during bright sunshine. Due to the benefits of adaptive modulation coding (AMC), this approach is accepted by both service providers and regulators. With progress in technology higher levels of modulation such as 1024 QAM and 2048 QAM are being supported in recent products, but they need larger bandwidth. Now, we realize how important the local regulations are. If our links are in areas of minimal interference, then highest throughput is likely.

14.3 Regulations to Promote Microwave Links

In recent years several regulations were brought in by the FCC and ITU to promote microwave links for two reasons: (a) Most link providers need a license, thereby requiring efforts to file and get an approval prior to building a tower. (b) Microwave links reduce latency to <1 ms. This is significant considering most IP networks today have 30 ms latency. In 5G, the ultra-reliable low latency communication (URLLC) avoids the use of traditional IP networks to reduce latency.

For example, in support of reason (a), FCC made major parts of 5–6 GHz as "unlicensed infrastructure", with IEEE 802.11 or related standards following the channel plan, as indicated earlier in Chap. 12, Sect. 12.5. This action indirectly reduces the number of microwave towers with 8 ft or 10 ft antennas in the 5–6 GHz band. In a coordinated action, FCC 12-97A1 Rule provides additional flexibility to broadcast auxiliary service and operational fixed microwave licensees. Released on August 3, 2012, it encouraged the use of smaller dishes in the 6, 8, and 23 GHz bands [7]. FCC allowed 2 ft (0.61 m) antenna dishes to be mounted on rooftops, thereby reducing installation and operational costs. This action required allowing wider channels and relaxation in the level of energy outside of the beam using the policy of "permit-but-disclose" so that other operators know where and how the beam is used.

A major impact of links to reduce latency in 5G with URLLC is expected with deployments along the roadside to support intelligent vehicular traffic. There are schemes to completely avoid the use of fiber and wired backhaul in order to provide real-time communication between vehicles. Such infrastructure is planned to reduce transit time. This was discussed in Chap. 11, Sect. 11.6.

Reduction in latency in the link between any two points using wireless is due to a direct line of sight. Any wired network (such as fiber, CAT6, etc.) will have wires in conduits laid that are brought to street ground level, laid through streets with bends, etc., and finally gets to the destination. This obviously increases the distance traveled by the wave. The second and more important aspect is due to routers and switches on an IP network. That introduces a considerable delay in routing the traffic. Both of these are avoided in a microwave link since it is "point-to-point".

The FCC regulation needs some explanation in terms of how dishes are viewed by the public and the operator. The general public may be concerned about the RF energy within the beam coming towards them. This is not an issue since the designer makes a special effort to keep the beam away from any obstacle—this is not only due to the Fresnel zone (discussed earlier) but also to focus the energy entirely on the counterpart dish at the other end.

14.3.1 Wind Load: Regulation

The operator on the other hand has to account for structural wind load, a major factor that increases cost. The larger the dish, the higher will be the wind load on the tower. If we assume the dish to be like an umbrella, everyone has experienced the pull that umbrella makes on the user and how difficult it is to hold up an umbrella when the wind blows into it. Since the microwave dish is of solid metal (not porous—

doesn't allow wind to pass through), the force in the direction of the wind is directly proportional to the area of the dish.

Tower manufacturers provide wind calculations into which "moment force" or the twisting of the support structure due to the antenna must be incorporated. Government regulations require that twist at the top (where microwave dishes are mounted) must be within 2° during a gust wind of 160 kmph lasting 3 s, for example. In order to calculate this, it is important to consider the forces acting on the antenna. Figure 14.9 shows parameters used in wind load calculations on a microwave dish antenna [8]. It assumes the direction of the wind that results in maximum moment force at the antenna. It indicates the resulting twisting moment "MT", on the dish which ultimately transfers to the tower. X and Y indicate offsets of the antenna due to the mounting bracket from the pipe where the antenna gets mounted.

To fully evaluate that variation of these forces, they are usually expressed in terms of coefficients in different directions:

C = Coefficient. C_A is the coefficient of frontal area (compression), C_S is the coefficient of shear force (side force that twists the dish mount or pole), and C_M is the coefficient of moment (moment due to dish support from the tower as a cantilever).

A = Frontal area in square feet/meters.
V = Wind velocity in mph/kmph.
D = Antenna diameter in feet/meters

$$C_A = \frac{F_A}{AV^2} \tag{14.8}$$

$$\frac{F_s}{AV^2} \tag{14.9}$$

$$\frac{M}{DAV^2} \tag{14.10}$$

The actual force on the antenna mounting pipe must incorporate the offset introduced by the antenna mount. These are shown in Fig. 14.9.

X = Offset of the mounting pipe in feet (meters).
Y = Distance on the reflector axis from the reflector vertex to the center of the mounting pipe in feet (meters).
X = Offset of the mounting pipe in feet (meters).
Y = Distance on the reflector axis from the reflector vertex to the center of the mounting pipe in feet (meters).

Although Fig. 14.9 shows a parabolic dish with Radome, these equations equally apply to other antennas as well, such as panel, dipole, Yagi, and other types. In each case, the area

Fig. 14.9 Wind load calculations on a microwave dish antenna

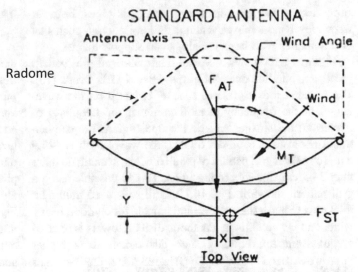

Figure 2: Positive Sign Convention

Microwave dish with wind force

will change depending on the dimensions and type of antenna. It is important to note that A is the "frontal area".

The total moment and forces exerted on the antenna mounting pipe can be determined by

$$F_A = F_A \qquad (14.11)$$

$$F_{ST} = F_S \qquad (14.12)$$

$$M_T = M + F_A(X) + F_S(Y) \qquad (14.13)$$

Using these equations, the coefficients C_A, C_S, and C_M can be calculated.

Owners of towers with antennas are guided by the standard ANSI/TIA-222-G which regulates the design criteria for all radio structures [9]. It shows how to calculate wind load, ice loads, wind on ice, seismic loading, by incorporating them in the terrain model, to improve the confidence of operators. This is an overarching standard that applies to all radio towers around the world (international agreement). It provides separate calculations for the axial force FA exerted on all appurtenances except the antenna wind load. Together, the two calculations provide guidelines of how the structure is designed to withstand forces due to all of these environmental factors. Let us return to our thought of link design using Fig. 14.5, bottom part. In Sect. 14.2.1 we came up with 0.2 W (+3 dBm) from transmitter. Considering the cable loss and insertion loss of duplexer, let us assume only 1 mW (0 dBm) enters the antenna. Due to field wind conditions antenna size was reduced to 4 ft, due to which instead

of 30 dB gain, the gain was only 27 dB. Then after amplification if we get +27 dBm (0.5 W) over the air, which is still sufficient (refer to Table 14.1). At this frequency even with a link of 40 km, the path loss will be 81 dB (assume 30 dB gain at receive antenna). Assume branch loss of another 3 dB and fade margin of 10 dB. After the entire path loss and receive antenna gain signal will be +27−(81+13) = −67 dBm which is a very good signal. Refer to MCS table in Chap. 12, Fig. 12.13 and it supports 64 QAM modulation with an effective throughput of about 200 Mbps. This is not an exact calculation, but just a quick check to validate the concept based on Fig. 14.5 bottom part. The real strength of microwave links are the huge antenna gains at transmit and receive ends. With larger dishes higher antenna gain of 40 dB (x10,000) are possible.

14.3.2 Grid Antennas: Reduce Wind Load

There are special efforts made to reduce the "Twist moment" force by making antennas with a grid metal pattern. Such grid antennas allow wind to pass through, reducing the twisting force (typically about 40% reduction). There are some limitations to such design: the primary limit is due to the wavelength of the operating band. In order to appreciate this, let us revisit Fig. 14.3 where the waveguide supporting half-wave was indicated on the righthand side. This is the "cut-off" or minimum frequency that can be supported. The design assumption for the grid pattern is that the wave is supported by the two adjacent metallic grid pipes. In lower frequency bands, the wavelength is long enough such that grid antennas with large openings can support them.

Figure 14.10 shows a typical dual-polarized grid pattern (grill) antenna that operates in the 4.8–6.5 GHz bands [10]. The antenna rods or frames spacing is based on the wavelength of the frequency band. The higher the frequency of operation, the smaller will be the opening. At the conventional microwave bands of 6 GHz, the wavelength is ∼4.9 cm. To make a grid pattern of parallel wires, spacing no more than 2.45 cm (for supporting ½ λ). This is possible using a grill pattern shown in Fig. 14.10. Grill rods need minimum thickness to make them strong and reliable for outdoor, heavy wind, and icy conditions. Parabolic dish on towers is open to the environment; building such dish antennas in higher microwave bands such as 11 GHz, 23 GHz, etc. becomes difficult since ½ λ will be 1.35 cm and 0.7 cm, respectively. At lower frequencies from 300 MHz up to 6500 MHz, ½ λ will be quite large (as shown earlier) and grid antennas offer an excellent opportunity for use in WAN/LAN, satellite, and other applications.

FCC has been opening up the 6 GHz bands and ITU recommends the use of both microwave and millimeter-wave links for TV video (needs <30 ms latency) and high-speed trading [11]. Further increase in telecommunication and data traffic has resulted in the global implementation of backhaul using microwave links. Globally more than 50% of all networks use microwave as backhaul links for the cellular, broadcast, and financial trading due to its low latency (<1 ms).

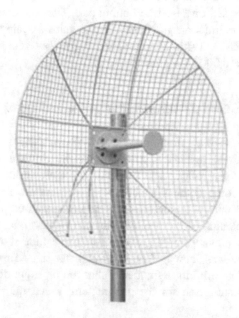

Fig. 14.10 Grid antenna—parabolic dish with grill for air passage

14.3.3 Radome: To Shield the Antenna

A Radome is a structural, weatherproof enclosure that protects any antenna. Radome is a covering material usually made of fiber-reinforced plastic or other material. Its primary purpose is to act as a cover to protect the metallic dish antenna (usually parabolic) and act as a shield from snow and ice. It can be flat or curved but is provided with thin heater elements embedded so that heating can be turned on during icy storms, allowing snow to melt. While accomplishing all this, it is important that Radome does not hinder the propagation of radio waves in both directions. Figure 14.9 shows a possible style of Radome (in dotted lines) that is curved and protruding. Such a curved Radome reduces the wind pressure (axial and twist forces) on the antenna. This and a variety of other curved or flat styles of radomes are offered by antenna vendors. It is up to the designer to choose the style based on local snow, wind, rain, and ice formation possibilities.

Most path calculations involve loss due to the insertion of Radome over the antenna. Usually, the attenuation or signal loss due to Radome should stay within 0.2 dB, although there are no specific regulations for such limits. Typically, large dish antennas tend to have a flat cover with clips to hold the Radome in place throughout the circumference of the antenna rim.

14.4 Emergency Deployments: Establishing Backhaul Links

Wireless systems come into prominence whenever there is an emergency such as a hurricane or other natural disaster. While satellite links are popular during disaster recovery, establishing links supporting long-term recovery needs microwave links. Quite often, the central distribution and command center is located in a place where supplies can arrive. It can be near a seaport or an airport depending on local terrain. Establishing links that pan out in different directions, quickly and reliably, brings microwave links to the fore.

After emergency, resuming other business and enterprise-related functions need the use of microwave links [12]. Fixed links using microwaves are definitely the best choice to resume data transmission quickly. Earthquakes, floods, landslides, and other natural disasters or even war zones require disaster recovery centers for uninterrupted business continuity for separate backup locations. It is necessary to incorporate microwave links as back up where important data must be stored and refreshed periodically. Microwave links are ideally suited to connect the main and back up centers since fibers can be (and usually are) damaged or cut off during disasters [13].

There are quite a few other arguments in favor of using microwave links during an emergency, including those from medical personnel attending to patients during an emergency. Almost every report seems to point to the possibility of fiber or cable being vulnerable during natural disasters.

14.5   Millimeter-Wave Link as an Alternative to Fiber

This brings us to a major and new topic of millimeter-wave links as an alternative to using fiber. Why is this important and how are millimeter-wave links different from traditional microwave links considered so far? "Millimeter wave", as the name suggests, has frequencies ranging from 30 GHz (wavelength 10 mm) to 300 GHz. For the topic here, it is important to consider two major bands allocated for links: (a) the unlicensed 60 GHz band and (b) the 70/80 GHz band which is licensed. Products have been available in these two bands for about two decades.

14.5.1   The 60 GHz Unlicensed Band

Millimeter waves are the subject of serious study and deployment in recent years due to the enormous bandwidth needed by 5G. But the study of millimeter waves at 60 GHz is not new. In 1897, Physicist J. C. Bose, demonstrated the use of 60 GHz, at the Royal College in London. For almost 100 years, the

research by Physicists continued, but engineering implementation began after a considerable period of lull.

In the early 1960s, the US military successfully used 60 GHz for secure communication. After the Cold War ended, in 2004 FCC made a historic release of millimeter-wave bands for public use. FCC allocated 7 GHz which is more than all of the spectrum allocations made thus far. The 60 GHz as unlicensed and the 70/80 GHz as licensed bands made a major impact on the backhaul network.

Figure 14.11 shows a photo of J. C. Bose's demonstration of millimeter waves; it was again unveiled during the commemoration of 100 years of millimeter-wave research [15]. The reference provides an overview of his discovery and further research on this frequency band.

An important property of millimeter waves is oxygen in the air absorbs considerable RF energy at 60 GHz, significantly increasing propagation loss. Figure 14.12 shows this property, with oxygen absorbing RF at 20 dB/km. Such additional absorption limits the link distance to ∼1 km. This property has some important implications; while it seems to be a limitation, it has some advantages. Due to severe absorption of RF, the link becomes inherently secure.

Figure 14.13 compares the 60 GHz spectrum to other unlicensed bands allocated earlier, including the popular 5 GHz UNII band. Unlicensed spectrum for microwave links at 24 GHz band, pales in comparison to massive spectrum availability in millimeter band of 60 GHz, with useful throughput of 1 Gbps comparable to fiber optic backhaul. 60 GHz links provide unique opportunities in major cities. Using high-rise buildings, it is possible to shoot

Fig. 14.11 J. C. Bose at the Royal Institution, London, 1897

Fig. 14.12 Atmospheric absorption of millimeter waves— oxygen peak at 60 GHz

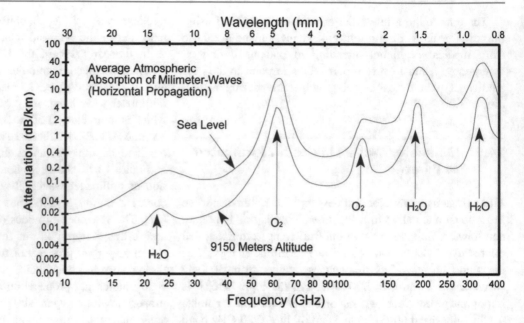

Fig. 14.13 FCC allocation for unlicensed broadbands—compare 60 GHz to others

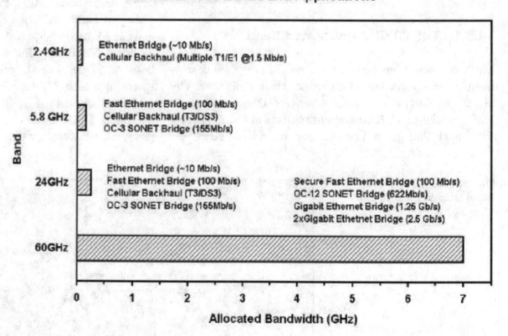

beams from rooftops to other rooftops while adhering to the rules of the Fresnel zone. A major advantage is the ease of installation and considerable reduction in latency (typically between 100 and 200 μs). The 60 GHz band offers a spectrum alone with a total bandwidth of 7 GHz [14]; that led to public interest in using these bands for point-to-point millimeter-wave links. Soon after 2004, suppliers who had built products for the defense industry started to release millimeter wave products into the commercial market.

Due to short wavelength, antennas tend to become smaller. Therefore, the largest antenna for millimeter wave is only 2 ft diameter (0.6 m) that provides a gain of over 40 dB (or 10,000). There is also another practical difference— coaxial cable cannot be used due to severe loss at these frequencies. Radio unit is directly integrated with the antenna and the RF circuit directly feeds into the antenna port using a small internal waveguide. The integrated unit only provides the user with a fiber port or a copper 1000baseT port and the RF part is not visible to the user.

Multiple vendors offer both the frequency division duplex (FDD) and time division duplex (TDD) versions of such 60 GHz link products which have been widely deployed in urban environments. FDD and TDD were earlier described in Chap. 6. Table 6.1 provided reference to how the two are different. Although its use in a rural environment is possible, distance limitation of 1 km restricts its use, but secure communication now becomes possible in the 60 GHz unlicensed band due to oxygen absorption of RF radiation. This can be viewed as an important advantage since outside the beam RF power drops off by over 80 dB (1/100 million), making it practically impossible to intercept the beam. Also, two antennas can be mounted on the same pole one below another, on both sides of the link. Both beams can travel close to each other (1–2 m apart), without causing interference to each other. In addition to natural security offered by oxygen absorption, vendors also offer 128-bit encryption as a standard feature.

14.5.2 The 70/80/90 GHz Licensed Band

The 70/80/90 GHz band was allocated by FCC as licensed bands for point-to-point links. They don't have the limitation of oxygen absorption, hence links up to 3 or 5 km can be set up [16]. Due to a larger bandwidth in this spectrum, it is possible to have 10 Gbps throughput offering a direct replacement to modern fiber optic backhaul for cellular networks. Both 71–76 and 81–86 GHz bands hold potential for the backhaul of mobile networks starting with 4G networks [17]. Products in these bands are readily available to support links with 10 Gbps throughput, the dishes are typically of 2 ft dia.

Millimeter-wave links have brought simpler deployment methods in microwave networks. Earlier, all microwave towers used to carry RF coaxial cable from the bottom of the tower (radio shelter room) to the top of the tower connecting directly to the antenna. Most modern towers now carry fiber or Ethernet (Cat 6) cable from the bottom to the top of the tower. This changes the cable length and weight limitation. Fiber cable is lightweight (6 kg for 100 m of 12 strands) and for 100 m it has practically no signal loss. RF coaxial cable is quite heavy (22 kg for 100 m for a single half-inch cable) and at least two such cables are needed which contributes to wind load on the tower. Also, coaxial cable loses about 30% of the signal in 100 m. In the new paradigm, on the tower is a digital remote radio head (RRH) or tower-mounted amplifier (TMA) which connects to this fiber/Ethernet cable.

Millimeter waves propagate in a straight line; therefore it absolutely needs a clear "line of sight", resulting in narrow beams of <3°. Reduced antenna size reduces wind load on the tower but has some deployment challenges. It is not easy to see a 1 ft or 2 ft (0.6 m diameter) dish mounted on a tower 3 or 4 km away. Using methods developed for millimeter-wave

alignment, smaller dish antennas now offer improved alignment kits. Some kits offer a laser beam accessory that shines a red dot on the other counterpart antenna, making it easy to align. The received RF signal is converted to DC voltage that allows a simple multimeter to indicate whether the far end is being aligned correctly. If alignment is in the proper direction, the detected DC voltage increases. Even before heavy microwave dishes get installed on the tower, using handheld signal sources, it is now possible to find out correct alignment (azimuth and elevation) [18].

14.6 The Remote End Tower: Design and Deployment

The concept of remote tower is as old as the microwave link itself. More often, traditional microwave links have repeaters located in a remote area. The remote end tower usually resides on hilltops, or in fields, and sometimes even in very harsh terrain. The service provider must analyze and evaluate how this tower will be monitored, serviced, and maintained by its crew [19].

Initial installation by the construction crew can be challenging. Carrying heavy equipment such as a tall crane to a remote hill site that has no roads may require a helicopter to drop off tower sections. For most countries, this is a major exercise, but it is performed as part of the site work. To appreciate isolated tower in a remote location, see Fig. 14.14. Unlike in urban areas, the construction vans or trucks normally used by tower maintenance provider cannot even reach such sites. It may be necessary to drop tower sections by helicopter. In terms of operation, primary concerns relate to electrical power and harsh conditions such as temperature and wind. The tower shown in Fig. 14.14 only uses diesel generators since solar panels are not useful in such cold conditions. But renewable energy such as solar panels or wind power or a combination may be needed for many sites. Inherent to such design would be the use of a battery with a DC voltage of 24 V or 48 V. Also, a diesel generator for standby power is considered essential since it is not easy to send maintenance personnel quickly if there are failures.

14.6.1 Monitoring the Remote Tower

To monitor different systems in and around the tower, a sensing system known as remote telemetry unit (RTU) is installed. It monitors parameters such as the temperature of critical units (tower-mounted amplifier, battery, and other locations on the tower). Voltages, temperature sensors, and other parameters are used as inputs and monitored by contact closures. Such units allow control such as system reboot or other remote control needed by the system. Another

Fig. 14.14 Remote site in harsh terrain—to bring the Internet to remote villages in Alaska

important consideration is how these measurements are communicated back to the master site.

There are two options to bring back such parameters from multiple sites: (a) to use the microwave link itself and (b) to use a separate RTU-based link. The first method known as "integral link" is simpler. It is included as part of the microwave link. But operators with multiple sites may prefer using a separate route. The second option, known as "RTU-based monitoring", is preferred by operators since it is not affected by the failure of the main microwave link. Often a separate spectrum (such as satellite feed) is used with an independent source of power to accomplish separate monitoring. In such cases, if there is a failure of the microwave transmitter at the remote site, still it can be observed at the main site. This is a major technology known as SCADA covered in Chap. 16.

In order to appreciate the harsh environments in which such towers operate, Fig. 14.14 shows a typical tower built in the remote region of Alaska [20], where there are no roads nearby. There are number of such locations on the globe where nothing other than microwave links would be a feasible method for telecommunication. At the same time, bringing heavy steel structures and erecting them at the sites

such as these are challenging, and is often made possible only by helicopters since cranes or other construction equipment cannot be brought to such locations by road.

In Fig. 14.14 the microwave antennas are not covered by ice and snow because there are heaters behind each Radome covering the antenna. Access to such sites (as can be imagined) is a challenge. To keep them operational and provide a crucial Internet link to remote villages is very important.

Microwave dishes and cellular towers are quite often the terms used together. Asia in particular has a major interest in microwave links since it is the primary means of cellular backhaul.

Terrains in Asia can be harsh in a different way. Instead of heavy snow seen in Fig. 14.14, Asia is home to heavy monsoon rain, major storms, and corrosion which could become major challenges. Despite such challenges, Asia continues to grow as the largest expansion in wireless communication, including microwave links as backhaul. Therefore, operators, vendors, and other users have formed a global entity known as "Towers Exchange" incorporating most Asian countries [21]. It offers a wealth of support and cooperation among tower operators and vendors in this largest growing market.

14.7 Conclusion

Microwave links are described as the arteries of modern-day wireless systems. This chapter focused on factors considered in its design, and deployment. Some of them are obvious such as there should be a line of sight between the two antennas. Others that may not be apparent are the curvature of the earth, objects such as mountains, buildings, and others that are not directly obstructing the path but high enough to cause reflections.

Microwave link operators are concerned about increasing the throughput; many schemes are available. Some of these are incorporated in design such as dual-polarization. While frequency selected directly relates to how far the link can operate, it can decide the capacity of the link. Other factors do play a role, such as whether there is an existing link that could potentially interfere. The regulatory authority may permit operation in a specific spectrum only depending on congestion, etc.

We conclude that two factors contribute to the superiority of microwave links: one is that it is a direct "line of sight" over the air and it is always the minimum distance with low latency and high reliability of communication. The second factor is that there are no routers or other network elements that introduce delay (typical of all IP networks). These two factors make every digital radio network provider prefer microwave to landline for voice, for broadcasting images, and other applications that need minimum latency.

Existing wired networks will continue to operate with fiber optic cable on the ground, and new networks often prefer the tower approach. Tower accommodates both the base station antennas and direct backhaul using microwave links. This winning combination was recognized by most Asian operators who have set up the "Tower Exchange" cooperative concept that can better serve a majority of the global population. Growth of towers inclusive of microwave/millimeter-wave links is favored globally by regulators and ITU alike.

This chapter supports and complements concepts of backhaul described earlier in Chap. 4. If wired fiber-optic networks exist, then microwave contributes by complementing them. If it does not exist, then microwave/millimeter-wave links become the main backhaul. Both of these network concepts are widely deployed across the globe. The Society of Cable and Telecom Engineers (SCTE) also offers training courses to those aspiring to qualify and work in the area of microwave systems.

Summary

Microwave links are terrestrial hops providing backhaul to all types of systems whether wired or wireless. They are in operation since the 1950s, but with the growth of digital wireless, all of them adopted digital modulation techniques with power control. The chapter started out by asking basic questions on what these links are and how are they designed.

Answer to these questions resulted in details of MW link design and parameters used to provide a reliable link. Other than the RF signal, factors such as wind load on the antenna and the tower it stands on were considered and reviewed. Several important considerations in microwave links were reviewed. They are Fresnel zone, Earth's curvature, polarization, terrain, climate, and frequency band. Other considerations that can affect the operation of links are wind load, snow, and ice. Some of these are addressed using active power control made possible in all digital microwave links.

There were several regulations and guidelines that supported and extended the use of smaller dishes and higher frequency bands. The historic change was the introduction of millimeter band for use in these point-to-point links. These bands directly support the maximum throughput rates expected by the cellular industry. In addition, they provide low latency that no wired network can. Adaptive modulation and coding (ACM) is one that was supported actively by regulators and operators since it allows a compromise solution. The compromise in terms of variable throughput rate due to weather conditions allows smaller antennas (simpler towers, including the option of mounting on rooftops), lower power limits over the air, and allows more links to co-exist.

Emphasis on Participation

I. What types of towers do you see in your neighborhood? Have you observed city blocks and microwave dishes on rooftops? Where are these links going? Make a formal map of locations and see if it is possible to make an informed guesswork of what the network looks like.

II. All microwave links must be registered with the local regulating authority (FCC in the USA). Find their website of registration to see how many microwave links already exist. Provide comments to the regulatory authority if you see something of concern (any dish even partly pointing towards an apartment complex).

Knowledge Gained

(i) Important aspects of microwave links were reviewed with details of all factors considered during the design

(ii) Factors related to structures, wind load due to large area of antenna surface were reviewed, with efforts to reduce wind load using Radome or using grid antennas

(iii) Millimeter-wave dishes and their impact on throughput and competition with fiber-optic cable were reviewed. Changes to lower frequency microwave links, such as the use of fiber from the radio room up the tower, and how they improve design were reviewed

(iv) Government regulations to encourage microwave links were reviewed with emphasis on reducing antenna size—modern design based on ACM and allowing higher levels of interference to accommodate more links in congested areas were considered.

Homework Problems

1. Make a list of at least 5 microwave links in your area (using regulator's website). Map them and observe whether they cross each other and how is interference avoided.
2. Review at least two remote sites with harsh conditions. Provide details of the frequency, size of dishes, and what these links are—what network they are connected to.

Project Ideas

(i) Using tools available online, design a microwave link to support your university buildings. What factors did you consider—list them.
(ii) In consultation with authorities design a millimeter wave link to carry most of the traffic from your university to the Internet provider point of presence. How efficient is this link? Consider both 60 GHz and the 70/80 GHz links and decide which would be better for your campus.

Out-of-Box Thinking

1. If a city were to use only microwave links/millimeter-wave links—what are the possible limitations. If city planners were to object, what are their likely observations?
2. If you were a tower operator, what are your likely plans to expand the network using millimeter/microwave links. Using the tower exchange website [21] makes the case of using only these links to extend Internet access to your favorite remote area. How many people does it serve?

References

1. Bray H (2002) Innovation and the communication revolution, from the Victorian pioneers to broadband Internet. In: The TD-2 story" by A. C. Dickieson, three-part article from the Bell laboratories record, p. 147. ISBN: 978-0-85296-218-3
2. Abdulrahman1, Yew OS, Shafikah N, Mohamud MM, Terrestrial microwave link design. In: Radar communication laboratory, Faculty of electrical engineering, Universiti Technogi, Malaysia.
    https://www.academia.edu/8699907/Terrestrial_Microwave_Link_Design
3. Oliver D (2014) Back to basics in microwave systems: polarization, 22 May 2014. Commscope.com, https://www.commscope.com/blog/2014/back-to-basics-in-microwave-systems-polarization/
4. Explanation of polarization angle: satellite signals limited, last amended 15 April 2018. https://www.satsig.net/polangle.htm
5. Planning a microwave link – Evans engineering solutions, Oct 10, 2012. Broadcasters Clinic, Middleton. http://evansengsolutions.com/wp-content/uploads/2014/06/Ben-Evans-Presentation.pdf
6. Horan S (2017) Introduction to PCM telemetering systems. In: 3rd edn, CRC Press. ISBN 13: 978-1-138-19670-4. Section 11.6 Atmospheric, Sun, Ground propagation effects
7. National Spectrum Managers Association, Recommendation WG 18.91.032, Automatic Transmit Power Control. https://nsma.org/wp-content/uploads/2016/05/WG18-91-032.pdf
8. Dependable, light weight parabolic grid antennas, Wind load information. http://markgridantenna.com/parabolic-grid-antennas/windload-information/
9. Ericsen J, ANSI/TIA-222-G Explained, Introduction, PE. https://www.towernx.com/downloads/TIA-222-G_Explained.pdf
10. Mimotik Dual polarized Grid antenna - MK-4865PG-28DP. https://mimotik.com/products/5ghz-28dbi-grid-dish-antenna
11. ITU Recommendation on Fixed Service use and future trends ITU-R F.2323–0 (11/2014). https://www.itu.int/dms_pub/itu-r/opb/rep/R-REP-F.2323-2014-PDF-E.pdf
12. Enterprise microwave backup – implementing disaster recovery connectivity/Ad-hoc networking, https://www.iskra.eu/en/Microwave-transmission-solutions/Enterprise-Microwave-Backup-Implementing-Disaster-Recovery-Connectivity/
13. Hegedus N (2010) Microwave links to reduce E 911 system's downtime, Times Herald Record, Dec 15, 2010. https://www.recordonline.com/article/20021128/News/311289989
14. Koh C, The benefits of 60GHz unlicensed communications, Director of Engineering, https://www.fcc.gov/file/14379/download
15. Emerson DT (1998) The work of Jagadish Chandra Bose: 100 years of MM wave research, National Radio Astronomy Observatory, 949 N. Cherry Avenue, Tucson, Arizona 8572, last revised Feb 1998. https://www.cv.nrao.edu/~demerson/bose/bose.html
16. Millimeter wave 70/80/90 GHz service, Rule part 47 C.F.R part 101. https://www.fcc.gov/wireless/bureau-divisions/broadband-division/millimeter-wave-708090-ghz-service/millimeter-wave-70
17. Review of the spectrum management approach in the 71–75GHz and 81–86GHz bands, OFCOM, 14 Oct 2013. https://www.ofcom.org.uk/__data/assets/pdf_file/0029/46775/condoc.pdf
18. Microw J (2015) Line of sight verification kit for microwave field engineers, Oct 14, 2015. https://www.microwavejournal.com/articles/25274-line-of-sight-verification-kit-for-microwave-field-engineers
19. A quick overview of microwave networks, Microwave system monitoring is a necessity, https://www.dpstele.com/network-monitoring/microwave/index.php
20. Nordrum A (2017) 109 microwave towers bring the internet to remote alaska villages. In: IEEE spectrum, 30 Nov 2017. https://spectrum.ieee.org/telecom/wireless/109-microwave-towers-bring-the-internet-to-remote-alaska-villages
21. A deep dive into Asian towers, Tower Exchange Asia Dossier 2018 https://www.towerxchange.com/wp-content/uploads/2018/10/TX_AsiaDossier_2018.pdf

Vehicular Technology

15

Abbreviations

PTC	Positive train control: A method to incorporate safety features in a long-distance train. Features include control of speed and following signal restrictions imposed on the track.
EMS	Emergency medical service: Part of the ambulance and health support system; its users are part of the public safety frequency band allocated by the regulator of each country
PTT or Push-To-Talk	is a feature where the handset has a button that is pushed before the user talks. Since it is typically a simplex (one-way, talk or listen), it is essential that the user keeps the push button held in a pressed position while talking. During this period, other users cannot talk and can only listen. Release of the button is essential so that the user now makes the channel free for others to use. In common terms this is known as "walkie-talkie", indicating push the button to talk each time and it is possible to walk around in the entire coverage area.
LMR	Land mobile radio is a term generally used to refer to public safety wireless network. Currently, it consists of 25 kHz narrow band system, with PTT (Push-To-Talk) voice support and limited data and text.
FDMA	Frequency division multiple access: User by analog LMR systems of P25 and others. It has one dedicated voice channel for transmit and another dedicated voice channel for receive.
TDMA	Time division multiple access: Used by digital LMR systems of TETRA and P25 phase two. TETRA provides four time slots while P25 provides two time slots in the 25 kHz bandwidth, thereby increasing the capacity of the system by four times (TETRA) or two times (P25) due to time sharing.
Talk group	Part of LMR, the talk group represents a set of users such as members of a police squad or an EMS squad or in a public transport system buses from a specific division or depot with its dispatcher.
PSTN	Public Service Telephone Network refers to the telephone system typically circuit-switched, but may consist of a hybrid with circuit-switched and packet-switched networks.
Simulcast	An LMR system in which all the sites operate on the same frequency channels. Control channels are separately designated but repeated at each tower, allowing users to seamlessly move around the area.
Multicast	An LMR system where specific channels are designated at each site, with a control channel embedded in the channel. The frequency used at one site can be reused at the alternate (adjacent to

TDMA frame structure

adjacent) sites, allowing higher capacity needed in urban areas. TDMA frame structure consists of hyperframe at the highest level which has 60 multi frames in it. Each multi frame contains 18 TDMA frames in which the last two are used for control.
The TDMA frame structure in TETRA consists of four slots. Slot is the lowest level in the frame structure. Each time slot consists of 510 modulating bits that contain information related to voice and data.

DMO

Direct mode operation refers to one mobile user talking to another without routing the call through infrastructure. This is allowed in TETRA and can be very useful in communications "on the scene of incidence" for police, fire, or other emergency responder teams.

OTAR

Over-The-Air-Rekeying is a process where the network provider can update mobiles and handsets using a specific high-security session. This may include software update, changing group, or other features as specified by the standard. Both P25 and TETRA support OTAR with their specific protocols.

FirstNet

A commercial term to denote that national public safety network using 4G LTE and later cellular technologies. It not only allocates 700 MHz public safety bands in North America but also supports all commercial bands of cellular such that public safety users get the full benefit of the latest technology in terms of features on the handset while retaining the Push-To-Talk feature.

LTE-R

The 4G and later cellular technologies that support rail network in different countries. It addresses features specific to the rail network that includes seamless handoff for high-speed trains traveling at 500 kmph (already implemented in some countries)

DSRC

Dedicated short-range communication is an unlicensed frequency band of 75 MHz (5.850–5.925 GHz). It is specially allocated to support vehicular automation, with way-side support on specific frequencies and on-board units on related frequencies.

WAVE

Wireless access in vehicular environments: It refers to the 802.11p standard that was specifically adapted for VANET (vehicular ad hoc network) using the ITS (intelligent transport system) band for autonomous vehicle communication.

V2I, V2V, and V2X

Vehicle to Internet refers to vehicular communication with way-side infrastructure, V2V refers to vehicle to vehicle communication, and V2X is vehicle to everything proposed by 5G cellular. Each of these has developed protocols that will be implemented in the vehicle and wayside.

CBTC

Communication-based train control refers to the technology that makes subway cars and light rail completely driverless and driven by way-side infrastructure and on-board radio system. It has been widely implemented in the 2.4 GHz band with TDMA protocol, quite different from the 802.11 standard.

CATC

Continuous automatic train control refers to the automation of high-speed rail traveling over great distances (about 1000 km) that include microcomputer-based interlocking systems, for extra safety to ensure trains follow the proper path at interlockings.

SAE

levels of autonomy: Society of Automotive Engineers have specified five levels of automation for vehicle starting with 1—fully manual and driven by human driver—to level 5—fully automated with no human driver.

URLLC	Ultra-reliable low latency communication is a standard that is part of 5G and aims to support V2X communication for autonomous vehicles.
LIDAR	Light detection and ranging is one of the technologies used in autonomous vehicles, particularly in reversing in narrow spaces, etc.
Farming automation	Farm automation (or smart farming) is a variety of tech innovations in traditional farming to optimize the food production process and improve quality. As of now, advanced farming technology can be an essential part of the farmer's daily work. These refer to driverless tractors, sprayer, and other farming equipment that move on same the track within ±3 cm.
Automated mining	There are two types of automated mining process and software automation, and the application of robotic technology to mining vehicles and equipment.

In the past decade momentum for electric vehicles with an emphasis on "connected vehicles" has gained public support. Prior to that, hybrid automobiles were introduced and worked well. There are other modes of transport such as rail, bus, public safety vehicles used by police, fire, emergency medical service (EMS), and control of vehicles that traverse difficult terrain. All of them use some form of radio communication. In addition, rail systems use control signals (generally known as positive train control—PTC). This Chapter will discuss all of them, providing an overview of vehicular technology using wireless communication that is essential to mobility.

15.1 Land Mobile Radio: Public Safety (Rail, Bus)

Life safety and emergency services form the quintessential support that all societies depend on. Whether it is ambulance, fire engines or police, or other vehicles with wireless systems, their use comes into its own during critical events that demand their presence. These systems, therefore, define "trunk-based services" which means a dedicated channel must be available to them all the time, independent of the cellular system used by the general public. Public safety systems mainly use voice with limited need for data, but due to the nature of public safety work they are termed "mission critical", which means life is at stake and reliability of this

system is crucial. All of them are based on "narrow band" technology (25 kHz bandwidth). Text messages are limited to driver receiving or sending "pre-canned" messages when a vehicle is not in motion; this is an essential safety feature.

Land mobile radio (LMR) predates the cellular system by several decades. The first deployment started in Detroit in 1928, soon after automobiles were introduced into the police force [1]. These early versions started in the 1920s using AM radio, followed later in 1933 by FM radio used by the state of Connecticut highway patrol. Current units operated by police and public safety still use FM radio but are gradually being replaced by digital radio technology based on TDMA. The system is "mission critical" as explained earlier and it has stayed that way for about a century.

There is no integration with other broadband services, as it occurred in cellular systems. There is another basic difference—the public cellular system discussed in Chap. 11 and also satellite system in Chap. 13 assume one-to-one conversation. There is a speaker and a listener and so long as two people are connected, the circuit is established. One user calling the other user and establishing a call or a session is the basic rule. In contrast, the public safety system typically uses "one-to-many". Usually, someone who is a dispatcher, or commander of a group, speaks to multiple end-users who are part of the group. There is no need for two-way conversation since the group members are listening. Conversations must be instantaneous—there is no time to waste on calling, ringing, and pick up (Figs. 11.3, 11.4, and 11.5). A simplex method of "Push-To-Talk" is natural for all these services. Duplex is available but rarely needed. Also, coverage is of concern and not capacity—this topic was discussed in Chap. 11. Therefore, tall towers with higher RF power are common.

Public safety systems are designed for coverage rather than capacity. The difference between coverage and capacity design principles was elaborated in Sect. 11.1.1. It was indicated that cellular systems are designed for capacity (for a large number of users), while LMR is designed for a large coverage area, typically a town, a municipal area, or a rural county. Due to its nature, LMR is shared by a variety of users who are segregated by "Talk groups" such that they can communicate within their own groups. Several such talk groups may share a common channel, which is why the concept of "trunk system" becomes important.

Trunk allocation protocols allow the system to control and prioritize a channel judiciously providing a system that allocates a channel fairly. Figure 15.1 shows a simple concept where multiple towers serve an area and public safety users—with mobile or portables accessing such a multi-site system. For simplicity, only two towers are shown, but multiple towers are common.

Figure 15.1 shows the basic scheme of how a public safety network operates. Each site may have either a base

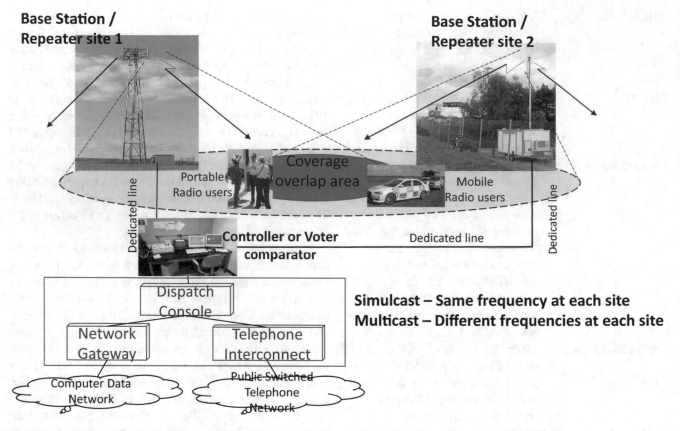

Base Station /
Repeater site 1

Base Station /
Repeater site 2

Dedicated line

Portable
Radio users

Coverage
overlap area

Mobile
Radio users

Dedicated line

Controller or Voter
comparator

Dedicated line

Simulcast – Same frequency at each site
Multicast – Different frequencies at each site

Dispatch
Console

Network
Gateway

Telephone
Interconnect

Computer Data
Network

Public Switched
Telephone
Network

Fig. 15.1 Typical two sites land mobile network—Simulcast and Multicast

station unit or it may have just a repeater located in the shelter shown next to the tower. A dedicated set of communication cables connects each base station site to a dispatch console from where it is connected to either a telephone interconnect panel or to a computer data network. From there it gets connected to the public network (PSTN or Internet). The operation is based on simplex—the user with a mobile radio or a portable radio will attempt to make a call using push to talk button (PTT) and waits until a channel is allocated, as explained below.

If the agency has many licensed frequency channels available to them, then they may prefer to use different frequencies at each site. In that case the scheme is known as multicast. When the user moves from one site to the next, there will be a handoff (similar to cellular handoff). Typically, the handoff is of 50 ms or less duration and the ear cannot discern the "click". Mostly the user may hear this handoff "click" if the duration of handoff exceeds 100 ms (not common).

Trunk-based allocation schemes are implemented at the controller or voter comparator block where a dispatcher or commander is connected by a dedicated line as shown. Algorithms at the controller are able to form a queue of callers who are waiting to be connected and judiciously allocates channels such that the users will not have to wait

for extended periods of time. Due to such efficient algorithms, trunk-based systems are popular and earlier methods of voting schemes (often known as "conventional scheme") to allocate channels are becoming less popular.

Figure 15.2 shows the difference in terms of how frequency channel is allocated in systems based on Simulcast and Multicast. In Simulcast scheme (shown on left), the seven sites operate using all channels F1-Fn. Users can move around the entire area since the same channels are everywhere.

In contrast, the Multicast scheme example shown on the right is made up of seven groups of channels. Groups G1, G2, G3, and G4 in the middle are used only once. But groups G5, G6, and G7 at the edges are reused at the alternate cell (adjacent to adjacent is known as alternate).

This is to avoid co-channel interference (energy present on the same channel). By reusing the same channel frequency two cells away, power radiating from the bottom right G7, for example, becomes so weak at the top left that it no longer interferes. Hence this channel can be reused at the transmitter at the top left site. Similarly, G5, G6, and G7 are judiciously reused. With this scheme users in ten cells are served by only seven-channel groups—due to "frequency reuse". This concept was borrowed from cellular systems, and "frequency reuse" was commonly used in 2G systems such as GSM.

Seven Site Simulcast LMR
Frequency Layout

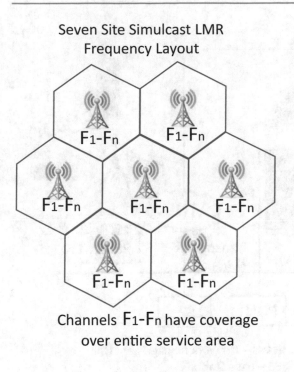

Channels F$_1$-F$_n$ have coverage
over entire service area

Ten Site Multicast or Zone type LMR
Frequency Layout

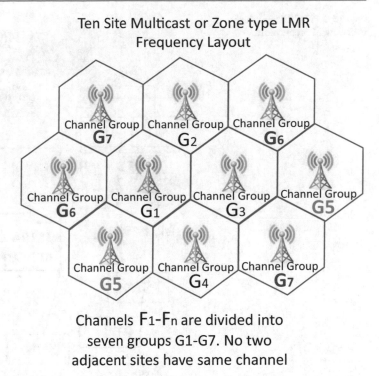

Channels F$_1$-F$_n$ are divided into
seven groups G1-G7. No two
adjacent sites have same channel

Fig. 15.2 Comparison of Simulcast and Multicast frequency layouts.

Trunk systems in Fig. 15.2 can be described in terms of how a control channel is operated. Let us consider an operation: User A wants to contact all of the units in the talk group [2]. The Push-to-Talk (PTT) key is pressed, which causes the radio to send a short burst of data to the control channel repeater. This data burst identifies the caller attributes and enters a channel request to the system controller. User A's radio then switches to receive mode to wait for a data response from the system controller. At the dispatch center, the system controller places User A in a queue and then attempts to select an available voice channel. When a voice channel becomes available, the system controller grants it to User A (according to queue). At the same time, it also sends a data message over the control channel switching all units in User A's talk group to the available voice channel. Only units in this talk group are automatically switched to the assigned channel. When User A starts talking, all members of that talk group hear the conversation. This preempts other users of that talk group from getting assigned to any other channel for the duration of the call, since they are in "receive" mode.

15.2   Key Digital LMR Systems

There are mainly two digital LMR standards for train, bus, as well as police, fire, and EMS. But there are many others that are also listed in [2]. The two major standards are

TETRA (Terrestrial Trunk Radio) and P25. TETRA is a digital TDMA technology similar to GSM used globally. P25 or "Project 25" developed by APCO (Association of Public Safety Communication Officers) is a hybrid system with phase 1 still using FDMA (analog) and phase 2 moving to TDMA technology but still supports only voice traffic. There are many differences between these two systems, which will be discussed in the later paragraphs.

15.2.1   TETRA (Terrestrial Trunk Radio) and P25 (Project 25)

There are two major standards for public safety—the TETRA system is used globally and the P25 system is used mainly in the USA. Figure 15.3 provides an overview of the TETRA system with standard interfaces shown as "I" interfaces.

Figure 15.3 shows the architecture of the TETRA system with multiple base stations and infrastructure, based on four-slot TDMA protocol. Its architecture has identified ISI (inter-system interfaces), in terms of the "I" interfaces I-1 through I-6 are shown in Fig. 15.3. Any of these units can be built by any vendor and it will work well with units by another vendor. In fact, TETRA standard has well-established laboratories with test methods to make sure that such compatibility is established and guaranteed. It verifies that systems with earlier versions of wired telephone

Fig. 15.3 TETRA standard—overview of architecture

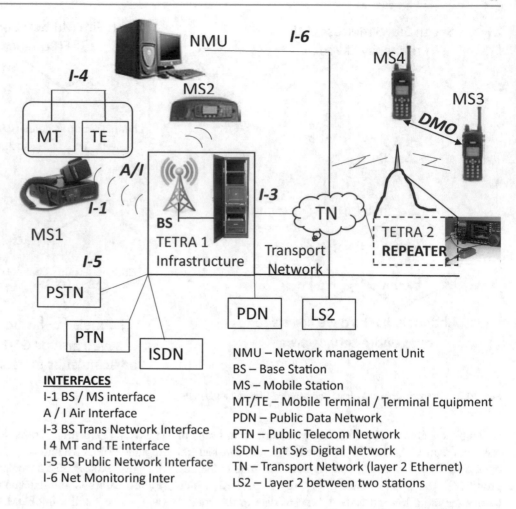

INTERFACES
I-1 BS / MS interface
A / I Air Interface
I-3 BS Trans Network Interface
I 4 MT and TE interface
I-5 BS Public Network Interface
I-6 Net Monitoring Inter

NMU – Network management Unit
BS – Base Station
MS – Mobile Station
MT/TE – Mobile Terminal / Terminal Equipment
PDN – Public Data Network
PTN – Public Telecom Network
ISDN – Int Sys Digital Network
TN – Transport Network (layer 2 Ethernet)
LS2 – Layer 2 between two stations

backhaul networks such as ISDN, PTN, and PSTN as well as PDN (packed data network such as the ATM / SONET, or the IP network) allow calls to support legacy handsets. These backhaul networks were described earlier in Chap. 4. All interfaces between TETRA base stations are based on a standard OSI model (such as layer 2 Ethernet), widely used in IP transport networks. Perhaps an important aspect of TETRA is that signaling is "multicast"—which means the control channel is embedded in each channel and the first slot always carries the control signal.

TETRA, somewhat like GSM (2G cellular standard), was started as digital-only radio technology with no intent for backward compatibility to analog. This approach was natural since Europe had multiple analog standards and backward compatibility was not useful. Based on TDMA with $\frac{\pi}{4}DQPSK$ modulation, the current version (Release 1) of ETSI (European Telecom Standards Institute) operates in all narrow band spectrum up to 1 GHz using only standard 25 kHz bandwidth.

Figure 15.4 features TETRA frame structure showing a hierarchy of hyperframe, multi-frame, and TDMA time slots. The control frame is embedded in the channel which is why each channel can be assigned and works independently with control derived from the central network controller. In the multi-frame, the last two (17 and 18) contain control frames as indicated. This allows smooth handoff, interworking between BS and a repeater. In the circuit mode (while interworking with earlier generation system) voice or data operation traffic from an 18-frame multi-frame length of time is compressed and conveyed within 17 TDMA frames. This allows the 18th frame to be used to control signaling without interrupting the flow of data. Besides the basic TDMA frame structure, there is a hyperframe imposed above the multi-frame structure, as shown. This is for the long repeat frame needed by encipherment synchronization.

Allocation of channels is similar to cellular allowing "channel re-use" in large urban areas. This concept was described earlier in Chap. 11, Sect. 2, where sector reuse was described. The architecture allows the use of repeaters as a second unit (shown earlier on the right in Fig. 15.3). Such smaller units mounted on towers, don't have full base station (BS) functionality, but the channel is assigned, and it works with a base station to serve small areas. The TDMA frame structure follows cellular design practices of frame,

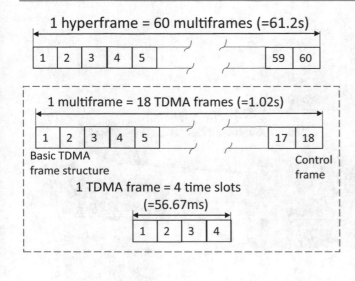

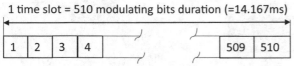

Fig. 15.4 TETRA—TDMA frame structure

multi-frame, and slot and voice use the standard block cipher mode (expects entire frame is received). This also allows standard encryption of the air link.

Due to separate air interface standard described in [2], DMO (Direct Mode Operation) shown in Fig. 15.3 on the right top is supported. This means one mobile/portable can talk to another without the need to go through infrastructure. This mode, for example, allows police or fire personnel to talk to each other at the scene of incidence so long as they are in the line of sight. Such features, as well as PEI (Peripheral Equipment Interface) allow a variety of devices (such as printer and other office products that operate over Ethernet) to interwork with mobiles/portables. Such features are an important reason why TETRA has become popular as an LMR technology.

The current release of TETRA supports the narrow band technology described so far. The second release known as Release 2 has a path forward moving systematically towards 4G LTE. The concept is that as more adjacent channels of 25 kHz are joined together higher modulation schemes can be implemented. The concept is based on adaptive modulation coding described in Sect. 14.2.7 of Chap. 14, where it was described in the context of microwave links. Larger bandwidth results in the use of a higher modulation scheme. Also, the higher the received signal level, the higher will be the modulation scheme. This combination results in higher throughput. This will be elaborated later in this chapter.

In terms of security, TETRA uses an authentication and encryption scheme fairly similar to GSM. It clearly

establishes the authenticity of the user and makes sure the uplink and downlink are both encrypted. If the infrastructure also happens to be a standard IP network, then "end-to-end encryption" can also be implemented.

Depending on the user agency higher levels of encryption are possible. "Over-The-Air-Re-keying" or OTAR is a standard method used in LMR to update encryption keys using the radio network. Major advantages of the encryption design are that encryption does not affect speech intelligibility nor does it affect the system's usable range. Both of these advantages are major improvements over encryption previously used in analog systems.

15.2.2 The APCO P25 System

P25 or Project 25 is a TIA (Telecom Institute of America) standard whose objective is to provide a transition plan from the current analog FDMA system to a new TDMA system. By definition, backward compatibility is required [3]. Therefore, both Simulcast and Multicast are supported. This concept is not new—in cellular a similar transition scheme using IS-136 was described in Sect. 11.2.3 under the category of channel usage. That scheme had a similar objective: to have interoperability between AMPS (analog) and IS-136 (TDMA digital). The transition of P25 is planned in three phases. Phase-1 with 12.5 kHz analog, digital, or hybrid modes. Phase-2 with two-slot TDMA, and phase 3 jointly being worked by ETSI, and TIA towards services with higher data throughput [4].

Backward compatibility has its advantages and limitations. The advantage is that the existing network is not disturbed, and a new technology network can be gradually phased in. The limitation is that older technology inhibits free evolution of the new technology. Whenever FDMA is used there is only one channel available—as compared with multiple "slots" on TDMA. There is an increase in capacity if TDMA is used, but it is possible only if there are multiple channels.

Figure 15.5 shows the general system model for P25 where RF subsystem is at the center having seven separately defined interfaces. The subsystems and the interfaces are:

- RF Sub-system (RFSS) Core Infrastructure
- Common Air Interface (Um) Radio to radio protocol
- Inter-System Interface (ISSI) RFSS to all other system interconnections (In progress)
- Telephone Interconnect Interface (E_t) PSTN to RFSS definition
- Network Management Interface (E_n) Network to RFSS definition (in progress)
- Data Host or Network Interface (E_d) Computer-aided dispatch to RFSS definition

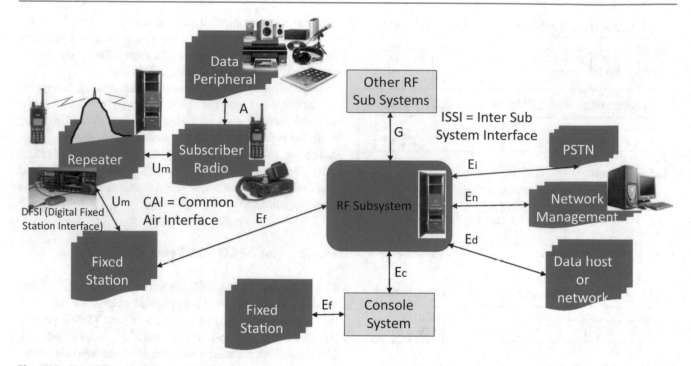

Fig. 15.5 The P25 general system model based on interfaces

- Data Peripheral Interface (A) Radio to Data Peripheral definition
- Fixed Station Interface (E_f) Base station to RFSS/Console Sub-system definition
- Console Sub-system Interface (E_c) Console to RFSS definition (in progress)

Since TIA publishes these as open interface standards, each of them (E_f or E_c or A or Um) is available to vendors who can build products to interface with. In that sense, they are similar in nature to TETRA although not identical. Standard IP network-based products do interface with P25 as defined in the TIA standard releases. The P25 was developed primarily for the North American market where considerable infrastructure (analog) exists; therefore backward compatibility is addressed at these interfaces.

Using a common control channel to address FDMA and TDMA allows P25 system to handle both legacy and new digital systems. Control channel signaling is an important part of how such transition can be implemented. P25 supports both Simulcast and Multicast methods of channel assignment. But Simulcast is commonly used since Phase 1 was based on the transition from analog systems that use Simulcast. Table 15.1 has a comparison of P25 and TETRA terms and usage.

Operations of TETRA and P25 are similar, but not identical. TETRA will interwork with a GSM cell phone, whereas P25 does not. There are other differences, particularly in the control channel due to differences inherent to frame structure;

some are worth mentioning. Security features of P25 have come under scrutiny and criticism by cryptographers, who claim that P25 design lacks robustness even though encryption is offered as an option [6]. While both standards look to 4G LTE as a future goal, the path towards broadband is slow. After all, narrow band technologies of 25 kHz cannot fathom bandwidth of 20 MHz used by LTE, which means 800 channels of narrow band—no public safety agency would have or can afford. But separate channel allocation in the 700 MHz band for public safety has been announced by the FCC (in the same band as cellular).

15.2.3   4G LTE Plans for P25 and TETRA

The cellular standards body 3GPP carefully evaluated the request from public safety to bring LMR into the cellular system. Due to high reliability and coverage requirements, it did so in several phases starting from Release 11 and carefully developing a plan over several years. With "Push-To-Talk or PTT" as primary means of voice communication, it evaluated multiple options to support data. Finally, release 17 of 4G LTE standard [6] will have full public safety communication as a mission critical system supported by a cellular network. With active support from NPSTC (National Public Safety Telecom Council), models of using broadband channels as a private network or hybrid with cellular or even by a third-party network provider are considered.

Table 15.1 Comparison of P25 and TETRA technologies

P25	TETRA	Comment
NAC code	MCC and MNC code	Equivalent
TGID (Talk group ID)	GSSI (Group short subscriber ID)	Equivalent
UID—User ID	ISSI—Individual Short Subscriber ID	Equivalent
ISSI	ISI Intersystem Interface	Equivalent
OTAR—Over-THE-AIR RE-KEYing	OTAK—Over-the-Air re-keying	Use of E2E encryption
Pre-programmed messages	Status messages	Equivalent
ESN (Electronic Serial #)	TEI Terminal Equipment Interface	Equivalent
Optimized for large area coverage (low density)	Optimized for high population density (higher capacity)	
7.2 + 4.8 = 12 kbps	Data throughput 28.8 kbps	P25 supports voice only
Simulcast and Multicast	Multicast only	
Two-slot TDMA	Four slot TDMA	
Future—Phase 3	TEDS—Tetra-Enhanced Data service*	Moving to 4G of 3GPP*

* Both P25 and TETRA are formally working with 3GPP towards evolution to 4G LTE

The US public safety team envisioned a first-responder network (FirstNet) dedicated to public safety. It has features that not only address the needs of police, fire, EMS, and others but provide a nationwide umbrella of common frequency bands. Based on this concept, the current model of networks having local coverage would be replaced by a national network with common bands. "FirstNet" will provide high-speed broadband services using modern cellular protocols, with features similar to modern smart phones [7]. Video or other broadband services that are familiar to cellular users would become part of mission critical network. Already handsets that operate on both P25 and FirstNet (4G LTE) are available in the market. When the public safety personnel encounter low signal, they can switch over to the cellular system and use the same handset to continue their communications with PTT (Push To Talk).

Economy of scale in cellular makes it possible to provide feature-rich portable and mobiles for public safety. Table 15.2 provides an overview of the current LTE and enhancement needed to support the rail industry which is

Table 15.2 System parameters of GSM-R, LTE, and LTE-R

Key	Parameter	GSM-R	LTE	LTE-R
1	Frequency	Uplink: 876–880 MHz downlink: 921–925 MHz	800 MHz, 1.8 GHz, 2.6 GHz	450 MHz, 800 MHz, 1.4 GHz, 1.8 GHz
2	Bandwidth	0.2 MHz	1.4–20 MHz	1.4–20 MHz
3	Modulation	GMSK	QPSK/M-QAM / OFDM	QPSK/16QAM
4	Cell range	8 km	1–5 km	4–12 km
5	Cell configuration	Single sector	Multi-sector	Single sector
6	Peak data rate, downlink/uplink	172/172 Kbps	100/50 Mbps	50/100 Mbps
7	Peak spectral efficiency	0.33 bps/Hz	16.32 bps/Hz	2.55 bps/Hz
8	Data transmission	Requires voice call connection	Packet switching	Packet switching (UDP data)
9	Packet retransmission	No (serial data)	Yes (IP packets)	Reduced (UDP packets)
10	MIMO	NO	2×2 and 4×4	2×2
11	Mobility	Max 500 kmph	Max 350 kmph	Max 500 kmph
12	Handover success rate	$\geq 99.5\%$	$\geq 99.5\%$	$\geq 99.9\%$
13	Handover procedure	Hard	Hard / Soft	Soft: no data loss
14	All IP (native)	No	Yes	Yes

moving from GSM-R towards LTE-R [8]. Many countries (USA, Korea, Japan, and others) have allocated bands for LTE-R initiative from the rail industry.

In order to modernize signaling and support communication within the railway network, LTE-R (LTE Rail) is seen as a new specification needed to replace the aging GSM-R. Key specifications of LTE-R shown in Table 15.2 indicate certain compromises and changes in PTT methods for convenient rail operations. Some stringent specifications that the rail industry imposes are speed—currently topping at about 500 kmph. Consequently, reliability of the network becomes crucial—quite often rail travels through harsh environments with no surrounding population or access. How to maintain the track-side communication network infrastructure and monitor them—such issues are being addressed with a pilot trial over a distance of 250 km in Korea, for example [7]. What are the effects of high speed on wireless communication and how does it affect design?

In Table 15.2, row 11 on mobility indicates the change in LTE specification—change from the current LTE speed of 350 kmph to 500kmph. This brings major challenges to radio—in terms of doppler and mobility-related issues already discussed in Chap. 11 under Sect. 11.1. High-speed rail needs track-side support—for example, how many handoffs are expected if towers are 8–12 km apart (cell range). Since rail travels at 500 kmph (~8 km in one minute), there would be 1 handoffs/minute, and each handoff must be reliable (line 12). This is shown in Table 15.2 as a distinct specification (line 13_ for handoff—"soft handoff

with no data loss". To implement this, a major rework of remote radio heads with seamless integration of redundancy of network becomes essential.

Such reliable networks are currently being tested in several countries [7, 8] and it is expected to result in adjustments to LTE-R specifications. Now let us consider enhanced railway services.

Figure 15.6 shows some services that LTE-R must address and also support the current GSM-R network. Such GSM services can be enhanced towards LTE-R as high-speed rail gets deployed. Current LTE-R trials indicate that rail control networks (known as positive train control—PTC) will continue with an extensive deployment of fiber or millimeter-wave links along the track side [7, 8]. Each additional service deployed enhances wireless infrastructure, making rail agencies pay close attention to 4G and 5G technology being deployed. These support services such as infrastructure monitoring and in station communication, railway IoT, mobile e-ticketing, interaction with passengers, and real-time remote monitoring, which are all shown in Fig. 15.6.

High-speed rail cars manufactured today are fitted with such wireless units to provide services shown as "in rail car wireless access" in Fig. 15.6. Connection between a rail car and the outside world is established by a combination of cellular and satellite links. WiFi will remain within rail car since it cannot support high-speed mobility [9]. Sophisticated three-axis stabilized platforms with antennas mounted on top of the rail car are used to support cellular radio

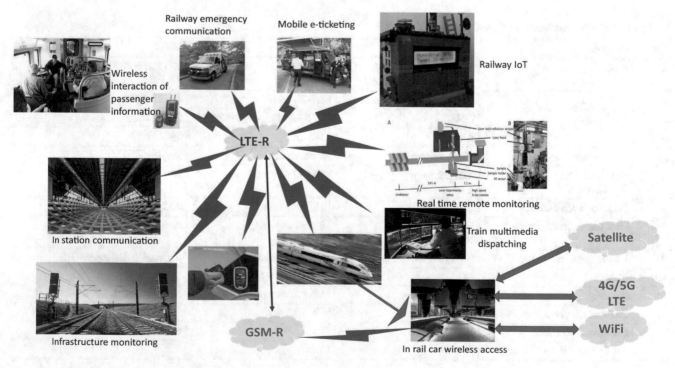

Fig. 15.6 LTE-R services and requirements structure

connection to remote radio heads (part of 4G / 5G network) located on towers. Such radio head units mounted on poles along track side are currently in service. Rail agencies find such deployment efficient and useful in the long term. Antennas on the stabilized platform also connect to satellites in areas where cellular services are not available. Cellular and satellite as mobility backhaul options were discussed in Chaps. 11 and 13 and their mobility features need enhancement to handle such high speeds.

15.3 Safety in Vehicles: DSRC and LAN in Vehicle

In the last decade, the most important change in road transport was ushered when safety measures were introduced in vehicles. It may be a simple warning to the driver or a swift application of brakes as the vehicle sensor notices danger ahead. Such features immediately brought public support to this new technology of DSRC (Dedicated Short-Range Communication). It addresses vehicle to vehicle communications with low latency. There is considerable information to be processed in real time to control the vehicle (independent of driver).

What is DSRC and how is it useful? DSRC is a wireless communication technology that enables vehicles to communicate with other vehicles on the road directly, without involving cellular or other infrastructure.

Figure 15.7 shows the concept of DSRC and how the system is used to alert driver or for vehicle control. This concept was introduced originally during 1999 to support ITS (Intelligent Transport System). To accommodate ITS the IEEE 802.11 standard was modified to 802.11p. Modification in basic structure is shown in Table 15.3, where some parameters were altered, as shown.

This modified standard 802.11p is known as Wireless Access in Vehicular Environments (WAVE). It was specially developed to adapt to VANETs (vehicular ad hoc network) requirements and support Intelligent Transport Systems (ITS). The IEEE 802.11p amendment allows the use of 5.9 GHz band (5.850–5.925) GHz with channel spacings of 20 MHz, 10 MHz, and 5 MHz and lays down the requirements for using this band in Europe and the USA.

The channel allocation in the ITS band is shown in Fig. 15.8. A total of 75 MHz bandwidth is split into four service channels (on either side of the control channel). Channel for public safety is separated from that for collision avoidance (instantaneous, low latency to support braking). In terms of physical layer challenges, quite a few are listed in [10]. These are five major considerations to this band, as indicated below (see 1 to 5):

1. *Effect of noise in bit and symbol energy*
2. *Effect of unused subcarriers on symbol energy*
3. *Rayleigh fading, frequency selective fading, delay spread*
4. *Doppler shift, channel variation, and channel estimation*
5. *Network coverage range and bit rate enhancement techniques*

While each of them is being addressed, new techniques and methods are being proposed and this is an active area of current research. In addition, the MAC layer has some current limitations due to the 802.11 format [11]. Some of the MAC layer limitations are:

(i) Sense and then transmit (no handshake)
    (a) Cannot deal with hidden terminal and congestion collapse.
    (b) High collision probability resulting in low reliability and efficiency.

Fig. 15.7 Concept of vehicle-to-vehicle communication (with alerts)

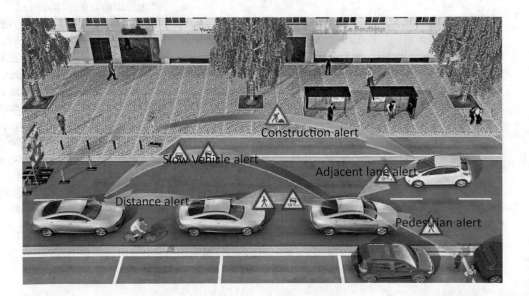

Table 15.3 Modifications to 802.11 standard to develop 802.11p

Parameters	802.11a	802.11p	Changes
Bit rate Mb/s	6, 9, 12, 18, 24, 36, 48, 54	3, 4.5, 6, 9, 12, 18, 24, 27	Half
Modulation modes	BPSK, QPSK, 16QAM, 64 QAM	BPSK, QPSK, 16QAM, 64 QAM	No Change
Code rate	½, 2/3, 3/4	½, 2/3, 3/4	No Change
Number of carriers	52	52	No Change
Symbol duration	4 μs	8 μs	Double
Guard time	0.8 μs	1.6 μs	Double
FFT period	3.2 μs	6.4 μs	Double
Preamble duration	16 μs	32 μs	Double
Subcarrier spacing	0.3125 MHz	0.15625 MHz	Half

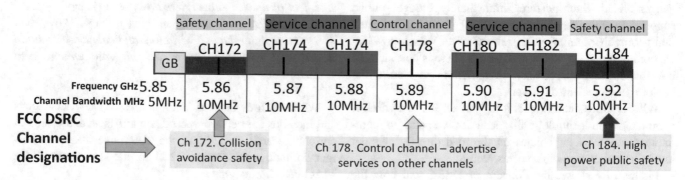

Fig. 15.8 Physical layer channel spacing for the ITS band

(ii) No feedback for beacon broadcasting
   (a) Cannot verify whether transmission failed.
   (b) Without confirmation, blind re-transmission can be applied but reduces efficiency

Research teams observed that compared to a professional driver, an autonomous vehicle generates almost a thousand times more data. It also encounters situations that are not easily resolved by machines. To address them, many field trials are under way by major automobiles makers. There is a realization that there are other parts of autonomous communication that must address issues related to how the enormous data can be transferred to the network. This is addressed in the next section, where both V2V (vehicle to vehicle) and V2I (vehicle to internet) are reviewed.

Consequently, there is a proposal to increase the allocated bandwidth by including some of the unlicensed spectrum in that range, as shown in Fig. 15.9.

The basis for this proposal consists of a series of interference measurements between 802.11AC channels and DSRC by the US Dept. of Transport [12]. This study supports a proposal for UNII-4 band (Unlicensed National Information Infrastructure) to co-exist with DSRC and possibly be used for vehicular infrastructure. This proposal is likely to be accepted, based on the fact that it may provide more bandwidth to even DSRC applications in the future.

Let us now consider a completely different topic—communication networks "inside the vehicle" for driver's convenience. Local networks operate within vehicles today that connect different devices, such as speakers, cell phones, or provide services such as WiFi or Bluetooth, within the vehicle [13]. There are many car vendors who have established their own standards and interfaces, but these are specific to their own models and not widely used across the industry. In order to provide common infrastructure, some neutral organization must provide standards that every

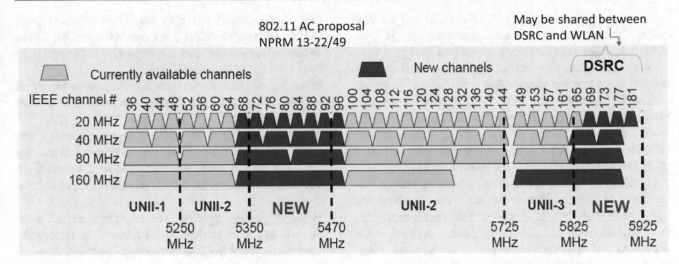

Fig. 15.9 DSRC and proposed U-NII-4 radio channels being considered by the FCC

Table 15.4 SAE classification of in vehicle networks

Network classification	Speed	Application
Class A	<10 kbps low speed	Convenience features—e.g., trunk release, electric mirror adjustment
Class B	10–125 kbps medium speed	General instrumentation transfer—e.g. Instruments, Power windows
Class C	125 kbps–1 Mbps high speed	Real-time control, e.g., power train, vehicle dynamics
Class D	Above 1 Mbps	Multimedia applications, e.g., Internet, Digital TV, Hard real-time critical functions, e.g., X-by-wire critical applications

vendor, component supplier, can adhere to. One such standard, provided by the Society of Automotive Engineers (SAE) classifies vehicular networks into four categories A, B, C, and D as indicated in Table 15.4.

Each of these was initially introduced by high-end car manufacturers. But standardization of protocols and interfaces allowed its use in all standard cars. The vehicle in general is a harsh environment for all electronics including wireless devices. Adverse conditions like mechanical vibration, temperature swings from –40 °C to + 80 °C, splashes from oil, petrol, and water, ice, strong electromagnetic fields (automotive field strengths can be > 200 V/m) are encountered.

In Table 15.4 Class D networks in particular are popular for wireless applications since these are accessible to the driver. In addition to entertainment (satellite TV streams to back seat passengers), audio and navigation aids are widely used. Traveling salesman may use devices like printer or office gadgets to support field work. For many electronic gadgets, the humble cigarette lighter socket in any vehicle has become the power source of choice. The basic design principle is "don't distract the driver". But driving itself may no longer be needed, as the next few sections reveal.

15.4   Vehicle to Vehicle, Vehicle to Way-Side Communications

While V2V is widely acknowledged, V2I (vehicle to infrastructure) which must be deployed on sides of the road has taken up an important role. Its role is to inform the vehicle about traffic conditions, how to re-route, and more recently traffic signals and pedestrian sense, to move towards autonomous—or self-driven car. While current drivers get used to this concept of V2V, there is still skepticism about V2I that allows the car to drive by itself over great distances.

There are valid reasons for such skepticism. There are concerns about weather conditions, poor visibility, slippery roads, and more—such as faulty traffic lights, manually operated junctions, etc. [14]. To support them, way-side infrastructure is essential. Such infrastructure must be not only aware of local construction but must inform the driver or even automate on-board unit on what to do. Many countries are already in the process of trying out extensive pilot projects on how to introduce way-side infrastructure and which real-world scenarios can be tested.

V2I infrastructure must support both human and machine vision. Current roadway infrastructure must change from analog messages designed for human eyes to digital messages designed such that technology in automated cars can interpret them and the surrounding environment. Response must be in real time, with redundancy to increase the confidence level of every road user. "Does the vehicle understand, and can it make critical driving decisions" are common concerns? Such redundancy is needed for increased safety and to enhance confidence in mobility [15].

Figure 15.10 shows some of the emerging special features in a car needed to improve roadways, mobility, and safety. Note that each of these must support humans and machines and will be elaborated further in Sect. 15.6. Since RADAR is a very important sensor in autonomous cars-function of Radar and application will be explained in detail in Chap. 18, under Sects. 18.3, 18.4 and 18.5. In that chapter, Radar and its applications are major topics. Even Fig. 15.10 is further expanded as Fig. 18.14 to provide more details as to what each sensor does etc. For now let us consider only the roadside infrastructure changes:

- Advanced road markings: Pavement marking must be visible to humans and also read by machines in any road condition.

- Pavement lane markings: they must work with automated vehicle sensors to detect lines outside the vision-based spectrum, improving lane detection and traffic safety in even extreme weather conditions.
- Smart signs: Directional signage that must be visible to humans and to machines that can be sensed irrespective of road condition.
- Retroreflective signs: To provide better readability that must offer accurate navigation and faster decision-making for both drivers and automated vehicle systems. In addition, smart signs must be compatible with traditional signage.
- **Wireless Communication**: There is a fundamental need for wireless communication that connects directly to vehicles, quickly helping to identify construction zones and potential safety hazards so that vehicle mobility and traffic flow can improve. Recent 5G standard URLLC (ultra-reliable low latency communication) was specifically developed to address such communication with a reliability of 99.9999% and latency of < 1 ms for an individual link and < 10 ms for end-to-end network. This was indicated in Sects. 11.6 and 11.8 and ushers a major change to current IP networks that operate the Internet.
- A DSRC multi-channel test tool is an independent multi-channel listening device that provides vehicle to

Fig. 15.10 Sensors on vehicle and special features recognition

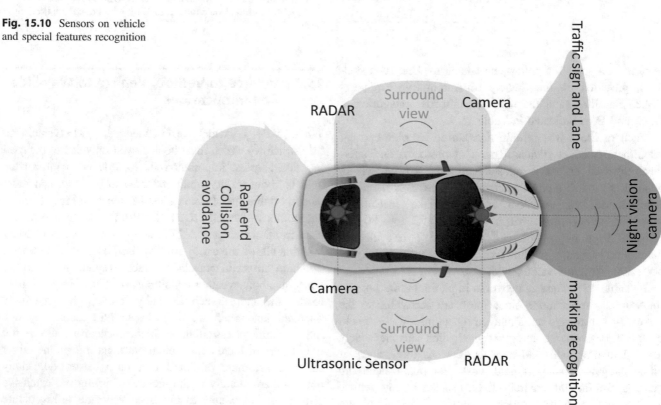

vehicle (V2V) and vehicle to infrastructure (V2I) communications. The test tool is an unbiased third-party resource, used to decode standards and connected vehicle requirements; SAE J2735, IEEE 802.11p, and IEEE1609. This will be elaborated further in Sect. 15.6.

15.5 Positive Train Control and CBTC

Trains with no human drivers have been accepted for quite some time now. Driverless monorail cars in airports or shuttle services over short distances are in service for many years. Normally terme "light rail" with three or four coaches to refer to shuttle trains over short distances of no more than tens of kilometers.

However, long-distance trains, particularly those that can attain very high speeds (up to or above 500 kmph or more), can become a challenge for such automation. There are major reasons why it is harder to deploy automation in trains.

1. Trains can travel at least three to four times faster than vehicles on the road (130 km on road, 500 km on rail)
2. The momentum of train (mass × velocity) is so large that it takes anywhere from 0.25 to 0.5 km before it can halt after applying brakes (based on average speed)
3. Depending on track conditions, it may bump into another train ahead or go past unmanned rail crossing without warning

4. Can move away to different section of rail (switch positions) or even derail moving on a curve at high speed.

15.5.1 PTC: Positive Train Control

Due to safety, the concept of positive train control (PTC) was introduced in high-speed rail. To understand PTC and its way-side infrastructure, some concepts of signaling are helpful, which is described now.

Figure 15.11 shows the concept of block signaling. Rail signaling is focused on safely moving trains such that they don't pass curves at excessive speed and stay on the track in a well-defined path (switching properly as the train moves in and out of stations). Signals are based on the concept of blocks that ensure safety.

A normal convention requires at least two open blocks between trains. In Fig. 15.11, at the top train 2 which is three blocks away from train 1 sees a green signal. This means the train is allowed to enter the block. At the bottom of figure, train 2 has entered the block and sees two yellow signals, indicating it can enter the next block "with caution". When it enters the next block there will be only one more block between trains. Therefore, caution is indicated showing driver must reduce speed and if necessary, apply brakes before entering the next signal. The next signal shows a single yellow indicating "slow down". The speed with which the train enters each block is determined by the gradient. If the gradient is downward, apply brakes immediately; if the gradient is upwards or uphill, then slow down naturally.

Fig. 15.11 Block signaling concept—for trains

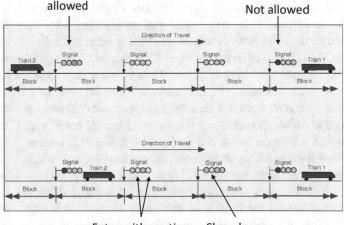

There is one more factor, which is the size of the train and how much it is loaded, which is momentum-related. Considering these factors, the speed of entry for each block is pre-determined using "maximum safety" scenario.

This can be translated into automation by using way-side equipment, which communicates with the trains using an automated mechanism. Using wireless signals, each way-side transponder can communicate with the on-board unit directing slow down or proceed. In addition, the color of the signal is translated using wireless communication so that the train unit "knows" whether the next signal is green or yellow, or red. Other situations such as a block with sharp curves may indicate the train to slow down before negotiating the curve. These rules are implemented even if there is a human driver on the train. Why? Because it is possible that the driver may be exhausted or did not pay attention to such signals. Then the system automatically applies brakes on the train slowing it down to negotiate a curve at the prescribed speed. This prevents the train from coming off the rails (derailing).

Automation of train movements using PTC (positive train control) started more as a safety measure, not with the intention of eliminating the human driver. It was shown that all train networks improved their safety records after implementing PTC. High-speed trains absolutely need such safety measures since accidents can result in major loss of human lives. For high-speed trains, the block size, therefore, becomes quite large (several kilometers) since braking may not result in train coming to a halt within the next few kilometers. Therefore, continuous automatic train control (CATC) is used by modern high-speed train systems that include microcomputer-based interlocking systems [16].

Rail infrastructure distances are often very long—several hundreds or even thousands of kilometers. Trans-Siberian rail network in Russia is 9,289 km (from Moscow to Vladivostok)—it is the longest rail route. Therefore, the frequency band used for wireless in PTC is often in the VHF band (217–220 MHz). This band is chosen to allow signals to propagate over long distances. Train receives such signals over great distances [17]. In situations where the train has to enter a tunnel, there will be a caution signal well before such entry. Although each block determines limits of speed with which a train can enter, but that can be varied with weather conditions and load on the train, its momentum, etc., which can be calculated by the ATC (automated train control).

Track circuits sense the movement of trains and know exactly where the train is at any given moment. It is also possible to convey this train location information via a wireless network, using way-side transponders. The wireless method is being increasingly preferred since the data throughput required is quite small, similar to voice circuit technology that was used during 2G. But the major focus in all signaling networks is on reliability and safety. The base

and locomotive radios can support data rates up to 32 kbps, while the way-side radio can support data rates up to 16 kbps [17]. Such data rates can be well supported by the VHF band reliably over great distances. This is key to safety and has been verified widely on various rail systems.

15.5.2 Communication-Based Train Control (CBTC)

PTC is a hybrid of traditional signaling systems integrated and expanded to accommodate modern signaling methods. However, CBTC is oriented towards communication network as the basic key to automation; in particular, wireless communication become the main thrust.

Communication-based train control (CBTC) is a system widely used in subway rail network where the block sizes reduce considerably. Although subway systems also use block signaling methods but quite often, they are deployed below the ground. There are many reasons for CBTC being implemented completely using wireless signals. The restriction includes lack of visibility in the tunnel, difficulty in laying ducts, and pathways for cable are important reasons for moving to wireless signals. Quite often the driver cannot see the train ahead, due to the curvature of the tunnel. Train speeds are no more than road vehicle speed, and there are no level crossings, making it better to implement CBTC.

Communication-based train control (CBTC) systems use the current generation of signaling technology adopted by most subways and other new modes of transportation (such as mono-rail or light rail).

CBTC systems integrate equipments that are installed on the trains and they are supported with way-side equipment using digital wireless communications as primary means, as indicated in Fig. 15.12. The system collects position information from trains and sends back control signals based on position. This significantly reduces the amount of way-side equipment and facilities. It also offers other benefits such as high traffic densities and the adoption of single driver or even unmanned trains [18]. The original concept strictly based on "fixed block control" is modified towards "moving block" between trains as shown in Fig. 15.12. This concept of "moving block" is based on ATP (automated train position) pattern that provides variable distance between two trains based on accurate position, speed, and related measurements. It still maintains a safe distance but is based on actual measurement data collected by the system as trains travel. The curve shown between the two trains is regularly calculated and used to determine "moving block". The entire system is operated by set of computers—Automatic Train Operation (ATO) that are networked with redundant feature set. In practice such trains operate mostly in automatic mode, needing human intervention only in areas where there is

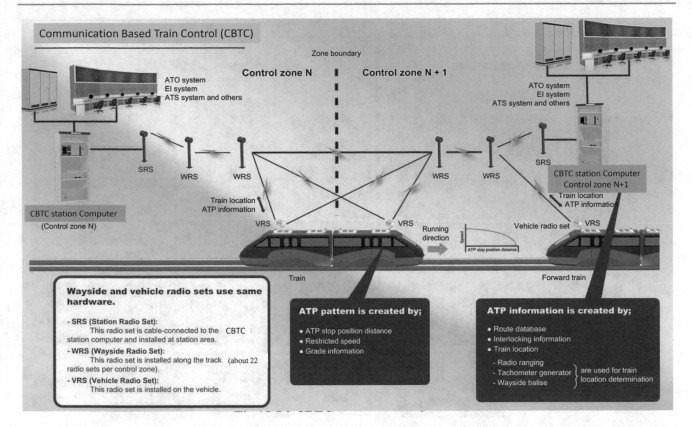

Fig. 15.12 CBTC system network concept

temporary change due to construction or wayside equipment problem.

The 2.4 GHz band (2400–2483 MHz) is unlicensed throughout the world. This band is a good choice for a uniform implementation of CBTC in subways of all major cities throughout the world. Due to higher frequency band, distances between WRS (way-side radio system) are reduced to no more than 200 m. But that is not a major limitation since signal blocks are quite small in most subway systems. The system is also robust enough to prevent interference from other wireless users and allows its use in mission critical situations. To avoid interference, CBTC does not use 802.11 protocol and it does not use the channel format used by WiFi standard. Instead, CBTC relies on and operates with full power allowed over the air (4 W of EIRP) in that unlicensed band (unlike cell phones or access points that operate with only 200 mW).

Many of these factors favor the deployment of CBTC that operates efficiently in the tunnel and subway environment. Quite often it follows the TDMA or the spread spectrum protocol for communication between multiple WRS and SRS units. Since VRS and WRS use the same hardware (as indicated in Fig. 15.12) the system offers flexibility in operational setup with software provisioning. Major challenges in implementing CBTC are operations-related questions such as "how does the on-board computer recognize construction crew in the tunnel"? These are usually resolved with a combination of manual and automated operations. For example, the agency may allow a driver to operate the train manually in the construction zone and revert to automated train control in other zones. With these excerpts on train automation, let us move to automated or driverless vehicles on the road, where considerable progress has been made in recent years.

15.6 Driverless Vehicles: Infrastructure Choices

The most exciting time of vehicle technology is just around the corner. "Driverless vehicles on the road" are eagerly awaited by many, and equally worrying to others. The advocates include all auto manufacturers and many in the younger generation, who believe that any vehicle must perform the task of transport on its own. Whether it is passengers or goods, the vehicle must be autonomous meaning self-driven. Given the rate at which auto industry is moving, it will not be a surprise if in the next 5–10 years a sizable portion of all autos on the road will be driverless.

There are three technology choices for building the infrastructure, one based on the DSRC technology that supports way-side communication. The second choice is to

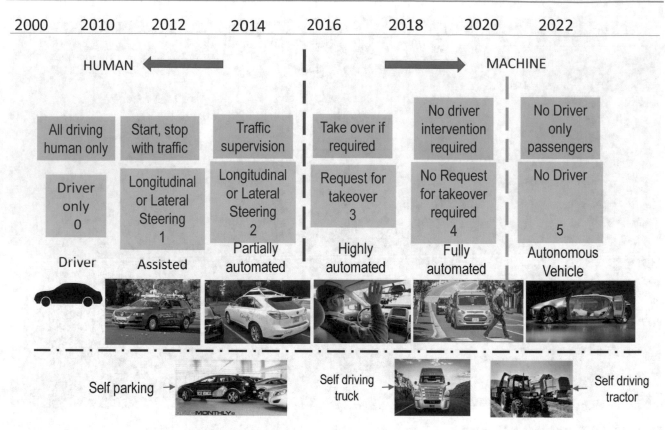

Fig. 15.13 Levels of autonomy defined by SAE (Society of Automobile Engineers)

use the cellular infrastructure and augment it with wayside. Debate on this topic is quite intense, with each side claiming its advantages. Cellular infrastructure is now supported by auto manufacturers who have named it "C-V2X", meaning vehicle to everything, using cellular infrastructure [19]. The third choice is a combination of cellular and satellite used by farming and mining industries. Before elaborating on these three technologies choices, certain common features may be reviewed.

Although personal transportation using cars was given priority by industry, other areas in which autonomous vehicles will be deployed include public transportation, delivery and cargo, and vehicles for farming and mining. All are being tested by different manufacturers. All of these applications have their own specific requirements. Different vehicles being built, sense their specific environment, process input and make decisions, and subsequently use it for automation. Five levels of autonomy are defined by SAE Society of Auto Engineers standard 3016.1.

Figure 15.13 shows different levels of automation that started about a decade ago towards partially automated vehicles built with the concept of trials to get inputs to further improve the design. These trials were very successful and DSRC technology largely helped move it forward to level 3. There are several cars that can cruise on their own

and perform well under normal condition, expecting driver intervention occasionally. From the beginning of 2020, auto manufacturers showed that cars have achieved level 4—that is they can perform very well under any traffic condition. Human driver has now become a standby observer who need not intervene.

Is level 5 possible to achieve or necessary? Current opinions are divided, but it requires considerable observation, as indicated earlier in the context of train automation. Compared to trains, vehicles on the road regularly encounter different scenario—the roads that they travel have some of the signs, signal lights, etc. that may remain common, but the traffic is never identical to the one day before, for example. Safety is of the highest priority; it is the very basis of why autonomous vehicles are being built; "they follow road rules" including signals and other directives. Unlike human driver who may try to sneak through an amber light, they will not.

Finally, autonomous vehicle development has extended to other transport vehicles such as trucks, tractors, and others as shown in the lower part of Fig. 15.13. Driverless trucks have been operating on roads for quite some time now. This could mean efficient distribution of goods and services that can operate safely through day and night on a well defined route. It is important to recall that accidents do

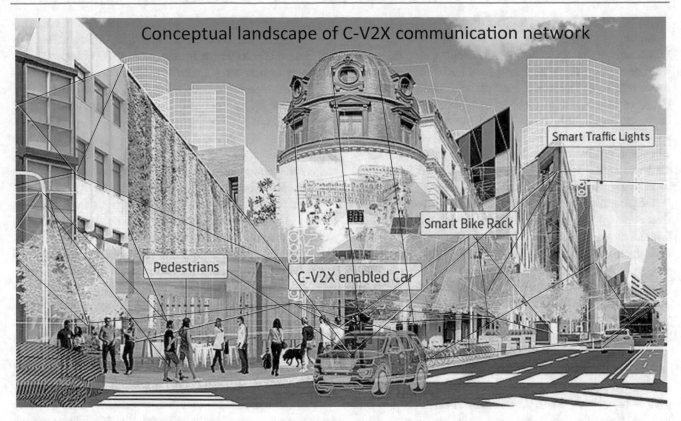

Fig. 15.14 Driverless vehicle with cellular "Vehicle-to-everything" links

happen due to drivers falling asleep, talking on cell phones, and other distractions that autonomous vehicles don't have. Conveniences such as self-parking of the car have become very popular, particularly in crowded cities where people spend considerable time and skill to do this routine task. Autonomous vehicles for agriculture, mining, etc., are further discussed in Sect. 15.6.3.

15.6.1 Cellular Infrastructure Choice C-V2X

Since many governments had advocated DSRC (Dedicated Short Range Communication) as the technology for autonomous vehicle, the choice of using cellular came as a surprise to many [19]. But it has a reality. DSRC requires extensive investment to convert all the roads in every country which typically gets done by governments. Without massive infrastructure investment everywhere, the autonomous vehicle must be built with a driver and manual as the primary option to navigate roads that don't have the infrastructure. In short, it can only be a hybrid, not a fully autonomous vehicle. In contrast, most of the globe already has cellular coverage, and sending messages related to vehicular infrastructure is an augmentation. Figure 15.14 shows the concept of such augmentation with a cellular network [19].

Although only some of the links are indicated by lines in the rendering on Fig. 15.14, these lines show the range of possible connections. These are needed locally between vehicles and smart traffic lights, vehicle and pedestrians, bicycle racks, and all possible movements on the road that must be regularly sensed and used for navigating the car, the bus (in the back), and even a car next to the bus. Such real-world scenarios were tested regularly by all major auto manufacturers. The major advantage of cellular comes from nearby buildings that connect to cellular towers and to the larger infrastructure of entire streets, multiple streets, and thereby entire city. This would allow the city to collect information on road conditions and a clear map of traffic conditions.

Proponents state that most vehicles already have a 4G/5G modem (including EMS—Emergency Medical Service), and pedestrians typically carry cell phones—all of them provide precise position information to the network regularly. These form the very basis of autonomous vehicles and how they can navigate streets whether it be in a city or suburb or even on highways. In fact, the 5G feature set of URLLC described earlier in Chap. 11, Sect. 11.6.1 shows major effort to support autonomous vehicles. Although it is early to say, the cellular network option seems to be receiving considerable support from all automobile vendors since 2018.

15.6.2 DSRC Network Choice V2X

DSRC was a well-thought-out technology with a frequency band allocated over a decade ago. A major part of autonomous vehicle testing was carried out with DSRC since most governments around the globe supported it. In addition to the dedicated bandwidth, modifications were made by 802.11 standards developing 802.11p dedicated exclusively to autonomous vehicle technology. This was described in detail, earlier in Sect. 15.3, where its advantages and the work done to augment way-side support system trials were discussed.

Despite multiple advantages gained over years of experience in trials of autonomous vehicles, there are some limitations of DSRC. One of them happens to be major investments needed from governments. The second one, though not obvious, is that cellular has long-standing experience in delivering wireless communication to mobility applications. These include the high-speed rail at over 500 kmph as well as rail safety-related standards such as LTE-R, which were described in this chapter under Sect. 15.2.3. The 802.11 standards, on the other hand, always had their strength in WiFi, which supports pedestrian speeds, but not high-speed mobility. This was discussed in Chap. 12, where the LAN standards including 802.16 WiMax were discussed, showing their limitations in supporting mobility. It is not clear whether the auto industry will continue its support to DSRC in some way. There are some indications that in cities where traffic signals directly interact with vehicles and pedestrians to regulate pedestrian crossing, etc., the DSRC technology may provide good support to autonomous vehicles.

15.6.3 Autonomous Vehicles for Farming, Mining, and Construction

An interesting aspect of wireless network infrastructure that looms large is in the area of autonomous vehicles for the agriculture and construction industry [21]. Due to vast expanses, the only wireless options to cover such large areas are through lower band of cellular (800 MHz or lower) or unlicensed 900 MHz (or licensed 450 MHz) bands. This is expected to become a major growth area, given that such tasks as indicated in Fig. 15.15 are repetitive, labor-intensive, and depend on weather conditions for all their operations.

Driverless tractors and autonomous spraying machines must operate on precise tracks and not go on useul plants or vegetables being grown on the farm. This may not seem apparent since humans did this, work manually, till now.

Considerable effort was made by various manufacturers around the world to build, test, and deploy autonomous vehicles that adorn countryside and in the mining fields. Data used by such equipment is classified in terms of GPS satellite receiver (used for location/position accuracy), terrain compensation to detect local conditions. There are several parameters to be considered in agriculture operations, but primarily the accuracy with which farm equipment (such as tractor) follows the same route, resulting in minimal damage to crops. Table 15.5 shows some of the current equipment capability utilized for both autonomous and manual farming, which is quite impressive.

Typical options for wireless receivers used in farm equipment involve directly using satellite GPS signal, which may take considerable time (half an hour) pull in time to get an accuracy of about 3 cm. Repeated farming operations mean that there will be time lost between each pass of the tractor.

The second option is to build a radio infrastructure around the farm area using either a 450 MHz licensed band or use the 900 MHz unlicensed (ISM) band. Since the radio infrastructure develops accuracy locally (but based on satellite signal and its own differential calculations), the pull-in time for each pass of tractor considerably reduces (just 1 min). This speeds up farm operation.

The third option is to use a commercial cellular signal which would involve a subscription and a cellular modem (4G modem, for example) but provides the same accuracy as

Autonomous mining truck   Driverless tractor   Autonomous Sprayer

Fig. 15.15 Autonomous farm equipment and mining truck

Table 15.5 Technology used by farm equipment

Wireless technology used	Horizontal pass to pass accuracy	Pull in time	Repeatability	Delivery method	Comment
Satellite radio receiver	±3 cm	<30 min	±3 cm in-season repeatability	Satellite	Radio operates using GPS
UHF receiver	±2.5 cm	<1 min	±2.5 cm long term	Radio	Radio uses 450 MHz or 900 MHz
Cellular modem	±2.5 cm	<1 min	±2.5 cm long term	Cellular	Commercial cellular

the second option. The advantage is that there is no need to build a radio network around the farm. In all cases the parameters available relate to received signal strength at the current location.

Satellite data received is measured in kilobytes—calculation of location accuracy starts once the connection is accepted and used by the receiver for each pass. Therefore, achieving accuracy depends on both GPS (Lat, Long) and terrain accuracy calculated by the receiver unit.

Licensed radio receivers used in the agricultural or mining industry operate on 450 MHz band using traditional 25 kHz or 12.5 kHz bandwidth, similar to those described in P25 and TETRA technologies described earlier in this chapter under Sect. 15.2. Data rates required are low, but repetitive accuracy is important. Alternatively, a 900 MHz unlicensed band can be used but it offers channels of 1 MHz bandwidth, with a frequency hopping mechanism. Both of these bands, as well as cellular bands of 700/800 MHz, offer enough signal range to cover large areas. The IoT infrastructure offered by cellular may play a major role in the expansion of this area of autonomous technology since many providers plan to continue GSM service in rural areas.

15.7 Conclusion

This chapter discussed various aspects of vehicular technology—concluding it as a major growth area in recent years. It discussed not only vehicles that travel on roads and rails, but others such as autonomous tractors for agriculture. These vehicles use communication systems that are well established over decades. Autonomous vehicles, in particular, plan on major network growth to accommodate driverless vehicles based on new technology that has evolved rapidly.

This may seem natural, but the number of sensors and data required, far exceed expectations. For example, an autonomous vehicle is expected to require 10 times the amount of data used by human drivers. The human ability to sense vehicular traffic, lights, road signs, and other cues must be replaced by sensors such as cameras, RADAR, or LIDAR for gauging distance, etc., with GPS location as the central key. This challenge took about a decade of development and now autonomous passenger vehicles are on the road. In spite of extensive testing, human concern to allow a vehicle to drive by itself is deep-rooted in human driving experience spanning over one century.

Land mobile radio, on the other hand, is experiencing a different turn since it always used narrow band technology with only human voice for about one century. Public safety communication now faces the challenge of moving towards broadband cellular network which is at least two decades ahead in technological development. Land mobile radio has two major standards—the P25 and TETRA, both of which incorporate in 2G, TDMA. They approached 3GPP (cellular standards body) on "how to move forward". This discussion resulted in Release 12 of the 3GPP which started roll out of PTT (Push-To-Talk) feature which is at the core of public safety communication. Over the next few years, 3GPP has more releases in 4G LTE that bring more features of public safety, culminating in release 17 which fully supports all features of public safety.

The rail industry followed a similar path to move towards commercial cellular. It has conducted multiple trials to bring 4G LTE features specific to rail, now known as LTE-R. This is in some ways an update to the earlier GSM-R standard. In a separate effort rail industry started to modernize signaling systems using wireless communication. Above-ground rail signals have taken a hybrid approach of PTC (positive train control) that uses some automation with some wireless features. The subway systems have moved fully towards wireless technology. CBTC (communication-based train control), as the name suggests, is completely automated and controls subway trains exclusively using wireless technology. Both signaling systems continue to have safety as the core feature.

All vehicular systems have made a deliberate attempt to move away from the existing wired communication infrastructure, towards wireless technology in a big way; this is the core message of this chapter.

Summary

This chapter started with a review of current land mobile radio used by public safety agencies such as police, fire, and EMS personnel. With trunk allocation for talk groups, it is the oldest form of a wireless telecommunication network that has operated for a century. After using narrow band technology for so long, current standards of digital radio such as P25 and TETRA were introduced recently. Their interest in moving from voice-centric system towards the 4G standard of LTE was discussed. This gradual move allows all public safety agencies to make changes and adjust to commercial broadband cellular networks. In terms of throughput requirement, they will continue to operate at very low data rates.

Autonomous vehicles of all types (car, train, truck, etc.) started out with the concept of improving safety and reducing accidents. The DSRC band was allocated exclusively for this effort and considerable design and trials were conducted over the past decade. Governments around the globe endorsed the use of DSRC, allowing vehicle to vehicle and vehicle to way-side (infrastructure) communications. But recently the auto industry indicated it prefers cellular.

Positive train control (PTC) has become a key word for safety in long-distance trains globally. Incorporating this technology to improve safe operations resulted in reducing accidents. Integrating modern digital communication with wireless signaling in the 217–220 MHz band has remained the focus of PTC, with a major effort to automate part of the signal system. Running at more than double the speed of any vehicle on the road, long-distance trains remain the fastest means of land travel that incorporated wireless network for increasing safety.

The subway system uses CBTC, which has fully moved towards wireless communication network as the basis for location, and control of the entire system. Operating in the 2.4 GHz but not using the 802.11 standard, CBTC has evolved as a signaling network that remains unaffected by the widespread use of WiFi in an urban environment.

Perhaps the major theme of this chapter is brought through choices of infrastructure for driverless cars and autonomous vehicles. After about a decade of work with DSRC major auto manufacturers have embraced cellular network as their choice, moving away from DSRC. Although the final decision is likely in the next few years, it seems cellular network has made a mark. Readily available way-side infrastructure, the fact that all smart phone users have location inherent in cell phone seems to have made an impact to move towards cellular.

In an interesting evolution of events, agricultural and mining operations seem to have gained an important foothold in the use of autonomous vehicles. Due to the vast areas they cover, infrastructure choices for communication are based on satellite and cellular technologies. It seems likely that these two areas that require considerable labor force and repetitive work may gain by the use of modern wireless technology.

This chapter overall brings to focus the use of wireless as a means of URLLC (ultra-reliable low latency communication) wireless systems. This 5G technology has proved to be superior to wired counterpart such as fiber optic communication. This is one of the three key technologies being brought about by 5G of cellular and is likely to support major growth in all vehicular technologies.

Emphasis on Participation

1. Where are vehicle automation trials in your area? What type of data are they collecting and how are they analyzed? Contact manufacturers and make a list of safety features used.
2. Are the trains of your area deploying CBTC or PTC? What additional safety measures do they plan to address? Make a list of new equipment and its functions.

Knowledge Gained

1. The most important knowledge gained in this chapter is all vehicles, whether on rail or road or even not on the road (on farms and mines) have taken deployment of a wireless network as the way to move forward.
2. Safety and reliability are two key features that have driven all manufacturers towards wireless systems.
3. Public safety communication is on its way to move towards 4G LTE, having experienced networks that sustained almost a century of public service (police, fire, EMS).
4. Vehicular technology does not necessarily need broadband, but they certainly need redundancy in network design to provide reliable means of wireless communication.

Homework Problems

1. DSRC network operates in the 5850–5925 MHz band that has signal coverage for about 200 m, assuming bow-type pattern. How many wayside radio units are needed for a city block of 5.5 km. Allow at least 25% additional units for overlap to account for unit failures.
   A high-speed train traveling at 250 kmph wants to use PTC operating at 220 MHz; if you plan to use transmitters every 10 km along this path, how much is the free path

signal loss? Use the free space path loss (FSPL) equation. Use transmitter gain of 6 dB and receiver gain of 3 dB:

$$FSPL = 20\,log_{10}(d) + 20\,log_{10}(f) + 20\,log_{10}(4\pi c) - Gt - GrFSPL$$

$$= 20\,log_{10}(d) + 20\,log_{10}(f) + 20\,log_{10}\left(\frac{4\pi}{c}\right) - Gt - Gr$$

where d is the distance in meters; f is the frequency; Gt is the transmitter gain; Gr is the receiver gain, and c is the speed of light.

Instead of using 220 MHz if the PTC designer uses 2400 MHz, how much is the path loss?

2. A driverless vehicle shown in Fig. 15.4 has five different sensors. What would you recommend as the angle of vision for each sensor and why? Will there be any interference for any of these sensors? List them and provide details of frequency bands and power levels used based on an industry survey.

Project Ideas

1. Using MATLAB simulation, create at least two traffic junctions with signal lights and traffic consisting of both manually driven cars and autonomous cars. List your assumptions. If all manual cars are converted to autonomous what is the traffic improvement during peak hours? Remove traffic lights completely and use autonomous sensors to regulate traffic.
2. Simulate a CBTC system for any major city subway system. Since CBTC uses the 2.4 GHz system globally, assume all lines use this band and create the infrastructure needed to regulate and monitor the entire system from a single command center.

Out-of-Box Thinking

Instead of PTC, superfast trains used CATC (continuous automated train control). If this system were to be automated fully like CBTC, what is the communication infrastructure needed? What frequency band would you propose and how many transmitters are needed along a 1000 km track? List a set of additional safety features you would incorporate to make sure accidents do not occur?

References

1. Noble DE (1962) The history of land-mobile radio communications. Proc IRE 50(5):1405–1414 (1962). https://doi.org/10.1109/JRPROC.1962.288119. https://ieeexplore.ieee.org/document/4066864

2. Booz AH (1999) Comparison of conventional and trunked systems, public safety wireless network, US department of Homeland security Library, May 1999. https://www.hsdl.org
3. Digital land mobile systems for dispatch traffic: Report ITU-R M2014–3, 11/2014 https://www.itu.int/dms_pub/itu-r/opb/rep/R-REP-M.2014-3-2016-PDF-E.pdf
4. Arcuri D, Davis A, Project 25 technology interest group white paper, P25 Trunking control channels, 09/14/17. http://www.project25.org/images/stories/ptig/White_Papers/P25_TDMA_Control_Channel_FINAL__170915.pdf
5. Training guide TG-001 P25 radio systems, by Daniels Electronics Ltd, Sept 2004. https://www.dvsinc.com/papers/p25_training_guide.pdf
6. Delivering public safety communications with LTE, July 2013. https://www.3gpp.org/news-events/1455-public-safety
7. Choi HY, Song Y, Kim Y-K (2014) Standards of future railway wireless communication in Korea, Korea Railroad Research Institute, Recent advances in computer engineering, communications and information technology. **ISBN: 978-960-474-361-2**. http://www.wseas.us/e-library/conferences/2014/Tenerife/INFORM/INFORM-50.pdf
8. He R, Ai B, Wang G, Guan K, Zhong Z, Molisch AF, Briso-Rodriguez C, Oestges C (2016) High-speed railway communications: from GSM-R to LTE-R. In: IEEE vehicular technology magazine, Sept 2016. https://doi.org/10.1109/MVT.2016.2564446 https://www.researchgate.net/publication/307518255_High-Speed_Railway_Communications_From_GSM-R_to_LTE-R/link/5a051b31aca2726b4c745d14/download
9. Communications on the move (COTM)—white paper Jul 2014 https://www.hughes.com/sites/hughes.com/files/2017-04/Comms-on-the-Move-Railways_H51432_HR_07-14-141.pdf
10. Abdelgader AMS, Lenan W (2014) The physical layer of 802.11p WAVE communication standard: The specification and challenges. In: Proceedings of the world congress on engineering and computer science 2014 Vol II, WCECS 2014, 22-24 October, 2014, San Francisco, USA. https://www.researchgate.net/publication/279474688_The_Physical_Layer_of_the_IEEE_80211_p_WAVE_Communication_Standard_The_Specifications_and_Challenges/link/56e68e9308ae98445c223940/download
11. Ni Y, Cai L, He J, Vinel A, Li Y, Mosavat-Jahromi H, Pan J (2020) Toward reliable and scalable internet-of-vehicles: performance analysis and resource management. Proc IEEE 108(2):324–340 (2020)
12. Vehicle-to-Vehicle Communications Research Project (V2V-CR), DSRC and Wi-Fi baseline cross-channel interference test and measurement report, Dec 2019. https://www.nhtsa.gov/sites/nhtsa.dot.gov/files/documents/v2v-cr_dsrc_wifi_baseline_cross-channel_interference_test_report_pre_final_dec_2019-121219-v1-tag.pdf
13. Leen G, Heffernan D, Dunne A (1999) Digital networks in the automotive vehicle, Dept. electronic and computer engineering, University of Limerick, Ireland. Donal.Heffernan@ul.ie, Gabriel.leen@ul.ie; PEI Technologies, Foundation Building, University of Limerick, Ireland. Alan.Dunne@ul.ie. https://www.researchgate.net/publication/3363624_Digital_networks_in_the_automotive_vehicle/link/02e7e520d193fdc09b000000/download
14. Vehicle to infrastructure—resources. Intelligent transportation systems, joint program office. U.S Department of Transportation. https://www.its.dot.gov/v2i/index.htm
15. What is Vehicle to Infrastructure (V2I) Technology? Everything RF, Nov 4, 2019. https://www.everythingrf.com/community/what-is-vehicle-to-infrastructure-v2i-technology
16. Hummer KH (1991) Operation control and signaling system for high-speed lines. Transp Res Record 1314. http://onlinepubs.trb.org/Onlinepubs/trr/1991/1314/1314-019.pdf

17. Bandara D, Abadie A, Melaragno T, Wijesekara D (2014) Providing wireless bandwidth for high-speed rail operations, George Mason University, 4400 University Drive, Fairfax VA 22030, USA. Proc Technol 16:186–191 (2014). www.sciencedirect.com

18. Railway Signal Systems, CBTC, Nippon signal systems, http://202.191.114.78/english/products/railway.html

19. Bigelow P (2019) A new connected-car battle: cellular versus DSRC, Automotive News, Feb 03, 2019. https://www.autonews.com/mobility-report/new-connected-car-battle-cellular-vs-dsrc

20. Efficiency Wins—2020 Autonomous vehicle technology report, Mar 06, 2020. https://efficiencywins.nexperia.com/innovation/2020-autonomous-vehicle-technology-report.html

21. Fortuna C (2019) Autonomous tractors, mining equipment and construction vehicles—oh, my: by Carolyn Fortuna, July22, 2019, CLEAN TECHNICA. https://cleantechnica.com/2019/07/22/autonomous-tractors-mining-equipment-and-construction-vehicles-oh-my/

Supervisory Control and Data Acquisition (SCADA)

16

Abbreviations

SCADA	Supervisory control and data acquisition refers to monitoring and control of devices, equipment, or system that may be in a remote location. It is a major thrust area for the expansion of narrow band IoT, which is being widely deployed using wireless technologies
PLC	Programmable logic controller represents an industrial computer that is ruggedized and adapted for the control of manufacturing processes such as assembly lines machines robotic devices or any activity that requires high reliability ease of programming and process fault diagnosis. In recent times it is implemented as a chipset that allows provisioning and monitoring and control of various devices using remote commands
WSN	Wireless sensor network indicates networks that are geographically dispersed, yet dedicated sensors to monitor and record the physical conditions of the environment in an industrial plant or other places. WSN typically forwards the collected data to a central location. WSNs can measure environmental conditions in the area such as temperature, sound, pollution levels, humidity, and wind
ZigBee	It is the IEEE 802.14 standard that supports personal area network (PAN). A large number of devices and equipment support ZigBee and have various applications
6 LoWPAN	It is a wireless personal area network based on IPv6. It is based on the concept that even the smallest low-power device should be able to send information over the IEEE 802.15.4 personal area network, based on the encapsulation and header compression defined by this standard in IETF
LoRa	Long Range is a wireless network protocol that allows low-throughput data over extended distances up to 5 km. It is a wireless platform connecting sensors to the Cloud and enabling real-time communication of data and analytics
IoT	Internet of Things is a term popularized by the large number of devices that are being connected to the Internet. The majority of the devices need narrow band connection with no more than a few hundred kbps data throughput and send data a few times a day or week. But broadband IoT uses much larger bandwidth and can support hundreds of real-time CCTV camera feeds that may prove to be a goon during emergency deployments
FOTA—Firmware over the air	It is a wireless-based method of updating or installing firmware that can change functions and operations of devices. This is similar to OTAP (over the air provisioning) where a special secure protocol is established between the server and remote device and updating is performed
WiFi HaLow	It is a wireless technology that operates below 1 GHz based on the

16

© Springer Nature Switzerland AG 2022
K. Raghunandan, *Introduction to Wireless Communications and Networks*,
Textbooks in Telecommunication Engineering, https://doi.org/10.1007/978-3-030-92188-0_16

802.11ah standard. It is also a designation for products incorporating IEEE 802.11ah technology. It augments WiFi by operating in the lower spectrum (below one GHz) to offer a longer range and lower power connectivity. WiFi HaLow meets the unique requirements for the Internet of Things (IoT) to enable a variety of use cases in industrial, agricultural, smart building, and smart city environments

Z-Wave   Z-Wave is a proprietary wireless technology with a focus to build better, smart homes. It has some unique benefits that make a Z-Wave-based smart home faster and safer. But due to the wide range of devices available, it can be used in SCADA as well

LPWAN   Low power wide area network can be used to support wireless sensors and related SCADA network, and LPWAN data rate ranges from 0.3 kbit/s to 50 kbit/s per channel, up to 10 km distance

WhiteFi   WhiteFi refers to wireless technology that operates in white spaces. White space is defined as a local TV channel that is unused in the area. Such unused channels in the low VHF/UHF bands offer good propagation characteristics over great distances, using the 802.11af standard

The concept of supervising machines/equipment, originates from human interest in knowing the status of a device, an equipment, or a system already built with the ability to provide measurements or data from a remote location to a centrally located operations center. Quite often such remote locations can be visited only periodically. In rare instances, it may not be practical to visit that location again.

The concept of SCADA (supervisory control and data acquisition) is based on setting up a method of monitoring, during the design phase. SCADA is a network architecture that consists of computers, network communication infrastructure, and diagnostic monitoring screens. It is not an after-thought. During the design phase, it is obvious to the designer what must be monitored. It is possible that more parameters may be added later based on how the system performs, but that is typically an enhancement; provision is already made for such enhancement, during design.

This chapter describes the design concepts, methods, and measurements used in SCADA and the application of wireless network to help monitor such locations. Initial deployments were mostly wired, and the objective was to bring signals up to a local monitoring point. Over a period of time the number of points monitored steadily grew, incorporating control (using an actuator) from a remote location. All of these will be described using a systematic approach starting with what is the current status and how wireless technology has started to change this field.

16.1 Current Deployments in the Process Control Industry

SCADA is often associated with the process control industry (chemicals), but it is not exclusive to that industry. SCADA systems are utilized in many other industries as well. Some examples include food and beverage production, manufacturing, transportation (rail, mass transit, and traffic control) asset tracking, telecommunications, mining, electric power, oil and gas, water and wastewater, buildings, and environment [1]. Specifics of SCADA system design may differ based on the industry. In addition, application software and system response may be matched to the specific needs of each industry. Some examples of the industrial sectors and what they monitor are indicated below.

In the electric power industry, monitoring of substations, generators, switchgear, begins with the objective to check how these systems behave based on demand. Under extreme weather conditions, discharge of surge current during lightning, thunderstorms, etc. are closely monitored. These are of concern to ensure the safety of equipment and personnel, hence monitoring them is a priority.

SCADA has existed in the oil and gas industry for a very long time, associating it with the processing of crude, by-products; flammability of petrochemicals is the prime concern since their safety relates to flashpoint at nominal temperatures. Monitoring of gas and oil lines is associated with the movement of crude and end products, both of which can catch fire. Wired SCADA systems are common in this process control industry mainly to avoid fire and maintain safe working conditions even during the storage. However, a wireless network is gradually getting introduced in these plants since it offers easy installation in existing plants.

Water and wastewater treatment are associated with every major town and municipal system. Since water quality must be guaranteed to the citizenry, townships monitor it regularly. Wastewater on the other hand requires processing treatment using a sewer system before it can be let out into the environment. It is important to establish that such wastewater is safe to be let out into the environment, hence wireless monitoring becomes important. Since there are multiple regulations associated with it, monitoring them at various locations spread over a large area becomes mandatory. It is common to see VHF wireless system spread over a wide area that helps monitor pump and discharge stations.

Figure 16.1 shows the generic architecture of a SCADA system where "plant" represents any of the industrial plants indicated in earlier examples. The plant consists of some type of line equipment involved in the process related to industry. It could be an electrical power utility (transformers, switchgear), waste-water sewage process, or chemical industry where food or chemicals are being processed. One of the key components in SCADA is programmable logic controller (PLC). This can be programmed to control certain process. For example, its logic maybe if the temperature reached "X" degree, then do "Y". It could also be if carbon monoxide level reaches "Z" then sound an alarm and shut down process. Since this logic can vary over time, in different field locations, it is "programmable" using short field programs implemented by a technician.

Each plant also incorporates a local router from where the alarms and control signals are routed to a central SCADA monitoring command center. At this center, alarms and parameters across multiple plants can be monitored and controlled. Some users may program PLC such that only specific actions related to safety are incorporated locally. All long-term monitoring, retaining the history of operations and other functions are implemented typically from the command center. In certain situations, the location may involve a remote site not easily accessible. Remote monitoring unit from such sites would only send information to the command center. PLC may not be used in such remote locations, and all control can be implemented from the command center. A detailed description of SCADA systems, their design, and operation is described in [2].

SCADA operations date back to 1960s starting with the electrical industry. Gradually it was incorporated in other industries during 1970s when the term SCADA emerged. The design objective was to have all parameters first converted to electrical signals before getting transmitted by wires to command center. This practice has continued, but the signal is now converted to digital signal using ADC (analog to digital converter), which is incorporated either as part of PLC (programmable logic controller) or the sensor itself has an integrated unit so that this Logic is accomplished and integrated into the network. There are five programming languages as per IEC standard 61,131–3: Ladder diagram, Sequential Function charts, Functional Block Diagram, Structured Text, and Instruction List [3]. Early networks operated on relays to turn something "on" or "off". It could also switch different voltage levels to the circuit. At the core of such relay used to be a coil or solenoid which acted as an electromagnet. PLC accomplished a similar function with logical gates but using a solid-state device.

Fig. 16.1 Generic architecture of SCADA system

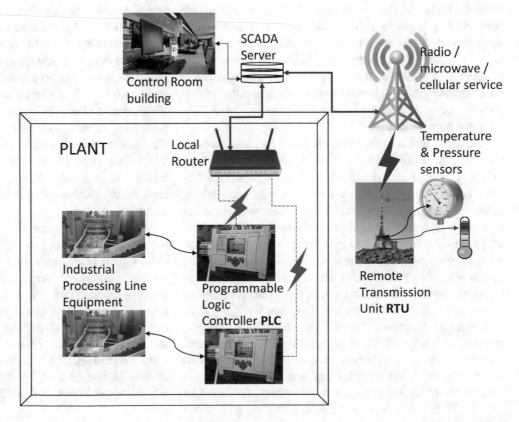

Fig. 16.2 PLC block diagram—with multiple input and output devices

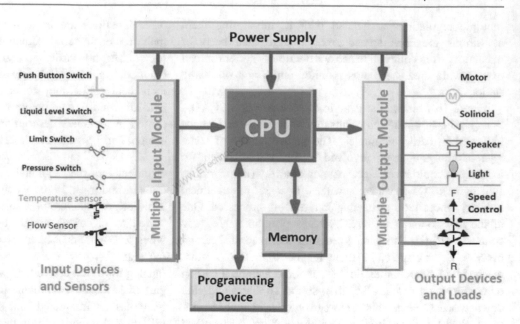

Figure 16.2 shows the system-level architecture of PLC where the input to PLC may consist of a variety of devices. It could be as simple as a push-button switch or complex sensors that may provide signals resulting from pressure, or liquid level or flow, or temperature variation. These are shown as input examples on the left. The logic built into PLC may result in outputs that control a motor or a solenoid or a speaker or even speed control of a stepper motor (shown on right). The biggest change in SCADA started when multiple inputs could be located in different places but connected by wireless links. Similarly, the output control devices could be far away from inputs and also connected by wireless links. This is implemented in water and wastewater system where pumps located in far-off locations work based on water flow. Flow control is used to regulate either piston pumps or centrifugal pumps commonly used for water/wastewater pumps.

Among output devices shown in Fig. 16.2, loudspeakers are normally used for announcement in manufacturing plants or venues where periodic announcements related to safety are necessary. Similarly, lights of different colors or blinking lights with siren are commonly used to alert workers about some type of hazard or to attract the attention of operators. Motors are often attached to pumps for water flow control. Another type of motor known as "stepper motor" can be turned around in 2° or 5° steps allowing adjustments similar to volume control knobs. This allows proportional control of fluids (gas or liquid). Solenoid, on the other hand, has only two extreme positions: "open" or "closed". It typically responds to pulses for "on" or "off" control. All of these controls are used in aircraft or process control plants. Also control solenoids are used as thrusters to control launch vehicle control systems or satellite orientation, for example.

Diagram in Fig. 16.2 raises another question—how can the wireless senor network (WSN) change the concept of PLC and thereby SCADA? The change occurs since cellular service is universally available (there are more people in the world that have cell phones, than those who have access to running water). Widespread availability of cellular signal brings three key factors together: sensing, computation, and communication. The power of wireless sensor networks lies in the ability to deploy large numbers of tiny nodes that can assemble and configure themselves [4].

Why is WSN (wireless sensor network) considered better than the existing SCADA system? One of the major limitations of current SCADA systems is that they are static, inflexible, and focus on a centralized architecture, as shown in Fig. 16.1, with a central control room. In contrast, the WSN decentralizes its functional elements that perform protection, control, and monitoring of the various signals. The sensor nodes themselves perform these functions in WSN. Due to its modified architecture, WSN can provide results with a detailed system-wide analysis of an event. It can alert the system coordinator within seconds of the event occurring.

WSN consists of a number of integrated sensor nodes—each of these nodes communicates among themselves using radio signals [5]. The number of sensors deployed depends on the quantity to sense, monitor, and understand the physical world. WSN acts as a bridge between the physical nodes and the virtual cloud network. The physical sensor nodes operate on small batteries (AA type or solar cells) that last for many years. These nodes are typically networked using the IEEE 802.15.4 standard as the base, with standards such as Zigbee, 6LoWPAN, "Wireless HART", etc., forming the upper layers. These standards address the primary

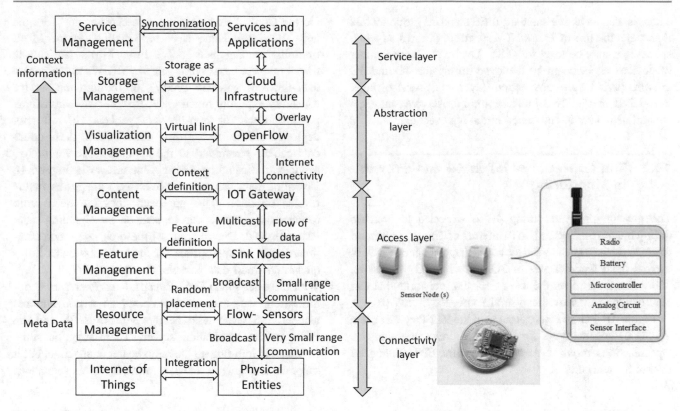

Fig. 16.3 Wireless sensor network architecture with sensor nodes

requirement of wireless sensor nodes—a long battery life. Security aspects of sensor nodes and how security features of these standards are being improved were addressed in Chap. 6 (wireless security).

Wireless nodes form part of the connectivity layer; due to their small size and low power, they operate over short distances of 100 m.

Access layer could be radio/microwave/cellular as indicated earlier in Fig. 16.1, where a radio tower symbol was used to indicate the connectivity layer. Different parts of PLC shown in Fig. 16.2 are part of the access layer radio; they are integrated into tiny sensor nodes. This is indicated in a de-centralized architecture of Fig. 16.3, where the size

of a tiny sensor node is shown in comparison to a small coin. The sensor node includes radio, battery, microcontroller, analog circuit, sensor interface, and antenna integrated within the package as shown on bottom right.

Different layers of the entire WSN are shown in Fig. 16.3, where various services of WSN and how it all gets integrated are illustrated. A typical flow sensor is shown next to the architecture diagram to indicate how small it is, but any type of sensor node can be used in practice.

Table 16.1 provides a comparison of WSN and WLAN (WiFi) with the traditional OSI model. The WLAN category could refer to a variety of IEEE standards such as Zigbee 802.15.4, 802.15.1 Bluetooth, 802.15.5 Mesh networks.

Table 16.1 Different layers of WSN, WLAN, and OSI

Wireless sensor network	WLAN (IEEE 802.1x)	OSI model
WSN application	Application programs	Application layer
WSN middleware	Middleware	Presentation layer
	Socket API	Sessions layer
WSN transport layer protocols	TCP/UDP	Transport layer
WSN routing protocols	IP	Network layer
Error control WSN MAC protocols	WLAN adapter & device driver WLAN MAC protocols	Data link layer
Transceiver	Transceiver	Physical layer

There is also an active standard IEEE standard group 802.22 that seeks the use of unused TV spectrum (known as white space) that may be used for WSN. Due to the large number of devices expected to be deployed during the 4G and 5G growth period, the abstraction layer and service layer introduced in Fig. 16.3 indicate multiple services may get streamlined. This is elaborated in the next section.

16.2   The Concepts of IoT in 4G and Growth in Monitoring

The growth observed during 4G is expected to increase during the 5G time period. IoT (Internet of Things) in general refers to devices such as sensor nodes that can operate well in narrow band technologies of 3G and even 2G. Therefore, 3GPP has focused on the use of existing networks that can readily support narrow band services needed by IoT [6].

Figure 16.4 shows user groups on the left; they can use a 2G/3G cellular network for IoT. It requires some changes. Optimizations required to support IoT at EPS (Enhanced Packet System) are:

- On the Control Plane, uplink data can transfer from the eNB (enhanced Node B or base station) to the MME (mobility management entity). From MME data can be transferred via the serving gateway (SGW) to the packet data network gateway (PGW), or through the service capability exposure function (SCEF). This data transfer is possible only for non-IP data packets (IP was not deployed during the 2G time frame). From these nodes data can be forwarded to the application server or IoT services (indicated by Orange line arrows in Fig. 16.4). Downlink data can be sent over the same paths in the reverse direction. This approach does not need radio bearers; data packets can be sent on the signaling radio bearer instead. Due to some of these changes such solutions would be appropriate for the transmission of infrequent and small data packets, common in IoT.

- With the User Plane EPS (Encapsulated Post Script) optimization, data can be transferred in a conventional manner, i.e., over radio bearers via the SGW and the PGW to the application server. There may be some overhead in building up the radio bearer connection, but it supports a sequence of data packets to be sent. This

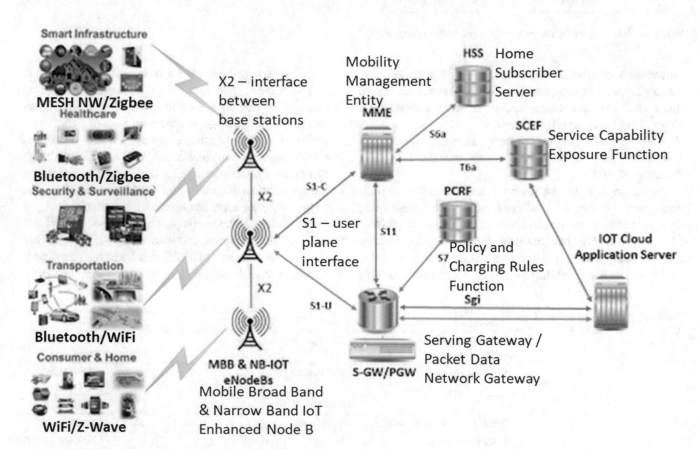

Fig. 16.4 Narrow band LTE architecture for IoT

approach requires support delivery of both IP and non-IP data packets with EPS.

- Another optimization approach could be to reduce signaling by guiding IoT devices to perform periodic location updates less frequently and by optimizing paging. Reducing signaling can help in avoiding overloads in massive device network.
- Subscriber data storage handling in the HSS (Home Subscriber Server) may need some optimization to support a large number of IoT devices.

The reason for taking this approach of retaining 2G and 3G networks is because it has dual benefits:

(a) 2G and 3G deployed in the last two decades can be re-utilized since the demand for voice calls has reduced (was common during 2G/3G). That infrastructure can be well utilized and integrated with 4G LTE base stations (known as eNodeB), using the smallest bandwidth of LTE 1.04 MHz.

(b) Sunset backhaul technologies of SONET/ATM discussed earlier in Chap. 4 (backhaul) are packet networks that did not use IP; they too can be re-utilized since they still operate well in most countries.

From Table 16.2 and Fig. 16.4 it seems clear that all non-cellular technologies can be used for local aggregation of traffic; for long distance it would be cellular. In addition, all non-cellular technologies use "unlicensed" bands and therefore may be less secure. Is that the case always? The answer to these questions is not that simple. There are phenomenal features that non-cellular technologies provide. Some of them are low-power, battery-operated devices that can be literally attached by tape, sticker, etc. They also closely work with applications that reside on smart phone. They can communicate with the cellular network to send alarms and messages over great distances. The only requirement is that a cellular signal of "reasonable strength" must be available in that area. What is "reasonable strength"? It means ~ 5 dB above the noise floor.

At this signal strength (reasonable level of ~ 5 dB above noise floor), even voice calls may not operate well. Then why is that acceptable to IoT? The answer is in the last column of Table 16.2. Let us elaborate on it; IoT/SCADA has one key requirement "low data rate". Therefore "narrow band 4G modem" is sufficient. IoT can tolerate periodic failure of messages—try and resend the message using multiple trials. After all, messages or alarms are needed a few times an hour or a few times a day or in some cases a few times a week. The message is critical but not urgent, which means real-time communication is not expected. Bluetooth or Zigbee algorithms are built on such "retry" efforts with their own security features already built into the sensor hardware. This was explained in Sect. 6.2 (security of Bluetooth, Zigbee devices) where security schemes for those were elaborated. Section 16.1 of this chapter discussed some features.

This is why 3GPP (cellular standards body) introduced new categories of 4G modems for narrow band—Cat 1, Cat M1, and Cat NB are for low throughput. Service providers charge lower rates since the device needs very limited service with the ability to tolerate lower signal levels. Simple narrow band categories only provide throughputs that can be as low as 250 kbps (one-hundredth the broadband data rate). Even such low data rates easily meet IoT requirements. The focus is on a large number of devices, each with a very low data rate. The counter-argument from non-cellular technologies, such as WiFi would be that it is capable of covering large distances of over 500 m, using directional antennas. In recent years, the 5.x GHz bands of UNII have a considerable spectrum, allowing monitoring CCTV video feeds along with alarms/actuators, for SCADA and IoT.

Table 16.2 IoT options—cellular vs non-cellular

Parameter	Non-cellular IoT					Cellular technology				
	Bluetooth	Zigbee	LoRa	SigFox	WiFi	GSM	eGSM IoT Rel 13	LTE Rel 8	LTE-M Rel 12	NB IoT Rel 13
LTE user device Cat	NA							Cat 1	Cat M1	Cat NB
Coverage distance	10 m	10–20 m	Few km	Few km	100 m	<35 km	<35 km	<100 km	<100 km	<35 km
Spectrum	Unlicensed					Licensed				
System bandwidth	1 MHz	2 MHz	<500 kHz	100 kHz	20 MHz	200 kHz	200 kHz	20 MHz	1.04 MHz	180 kHz
Max data rate	20 Mbps	250 Kbps	50 Kbps	100 Kbps	100 Mbps	<500 Kbps	<140 kbps	<10 Mbps	<1 Mbps	DL < 170 Kbps
								<5 Mbps		UL < 250 Kbps

16.3 Devices, Deployments, and Use of Narrow Band

This brings us to devices at the heart of narrow band IoT, where broadband may have a role.

Figure 16.5 is an example of how WLAN technology can be deployed in a petro-chemical facility. Using 60/80 GHz links as backhaul, local throughput rates easily compare to fiberoptic links (10 Gbps throughput with AES256 encryption). At the same time, network security is achieved using geometry/trigonometry described earlier in Sect. 6.4.4, which showed how to deploy systems with highly directional antennas. Point-to-multi-point (PTMP or P2MP) links use 802.11 technology but instead of using AP (access point) they use a base station. The base station (BS) communicates with multiple subscriber units (SU) using TDMA protocol. Only SU with its credentials pre-recorded in the base station's database can talk to that base station. Any SU even

from the same vendor cannot communicate with the BS unless pre-registered with that BS. An unlicensed band doesn't mean there is no security.

Since the PTMP base stations are strategically located within the facility, it is possible to set up redundant paths for the signal, creating a highly reliable, secure network using 802.11 technology and millimetric wireless links for backhaul. Such networks can be set up in difficult terrain such as industrial complex, rail yards, large open areas, where the deployment of fiber-optic conduits is prohibitively expensive and difficult. Devices that integrate with one of the wireless LAN technologies are now quite common. Integration of WLAN or cellular at each location brings changes in the IoT industry. At the same time, safety-related monitoring using CCTV cameras and local recording of video from camera allows records for verification if an incident occurred.

To deploy narrow band sensors along with broadband IoT (petrochemical example that was just described), integration

Fig. 16.5 Typical example of WLAN deployment over a large facility

can be accomplished at the cloud as indicated in Fig. 16.3. Such a cloud platform, known as "Twilio", is gaining widespread support [7]. This platform provides two types of development support:

- Broadband IoT—the type with cameras, as in a typical petro-chemical facility
- Massive IoT—very large number of sensors but all use narrow band IoT.

Broadband IoT uses super SIM card (for business users) that allows such systems to request high throughput rates—something that 4G LTE is capable of. Typically, 20 MHz spectrum is granted to such applications that can support CCTV camera traffic.

Narrow band IoT service, on the other hand, would be requested by a narrow band SIM card. Depending on the category of modem used, one of the services listed on the right five columns of Table 16.2 would be provided. Depending on local conditions cellular service from either GSM network or 4G LTE narrow band service will be provided to the modem. Such flexibility to operate on 2G, 4G, or even 5G network makes the platform very useful, which is why "Twilio" has become popular. It is important to note that this platform is already supported in about 140 countries since many cellular providers find this strategy very useful.

When there is such a wide gap between broadband and narrow band in terms of network usage, how are the two differentiated in terms of cost, throughput, and other factors? Massive IoT relies on low cost per sensor, while broadband IoT would carry high throughput from a few typically CCTV cameras. This can be clearly differentiated at the cloud and therefore, metering such service takes place at the cloud [8]. Applications such as monitoring smart grid depend on such advanced metering infrastructure (AMI). Factors like reliability, latency, security scheme, battery life of devices used are considered in the design of AMI. Such features are discussed in the next section.

16.4 Network Monitoring System (NMS)

Earlier in this chapter, central monitoring of SCADA was discussed in Fig. 16.1, where control room was shown as the monitoring system. Gradually as IoT evolves it is seen as a replacement to conventional SCADA. There are some reasons—IoT deals with SCADA and many more that have newly arrived due to expansion of the Internet and cloud management of services [9].

Mobile devices lead this expansion—being small, but they are present in very large numbers. They may not need mobility but use wireless functionality. In addition, the use of social media and other platforms led to a new definition of network management. It is no longer a centrally located and TCP/IP-based MIBs service that it was for many decades. It is a cloud-based service whose location is not important since it serves the customer in any case.

With this perspective, network management system (NMS) and its monitoring center in particular become the responsibility of the cloud server. For those developing applications for IoT, setting up a common platform became a central issue. How does a network designer define a common platform that supports various industries? In 2012, when the term IoT was not in use, the commonly used term was M2M or machine to machine communication.

This common platform was brought about due to efforts by many standard bodies that jointly decided in 2012, to work on M2M (machine to machine) communication. Seven standards bodies—TIA and ATSI from the USA, ARIB and TTC from Japan, CCSA from China, ETSI from Europe, and TTA from Korea—jointly decided that they would work on a unified, global standard known as "oneM2M" partnership project (http://www.onem2m.org). All these standard bodies decided to transfer their work done thus far to OneM2M project and stop individual work on the IoT application services layer. The reason for commonality was to achieve globally acceptable cloud architecture for IoT as seen in Fig. 16.6, known as oM2M domains.

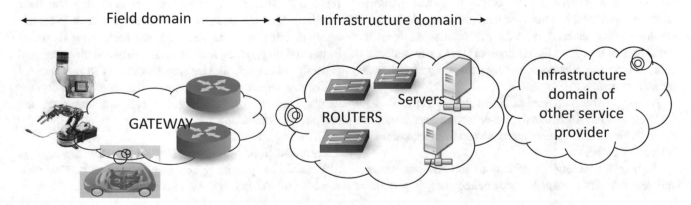

Fig. 16.6 OneM2M domains—IoT standard view

Fig. 16.7 Common IoT services platform functions

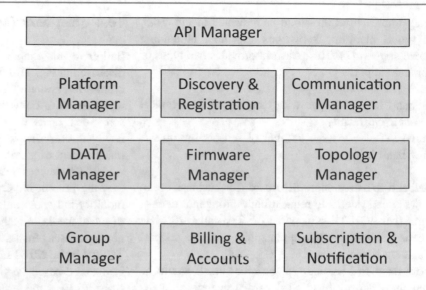

In the "oM2M standard" model the entire IoT is considered to be made up of two domains: the field domain and the infrastructure domain. The field domain includes things (sensors, actuators, etc.), including gateways. The range of devices can vary from smart sensors on a chip with a battery that lasts almost 10 years (measuring voltage, pressure, temperature, or other parameters) to robots that work on the floor of an industrial plant, to driverless vehicles. With such varied industries where each of the sensors or actuators may perform very different functions, all are included in the field domain.

The infrastructure domain consists of communication networks (aggregation, core) as well as data centers (routers, servers). It is essentially a data processing domain. From a functional perspective, each of these domains includes three flavors of entities—an application entity, a common service entity, and a network service entity. Gradually this model of two domains evolved into a common service platform, which is equivalent to a network management system used at the monitoring center. But with more functions from the field domain and literally millions of devices to support, a wide range of applications and services are grouped under infrastructure.

Although the services platform seems generic with functions widely distributed, it has well-defined functions due to efforts made by standard bodies to bring commonality across the globe.

With such a variety of applications and services, it seems apparent that API manager may be defined only in generic or

common terms. Other functions also seem distributed, based on various types of industries that use sensors and devices. With such open architecture, platform functions do not constraint the function of the blocks; this is why the relationship between them is not indicated in Fig. 16.7. The functions of each of these blocks will be described now in brief.

Platform Manager: is responsible for managing the internal modules and interfaces. He works with the communication manager and element manager to monitor, configure, troubleshoot, and upgrade the services platform modules. The platform manager may be a physical system/server, or virtual system with functions distributed among the common management components.

Discovery, Entities, Services, and Location: Discovery is the process of identifying and transferring information regarding existing IoT entities and/or resources and their location. Accurate discovery is essential for most IoT management tasks, such as asset management, network monitoring and diagnostics, fault analysis, network planning, and capacity planning. Device registration is a process of delivering device information to management entity (or another server) so that devices can communicate and exchange information.

Communication Manager: Communication manager is responsible for providing communications with other platform functions, applications, and devices.

Data Manager: Data manager's tasks include collecting, storing, and exchanging information among various platform entities. Data retrieval followed by repository is another key functional requirement for the IoT service platform.

Firmware Manager/Element Manager: Perhaps the central and most important one is the element manager. Its function is central to NMS (Network Management Services). Its function encompasses monitoring IoT sensors, actuators, gateways, and all devices residing within platform boundaries. Element manager also handles all security-related functions.

Firmware manager has come to the fore since a variety of devices on different platforms can update firmware over the air (FOTA). This is similar to OTAP (Over-The-Air-Provisioning) discussed earlier in the chapter on cellular communication and security. Mobile devices need not be brought to some central location to update their firmware or key assignment, it can be done over the air, using a secure protocol.

Topology Manager: Network topology refers to the physical or logical connection and layout of devices, actuators, gateway, and other elements that are in the network. Physical connections relate to positional locations on a map for example, while logical connections show virtual data, data flow, layers, and related information.

Group Manager: Unlike traditional networks, IoT tends to have a large number of devices. At the extreme it could be in thousands of devices in one location. Group Manager supports requests from subgroup that manage their devices. Group Manager will have the ability to create, update, retrieve, or delete sub-groups or groups that report to Group Manager.

Billing and Accounting: This will be responsible to calculate and report charges based on usage and billing policies. It also handles different charging models including real-time credit control and related functions.

Subscription and Notification: Allows authorized devices to subscribe to services they are entitled to. It also sends notification related to security alert or notice to the subscriber.

API Manager: The function of API manager is to provide information to the communication manager related to the IoT network, so the communication manager can include that information to determine proper communication handling.

With the evolution of this common standard, the new OM2M which implements these functions is carried out under Eclipse OM2M project. (https://www.eclipse.org/om2m/). It is important to note that this is a continuously evolving field and more activity is expected.

16.5 Standards of ZigBee and 802.11af

There are multiple technologies that support IoT devices. Zigbee, LoRa, LTE-M, Wi-Fi 802.11ah (HaLow), and Z-Wave are all standards created by different standard bodies specifically for IoT and they compete for a slot in this fast-growing sphere.

Perhaps the biggest competition to get a share of the IoT market also belongs to IEEE standards. There are some such as Zigbee IEEE 802.15.4 that were established earlier, but in recent times the field has expanded. Recent standards of IEEE emerged to serve specific distances from the wireless node. The concept of signal attenuation due to frequency and distance was explained earlier in Chap. 3, where expanding volume of space and how it attenuates the signal were explained. Note that frequency is an important component of path loss. A major requirement of IoT is "extremely low power" from the wireless node, expecting a small battery to last over 10 years. Low power requirement forces designers to look for alternatives based on "lower frequency" to help signal propagate farther or reduce attenuation by operating at a lower frequency. Such technologies are called LPWAN— low power wide area network.

Figure 16.8 shows typical distances signal can travel from a low-power wireless source. These are the maximum distances signal can propagate and still be useful at the receiving end [10].

It is important to observe cellular standard LTE-M as one of the standards in the 600–1900 MHz. Earlier in Sect. 16.2, Table 16.2 showed five cellular standards—two from the erstwhile 2G GSM standard and three from the recent 4G LTE directly vying for the IoT market. Cellular system operates over a wide range of frequencies with 5G moving

Fig. 16.8 Wireless devices and operating standards

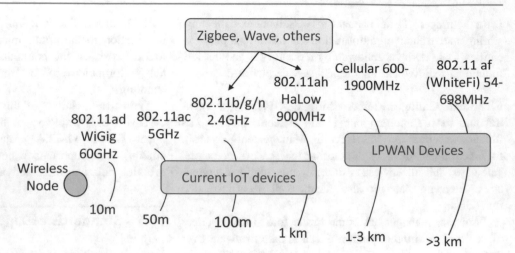

well into the 60 GHz and beyond. In addition, cellular bands now operate in the 400 MHz band, which is not indicated here. In essence, cellular covers the entire spectrum of frequencies indicated in Fig. 16.8.

Let us consider the individual IEEE standards. The 802.11ad is considered broadband IoT, usually at home or office, where it can be visualized as a very short wireless link. Prior to this standard's introduction, the 60 GHz band was unlicensed and deployed in 2006 as a millimetric wave link, which was described in Sect. 14.5.1. It served well as an alternative to fiber-optic link with significantly lower latency (usually <200 μs) and 1 Gbps data rate. Since WiGig (802.11ad) Gigabit WiFi operates over a short distance, it provides a 7 Gbps throughput rate and can support hundreds of video links at home or office. It can be viewed as a broadband version of Bluetooth. WiFi with the current 802.11n and 802.11ac versions operates in all 5 GHz (UNII) and 2.4 GHz bands. This was extensively dealt with in Chap. 12, with a focus on broadband IoT. Figure 16.5 described an example that used these bands with 60 GHz links backhaul.

Major IoT standards focus on narrow band that operates in 2.4 GHz and 900 MHz as LPWAN—low power wide area network (Zigbee or 802.11ah are good examples). For long ranges, 802.11af (WhiteFi) and LoRa (Long Range) provide better coverage at distances of over 3 km. LoRa is a technology focusing on supporting a large number of devices spread over a large area (a circle of 6 km or more). It has categorized different entities as follows [11]:

- LoRa end devices: These are end devices (sensors) that perform sensing and actuation
- LoRa gateway: Similar to an IEEE 80211ah gateway or a ZigBee/6Lowpan (personal area network operated with IPv6) coordinator, a LoRa gateway receives communications from LoRa end devices and offers Internet backhaul functionality. LoRa gateways are projected to be housed with cellular base stations
- LoRa server: It is a network server that manages the LoRa network including packets filtration, data rate adaptation among many other network management and control capabilities
- LoRa remote server/cloud system: Provides high-level application services such as collecting and processing the data gathered by end devices, performing data analytics, and running IoT applications.

The system is expected to operate in the white spaces of TV band (where TV bands may not be in use). Given that it can operate in lower VHF and UHF bands, a considerable area can be covered and the technology holds promise. In addition, LoRa considers locations and services that only expect very low data rates with three different quality of service categories:

(1) **Class A devices**: Class A is the lowest power end-device as it is more energy-efficient. Class A devices spend most of their time in sleep mode. They only wake at a scheduled time or when they are ready to transmit data (event-driven). Thus, communications

Table 16.3 Security risks to WBAN and corresponding security requirements

Attack assumptions	Risks to WBAN	Security requirements
Computational capabilities	Data modification	Data integrity
	Impersonation	Authentication
Listening capabilities	Eavesdropping	Encryption
Broadcast capabilities	Replaying	Freshness protection

from the server are only possible during the scheduled uplink of an end device.

(2) **Class B devices**: Class B devices receive a time-synchronized beacon from the LoRa gateway which allows them to open extra receive windows at scheduled times. This allows the server to determine the time an end-device can receive data.

(3) **Class C devices**: Class C end-devices are able to receive data at any time, except when transmitting data, as they are always listening. Obviously, this is the lowest energy-efficient among all classes of devices.

16.6 IoT and Impact on Safety and Public View

With the proliferation of wireless devices, general concern about public health and safety will become an important issue. Public concern can be visualized in two categories: (A) public health due to radiation and (B) annoyance or misgivings with device malfunction.

From the operator's point of view, their concern could be to maintain centralized facility and how to handle such a large amount of data that grows over time. They are turning to IIoT or industrial IoT where cloud computing can largely perform the task of storage and upgrade [12]. There is even a thought that such cloud-based solution may eliminate the traditional SCADA layer and allow easier maintenance and upgrade using the cloud server.

16.6.1 Public Health: Device Proximity

IoT devices may operate close to children and adults. With a very large number of these devices in each household, this becomes a genuine concern. Although WiFi operates in most homes with client devices that can add up to tens of devices, IoT could substantially increase that number. However, there is one major factor that is in favor of IoT—"low power". Device power is so low that all of those devices combined may not produce the equivalent of one WiFi access point. Therefore, there will be a considerable effort by vendors to point out that when tiny battery on such devices lasts several years it indicates it will not be harmful to living beings. To an extent that argument is correct—but without a doubt, this concern will remain. In dense housing areas with multiple dwellings, signal noise floor could be high and such arguments may continue. In particular, the health industry has concerns about wireless sensors, not from a radiation perspective, but wireless body area networks (WBAN) [13]. This includes both the physical level of security with direct access to the sensor devices and the network security. These risks are summarized in Table 16.3 [13].

Some of these security concerns are already addressed in the WSN (wireless sensor network) as explained earlier. However, medical personnel have indicated their security requirement as shown in the right column of Table 16.3, where authentication and encryption get addressed as part of the design in WSN. The other two requirements, such as integrity and protection, are part of the security scheme in cellular network, but in WSN it needs specific attention when the number of sensors continues to grow, in the hospital environment.

16.6.2 Sensor Malfunction Leading to Concerns

The second concern has not been well addressed in standards, which is what happens when some of these sensors malfunction. This can be a simple problem such as the device stopped working, which can mean short-term inconvenience to users. It could be due to a low battery or some other failure in the device [14].

More serious concerns emerge when the sensor device reports the wrong value. It can indicate a temperature that is much higher or lower than the nominal value, when the true value may be within the normal range. Any parameter, if reported to be outside the nominal values can raise concerns and alarm. In course of time such occurrences lead to a lack of public trust. This category is expected to loom large as a percentage of sensor/actuators malfunction. If an actuator was expected to turn off something and it does not turn off that function, but reports as if it was turned off, that can become a crucial situation.

16.7 Conclusion

This chapter described one of the traditional technologies—widely known in process control and other industries as "SCADA". This technology for monitoring and control remained largely unaffected until recently. It started to change with concepts of "machine-to-machine" communication that began with 4G cellular and has now evolved into a major area known as IoT or "Internet of Things". In addition to the traditional SCADA, it now involves everything from home networks, security and monitoring in health care industry, transport, and a wide range of sectors.

The revolution is so powerful that separate service providers to monitor a wide range of industries is steadily evolving. Cloud computing and cloud-based services allow limitless possibilities—a view shared by many who believe that IoT devices will be 10 times larger than the number of cell phones globally.

With the global expansion of IoT, many standards bodies joined together to offer a high-level view of how this field should be perceived. With generic definitions they hope to achieve a level of commonality that has become possible by separating the sensor devices from the service infrastructure. In terms of technology, there is a wide range of standards mostly from IEEE WLAN family and from 3GPP that is mostly based on cellular. It is likely that a hybrid of these two has evolved and will remain so in view of the advantages of each.

In terms of throughput requirement are two types of devices—the first one requiring narrow band or very low throughput using very low power has emerged as the biggest. The smaller part is occupied by devices such as camera that need broadband service which can be served using existing cellular and WiFi technologies. Narrow band devices are expected to grow into this huge IoT sphere of integrated sensors making up billions of devices to serve global services.

Summary

The chapter started with a classical view of SCADA and how it started as a monitoring activity in the process chemical and other industries where safety was an important part of the operation. PLC (programmable logic controller) is an important element in all SCADA systems; its functionality and integration in the network were discussed.

Some of the industries where the current deployment of SCADA has existed for decades were discussed to appreciate a better understanding of harsh environments in which sensors operate. Quite often satellite and microwave links were used in such systems, but local monitoring always favored wired network. However, with increased deployment of cellular and its reliability steadily improving, the scene shifted in favor of considering cellular and WLAN services.

Realizing that most SCADA needed very little bandwidth with periodic throughputs of no more than a few hundred bytes, wireless industry quickly moved in providing a range of options from cellular such as 2G, 3G, and special flavors of LTE that offered low throughput. Similarly, the WLAN standards also offered resilient standards such as Zigbee that could operate well to support SCADA. With the grand opening of this area into individual homes and small office locations, the term machine to machine communication became popular.

With the advent of 4G LTE and 802.11ah and 802.11af, the term IoT (Internet of Things) became popular. It not only incorporated SCADA but considerably expanded it, bringing a wide range of technologies and applications. Projections of 10 billion or more devices in IoT led to a considerable thinking in standards, and a new architecture to support different layers was considered. This chapter explored those layers and how their generic definitions lend to bring a common service provider layer that essentially replaces the NMS (network monitoring system) as a service across multiple industries, business, and residential sectors.

Finally, with sensors occupying every home, office, and public area, concern for public health and safety were briefly touched upon. Such concerns also extend to undue stress based on malfunctioning of sensors and actuators. This is projected as an area where considerable work is needed to assure citizens and at the same time set clear protocols on how to recognize and handle malfunctioning of devices.

Emphasis on participation

Contact a local petro-chemical or process industry or service provider and discuss options of streamlining their service with either Zigbee or cellular network. Discuss topics such as:

Are there specific advantages in moving to a wireless network?

What are the concerns of traditional SCADA operators?

Contact local child daycare centers. Discuss the use of child monitoring and other devices. Why do they prefer using them?

In the University/College sports or Gym department do they encourage the use of body monitoring sensors? What type of sensors would they prefer or not use? Why?

Knowledge gained

With traditional SCADA that remained for about five decades—the field is poised for explosive growth.

It is perceived that device technology, particularly battery-operated devices, led to this explosive growth. From process chemical and other heavy industries, this technology has become universal operating in all areas, starting with baby monitors to sports and athletic sensors, home security, office, and public area monitoring. Most of these sensors are also centrally monitored—offered to all as a monthly service.

In almost all new deployments, the wired option is not preferred. For example, home security using wireless technology such as Z-wave is more popular than a wired system that is centrally monitored.

Homework problems

1. Using Fig. 16.8 assume a device that operates at 2450 MHz and the device output 10 mW. Assume the receiver needs a minimum level of −74 dBm to operate with guaranteed 100 Kbps throughput. The same device also provides an option to operate at 915 MHz. How far would the signal travel in each case reliably?
2. Choose three types of sensors shown on the left side in Fig. 16.2. If each device malfunctioned, what would be the effect on the actuators shown on the right side?
   (i) Flow sensor indicates a decrease in flow rate from 10 Gallon/minute to 2 Gallon/minute, expecting the motor to increase speed. But flow rate is actually normal, the reported decrease is due to malfunction. How would you verify this? Hint: there are other

sensors connected to the stream, before and after this sensor
   (ii) At a separate site, the pressure switch malfunctions and indicates an increase in pressure on the gas line, expecting the solenoid to open up the gas line. There are two more pressure switches after the gas line branches into two. How would you verify the malfunction?

Project ideas

1. Projects based on the programming language IEC 61,131–3 are described in 16.2. Using the video in Reference 3, https://www.youtube.com/watch?v=Qf32qtHfowQ use the sequential function chart and develop a PLC program for a process control flow, as given below.

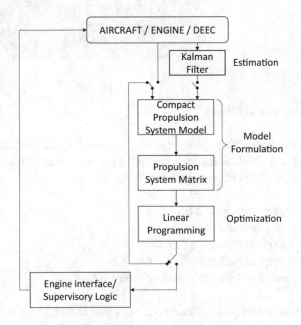

2. Using the same video, another project using ladder diagram creates programs to show fault, as shown in the diagram below.

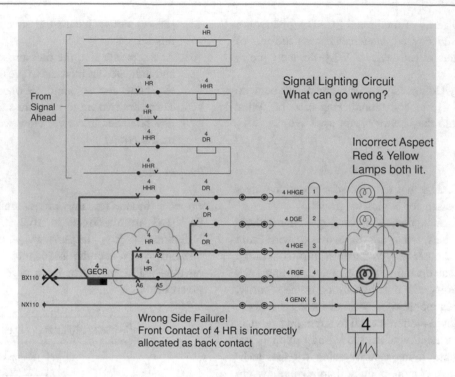

Out-of-box thinking

Narrow band usage is the primary assumption for IoT. If there are 120 locations reporting to one master service center, each location reports the following:

1. Temperature over 2 h period
2. Flow rate twice a day
3. Voltage fluctuation every 8 h
4. Pressure switch reporting every 3 h
5. Liquid level switch reporting every 6 h.

How would you allocate time such that there is no clash of signals reported from these locations? All of these SCADA monitoring functions are needed daily and a historic record must be maintained. Create links between flow rate, pressure switch, and liquid level and explain the logic of monitoring such a system.

References

1. SCADA Technology summit, a digital conference experience. https://www.scadatechsummit.com/
2. Department of the Army, TM 5–601, Supervisory Control and Data Acquisition (SCADA) Systems for Command, Control, Communications, Computer, Intelligence, Surveillance, and Reconnaissance (C4ISR) Facilities, 21 January 2006. https://www.wbdg.org/ffc/army-coe/technical-manuals-tm/tm-5-601
3. What are the Most Popular PLC Programming Languages? https://www.youtube.com/watch?v=Qf32qtHfowQ
4. Amarawardhana C, Dayananada KS, Porawagama H, Gamage C (2009) Case study of WSN as a replacement for SCADA. https://www.researchgate.net/publication/241166910_Case_study_of_WSN_as_a_replacement_for_SCADA
5. Zennaro M (2012) Introduction to wireless sensor networks. ICPT, Trieste, Italy. http://wireless.ictp.it/wp-content/uploads/2012/02/Zennaro.pdf
6. Lutra M, Atri R, Sadeghian M, Malik S, Rekhi P (2016) Long term evolution for IoT (Narrow band LTE—Cellular IoT), a short note on design, technology and applications. https://www.academia.edu/29092140/Long_Term_Evolution_for_IoT_Narrow_Band_LTE-Cellular_IOT
7. Twilio Programmable wireless: IoT developer kits. https://www.twilio.com/docs/iot/wireless/twilio-developer-kits
8. Wan L, Zhang Z, Wang J (2019) Demonstrability of Narrowband Internet of Things technology in advanced metering infrastructure. EURASIP J Wireless Commun Netw 2. https://link.springer.com/journal/13638
9. Hayes A, Salam S (2019) Internet of Things—from hype to reality. 2nd edn. Chapters 5.4.2 and 7.1. ISBN 978–3–319–99515–1 ISBN 978–3–319–99516–8 (eBook). https://doi.org/10.1007/978-3-319-99516-8
10. Frenzel L (2017) What's the Difference Between IEEE 802.11ah and 802.11af in the IoT? Electronic Design Mag. https://www.electronicdesign.com/industrial-automation/article/21805297/whats-the-difference-between-ieee-80211ah-and-80211af-in-the-iot
11. Elkhodr M, Shahrestani S, Cheung H (2016) Emerging wireless technologies in the Internet of Things: a comparative study. School

of Computing Engineering and Mathematics, Western Sydney University, Sydney. Australia, International Journal of Wireless & Mobile Networks (IJWMN) 8(5) https://doi.org/10.5121/ijwmn.2016.8505 https://arxiv.org/ftp/arxiv/papers/1611/1611.00861.pdf

12. A cloud based alternative to traditional SCADA, IIoT magazine. https://iiot-world.com/industrial-iot/digital-disruption/cloud-based-alternative-to-traditional-scada/

13. Al Ameen M, Liu J, Kwak K Security and privacy issues in wireless sensor networks for healthcare applications. J Med Syst https://doi.org/10.1007/s10916-010-9449-4

14. Internet of Things: Limitless connections and ways to fail, Insights, Oct 2018. https://www.mmc.com/content/dam/mmc-web/insights/publications/2018/dec/IoT–Limitless-Connections-and-Ways-to-Fail/Internet-of-Things_%20Limitless%20Connections.pdf

Antennas, Transmission Lines, Matching Networks

17

Abbreviations

Antenna	In wireless communication antenna is the element that is between radio waves propagating and the conducting wires (transmission line). An antenna can send (transmit) or receive electro-magnetic waves or do both
Waves and fields	These two terms refer to the electric energy represented by waves and magnetic energy represented by fields. But these two always go together forming the electro-magnetic waves and fields. One doesn't exist without the other. But since electric field or energy is more dominant that is why mostly they are called radio waves
Maxwell's equations	Attributed to the physicist Maxwell, the four equations along with Lorenz's force equation completely represent the electro-magnetic waves. Although these are highly mathematical, they describe the nature of electro-magnetic energy that is not only used in wireless but also in fiber optics and every endeavor related to light
VSWR	Voltage standing wave ratio represents maximum voltage to the minimum voltage on the transmission line. The ideal VSWR is 1, which means all the power was absorbed and nothing was returned. All field values will be > 1, which means at least a small part of the energy gets returned
Reflection coefficient	It is the ratio of reflected voltage of current to the forward voltage or current vector, as a number. It is related to VSWR $= \frac{1+\Gamma}{1-\Gamma}$, where Γ is the reflection coefficient
Return loss	Return loss is the measure of how much of the signal is lost when it is reflected back to the source. It is usually expressed in dB relative to a short circuit and related to VSWR. Nominal VSWR (close to ideal) of 1.01 provides a return loss of 45.064 dB, indicating less than 1/40000 of the forward energy was returned (negligible)
Coaxial transmission line	Coaxial transmission line to carry RF signal consists of an inner conductor and an outer conductor with a solid polyethylene (PE) or solid Teflon (PTFE —"Poly-Tetra-Fluoro-Ethylene" (PTFE, or commonly known as Teflon) insulator as a dielectric to keep losses to a minimum. While cable diameter is an important parameter, the distance between the inner and outer conductor and how strictly it is maintained is the key
Waveguide	A waveguide is an electro-magnetic transmission line used in microwave communications, broadcasting, and radar installations. A waveguide consists of a

© Springer Nature Switzerland AG 2022
K. Raghunandan, *Introduction to Wireless Communications and Networks*,
Textbooks in Telecommunication Engineering, https://doi.org/10.1007/978-3-030-92188-0_17

	rectangular or circular cross-section of a hollow metallic tube. The electro-magnetic field propagates lengthwise along that tube. Waveguides are often used with horn antenna or dish antenna and relative to coaxial cable can carry more RF power, with lower loss
Bending radius	The term bending radius applies to RF coaxial cable, where bending the cable deteriorates the performance of the system. The cable is therefore made semi-rigid and the bending radius is specified as the "tightest bend allowed". The general rule applied is "5 times the diameter of the cable" for tight bend and "10 times the diameter of the cable" for nominal bend
Coaxial connectors	There are various types of RF connectors developed over the past 80 years—all of them have one important criterion—distance between the inner and outer conductor and how strictly it is maintained is the key to connector performance
Radiating cable	It is a coaxial cable with holes punched into the outer conductor to carefully allow RF radiation to escape, therefore making it into a long radiating element. Unlike antenna it allows RF energy to leak only along its length. Due to this design, it is often known as the leaky cable used in tunnels, mines, and other restricted environments. Other than this feature, it looks like and works similar to a coaxial transmission line
Antenna gain	Antenna gain consists of three separate entities: Gain, realized gain, and directivity. Only the gain is defined by the IEEE as the ratio of the power produced by the antenna from a far-field source on the antenna's beam axis to the power produced by a hypothetical lossless isotropic antenna, which

	is equally sensitive to signals from all directions. In practice, such an isotropic antenna is used as a theoretical reference and practical antennas indicated their gain in dBi, or dB with reference to isotropic. Realized gain includes antenna efficiency and losses showing how much is the "actual" gain. Directivity is due to the orientation of the beam in a specific direction, thereby increasing energy in that direction, with no useful radiation in any other direction
EIRP (effective isotropic radiated power)	EIRP or the effective isotropic radiated power is the total power radiated by a hypothetical isotropic antenna in a single direction. ERP is effective radiated power is the total power radiated by an actual antenna relative to a half-wave dipole rather than a theoretical isotropic antenna
Beam width of antenna	Using the radiation pattern, the point downs the propagation path, where the energy reduces to half of the original (starting) power, and that point is used to measure the beam width. Typically, this is indicated by the antenna vendor, showing the beam width in degrees. Beam width is specified separately for the elevation angle and the azimuth angle
Polarization	Polarization normally refers to the orientation of the electric field. Vertical polarization is commonly used in cellular and land mobile systems, while horizontal polarization is used in aircraft communication. Satellite systems often use the third one which is circular polarization
Cross-polarization	If the antenna element is set to 45° instead of the usual 90°, then energy is partly available in both vertical and horizontal directions. This is deliberately done in urban areas where most users have no direct line of sight to the antenna.

	In such cases cross-polarization allows better communication serving users		induced in the secondary device (antenna) due to current in the primary (transmission line)
FSPL (Free space path loss)	IEEE Std 145–1993 defines "free space loss" as "The loss between two isotropic radiators in free space, expressed as a power ratio". It does not include any power loss in the antennas themselves due to imperfections such as resistance. Free space loss increases with the square of the distance between the antennas because the radio waves spread out by the inverse square law and decreases with the square of the wavelength of the radio waves	Electrically small antennas	Generally, antennas that are about one-tenth the size of operating wavelength are considered electrically small. Devices that operate in the VHF or lower bands where the wavelength is quite long, use this concept while building small devices. There are some challenges including negative gain, but it may be possible to build a good matching network to compensate for this
Radiation pattern	The radiation pattern of an antenna is a three-dimensional model of the RF energy propagated from the antenna. Since 3D view is not easy to be effectively displayed, vendors usually provide in their datasheets, the elevation pattern and azimuth pattern as separate plots	Hidden antenna	Antenna housed inside a casing and not visible to the user. This includes antennas inside a cell phone, but a number of battery-operated wireless devices also incorporate antenna inside the plastic housing, with the main intent of keeping the devices free from encumbrances
Antenna impedance	Antenna impedance comprises three parts: impedance of the device, self-impedance, and mutual impedance. Impedance of the antenna as a passive device is related to the electric and magnetic fields and is defined at the terminal point of the antenna system as the ratio of voltage or current across the particular terminal. Self-impedance is the input impedance of the antenna in the absence of all other elements. Mutual impedance—if we view the transmission line that feeds as one device and the antenna as the other device, there is a mutual impedance between these two. By viewing the transmission line as the primary transformer and the antenna as the secondary transformer, mutual impedance can be defined as the voltage	EMI/EMC	Electro-magnetic interference usually relates to any unwanted, spurious, conducted, or radiated signal of electrical origin that can cause unacceptable degradation in system or equipment performance. EMC or electro-magnetic compatibility is the ability of systems to function as designed, without malfunction or unacceptable degradation of performance due to EMI from others within the operational environment
		RFFE	Radio frequency front-end circuitry refers to circuits right behind the antenna that become part of the antenna assembly. This can include multiple layered integrated circuit packages. It can also be a part of the smart phone or wireless device that includes RF and related portions that are designed to be sensitive to interact with radio waves outside the plastic enclosure of such devices

In all wireless systems there is an antenna which is sometimes known as an "Aerial". Any antenna will either transmit or receive, or it may do both. Without this key element, there can be no wireless communication. In all wireless communications "transmission line" is another key component; it may take the form of two wires or more commonly a coaxial cable or a waveguide. Both of these are passive elements; there is some form of voltage or current in them. This chapter deals with these two passive elements, their important features, and how they help any wireless network.

If they are passive elements why should we consider them in such detail? The answer is "these two key elements together transform the world from wired to wireless". The care given to their design makes the wireless unit work or not work. Even in the best cases, how the transmission line is matched to the antenna directly shows how much electro-magnetic energy leaves the antenna and goes into the air and how much gets returned (rejected by the antenna could be another way of stating it). Therefore, antennas that are extremely well designed with minimum rejection could cost almost ten times a nominally designed antenna. The key factors of good design must be analyzed, which is the important focus of this chapter.

17.1 Electro-Magnetics and Fields

In Chap. 2, initial aspects of how to perceive an electro-magnetic wave, their nature, and how they travel through free space medium of air or vacuum were reviewed. Let us consider them in some detail now. For most physicists and engineers, the study of electro-magnetics and fields relate to classical equations by James Clark Maxwell. In the 1860s he put together extensive work done by other physicists to provide a complete understanding of electro-magnetic waves and fields. Maxwell's framework consists of four equations. These equations along with Lorenz's force equation provide a complete understanding of electro-magnetic waves and fields [1].

Waves and fields are three-dimensional. Their study involves complex mathematics. It is better to learn them initially through a visual model of waves shown in a classic video lesson from the 1950s—an example is shown in [2]. The analytical approach followed by physicists involves mathematical quantities such as "curl" and "divergence"; these are explained using 3D figures with forces indicated by "vectors" that have magnitude and direction. Derivations of these equations are addressed by physicists at graduate-level courses and are well beyond the scope of this book.

17.1.1 Maxwell's Equations

Maxwell's equations describe how electric charges and electric currents create electric and magnetic fields. These four equations describe how an electric charge can generate a magnetic field and vice versa. Any charged particle also creates a force. The force per unit charge is termed as a "field". The fields could be stationary or moving; therefore their study branch off into electrostatics and electrodynamics. Maxwell's equations and the Lorentz force equation offer all the tools one needs to calculate the motion of classical particles in electric and magnetic fields (dynamics). They are shown in Fig. 17.1.

Maxwell's first equation shows how to calculate the electric field created by a charge. His second equation explains how to calculate the magnetic field and confirms that any magnet will always have two poles (north and south poles). The third and fourth equations of Maxwell deal with fields, showing how fields "circulate" around their sources. Magnetic fields "circulate" around electric currents and time-varying electric fields. Further work by others expanded Maxwell's equations.

Ampère's law with Maxwell's extension showed how currents are formed in closed surfaces, while Faraday's law dealt with electric fields that "circulate" around time-varying magnetic fields. Finally, Lorentz force is a combination of the electric and magnetic force on a point charge due to electro-magnetic fields. Lorentz equation indicates the electro-magnetic force on a charge "q", which is a combination of a force in the direction of the electric field $\mathbf{E}$ proportional to the magnitude of the field and the quantity of charge, and a force at right angles to the magnetic field $\mathbf{B}$ and the velocity v of the charge, proportional to the magnitude of the field, the charge, and the velocity.

Variations on this basic formula describe the magnetic force on a current-carrying wire (known as Laplace force), the electromotive force in a wire loop moving through a magnetic field (an aspect of Faraday's law of induction), and the force on a moving charged particle. Lorentz force equation is useful in electro-magnetics that uses movement such as a solenoid or electric motor. Figure 17.1 shows Maxwell's equations, followed by the Lorentz force equation. Together they form the basis of understanding in all electrical systems—both wired and wireless. Detailed analysis of these equations is beyond the scope of this book. They are typically studied at graduate level courses in Physics. We will focus on the practical aspects of how fields and electromagnetics operate in transmission lines (cables) and how they propagate through antenna.

MAXWELL'S EQUATIONS

Differential form

$$\nabla \cdot \mathbf{E} = \frac{\rho}{\varepsilon_0}$$

Gauss's Law: The relationship between electric field and electric charge. The electric field's mapping is equal to the charge density divided by the permittivity of free space.

$$\nabla \cdot \mathbf{B} = 0$$

Gauss's Law for magnetism: There is no such thing as a magnetic monopole. The net magnetic flux out of any closed surface is zero

$$\nabla \times \mathbf{E} = -\frac{\partial \mathbf{B}}{\partial t}$$

Electric field can be created by changing magnetic field

$$\nabla \times \mathbf{B} = \mu_0 J + \mu_o \varepsilon_0 \frac{\partial \mathbf{E}}{\partial t}$$

Magnetic field can be created by changing Electric field

Integral form

$$\oint_A \mathbf{E} \cdot \mathrm{d}a = \frac{Q}{\varepsilon_0} \qquad \textbf{...Eq 17.1}$$

$$\oint_A \mathbf{B} \cdot \mathrm{d}a = 0 \qquad \textbf{...Eq 17.2}$$

$$\oint_L \mathbf{E} \cdot \mathrm{d}l = -\frac{\partial \mathbf{B}}{\partial t} \oint_A \mathbf{B} \cdot \mathrm{d}a$$
$$\textbf{...Eq 17.3}$$

$$\oint_L \mathbf{B} \cdot \mathrm{d}l = \mu_0 \mathbf{I} + \mu_o \varepsilon_0 \frac{\partial}{\partial t} \oint_A \mathbf{E} \cdot \mathrm{d}a$$
$$\textbf{...Eq 17.4}$$

$$F = qE + qv \times B$$

...Eq 17.5

The force exerted on a **charged particle** q moving with **velocity** v through an electric field E and **magnetic field** B. The entire electromagnetic force **F** on the **charged particle** is called the Lorentz force

LORENTZ FORCE EQUATION

Fig. 17.1 Maxwell's equations and Lorentz force equation

Figure 17.1 shows these four equations by Maxwell, and the fifth one is Lorentz force equation, which are useful in understanding how electro-magnetic waves travel through transmission lines. If the transmission line is a coaxial cable, then it is supported by an internal structure of an insulating (dielectric) material to maintain uniform spacing between the inner center conductor and the outer shield. There are coaxial cables where the dielectric happens to be air. Transmission lines could also be made of waveguides that typically use only air as dielectric.

Their dielectric losses depend on the material used and losses increase in the following order:

(1) Ideal dielectric (no loss)
(2) Vacuum
(3) Air
(4) Polytetrafluoroethylene (PTFE)
(5) Polyethylene foam
(6) Solid polyethylene

The μ (permeability of the medium) and ε (permittivity of dielectric) indicated in the first and fourth equations of Maxwell mathematically show how the field and transmission are affected.

How do the antenna and the transmission line work together? Let us consider an ideal situation where the source (base station) sends, say, 100 W to the antenna on the tower. Under ideal conditions there would be no loss of power through the coaxial cable and the antenna would receive 100 W and transmit it over the air. But a practical system indicated in Fig. 17.2 shows the limitations.

The upper part of Fig. 17.2 shows the ideal connection where everything is expected to be purely resistive. Under practical conditions, the transmission cable will have some loss, and the connector will also have some loss. The antenna will not see the transmission line as ideal 50 Ω, therefore most of the 100 W power is received but part of it gets returned back towards the base station. This is shown in the lower part of Fig. 17.2 as the forward wave (thick blue arrow) and return wave (thin red arrow). How does this loss occur? Coaxial cable consists of two conductors (center and outer) in parallel that forms a capacitor. The value of this capacitance is distributed (it is C/meter) and it results in reactance $X_c = \frac{1}{2\pi fc}$. In addition, the cable has distributed inductance (it is L/meter), resulting in reactance $X_L = 2\pi fL$. Capacitive and inductive reactance (X_c, X_L) are complex numbers that are together expressed as impedance Z.

Fig. 17.2 VSWR concept—with RF source, transmission line, and antenna

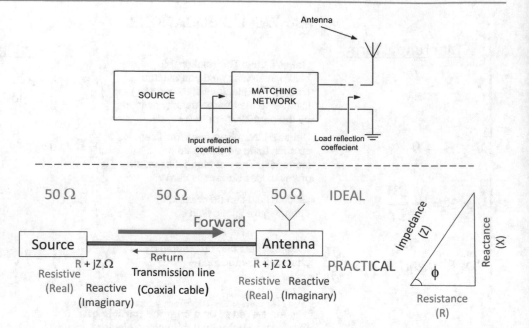

Since reactance (capacitance and inductance) varies with the frequency of operation, impedance is not a fixed value. It consists of the real part "R" (shown in blue) which is resistance and imaginary part X (shown in red) which is reactance. Together they are known as impedance $Z = (R \pm jX)\ \Omega$, where $j = \sqrt{-1}$ is an imaginary number. For a series RLC circuit, the relationship will be $Z = \sqrt{R^2 + (X_L + X_C)^2}$. It will have a magnitude less than 50 Ω, with phase difference $\emptyset = arctan\frac{X}{R}$. Both magnitude and phase are not fixed values and change with frequency "f", as already explained. This is why the word "practical" is split in blue and red, showing that it has a resistive and reactive part in all practical systems, expressed mathematically by the triangle.

Let us assume that at some frequency jX part has a value equivalent of 0.7 Ω, then expressed as $Z = (50-0.7) = 49.3$ Ω. In order to compensate, a matching circuit with value equivalent of $Z = (50+0.7)$ is needed so that the line appears to be purely resistive with the value of $Z = 50\ \Omega$. Such a matching circuit must be inserted before the antenna. The difficulty in designing it for a fixed value is, the Z works only at that frequency. If the frequency is changed there will be a mismatch again in the reactance and its phase. Therefore, such components are designed over a narrow band of frequencies and work well only over that frequency band. Matching circuits for antennas are based on using complex numbers described earlier, and will be discussed further in Sect. 17.1.4.

Two important factors are used by engineers/physicists to address transmission lines and load (antenna). The first factor is known as VSWR (voltage standing wave ratio) and

the second factor is return loss (RL). Both VSWR (voltage standing wave ratio) and return loss are a measure of the same parameter. That is the amount of signal reflected by a connector. VSWR is the major factor contributing to the total signal efficiency of the connector. Return loss (RL) is the portion of a signal that is lost due to a reflection of power at the location of transmission line discontinuity. In Fig. 17.2, there are two such discontinuities: one where the cable is connected to the source and the second where the cable is connected to the antenna.

VSWR is mathematically defined as

$$VSWR = \frac{V_{Max}}{V_{min}} \qquad (17.6)$$

In terms of impedance,

$$VSWR = \frac{(1+\Gamma)}{(1-\Gamma)} \qquad (17.7)$$

where Γ (gamma) is the voltage reflection coefficient that can be expressed as

$$\Gamma = \frac{(Z_L - Z_o)}{(Z_L + Z_o)} \qquad (17.8)$$

where Z_L is the load impedance antenna. Z_o is the source impedance, also shown in Fig. 17.2. In an ideal case since maximum and minimum voltages are the same, VSWR = 1.0. In real systems the number is likely to be between 1.01 and 2, or even more than 2.

Return loss is similar to VSWR and is generally preferred in the cable industry. Return loss is a logarithmic measurement. It is very useful when displaying very small reflections. Return loss is defined as the logarithmic ratio of forward power to return power. Return loss is normally calculated in terms of incident power and reflected power as follows:

$$RL = 10 \log_{10}\left(\frac{P_i}{P_r}\right) \qquad (17.9)$$

Since the reflection coefficient Γ is the ratio of the forward and reflected voltages, and power is proportional to the square of the voltage, return loss (RL) can be expressed in dB as

$$RL = 20 \log_{10}\left(\frac{P_i}{P_r}\right) \qquad (17.10)$$

There are tables available that provide equivalent values of RL and VSWR, although they can be calculated. RL of 14 dB will be equivalent to VSWR of 1.5, which means if 100 W of power was sent to the antenna (load), 4 W would get returned. This is considered fair performance. If the user chooses RL = 20 dB that would mean when 100 W is sent only 1 W is reflected. This would be an excellent performance where VSWR would be 1.22. Such high performance is needed in systems where special design effort is made to meet such stringent specifications. That happens more often in space systems or military networks where the power is at a premium.

What is the difference between these two factors and when should each be used? VSWR is the ratio of the voltage applied to voltage reflected. VSWR is generally preferred in the connector industry since it provides a measure of how good the connector is (at the point of discontinuity). VSWR of 1.5 or less is preferred, but a VSWR of 1.03, for example, can be difficult to achieve [3]. Since it is a linear measurement, it can be useful when displaying larger reflections due to the fact that small differences are not compressed as they are in a logarithmic measurement. Return loss RL, on the other hand, is preferred by the cable industry since it expresses how much of the signal got reflected. It is a measure of the quality of the cable loss due to dielectric, capacitance, and inductance of the conductor [4].

Discussion thus far summarizes into an important parameter of transmission line—its "characteristic impedance", which will be reviewed in the next section. Its importance lies in the fact that this parameter is the key to "matching network" which is the important part that "connects" a transmission line to the antenna or load.

17.1.2 Types of Transmission Line: Coaxial or Waveguide

The concept of coaxial cable and waveguide was discussed in Chap. 2, Sect. 2.2 and Chap. 14, Sect. 14.2 Link design. The power line cables discussed in Chap. 2 are useful at lower frequencies. Waveguides are limited by the physical width (rectangular configuration) as discussed in Sect. 14.2. Waveguides had limited usage in applications until 5G introduced millimeter wave bands. At such high frequencies, coaxial cable losses are severe. An example provided in Chap. 11 shows a meter-long jumper cable in Fig. 17.2 used to connect the last leg from the tower antenna to the thicker coaxial cable, can lose up to 75% of the power at 67 GHz.

All the theories described so far also apply to the waveguide structures (skin depth effect, etc.).

In terms of cost and power capabilities, there are a few differences. Cables are cheaper; metallic structures are always more expensive. The power capabilities are higher in waveguides (lower losses too), but they are quite bulky. Waveguides also have smaller bandwidth, however, at frequencies beyond around 60 GHz, the signal cannot be well contained in the cable. For example, VNA (vector network analyzer) used for measurements at 60 GHz has waveguide ports, not cable connectors [5]. Therefore, waveguides become useful in millimeter wave bands.

At millimeter and higher bands, the connection of RF signal to any circuit or antenna uses a waveguide that is short, integrated into the circuit itself, and not flexible. There are a multitude of applications for such extremely small waveguides that typically fit into the integrated radio head connecting directly to the RF chip. This has now become such an important topic for all of 5G and millimeter-wave frequency applications that there is a separate textbook to cover the topic [5]. Certain aspects of this will be discussed later in this chapter.

What are the types of coaxial cables and where are they used? Chart in Table 17.1 shows a wide range of connectors used with coaxial cables [6]. Why are there so many connector types with different cables? To answer this question, some background of coaxial cable is needed. With the increase in frequency "how the signal is carried" becomes important for efficient transmission of a signal. At audio frequencies either twin wire (balanced) or coaxial (unbalanced) cable is used. Twin wire has low loss, but is prone to external influence from the coupling of electric or magnetic fields. Coaxial has better shielding from external signals, but higher losses. With an increase in frequency the "skin effect" becomes prominent. It means the depth through which the RF current flows inside the conductor decreases and localizes to the skin (outside surface) of the conductor. In contrast

Table 17.1 Different types of RF connectors for various frequency ranges

Connector name	Frequency range GHz	Impedance Z ohms	Year developed	Mating torque in lbs	Outer shield inside diameter (mm)	Center conductor diameter (mm)	Center conductor contact depth tolerance
BNC	DC-4	50, 75	1944	NA			0.186/0.206
TNC	DC-4	50, 75	1956	12			0.186/0.206
14 mm or GR900		50, 75	1962	12	14.2875	6.204	
7 mm or APC-7	DC-18	50	1964	12	7.000	3.040	−0.002/−0.002
N	DC-12.4	50, 75	1942	12	7.000	3.040	
Precision N	DC-18	50, 75	1965	12	7.000	3.040	M0.021/0.230 F0.187/0.207
3.5 mm/APC3.5	DC-34	50,75	1976	8	3.500	1520	0/−0.003
SSMA	DC-26	50	1960				
SMC	DC 10	50					
SMB	DC-4	50,75		NA			
SMA	DC-18	50, 75	1958	5	3.5	1520	0/−0.01
2.92 or K	DC-40	50	1974–1983	5–8	2.920	1.268	0/−0.003
2.4 mm	DC-50	50	1986	8	2.400	1.042	0/−0.002
1.85 mm or V	DC-65	50	1989	8	1.850	0.803	0/−0.002
1 mm	DC-110	50	1990's	3	1.000	0.434	

at DC and AC (60 Hz), the current flows through the conductor's entire cross-section. Skin depth δ is defined as

$$\delta = \sqrt{\frac{\rho}{\pi F \mu_r \mu_0 \sigma}} m \qquad (17.11)$$

where ρ = resistivity = $1/\sigma$ in Ω/m,

σ = electrical conductivity in m/Ω,

F = frequency in Hz,

μ_r = permeability of material relative to μ_0, which is a permeability constant defined below.

μ_0 = free space permeability constant = $4\pi \times 10^{-7}$ H·m^{-1} = $1.2566370614 \times 10^{-6}$ H·m^{-1} or N A^{-2}

$$\mu = \mu_o \mu_r \qquad (17.12)$$

These equations are used in the design of coaxial cables and connectors. Table 17.1. shows different types of RF connectors and their specifications. There are some important observations to be made. The table progressively moves towards smaller cables built in later years. But most of them have tight tolerances between the inner and outer conductors and maintain 50 Ω as their standard impedance. Only some of them offer a choice of 75 Ω.

In Table 17.1 observe that cables do operate well into the millimeter-wave band (all the way to 110 GHz), but their use may be limited. Short cables within circuit boards or very short cable distances to connect internally in equipment

(typically in tens of cm). For very long coaxial cables used in outdoor towers, "characteristic impedance" plays an important role in terms of signal transmission quality. The characteristic impedance of coaxial cables in ohms is

$$Z = \frac{59.9586}{\sqrt{\varepsilon}} ln\frac{D}{d} = \frac{138}{\sqrt{\varepsilon}} log_{10}\frac{D}{d} \Omega \qquad (17.13)$$

where D = outer diameter (any unit of length, but must use the same units as "d", d = inner diameter (any unit of length, but must be same as "D").

ε_0 = free space permittivity in a vacuum = $8.8541878176 \times 10^{-12}$ F/m.

ε_r = relative permittivity or dielectric constant.

$\varepsilon_0 = \varepsilon_0.\varepsilon_r$.

Another interpretation of the characteristic impedance is

$$z = \sqrt{\frac{L}{C}} \qquad (17.14)$$

where L = inductance/unit length and C = capacitance/unit length. Typical coaxial impedances are 50, 75, and 93 Ω. Among them 50 Ω is the most common. A typical question

Calculating Bending Radius

Cable Diameter	Minimum (5 X OD) Bending Radius	Optimum (10 X OD) bending radius
½ inch	2 ½ inches	5 inches
7/8 inch	4 ½ inches	8 ¼ inches
1 ¼ inch	6 ¼ inches	12 ½ inches
1 (5/8) inch	9 ½ inches	16 ½ inches

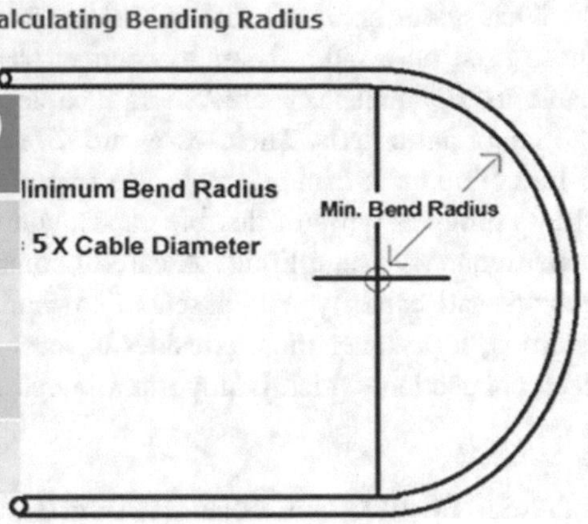

Minimum Bend Radius

5 X Cable Diameter

Min. Bend Radius

Fig. 17.3 Bending radius of coaxial cables—minimum and optimum

by many is "why 50 Ω"? To answer this question, the best coaxial cable impedances in high-power, high-voltage, and low-attenuation applications were experimentally determined at AT & T Bell Laboratories in 1929 to be 30, 60 Ω, and 77 Ω, respectively [7]. For a coaxial cable with air dielectric and a shield of a given inner diameter, the attenuation is minimized by choosing the diameter of the inner conductor to give a characteristic impedance of 76.7 Ω. When more common dielectrics were considered, the best-loss impedance dropped down to a value between 52 and 64 Ω. Maximum power handling is achieved at 30 Ω. The arithmetic mean between 30 and 77 Ω is 53.5 Ω; the geometric mean is 48 Ω.

The selection of 50 Ω as a compromise between power-handling capability and attenuation is, in general, cited as the reason for the number indicated in Fig. 17.2 which is considered ideal. Many of these cables are not very flexible. Their flexibility is defined by "bending radius", shown in Fig. 17.3. In outdoor locations with long cable run, commercially available cables have outer diameters that are typically 1/2, 7/8, 1 1/4 inch radius and 1 5/8 inch, as shown in Fig. 17.3 They have a minimum bending radius of about 5 X OD (very sharp bend) and nominal (10 X OD) which is normally allowed. While bending is allowed, good practice prescribes no more than two tight (sharp) bends with minimum bending radius, in the entire run. RF cables must be laid out carefully to avoid bends and minimize loss.

Fig. 17.4 Different types of flexible RF cables and connectors

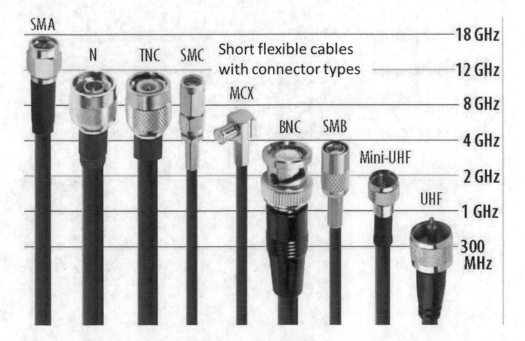

What about short flexible cables used in equipment or those used on a radio tower to connect from a bulky long cable to the antenna? Here "short flexible" cable becomes the important criteria. Their losses are severe; hence flexible cables of no more than one meter are prescribed. Figure 17.4 shows different types of flexible cables with connectors and their frequency range limits. A careful combination of connectors and cable type is essential to minimize losses. In addition, a designer must consider at least 0.5 dB loss for each connection, which is important increases path loss.

17.1.3 Radiating Cable: Transmission Line Designed to Be Lossy

Radiating cable is a coaxial cable in which outer conductor is deliberately punched out selectively such that RF energy can enter and leave the cable through openings. This serves as an antenna for bi-directional communication in enclosed spaces such as tunnels, mines, and long corridors or even factory shop floors where such cable can be placed near or close to equipment racks [8]. It is used also in server racks in large storage areas or within multi-storied office buildings.

Figure 17.5 shows an example of slots punched in the outer conductor to make radiating cables [9]. Each type of slot offers certain unique features that support different applications.

Due to its special design, radiating cable is initially fed with higher RF power such that it can radiate (or leak) through the entire length of the cable. Because of this property, the term "leaky cable" is often used for this unique cable type. Radiating cable has "only loss no gain" since it spreads radiation along its length, which means signal linearly decreases along the cable length. In essence, radiating cable works as an antenna throughout its length but radiates higher power initially and gradually reduces its radiation, as power is lost due to leakage along its path. From outside physical appearance it looks just like coaxial cable but only if the outer sheath is peeled, then punched-out holes in the outer conductor get revealed, as shown in Fig. 17.5.

There are five parameters to be considered when using all coaxial cables [9]. Only the coupling loss is specific to radiating cable. Figure 17.6 illustrates some of them. At the top in Fig. 17.6a is an illustration of differences between the normal coaxial cable and radiating cable; it shows how radiating cable transmits to the devices around it and

Fig. 17.5 Example of different slot configurations for radiating cable

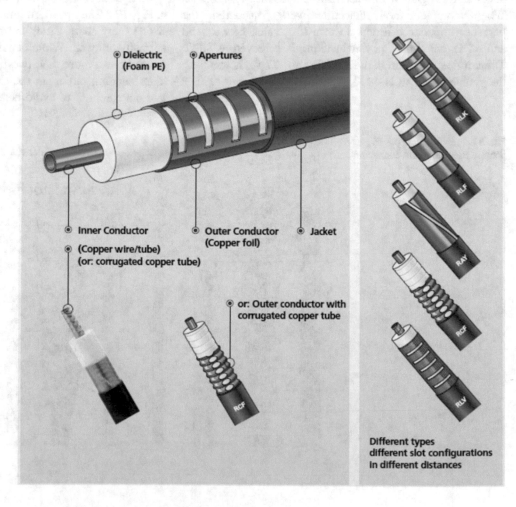

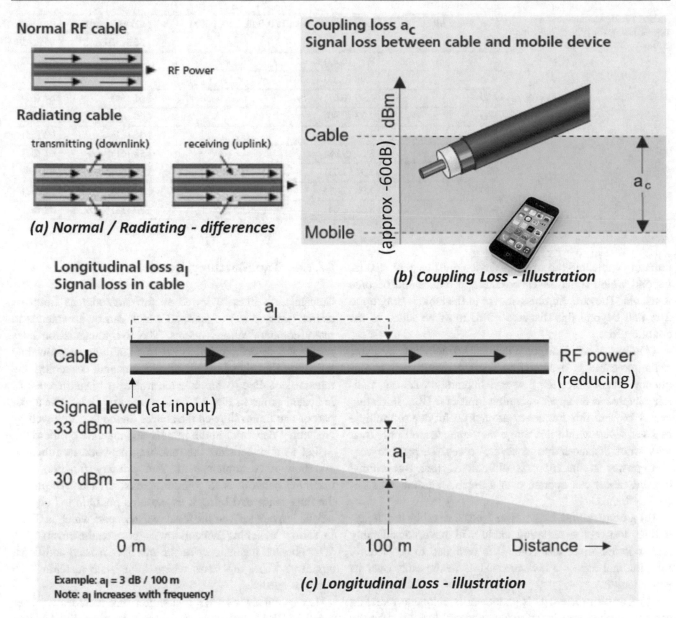

Fig. 17.6 Key parameters of radiating cable illustrated

receives signals from the devices (smart phone for example) through these slots along the way. Therefore, the radiating cable can be thought of as a small base station which is distributed in nature.

Figure 17.6b shows on the right, coupling loss a_c which is constant but unique to only radiating cable. Although data sheets for each cable will indicate measured values at a distance of 2 m from the radiating cable, coupling loss in general is about 60 dB. It is an important loss that is absent in normal coaxial cables. The bottom of Fig. 17.6c shows the longitudinal loss a_j for a leaky cable, which depends on the frequency and drops to half the original value (3 dB) within 100 m. It continuously loses signal level along its

length. Longitudinal loss for radiating cable is much higher compared to normal coaxial cable.

In order to illustrate losses, Table 17.2 shows typical values a_j and a_c for a specific radiating cable. For radiating cable coupling loss forms bulk of system loss; it is far more than the longitudinal loss.

In systems using normal coaxial cable there is a longitudinal loss but there is antenna gain effectively providing higher (sometimes much higher) radiated power. Signal loss over the air depends on the frequency described by FSPL (free space path loss) in Sect. 2.2.1 Hence coupling loss, in this case, could be considered as an equivalent of FSPL, but the difference is that coupling loss occurs over just a few meters from the cable. Typically, the distance between

Table 17.2 Attenuation and power rating for typical radiating cable

Frequency MHz	Longitudinal loss dB/100 m (dB/100ft)	Coupling loss	
		50%, dB	95%, dB
75	0.74 (0.23)	53 (56)	64 (67)
150	1.08 (0.33)	60 (63)	68 (71)
450	1.99 (0.61)	61 (64)	64 (67)
800	2.93 (0.90)	60 (63)	64 (67)
870	3.13 (0.95)	58 (61)	62 (65)
900	3.21 (0.98)	58 (61)	61 (64)
960	3.37 (1,03)	57 (60)	61 (64)
1800	7.98 (2.43)	53 (56)	59 (62)
1900	8.65 (2.64)	50 (53)	58 (61)

radiating cable and the wireless device must be kept <10 m, beyond which signal levels continue to reduce and become unusable. The coupling loss shown in the table is only up to 2 m and beyond this distance FSPL must be added to the coupling loss.

(1) **Operating Range**: Like in all wireless systems, radiating cable is designed to operate over a certain frequency range. Since RF power deliberately leaks out, radiating cable has an upper operating limit of 6 GHz. Radiating cables beyond this frequency are not useful due to multiple reasons. Even within this range they operate well only over very short distances due to severe losses. To provide very high power at the input is difficult as well. But normal coaxial cables can operate over a much wider range of frequency bands.

(2) **Longitudinal Loss**: Every coaxial cable has longitudinal loss but a radiating cable will have considerably higher longitudinal loss since it is designed to be "lossy". Longitudinal loss is a measure of loss in the cable over its entire length.

(3) **Coupling Loss**: This parameter is unique to radiating cable. Coupling loss is defined as a signal loss between the cable and a test receiver at a distance of 2 m. It is important to note that 2 m is the reference distance, although signal will be available beyond this point but energy will reduce per FSPL above and beyond coupling loss, as can be expected.

(4) **System Loss**: This is the sum of longitudinal loss and coupling loss for radiating cable. For normal cable it is predominantly longitudinal and connector losses combined.

(5) **Reception Probability**: This parameter is the same for all cables. But the difference is it must be considered only around the conductor for a radiating cable.

In conclusion, the radiating cable not only acts as a cable but also as an antenna. All its characteristics listed above are applicable to a traditional coaxial cable as well.

17.1.4 The Matching Network

Transmission lines reviewed so far carry signals from the source (transmitter) to the load which can be an antenna in many communication systems. The five transmission lines parameters described above refer to all of them and how they influence the signal reaching the antenna. Matching the transmission line to the antenna makes a big difference as indicated earlier in Fig. 17.2 and discussion there. The upper part of that figure showed the source on the left, followed by "matching network" in the middle and antenna on the right.

Let us now consider the matching network, its function, and how it is implemented. The primary function of a matching network is to counter against the reactive part of the impedance and bring it as close as possible to "characteristic impedance", or the ideal resistive part which is 50 Ω. In short, "matching network must correct the mismatch". Why do such mismatches occur and does it need additional circuitry? There are cases where impedance matching circuitry is needed.

Let us consider an RF transmitter that includes a power amplifier (PA) and it must connect an antenna. The PA manufacturer may design the PA for 50 Ω output impedance, but the impedance of the antenna it connects to changes according to its physical characteristics as well as the characteristics of the surrounding materials [10].

Antenna impedance is not constant relative to the signal frequency. A manufacturer may design an antenna for 50 Ω impedance at one specific frequency, but it may result in a nontrivial mismatch when the antenna is used at a different frequency. Let us consider a plot shown in Fig. 17.7, where a ceramic surface-mount antenna is intended for use in the 2.4–2.5 GHz band. Its return loss is shown on the Y-axis. It is a ratio of reflected power to incident power. The antenna measurement indicates significant change as the signal frequency moves away from the center point of 2.45 GHz. At

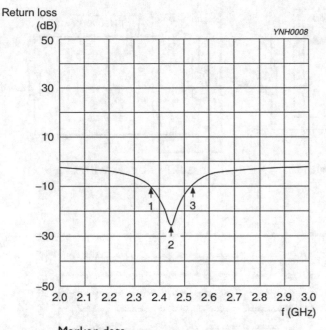

Marker data
1.   2.37GHz, -10dB
2.   2.45GHz, -24.9dB
3.   2.54GHz, -10dB

Fig. 17.7 Return loss of antenna in the 2.4 GHz ISM band

the center point (points 1 and 3) the reflection coefficient is excellent. This means at the center frequency of the band (2.400–2.483) antenna reflects less than 0.01 of the signal. But outside the band, at 2.37 GHz (80 MHz below) and 2.54 GHz (90 MHz above) the reflection coefficient deteriorates to −10 dB, indicating 10% of the signal gets reflected. The objective of this curve is to show how a matching circuit is useful to match the antenna reactance over a narrow band of frequencies.

If the RF circuit contains components that do not have matched impedances, there are two options for the network designer:

- Modify one of the components
- Add circuitry that corrects the mismatch.

In recent decades the first option is no longer practical. It is difficult to adjust impedance by physically modifying an integrated circuit (PA) or a manufactured coaxial cable. But the second option is possible to implement. Such an additional circuit is known as a "matching network or an impedance transformer". Both names are useful in understanding the fundamental concept: It means a matching network enables proper impedance matching by transforming the impedance relationship between source and load.

The design of matching networks is not always simple, and it's not something that is discussed in textbooks. But we can consider some basic principles and take a look at a fairly straightforward example. Here are some salient points to bear in mind:

A matching network is connected between a source and a load, and its circuitry is designed such that it transfers almost all power to the load while presenting an input impedance that is equal to the complex conjugate of the source output impedance. Alternatively, a matching network is used for transforming the output impedance of the source such that it is equal to the complex conjugate of the load impedance, which is another useful method to implement.

In reality, circuits with source impedance may have no imaginary part, and there is no need to examine its complex conjugate. The load impedance must be made equal to the source impedance. The complex conjugate becomes irrelevant when the impedance of both source and the antenna is purely real (not common). Typical matching networks (referred to as "lossless" networks) use only reactive components which means components that store energy rather than dissipate energy. This characteristic is helpful for the purpose of a matching network. Its function is to enable maximum power transfer from source to load. If the matching network contains components that dissipate energy, it consumed some of the power that could have been delivered to the load. Thus, matching networks use capacitors and inductors, but will not resistors since they dissipate energy.

A wideband matching network is a design compromise since the matching network comprises reactive components. The impedance of inductors increases with frequency while for capacitors it decreases with an increase in operating frequency. Hence, changing the frequency of the signals passing through the matching network makes it tricky to balance them.

Figure 17.8 shows a combination of passive circuits using L and C—the first of which shown in (a) is in the shape of a simple L network. This network performs well over narrow bandwidth of 10–50 MHz but does not have extensive matching capabilities. The second network shown in (b) is a π network that uses capacitors C1 and C2 in the vertical arms. It may be viewed as the equivalent of back-to-back L network. This will work well at mid-range bandwidth since both C1 and C2 can be seen as by-pass capacitors at higher frequencies. By cascading multiple sections of the L, it is possible to obtain wide band matching, but the Q factor (Quality factor) will decrease. That is a compromise the designer must make. The third option is to use inductors in the straight arm, as shown in (c) with capacitors C1 and C2 in vertical. It forms a T network for matching reactance. The concept is to select them such that they are in resonance with the load impedance.

Figure 17.9 shows two methods of canceling the load reactance. They are indicated on the left in Fig. 17.9a. Their

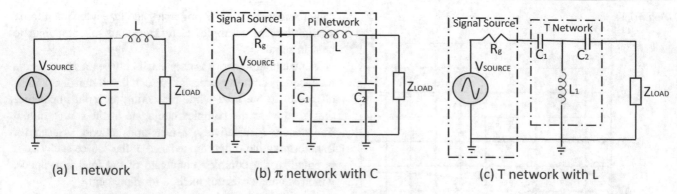

(a) L network   (b) π network with C   (c) T network with L

Fig. 17.8 Matching networks with L and Pi network arrangements

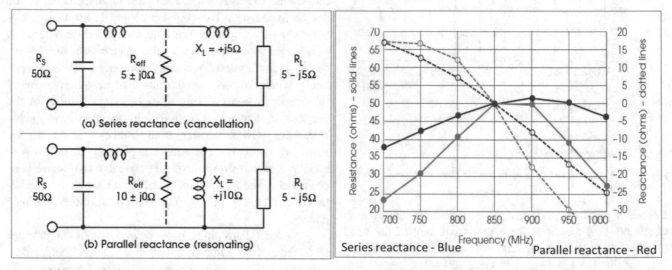

(a) Two methods to cancel load reactance   (b) Impedance at RS port for the two matching networks shown in (a)

Fig. 17.9 Two methods to cancel load reactance—their results

response is shown on the right in Fig. 17.9b with series reactance measurements in blue curve, and parallel reactance circuit response shown in red curve [10]. In the plot on the right, dotted lines indicate "before matching", and solid lines indicate "after matching".

Considering the matching circuits in (a), note that the effective load R_{eff} is 5 ohms for the series canceling method (shown by blue lines), but R_{eff} is 10 ohms for the resonating inductor method of (b) shown by red lines. This reduces the magnitude of the due to higher impedance offered transformation by the L-network, but it results in a matching bandwidth. This series circuit has significant variation in impedance away from the center frequency, but it will be very effective from 850 to 900 MHz, in matching the load impedance of 50 Ω.

Although we will not review details of many matching network designs, it is important to note that in each case capacitors or inductors can use variable elements (variable capacitor or inductor), allowing better matching in the field,

and over a select frequency band of interest. Such adjustments in the field are not commonly done, but certain networks that need the highest performance from the antenna may prefer to provide that option.

Before we consider a discussion on antenna, it is important to relate it to the matching network and how the waves radiating in/out of the antennas get affected and countered by matching.

17.1.5 Antenna: How It Connects to the Transmission Line

In Chap. 2, Sect. 2.2 the concept of how the antenna radiates was considered. In Chap. 13 antennas for satellite communication and VSAT terminals were reviewed (Sect. 13.3). Chap. 14 elaborated on parabolic dishes used for microwave communication, in Sects. 14.2 and 14.6. Since these are used in different applications, is it necessary that each of them needs a matching network? The answer is yes, but sometimes it may be incorporated as part of the RF front-end.

What are the key parameters of an antenna? The key parameters of antennas are (a) gain, (b) beam width, (c) polarization, (d) bandwidth, (e) radiation pattern, (f) efficiency, and (g) impedance. Let us consider each of them to understand their importance.

(a) **Gain**: Since the antenna is a passive element, its gain cannot be measured using the conventional definition, which is Gain = output/input. There are three aspects: gain, realized gain, and directivity. In electro-magnetics, an antenna's power gain or simply "Gain" is a key performance number which combines the antenna's directivity and electrical efficiency. In a transmitting antenna, the gain describes how well the antenna converts input power into radio waves radiate in all directions. This must be clarified using the definition for effective radiated power (ERP).

Effective radiated power (ERP), which is synonymous with equivalent radiated power, is an IEEE standardized definition of directional radio frequency (RF) power, such as that emitted by a radio transmitter. An alternate parameter that measures the same thing is effective isotropic radiated power (EIRP). EIRP is the hypothetical power that would have to be radiated by an isotropic antenna to give the same ("equivalent") signal strength as the actual source antenna in the direction of the antenna's strongest beam. The difference between EIRP and ERP is that ERP compares the actual antenna to a half-wave dipole antenna, while EIRP compares it to a theoretical isotropic antenna. Since a half-wave dipole antenna has a gain of 1.64 (or 2.15 dB) compared to an isotropic radiator, if ERP and EIRP are expressed in watts, their relation is:

$$EIRP\ (W) = 1.64\ ERP\ (W) \qquad (17.15)$$

If they are expressed in decibels then

$$EIRP\ (dB) = ERP\ (dB) + 2.15 \qquad (17.16)$$

In Fig. 17.10, Gain of "directional antenna dish" shown within the isotropic sphere (very small parabolic dish located at the center of Isotropic sphere in grey) is known as EIRP—equivalent isotropic radiated power and its power is shown on the right in Fig. 17.10. Gain using EIRP is expressed in dBi or dB isotropic as indicated, which is the additional part starting from the edge of the isotropic sphere (in grey) to the 3 dB beam width point. The directional antenna uses the property of "directivity"—means energy is propagated in a specific direction (the squeezed yellowish green sphere that extends towards the right side of the gray

sphere), directing the same energy into a narrow beam. Comparison of energy in a particular direction, taking an ideal isotropic sphere as reference (equal radiation in all directions) defines directivity. The concept of elevation and azimuth of an antenna is three-dimensional in nature. Hence, the bottom of Fig. 17.10 is a 3-axis diagram which represents the antenna in the middle with other angles defined on the surface of the Earth (grey sheet). Azimuth and elevation are also measures used to identify the position of a satellite flying overhead. Azimuth indicates what direction to face and elevation indicates how high up in the sky to look. Both are measured in degrees. Azimuth varies from 0 to 360°. It starts with North at 0°. When the antenna is turned on its own axis to the right (in a clockwise direction) it initially faces East (which is 90°), then South (which is 180°), then West (which is 270°), and finally return to North (which is 360° and also 0°). If the azimuth of the satellite is, say, 45°, which means the satellite is northeast. Elevation is also measured in degrees. A satellite just barely rising over your horizon would be at 0° elevation, whereas a satellite (shown by yellow circle) directly overhead would be at 90° elevation (a.k.a., "the zenith").

Earlier, in Chap. 14, it was indicated that satellite antennas will have an elevation of 6° or higher, whereas terrestrial (including microwave) antennas will have an elevation angle of 5° or less. This is to minimize interference between satellite and terrestrial antennas. In the diagram shown in the lower part of Fig. 17.10, the yellow circle represents the satellite. It has an azimuth of about 200° (southwest of the observer) and an elevation of about 60° (about two-thirds of the way up in the sky). Although it illustrates a satellite antenna, the angles and conventions shown are equally applicable to all terrestrial antennas as well. Now let us return to our discussion on antenna gain and how practical antennas are used in wireless systems. In Fig. 17.10 the Gain in dBi is shown from the edge of the grey sphere, till the 3 dB point. This is the extra distance "gained" due to directing the beam in a specific direction.

To obtain higher gain using "directivity": By squeezing RF energy as a beam in a particular direction, "higher power is observed within that narrow beam". Using such a directional beam, the radiated power (EIRP) within that beam can be considerable. An isotropic sphere may have only 1 W but a directional beam can carry 10 times (Gain 10 dBi) of that power. Using very narrow beam (typically less than 10°) beams can even carry 10,000 times (Gain 40 dBi) the original input signal. Such intense power allows the beam to propagate farther as needed in satellite systems for example. Microwave links carry signals 40 km or even 60 km after path loss. But there are many antennas with a gain close to isotropic sphere and these are known as omnidirectional antennas or dipoles. Their radiation pattern will not be an ideal sphere, but somewhat like that of an apple or

Fig. 17.10 Gain of antenna defined in comparison to isotropic antenna

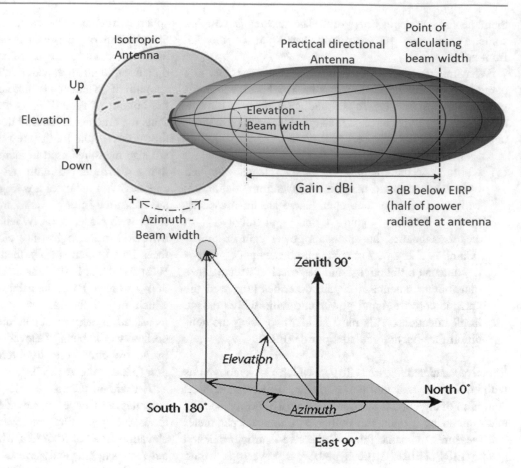

compressed doughnut. A generic formula for practical antenna gain is

$$\text{Gain} = \frac{4\pi A_e}{\lambda^2} = \frac{4\pi f^2 A_e}{C^2} \qquad (17.17)$$

where G is antenna gain, A_e is effective area, F is carrier frequency, C is the speed of light ($3 \times f10^8$ m/s), and λ is carrier wavelength.

(b) **Beam Width**: How is the beam width specified? In Fig. 17.10, when the beam propagates through its path, its power reduces due to path loss. When this power reduces to half the initial value, that point is taken as a reference to indicate beam width. This is indicated in Fig. 17.10. In theory, it is also known as the 3 dB beamwidth.

For example, if EIRP was 40 dBm (10 W) at the starting point of the beam (center of isotropic sphere), the point where its power reduces by 3 dB which is 37 dBm (5 W), that point is taken as a reference to calculate beam width as shown. In Fig. 17.10, beam width refers to elevation (vertical or up/down at antenna). The azimuth beam width would

be in the horizontal direction (width or measurement referencing true north as 0° movement of either West or East at the antenna). When the observer should be directly on top of the antenna and observe the beam on its own axis (shown in the bottom part of Fig. 17.10). Both of these beam widths are important—elevation angle indicated by the figure on the top and the bottom figure illustrating how the azimuth is measured. Azimuth is always measured in degrees with true North as reference (zero).

Both of them together allow designers to shape the beam and design antennas accordingly. In modern antennas, beam shaping is done by signal processing which will be discussed in later sections of this chapter. The generic term used for such elevation adjustment is "electrical tilt". Many antenna vendors also provide the "mechanical tilt" which can be done using an antenna mount that has bolts and notches for adjustment. Although the beam shown here is symmetrical (to the reference line), it need not be that way. For example, if an antenna is located on top of a hill, the beam may point downward. For satellite earth stations, the antenna will point upward. In each case, antenna elevation angle with beam width, as well as azimuth beam width, is specified at each location. This provides a clear bearing on where the radiation is propagating.

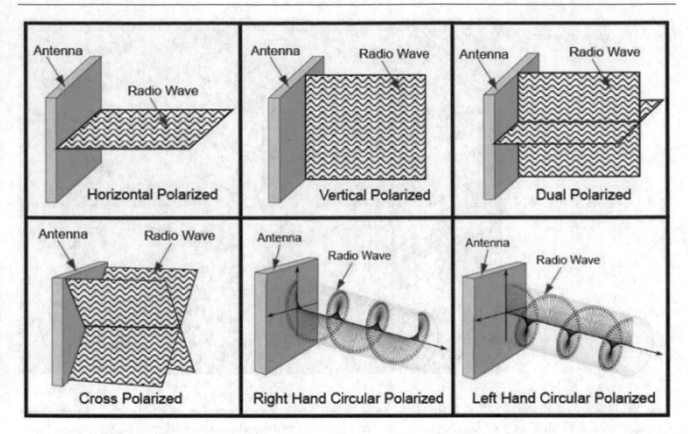

Fig. 17.11 Different types of polarization of antenna

(c) **Polatization**: Polarization of antenna was explained in Chap. 14, in Sect. 14.2.2 with a detailed description of dual-polarization in Sect. 14.2.3. Polarization refers to one dominant field, which is assumed to be an electric field. If the electrical field is oriented vertically, then it is termed "vertical polarization" which is widely used, including all cellular systems. Six types of polarization methods are indicated in Fig. 17.11, and other combinations are possible [12].

Antenna polarization is similar to the polarization of light. It concerns the transmission and reception of electro-magnetic radiation based on its orientation. With light polarization, a film or glass will appear darker, since it blocks orthogonally polarized light, but allows correctly polarized light to pass through. Similarly, in an antenna, polarization determines how EM radiation is transmitted or received by that antenna. Typically, RF antennas are either linearly or circularly polarized. Linearly polarized antenna is either vertically polarized or horizontally polarized, while circularly polarized antenna is either left-hand or right-hand circularly polarized. Cross-polarization was briefly discussed in Chap. 14, Sect. 14.2.3 showing XPD, the cross-polarization discriminator. Polarization of the waves are shown in

Fig. 17.11 for six different cases. Cross-polarization happens when the waves are at 45° angle rather than 90° (orthogonal). Sometimes this is helpful when the designer prefers to get at least part of the signal to the receiver. This can help improve coverage in urban environments. By slanting one of the antenna elements 45° to the left and the other 45° to the right, improved equality in received signal levels can be achieved.

(d) **Bandwidth**: The bandwidth of an antenna refers to the range of frequencies over which the antenna can operate. It is connected to the reflection coefficient (in scattered parameters, usually S11 below 10 dB antenna is considered good because 90% of energy being used). Scattered parameters or S-parameters are discussed in detail in in Chap. 19. There is a separate discussion that relates to percentage bandwidth, which is defined as

$$\text{Bandwidth (in \%)} = 100 \times \frac{F_H - F_L}{F_C} \qquad (17.18)$$

where F_H is the highest frequency it operates at, F_L is the lowest frequency it operates, and F_C is the carrier frequency.

Fig. 17.12 Measurement of
antenna prototype in an anechoic
chamber

If we consider typical example of 850 MHz as the carrier
frequency with 860 MHz as the highest frequency and
840 MHz as the lowest frequency, then BW in percent is
calculated as $100 \times \frac{860-840}{850} = 23.5\%$. The value of this
approach is that percentage directly relates to the carrier
frequency. At higher frequencies the BW or window
(20 MHz in this example) increases considerably. If we need
the same 20 MHz in the PCS band at 1900 MHz (100 ×
) = 1.05%. At higher carrier frequency the same BW in
percentage seems smaller. This is a point of discussion to
progress towards higher bands. It will be discussed towards
the end of this chapter.

To provide examples in lower frequency bands, consider
the lower frequency bands of AM and FM used for broadcast. The amplitude modulated (AM radio) carrier frequencies are in the frequency range 535–1605 kHz. Carrier
frequencies of 540–1600 kHz are assigned at 10 kHz intervals. At 545 kHz, the wavelength is about 500 m, where
even half wavelength is 250 m. To build sizes that can
handle kW of power, entire towers themselves make up the
antenna. These are known as live towers (tall towers painted
red and white as an indication) and are generally located in
unpopulated areas.

Even then about 4 towers of 65 m must be connected in
series to make up just one antenna and that too at half
wavelength! Whether to use ½ or ¼ or other wavelengths
involves factors such as gain, pattern, coverage, etc., and
will not be discussed here.

(e) **Radiation pattern**: One of the important characteristics
of any antenna is its radiation pattern. It provides a good
preview of how RF power from the antenna radiates in
three-dimensional (3D) space, as shown in the measurement setup illustrated in Fig. 17.12.

To characterize antenna radiation pattern requires elaborate arrangement, which has been somewhat simplified in
recent years. For effective antenna radiation pattern measurements, instruments such as vector network analyzer
(VNA) are used in an anechoic chamber with two antenna
masts, one for transmit and another for receive, each capable
of rotating 360° [13]. In an anechoic chamber, all walls are
covered by cones of foam that are treated such that no radio
wave reflection bounces back from any wall (no echo or
anechoic). It not only allows measurements at various distances in the horizontal and vertical planes but also allows
rotation in the third axis (roll or azimuth). Since it is not easy
to describe details, readers are strongly encouraged to view
the video of how measurements are made in an anechoic
chamber, see [14].

To elaborate, anechoic chambers support measurement of
all antenna characteristics such as gain, beam width, polarization, bandwidth, radiation pattern, efficiency, and
impedance.

The radiation pattern of a typical dipole antenna is shown
in Fig. 17.13, where the elevation pattern and azimuth pattern are shown at the top of the figure [15]. Such 2D patterns
are provided in any antenna data sheet. Elevation angle

Fig. 17.13 Radiation pattern of a typical dipole antenna

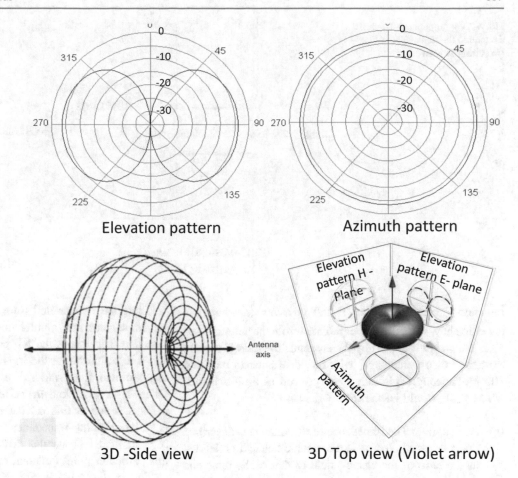

patterns are shown with the center of the antenna as a reference. For azimuth, true North is always taken as a reference. To visualize the antenna 3-D pattern, observe the lower diagrams. Elevation patterns are measured in an anechoic chamber (shown in Fig. 17.12) with viewing angles along the green arrow and red arrow. Both are projected on the sides opposite to the arrow as cut sections on the H plane and E plane. In this case, they happen to be identical since a dipole pattern is symmetrical in both planes. In most other cases, they are different, as measurements in the x, y, and z axis are made using set up shown in Fig. 17.12.

The azimuth pattern (on the right) is viewed from the violet arrow as the axis (from the top). Observe that the pattern is a circle (looking from the top). Antenna manufacturers normally provide 2D patterns as part of their specification sheets for any antenna. There is a tendency to regard the 2D polar plots as a distance-covered plot, which it isn't. What the plot indicates is that moving from the antenna located at the center, the relative readings of amplitude or power in the radiated field will fit these patterns. The power coming from the antenna can vary, which is why only relative numbers in dB are indicated.

In an isotropic antenna, the pattern would be a circle in both horizontal and vertical planes. Since Fig. 17.13 shows a dipole antenna, the pattern is doughnut-shaped and the 3D images below show the doughnut pattern, viewed from the side and from the top. On the bottom right the pattern in "green" is the elevation pattern in the H plane, the plot in "red" is elevation pattern in the E-plane, which is orthogonal to the "green". In this case they are symmetrical resulting in the doughnut shaped pattern. Looking from the top "violet" arrow indicates Azimuth, which is a circle in this case (shown as azimuth circular pattern). It is very important for any antenna designer to get this 3D view of what the radiation pattern looks like. The objective of Figs. 17.12 and 17.13 is to impress upon readers on how the 2D patterns are generated by the antenna manufacturer and presented to users. Designer should consider them mentally as a 3D model of the radiation pattern while choosing radio coverage of an area.

(f) **Efficiency**: Most antennas support transmitters (or radiators) and receivers (or acceptors)—this means the same antenna can "send" and "receive" information, although these may be at slightly different frequencies.

Fig. 17.14 Impedance by the antenna to the transmission line—two-terminal network

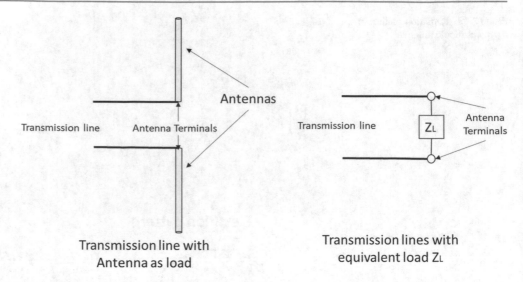

In general, antenna efficiency is defined as $\eta = \frac{P_r}{P_a}$, where η is the efficiency, P_r is the power radiated from the antenna, and P_a is the power accepted by the antenna. The ratio is usually expressed as a percentage, with the ideal antenna reaching 100% efficiency. This may be expressed as a ratio in dB, where 0 dB would be the ideal antenna.

(g) **Impedance:** The impedance of the antenna is described using terms of electric and magnetic fields; it is defined as the ratio of the electric field to that of the magnetic field at specific points.

There are various methods used to calculate the impedance of an antenna. They are:

1. Boundary value method
2. Poynting vector method
3. Transmission line method

Quite often, the boundary value method is used for impedance calculation. In this method, the impedance is obtained by applying the boundary conditions that represent the absence of the tangential electric field component at the conducting surface. This determines the ratio of applied emf to that of the current, i.e., impedance.

In the Poynting vector method, the Poynting vector, i.e., the power density is integrated over a closed surface. Generally, a sphere of extremely large radius R is considered, where R must be greater than or equal to $2L^2/\lambda$. Here L represents the largest dimension of the antenna which is considered.

In the transmission line method, the antenna is assumed to be a transmission line and this method is most convenient when applied to biconical antennas (broadband dipole antennas).

There are two related impedances for the antenna—the self-impedance and mutual impedance [16]. The transmission line that carries the RF power has its own impedance and it is connected at the input terminal of the antenna—at this terminal it is said to be "antenna input impedance". Often it is also known as feed point impedance as the antenna is fed at this particular point. Another name used is "driving point impedance" since the transmission line carrying the RF power that drives the antenna is connected at this particular point (antenna terminals).

This is illustrated in Fig. 17.14 with a physical connection shown on the left and represented as a two-terminal network on the right. Earlier, in the matching network, care was taken to minimize and eliminate the resistive part since it dissipates power. Since no heat loss is associated with it and it is placed away from the ground and other objects, then the terminal impedance (Z_L) will be equal to self-impedance (Z_{11}) of the antenna. In such a case, the self-impedance will be given as

$$Z_{11} = R_{11} + jX_{11}$$

where R_{11} denotes the radiation resistance; X_{11} represents the self-reactance.

For a thin linear half-wave center-fed antenna, the self-impedance of the antenna is given by $Z_{11} = 73 + j42.45\Omega$.

Note that self-impedance is a complex quantity where radiation resistance or self-resistance is the real part while self-reactance is the imaginary part. Self-impedance is always positive and it is important to remember that "self-impedance is the input impedance of the antenna in the absence of all other elements".

The transmission line and the antenna—each of them have their impedance. These two can be thought of as a transformer where primary is the transmission line and secondary is the antenna. Therefore, the mutual impedance between them must be reviewed. Similar to the coupled circuits, multiple antennas can be coupled to form an antenna array. But in an antenna array, the input impedance not only depends on the self-impedance of the single antenna but also the mutual impedance of other antennas as well. Therefore, it is important to define mutual impedance, which is given by

$$Z_{21} = -\frac{V_{21}}{I_1}$$

V_{21} represents the voltage in the secondary (antenna), due to current I_1 flowing in primary (transmission lines). According to the reciprocity theorem two mutual impedances are equal.

$$Z_{12} = -\frac{V_{12}}{I_2} = Z_{21} = -\frac{V_{21}}{I_1} = Z_m$$

$$Z_m = \frac{V_{12}}{I_2} = \frac{V_{21}}{I_1}$$

Therefore, by knowing V_{12} and I_2 or V_{21} and I_1, Z_m can be calculated.

To get maximum power transfer from a transmitter to an antenna, the RF amplifier output impedance and the antenna impedance must be known so that a matching network can be designed. For low-power transmitters with radiated powers of tens of microwatts, close matching is not critical. For a radiated power of 10 milliwatts and battery-operated devices, proper matching can save battery energy. By increasing efficiency, transmitter design can be simplified. In all cases, the matching network certainly helps to transfer maximum power to/from an antenna.

17.2 Types of Antenna: Impact of MMIC (Reflector Antennas)

Types of antenna can be described along with examples and applications since there are too many of them to elaborate on. Some of them were discussed in other chapters such as 11, 14, and 15. A brief overview of the antenna types is indicated here with some examples. Table 17.3 illustrates different types of antennas and their applications [11].

The intent here is to illustrate a few types of antenna with typical characteristics showing the basis of their design and

construction [17]. Only two types of antennas are considered at random to show the theoretical and engineering aspects. The wire antennas and the traveling wave antenna are used to illustrate design and theory aspects of how they are built. The topic of MMIC (monolithic microwave integrated circuit) will be covered towards the end of this section.

17.2.1 Wire and Aperture Antennas

The simplest of the antennas are constructed out of metallic wires, strips, or plates. They can be hung and connected through different means but offer valuable experience in understanding. They are not only straightforward to use but often form the core of amateur radio users and are widely deployed in all continents. Fig. 17.15 shows a typical wire from the house using a pole, connected to a pole, but carefully insulated—this type of antenna is often used by amateur radio users. Observe that the actual part of wire used as an antenna, is carefully isolated from the house as well as the outside pole. Effective length of this wire antenna is between the two insulators shown.

17.2.2 Long Wire Antenna

Long wire antenna is literally a long piece of wire that is well over one wavelength long. Particularly used by amateur radio hams for HF communication, they take reference of the lowest frequency they expect to use and keep the long wire at least one wavelength long. The general assumption is all higher frequencies can be supported since lengths become shorter when the operating frequency is increased [18]. Figure 17.15 can also be used as a long wire antenna, by choosing proper length between the two insulators.

The construction of such a wire antenna is straightforward. An appropriate length of wire is first secured to a post, usually in a house, and cable is extended up to a cable insulation block.

From there antenna wire is run up to a pole that has a pulley secured at the top. Even at the other end, an insulation block is used. The secure short cable at the pole is run over a pulley and then tightened. An extra piece of cable is left hanging. Shorter wires include dipoles of various configurations and lengths based on the wavelength of operation λ. Their length could be one-fourth of the wavelength or half the wavelength. In some cases, it could be shorter than one-fourth wavelength. Further enhancement can be made using "ground plane", which is a metal plate or flat metallic sheet. Next let us consider another popular type of wire antenna—the dipole, which can take different physical forms as shown in Fig. 17.16.

Fig. 17.15 Typical end fed wire or long wire antenna

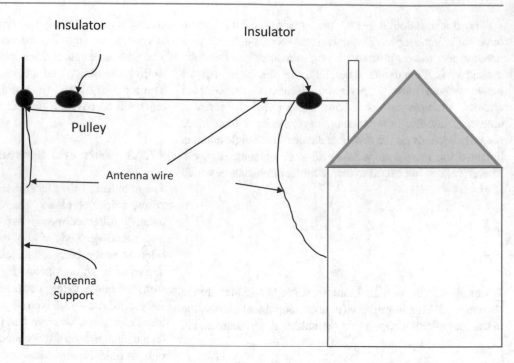

Table 17.3 Types of antenna and their application areas

Type of antenna	Examples	Applications
Wire antennas	Dipole, monopole, helix, loop antenna	Personal applications, buildings, ships, automobiles, spacecrafts
Log periodic antennas	Bow tie, log-periodic, log-periodic dipole array	Narrow beam high gain in tunnels, congested areas that need highly directed beam patterns
Aperture antennas	Waveguide (opening) horn antenna, slot antenna	Flush mounted applications, aircraft, spacecrafts
Reflector antennas	Parabolic reflectors, corner reflectors, flat plate reflector	Used in very high-frequency applications (microwave, millimeter-wave)
Lens antennas	Convex plane, concave plane, convex-convex and concave-concave lenses	Used in very high-frequency applications (microwave, millimeter-wave)
Microstrip antennas	Circularly shaped, rectangular, metallic patch above the ground plane in small packages, quarter-wave patch	Aircraft, Spacecraft, Satellites, Missiles, Cars, Mobile phones, and Consumer devices
Traveling wave antennas	Long wire, helical wire, spiral	Broadband applications, Radio, GPS, Satellite communication, Mariner and other Spacecraft, Aircraft, missiles. RFID devices implanted in the human body
Array antennas (set of multiple connected antennas)	Yagi-Uda, microstrip patch array, aperture array, slotted waveguide array	Used in very high-frequency applications, in particular where control pattern must be controlled. They are used in all frequency bands. (like huge military radars, etc.)

Fig. 17.16 Dipole antenna connected to the ground plane

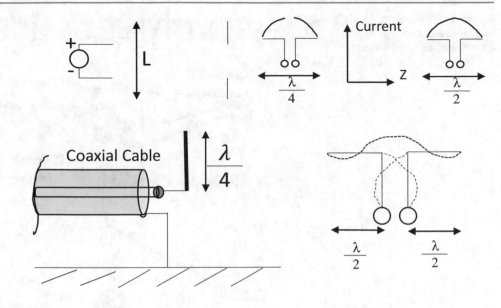

(a) Earth ground or conducting plane   (b) One Wavelength Dipole

Figure 17.16 shows in (a) how a typical quarter-wave dipole is connected to the ground plane as a conducting plane.

Here the ground structure electrically serves as the other half of the $\lambda/4$ antenna making the combined unit equivalent to $\lambda/2$ dipole. If the ground plane is adequately sized and conductive, the performance of the ground plane becomes a vertically mounted $\lambda/2$ dipole. This is useful in places where there is insufficient space to accommodate half-wavelength dipole. It is also useful in outdoor applications where the ground plane can serve as a base already exists. For example, an antenna mounted on a vehicle (car, ship, or aircraft) —the antenna uses the vehicle's body surface as a ground plane [19].

Since dipoles of various lengths were discussed, inquisitive thinkers would wonder why one full wavelength dipole is not used? Indeed Fig. 17.16b on the right shows the answer, which is interesting—the voltage pattern that each half wavelength induces, is shown by dotted lines. In the case of a full wavelength dipole, its positive charges and negative charges exist at the same time and phase, therefore cancel out each other as shown in the figure on the right. The induced charges make no further attempt to radiate since they get canceled. The output radiation will be zero for a full-wave transmission dipole [11].

Waveguide antenna is an elegant design of aperture antennas. The basis for antenna aperture was discussed in Sects. 2.2 and 13.3 (VSAT). The transition of a waveguide into an aperture was indicated in Fig. 2.2. Dimensions of the waveguide are key to horn antenna and how they relate to

different modes of propagation were discussed in Chap. 14 and Sects. 14.2.2 and 14.2.3. The use of waveguide as a transmission line and its advantages were reviewed and it was indicated that waveguide can operate up to 110 GHz with lower loss compared to coaxial cable.

Also, the waveguide has the advantage of operating in either electric and magnetic modes (TE and TM modes). This distinction and how the two modes when combined helps to increase throughput in microwave links were discussed in Chap. 14, Sects. 14.2.2 and 14.2.3.

In terms of the horn antenna configuration, there are four types, and Fig. 17.17 shows the four configurations [11]. All of these are used depending on the system design. The width of the waveguide will be typically half wavelength, therefore its size decreases as frequency increases. The sectoral horn, is a typical example of small aperture antenna to send out or receive either horizontally polarized (H plane) waves, or vertically polarized waves. The pyramidal horn, although similar to sectoral horn, has an expanding waveguide pattern. Depending to the length of the pyramid, its opening can be large and may be used for low frequency operation. The conical horn supports a circularly polarized wave, often used in satellite communication, for example. Although these horn antennas can be used for transmission or receiving, the term "aperture" is defined assuming it as a receive antenna.

Relationship between effective area (or aperture area) and gain is given by

$$G = \frac{4\pi A_e}{\lambda^2} = \frac{4\pi f^2 A_e}{C^2} \qquad (17.19)$$

Fig. 17.17 Different
configurations of waveguide
aperture leading to the antenna

DIFFERENT TYPES OF HORN ANTENNA

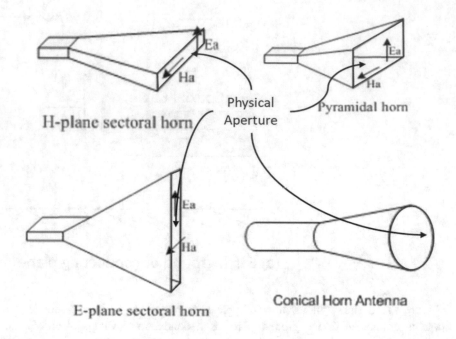

where
  G is antenna gain
  A_e is the effective area
  f is carrier frequency
  λ is the wavelength
  C is the speed of light (3×10^8 m/s).

The effective area or aperture area A_e is where most of the energy from the antenna is focused. Gain of the antenna also depends on the square of the frequency or inversely square of the wavelength.

Note that antenna aperture is a measure of what is presented to the antenna as a receiving element. The term "aperture" is associated with the receiving area of the antenna; this shows the electro-magnetic power capturing characteristics of receiving antennas. An aperture is classified into:

- Effective aperture
- Scattering aperture
- Loss aperture
- Collecting aperture
- Physical aperture

The effective aperture of an antenna was referenced in Eq. 17.19. It is the region of the receiving antenna which effectively collects the electro-magnetic energy from the radiated wave present in the overall region of the antenna. In other words, the larger the extracting region of the antenna, the more efficient it is. It is defined as the ratio of power

received by the antenna to the average power density of the incident wave.

$$Effective\ Aperture\ A_e = \frac{Power\ Received}{Average\ Power\ density}; \ or\ A_e$$
$$= \frac{W}{P}$$

$$(17.20)$$

Figure 17.17 points to the physical aperture as an example. The physical aperture is nothing but the cross-sectional area of the horn or in a parabolic reflector area of the reflector. For large antennas the cross-sectional area plays an important role in its efficiency. Such details are important for the antenna designer in terms of how it is constructed, measured, how much voltage is induced, etc. Although more details on the other antenna apertures listed are available in the literature, it would be beyond the scope of this book and will not be discussed here.

17.2.3 Traveling Wave Antennas

The only other type of antenna we will touch upon here is the traveling wave antenna. It offers radiation that is circularly polarized which is less prone to multipath effects (typically used in satellite systems). They offer higher antenna gain, and some of the most elegant are the "traveling wave" antenna types that mimic the wave being launched.

Fig. 17.18 Radiation patterns of the same size helical antenna in normal and axial modes

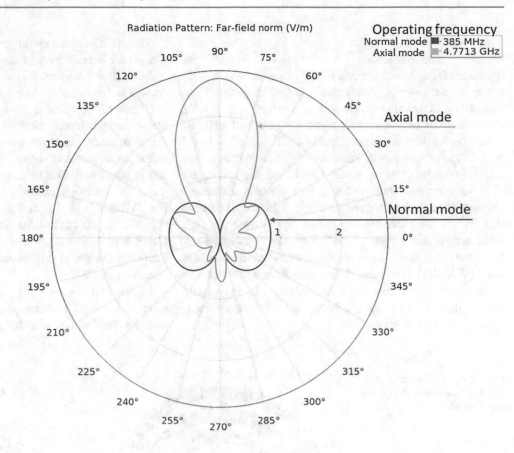

- Helical wire antenna
- Spiral antenna

Let us consider each one in order.

17.2.3.1   Helical Wire Antenna

Helical wire antennas are built with a base plate acting as a ground plane, with a central support rod and coil of wire that winds around it. Its smaller version would look like a spring coil [20]. Helical antenna operates in two modes:

(a) Normal mode—operating with linear polarization
(b) Axial mode—operating with circular polarization

How are the two distinguished? The basic difference is if the circumference of the helix is significantly less than the operating wavelength and the pitch of the helix is significantly less than the quarter wavelength. Therefore, helical wire antennas operating in the normal mode tend to be very small and can fit into small areas including chips that are embedded in the human body. In the normal mode the far-field radiation pattern is orthogonal to the direction of the helix and similar to that of the dipole antenna discussed earlier.

In the axial or end fire mode, the antenna radiates circularly polarized waves, in the direction of helix. In this

mode the circumference of the helix is comparable to the wavelength of operation. Therefore, only higher frequency bands tend to use the axial mode of operation. In the axial mode, the radiation pattern is highly directional since the helical wires (each pitch) act as an array, pushing radiation much farther in a specific direction.

Double helix—advantages: In the normal mode input impedance of the antenna tends to be lower. By adding a second helix and shorting it to the ground, the double helix exhibits impedance that is four times higher compared to a dipole of the same size. Therefore, double helix antenna allows much easier impedance matching to typical 50 Ω coaxial cable [20].

To provide a comparison of the radiation pattern in normal mode and axial mode for a helical antenna of the same size, Fig. 17.18 indicates the two patterns for helical antenna [20].

Figure 17.18 shows the sharp contrast between the "normal" mode and the "axial" mode of operation. Note that in the axial mode (or end fire mode), the operating frequency shifts by over 10 times that of the normal mode. Helical antennas are typically easier to build, and also they provide an opportunity for a good impedance match and operate well.

Spiral antenna

Spiral antennas belong to the class of frequency-independent antennas. They operate over a wide range of frequencies. Polarization, radiation pattern, and impedance of spiral antennas remain constant over a wide bandwidth. These antennas are circularly polarized with low gain. Spiral antennas are classified into different types based on the type of spiral used. Some examples are Archimedean spiral, logarithmic spiral, square spiral, and star spiral, etc. The Archimedean spiral antenna is a popular frequency-independent antenna. It is shown in Fig. 17.19a and observe that "r_0" is the inner starting radius (spiral does not start from centre, but offset by "r_0", and the outer radius is "r". Many wideband array designs with variable element sizes (WAVES) have used the Archimedean spiral antenna as the radiating element. The Archimedean spiral is typically backed by a lossy cavity to achieve wider frequency bandwidths of 9:1 or greater.

The Archimedean spiral antenna radiates from a region where the circumference of the spiral equals one wavelength. This is called the active region of the spiral. Since it has two arms it is important to balance them in terms of radiation. Each arm of the spiral is fed by signals that are 180° out of phase, so when the circumference of the spiral is one wavelength the currents are complementary at opposite points on each arm of the spiral. Therefore, the currents will add in phase in the far field, making it similar to a dipole in terms of radiation. Details of the feed are shown in Fig. 17.19b and its radiation pattern is indicated in Fig. 17.19c. To control the two feeds a BALUN (Balance Unbalance) circuit is added such that the feed carefully maintains the phase difference. Details of Balun design are described in [21].

Figure 17.19 shows the construction of Archimedean spiral, whose geometrical property is that the spiral has constant line width; also its arm radius is still "partially

Fig. 17.19 Construction of Archimedean spiral and typical radiation pattern

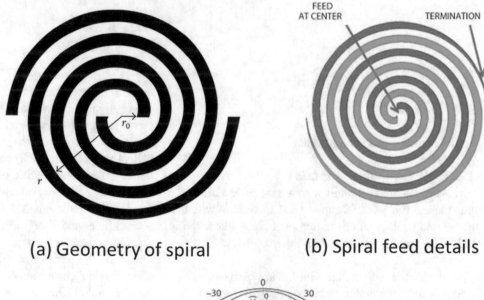

(a) Geometry of spiral

(b) Spiral feed details

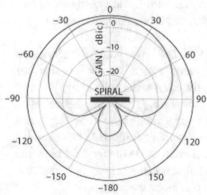

(c) Archimedean Spiral – typical radiation pattern

self-scalable" and "self-complementary". This means its radius is increasing linearly with angles, but the line width does not increase and remains constant. The Archimedean spiral is considered nearly frequency independent, and if the spiral arm line width is sufficiently small spiral antenna is built on a flat or planar surface; therefore is it possible to use this surface as a ground plane? Yes, this is possible but not desirable. One technique to convert a bi-directional to a unidirectional pattern is to use a cavity. A spiral antenna without a cavity housing will radiate in a bi-directional pattern on both sides of the spiral plane. A bi-directional pattern antenna is more sensitive to installation effects such as scattering from the mounting surface. Because of this limitation, unidirectional patterns are more desirable, which helps in radiating in front but not radiate behind the planar surface.

Although spiral antenna can be built in various sizes, there is a minimum diameter that must be maintained, to avoid short circuit between spiral elements [21]. The minimum size of the spiral antenna is established based on Eq. 17.21

$$Diameter = (1.2)\frac{C}{F_{Low}}\left(\frac{1}{\pi}\right) \quad (17.21)$$

where F_{Low} is the lowest operating frequency of operation, and C is the speed of light.

The lower part of Fig. 17.19 shows the spiral and its radiation pattern. Observe that most of the radiation is towards the front of the planar surface on which spiral is placed or imprinted. Spiral antenna is popular in applications that have space constraints and can use circularly

polarized radiation. This brings us to the next topic which moves towards "antennas within chipsets".

17.2.4 Impact of MMIC

With the advent of smaller devices where antenna is contained within chipsets, its formation and ability to interface with monolithic microwave integrated circuit (MMIC) became pivotal to support broadband communication system, particularly in the millimeter-wave band. A fully integrated system with MMIC and antenna is described in [22]. Such systems are typically used by security personnel at various locations such as airport security to screen passengers, etc.

The basic concept of how such a system works is shown in Fig. 17.20, where passengers or others to be screened by security personnel are shown as the target. The objective of the system is to create a 3D image of the person to verify that person does not have weapons or explosive material. To create a fine image the scanning speed must be very high such that resolution in terms of wavelength is very small (typically in millimeters). Typical numbers indicated are about 10 ms per pixel. From the transmit antenna (horn type), millimeter waves (shown in red) are reflected towards the person using a large reflector (primary reflector).

During the quick scan, multiple rays are reflected back (shown in green) to a sub-reflector, which then sends them to multi-pixel MMIC where it is processed and a 3D image gets generated. Although Fig. 17.20 shows these as separate

Fig. 17.20 MMIC-based security scanning system with multiple antennas

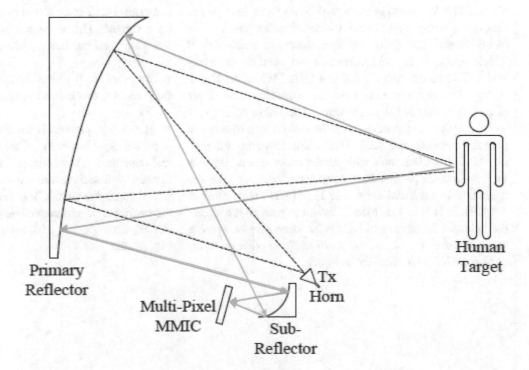

Fig. 17.21 Inside view of integrated quasi-optical HEB-MMIC receiver

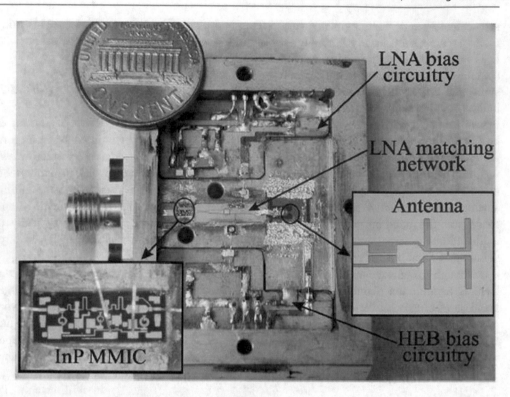

components, at millimeter-wave band these devices are so small that they must be integrated into the chip that consists of DHBT—double hetero-junction bipolar transistor. Although the image of this particular application is not shown, an earlier version of similar technology using MMIC and antenna is shown in Fig. 17.21.

Here HEMT (high electron mobility transistor) MMIC LNAs that provide outstanding noise performance are shown. The low noise amplifiers (LNAs) were discussed in Chap. 13, where Sect. 13.3.1 discussed noise power P_N = 4kTB (watts) indicating the importance of bandwidth B which is large (Eq. 13.5). Here is an example of bandwidth B being extremely large at 4 GHz [23]. In the chip it is important to observe the LNA and matching network are placed right next to the slot antenna. The authors claim very low dc-power consumption, with the additional advantage of reduced physical size [23]. HEB (Hot Electron Bolometric) MMIC is often built with semiconductor material such as InP—Indium phosphide, for better performance. The slot antenna (enlarged and shown in Fig. 17.21) was tested at 1.4 THz. It is clear that MMIC has an impact on the size of the antenna. Its placement within the same chip or housing keeps it extremely close (within 20 mm), in addition to the need for LNA and matching network.

17.2.5 Stripline and 3D Printed Antennas

With the advent of MMIC and antennas operating above 1 GHz, there is another manufacturing function that brought about change in how antennas are built. It is not easy to build the antenna mechanically and connect it to associated circuitry when dimensions become small. It is better to use 3D printer with conducting ink to layout the matching circuit component and build the antenna by directly printing them on a surface. This approach is known as "additive manufacturing" and has certain advantages [24]:

a It can reduce the time and material costs in a design cycle
b It enables on-demand printing of customized parts

The design process is somewhat different. Several existing tools and sites are available to help customize mechanical structures. This concept can be expanded into the RF domain with software that uses a high-level design parameter to create the circuit. The same tool can be used to model the performance and create computer-assisted manufacturing (CAM) files. Details of the CAD design are illustrated in Fig. 17.22a.

Fig. 17.22 3D printout of **(a)** stripline BALUN and **(b)** monopole array antenna

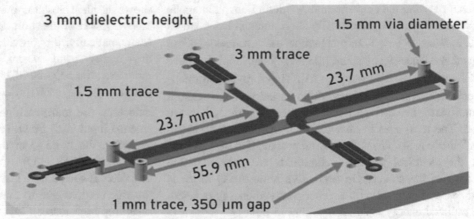

(a) CAD view of a strip line Marchand Balun with two strip line inner layers (green and red) and a top layer coplanar waveguide section (blue)

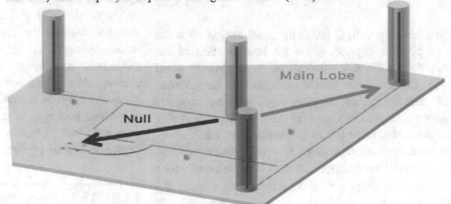

(b) CAD view of a four-monopole array with an integrated strip line beamforming network

By carefully using this process, circuit design can be updated or customized after the initial development. A computer-assisted design (CAD) tool can be used to further modify the structure to customize the mechanical interface. A machine tool pathing code (a slicer) could be used to translate the CAM files into a format the 3D printer can recognize and build.

Figure 17.22b shows the actual printed design of a monopole array with an integrated beam-forming network. Height of the dielectric base was 4 mm. The array was designed as an ultra-light directional antenna to be used as part of WiFi on a UAV (unmanned aerial vehicle) to operate camera traffic. Each monopole 31 mm tall is a quarter wavelength (WiFi band 2400–2483 MHz).

Such low-profile low-weight designs are possible using 3D-printed antennas. In this case the monopole array is used as part of the directional antenna system to extend the range of operation. A Radome is specifically used to illustrate the process that is used for each component. There are some practical limits. For example, Voxel8 Standard Silver ink, a room-temperature-curing silver conductive ink was used for these prints. Its DC conductivity is 3.45 MS/m (mega

siemens per meter), which is quite different from pure silver that has a conductivity of 61 MS/m.

When designing printed circuit boards, it is common to consider the trace and space tolerances. Accuracy to which designer can maintain the desired trace width and the gap between traces is now limited by the 3D printer. The 3D printer deposits ink with a 0.25 mm nozzle, and it should be expected that small traces might not print as designed. The vendor recommends 0.5-mm lines for general prints (two passes); however, RF circuits generally require more design flexibility [24]. With current technology, circuits in the millimeter-wave band can be built that could operate well into the bands considered for 5G. Some of the later technologies currently being ventured that are well into terahertz may find this as a limitation.

17.2.6   Electrically Small Antennas

Electrically small antennas are very short, much smaller than the wavelength they are expected to operate on. Despite this limitation they have become popular and are widely used.

Instead of using the conventional λ/2 definition, an alternate approach is used. An electrically short antenna has a length of 2 h, where $\frac{2\pi h}{\lambda} \ll 1$. In practical terms, antennas of $\frac{\lambda}{10}$ are considered "electrically small". This includes loops that have a diameter of $\frac{\lambda}{10}$, for example [25]. They provide an omni-directional radiation pattern with no variation in the azimuth plane. Also the conventional "propagating mode" is not used. The term used to describe their operation is "evanescent" mode. This means there is no propagation of the wave, but the electrical and magnetic fields are present in the vicinity of the source. On a positive note, the Q factor is high. Since Q is a ratio of the energy stored/energy radiated, the radiated energy is small resulting in high Q.

$$Q \propto \frac{1}{k^3 r^3} \quad k = \frac{2\pi}{\lambda} \tag{17.22}$$

In practical terms, the Q factor for small antennas is in the range of 50–200, depending on the configuration of the design. The maximum bandwidth of an electrically small antenna is limited by the sphere of radius r of the antenna.

With such limitations, why are small antennas popular? The answer emerges from some of the VHF implemented at lower frequencies that have long wavelengths. Such wavelengths cannot be accommodated in small units. A wireless remote controller uses either 315 or 433 MHz band where the wavelength is about 1 m. Even a monopole antenna of $\frac{\lambda}{4}$ would be 17 cm long, making it a loop of that diameter will be large as well. Similarly, many RFID devices operate in bands below 1 GHz where the wavelength is of similar order. Miniature GPS antennas have become possible due to this principle. It, therefore, makes sense to use electrically small antennas.

There are some challenges to building and using small antennas. They are:

- Impedance matching
- Insertion loss is high and may result in heating
- Small radiation aperture with low radiation efficiency

Although they seem formidable, there are well-planned methods to address them. Matching network: Due to higher loss, it is important to use high X_L or X_C in the matching network. In a traditional matching network X_L is used as the primary reactance, whereas in this case, inductor of significant value will be needed. Figure 17.23 shows a typical example of how to design a matching network at 100 MHz, considering a practical small antenna dipole with Q = 100 [25].

Since capacitors generally have higher Q compared to L, it is preferred that X_C can be used as an important element for reactance matching. Let us consider Fig. 17.23a which shows an ideal, lossless matching network to transform the 1.96–j1758 Ω of the short dipole to 50 Ω system impedance. Mathematically, this provides a proper matching but results in a narrow band system. Figure 17.23b shows practical inductors with a Q of 100, each inductor will have a resistive loss of $\frac{X_L}{Q}$, or 879/100 = 8.79 Ω. Since there are two inductors, the total additional resistance in series with the antenna input is 17.58 Ω. Ignoring the smaller loss from the capacitor, the finite Q of the inductors results in a loss of 20log[1.96/(17.58 + 1.96)] = 21 dB. One approach would be to accept such a high loss and some designers incorporate the antenna loss due to matching, into their path equation, accepting high losses [25].

This is not necessary, as a matching network (like the one shown in Sect. 17.2.5) using stripline and printed circuit, may be better and cost-effective to implement. Small antennas are usually housed in small enclosures that are battery-operated. Therefore, it is important that losses are minimized and range is extended—this can be better accomplished with a carefully designed matching network. Miniature air-core inductors with Q = 100 are available in very small packages of no more than 4 × 2 mm. Matching networks will continue to shrink in size and provide improved overall performance.

17.3   The Smart Phone and Hidden Antenna

Let us consider the smart phone and take a closer look—how many antennas does it have? First, the cellular mobility network LTE requires a minimum of two antennas one to support the low band (600–800 MHz) another for the high band (1800–2100 MHz). There could be two more for cellular to accommodate 2 × 2 MIMO which is part of LTE standard. The WiFi network also requires two antennas (MIMO and 2.4/5.x GHz bands). GPS requires a separate antenna. Bluetooth also requires a separate antenna. That brings the total to a minimum of six antennas. It also has the challenge of staying "hidden" within the phone housing (antenna does not project out of plastic case). Now 5G already has allocated bands in the 28 GHz and upper bands for which additional antennas must be incorporated. "Tap and go" is the other popular feature which needs an NFC (near field communication) antenna. NFC was discussed earlier in Chap. 3, Sect. 3.6. This requires the near field antenna operating in an entirely different frequency band.

In terms of design and performance, so many antennas packed within close proximity of the same smart phone package pose some interesting challenges to the designer [26].

Table 17.4 provides an overview of what cell phone designers encounter as essential to every smart phone. There

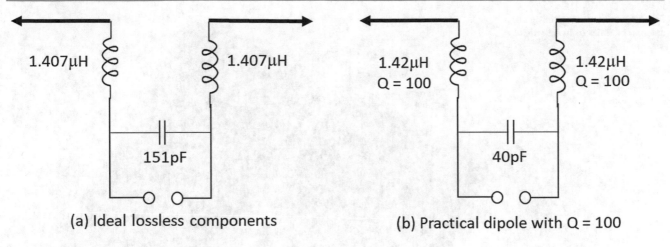

Fig. 17.23 Short dipole matching—small antennas

Table 17.4 Number of antennas in Smart phones

Function	Frequency bands (MHz)	Possible options	Number of antennas
Cellular service	600–900, 1800–2100 MHz, 2500–6 GHz, 24–28 GHz	Combine 2100 with WiFi at 2400 and 5.xGHz band	Minimum 5 (if WiFi combined), Normal 7
WiFi service	2400, 5.x to 6 GHz	Combine with Bluetooth at 2400	Minimum 2, Normal 3
Bluetooth	2400	Combine with WiFi	Minimum 1 Normal 2
Near field comm (Tap and go)	13.56 MHz	Not combined. Operates if near field	Normal 1
GPS, GNSS satellite	1207.x MHz and 1575.x	Can use one antenna for both bands	Minimum 1, Normal 2
Mid band cellular	3.7–4 GHz (new)	Cannot combine with WiFi	Minimum 1
High band (Upper)	37–47 GHz (new)	Expected later–may have multiple bands	Minimum 1, Normal 3

are 8–12 antennas and their number is likely to grow, due to newer frequency band licenses. A sensible solution would be to consider a new approach to antenna design. Instead of creating antennas as individual elements and allowing smart phone designers to choose and after that ponder on where to fit them within phone unit, it may be better to design a suite of antennas that work well together as a system and then place that system carefully in the mobile device as a single cohesive unit [27]. What are the constraints? They are:

- Antenna should stay at the edge of the unit, preferably two at opposite ends.
- Antennas should be as far from each other as possible to avoid mutual interference.
- Avoid edges where the hand holds the smart phone (that reduces signal to/from phone).
- In addition, laptops also use WiFi and Bluetooth—they have an additional constraint that they are made of metal

or carbon—both of these do not allow signals to enter/leave, except at hinges or plastic screen bezels.
- Since cell phones and laptops have also become slimmer, antennas must be carefully inserted and shaped to follow the edge/hinges.
- Diversity—two separate antennas for the same band—is important to avoid signal loss.

Such constraints limit options of where and how antennas are placed within the smart phone. Just to provide an example Fig. 17.24 shows (photo by "iFixit") of a modern cell phone.

It is worth noting that NFC antenna operates at a lower frequency and therefore larger wavelength. It is fitted at the back edge since tapping happens only with the flat side of the phone.

To address diversity, it may be possible to select one antenna at a time which has the best signal reception. This is

Fig. 17.24 Modern 5G design of antenna modules to fit into smart phone

easier with WiFi since mobility and speed at which the user travels is not a factor for WiFi. But cellular which must handle mobility may be forced to use diversity for better reception during mobility. Also, cellular signals use vertical polarization, while GPS uses circular polarization. To an extent, such factors reduce mutual interference between antennas.

Figure 17.25 shows one of the solutions by Qualcomm to provide antenna module for smart phone. All antennas within the module are designed carefully allowing designers to fit them within smart phone [28]. The role of RF front end (includes antenna module) has changed during 5G. The sheer complexity indicated in Table 17.4 dictates that RF must take an active partner role to support smart phones with many functions and diverse wireless communication technologies. Details of how antennas are separated within the module and integrated into the data stream at baseband are described in [28]. In general terms, this is known as "modem-to-antenna solution" or RFFE (radio frequency

front end). Moving into the new design cycle in 2020, the second-generation 5G solutions are expected to feature tighter component integration between 4G LTE and 5G NR, with 5G devices that combine both sub-6GH and "mmWave" in the same RFFE.

Deployment of cellular service in the millimeter-wave band will introduce another generation of RFFF that is likely to make integration of antennas into the RF front end even more important since losses at those frequencies are severe and careful design is essential.

17.4 EMI/EMC Issues With Embedded Devices

Not only the smart phone but a variety of miniature consumer devices have embedded "Radio Frequency devices and antenna" inside plastic housing. How do they radiate and what are interferences due to radiation from other

Fig. 17.25 Inner details of antenna placement within a modern smart phone

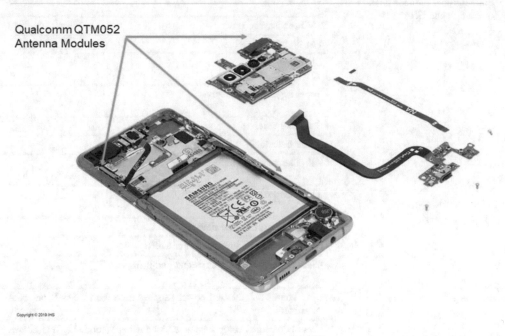

Qualcomm QTM052
Antenna Modules

devices? What are the critical issues for embedded devices? If the embedded device radiates that must not affect other devices in the vicinity (such as pacemakers, etc.), which is called EMI. EMI or electro-magnetic interference relates to any unwanted, spurious, conducted, or radiated signal of electrical origin that can cause unacceptable degradation in system or equipment performance. Electro-magnetic compatibility (EMC) is the ability of systems to function as designed, without malfunction or unacceptable degradation of performance due to EMI from others within the operational environment [29]. How do we know whether an embedded device is a radiating device? There are two possibilities:

- Intentional radiator—this is indicated by the manufacturer with its frequency specified
- Unintentional radiator (also see Table 17.5).

For intentional radiators EMI level radiated by the device should be within limits specified by the national regulator such that nearby systems continue to function properly. To limit the susceptibility of the device due to outside radiation (EMC), shielding is used as the first line of defense.

Figure 17.26 shows a typical shield—on the left is the outside world and on its right is inside of the device being shielded [30]. Since embedded devices are small, shielding to control EMC with a traditional shield using the metal of a certain thickness may not be practical. Device vendors rely on other means to control EMC. For example, the use of metal as casing, or if it has a plastic housing then they may

spray some form of conductive coating (conductive paint or vacuum deposition). Such coating may even include carbon deposit (a conductor) or making the device with carbon fiber or some conductive composite. For unintentional radiators, Table 17.5 shows regulations by the FCC for devices to operate in the USA [31].

Observe that Table 17.5 shows a variety of devices whose primary function is not transmitting RF but they do so "unintentionally". Shielding effectiveness (SE) in dB is a measure of how well the conductive housing attenuates electro-magnetic fields.

$$SE_{dB} = 20\log_{10}\frac{E_{inside}}{E_{outside}} \qquad (17.23)$$

Theoretical SE of homogeneous material is given by three factors:

(a) Reflective losses, R
(b) Absorption losses, A and
(c) Secondary reflective losses, B (ignore if A > 8 dB).

In that case it simplifies to, SE = R + A + B → SE = R + A.

There is the other dilemma—the embedded device must communicate. Hence its RF radiation must escape from its enclosure, but the device itself must be shielded from radiation from other devices within its vicinity. To address this conflict, the radiating antenna must be kept at the outer layer

Table 17.5 FCC regulations for unintentional radiator devices to operate in the USA

Type of device	Equipment authorization required
TV broadcast receiver	Verification
FM broadcast receiver	Verification
CB receiver	Declaration of conformity or certification
Super regenerative receiver	Declaration of conformity or certification
Scanning receiver	Certification
Radar detector	Certification
All other receivers subject to part 15	Declaration of conformity or certification
TV interface device	Declaration of conformity or certification
Cable system terminal device	Declaration of conformity
Stand-alone cable input selector switch	Verification
Class B personal computers and peripherals	Declaration of conformity or certification
CPU boards and internal power supplies used with Class B personal computers	Declaration of conformity or certification
Class B personal computers assembled using authorized CPU boards or power supplies	Declaration of conformity
Class B external switching power supplies	Verification
Other Class B digital devices and peripherals	Verification
Class A digital devices, peripherals, and external switching power supplies	Verification
Access broadband over power line (Access BPL)	Certification
All other devices	Verification

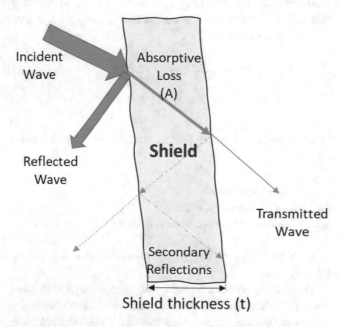

Fig. 17.26 Shielding and its effectiveness

such that it can radiate (both transmit and receive). The smaller the antenna for a given operating wavelength, the less efficient it will be. An excellent guideline of how to incorporate such an antenna in an embedded device is available in [32].

To illustrate this with an example, consider a typical frequency range below 1 GHz, where many 4G cellular bands (low bands) operate. Sub-gigahertz antennas need a minimum of 100 mm as the ground plane. Yet, many IoT devices today are around 50 mm in length or smaller. Quarter-wave antenna at 916 MHz would be ~ 82 mm long, and this will not fit. One approach is to use multi-layer PCB (printed circuit board). Another possibility is to extend beyond the device—for example, an electronic watch can have a ground plane extending into the watch strap.

Figure 17.27 shows a small PCB that operates in the LTE band but at one band at a time. It is possible to use one antenna to span the entire 4G low band from 698 to 2690 MHz but use a tuning circuit with switch bands and operate only one band at a time [32]. Such arrangement is

Fig. 17.27 Example of active tuning circuit for 4G embedded device

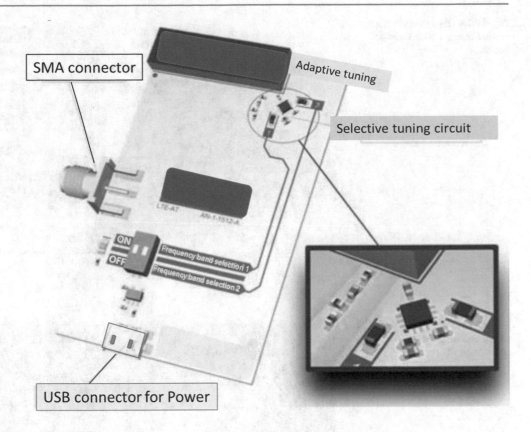

SMA connector

Adaptive tuning

Selective tuning circuit

ON
OFF
Frequency band selection 1
Frequency band selection 2

USB connector for Power

shown in Fig. 17.27 and it allows the circuit to better match the antenna and selects the band that operates well at a given location.

There are other issues—the antenna must be located such that it is not blocked by the user. For example, if it is a handheld device like a cell phone, all-natural ways of hand-holding cell phone must be explored. Those locations must be avoided for antenna layout, an example was shown in Fig. 17.25, where the antenna and entire RF front end (RFFE) system was designed separately by working closely with the phone manufacturer.

17.5 Opportunities for Millimeter-Wave Antennas

Millimeter-wave is becoming popular starting with 5G and antenna size at these higher bands is very small, since full wavelength itself is in the order of millimeters. But conduction loss is so high that the use of cable or waveguide has to become part of the circuitry. This is the new design approach where the antenna and its connection using cable or waveguide directly interfaces with an integrated circuit chip. A typical service provider may only see the antenna as part of the RF circuitry. In earlier sections of 17.3 and 17.4, only the handset and embedded devices were addressed. Now we address the base station and outdoor antennas.

At the base station, a whole new evolution is occurring due to the following factors:

- Massive MIMO, which means at least 8×8 antenna array at the base station as shown by matrix beam in Fig. 17.28 (left top)
- RFFE—RF front-end circuitry right behind the antenna panel as an array
- Beam steering—multiple beams that could be following a user as they move or it could be fixed beam for residents in an apartment block

The concept of RF front end is brought out in Fig. 17.28 where 6×6 matrix panel antenna (for simplicity 6×6 panel is shown in the figure) serves fixed customers in residences and apartments and mobility services with beam steering.

Fixed wireless service will have services of at least 1 Gbps for each customer, while mobility may get supported with lower throughput levels of 100 Mbps or higher. The antenna shown as a panel is an "active antenna system" with RF front end directly mounted behind the panel antenna. Throughout this chapter antenna was addressed as a "passive element", which is true. But the term "active antenna" is meant to indicate that active RF equipment is "integrated into" the antenna panel.

Fig. 17.28 End-to-end cellular service concept with multiple beams

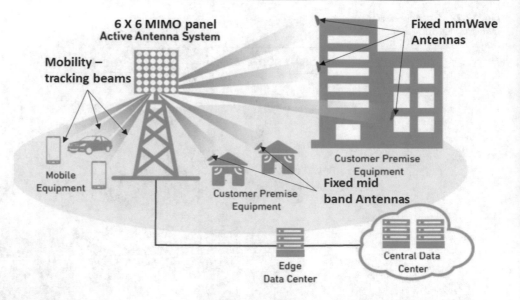

Fig. 17.29 Active antenna panel with a radiating element in front and ICs on the back

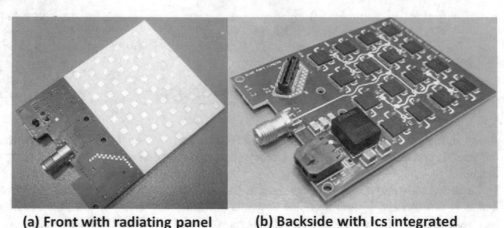

(a) Front with radiating panel   **(b) Backside with Ics integrated**

An example of such an active antenna with integrated circuits on the back is shown in Fig. 17.29. Many radiating panels such as the one shown here must be placed in plastic or hardened fiber outer casing so that they can be mounted on a typical base station tower to form a 5G network with different services as indicated earlier in Fig. 17.28.

Requirements of upcoming 5G millimeter-wave communications, such as high antenna gain and beamforming agility, are addressed by designing, manufacturing, and measuring a compact and scalable phased array antenna for a 28 GHz access network [32]. The panel shown in Fig. 17.29 is based on a multi-layer printed circuit board (PCB) with 64 radiating elements (8 × 8) and 16 commercial quad-core monolithic microwave integrated circuits (MMICs) providing independent 5-bit phase and amplitude controls for each radiating element. The focus is on measured radiation patterns, which show good characteristics (beam width, side lobe level, cross-polar discrimination, beam steering range) and correspondence with theory, even without calibration. The antenna reaches effective isotropic radiated power

(EIRP) up to 48 dBm. Measurements with real modulated signals confirm the feasibility of the antenna for multi-Gbps communications.

The future of antennas is moving from a passive element to an integrated part of the RF front end right behind the antenna itself. This change is mainly due to multiplicity of wireless devices such as massive IoT, low latency, and enhanced mobile broadband. The overall growth in wireless has brought the antenna to a point where it is often integrated by a 3D printer that uses conductive ink to form an antenna in 3D to satisfy the coverage and functionality requirements of different industries.

17.6   Conclusions

This chapter on antennas covered a whole range of topics—all related to how waves propagate and antenna interfaces to a modern wireless network using a transmission line. Starting with the famous Maxwell's equations, basic definitions such as VSWR and other antennas characteristics were described. Types of transmission lines and their characteristics were described. Antenna is the key element that moves EM waves from the wired medium of transmission line to the wireless medium of air/vacuum. This was stressed using its key parameters to show how this becomes possible in an efficient manner.

Different types of antennas have always featured throughout the wireless network. Irrespective of distance or power levels antenna is one element that absolutely must exist. Whether a coaxial transmission line terminates in the antenna, or a waveguide opens out into the aperture of a horn antenna, the result is still radiating wave—a function which antenna is expected to perform.

There were aspects such as how to find the resonant frequency and band over which it operates well. There were other aspects such as ground plane which allows a quarter wavelength to operate as a half-wave. The use of ground plane often becomes essential when operating at lower frequencies and wavelength becomes long. Ground plane helps by reducing the length of the antenna to quarter and supplement the other half (virtual length of ground plane). This is true even in recent 5G designs where the same antenna must operate over a wide range of frequency bands.

For every wireless enthusiast—antenna is one element they will always encounter. It becomes the window through which radio waves are sent and received. The importance of this element along with transmission line is at the very core of everything related to wireless communication.

Summary

This chapter on antennas started out rightly with Maxwell's equations that describe electro-magnetic waves whether they are on a conductor or in the air or vacuum. Rightly antenna is one element that allows these waves to move from conductor to air when transmitting or the other way around when receiving. With that introduction, the topic of transmission line was touched on both coaxial cable and waveguide. It described the conditions necessary to allow the wave to move along this conducting medium and how the antenna must interface to it.

This was followed by key parameters of the antenna, what they mean and how they affect the quality of sending the waves over the air. A key part of this understanding included the physical motion of waves using video—that dates back to Bell Labs vision in the 1950s on how waves of different types behave. Radiation pattern is perhaps the most vital characteristic of an antenna. How this is measured to create a 3D pattern and how the data sheets typically portray radiation in the elevation and azimuth diagrams were described with examples. The video in reference [14] provides a view of how this complex radiation pattern gets generated.

This was followed by types of antennas—by listing most types and elaborating on a few of them. The intent was to describe how the antenna design process works, how to control the radiation patterns, and how are they visualized during design. The direction antenna fed makes a considerable difference—this was highlighted using a helical antenna where the pattern becomes completely different depending on the type of feed. One of the unique types is the radiating cable that works as an antenna but is actually a coaxial cable with holes punched in its outer conductor. It radiates only around the cable but generally will not radiate far—meant for propagation within tunnels.

Since the late 1960s monolithic microwave integrated circuit (MMIC) started to have a great impact on the design of radio communication, particularly in the upper microwave bands. Design based on MMIC and stripline with an antenna printed along with feed and balancing circuit was described to show how far the technology has moved. This process of printing antennas lends itself to customized characteristics and patterns.

Perhaps the biggest driver for an antenna in recent times has been the smart phone. Requiring multiple antennas within a small area has forced RF engineers to offer complete solutions known as RFFE (RF front end) that takes care of all antennas and their interconnections. Such solutions provide a better product that works well without compromising the quality of signal and performance.

With the integration of multiple technologies in compact packages, the issue of interference from nearby devices and how the device could become an interferer were briefly reviewed. Millimeter-wave band has ushered a new role to antennas. For keeping components small and accurate in terms of beam forming, antenna has become closely integrated with active components in the form of matrix. The modern panel antenna is a passive element only on one side but acts as the base on which multiple layers of integrated circuits work as RFFE. Such antenna matrix stacked together is expected to provide 5G services in every neighborhood.

Emphasis on participation

(1) Using Matlab write a program to create a 3D pattern for the following:
   a. Dipole with half-wavelength
   b. Dipole with quarter-wavelength
   c. Dipole with quarter-wavelength but connected to ground plane
   d. Basic design of Yagi-Uda antenna.
(2) Long wire antenna—using cable derive taps of three different wavelengths—check the reception of local

AM and FM stations using antennas at different taps—record readings

(3) Contact your local radio / TV station—study their transmission site. Record the following
   a. what type of antenna do they use?
   b. Over which frequency bands does it operate?

(4) Using a plexy-glass or hard plastic surface draw out a spiral antenna made of some metal (copper or aluminum preferably). Connect as Archimedean spiral and measure its radiation.

Knowledge gained

I. Basic concept of electro-magnetic waves and Maxwell's equations

II. Types of transmission lines—parallel wires, coaxial, and waveguide

III. Key parameters that define antenna characteristics. How they are measured, and why they are needed.

IV. Different types of antennas—how they are designed, what are the radiation patterns and other features that distinguish them.

V. Monolithic microwave integrated circuits—modern chipsets that extended key concepts such as miniaturization, operating at very high frequencies (close to the optical band).

VI. Stripline with 3D printed antennas using metallic ink—advantages of building antennas from equations.

VII. Smart phone and transformation in hidden antenna design with RFFE.

VIII. EMI and EMC issues with embedded small wireless devices—how to limit interference to other devices in the vicinity and build immunity to reduce interference from other devices.

Homework problems

(1) Certain cable has inductance L = 5.005 mH per meter and Capacitance of 2 μ F per meter. Using the equation, find out its characteristic impedance Z (*Eq.* 17.14).

(2) A coaxial cable is connected to an antenna. Using power meter, the incident power is measured as 950 mW. The reflected power is measured as 50 mW. What is the VSWR? (*Eq.* 17.9).

(3) Certain antenna has effective area Ae = 0.25 sq.m and operates at 600 MHz. Determine its gain (*Eq.* 17.19).

(4) A spiral antenna must be designed to operate up to the lowest frequency of 300 MHz. Determine the diameter of the spiral (*Eq.* 17.21).

(5) A radiating cable operating at 1900 MHz is to be used for a corridor 500 m long. The maximum cellular signal available is +30 dBm. Calculate system losses and indicate the signal available at (a) 2 m away from the cable, (b) at the far end of the corridor.

Project ideas

(a) At the 900 MHz ISM band the maximum signal allowed to be radiated is +36 dBm. Design a hospital signal network along the corridors and in patient rooms—for equipment operating in this band. Use data sheets from commercially available medical devices.

(b) Using metallic paint, design a spiral antenna operating from at least 800 MHz and measure its performance over the 800–5800 MHz band. A second project team to build a helical antenna operating in normal mode or axial mode over the same frequency range—which one is recommended for a small medical device?

(c) Using general guidelines provided in the link below, construct Yagi-Uda antenna for a point-to-point link (both sides using the same antenna) operating in the 902–928 ISM band or 2.4 GHz for a signal and control link that needs data rates in the range of 200 Kbps to 200 Mbps throughput. https://www.electronics-notes.com/articles/antennas-propagation/yagi-uda-antenna-aerial/gain-directivity.php.

Out-of-box thinking

Using Matlab calculate and build a plastic/clay model of an antenna of your choice. If this were coated with metallic paint, would it be possible to use it as a 3D-printed antenna? What are the advantages and limitations?

The use of Yagi-Uda antenna for TV reception has been common for almost five decades. Is it possible to use Yagi-Uda antenna for cellular bands of 800/900 MHz and 1800/1900 MHz, instead of the current panel antennas on towers or in tunnels? What are the advantages of such a design?

References

1. Maxwell's equations explained in 39 minutes (+ Divergence, Stoke's theorem), by Universal denker. Retrieved October 5, 2019, https://www.youtube.com/watch?v=hJD8ywGrXks
2. AT&T archives: Similarities of wave behavior (Bonus Edition). Retrieved April 3, 2012, https://www.youtube.com/watch?v=DovunOxlY1k
3. Coaxial cable calculator: definitions of coax cable parameters and formulas used for calculation
4. Oliver D (2014) Back to basics in microwave systems, Return loss and VSWR. CommScope Blog
5. Lioubtchenko D, Tretyakov S, Dudorov S (2004) Millimeter wave waveguides. Helsinki University of Technology, Kluwer Academic Publishers, Springer Science and Business Media, Finland.

ebook ISBN 0–306–48724–1 and print ISBN 1–4020–7531–6. 2004

6. RF connectors and cables, By Joe Cahak, June 9, 2014RF Café. https://www.rfcafe.com/references/articles/Joe-Cahak/rf-connectors-cables-joe-cahak-6-2014.htm

7. Thiele B, Hardin S (2010) Choosing the optimal high frequency cable. Microwave J https://www.microwavejournal.com/articles/9169-choosing-the-optimal-high-frequency-coaxial-cable and https://electronics360.globalspec.com/article/15538/practical-overview-of-antenna-parameters

8. It's not a Cable, It's an Antenna! By Keith Blodorn, ProSoft Whitepaper—Radiating Cable 2015. http://www.prosoft-technology.com/var/plain_site/storage/original/application/1afac1ab3e938a0d71856ff36e7dfab4.pdf//clickId=81880

9. Radiaflex brochure 2012–03. Radio Frequency Systems, http://www.rfsworld.com/userfiles/brochures/2012-october/RADIAFLEX_BROCHURE_2012-03.pdf

10. Breed G (2008) Improving the bandwidth of simple matching circuits, Editorial Director, High Frequency Electronics, https://www.highfrequencyelectronics.com/Mar08/HFE0308_Tutorial.pdf

11. Antenna Theory—basic parameters, Tutorial Point, https://www.tutorialspoint.com/antenna_theory/antenna_theory_basic_parameters.htm

12. What is Antenna Polarization, and Why Does it Matter? By Peter Mcneil, Feb 22, 2018. Pasternack Blog. https://blog.pasternack.com/antennas/antenna-polarization-matter/

13. Perform cost effective antenna radiation measurements, Miroslav Joler, Faculty of Engineering, University of Rijeka, Rijeka 51000, Croatia. August 21, 2018. Microwave and RF. https://www.mwrf.com/technologies/test-measurement/article/21849337/perform-costeffective-antenna-radiation-measurements

14. Radiation Pattern Measurements in an RF Anechoic Chamber, by Limpkin, 15 Oct 2018. https://www.limpkin.fr/index.php/post/2018/10/15/Radiation-Pattern-Measurements-in-an-Anechoic-Chamber

15. Miron DB (2006) Small antenna design. Elsevier publications Chapter 1. https://www.sciencedirect.com/topics/computer-science/radiation-pattern, https://doi.org/10.1016/B978-0-7506-7861-2.X5000-4

16. Antenna Impedance, Antenna and wave propagation by Roshni Y, https://electronicsdesk.com/antenna-impedance.html

17. Different types of antennas and characteristics of antennas. Electronics Hub.

18. Freznel L (2013) What is the difference between a dipole and a ground plane antenna? Electronic Design Mag https://www.electronicdesign.com/home/whitepaper/21803011/whats-the-difference-between-a-dipole-and-a-ground-plane-antenna-pdf-download

19. Long wire: End fed wire antenna, Electronics notes. https://www.electronics-notes.com/articles/antennas-propagation/end-fed-wire-antenna/end-fed-long-wire-antenna-basics.php

20. Forrister T, Blog C (2018) Analyzing operating mode for helical antennas, https://www.comsol.com/blogs/analyzing-operating-mode-options-for-helical-antennas/

21. Lam T, Bidwell R, Blee S (2013) Spiral antenna design considerations. Microwave J https://www.microwavejournal.com/articles/18897-spiral-antenna-design-considerations

22. Karandikar Y (2012) Integration of Planar Antennas with MMIC Active Frontends for THz Imaging Applications, Ph.D. thesis, Microwave Electronics Laboratory, Department of Microtechnology and Nanoscience (MC2), Chalmers University of Technology Goteborg, Sweden, http://publications.lib.chalmers.se/records/fulltext/156421.pdf

23. Rodriguez-Morales F, Yngvesson KS, Zannoni R, Gerecht E, Gu D, Zhao X, Nicholson J (2006) Development of Integrated HEB/MMIC receivers for near-range terahertz imaging. J. IEEE Trans Microwave Theory Techn 54(6) https://authors.library.caltech.edu/5997/1/RODieeetmtt06.pdf

24. Kiesel G, Bowden P, Cook K, Habib M, Marsh J, Reid D, Baker B (2020) Practical 3D Printing of Antennas and RF Electronics. Aerospace & Defense Technology Magazine, Atlanta, GA. https://www.aerodefensetech.com/component/content/article/adt/features/articles/37095

25. Reed G (2007) Basic principles of electrically small antennas, Editorial director. High Freq Electronhttps://www.highfrequencyelectronics.com/Feb07/HFE0207_tutorial.pdf

26. Valkonen R (2018) Compact-28-GHz-phased-array-antenna-for-5G-access. in Engineering, 2018 IEEE/MTT-S International Microwave Symposium —IMS. https://www.semanticscholar.org/paper/Compact-28-GHz-phased-array-antenna-for-5G-access-Valkonen/810f75eb78f31c34b0743bce7931b63db640412b

27. Hu S, Tanner D (2018) Building smartphone antennas that play nice together. IEEE Spectrum Magazine https://spectrum.ieee.org/telecom/wireless/building-smartphone-antennas-that-play-nice-together

28. Lam W (2019) In 5G smart phone designs, RF front end graduates from traditional supporting role to co-star with modem. Principal Analyst, Mobile devices and networks. https://omdia.tech.informa.com/OM004560/In-5G-smartphone-designs-RF-FrontEnd-graduates-from-traditional-supporting-role-to-costar-with-modem

29. Moe A, Raghunandan K (2016) Radiation Safety and EMC Compliance. in IEEE NJ coast AP/EMC/VTS chapter talk. Rutgers University, Student Center, The Cove, Busch Campus, East Brunswick, NJ

30. Colotti J (2006) EMC DESIGN FUNDAMENTALS. in 2006 IEEE Long Island Systems Applications and Technology Conference. Long Island, NY, pp 1–2. https://doi.org/10.1109/LISAT.2006.4302648

31. Electronic Code of Federal Regulations (e-CFR).

32. Newman C, Antenova TC Mission Impossible? How can an embedded antenna perform brilliantly within an IoT device with a very small PCB? Hatfield UK. www.embeded-world.eu, https://blog.antenova.com/hubfs/mission-impossible_antenova-paper.pdf

Abbreviations

RADAR	The word RADAR is an abbreviation for "RAdio Detection And Ranging". Traditionally came from the military, but currently in use at all airports and in driverless cars. It is a technology set for considerable growth, with new frequency allocations in the millimetric band
Radar pulse	Usually associated with the outgoing RF transmission, typical Radar pulse has energy in the kW or MW range, but over a very short duration of microseconds. It was originally intended to reach and return from target hundreds of km away. But current transponder Radars do not need that level of energy since the aircraft or ship will respond using their own transmitter (not just reflected)
Duplexer	An RF device that combines as well as separates the transmit and receive signals to send them to a common antenna. If transmit and receive are at the same frequency then it works as a switch using clock synchronization. If transmit and receive are at separate frequencies then it uses filters to isolate them to minimize interference
Transponder	A device that receives the signal, frequency shifts it before sending back. It is abbreviated as "transmit + respond" or the one that responds to transmitted signal by receiving, frequency shifting, and responding with its own identity. When used in satellites, these units help reduce the noise associated with the transmit path (uplink noise)
Radar range	This is the distance between the Radar antenna and the target (in the sky or a ship). It is measured in terms of time taken by electromagnetic wave to travel up to and come back from the target. If both the Radar and the target are moving, this calculation becomes complex. The simplest case is when the Radar is stationary and the target is also stationary or not moving at high speed so that transmit path and receive path can be considered as equal
Radar cross section	This is the cross section of the target that is available for signal reflection. In large

© Springer Nature Switzerland AG 2022
K. Raghunandan, *Introduction to Wireless Communications and Networks*,
Textbooks in Telecommunication Engineering, https://doi.org/10.1007/978-3-030-92188-0_18

	commercial aircraft this area may be large, but small military jets may offer minimal area, and with stealth technology the reflected echo, may not even come back		screen, the VRM appears as a circle that is centered on the present location of the ship or boat, and the EBL appears as a line that begins at the present location of ship or boat and intersects the VRM. The point of intersection is the target of the VRM and the EBL
Radar bands	The frequency band designations used by the Radar community are different from conventional radio community. Radar community uses only letters starting from A to O, while radio community uses key abbreviations such as HF, VHF, and UHF	TCPA and CPA	Time to Closest Point of Approach and Closest Point of Approach are warning alarms to ship / boat that may be based on radar's ability to see beyond the horizon. By plotting them on the radar screen (based on navigation map) the system is able to warn to steer clear of such points and avoid collision with such points
Transponder beacon	Radar transponder typically identifies itself at 1040 MHz and expects the commercial aircraft to respond with its identity using the same band, providing Lat, Long, and Altitude, that immensely helps ground-based navigators to guide the aircraft to the landing runway strip	SAR	Synthetic Aperture Radar: This uses a clever concept whereby, a sequence of acquisitions from a shorter antenna are combined to simulate (or synthesize) the equivalent of a much larger antenna, thus providing pictures of much higher resolution. This concept is particularly useful for a Radar on satellite that looks to the earth but is able to get a much better resolution— something that could be achieved by real antenna that would have been several kilometers long
Meteorological OTH radar	Used in coastal areas, Over-The-Horizon Radar helps in identifying high-speed winds, waves, and tornados over the sea. It is also used to identify smugglers operating over vast land areas		
COLREGs (Convention on the international regulations for preventing collisions at Sea)	It is an agency set up by the International Maritime Organization (IMO) that "mandates" that all sea going vessels of 3000 tons or more MUST HAVE a X-band Radar to identify other vessels or fixed posts (such as rocks) to avoid collision at sea	SRTM	Shuttle Radar Topography Mission: It was a mission by space shuttle to photograph and catalog different parts of the land area on earth, that led to 3-dimensional model or 80% of earth land mass based on the concept of phased array antenna, it offers an opportunity to improve both detection and resolution of target
VRM variable range marker and EBL	The variable range marker (VRM) and the electronic bearing line (EBL) measure the distance and bearing from a ship or boat to a target object. On the Radar	MIMO radar	

Mono-static MIMO radar	In a MIMO radar system, each radiator has its own arbitrary waveform generator. Therefore, each radiator uses an individual waveform. This individual waveform serves as the basis for an assignment of the echo signals to that source	MRR (Medium Range Radar)	Medium Range Radar (MRR)—detection of objects 1–60 m from the vehicle—height detection which reliably classifies objects and brake safely, even when the object is stationary (parked object–is it a truck or small box, is it a child running across, etc.)
Bistatic MIMO radar	In a Bistatic MIMO Radar system each radar antenna views the target from a different aspect angle. Hence, the target provides a different radar cross section for each radar antenna. This requires more complex target models for radar data processing	FMCW radar	Frequency Modulated Continuous Wave is a new concept in Radar that uses "chirp". Radar operating at 80GHz with 4GHz bandwidth is able discern objects that are only a few Cm apart. The unique feature of this Radar is it recognize objects based on both Frequency and phase, allowing a unique signature even for someone standing next to a stationary car for example
Doppler radar	Unlike conventional Radar, the Doppler Radar not only revolves on its own axis, but also changes its elevation angle, to collect data from space around it. It transmits for very short period of 1.56 μs but listens for a very long time 998.43 μs. During its long "listening period" it listens to responses even from small rain drops, to build a detailed 3D model around it, helping meteorologists in weather predictions	Chirp	Chirp represents a signal that increases or decreases with time. Usually, such sweep signal of chirp is associated with Radar to form signature of very small objects including drones flying at considerable distance. When applied to Doppler Radar, it is able to discern small flying objects such as snow or debris from wind, helping predict weather more accurately
ADAS	Advanced Driver Assistance System. This is a support system human driver in autonomous vehicles of level 3 and higher. It can support functions such as parking, lane change assistance) as part of the standard adopted by the Society of Automobile Engineers (SAE)	Long range radar (LRR)	detecting objects up to 250 m from the vehicle. Ability to accurately detect small objects or something crossing the road further down (slow moving vehicle further down, allows gradual braking and speed regulation)
SRR (Short-Range Radar)	Used in autonomous vehicles it operates in the range of 1–10m from the vehicle including Blind Spot Detection (BSD) and Lane Change Assistance (LCA)		

The word RADAR is an abbreviation for "RAdio Detection And Ranging". Analyzing it further—what does RAdio do —it aids in detection, what does it detect—any object farther away that cannot be seen by the eye or camera for example. What is ranging—it is an estimate of distance adjudged by calculation based on radio wave propagation. With this explanation Radio is at the heart of such a system. The concept of RADAR therefore is for detection, it is not for communication with someone.

Therefore, the signal which must be sent as a powerful pulse must be received within a very short time duration. The power of the pulse must be sufficient for it to travel hundreds of kilometers, get reflected by a metallic or some other object and that reflected signal must get back to originating antenna. Since estimation of range is the objective, there is no reason to have continuous communication. It is sufficient to send stream of pulses that travel up to that object, get reflected and return, then the range is calculated. But over that distance since signal gets dispersed it is important to also keep the beam somewhat narrow, such that energy is not wasted in all directions.

This is the objective of RADAR—but as always, details of how it is designed, how it operates despite clouds or other obstructions in its path, how is it affected by mobility, etc., are aspects that will be addressed in this chapter. Its use as a technology is now observed by the public in daily life particularly as Doppler Radar for weather images that has brought considerable interest on "RADAR" in recent years.

18.1 A Brief Overview of RADAR

Radar as a technology started its application in the military well before the second world war. Radar was the outcome of research during the mid- and late-1930s by scientists and engineers in nine countries: United States, Great Britain, Germany, Holland, France, Italy, USSR, Japan, and Switzerland. Each country believed that this was its own development and held the technology in highest secrecy. Great Britain gave the basics later to four advanced Commonwealth nations: Australia, Canada, New Zealand, and South Africa, and indigenous systems emerged in each.

Why such a great secrecy? In each case, the use of Radar was to detect enemy aircraft over great distances. That would offer distinct advantages in terms of directing weapons (anti-aircraft guns for example) towards the aircraft. Radar was also used in an offensive strategy by giving pilots the ability to attack targets at night during inclement weather etc. Over the years, this technology progressed considerably ushering civilian uses and wide range of other applications that have made it one of the very useful wireless technologies to help the general public [1].

Figure 18.1 shows the concept of how a RADAR operates. Its antenna is looking up for objects in the sky. It continuously rotates on its own base sending out stream of pulses in all directions around it (shown by black pulse stream). The green circle on the bottom right corner, depicts the Radar scan, with its antenna located at the center of the circle. Concentric circles of increasing radius indicate "range" or distance from the Radar antenna to the object (Aircraft in this example). Once the RF pulse energy touches any object in the sky (aircraft), it gets reflected and an echo or reflection, returns towards the sending antenna (return is shown by red pulse stream). The time taken by outgoing pulse to travel up to the aircraft and time taken to return from the aircraft back to the antenna are measured and calibrated as distance. Since pulses travel at the speed of light, actual location of the aircraft at that instant is indicated by "blips" on the green circle (bottom right).

In Fig. 18.1 at the bottom right is a green circle, which represents a typical Radar screen and the spinning action of antenna is shown by the hue in the clockwise direction, on the green circle. Since antenna is located at the center, repetitive scans provide updates in terms about direction and location of the aircraft and how it is moving. Further sophistications, such as its altitude, estimated speed, are all calculated and displayed on the Radar operator's screen, but those estimates require some understanding of Radar equations.

Radar equations consider the upstream pulses shown in Fig. 18.1, their power density, and the reflected pulses (red stream) from the aircraft and their power density. They also consider other factors such as beam width, aperture of the antenna, and the surface of aircraft (or another object being detected). These factors form the basis of calculations in the Radar equations.

Before progressing to the equation, it may help to review the functional block diagram of a Radar system. Figure 18.2 shows the block diagram of a Radar system [2]. It consists of two separate sections—the transmission section at the top and receiver section at the bottom.

The two sections are connected to the radar antenna using a "Duplexer", which is located behind the antenna on top left. Duplexer is a device that has three ports indicated by dots—Transmit (TX), Receive (RX), and Antenna (ANT). Why do we need this Duplexer? Since the outgoing pulse is powerful (almost as if Radar shouts out during each pulse) but the receiver gets weak signal (echo is coming back as if it is a whisper), it is important that the two are well isolated. The need for isolation is because Radar's transmit and receive frequencies are typically very close and can only reasonably be separated through use of the duplexer. The Radar has to first send out a powerful pulse and then wait for the weak echo from any object in the sky—like an aircraft. To hear that echo, it must open up its receiver with big ear in the form of "Low Noise RF Amplifier (LNA)" seen as the first block in receiver section. LNA is specially designed to receive weak signals.

Since this transmit and receive happens repeatedly, the "Duplexer" acts as a channeling device allowing loud talk (sending out pulse) and then listen intently (receiver). It

Fig. 18.1 Concept of radar's detection and ranging

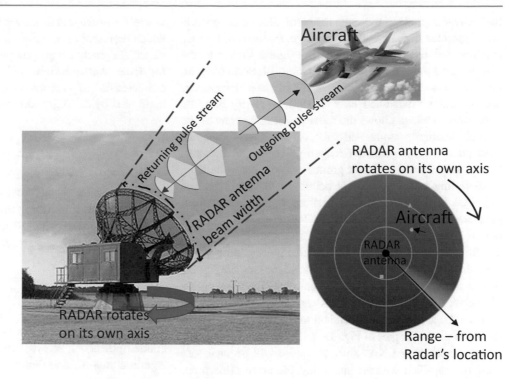

Fig. 18.2 Functional block diagram of radar system

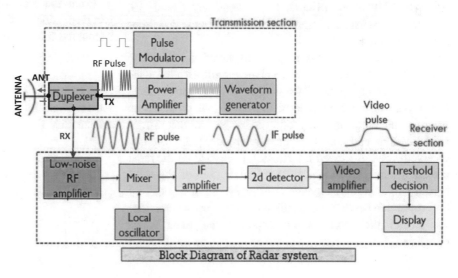

functions as a switch which allows powerful transmit and weak receive signal to work separately but using a common antenna. Either switched systems such as Duplexer or devices such as the "magnetic circulators" can be used to create such isolation between the outgoing and incoming signals. Although the Duplexer and Magnetic circulator operate on different principles, those terms are often used interchangeably as a common simplification, but it is not always accurate. We know that return frequency will be almost the same as the one transmitted. Therefore, in addition to the isolation, timing must be synchronized with the pulse modulator, which is shown in the transmission section. Such control of transmit is separated from receive function

and duplexer acts as a switch, allowing either transmit or receive at any given time.

If the Radar operates in "transponder mode" using beacon, then the transmit and receiver frequencies are distinctly different. Then duplexer is designed with different frequencies for transmitting and receiving. Unlike the pulse the "beacon acts as a broadcast message" identifying the radar location and expects the aircraft to respond with a message. The aircraft or object being detected receives the beacon pulse, does a frequency shift (transponder = transmit + respond), and returns the beacon as a response pulse. The response pulse stream contains information such as altitude and the aircraft ID. This concept was introduced by the military as "friend or foe" where aircrafts of the same nation or alliance understood and responded to the broadcast message. In the radar system, transmit and receive ports will be designed similar to the conventional duplexer explained in Chap. 14, Sect. 14.2 and later in Chap. 19, Sect. 19.7.3. Details of radar transponder beacon will be explained later in 18.2, to clarify how the transponder beacon allows the radar to get a 3D view of the aircraft (azimuth angle, distance, and altitude (height). Transponder mode primarily helped all commercial aircraft to take help from any airport ground-based radar system.

Reverting to the functional details in the upper part of Fig. 18.2 is the transmission section. It has a waveform generator that produces continuous RF waveform. But it is pulsed using a pulse generator. The output of power amplifier consists of short but powerful pulses that get sent over the air, towards the target. This is the outgoing pulse stream shown in Fig. 18.1 (in grey color). When these pulses hit target, they return (red pulses) and are received by the same antenna, but duplexer sends it to the receive section, as shown in the lower part of Fig. 18.2. The receiver uses "Low Noise Amplifier or LNA" since the pulse will be weak and must be amplified without increasing the noise. This function of LNA was discussed in detail in Chap. 13, Sect. 13.3.1, where its importance in satellite earth stations was emphasized.

Since the received pulse must be displayed as a "blip" on the video screen, rest of the receiver circuitry follows the video processing path. This design is like conventional heterodyne receiver, with local oscillator that produces an Intermediate Frequency (IF) followed by detector that assigns azimuth angle and scaled distance (references pulse sent and received). The processed information is aligned on the radar screen where scan shown on the right bottom of Fig. 18.1 (green circle) indicates location of the aircraft or object.

To provide better understanding of how these calculations are made, the next section explains the radar range equations.

18.1.1 The Radar Range Equation

With this basic understanding, an isotropic radiator is considered initially for the calculations [3]. Isotropic radiator that was already discussed in Chap. 17, and Fig. 17.2 shows unity sphere radiation. In general terms, area of the sphere of radius R is given by A = 4 πR^2. Substituting "non-directional power density S_u" or "Uniform power density" is given by $\frac{Power}{Area}$:

$$S_u = \frac{P_s}{4\pi R_1^2} \qquad (18.1)$$

where P_s—transmitted power in watts, and here we use R_1 which represents the "range from transmitted antenna to the target". In practice high gain parabolic dish antennas are used for Radar. Antenna Gain "G" was described as an important characteristic of antenna in Chap. 17. This gain must be multiplied by S_u to get "directional power density", S_g.

$$S_g = S_u \cdot G = \frac{P_s}{4\pi R_1^2} \qquad (18.2)$$

The reflected power P_r, depends on the surface area of the target, known as "Radar cross section σ". This power travels through the same path and gets to the radar antenna receiver.

Reflected power which is an "echo" is indicated as S_e at the Radar receiver, with range R_2

$$S_e = \frac{P_s}{4\pi R_2^2} \qquad (18.3)$$

Figure 18.3 shows the relationship between uniform power density S_u, Radar Cross section σ, Ranges R_1 and R_2, emitted power P_e, received power P_r, signal from echo S_e, Aperture area A_w and directional power density S_g.

From the antenna only part of this "echo" enters the receiver, since effective antenna aperture A_w must be accounted for. Therefore,

$$P_e = S_e \cdot A_w \qquad (18.4)$$

A_w in turn is a product of A the geometric area of the Antenna and its efficiency K_a; by substituting for A_w this will modify Eq. 18.4 as

$$P_e = S_e \cdot A \cdot K_a \qquad (18.5)$$

By substituting for S_e from Eq. 18.3 we get power density at the receiver.

$$P_e = \frac{P_r}{4\pi R_2^2} \cdot A \cdot K_a \qquad (18.6)$$

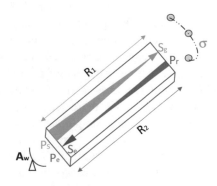

Fig. 18.3 Relationship between Radar equation parameters

However, we note that P_r the power received at the receiver unit will have been multiplied by the gain of the antenna G and the cross-sectional area σ which is actual area offered by the targeted Aircraft. If it is large commercial airliner, that target area is large. A two- seater sports aircraft will have a much smaller cross-sectional area offered to the pulse. In general terms this is described as the "radar cross section".

$$P_r = \frac{P_s}{4\pi R_1^2} \cdot G \cdot \sigma \qquad (18.7)$$

where P_r is the reflected power in watts, and σ is the radar cross section in m^2; this will be the same as S_e the power density at receiver since the transmit and receive paths are the same. For clarity of signal path the term R_1 and the term R_2 were used to indicate "range from receiver antenna, range from the target". But in practice the range is same and can be substituted by R. Hence range $R_1 = R_2 = R$. The power density of received signal is affected by another factor —"aperture of the antenna A_w" which is a product of Antenna's geometric area A and efficiency K_a.

Equation 18.4 showed power density at the receiver site $P_e = P_e = S_e \cdot A_w$, after substitutions from three equations above, we get

$$P_e = \frac{P_s \cdot G \cdot \sigma}{(4\pi)^2 R^4} \cdot A \cdot K_a \qquad (18.8)$$

Although determination of transmit and receive may seem redundant, they may become important if the Radar is on a moving platform (such as car, plane or ship) and the target is moving as well (aircraft) depending on the application. This will become relevant in situations to be discussed later. For now, we will proceed with simplified "fixed radar" and a "fixed target" situation.

In Eq. 18.2 the term for antenna gain G can be substituted using **Eq. 17.3**, from the Chap. 17.

The equation for gain G is given by

$$G = \frac{4\pi A \cdot K_a}{\lambda^2} \qquad (18.9)$$

where Antenna's geometric area is A and its efficiency K_a. For given radar equipment of different sizes the parameters (P_s, G, λ) can be regarded as constant since these parameters vary only within narrow ranges. The radar cross section, on the other hand, varies considerably. But for practical purposes, it is assumed to be 1 m^2.

After substitution and solving, we get the classic Radar equation for the range R:

$$R = \sqrt[4]{\frac{P_s G^2 \lambda^2 \sigma}{P_{e_{min}}(4\pi)^2}} \qquad (18.10)$$

Here $P_{e_{min}}$ indicates the minimum power received by the radar antenna [3]. With these parameters the Radar station can obtain azimuth direction and scaled distance of aircraft. Early Radars usually operated in the UHF or SHF (Ultra or Super High Frequency) band to maximize range and keep the size of antenna within reasonable limits. Over the past six decades a wide range of frequency bands have resulted in radar applications that are quite widespread. The frequency bands and applications will be depicted as part of an overview in Fig. 18.4 in the next sub-section.

18.1.2 Radar Transponder Beacon

All Radars do not work only on reflected signals in stealth mode. There is a good reason to identify and operate cooperatively with aircraft or ships as seen in ATC (Air Traffic Control) for example. A radar beacon transponder, or simply, a Transponder (**Trans**mit-Res**ponder**), uses the concept of broadcast and expects a response from the aircraft. This method provides positive identification and location of an aircraft on the radar screens of ATC. The methodology is akin to cellular systems where the base stations broadcast beacons all the time and expect cell phones in the vicinity to respond. In this case the ATC radar sends out beacon all the time and expects aircraft in the vicinity to respond.

For each aircraft equipped with an altitude encoder, the transponder on aircraft provides the pressure altitude of the aircraft. It can be displayed adjacent to the "on-screen blip" that represents the aircraft. The Secondary Surveillance Radar (SSR) identifies the aircraft and communicates with the pilot [4]. For example, a pulse can be sent from the aircraft at 1030 MHz and it contains one of the digital octal codes. There are a total of 4096 codes, the pilot can select any one of these codes and send back a reply at 1090 MHz to the SSR.

The third dimension "altitude" is incorporated using such inputs from an altitude encoder that codes the altitude based on barometric pressure [4]. By pre-assigning an octal code (chosen from the 4096 codes) and using the IDENT (identification button) the system can uniquely identify each aircraft and their attitude (height above ground). This third dimension immensely helps navigators at the control tower to carefully direct each pilot to navigate and land at the airstrip.

Figure 18.4 provides an overview of frequency bands and applications, indicated in green is the convention for Radar. The bands indicated in red are the conventional bands used in all wireless communication systems. Each application is briefly described below based on the bands [3].

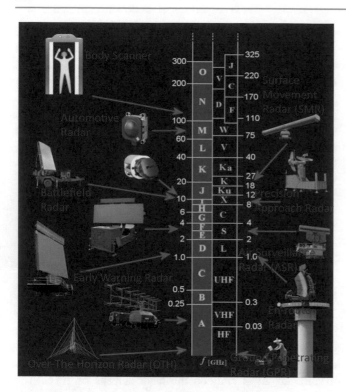

Fig. 18.4 Radar frequency bands and some applications

HF band up to 30 MHz (Radar band A)

OTH—Over the Horizon Radar (OTH): Meteorological OTH is used near coastal areas to measure, waves, wind direction, wind speeds and Tsunamis, but no longer used for reconnaissance. They are also used to monitor drug trafficking illegal migration activities of smugglers.

Ground Penetration Radar (GPR): Used to identify objects below ground—used for geological survey and to identify objects buried under the earth surface, such as pipes, cavities, archeological objects in construction sites particularly in urban areas. Quite often these radars use pulses much shorter than wavelength of the antenna used.

VHF band 30–300 MHz and (Radar band B)

EWR—Early Warning Radar: It served important function during WW-II, but was gradually replaced by reconnaissance satellites and drone that can travel great distances, without detection.

Non-Stealth (conventional) Radar: Not common in recent years—these Radars detect hardly visible stealth bomber and other aircraft operating usually under night cover.

UHF band 300 MHz–1000 MHz (Radar band C)

En-route Radar: Operating in the L-band (around 1 GHz) it is used for aircraft either on way or leaving the area—range is up to 450 km.

Medium Extended Air Defense (MEADS) Radar: Operating in L-band (below 1000 MHz) this can operate well in any weather and maintain considerable range ($\sim$500 km) with great accuracy.

1000 MHz–3 GHz (Radar E and F bands)

Air Surveillance Radar (ASR): Used to identify position of aircraft within the aerodrome area and can also normally display weather conditions around the area. It offers good compromise between range and accuracy. Usually, operation in the sub 2 GHz frequency bands.

3 GHz–10 GHz (Radar G and I bands)

Battlefield Radar: In order to keep its range short and decrease size of antenna for mobility, this operates in the G and I band.

Precision Approach Radar: Operating fairly close to the airstrip, it allows navigators to position aircraft on the approach path to landing. Any deviation from this path would result in navigator talking to pilot for precision guidance to the runway landing zone.

10 GHz–30 GHz (Radar J and K bands)

Surface Movement Radar (SMR): The most common radar used in airfield to monitor aircraft movement as well as vehicular movement within the airfield. It provides good accuracy but operates over distance of no more than 10 km.

Airborne Radar: In order to keep its size small, airborne radar used on aircraft operate around 10 GHz. Even though operating power is quite high, the range is limited to a few hundred kilometers, which is still sufficient to indicate aircrafts within the general region.

30– >120 GHz (Radar bands M and N)

Automotive Radar: A major emerging application for smart cars, it operates in the millimeter wave bands of 60 GHz or higher. Due to large number of vehicles expected on the road, this band is helpful since signal will attenuate quickly reducing range to about 100 m or less.

Body Scanner: Used in airport security it operates in 120 GHz band, using low power. Dry clothing is permeable but body's moisture stops its penetration to only a few mm making it safe and causes no ill effect on human body.

From the above common types, logic seems to indicate co-relation between frequency band used and the objects and surrounding distances. All of these relate to earth and its surroundings. In terms of objects in space, general approach seems to be that Radar is used on a satellite to observe objects on the earth. Tracking satellites uses a different technology—not discussed here.

In terms of operation, these are the two basic types of Radar—one based on reflected signal and the other based on transponder beacon interrogation. But depending on the frequency band and the pulse width used, it is possible to use Radar in a variety of applications. There are Radar Satellites whose main application is to scan features on the earth are report on a periodic basis. The number of applications continue to grow—there are well identified band allocations for Radar use. In international applications such as commercial aircraft used for International flights, such standardized frequency bands are very useful. Such systems are used in other applications such as ships and other vehicles.

18.1.3 Ship Borne Radar Systems

There are certain capabilities and limitations for Radars used on a ship. The first and obvious difference is that Radar is mounted on a moving platform—the ship. Therefore, it is important that this is mounted on a gimbled platform so as to avoid ship movements getting transferred on to the Radar platform. This is similar to the 3-axis stabilized platform indicated in Chap. 13, where VSAT satellite terminals were described with Fig. 13.13. The Radar antenna typically could be located between 10 and 20 m above the ship deck [5]. But use of Radar on a ship is considered "very important". The fundamental reason is safety—in terms of seeing other vessels at a distance and manage steering (navigating) clear of danger.

The International Marine Organization (IMO) set out Convention on the International Regulations for Preventing Collisions at Sea (COLREGs) in 1972. These regulations are safety related to avoid collisions at sea, and "mandatory for all vessels over 3000 tons" gross weight. COLREGs stipulate that it is compulsory to have marine radar in the x-band (10 GHz) and another secondary S-band (3 GHz) radar as a navigation aid. The x-band radar is commonly used for better resolution during normal weather conditions. The S-band radar is used especially during rain or fog conditions. It is also well suited for identification and tracking. These bands are similar to those used by aircraft (see Fig. 18.4).

Tracking devices on ships are mandated by COLREGS to Prevent Collisions at Sea. Terms commonly determined using marine radar are—Electronic Bearing Line (EBL), Variable Range Marker (VRM); these two are used to set the target on bearing and help if the vessel drifts. Time to Closest Point of Approach (TCPA), Closest Point of Approach (CPA) to be verified if there is any vessel within the vicinity of radar. To avoid collision, change of course is needed well in advance (time is of essence), since ships cannot turn around quickly. Some guidelines for boats are available at [6].

Over the years, safety related courses brought considerable education to all sea going vessels. This education was so effective that even smaller boats have started the use of Radar for safety and follow those regulations [7].

18.1.4 Satellite-Based Radar Systems

Satellite-based radar systems commonly use Synthetic Aperture Radar (SAR) which is a technique used by first sending microwave signal, receive backscatter and based on that information, develop model of the target area [8]. This uses a clever concept whereby, a sequence of acquisitions from a shorter antenna are combined to simulate (or synthesize) the equivalent of a much larger antenna, thus providing pictures of much higher resolution. This concept is particularly useful for a Radar on satellite that looks to the earth but is able to get a much better resolution—something that could be achieved by real antenna that would have been several kilometers long. Its basic measurements are intensity (or amplitude) and phase of the backscattered signal, sampled in "time bins along the azimuth" (along track direction of the sensor antenna) and range (across track or perpendicular to the direction of the sensor antenna). There have been altogether 87 spaceborne remote sensing platforms and sensors to build models. Some of the well-known platforms are

ERS-1/2–*European Remote Sensing Satellite 1 and 2*
JERS-1–*Japanese Earth Resources Satellite–1*
Light-SAR–A JPL lead US project.
RADARSAT–Commercial, very similar to ERS.
SEISM–*Solid Earth Interferometric Spaceborne Mission, a French concept.*
SIR-C/X-SAR–Shuttle-borne Imaging Radar
SRTM–*Shuttle Radar Topography Mission.*

Major impact was brought by SRTM. In eleven days, it collected about 18 Terabytes of radar measurements data that allowed scientists to virtually reconstruct a 3-dimensional model of 80% of the Earth's continental area. The radar images collected were converted to Digital Elevation Models (DEMs) spanning the globe between 60° North and 58° South. The `virtual Earth' was reconstructed as a mesh of 30 m spacing and accompanied for each point by a measure of the reflected energy of the radar signal, the intensity image. These data points became an important reference for comparison and correlations with earlier and future satellites for other *Earth Observation* (EO) data. SRTM became a valuable asset for many applications ranging from geology, tectonics, hydrology, cartography, to navigation and communications.

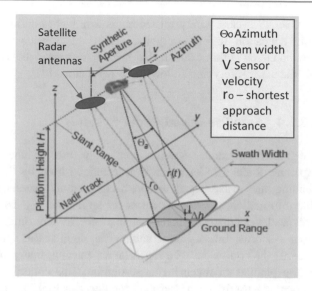

Fig. 18.5 Illustration of SAR satellite geometry

Figure 18.5 shows the basic geometry of satellite in orbit and various angles used for backscatter measurements. Earlier, while describing remote sensing satellites in Chap. 13, Fig. 13.8. indicated "polar orbit" that allows the satellite swath to cover the entire earth surface. Some of the terms indicated in Fig. 18.5 such as swath width refer to resolution of the image. Many remote sensing satellites are in this orbit and carry radar units with parabolic antenna oriented towards the earth as shown in the figure. Typical Sun-synchronous orbits around Earth are about 600–800 km in altitude, with orbital periods in the 96–100-min, and inclinations of around 98°. The radar system transmits electromagnetic pulses with high power and receives the echoes of the backscattered signal from the earth in a sequential manner. Typical values for the pulse

repetition can vary from a few hundred to a few thousand hertz for airborne and spaceborne systems, respectively. The swath width varies from a few kilometers to 20 km in the airborne case and from 30 to 500 km in the spaceborne case. This is used in turn to build 3D models and 4D models (with time) of the earth [9].

18.1.5 Concept of MIMO in SAR

There are several techniques to improve quality of image as well as resolution. Recent research has focused on MIMO where images coming back from the earth could be used by multiple receivers. A receiver with "n" sub-apertures allows simultaneous sampling of the arriving wavefronts with "n" phase centers. The effective number of phase centers can be increased by using additional transmitters. Such extension was suggested for resolution improvement in the context of a forward-looking imaging radar experiment on a helicopter and later elaborated in more detail for a 3D imaging system on a UAV. In a SAR system, possible benefits of using multiple transmitters are as follows: increase of the coverage area, reduce ambiguity in range and azimuth, provide additional baselines for interferometric or even tomographic applications.

In summary, satellite-based Radar has major accomplishments and continues to evolve. The ECM (Essential Climate Variable) that were agreed upon by IPCC (Inter Governmental Panel for Climate Change) and are listed in Table 18.1

In the Table 18.1, items listed by * indicate—parameters observed by SAR. MIMO radar based on the concept of phased array antenna, it offers an opportunity to improve both detection and resolution of target. Each transmit antenna radiates an arbitrary waveform independent of the

Table 18.1 SAR imagery of ECM agreed by IPCC	Features	Imagery "focus areas"
	Land	Multi-purpose land surface imagery, soil type, land cover*, Earth topography (elevation and surface model), lake levels*, subsidence, landslides, erosion, earthquake and volcano monitoring, disaster monitoring, mitigation and assessment, flooding monitoring, coherent change detection, urban and infrastructure planning, road traffic monitoring, soil moisture*, wetlands monitoring, permafrost and seasonally frozen ground
	Vegetation	Vegetation type, forest biomass*, forest biomass change, biodiversity, forest profile, forest height, fire disturbance and monitoring, crop classification, crop height, crop biomass, deforestation, and forest degradation
	Ocean	Multi-purpose ocean imagery, sea state*, ocean currents*, wind speed and vector over sea surface, bathymetry at coastal zones, wave height, ocean wavelength, wave direction, oil spill cover, ship monitoring
	Sea Ice	Sea-ice cover and extent, sea-ice type, sea-ice thickness, iceberg cover and movement, ship route optimization
	Snow and Land Ice	Snow cover*, ice and glacier cover*, snow melting status (wet/dry), snow water equivalent, glacier motion and dynamics, glacier topography

other transmitting antennas. Each receiving antenna can receive these signals. The advantage is that due to the different waveforms, the echo signals can be re-assigned to the single transmitter. An antenna field of N transmitters and a field of K receivers mathematically results in a virtual field of K·N elements with an enlarged size of a virtual aperture. MIMO radar systems can be used to improve the spatial resolution, and they provide a substantial improvement in immunity to interference. By improving the signal-to-noise ratio, the probability of detection of the targets increases.

The MIMO radar systems can be classified into two categories:

- MIMO radar with collocated antennas (known as "Mono-Static" MIMO). The target is a point target as in traditional radar systems.
- MIMO radar with widely separated antennas (known as "distributed" or "Bi-Static" MIMO). The target is regarded by each antenna from another aspect angle.

"Mono-Static" MIMO

When radar antennas are collocated, the transmitting antennas are close enough such that the target radar cross sections (RCS) observed by each transmitting antenna elements are identical. This concept is similar to phased array antennas in which each radiator has its own transceiver module and its own A/D converter. However, in a phased array antenna, each radiator only transmits (possibly time-shifted) a copy of a transmission signal, which is generated in a central waveform generator. In a MIMO radar system, each radiator has its own arbitrary waveform generator. Therefore, each radiator uses an individual waveform. This individual waveform serves as the basis for an assignment of the echo signals to that source.

"Bi-Static" MIMO

In the distributed arrangement of the antennas, the radar data processing becomes more complex. In contrast to "Mono-Static" MIMO, each radar antenna views the target from a different aspect angle. Hence, the target provides a different radar cross section for each radar antenna. This requires much more complex target models for radar data processing.

18.2   Doppler Radar Accuracy and Safety

"Doppler Radar" for the general public represents the "weather radar". But Doppler effect was explained earlier in Chap. 3, Sect. 3.2 as a major issue to be addressed in mobility. In the context of mobility doppler compensation techniques are used to counter doppler shift—an essential effect of any wireless systems to support users who are on the move (such as cellular, satellite). How does doppler radar encounter mobility and use it for weather forecast? Its focus is to study the atmosphere and to report back changes including oncoming storm, rain, tornado, etc. Doppler radar targets clouds, rain drops, and objects in atmosphere that are on the move and approach the region. Doppler radar is a long-range radar—sending out very powerful radio pulses changing both the elevation angle and azimuth [10].

In short, the weather radar tries to build a 3D image of all movements of moisture around it. They send very short pulses and wait for a long time. Major difference in approach is due to the fact that atmosphere in 3 D means a huge space. With no specific target if pulses must be sent—looking for some object such as cloud, rain drop, snow, or ice, then the power of signal has to be extremely high. Since multiple objects from different directions and different distances respond to this pulse, it takes a long time to receive all of them, align them in terms of their location in three-dimensional space. By repeated measurements movement of each of these hundreds or even thousands of objects must be studied before making up a model and then based on it make predictions.

Upper portion of Fig. 18.6 shows how the Doppler Radar operates. Note that it not only turns on its own axis, but also changes its elevation up and down. By repeating rotations on its own axis AND at different look angles (elevation), it gradually builds up a 3D image around itself. That image is based on back scatter received from every small object around it. Fixed objects (buildings or other) or flying objects (aircraft) are distractions, that it must be removed since they don't contribute to the atmospheric change.

It operates by sending high-power pulse transmitted at regular intervals, followed by a long period of "listening". Doppler radar sends one pulse with average power of 450,000 watts or 450 kWs, over very short time of 1.157 μs, followed by a long period of listening which is 998.53 μs. Why such a long period of listening? The emitted wave can travel quite far and gets scattered by even a small rain drop. Also, some of the tornados may be touching the ground, therefore at very low "look angles" the Radar has to filter out a lot of noise, before building an image, which requires considerable time for processing.

Based on the phase and frequency of received signal, the Radar gradually makes up a 3D model of the atmosphere around it. Lower part of the block diagram in Fig. 18.6 shows a fully "coherent radar". All frequencies and clock pulses are derived from the highly stable frequency of a master oscillator and these signals therefore have a phase reference that is stable and the units are synchronized with each other [11]. By many frequency multiplications and mixing of the intermediate results, the frequency of STALO (Stable Local Oscillator) is generated. It is chosen so that it is higher than the transmission frequency with a separation of the used intermediate frequencies (IF).

Fig. 18.6 Block diagram of doppler radar with pulse pattern

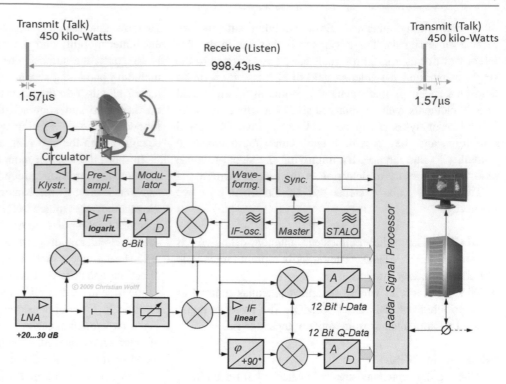

For weather radar, the echo signal is "measured, not just detected". This principle is indicated in the block diagram of receiver (left bottom). The receiver must be protected against severe interference from fixed targets. This requires a large dynamic range, excellent sensitivity, and very good accuracy. To achieve these a logarithmic amplifier is used, with a subsequent analog to digital converter (ADC) to generate a control signal for a regulated attenuator (bottom right). In addition, LNA (Low Noise Amplifier) is used to boost the incoming signal level. On the top left is the transmitter chain —in order to achieve very high power, multi cavity Klystron is used that enables a high pulse power during a short pulse duration. To separate the very high-power RF pulse (transmit) from the weak reflections received, an isolation filter is included which is placed directly at the antenna.

Dual-Polarization

Doppler radar uses another enhanced feature known as dual polarization of the radar pulse. This means the pulse is sent with both vertical and horizontal polarizations. Concept of a "dual polarized antenna" was discussed in Chap. 14 and Fig. 14.14 shows a dual-polarization feed to the antenna. The `dual-polarization' for Doppler Radar, must be further supported by software and hardware attachment to the antenna dish since it brings a lot more information than the conventional two-dimensional picture. Signal processing unit is shown in Fig. 18.6 (on the right side bottom showing

two signals, one of them with 90 degree offset). It must process the vertically and horizontally oriented signals separately and then combine them to provide 3D image of the weather condition.

Dual-polarized radar helps weather forecasters to clearly identify rain, hail, snow, the rain/snow line, and ice pellets. Such details improve forecasts for all types of weather. Another important benefit of dual polarization is that it clearly detects airborne tornado debris (the debris ball). Such details allow forecasters to confirm a tornado is on the ground and is causing damage—this helps them to confidently warn communities in its path. This feature is especially helpful at night when ground spotters are unable to see the tornado.

Doppler radar is also located in the nose cone of aircraft to help weather ahead. Three types of weather-related equipment are used by aircraft during international flights [12].

1. Actual on-board radar for detecting and displaying weather activity.
2. Lightning detectors; and
3. Satellite or other sources—weather radar information that is uploaded to the aircraft from an outside source.

Severe turbulence, wind shear, and hail are of major concern to the pilot. Although hail provides a return on weather radar, wind shear, and turbulence have to be interpreted from the movement of any precipitation that is detected. An alert is

Table 18.2 Comparison between surveillance radar and weather radar

Characteristic	PSR (Pulsed surveillance radar)	Weather radar
Frequency	L, S-band	S, C and X-band (+L-band)
Doppler	Yes	Yes
Scanning	Azimuth or Elevation	Azimuth and Elevation
Processing	Complex & real-time	Very complex, not time-critical
Polarization	Linear and Circular	Dual (vertical and horizontal)
Peak power	Various (kW - Mw)	Various (kW–Mw)
Processing	In phase and Quadrature (I and Q)	In phase and Quadrature (I and Q)
"Picture" update	6–12 s	5–15 min
Clutter processing	Yes (but weather is clutter)	Yes (but aircraft are clutter)
Antenna size	Larger (longer wavelength)	Smaller (shorter wavelength)

annunciated by the pilot to passengers on board, if this condition occurs on a weather radar system. Dry air turbulence is not detectable. Table 18.2. underlines the complexity of weather radar compared to PSR (Pulsed Surveillance Radar).

18.3 RADAR on Aircraft Versus in Vehicles on Road—Comparison

From Fig. 18.2 showing band allocation for different applications, it is clear that most aircraft related Radars operate below the millimeter wave band, while vehicles on the road operate in the millimeter wave band. Reasons for this is due to the following—range of radars for vehicles on the road will be much shorter. At the same time, it is necessary for the radar on a vehicle to differentiate between multiple objects not far from each other. The number of vehicles on the road is considerably higher than aircrafts in the sky. Initial intent of introducing radar in cars and trucks was to emphasize its function as safety tool that assists the driver. Later versions brought automatic braking to boost confidence of passengers, because radar can often sense danger quickly which assists the system to act immediately [13].

The general term for such systems is ADAS (Advanced Driver Assistance Systems) of which Radar happens to be one of the key technologies. Utilizing millimeter wave bands allocated in 24 and 77 GHz they were able to observe full 360 degrees around the vehicle. In order to perform such observation Radars were categorized into three zones based on range.

Three types of Radar are incorporated in ADAS:

- Short-Range Radar (SRR)—1–20 m from the vehicle including Blind Spot Detection (BSD) and Lane Change Assistance (LCA).
- Medium Range Radar (MRR)—detection of objects 1–60 m from the vehicle—height detection which reliably classifies objects and brake safely, even when the object

is stationary (parked object—is it a truck or small box, is it a child running across, etc.).
- Long Range Radar (LRR)—detecting objects up to 250 m from the vehicle. Ability to accurately detect small objects or something crossing the road further down (slow moving vehicle further down, allows gradual braking and speed regulation).

In Chap. 15, Sect. 15.4 covered many sensors used in automobiles (including radar) and their functions. The focus in Chap. 15 was on allocation of frequency band for communication between the vehicle and outside world and how it can be enhanced to driverless vehicles. This will be discussed later in Sect. 18.5. The focus here is in terms of how the radar senses objects and provides accurate information into ADAS system leading to safety. Driverless cars are emerging and the technology of radar (and Lidar) enables safety aspects which will be discussed now. The next section will focus on the need to use millimeter wave band.

Since Radar, Lidar, and other sensor devices must be designed using an established requirement, it is useful to reference an industry standard known as Society of Automotive Engineers (SAE) International's J3016. This standard lists Taxonomy and Definitions for Terms related to On-Road Motor Vehicles. Automated Driving Systems standard is a classification system designed to provide a common terminology for automated driving. This standard defines 5 levels, which are steps with increasing level of automation in automobiles. Bullet points under Sect. 15.6 in Chap. 15 indicated the 5 levels in Fig. 15.13—Current technology is around level 3–4, depending on the vehicle and infrastructure supporting it. All of auto industry has accepted and follows these guidelines while developing vehicles used on the roads today. The levels indicate dynamics of early detection and appropriate actions—time for such dynamic situations are measured using field trials / scenarios. Actions are described in the next Sect. 18.3.1, corresponding to levels indicated earlier in Sect. 15.6 and Fig. 15.13.

18.3.1   SAE International's Levels of Driving Automation for On-Road Vehicle

- Level 0 (no automation): In these vehicles, the driver is in full control at all times.
- Level 1 (function-specific automation): The vehicle takes control of one more vehicle functions, such as dynamic stability control systems. Most modern vehicles fall into this category.
- Level 2 (combined function automation): This involves automation of at least two primary functions. For example, some high-end vehicles offer active cruise control and lane keeping, working in conjunction, which would classify them as level two.
- Level 3 (limited self-driving automation): The vehicle is capable of full self-driving operation under certain conditions, and the driver is expected to be available to take over control if needed.
- Level 4 (full self-driving automation): The vehicle is in full control at all times and is capable of operation, even without a driver present.
- Level 5 (Full automation): The full-time performance by an automated driving system of all aspects of the dynamic driving task under all roadway and environmental conditions that can be managed by a human driver.

Current leading models of vehicles do perform tasks at level 3 such as—"the car takes over driving in certain circumstances, but the driver must be ready to resume control". Take for example highway driving. An important semi-autonomous feature for highway driving is adaptive cruise control, which changes speeds automatically to keep pace with traffic and road rules (speed limits). Another is lane-keep assistance, which uses a front or rear camera to keep the car centered in its lane and maintain safe distance from other drivers. Park assist takes full control during parking in crowded parking lots and garages, and in-cabin driver monitoring can detect an incapacitated driver and initiate a maneuver to pull the car over and stop it safely [14].

Figure 18.7 shows typical interfaces needed in a vehicle to achieve performances indicated in the bullet points from level 3 to 5. Interfaces that work towards level 4 and 5 are in progress in terms of adapting to local road regulations, policy directives for that region. However, trajectory management must be checked particularly under adverse conditions such as snow, heavy rain, and wind. Many of the current models do well in areas with dry weather, even poor visibility conditions posed by inclement weather, does not affect Radar, but it affects cameras.

Figure 18.7 shows that four types of sensors are considered for all autonomous vehicles (RADAR, LiDAR,

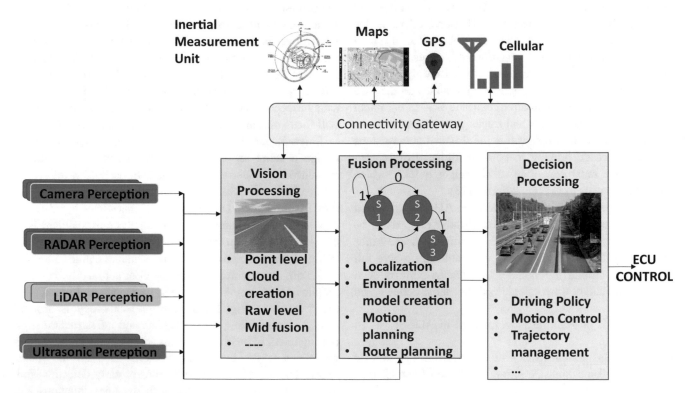

Fig. 18.7 Typical level 3 to 5 system interface

Ultrasonic, and Camera) indicated as inputs on the left side of the figure. These four sensors provide high-speed updates that are first processed by the vision processing unit that acts like eyes of the driver. Inputs from the vision processing unit are then further processed by fusion processing unit, whose function is to collectively use the sensor data with the road view and process it based on local road rules and logic. It also gets inputs from IMU (Inertial Measurement Unit) which helps sensors to stay aligned towards the road on which vehicle is traveling, local maps, GPS (current location with Lat, Long updates), and cellular signals (local traffic update, congestion, alternate routes, etc.). Fusion also enables in overcoming the drawbacks of individual sensor solutions and can provide some level of redundancy.

Finally, the decision processing unit will process all of these to decide whether to change lanes, regulate speed (accelerate or brake), follow speed limits, etc. Its output is used by Electronic Control Unit (ECU) of the vehicle to control acceleration, braking, steering, and other indicator signals that every vehicle must use. In essence it makes the driving related decisions that a good, well-trained driver makes.

Given the overall complexity of how driverless vehicle must operate, there are multiple and complex measurements that RADAR will make. It needs to scan full 360 degrees around the vehicle. The reason to operate it at millimeter wave band, objects, and signs it must discern, and how quickly it must convey information to decision unit will be described in the next section.

18.4   Use of Millimeter Waves in Vehicles

How are the objects viewed by Radar, what are the parameters needed by the vision and fusion processing units, and how is the decision made based on Radar observations? These and related questions are addressed now. Earlier it was noted that large number of vehicles on the road and proximity between them must be carefully addressed. To discern objects so close to each other (such as a person standing next to a car) raises the question don't we need millimeter wave band? The answer may be "yes", but what is so unique about millimeter waves? Let us begin with object resolution on the road and distances. What the driver can see must be evaluated with machine vision and decisions made. This means not only visual clues but quite a few other factors that human driver knows by "intuition" that are not easy to write down as logic.

Earlier in Sect. 18.1, Figs. 18.1 and 18.2 indicated that the range distance from radar to the object is decided by measuring the time taken for a pulse to touch the object and return. It is based on "c" the speed of light and how much

time RF signal takes to reach the object and return. Radar equations provide clues of how range is calculated.

18.4.1   FM Continuous Wave (FMCW)

Since objects such as vehicles are expected to be in close range, slightly different approach to detection is needed. Data must differentiate between two adjacent objects—it may be vehicle or some other stationary object on the road next to some other object that is moving. Range being small, there is very little time to calculate range and act on it quickly or in real time. Instead of a pulse modulated signal, an FMCW (Frequency Modulated Continuous Wave) signal called "chirp" is considered. This "chirp" has properties that are helpful in separating multiple objects at close range and also determine if they are stationary or moving.

An FMCW uses stable frequency source that is frequency modulated over a very wide frequency range—the range needed is so large that it is necessary to operate at extremely high frequency bands that offer such a wide bandwidth (4 GHz in this example). Hence, millimeter wave band of 77 GHz has been allocated by regulators and globally accepted. Such a "chirp" offers better resolution at short distances [15]. To elaborate further let us consider FMCW signal and its properties.

In a chirp signal, FM or Frequency Modulation is used as a ramp on a CW (Continuous Wave) resulting in a frequency that linearly increases with time. The upper part of Fig. 18.7 shows such a signal with increasing frequency over a period of time "t".

To identify an object (stationary or moving) the chirp signal is generated, starting from a very accurate synthesizer whose output is sent out from the Radar transmitter. When this signal is reflected from another vehicle or stationary object, it arrives with the change in frequency and change in phase. The transmitted signal and the received signal are combined in a mixer whose output is known as "IF signal". The IF (Intermediate Frequency) signal is made up of the difference in frequency between the transmitted and the received signal. The IF signal also contains the phase difference between the transmitted and received signal.

Using these two differences, the position of the vehicle can be calculated, since the Radar produces a constant frequency of S2d/c. The slope S, indicated in Fig. 18.8, shows 100 MHz/μ s. If the distance to an object is 300 m then it is a constant tone at 200 MHz, indicated by the f to t plot (frequency of Y-axis and time on the X-axis) in the lower section of Fig. 18.8.

Details of implementing FMCW are illustrated in Fig. 18.9 where the transmit and receive antennas are shown separately although in reality it is the same antenna. A stable

Fig. 18.8 Concept of Chirp used
in FMCW radar

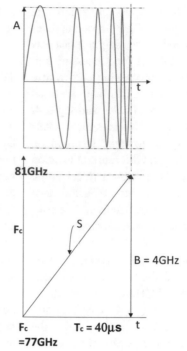

An FMCW radar transmits a
signal called a "chirp". A chirp
is a sinusoid whose frequency
increases linearly with time,
as shown in the Amplitude vs
time (or 'A-t' plot) here.

A frequency vs time plot (or 'f-t
plot') is a convenient way to
represent a chirp.
• A chirp is characterized by a
start frequency (fc),
Bandwidth(B) and duration (Tc).
•The Slope (S) of the chirp
defines the rate at which the
chirp ramps up. In this example
the chirp is sweeping a
bandwidth of 4GHz in 40us
which corresponds to a Slope of
100MHz/μs

Fig. 18.9 Concept of radar using
chirp to find objects

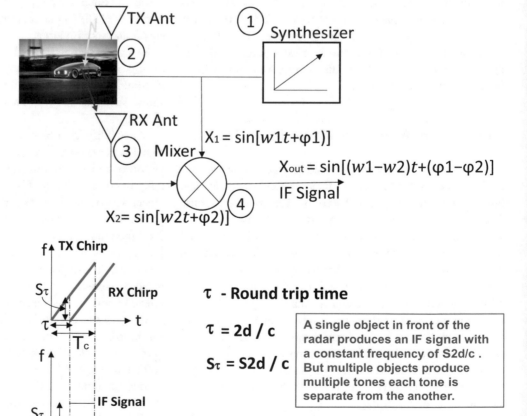

τ - **Round trip time**

$\tau = 2d / c$

$S_\tau = S2d / c$

A single object in front of the
radar produces an IF signal with
a constant frequency of S2d/c .
But multiple objects produce
multiple tones each tone is
separate from the another.

synthesizer is connected to both providing a common source for measurement. When the Radar signal touches the target and return echo is received, a tone is produced as IF signal. A single object provides a single tone. If there is more than one object in the vicinity, more tones are generated. This is shown in the lower part of Fig. 18.9 where two objects are detected and uniquely identified. Advantage of using the mixer is that the "frequency and phase differences" are obtained, separately for each of the objects [15]. That uniquely identifies each object.

In Fig. 18.9 the upper part shows the block diagram concept. This FMCW radar measures the phase, angle of arrival, velocity and helps precise location of objects, using the following:

1. A synthesizer generates a chirp (X1). $X1 = \sin[(w1t + \phi1)]$.
2. The chirp is transmitted by the TX antenna (Yellow line).
3. The chirp is reflected off an object and the reflected chirp is received at the RX antenna (X2) shown as Red line. $X2 = \sin[(w2t + \phi2)]$.
4. The RX signal and TX signal are "mixed" and the resulting signal is called "IF signal" (X_{out}). $X_{out} = [(\sin w1 - w2)t + (\phi1 - \phi2)]$.
5. Mixer is a 3-port device with 2 inputs and 1 output.
6. For two sinusoids $\times 1$ and $\times 2$ are sent to the two input ports, the output X_{out} is a sinusoid which has two parts—the frequency part and the phase part.
   (a) Instantaneous frequency is equal to the difference of the instantaneous frequencies of the two input sinusoids.
   (b) Phase is equal to the difference of the phase of the two input sinusoids.

Lower part of the figure shows the chirp with tones generated. Note that round trip time τ is typically a small fraction of the total chirp time Tc $\rightarrow$ non-overlapping segment of the TX chirp is usually negligible. For example: a radar with a max distance of 300 m and $T_C = 40\mu s$. $\frac{\tau}{Tc} = 5\%$.

18.4.2 Use of DSP (Digital Signal Processors) in FMCW

To provide a faster method to calculate objects and distances, let us consider FFT—Fast Fourier Transform, which has the property of converting time domain signal to frequency domain signal. Concept of FFT was introduced in Chap. 2, with some details of FFT in 2.4.2 and Sect. 2.5 elaborated on DSP. In this case, FFT helps Radar to detect multiple objects and update their positions quickly (in real time) under highway speed driving conditions.

Figure 18.10 shows the concept of observation window. On the left side of the figure are signals in time domain and on the right are the FFT transform of those signals in frequency domain. The top left has two signals, one in blue and the other in red. The blue signal has two and half cycles completed during time interval T, while the red signal completes two cycles. The half cycle difference between the two signals cannot be discerned by the FFT—therefore it shows as only one signal in the frequency domain on the top right [15].

If the window of observation is increased to 2 T then the two signals differ by one cycle. This is because "Blue" completes five cycles while red signal completes four cycles. This one full cycle difference is sufficient for FFT to discern the difference and shows as two distinct signals in the frequency domain on the right bottom.

To generalize this concept, longer the window of observation T, better will be the resolution of signals. The difficulty is that time interval T cannot be kept too long under highway driving conditions. If time interval T is to be kept short, then the signal wavelength must be kept very short. In this case 77–81 GHz (indicated in Fig. 18.8) provides two full cycles in about 25 Ps which is extremely short interval of time. In fact, it may be practical to keep the window open longer for even ten full cycles to get much better resolution. "This is where millimeter wave band becomes extremely useful—window of time to observe is no longer a major limitation". The key feature in millimeter wave band is that frequency is very high and therefore the time needed to complete one cycle is very short: time $T = \frac{1}{f}$. Next, let us consider the objective of detecting objects that are fairly close by.

Earlier in Fig. 18.9 it was observed that Rx chirp is based on the returned signal. The Rx chirp consisted of both frequency and phase that were different from what was transmitted. At the mixer the output signal was $X_{out} = [(\sin(w_1 - w_2)t + (\phi1 - \phi2)]$ which indicates the difference between transmitted and received signal. When multiple objects are present in close vicinity (for example, three objects in close vicinity), the difference in frequency $(w1 - w2)\ t$ and the difference in phase $(\phi1 - \phi2)$ will

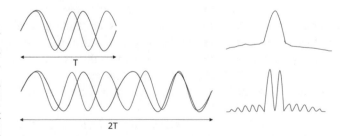

Fig. 18.10 Window of time interval to resolve Radar signals

Fig. 18.11 Multiple objects in close vicinity detected by Radar

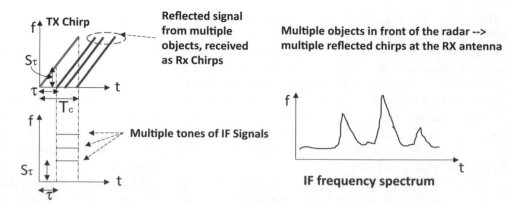

become multiple. To discern three objects, in addition to $(w1 - w2)$, there will be $(w1 - w3)$ and $(w1 - w4)$. Similarly, the phase differences $(\phi1 - \phi2)$, $(\phi1 - \phi3)$, and $(\phi1 - \phi4)$ will be available at X_{out} and they must be distinct.

Figure 18.11 shows such a case where three objects reflect the Radar signal back and receiver indicates three separate chirp signals. Since time window of observation is not a constraint, the frequency components received and their phase differences need to be resolved. The two objects can be resolved by increasing the length of the IF signal. Note that this also proportionally increases the bandwidth. Thus intuitively: Greater the Bandwidth = > better the resolution. Earlier, in Fig. 18.8 it was indicated that the bandwidth in the 77 GHz band is 4 GHz.

For two objects separated by distance Δ_d, difference in their IF frequencies is $\Delta f = S2\Delta dc$. Since the observation interval is Tc, this means that

$$\Delta f = \frac{1}{T_c} \Rightarrow s2 \cdot \frac{\Delta d}{c} > \frac{1}{T_c} \Rightarrow \Delta_d > \frac{c}{2ST_c} \Rightarrow \frac{c}{2B} \quad (18.11)$$

The Range Resolution (d_{res}) depends only on the Bandwidth swept by the chirp:

$$d_{res} = \frac{c}{2B} \quad (18.12)$$

For the 77–81 GHz the bandwidth B = 4 GHz. Hence, the range resolution is 0.0375 m or 3.75 Cm. The radar and its ability to detect objects during high mobility and detecting objects that are close to each other (during backup for parking) have all become possible due to use of millimeter waves along with FFT, it offers some challenges in terms of computation. Part of the challenge is that the world is analog and signals must be converted to digital for processing. Analog to Digital converters that operate at such high switching speeds are needed, which will be discussed later in the next Sect. 18.5.

The allocated frequency band is from 77 up to 81 GHz provides a large bandwidth of about 4 GHz. The velocity resolution and accuracy are inversely proportional to the

radio frequency (RF). Thus, a higher RF frequency leads to a better (smaller) velocity resolution and accuracy. Compared to 24 GHz sensors, the 77 GHz sensor improves velocity resolution and accuracy by a factor of 3. For automotive park-assist applications, velocity resolution and accuracy are critical, due to the need to accurately maneuver the vehicle at the slow speeds during parking [16].

Figure 18.12 shows a representative 2-D fast Fourier transform (FFT) range-velocity image of a point object at 1 m and depicts the improved resolution of the 2-D image obtained with 77 GHz. Image on the left using 24 GHz, cannot resolve objects well (smudged colors), while image on the right clearly shows two objects 1m apart. It is important to note the "relative speed" compared to the automobile's own velocity. It means objects are stationary on way side, as shown in Fig. 18.12. Recent research with improved algorithms uses radar for pedestrian detection and advanced object classification. Due to higher resolution "micro-Doppler signature" has become available from the 77 GHz sensor. The increased precision of velocity measurements helps other industrial applications as well. Consider a process chemical plant where liquid level must be measured. High range resolution minimizes dead zone at bottom of the tank. Level sensing can be performed accurately in tanks filled with liquid—the level can be sensed till "the last drop"—unlike mechanical or other sensors that have a "dead zone" and will not sense once liquid level falls below a certain depth, since the resolution will be poor [16].

18.5 Integrating Radar with Wayside Network

Radar and other sensor devices must get integrated into self-driven vehicle network. Also, wayside network may communicate outside the vehicle—this is a major topic in recent years. It was indicated earlier that there are four major types of sensors on autonomous vehicles. Among them

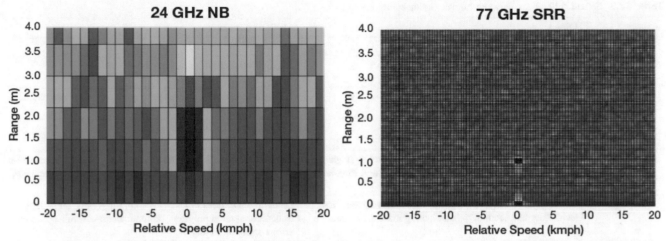

Representation of 2-D FFT image of a point object – comparison between 24 GHz and 77GHz

Fig. 18.12 Resolution of FFT image of an object–at 24 and 77 GHz

Radar, lidar, ultrasonic sensors are three very useful inputs in overcoming ambient light needed by the fourth sensor which is camera. Depth information, i.e., distance to objects, can be measured effectively to retrieve 3D information with these three sensors, and they are not affected by ambient lighting conditions [16]. Lidar uses Electro Magnetic waves in the infrared band of the spectrum and helps detect objects better if they are less than 100 m away. With this information, let us review how they get integrated into the wayside network.

18.5.1 Human Driver to Autonomous Vehicle— Motivation

A major finding from study comparing human driver to machine driving, states that "even though human drivers are still better at reasoning in general, the perception capability of ADSs (Advanced Driving System) with sensor-fusion can exceed humans, especially in degraded conditions such as insufficient illumination [17].

To keep up with real-time driving conditions, the concept of YOLO (You Only Look Once) is introduced to provide images of reasonably good quality. YOLO limits post processing to accommodate real-time driving usage—to provide scenes of acceptable quality [17]. Human drivers have handled automobiles for a century, yet there are some aspects humans are not very good at—resulting in accidents, for example, stress and other factors during driving [18]. Table 18.3 shows salient differences between human and machine driving abilities.

This table shows that human driver, under adverse weather conditions is not safe. In addition, there are other factors related to human health conditions that contribute to accidents. Machine and ADS were introduced with focus

towards safety and avoid tenuous situations—examples include immediate braking during unanticipated situations (child runs across or vehicle ahead breaks suddenly), entire line of vehicles entering next lane due to accident ahead—exercising caution while changing lanes and others while following local speed regulations.

To overcome limitations such as ambiguity, perception, a number of algorithms were developed based on a wide range of sensors. This led to the concept of fusion—which means to take an average or best of the inputs, etc. based on situation. Other compensatory methods including laying out the route well before hand and not depend on GPS signal all the time. Also, knowing local road regulations beforehand and applying them appropriately throughout the route—minimizes ambiguity. Some of these features are made available to human drivers—to make them appreciate how a machine understands such information and truly follows them during travel.

18.5.2 Sensors—Comparative Strengths

LIDAR, Radar, ultrasonic sensors, and cameras—each of them has their own niche sets of benefits and disadvantages. Highly or fully autonomous vehicles (levels 4 and 5) typically use multiple sensor technologies to create an accurate long-range and short-range map of a vehicle's surroundings under a range of weather and lighting conditions. Sensor technologies complement each other and have sufficient overlap to increase redundancy and improve safety. Sensor fusion is the concept of using multiple sensor technologies to generate an accurate and reliable map of the environment around a vehicle. Ultrasonic waves suffer from strong attenuation in air beyond a few meters; Ultrasonic sensors

Table 18.3 Salient differences between human and machine driving

Aspect	Human	Machine/computer
Speed	Relatively slow	Fast
Power output	Relatively weak, variable control	High power, smooth and accurate control
Consistency	Variable, fatigue plays a role, especially for highly repetitive and routine tasks	Highly consistent and repeatable, especially for tasks requiring constant vigilance
Information processing	Generally single channel	Multichannel, simultaneous operations
Memory	Best for recalling/understanding principles and strategies, with flexibility and creativity when needed, high long-term memory capacity	Best for precise, formal information recall, and for information requiring restricted access, high short-term memory capacity, ability to erase information after use
Reasoning	Inductive and handles ambiguity well, relatively easy to teach, slow but accurate results, with good error correction ability	Deductive and does *not* handle ambiguity well, potentially difficult or slow to program, fast and accurate results, with poor error correction ability
Sensing	Large, dynamic ranges for each sensor, multifunction, able to apply judgement, especially to complex or ambiguous patterns	Superior at measuring or quantifying signals, poor pattern recognition (especially for complex and/or ambiguous patterns), able to detect stimuli beyond human sensing abilities (e.g., infrared)
Perception	Better at handling high variability or alternative interpretations, vulnerable to effects of signal noise or clutter	Worse at handling high variability or alternative interpretations, also vulnerable to effects of signal noise or clutter

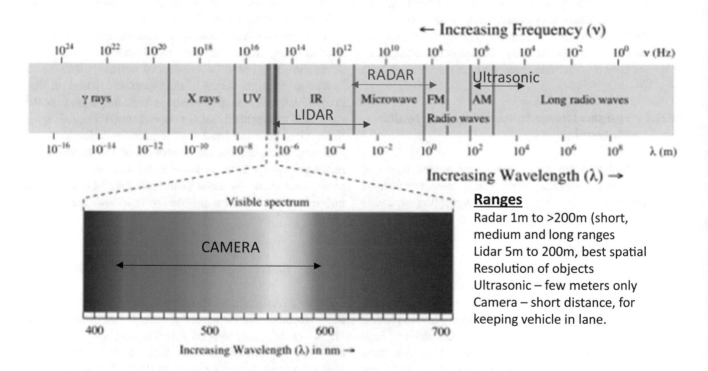

Fig. 18.13 Various sensors—range and resolution

are primarily used for object detection within a few meters. Typical applications are during parking where they are helpful in finding adjacent vehicles, curb, and other limits that help in parking the vehicle.

Figure 18.13 shows the frequency bands and ranges over which the four sensors operate.

Each sensor has special features and operates in specific frequency bands that do not interfere with each other.

Designers tend to use camera for keeping vehicle in lane or follow road signs since it can see color vividly, using good ambient light (daytime is better). Lidar on the other hand provides excellent spatial resolution of 3D objects since it operates well beyond the Infra-Red band (from 905 to 1550 nm).

Radar operates over a wider frequency range and operates well even under adverse weather conditions. Due to this feature Radar is used for two separate functions. One is for the normal range in front of the vehicle that helps in immediate inputs to the decision-making process. The other feature is "long range" radar which can see objects such as trucks or other obstructions such as parked vehicle on the road (due to construction or an emergency, for example) that helps to maneuver the vehicle either by shifting lane or reduce speed or both. This specialty of Radar is illustrated in Fig. 18.14 with relevant comments.

Ultrasonic is based on sensing using sound (beyond human hearing)—it has strength in resolving objects around the vehicle within its lane. It is, therefore, very useful if there is another vehicle next to it on either side during highway drive, or it can help during self-parking to know objects within the immediate vicinity.

Figure 18.14 shows the concept of autonomous vehicle with different sensors [19]. Upper part of the figure shows the four sensors, their capabilities and the range of operation. Note that Radar is indicated twice, once for the short range and again for long range. Radar is one sensor that has three separate ranges as indicated earlier—short, medium, and

long range. Radar provides information such as distance, angle, and velocity both in the front and in the back.

Post processing of data in real time to enable highly automated vehicle (Level 5), is still a challenge. It is actively being worked by various research and industry experts. Since it assumes "no driver present", it is important to first try algorithms on vehicles that operate on known routes. That provides the basis to pre-establish (or use as a reference) the entire route, with clearly known signs, exits, number of lanes, etc. That helps the driverless vehicle to "know" everything about that route, allowing it to focus on current traffic conditions which must be recognized, processed, and used in real time to drive the vehicle. Since road conditions vary considerably even within the same route, a wide variety of road conditions were studied over the past decade, the result is being applied to update the vehicle's capability.

Finally, "communication with network" is all around the vehicle. This concept is used to define the "connected vehicle". What is the difference between an autonomous vehicle and a connected vehicle? An autonomous vehicle can drive without a human driver. But a connected vehicle is not only autonomous, but also communicates with other vehicles on the road in its vicinity. That helps other vehicles to know the intent of this connected vehicle. Is it planning to take the next exit—does that mean it will change lane? Does it plan on moving to a slower lane since its passengers requested, etc. "Connected vehicle" will also communicate with the Internet to know about "weather at the destination", whether there are restaurants or other facilities within the vicinity and can show that to passengers. Connected vehicle can also provide a lot of other information that typical traveling passengers would look for. Online connection may be needed for other purposes as described in the next section.

18.5.3 Use of URLLC Standard for Network Connection

Earlier in Chap. 11, Sects. 11.6 and 11.8 indicated that work in 5G standard includes URLLC (Ultra Reliable Low Latency Communication) particularly to support "Connected vehicles". What is unique about this part of 5G standard is that it considerably challenges current network technologies. The accepted standard of five 9 s reliability 99.999% for any network, but it is now taken a step further—improve it to six 9 s reliability 99.9999%. This is not a small task but given the importance of driver safety on roads is now being assigned to a machine—demands such reliability. The second part is "Low Latency"—what does this mean? The current IP network has latency in the range of 30 ms because it is based on Ethernet and routing the signal through the

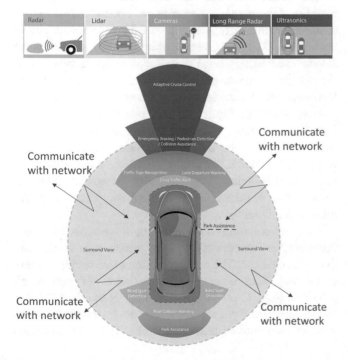

Fig. 18.14 Vehicle sensor with range and limitations

"best route". Routers and switches that constantly provide best route search, introduce delay to accomplish this task. But "Low Latency" defined by this standard is 1 mS or less. End-to-End delay tolerated by this standard is 10 mS and compare this to current IP network technology which uses end-to-end delay of 100 ms.

Such dramatic increase in reliability combined with decrease in latency has become a major challenge. Yes, over the air "from vehicle to any network node nearby" the signal travels at the speed of light. If this happens to be a cellular network then security is much better than WiFi or other wired networks. But other wayside networks cannot continue its current topology of routers and switches. It must offer a separate dedicated link to support such "quick, low latency, real-time turn around" for data needed by the vehicle. There is considerable work in this direction and 5G standard URLLC has set a higher bar for all Ethernet-based IP networks.

The amount of information in terms of throughput is not expected to very high, it is in the range of 50–10 Mbps, as indicated earlier in Chap. 11, and in Table 11.6.1. But the turnaround time (latency) from the network and the reliability are key factors. How does this relate to the four sensors including Radar? Each of the sensors—after providing output, must continue their real-time sensing, detection, angle, distance, and velocity numbers. Behind these sensors the fusion and later decision processing work with high-speed ADC (A to D converters) and DSP as explained earlier. Many of these results are used locally at the vehicle from decision processing to actuators that drive the vehicle. However, information related to its route plan—if there was a route change announced on the Internet, or a weather update that might change the route, are all needed quickly.

On the other hand, if there is an accident ahead, resulting in piling up of vehicles in the lane, how to slow down (break) carefully must be communicated immediately to vehicle ahead and vehicle behind. Long range radar may be put to work in such situations. These are situations that need immediate attention of the wayside network and quick turnaround. Such "communication from the vehicle" is known by the generic term V2X communication. V2X (Vehicle to everything) is a key technology enabler to enhance the vehicle's perception of its surroundings. It enables the vehicle to communicate with other vehicles and its surrounding environment. The concept involves overall ability of vehicles being able to exchange messages with each other (V2V or Vehicle to Vehicle), with the network (V2N or Vehicle to Network), and pedestrians (V2P or Vehicle to Pedestrians that uses short-range radar), or the infrastructure (V2I or Vehicle to Infrastructure which

includes the Internet). There are potential lifesaving benefits as V2X communication brings a new level of situational awareness leading to increased road safety as well as traffic efficiency [20].

Figure 18.15 shows a typical scenario where multiple vehicles are at a traffic junction—camera, short-range Radar and ultrasonic sensor may be active during such a scenario actively using V2X communication with the signals as well as pedestrians.

In Chap. 15, it was indicated that much work was done using WiFi for almost a decade. Then the industry changed course and moved in favor of cellular network to support V2X. Proponents of C-V2X (Cellular V2X) highlighted advantages of a cellular-based system over an ad hoc WiFi-based system (better security, congestion control, reliability, etc.,). The ability to leverage the existing cellular infrastructure (V2N) and 3GPP-R14 (release 14 of the cellular standard 3GPP) enhanced the "PC5 Side link" interface for Direct Communications allowing V2V communications between vehicles. Finally, as a part of the 3GPP standards family, C-V2X offered an evolution path to 5G.

Due to the work done by 802.11p and now major drive from the cellular 3GPP standards body, there are two distinct technologies to be considered for Cooperative ITS (C-ITS): WiFi-based 802.11p and C-V2X. While there is a debate on which technology is better suited for V2X, there are also some interest to make both technologies coexist to leverage the best of both since they share a common goal, i.e., realizing a safer and more efficient mobility for all.

The OPEN (One Pair Ethernet) Alliance was formed to establish Automotive Ethernet as an open standard and to encourage wide scale adoption of Ethernet connectivity in automotive [21]. The Ethernet, although very popular in the Internet Technology network industry, is a relatively new to "in-vehicle point-to-point communication technology". Currently many vehicle manufacturers use their own proprietary standard. Open Alliance based on Ethernet was introduced to address the automotive increasing need of bandwidth. It is expected to reduce cost, weight, and complexity as compared to existing in-vehicle wired communication technologies. The modular setup of Ethernet transceivers, switches, and controllers support such scalability and flexibility.

Ethernet communication network is seen to be a key infrastructure component for future autonomous driving and the connected car. The trend for automotive wiring harness is moving from heterogeneous networks of proprietary protocols (e.g., CAN—Controller Area Network, MOST—Media Oriented Systems Transport) to hierarchical homogeneous automotive Ethernet networks. The IEEE standards

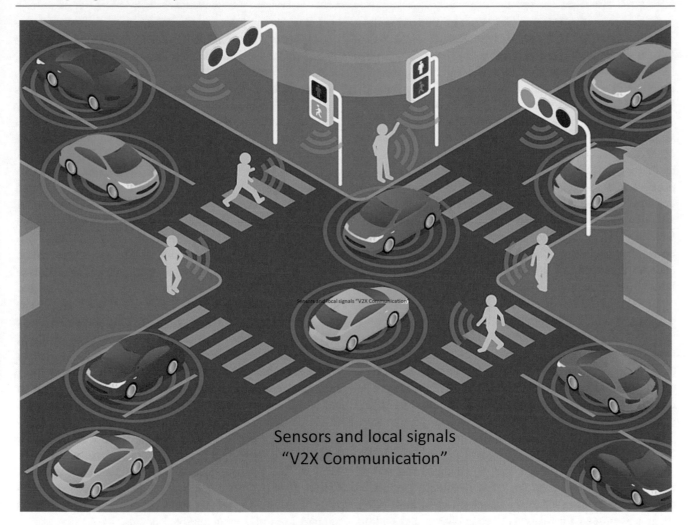

Sensors and local signals
"V2X Communication"

Fig. 18.15 Sensors and local signals at traffic junction communicate using URLLC

association is currently working to add Time Sensitive Networking (TSN) features to the existing standards for 802.1 and 802.3 Ethernet to provide deterministic performance, which is vital for real time, mission critical applications. With the inclusion of the TSN feature, Ethernet is poised to become the core network backbone meeting the complex demands of future vehicles [21].

For a glimpse of current technology in autonomous and connected vehicles, it is worth visiting the technology page of Tesla, one of the popular automated vehicles [22].

18.6 Conclusion

It would be appropriate to describe "Radar" as the ever-young wireless technology admired equally by centenarians (who participated in the second world war) and teenagers who love the concept of Radar in a driverless car. Imagine a Radar conference—it would be where grand old daddy and college going youngster would feel at ease—

since Radar has done what it always did "to measure distance, angle, and velocity" of the object unseen.

Starting out with the concepts used during WW-II, the chapter overviewed the "Radar equation", its relevance, followed by Transponder (**Trans**mitter-Res**ponder**) to not only identify but move a step further—communicating and helping aircraft land safely at the airstrip. Use of Radar as a ship-based sensor was introduced for safety reasons. Its use as a helpful tool quickly led to maritime organization approving it as mandatory for all ocean-going vessels.

From the earth-based vehicles—moving into space and stay on a satellite was a major jump for Radar. Its usefulness in space was underlined by the space shuttle that established it as a major tool to study the earth and exploit its abundant resources. Synthetic Aperture Radar or SAR is a very useful technique used by satellites to study the backscatter from earth and make use of amplitude and phase of the information obtained to form a 3D model of earth's ever-changing atmosphere and long-term environment.

The concept of MIMO—widely used in WiFi and cellular systems—is also used by Radar. But Radar uses MIMO to look at the same object from various angles to form a good 3D model. At the same time, it is also used to improve the accuracy in terms of location and resolution. This approach has provided pointers to more research work in the area.

In recent times Radar moved further into public arena of vehicles on the road. It challenges the human driver and often does better by showing objects unseen by human eye, and far beyond the visible range of the driver. It brought the possibility of replacing the human driver—allowing global population of vehicle drivers to get away from the mundane task of driving. Use of newer technologies including millimeter wave allowed better resolution within the driving range. The concept of chirp—applied to millimeter waves, allowed many other applications switch as more accurate liquid level sensors among others.

To identify and help global population confronts everyday "the weather"—Radar adopted an interesting strategy—to send a powerful pulse and take the backscatter response from the environment around and study it for a "long time" to develop a model of dynamic environment around the radar site. This method of pulsed radar signal anchored its role as weather radar—that is so popular in airports and shipyards. It is now popular with the general population as "doppler radar" and everyone observes on TV or smartphone of what id a "weather map" or "radar map".

Effectively, it is perhaps best to conclude that, of all concepts in wireless technology this is perhaps the longest running and most successful technology. It is most often, silent and unseen by many yet accomplishes and endears itself to all—the silent giant that tirelessly works and becomes the most wanted technology for all generations to come.

Summary

The chapter started with a quick overview of how RADAR was independently developed by multiple countries before the second world war. This was followed by Radar equation which provides the relationship between Range and power sent out, power received, antenna gain, antenna cross section. This provides a 2-D view of Azimuth distance (Range) from the Radar site. For quite some time it remained a one-way communication system with Radar site silently seeking out objects (aircraft) without anyone in the aircraft knowing whether they are being tracked. Radars operate over a wide range of band starting from a few MHz (UHF or SHF) all the way into the millimeter wave band. These bands and their applications were reviewed.

Later, the ATC (Air Traffic Control) established the concept of bi-directional communication. It clearly broadcast which Radar it is, its location and sends it out as beacon, seeking information from the aircraft in terms of its altitude. This provides a 3D view of aircraft location that includes its height. ATC systems are used by all civilian airports today providing support to aircraft to safely land at the runway. In addition, the system also tracks other aircraft and vehicles in the airfield.

Use of Radar in ships was a natural extension and its location at a high mast on the ship deck with a gimbled platform made sure that Radar could look out for other ships, aircraft or while approaching a port buoys and at open sea look for ships that are beyond the horizon. Its usefulness was recognized by the International Marine Organization (IMO) which has set out regulations known as COLREGs in 1972. These regulations are mandatory for all major sea going vessels. In recent years, even small boat owners use Radar to navigate through crowded channels avoid buoys and other objects that are used as warning of shallow water.

Synthetic Aperture Radar (SAR) is a modified version of Radar mounted on satellites which move in polar or other near-earth orbits with SAR looking towards the earth. It sends out pulses and receives backscatter from the earth. Since the satellite move around the earth it collects information from every part of the globe. Then it analyzes the backscatter extracting information about various resources such as forest cover, mineral deposits, and others from Land, Ocean floor etc. Such information has long-term implications about environmental pollution, global warming and has support from major international bodies.

The concept of MIMO used in WiFi and Cellular is used in an entirely different way. Instead of focusing on increasing network throughput, it uses multiple streams as beams to observe the same object from various angles, to provide more accurate 3D information about that object.

Perhaps the most popular form of Radar image that general public is familiar on TV screens is the "weather radar" map showing movement of rain, storm, and related images. Weather prediction uses "doppler radar" that is unique in the sense it mostly listens. In about 1000 microseconds it only transmits a powerful pulse for less than 2 microseconds, and for the other 998 microseconds it listens analyzing information from various objects in the sky and land surface. Backscatter received from even small objects like rain drop to large mass of snow, ice pellets, and storm— are analyzed. Dual-polarized antenna sends pulses in vertical and horizontal polarization, to collect more details in three dimensions indicating not only where such storms are, but in addition provide details of in which direction it is moving, its speed, etc.

Use of Radar in automobiles started with certain vehicle safety related features—such as quick braking in case an object or person running across the driving lane. It gained acceptance and moved to serve other safety features such as cautioning drivers about other vehicles passing on either side of the vehicle. Based on its performance the Society of Automotive Engineers (SAE) established levels of automation towards "driverless" vehicles.

With extensive work supporting autonomous vehicle brought the concept of using millimetric wave bands for Radar. This band offers considerable bandwidth (about 4 GHz bandwidth) allowing a different type of Radar known as FMCW (Frequency Modulated Continuous Wave). This concept known as "chirp" was described showing why it is able to differentiate objects that are in close range and are fairly close to each other. Such resolution at short range is very important for automobiles traveling along highway and need to observe and measure multiple objects in real time.

Finally, to move from autonomous vehicle (self-driven) to a "connected vehicle" four major sensors are needed. Radar (short, medium, and long range), Lidar (Light detection and ranging), Ultrasonic (uses sound for analyzing nearby objects during self-parking, etc.), and camera which can see color and road signs. How to integrate them, analyze and then communicate with other vehicles on the road, was the final topic of this chapter. Radar plays an important role in this effort, providing information about object that are either near the vehicle or much farther beyond visible range. Communication methods include radar sensing traffic intersections and provide information to other vehicles and pedestrians within the vicinity.

This chapter covered a wide range of applications supported by Radar for almost a century of its existence. While its basic functions remain the same—"detection of objects" applications demanded modifications to its design, including features that need considerable data processing. Operating over a very wide range of frequency bands, radar has remained an important tool that supports an ever-growing range of new trends that rely on wireless technology.

Emphasis on participation

(a) Where is the Doppler Radar system located in your area? Contact the local meteorologist and find out what weather patterns can they predict? How does the national weather service work—does it have multiple locations with local forecast?

(b) If there is an airport or shipping port in your area, request a visit to the radar operations office and note down how good is the accuracy of their system? How many aircraft can they handle simultaneously and track them?

(c) If there is a military museum in your area, look for exhibits of Radar from different times. Which is the oldest and most recent military radar and which force uses it—Army, Navy, or Air Force?

(d) Is there a body scanner radar in your nearby airport or other facilities? Who was the manufacturer and what frequency band does it use?

Knowledge gained

A. Concept of Radar—how it operates, frequency bands, and applications.

B. Radar equation—parameters and concepts used in detection of objects.

C. What are the different types of radar—how does the Radar adapt to different applications, in terms of post processing, hardware changes, range, resolution, and accuracy.

D. Radar on a fixed platform and rotating, Radar on a moving platform—aircraft, ship, satellite, and dynamics of estimation.

E. Differences between detection of objects in the sky / sea, Versus from a moving vehicle detection of objects stationary or moving on crowded streets.

F. The "Doppler Radar"—used as a tool in studying atmosphere around it—limitations such as why it cannot detection dry air storms.

G. Concept of SAR (Synthetic Aperture Radar)—its use on satellites.

H. Concept of FMCW with chirp, its application in autonomous vehicles.

Homework problems

(1) simulation software notebook, first run (1) radarsimpy. Transmitter (2) radar.simpy.Receiver and (3) radarsimpy. Radar for example see: https://github.com/rookiepeng/radar-simulation-notebooks

(2) Simulate and understand Radar parameters and compare them with the Radar equations. What are the range of power levels, antenna gain that you can use? Check with practical radar systems indicated in ref. [2].

(3) Minimum power received by radar is 10 W to operate well in the 300 MHz band. If transmitted power is 2000 watts and cross-sectional area of antenna is 1 sq. meter, what is the range of the Radar whose antenna whose gain is 30 dB?

(4) The 77 GHz band offers 4 GHz of bandwidth for FMCW Radar? There are other bands such as 90 GHz

where the bandwidth could be 6 GHz. All other features being same, what will happen to its range, resolution?

Project ideas

Use Radar demonstration kits see: https://www.pasternack.com/pages/RF-Microwave-and-Millimeter-Wave-Products/radar-demonstration-kit.html

Build and operate a radar system in the following modes:

- CW,
- FMCW,
- Doppler and other modes: Radar is quite advanced now. Multifunctional radar, OFDM radar, Passive Radar are also proposed.

(a) What is the range achieved in 2.4 GHz band? Can it be operated in other bands?

(b) Can this be built as a SAR (Synthetic Aperture Radar) —what software can you search interface?

Out of the box thinking

Since many boats currently use ship's radar systems—is it possible to use Radar system to navigate river rafting using conventional or FMCW radar used in cars?

What options does a hang glider have in terms of navigating and getting support from a ground-based Radar—is it possible for such a glider to use an airport-based ATC and travel longer distances?

References

1. Foley S (2011) World War II technology that changed warfare–radar and bombsights. Johnson & Wales University–Providence https://scholarsarchive.jwu.edu/cgi/viewcontent.cgi?article=1011&context=ac_symposium
2. Roshni Y Electronics desk, radar system https://electronicsdesk.com/radar-system.html
3. Wolff C Radar tutorial—radar equation. https://www.radartutorial.eu/01.basics/The%20Radar%20Range%20Equation.en.html
4. Aircraft radar beacon transponder, https://www.aircraftsystemstech.com/2017/05/radar-beacon-transponder.html
5. Capabilities and limitations of ship borne radar, chapter III, section D. https://www.ibiblio.org/hyperwar/USN/ref/RADONEA/COMINCH-P-08-03.html
6. Bhattacharjee S (2020) Marine Radars and their use in shipping industry. https://www.marineinsight.com/marine-navigation/marine-radars-and-their-use-in-the-shipping-industry/
7. Englert K (2017) Low visibility radar tips. Boating magazine https://www.boatingmag.com/gear/low-visibility-radar-tips/
8. Inggs MR, Lord RT Applications of satellite imaging radar. Department of Electrical Engineering, University of Cape Town, South Africa, Private Bag, 7701 Rondebosch http://www.rrsg.uct.ac.za/applications/applications.html
9. Moreira A, Prats-Iraola P, Younis M, Krieger G, Hajnsek, I., & Papathanassiou, K. P. (2013) A tutorial on synthetic aperture radar. In: Microwaves and Radar Institute of the German Aerospace Center (DLR), Germany. in March 2013 IEEE Geoscience and remote sensing magazine. https://elib.dlr.de/82313/
10. National weather service—how radar works: Doppler Radar. https://www.weather.gov/jetstream/how
11. Wolff C The technical principal of weather radar, Radar tutorialhttps://www.radartutorial.eu/15.weather/wr04.en.html
12. Aircraft weather radar https://www.aircraftsystemstech.com/2017/05/weather-radar.html
13. Browne J (2019) Driver safety with automotive Radar systems. Microwave and RF magazine https://www.mwrf.com/print/content/21849983
14. Sagar R (2017) Making cars safer through technology innovation. ADAS automotive processors, Texas Instruments https://ti.com/lit/sszy009
15. Rao P Introduction to mmWave sensing, FMCW Radars. TI Training Texas Instruments https://training.ti.com/sites/default/files/docs/mmwaveSensing-FMCW-offlineviewing_4.pdf
16. Ramasubramanian K, Ramaiah K Moving from legacy 24GHz to state of the art 77GHz Radar. Texas instrumentshttps://www.ti.com/lit/wp/spry312/spry312.pdf?ts=1608490693068
17. Yurtsever E, Lambert J, Carballo A, Takeda K A survey of autonomous driving: common practices and emerging technologieshttps://arxiv.org/pdf/1906.05113.pdf
18. Schoettle B (2017) Sensor fusion: A comparison of sensing capabilities of human drivers and highly automated vehicles. University of Michigan,Sustainable Worldwide Transportation, Tech. Rep. SWT-2017–12
19. How to see the invisible infrared world with your mobile camera, blog by Ken–Ken tips April 4, 2010 https://kenstechtips.com/index.php/how-to-see-the-invisible-infrared-world-using-your-mobile-phone-camera
20. Tuveri G (2017) Technologies for connected and autonomous driving. Keysight Technologies, Tech Time, Electronics and Technology News, Israel https://techtime.news/2017/11/29/connected-and-autonomous-driving/
21. Open signal—standards body to enable adaption of Ethernet based automotive connectivity http://opensig.org/home/
22. Tesla motors. Autopilot press kit https://www.tesla.com/presskit/

Abbreviations

Meterlogy	The scientific study of measuremen. It establishes a common understanding of units, Crucial in linking human activities
National standards laboratory	The repository in each country where standards of different units are kept, regularly measured, and verified with standards of other countries. The objective is to establish a common basis for all measurements that are internationally agreed upon
Precision and accuracy	In a set of measurements, accuracy is closeness of the measurements to a specific value, while precision is the closeness of the measurements to each other
Cesium beam standard	A primary standard based on the fundamental property of Cesium 133 atom. By definition, radiation produced by the transition between the two hyperfine ground states of cesium (in the absence of external influences such as the Earth's magnetic field) has a frequency, $\Delta \nu_{Cs}$, of exactly 9192631770 Hz
Crystal oscillator	Refers to an oscillator built around a quartz crystal whose frequency of oscillation is quite accurate
Short-term stability	Short-term RMS (Root Mean Square) average of frequency variations or time domain stability is a measure of the frequency or phase noise. It is measured in terms of variation over seconds and minutes. Variations over this time period are due to temperature variations, shock, and environmental factors, some of which can be controlled. But it is inherent to the crystal and well documented
Long-term stability	Long-term stability is generally referred in days, weeks, and months. Variations over this time period are mainly due to aging of the crystal. Aging is a constant and linear process for which there is no control
OCXO or TCXO	Oven Controlled Crystal Oscillator or Temperature Controlled Crystal Oscillator, indicate oscillator is kept in a controlled enclosure in terms of temperature, shock, and vibration to control short-term stability. It can improve accuracy from 1×10^{-6} to $1 \times^{-8}$, offering a considerable improvement in stability. Often these are located within a precision instrument or located externally with feed as

© Springer Nature Switzerland AG 2022
K. Raghunandan, *Introduction to Wireless Communications and Networks*,
Textbooks in Telecommunication Engineering, https://doi.org/10.1007/978-3-030-92188-0_19

	an external reference to measuring equipment		conductor continues through the instrument
MEMS (Micro Electro-Mechanical Systems)	MEMS oscillators offer frequency stability with excellent environmental tolerance useful for rugged applications such as automobile "under the Hood" applications. They are fairly accurate 1ppm to 0.1 ppm accuracy but can tolerate much wider temperature range. They also lower frequency bands and can be fitted directly in an Integrated Circuit	Load matching	The process of measuring the forward and return waves. Therefore, load matching often implies changing the cable length to match "half wavelength" for the frequency at which the system operates
		Impedance	The electrical terms, the elements that "impedes" the flow of current is known as impedance, which refers to the three basic elements, Resistor, Inductor, and Capacitor
Calorimeter	Calorimeter is a process or device to measure heat— usually heat absorbed by some known substance such as water. Absorbed dose is the energy imparted per unit mass by radiation to an irradiated target. The SI unit of absorbed dose is the gray (abbreviated G_y), equivalent to J/kg, and it is realized most directly via calorimetry. Calorimeter is used as the primary standard for RF power or in medical science for similar heat measurements	Reactance	The electrical terms, the elements that "reacts" to the frequency. This does not include Resistor since does not "react" to frequency, but it includes Inductor which "reacts" linearly to the frequency. The term also applies to a capacitor which "reacts" inversely to the frequency of operation
		Quality factor Q	Quality factor is an important measurement in AC circuits. It is the ratio of reactance to resistance. Its reciprocal is known as dissipation factor D. In practice Q is used for inductors and D is used for capacitors
Thermistor mount	"Thermally sensitive Resistor" is Thermistor. Thermistor mount includes a carbon absorption material that receives the RF power allowing the Thermistor to measure it. Thermistors normally have negative temperature coefficient, i.e., their resistance decreases as the temperature raises. Often, they are kept in a bridge configuration to provide better accuracy	Josephson Junction	This Noble Prize-winning work states that "at a junction of two superconductors, a current will flow even if there is no drop in voltage. This is used as an excellent voltage reference standard that in turn provides impedance reference for capacitors providing an essential primary standard of impedance both at DC and AC
Thruline power meter	An instrument through which the RF coaxial line passes through. Only the outer conductor is opened into a metal housing, to tap RF power and measure, the central	Parasitic for inductor and capacitor	Any circuit element, always includes "parasitic" which includes inductance due to the leads connecting that device, capacitance between the leads and other metallic leads, etc.,

Impedance measurement methods	All practical systems carefully consider parasitic as part of measuring methods/procedure Depending on the frequency and the type of circuit (series or parallel) different measurement approaches are used for impedance measurements. Several dependencies, such as frequency, signal level, and temperature, that alter the measured value must be considered before selecting a measurement method	VSWR	Voltage Standing Wave Ratio is defined as the ratio of the maximum to minimum voltage on a loss-less line. The resulting ratio is 1 for an ideal line, where load impedance Z_L is the same as characteristic impedance Z_O. Most practical systems have values of 1.1 or more. For all well-balanced systems 1.3–1.5 can be expected. VSWR is related to the Reflection Coefficient Γ but considers only the magnitude of Γ, and not its phase
S-parameters	Scattering parameters refer to the RF circuit as a 2-port device with input port (port 1) and output port (port 2). All signals are considered in terms of waves at these two ports. The source impedance Z_S and the load impedance Z_L are considered in relation to the waves arriving or reflected from these two ports	Reflection coefficient Γ	It is the ratio of the reflected wave to the incident wave, both expressed as phasors. Therefore, reflection coefficient provides information about both the magnitude and phase and is inversely related to VSWR
Vector network analyzer (VNA)	VNA is an instrument that measures the network parameters of electrical networks. They commonly measure s-parameters because reflection and transmission of electrical networks are easy to measure at high frequencies. Network analyzers are often used to characterize two-port networks such as amplifiers and filters, but they can be used on networks with an arbitrary number of ports	Smith chart	The Smith chart is a graphical tool for determination of the reflection coefficient and impedance along a transmission line. It is a polar graph of normalized line impedance in the complex reflection coefficient plane. Its main objective is to help solve problems with transmission lines and matching circuits
Return loss	Return Loss is the portion of a signal that is lost due to a reflection of power at a line discontinuity (such as an antenna or a load). It is a logarithmic measurement expressed in dB and is useful to indicate very small reflections, encountered during load matching. It is the magnitude component of signal reflection	Noise power p_N	Noise power mainly addresses the noise received or observed in a system operating with bandwidth B. It also depends on the temperature at which it is operated, resulting in the classic equation $P_N=kTB$, where k is Boltzmann's constant, T is the temperature in degree Kelvin and B the bandwidth of the system
		Noise Figure F	Noise figure is a value which gets added to the level of noise produced when a signal passes through an active device. For example, an amplifier would not only amplify signal but noise as well. If that amplifier

has a low noise figure, then the noise added to the output will be less. It is important to carefully evaluate the Noise Figure of the amplifier (particularly in satellite systems where amplifier gain can be very large)

ADC (Analog to Digital Converter)   A device that takes analog signal as input, converts it into a set of bits, that is convenient to use in digital systems. The number of bits produced as well as the how often it is sampled (clocking rate) can be controlled by design, but both of them have a direct impact on noise, increasing the noise floor

DAC (Digital to Analog Converter)   A device that takes digital (bits) as input and coverts it into an analog output. The number of bits produced as well as the sampling rate can be controlled by design, but the DAC usually tries to match the ADC used by the system (same number of bits and clocking rate)

Quantization noise   In an ADC (Analog to Digital Converter) the process of converting analog signal into bits involves the process of "quantization" which needs referencing to known voltage level V. This process is repeated for each bit. The error that can result is 1/2V uncertainty for each bit, known as "quantization error". For example, if the input is a sinusoidal wave then for each bit of ADC, it is known that there will be a 6.02dB of quantization noise added for each bit

Aperture jitter of ADC   Aperture jitter is the sample-to-sample variation during the encoding process. It has three distinct effects on system performance. First, it can increase system noise.

Second, it can contribute to the uncertainty in the actual phase of the sampled signal itself, giving rise to increases in Error Vector Magnitude (EVM). Third, it can increase Inter-Symbol Interference (ISI)

Oversampling of ADC   Oversampling is the process of sampling a signal at a sampling frequency significantly higher than the Nyquist rate. While the Nyquist rate is defined as twice the frequency of signal variation. Oversampling often involves rates much higher, like 10 times or even more. Oversampling increases the system gain, thereby improving the SNR (Signal to Noise Ratio)

EVM   Error Vector Magnitude is a design parameter of a transmitter, indicative of the quality of modulation, sometime also called as Relative Constellation Error (RCE). EVM has two important components (magnitude and phase) whose measurements are needed to determine modulation quality

OOS   Origin Offset Suppression or Origin Offset Error is a design parameter indicative of the quality of modulation, indicative of how well the constellation's origin is controlled by the transmitter. Origin offset is derived as part of the modulation accuracy measurement. This is a measure of the DC offset in the I and Q paths of the transmitter and is expressed in dB (as a ratio of nominal signal vector magnitude). Frequency error is also derived from this measurement

Receiver sensitivity   Sensitivity of a receiver is normally taken as the minimum input signal (S_{min}) required to produce a specified

E_b/N_o. Energy per bit to the spectral noise density

Physical cell ID (PCI)

PSS

SSS

SSB beam forming in 5G

output signal having a specified signal-to-noise (S/N) ratio. Sensitivity is defined as the minimum signal-to-noise ratio that the receiver can detect, multiplied by the mean noise power (kTB), and noise figure of the receiver

In digital communication E_b/N_o is the normalized Signal-to-Noise Ratio (SNR) measure, sometimes indicated as SNR per bit. It allows a fair comparison of different digital systems without taking into bandwidth used in each

This is a number used to identify each cell indicated in the physical layer of LTE systems of 4G and 5G. In 5G there are 1008 such cell IDs are available. Due to large number of cells used, reuse of PCI may be necessary with several cells using the same PCI but not at the same location. Specific rules and algorithms govern cell planning resulting in methodic distribution of PCI. The PCI is represented using PSS and SSS signals in the physical layer

Primary Synchronization Signal is a physical layer signal used by the handset (4G or 5G cell phone) to synchronize with a specific cell site. It has a range of 1, 2, 3

Secondary Synchronization Signal is a physical layer signal used by the handset (4G or 5G cell phone) to synchronize with a specific cell site. It has a range of 1–335

Synchronization Signal Block (SSB) is used by each cell site in 5G. SSB beams are static, or semi-static, always pointing to the same direction. They form a grid of beams covering the whole cell area. The UE

PIM (Passive inter-modulation)

DTF (Distance-T- Fault)

Lightening arrester or surge arrester

The gas discharge tube (GDT, Gas Capsule)

Directional coupler

searches for and measures the beams, maintaining a set of candidate beams that it can go to. The candidate set of beams may contain beams from multiple cells

PIM is interference resulting from the nonlinear mixing of two or more frequencies in a passive circuit. If the interference coincides with an LTE or 5G network's uplink receive frequencies, it can adversely affect network performance and throughput. All cellular providers make it mandatory to conduct PIM tests at each site and also for any indoor cellular coverage area

An important measurement for either cable failure, or PIM or any related measurement in the field. Useful for diagnosis and correcting faults or used in planning

A surge arrester is a protective device for limiting voltage on equipment by discharging or bypassing surge current coming from lightening. An arrester does not absorb lightning or stop lightning. It diverts the lightning current to ground, limits the voltage build up and protects the equipment installed in parallel

GDT is a tube filled with inert gas and two electrodes that are kept within arcing distance at a higher voltage. During lightening there will be an arc that allows high current to pass through to ground. It doesn't affect performance of GDT during subsequent lightening. However, GDT must be replaced every 5 years, to retain reliable performance

Directional couplers are four-port circuits where one port is isolated from the input

	port. They are passive reciprocal networks. All four ports are closely matched, and the circuit has very little loss. Directional couplers can be realized in microstrip, or in strip line form, coaxial, or in the form of waveguide. They are used for sampling a signal, sometimes both the incident and reflected waves
Duplexer	Duplexer" is a three-port device with Transmit, Receive, and Antenna ports. Many antennas on cell towers do transmit and receive signals. Such systems are based on FDD—Frequency Division Duplex. Duplexers are also used in all cell phone which use a common antenna for transmit and receive
Diplexer	"Diplexer" is a three-port device, but it is used for "combining" two independent frequencies from two distinct signal sources and sending it to the same antenna. The two frequencies must be on separate bands such that they remain distinct. In order to achieve this, two distinct filters are used to keep signal from the sources
Splitter	The splitter as the term indicates, only splits signals. The input signal P1 is split into two paths P2 and P3 therefore each of them provides only half the power of P1. Splitter can be designed either using resistors or hybrid elements but operates over a specified frequency band
Balun transformers	Balun" is shortened form of "Balance-Unbalance". It is an important element used in television and other RF circuits where the two sides may not be balanced. Balun transformer is primarily used to bring about balance in an unbalanced
	circuit. In an application like TV, the 300Ω ribbon cable (balanced) from an antenna, must be attached to coaxial cable of 75Ω (unbalanced or single ended). Balun transformer in this case uses 4:1 ratio in coil winding
Circulator or Isolator	An RF Circulator is a non-reciprocal ferrite device which consists of 3 or more ports. Due to ferrite material placed inside circulator or an isolator, its strong magnetic interacts with the incoming signal and moves the signal to the adjacent port. So long as this port has little mismatch and good impedance with VSWR closer to 1.0, these devices operate well
Attenuator	An attenuator is a two-port resistive network designed to weaken or "attenuate" (as the name suggests) the power being supplied by a source to a level that is suitable for the connected load. A passive attenuator reduces the amount of power being delivered to the connected load by either a single unit of fixed value, or a variable unit or in a series of known switchable steps

Any term in science or engineering is better understood if it can be quantified and measured. Every measurement provides a better understanding and the units used in measurement convey fundamental understanding of that term. Such thought process is essential during conceptual design, building of any device, or its construction and deployment in the field, followed by a scheme to maintain the system in a state of good repair. The science of measurement dates back to human evolution, when there were perhaps no physical instruments, yet the numbers indicated in such predictions by ancient civilizations are quite accurate [1]. It subscribes to the view that theoretical understanding can lead to fairly accurate predictions—which is the method used even today but substantiated by "Metrology"—the scientific study of measurements.

In the context of wireless communication measurement offers some challenges, which is the basis for this chapter. There is a general observation by many that any RF measurement is difficult—at the same time it is essential. The focus on practical measurements relates to theoretical understanding and use of wireless equipment in real environment. At times, it is necessary to establish laboratory measurements where more sophisticated instrumentation is needed. Whenever a new standard is established—(5G or WiFi 6, etc.), the first team that clearly establishes the concept is "Metrology team" who worry about—how can this be measured and what are the challenges. They bring reality to what was conceptual, can it be built and used? Each nation supports "National Standards Laboratory", where such thoughts take shape.

Standards hierarchy

Standards of measurement are based on a hierarchy of three levels. Primary standards are the highest, usually maintained by each nation as their national standards laboratory. Physicists, Chemists, and other scientists strive to establish cutting edge methods to improve accuracy and achieve highest levels of precision, periodically comparing them to other international standards.

This is followed by secondary standards of regional bodies or agencies such as the military, space, and other bodies who need to maintain high accuracy standards to calibrate their own instruments. The secondary standards are annually referenced to the primary national standards.

The working standards are operated by all field organizations who reference them to the secondary standards. These are typically used to make actual measurements on field equipment fairly regularly (weekly, monthly, or as required) during design, installation, and operation. Thumb rule for referencing to a higher standard is—the higher standard should have one order of magnitude higher in accuracy. This means ten times more accurate than the one being compared to. This is to make sure that if there are adjustments to be made during annual calibration, it is always traceable ultimately to the national standard.

Precision and accuracy—difference

These two terms mean quite different things that could be explained with an example. Consider an instrument that measures with an accuracy of 0.5%. When the instrument is used to measure a true value of 100, it can read 99.5 or 100.5 and is still considered accurate so long as readings are within that range. However, if the same instrument when used ten times and measures 100.4, 100.5, 100.5, 100.5, 100.5, 100.4, 100.5, 100.5, 100.4, 100.5—then it is precise and accurate. What precision means is "togetherness of multiple readings". In the earlier case, the instrument is accurate, but not precise since readings can vary over the entire range of 99.5–100.5 which is 1% of 100.

Engineers often look for precision—since inaccuracy if already known, can be adjusted, but consistency in readings brings confidence in the measurement. The process of such "adjustment" from true value is known as "calibration" which is performed against known or established standard which is at least 10 times more accurate and precise. That means comparing it with a standard that varies only by 0.05% and more precise. Therefore, higher standard must be within 99.95 and 100.05 as well as precise. It means the reference may vary around 99.95 or 99.96 or may be around 100.05 or 100.04. This "togetherness of readings" is also indicated by another term—"stability". Stability is based on variety of physical controls in the environment such as minimum variation in temperature, no shock or vibration to the instrument, moisture control (preferably dry). This is why accuracy and precision are often quoted under "lab conditions".

There is another important feature that Physicists often point out. "Does the act of measurement itself alter the value?". Initially this may seem intriguing but let us consider it carefully. If we measure voltage between two terminals, the measuring instrument will consume a small current. No matter how small, it does drop the voltage ever so slightly. Therefore 10.000 V may actually measure 99.999 only because of such a small current is used by the voltmeter. Then how should the ideal measurement be performed? "By comparison"—that means use a bridge, use another precise voltage source to oppose this unit under test. When no current flows, meter reads "null or zero", and the "bridge is in balance" and the precise source provides the true value of voltage.

In terms of mass when a weighing machine is used it is inherently not the "true value". The ideal method would be to use the good old traditional balance with two arms—one with standard weights and the other arm with object being weighed. It is also used to measure gold and gems since the standard weight precisely indicates the "true weight" of gold placed on the other arm.

In practice, instruments that measure are quite precise and "fairly accurate" but operate very fast. It may perform hundreds of measurements in a minute, some even faster. The compromise is that their measurement is close to "true value" but not exact. Therefore, the traditional balance, or the bridge method of "true value measurement" are now confined to national standards or regional calibration laboratories. Fast measuring machines have become working standards in the field.

19.1 Frequency and Time

Frequency and time are directly related to each other since one is the inverse of the other. Of the three fundamental units in Physics—length, mass, and time, "time" is the most accurately measured quantity in the world. That might seem like a tall statement, but the accuracy of a primary standard of time/frequency is 1 part in 10^{-16}. An atomic clock referenced to the Cesium beam is kept at most national standard laboratories to provide accuracy of better than 1 s in about 1000 million years. How do these clocks manage to keep time so accurately? What is the principle? These clocks are based on fundamental property of matter thanks to Nobel prize winning work by Prof. Norman F. Ramsey, whose concept is explained now, using Fig. 19.1 as reference.

All atoms can be in different energy states, called hyperfine levels or hyperfine structure in scientific terms. In the element "Cesium", there are a pair of energy states, state A and state B, respectively. Atoms are heated and bundled into a beam. A magnetic field removes atoms in state B so only state A atoms remain. This is a crucial step. Next, the state A atoms pass into a cavity resonator. Inside the resonator these atoms are pelted with microwave radiation, which causes some of those atoms to change state, moving to state B. The atoms are passed behind the resonator, and another magnetic field extracts atoms that remained in state A.

Next, a detection system collects and counts the atoms left in state B. This is where things get complicated. The microwave radiation's frequency will determine the total percentage of atoms that change their state after passing through the cavity resonator. If the radiation frequency is closer to that of the atom's oscillation, then more will change state [2].

This is how time is determined using an atomic clock. The clock is tuned—or its microwave frequency is tuned—to match the oscillation of the atoms and then the outcome is measured. After 9,192,631,770 oscillations have occurred, it can be counted as a whole second. Thanks to Prof. Norman Ramsey's groundbreaking work (that led to his Nobel

prize) he proved this was a fundamental property. This has been accepted as the primary standard, which needs no further calibration (being a fundamental property in Physics). Since 1967 one second has been defined as the time during which the cesium atom makes exactly 9,192,631,770 oscillations.

Figure 19.1 is an illustration of how Cesium beam primary standard is set up. The complete system configuration and its measurements are beyond the scope of this book. Readers interested must refer to a brief introduction at [3]. Current generation of primary standard beyond NIST-7 has reached accuracies of 1 part in 10^{-16}. This means 1 MHz signal will have an accuracy of 1.000000000000000x where the very last digit or 16th digit x will only vary by +/− 1. Work is in progress to improve this even further. In short, no other physical quantity has been measured to this accuracy.

Another fundamental unit "Length" is now redefined using "time taken by light" to travel a distance of 1 m. In turn that "Time taken" is referenced to this Cesium beam primary standard. Therefore "Length" is defined in terms of "Time". Now, the last remaining fundamental unit of mass "kilogram" is likely to get redefined again referencing it to "TIME" [4]. Why is this shift so dramatic? Recall that students the world over have studied fundamental measurements in Physics consisting of three basic units: Length, Mass, and Time. Now the enormous accuracy of "Time" has redefined "length" and now will redefine "Mass", both referenced to "Time".

The most important application of this "Time" standard is the GPS clock that everyone has come to appreciate. GPS clock transmissions (used globally) has such an atomic clock from which its transmissions are derived [5].

Fig. 19.1 Cesium beam primary frequency standard at NIST

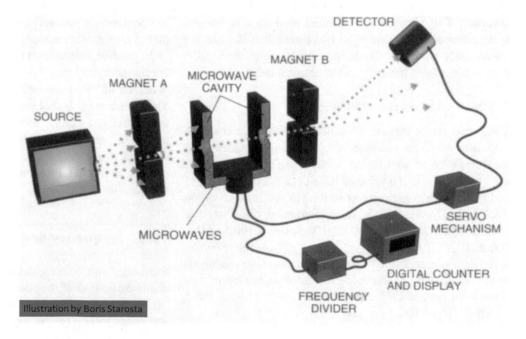

19.1.1 Secondary Standards of Time/Frequency

Secondary standards of frequency/time are Rubidium frequency standard, based on principle similar to Cesium but with a lower accuracy of 1×10^{-12}. These are used typically at satellite earth stations where regular synchronization is required among all earth stations in that region.

Quite often working standards are also fairly accurate and provide accuracy of 1×10^{-11} if they are frequently calibrated, or 1×10^{-10} if they are annually calibrated. These are TCXO or Temperature Controlled Crystal Oscillators that are usually kept in a sealed unit with temperature control by placing heating element in an oven like package. Reason for temperature control is—crystal oscillators are prone to two types of drift. The short-term drift and the long-term drift, which will be addressed now.

19.1.2 Crystal Oscillator—The Reference

So far, the crystal oscillator was mentioned as part of the primary, secondary, and working standards. Originally built over a century ago (1918, by Bell Labs) it is used in a wide range of cellular devices, short-range wireless modules, wearable devices, and automotive multimedia devices. It is worth learning the rudimentary principles of this electro-mechanical device. Quartz crystal which is made out of silica exhibits a unique property. When mechanical pressure is exerted on one face of the crystal, on its orthogonal face voltage is generated. If AC signal is applied on that face, then in orthogonal face it vibrates. This property known as "piezo electric effect", which is fundamental to its oscillation and stability [5]. The frequency of vibration is based on the type of cut, size, and shape of the crystal. But once cut, it operates within a narrow range of frequencies, providing stable operation. There is a general agreement that any reasonable crystal oscillator provides accuracy of 1×10^{-6} or 1 ppm (one-part per million), which means an accuracy of ± 1 Hz at 1 MHz; in terms of time, this results in 1 s difference in every 11 days, which is acceptable for most watches. This is why the watch industry moved away from mechanical clocks to quartz crystal clocks, even though some retain the traditional look of mechanical clock. This revolution of abandoning the old mechanical clock with main spring, winding it daily, etc., changed the concept of clocks. In the wireless industry, for example, a smartphone, or WiFi or any consumer product—just relies on crystal clock—corrected by GPS almost daily through a cellular base station or other means. They are orders of magnitude more accurate than any mechanical watch.

Synthesizers, vector network analyzers, frequency counters, spectrum analyzers, and many other instruments must accurately create or measure frequency. Their accuracy is affected by time, temperature, frequency, and even gravity and therefore they reference a crystal oscillator [6].

It was mentioned earlier that environmental factors do affect stability. Crystal oscillator is not an exception and the following uncertainties exist that are related to passage of time.

Time-based uncertainties are grouped into the following categories:

(1) Temperature effects
(2) Uncertainty with respect to time
  (a) Short-term stability
  (b) Long-term Aging
(3) Mechanical effects
  (a) Shock and vibration
  (b) Gravity

Some of these can be controlled.

Figure 19.2 shows the long-term aging feature of a crystal oscillator (graph on the left). Small variations superimposed on the aging curve indicates effect of short-term stability [7]. Short-term RMS (Root Mean Square) average of frequency variations or time domain stability is a measure of the frequency or phase noise. This is shown on the right, where it is the average over various time periods τ. This is specified as the standard deviation of the fractional frequency fluctuations for a specific averaging time (usually short durations of milli-seconds or seconds).

Crystal oscillators can be used over a wide temperature range (-50 to $+ 90$ °C), but the frequency variation would be in the 40 ppm range. This may be acceptable to many consumer products that do not need a highly stable source. For more professional and industrial applications, it is necessary to limit variation due to temperature. If it is for a laboratory environment, it can be compensated over a limited temperature range available in packages called TCXO or Temperature Controlled Crystal Oscillator [7]. Such packages specify operation over a limited range such as 0–70 °C. The better way would be to place the entire crystal oscillator circuitry in a small oven, with a heating element inside to maintain the package at constant temperature. The package including the oven is called OCXO—Oven Controlled Crystal Oscillator. OCXO provides better accuracy at specific temperature for reference sources used in many instruments, since outside/ambient temperature variation is no longer a major issue. Military and space applications typically would use such a package.

Shock, Vibration, Gravity all contribute to variations in the crystal oscillator and its performance.

Figure 19.3 shows the overall effect due to multiple factors [6]. Striking a crystal oscillator (shock shown in the figure) places a sudden stress on the crystal by temporarily

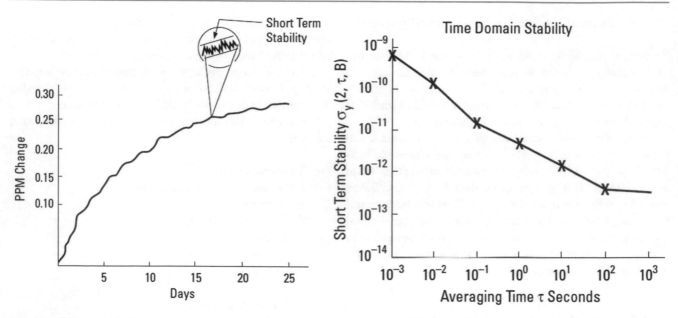

Fig. 19.2 Long-term aging and short-term stability of quartz oscillator

Fig. 19.3 Graphic representation
of environmental effects

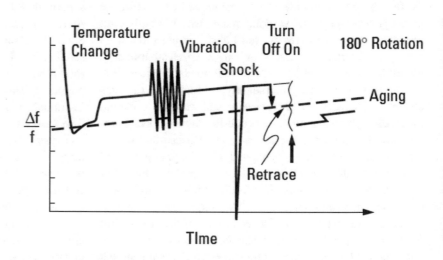

deforming the mounting structure. Shock can result in change of frequency by 1×10^{-9} per G. The frequency of an AT cut crystal (one type of crystal cut), will change by 1×10^{-9} with a variation in drive of one microwatt. Drive level requirements vary depending upon the crystal impedance. Excessive drive level may cause mechanical vibrations to exceed the quartz elastic limits resulting in a fracture.

Depending on applications vibration can be significant for certain applications [8]. The phase of vibration modulated signal is

$$\phi(t) = 2\pi f t = \left(\frac{\nabla f}{fv}\right) \sin(2\pi f_v t) \qquad (19.1)$$

When the oscillator is subjected to sinusoidal vibration, the peak phase excursion is

$$\nabla \phi_{peak} = \frac{\nabla f}{f_v} = \frac{(\bar{\Gamma} \cdot \bar{A}) f_o}{f_v} \qquad (19.2)$$

Example: If a 10 MHz, 1×10^{-9} per G oscillator is subjected to a 10 Hz sinusoidal vibration of amplitude 1 g, the peak vibration induced phase excusion is 1×10^{-3} radian. If this oscillator is used as the reference oscillator in 10 GHz radar system, the peak phase excursion at 10 GHz will be 1 rad. Such a large phase excursion can be catastrophic to the performance of many systems such as those that employ phase lock loops (PLL) or Phase Shift Keying (PSK) [8].

RETRACE: When an oscillator is turned off and then back on, it will not necessarily start at the same frequency at which it had been operating. Eventually the oscillator will begin to age at its previous rate but will most likely be offset slightly from its original frequency. A typical retrace offset may be in the order of 1×10^{-8} (shown as "Off/On" in Fig. 19.3). Operating in relatively higher frequency bands from a few hundred kHz to several hundred MHz, crystal oscillator continues to play a major role in many wireless applications. Typical offsets for 180° of rotation may be about 1×10^{-9} which is 1×10^{-9} per G.

Due to their history spanning over a century, their characteristics are widely known and understood. Often instruments use external crystal oscillator as references. This allows such units to use TCXO yet use a secondary standard as reference, while performing high accuracy measurements in the upper frequency bands.

19.1.3 MEMS—Micro Electro-Mechanical Systems (Oscillators, Timers)

For different types of consumer product applications, that do not need the accuracy of quartz crystal oscillators, MEMS oscillator could be an alternative. Its accuracy ranges from typically 0.1 ppm, with more precise versions with stability of better than 0.01 ppm. It provides better stability over operating temperature range. It is fair to classify them as devices of reasonable stability, for rugged applications such as "automobile electronics under the Hud" (regular vibration and temperature change) [9].

They offer advantages in terms of operating in the lower frequency band from 1 Hz to several hundred kHz all the way up to hundreds of MHz [10]. They are programmable in frequency and directly fit into the dye of an IC (Integrated Circuit). These features make them suitable for products such as toys, wearables, solar panels, IoT (Internet of Things) micro satellites, and others. Due to considerably lower power consumption (about 1 μ A), it allows sleep mode in devices such as IoT that operate for short intervals a few times each day using small button type battery. They can support time synchronization and jitter clean up functions in networks—for example 5G and IEEE 1588 synchronization applications.

Overall, MEMS oscillators are emerging to support a variety of low-cost applications that are expected to grow. It is likely to complement the crystal oscillator in many areas due to different approach in packaging. For example, MEMS oscillators with single and multioutput provide options of incorporating spread spectrum in automotive, network jitter control, and industrial applications [10]. For Ethernet-based network applications PSE ICs (Power Sourcing Equipment

Integrated circuits) are used in a variety of applications, such as Enterprise, SMB, and industrial Ethernet switches, Gateways, Routers, PD-PSE daisy chain applications.

19.2 RF Power

RF Power was briefly touched upon at the beginning of this book in Chap. 2, Sect. 2.1.2. Since RF power is a major issue addressed in terms of its effect on human body, it was considered again in Chap. 7, Sect. 7.2. In later chapters of 11 through 18 in various contexts RF power was addressed. In short, RF power and how it is measured is fundamental to all of wireless technology. Therefore, this review starts from DC (Direct Current) and how power is accurately measured at DC and move on to how it becomes progressively difficult to measure power in AC (Alternating Current) as the frequency increases, which is why it references to DC.

19.2.1 Primary Standard—RF Power

Earlier, in Chap. 2, Sect. 2.1.2 indicated that $P = I^2 R$ is the power that can be accurately measured at DC. In AC there are two issues—R becomes a complex number $Z = R \pm jX$ where, if the complex number for an inductor is $X_L = 2\pi f L$, and if it is capacitive $X_C = \frac{1}{2\pi f C}$. For transmission lines, it is a combination of both. With increase in frequency X_L increases and X_C decreases. The transmission line does not have a fixed L or a fixed C, it is distributed over its entire length. Hence, **Primary Standard of RF Power** starts with Calorimeter, which is contained in a double insulated chamber as shown in Fig. 19.4.

To compensate for power absorption in the cable, a micro-calorimeter as shown in Fig. 19.4 is used as primary standard. It consists of a thermally well-insulated receptacle that reduces losses due to heating. Losses along the feeding transmission line are first determined. A thermistor is used as load and the conversion losses in thermistor and their relationship to the absorbed RF power are determined [11]. The temperature increase generated by the power sensor within the calorimeter is determined by finding the difference in reading—with the RF power switched off and then switched on. The measurement is performed using an electric thermometer made up by series-connected thermoelements (thermopile) and is referencing it to a second passive thermistor sensor.

When the RF power is switched off, the heating is caused exclusively by the DC power that raises the thermistor to the nominal temperature. When the RF power is switched on, additional heating is caused due to RF power absorption [11]. The change in temperature and DC voltage that occurs

Fig. 19.4 RF power primary
standard using calorimeter

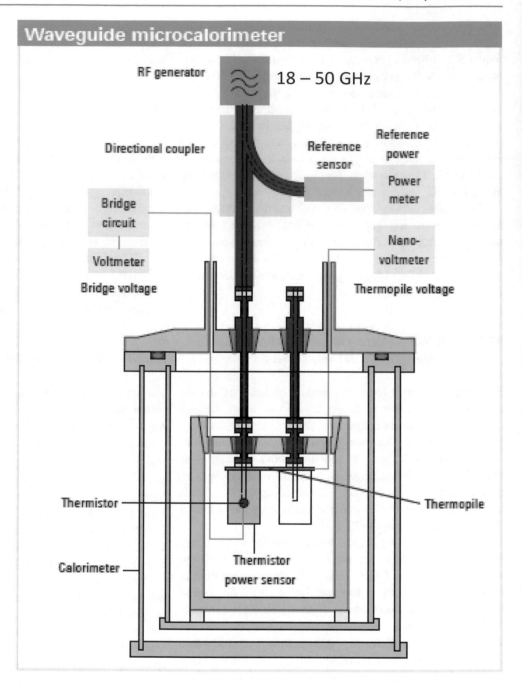

after applying the RF power yields the effective efficiency as
follows:

$$\text{Effective Efficiency} = \frac{DC\ substitution\ power}{Total\ absorbed\ RF\ Power} \quad (19.3)$$

Once this efficiency is known, displayed result of thermal
power meter can be corrected as a function of frequency.
This calibration factor is measured and noted at different
frequencies (from 10 MHz to 50 GHz). These measurements
are slow since it takes time to allow of thermal equilibrium;
it could be from 60 to 90 min for each frequency

measurement. Altogether about 40 such measurements are
made, which takes several days. At lower frequencies (up to
18 GHz) coaxial cable is used. At higher frequencies (18–
50 GHz) waveguides are used [12]. In each case the measurement methods are similar.

19.2.2 Secondary Standards of RF Power

These are based on the absorption type of thermistor power
sensors. With these it is possible to reference to the primary

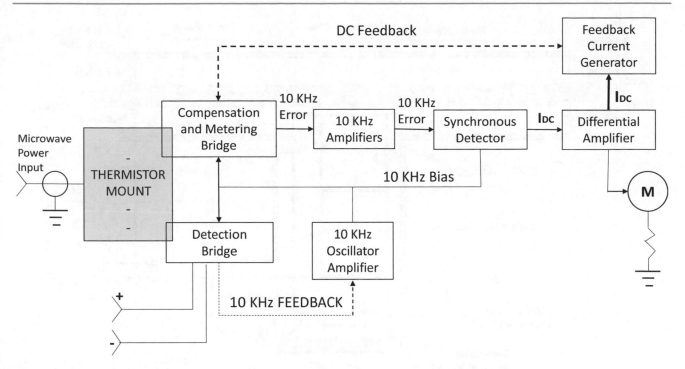

Fig. 19.5 RF power meter using Thermistor Mount

standard and they are essentially "transfer standards" used by regional calibration centers.

Figure 19.5 shows an RF power meter that indirectly measures microwave frequency power by using two bridge circuits—(1) The detection bridge and (2) the compensation and metering bridge [12]. The detection bridge incorporates a 10 kHz oscillator in which the amplitude is determined by the amount of heating of the thermistors in that bridge. This heating is the result by microwave power. The compensation and metering bridge contain thermistors that are caused by the same microwave power heating as those of the detection bridge. An imbalance in the metering bridge produces a 10-kHz error signal.

This error signal, plus the 10 kHz bias that is taken directly from the 10-kHz "Oscillator Amplifier" are mixed in the "Synchronous Detector". The synchronous detector produces a DC current (I_{dc}) proportional to the 10-kHz error signal. The I_{dc} error signal is fed-back to the compensation and metering bridge, where it substitutes for the 10-kHz power in heating the thermistor and drives the bridge towards a state of balance. The DC_{output} of the synchronous detector also operates the meter circuit for a visual indication of power. Such indirect procedure still references RF power to DC, thereby obtaining accurate result [12].

19.2.3   Working Standard of RF Power

It is not essential to absorb the RF power into a Thermistor or other device to make this measurement—the compromise solution is to "sample or tap into the transmission line", which is an interesting and widely used concept, explained now.

Figure 19.6 shows schematic layout of a RF power meter commonly known as the "Thruline meter" [13]. The advantage of this type of RF power meter is that it can be used in any environment where RF power is being sent to a load/or device such as antenna. The easiest way to visualize "Thruline" operation is from a traveling wave viewpoint. In transmission lines the voltages, currents, standing waves, etc., on any uniform line section are due to the interaction of two traveling waves: The Forward Wave and the Reverse Wave. In Fig. 19.6 as illustrated, the forward wave (and its power) travels from the source to the load. It has RF voltage E_f and current I_f in phase, will be described by

$$\frac{E_f}{I_f} = Z_0 \qquad (19.4)$$

The reflected wave (and its power) originates due to reflection at the load and travels from the load back to the source. It has an RF voltage E_r and current I_r in phase, hence

$$\frac{E_r}{I_r} = Z_0 \qquad (19.5)$$

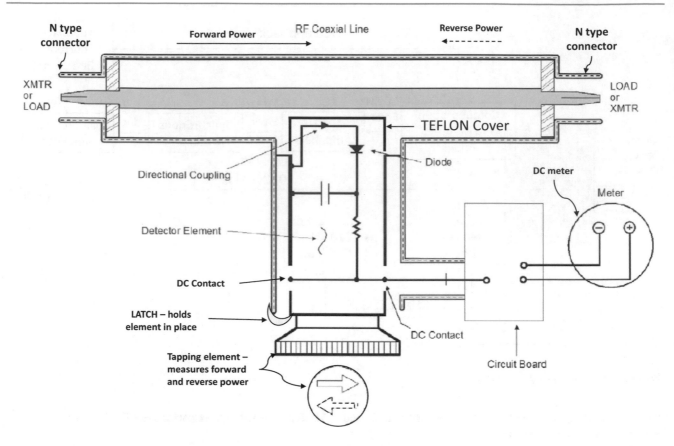

Fig. 19.6 Working RF power meter–"Thruline" type

Each wave is mathematically simple and has a constant power:

$$W_f = \text{Watts Forward} = \frac{E_f{}^2}{Z_0} = I_f{}^2 Z_0 = E_f I_f$$

$$W_r = \text{Watts Reflected} = \frac{E_r^2}{Z_0} = I_r^2 Z_0 = E_r I_r$$

Z_0 is the characteristic impedance of a uniform line section. For ideal lines it is usually a pure resistance of 50 Ω. The RF circuit of Fig. 19.6 is a length of uniform air line with $Z_0 = 50$ Ω.

The center conductor continues through the instrument (Thruline). The tapping element is carefully designed such that the diode and capacitor are placed inside a "Teflon cover" which is fully closed. There are different types of elements for various frequency ranges and modulation types. Such an element is inserted into the cavity of the instrument as shown and a latch is applied to keep it in place. The key to such design lies in the capacitor, resistor, and diode circuit that retains transmission line properties such that impedance of 50 Ω is maintained throughout that frequency range.

19.2.3.1 Load Matching

One of the very useful applications of the "Thruline meter" is for load matching, as described now. Figure 19.7 shows a graph of frequency versus Cable length (in inches), used for load matching. X-axis at the bottom is from 100 to 700 MHz and the corresponding cable length in inches is shown on the Y-axis (left). Top portion of the graph continues with frequency from 300 to 1000 MHz shown in parenthesis. Its corresponding cable length in inches is shown on the right also in parenthesis. How this graph is used will be explained in the next few paragraphs.

When the "Thruline meter" is used to tune a load to a transmitter and a good match is obtained, then removing it will not change the match quality. A good 50 Ω load can terminate a 50 Ω transmission line of any length without altering conditions at the transmitter. Thruline wattmeter adds short additional length (about 4 inches) of 50-Ω line in series with the measurement.

When the load is not well matched, e.g., an antenna with a VSWR of >1.5 or 2.0, then line length between the load and the transmitter will transform the load impedance as observed at the transmitter. Removing the wattmeter shortens the total line length by four inches plus two connectors. Cables are reconnected through an extending connector

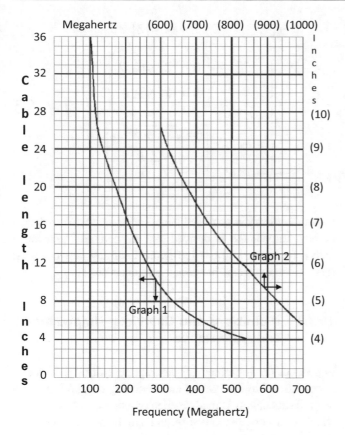

Fig. 19.7 Graph used for cable length/wavelength matching

(N-female). This is still not significant at low frequencies where ~ 10–12.5 cm (4–5 inches) is a fraction of a wavelength. At higher frequencies, the frequency or power output of the transmitter may be affected. For quick reference, at 300 MHz—wavelength is 1 m and half wavelength is 50Cm (~ 16 inches).

Transmission line theory indicates that "if the line length changes by exactly ½ wavelength, the impedance at the transmitter is unchanged". To have identical match quality with Thruline meter in or out of the circuit, it is necessary to insert or remove ½ wavelength worth of line (including the wattmeter). To do this, if we use a length of cable which, when added to the wattmeter, equals ½ wavelength at the frequency of interest. If more than one frequency is involved, a separate cable length is required for each. See Fig. 19.7 for sample cable lengths. For the first curve on the left Graph-1 applies (in the range 100–550 MHz). For the higher frequency band with scale on the right, Graph 2 applies (in the range 600–1000 MHz).

Observe that the directional coupling sensor is placed "inside the transmission line" enclosure. This allows it to couple inductively and capacitively to tap small part of the RF wave traveling through the coaxial line. The element is carefully designed such that by just turning the element by

180 ° (arrow in the reverse direction) the reflected power is measured by the same meter. For example, if the base station power was input at the left and the right connects to the antenna on the tower, simple measurements of forward and reflected power are possible, by turning the element to face forward or reverse. This is one of the key reasons why this instrument (Thruline) has remained popular since second World War.

Are there any limitations? Yes, in terms of accuracy, typical secondary standard meters provide ±2% while this "Thruline" meter will offer about ±5%. It is less accurate, but not significantly. Are there other precautions? Yes, the element that plugs in (bottom) is quite sensitive to shock and vibration—it must be typically carried in a well-cushioned box. Except that, instrument is rugged and portable—used by the military throughout the world, for many decades now.

19.3 Impedance

The term "Impedance" is used to mean something that "impedes" the flow of current. Impedance is made up of three distinct passive elements—Resistor R, Inductor L, and Capacitor C. Each of them has distinct type of response to frequency. Resistor response is not based on frequency, it follows Ohm's law V = IR. Whether it is DC or AC its response doesn't change, it still offers resistance. In Sect. 19.2.1, we observed that primary power standard starts with DC, which is used as a reference to derive power in AC for which we used the equation $P = I^2R$.

An inductor "reacts" to change in frequency, and its "Reactance" is $X_L = 2\pi fL$. At DC since f = 0, its reactance is $X_L = 0$ which means it acts as a "short circuit". But as frequency increases, its "reactance" increases linearly, becoming almost an "open circuit" at very high frequencies. In contrast, Capacitor "reacts" to change in frequency, but its "Reactance" is $X_C = \frac{1}{2\pi fC}$. At DC since f = 0, $X_C = \infty$. Capacitor acts as an "open circuit". But as frequency increases, its "reactance" decreases linearly. At very high frequencies, $X_C = 0$, it acts as a "short circuit".

The opposing "reactance" combinations of L and C are balanced carefully in all wireless systems. They are key design parameters to develop network of filters, feedback elements, etc., based on frequency band of interest. In quantitative terms impedance is still expressed as "ohms" but unlike the resistor, impedance is a vector. In impedance "ohms" represents only the magnitude, but the phase angle associated with it, becomes equally important for reactance. This leads to the condition that primary standards for resistor exist and to a limited extent also for capacitor since their value (magnitude) can be measured at DC.

Fig. 19.8 Definitions of impedance, reactance

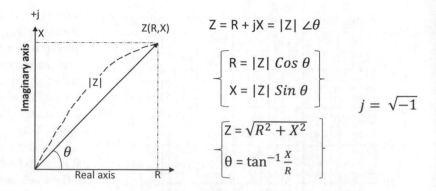

$$Z = R + jX = |Z| \angle \theta$$

$$\left.\begin{array}{l} R = |Z|\ Cos\ \theta \\ X = |Z|\ Sin\ \theta \end{array}\right\}$$

$$j = \sqrt{-1}$$

$$\left.\begin{array}{l} Z = \sqrt{R^2 + X^2} \\ \theta = \tan^{-1}\dfrac{X}{R} \end{array}\right\}$$

In recent years NIST (National Institute of Science and Technology—national standards laboratory of USA) has taken the approach of Quantum Hall Resistance (QHR) based on Josephson Junction [14]. Josephson Junction is a 1973 Nobel prize winning work by Brian Josephson (this work was performed when he was a Ph.D. student). He discovered that a supercurrent could tunnel through a thin barrier. He predicted that ``at a junction of two superconductors, a current will flow even if there is no drop in voltage. He also predicted that when there is a voltage drop, the current should oscillate at a frequency related to the drop in voltage; and that there is a dependence on any magnetic field'. This became well known as "the Josephson effect" and the junction as a "Josephson junction". How Josephson junction works at superconductivity (temperature near absolute zero) is explained in [15].

Reverting back to the topic of primary standards for impedance—development of wideband impedance measurement services requires reference standards that can be characterized over the impedance and frequency ranges of interest. NIST has developed a system to characterize commercial four-terminal-pair capacitance standards from 1 pF to 1 nF over the frequency range from 1 kHz to 10 MHz A bootstrapping technique using an LCR meter and an inductive voltage divider can extend the characterization to higher valued capacitance standards up to 10 μF [14].

To develop primary standard measurements for "impedance" requires a highly stable voltage standard as well. Such a standard based on "Josephson Junction" is available at several national laboratories now [16]. The primary standards use either DC or AC within a limited frequency range. Therefore, at frequencies used by RF, it is important to calibrate against these standards.

Impedance is basic to any circuit or device where AC (Alternating Current) flows. At higher frequencies used in wireless communication, impedance plays a major role, since it is one of three (frequency, power, impedance) fundamental entities that control any wireless application. Characteristic impedance Z_0 is a term often used throughout this book to describe antennas, transmission lines, etc. Only

a short summary is provided here. Those interested in details of how impedance is derived and mathematics associated with it for active and passive elements of any network must refer to the excellent handbook referenced in [17].

19.3.1 Definitions and Overview—Impedance and Admittance

Impedance (Z) is defined as "the total opposition a device or circuit offers" to the flow of an alternating current (AC) at a given frequency. It is mathematically represented as a vector which is a complex quantity and graphically shown on a vector plane [17].

Figure 19.8 shows the basic definition and relationships between Impedance Z, the Reactance (Reaction from Inductor or Capacitor) and Resistance R. These are three basic passive elements that reduce the flow of current in any AC circuit. Impedance Z is a vector, it has magnitude |Z| and direction indicated by angle θ. The imaginary number "j" is used in electrical engineering instead of the mathematical convention "$i = \sqrt{-1}$", since "the term i" is used for current.

The term X-reactance can take two forms depending on the two basic circuit components, Inductance, and Capacitance. In specific terms $X_L = 2\pi fL$ is the expression for inductive reactance. Similarly, $X_c = \frac{1}{2\pi fc}$ is the expression for capacitive reactance. Reactance X as a combination of these two, along with Resistance R (unaffected by frequency), must be vectorially added based on the basic rules for series or parallel circuits. Also, the Greek letter Omega "ω" is used to indicate angular frequency in radians/sec, therefore,

$$\omega = 2\pi f \qquad (19.6)$$

Inverse terms for impedance and reactance are shown for series and parallel circuits. Following are inverse relationships:

$$\text{Inverse of Impedance Z is Admittance Y}; Z = \frac{1}{Y} \qquad (19.7)$$

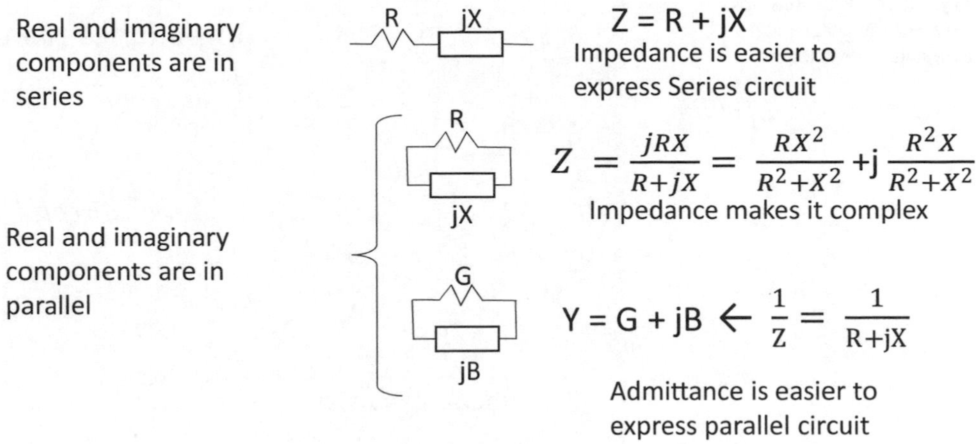

Fig. 19.9 Expressions for series and parallel combinations

Real and imaginary components are in series

$$Z = R + jX$$
Impedance is easier to express Series circuit

Real and imaginary components are in parallel

$$Z = \frac{jRX}{R+jX} = \frac{RX^2}{R^2+X^2} + j\frac{R^2X}{R^2+X^2}$$
Impedance makes it complex

$$Y = G + jB \leftarrow \frac{1}{Z} = \frac{1}{R+jX}$$

Admittance is easier to express parallel circuit

Inverse of ResistanceR is Conductance $G; R = \frac{1}{G}$ (19.8)

Inverse of ReactanceX is Susceptance $B; X = \frac{1}{B}$ (19.9)

These inverse terms are used in Fig. 19.9 to express series and parallel combinations.

Methods of measurement for both Impedance and Admittance have their respective advantages depending on whether the circuit is in series or parallel configuration.

There are two more units, often measured to characterize devices/circuits. The first one is the Quality factor Q defined as $Q = \frac{X}{R} = \frac{B}{G}$ and the second is its reciprocal, known as Dissipation factor D. In practice the Quality factor Q is used for Inductors while its reciprocal Dissipation factor $D = \frac{1}{Q}$ is used for capacitors.

$$Q = \frac{1}{D} = \frac{1}{\tan \partial} = \frac{X_L}{R} = \frac{-X_C}{R} = \frac{-B_L}{G} = \frac{B_C}{G} \quad (19.10)$$

Since series and parallel circuits using the three passive elements (Resistor, Inductor, and Capacitor) are quite common in all circuits, it is appropriate to view them in both the Impedance and Admittance planes. These are typically indicated in most measurements—it is relevant to reference them here. Observe that depending on the type of plane, the lead and lag of R and X vectors change [17].

Figure 19.10 shows the Impedance and Admittance planes with the RL and RC components in series and parallel circuits. Since Resistance is always positive, only the right-hand planes (first or fourth quadrant of Cartesian system) are used. Representation of "series circuits in the Impedance plane" is easier while representation of "parallel circuits in the admittance plane" is easier. Note that the angle ∂ is a complement of angle θ in each of the planes. Due to basic property of Inductance, j X_L vector is positive and always represented in the first quadrant of the cartesian coordinate shown as "vector plane". In contrast, the property

of Capacitance is j X_C negative is represented by the fourth quadrant of the Cartesian coordinate system. Since Resistance R is always positive only the right half of Cartesian Coordinate is used. These representations follow standard conventions used in all of electrical engineering, where relationships between the voltage and current vectors are indicated to be leading and lagging of the current vector with respect to voltage.

Upper part of Fig. 19.10 is the Impedance plane showing series circuits—resistor in series with inductor (inductive vector) and resistor in series with capacitor (Capacitive vector). Lower part of Fig. 19.10 is the Admittance plane showing parallel circuits—resistor in parallel with inductor (Inductive vector) and resistor in parallel with capacitor (Capacitive Vector). Observe Inductive vectors are shifted—the upper Impedance plane shows the first quadrant and the Capacitance vector shows the fourth quadrant. The difference between the Impedance plane and the Admittance plane is—they are inverse of each other, as explained earlier. First quadrant (positive reactance) and fourth quadrant (negative reactance). Beyond these basics it is important to note that there are no pure R, L, C components. Each element involves "parasitic". Parasitic capacitance and parasitic inductance in any device is inherent to the way it is built. For example, a capacitor, needs metallic leads or contacts to connect, that results in parasitic inductance and resistance. Similarly, an inductor coil will also have capacitance.

Parasitic of a typical component (capacitor) and its electrical equivalent circuit are shown on the right in Fig. 19.11. The reason why parasitic must be considered for any device is because every physical device contains them it is indicated in actual measurements [17]. Unless parasitic are properly accounted for, the measured value will be erroneous. Laboratory measurements in particular focus on such details since system providers use it as part of their design. Also, contractors use them in field instruments at site to verify, before inserting matching circuits.

Fig. 19.10 Relationships between Impedance and admittance parameters

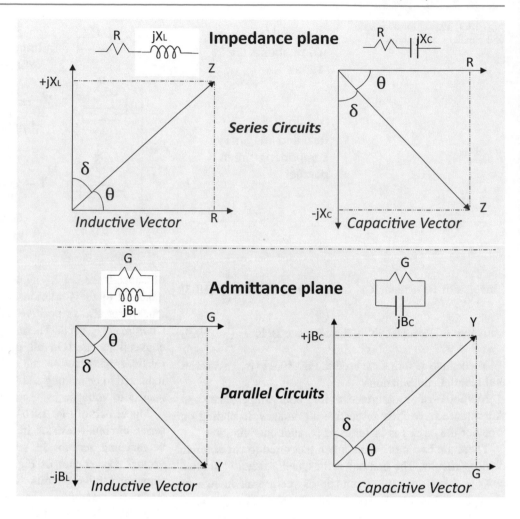

Fig. 19.11 Capacitor (component) with parasitic represented by equivalent circuit

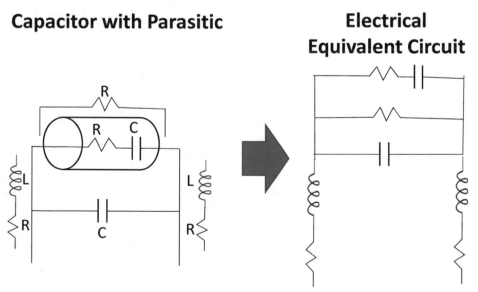

There are several dependencies such as frequency, signal level, and temperature that alter the measured value—these are explained in detail at [17]. Those interested in precision measurement in the laboratory, how components built from different materials behave, etc., will find it very interesting and useful to review [17].

19.3.2   Impedance Measurement Methods

Based on earlier discussions, it can be expected that measurement of impedance starts with a balanced bridge. This is true—however, there are other precision methods to measure impedance over higher frequency ranges. About six laboratory-based methods are used for precision measurements of impedance. These will be briefly reviewed using a table. Detailed methods of each are beyond the range of this book but are well described in [17].

Table 19.1 shows the six methods of measuring impedance, their merits, and limitations. Observe that frequency range plays an important role in each method used.

From Table 19.1 it is seen that RF-IV and Network analysis methods operate in the RF frequency range where cables, connectors, and other wireless devices are used. We shall review only those two methods. In addition to impedance both methods also measure Q the quality factor and D the dissipation factor.

Table 19.1 Common impedance measurement methods

Test method	Advantages	Disadvantages	Applicable frequency range	Common applications	Comments
Bridge Method	High accuracy (0.1% typ.) Wide frequency Coverage—by using different types of bridges	Needs to be manually balanced. Narrow frequency coverage with a single instrument	DC to 300 MHz	Standard Laboratory test	Depending on the frequency range—users select bridges and test devices over the required frequency band
Resonant Method	Good Q accuracy up to high Q	Needs to be tuned to resonance. Low impedance measurement accuracy	10 kHz to 70 MHz	High Q device measurement	This test focuses on obtaining Q value—not commonly used for Impedance
I-V method	Grounded device measurement suitable to probe type test needs	Operating frequency range is limited by transformer used in probe	10 kHz– 100 MHz	Grounded device measurement	Specific locations on a circuit board or device connection
RF I-V method	High accuracy (1% typ.) and wide impedance range at high frequencies	Operating frequency range is limited by transformer used in test head	1 MHz– 3 GHz	RF component measurement	Laboratory measurement and product quality tests
Network analysis method	Wide frequency coverage from Low to RF. Good accuracy when the unknown impedance is close to characteristic impedance	Recalibration required when the measurement frequency is changed. Narrow impedance measurement range	5 Hz and above	RF component measurement	Laboratory method or field installation test for devices—for example, if device is close to 50 ohms, useful to check load balance circuits
Auto balancing method	Wide frequency coverage from LF to HF. High accuracy over a wide impedance measurement range. Grounded device measurement	High frequency range not available	20 Hz– 120 MHz	Generic component measurement	Quality check on production line

19.3.3 RF I-V Method

RF I-V method is based on the principle of measuring the voltage and current through an unknown device—DUT (Device Under Test). Modern RF devices include RF chips that operate well into the GHz bands of cellular systems. The RF I-V method uses a simple technique—it involves measuring ratio of two voltages using a resistor R of stable value [18].

Advantage of ratio is that minor deviation in voltages will not become part of measurement, only the ratio of two voltage matter. Figure 19.12 shows the unknown DUT (Device Under Test) connected to the primary of a transformer. Secondary of the transformer picks up the current passing through DUT. The transformer can use equal number of turns in primary and secondary such that it retains a 1:1 translation of current through DUT. Resistor R of stable value is used across the secondary. By measuring voltage drop across R, current I, passing through DUT can be accurately measured. Unity gain amplifier isolates vector voltmeter V_2 in the secondary circuit. The source has a range from 1 MHz to 3 GHz, hence a wide range of devices to be measured [18].

In performing RF impedance measurements, two sources of error can affect stability:

(i) Changes in environmental temperature significantly influence the measurement of vector impedance at high frequencies.

(ii) Change in frequency channels (for a device covering multiple frequency bands) also affect stability of measurements. Known as "tracking errors", these occur

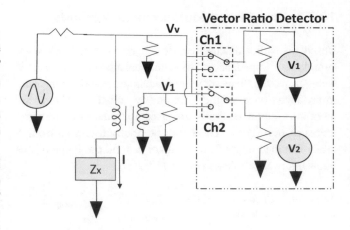

Fig. 19.13 RF-IV measurement with vector ratio detector

when relative gain and phase of the two channels vary after calibration is performed at each band.

Both errors can be eliminated by modifying the ratio measurement technique, shown in Fig. 19.13 where the input signals V_v and V_i are multiplexed by the input switch. Each signal is measured alternately with two-channel VRD (Vector Ratio Detector) circuits in each measurement cycle. As a result of the vector voltage ratio calculation, tracking errors are canceled [18].

This makes it possible to obtain stable impedance measurement results. An additional advantage of the RF I-V method is in the evaluation of temperature characteristics. Due to comparison made in each measurement cycle, this method ensures variations due to the temperature changes are eliminated, increasing accuracy of measurement (about $\pm 0.8\%$).

Fig. 19.12 RF I-V method—Impedance measurement

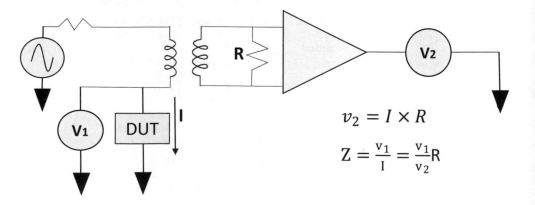

$$v_2 = I \times R$$

$$Z = \frac{v_1}{I} = \frac{v_1}{v_2}R$$

19.3.4 Network Analysis Method

Network Analyzers are used to evaluate electrical network components—not computer networks. Therefore, it will not involve routers, bridges, switches but involves high frequency RF networks. Network Analyzers are used to measure "Return Loss" described earlier in Chap. 17 using Eq. 17.8. In short, return loss is the complement of VSWR (Voltage Standing Wave Ratio). Network Analyzer's calibration and compensation are important processes which play a critical role in RF component testing. The conventional method to calibrate a network analyzer uses three conditions at the output port:

(i)   with output open,
(ii)   output short circuited,
(iii) output has load calibration.

To perform this three-step method of calibration, the open, short, and load reference terminations are connected one at a time to the test port and then each termination is measured [19]. The calibration data is then stored in instrument memory and is used to calculate and remove instrument errors. Such modern Network analyzers can measure and display a device under test's (DUT) with its magnitude and phase information across a frequency range, which characterizes a DUT in terms of scattering parameters, or S-parameters.

At this stage a brief review of scattering parameters is useful. Figure 19.14 shows basic definition of the "Scattering parameters" or S-parameters [20]. Signals are linked to terms that designate two ports—the input port, which is port 1 and the out port which is port 2. All signals are considered in terms of waves at these two ports. The source impedance Z_S and the load impedance Z_L are considered in relation to the waves arriving or reflected from these two ports. Left half of Fig. 19.14 provides the reference two-port circuit and definition of terms, below the circuit. The right half of shows two conditions, Fig 19.14b replaces the load Z_L with characteristic impedance of 50 Ω, while Fig. 19.14c replaces source impedance Z_S with 50 Ω.

Figure 19.14a shows an illustration of a two-port network. Recall that 50-Ω impedances are typical of most RF/microwave applications. The source impedance Z_S and load impedance Z_L shown in Fig. 19.14a on the left are assumed to be 50 Ω. The Cable-television (CATV) applications are the main exception since they operate in a 75-Ω environment. Definitions of are based on the definitions of a_1, b_1, a_2, b_2 noted directly below Fig. 19.14a. On the right half of Fig. 19.14, upper portion shows Fig. 19.14b where Port 2 of the DUT is terminated in a 50-Ω load, thus setting "a_2" to be equal to zero.

In Fig. 19.14b, a_1 is applied to Port 1 of the DUT. A portion of this incident wave is transmitted through the DUT and exits through Port 2, thereby resulting in b_2. A portion of the incident wave is also reflected back to the

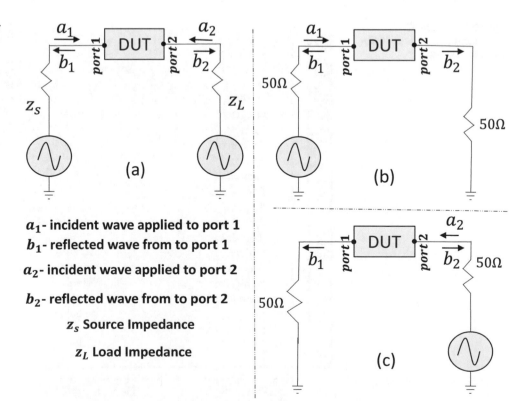

Fig. 19.14 Scattering parameter conditions and definitions

a_1- incident wave applied to port 1
b_1- reflected wave from to port 1
a_2- incident wave applied to port 2
b_2- reflected wave from to port 2
z_s Source Impedance
z_L Load Impedance

source, thereby resulting in b_1. Now, two S-parameters, S_{11} and S_{21}, are mathematically defined as follows:

$$s_{11} := \frac{b_1}{a_1}\bigg|_{a_2=0} \qquad (19.11)$$

$$s_{21} := \frac{b_2}{a_1}\bigg|_{a_2=0} \qquad (19.12)$$

In Fig. 19.14c on the right Port 1 of the DUT is terminated in a 50-Ω load, thus setting a_1 to be equal to zero. In Fig. 19.14c, a_2 is now applied to Port 2 of the DUT. A portion of this incident wave is transmitted through the DUT and exits through Port 1, thereby resulting in b_2. A portion of the incident wave is also reflected, thereby resulting in b_1. Now, the two remaining S-parameters, S_{12} and S_{22}, can be mathematically defined as follows:

$$s_{12} := \frac{b_1}{a_2}\bigg|_{a_1=0} \qquad (19.13)$$

$$s_{22} := \frac{b_2}{a_2}\bigg|_{a_1=0} \qquad (19.14)$$

A two-port network therefore has four S-parameter elements: S_{11}, S_{21}, S_{12}, and S_{22}. S_{11} and S_{22} are known as reflection coefficients while S_{21}, S_{12} are known as transmission coefficients.

S-parameters contain both magnitude and phase information. Magnitude is typically expressed in decibels (dB). This is mathematically defined as

$$S_{11}(dB) = 20\log_{10}|S_{11}| \qquad (19.15)$$

$$S_{12}(dB) = 20\log_{10}|S_{12}| \qquad (19.16)$$

$$S_{21}(dB) = 20\log_{10}|S_{21}| \qquad (19.17)$$

$$S_{22}(dB) = 20\log_{10}|S_{22}| \qquad (19.18)$$

Many of the devices listed in Fig. 19.14 were discussed in various chapters. This list of devices in Fig. 19.14 elaborates various levels of integration on the Y-axis and types of devices on the X-axis, to show readers the wide range of devices (active and passive) that "RF Network Analyzer" can test [20]. It is indeed an essential test equipment to establish any stable component in an RF network. It tests both passive and active devices. It is capable of conducting tests on single devices as well as highly integrated circuits. In Eqs. 19.15 and 19.18, the dB representations of S_{11} and S_{22} are known as return loss, which is the difference in dB between the reflected signal and the incident signal. Thus, a return loss of −20 dB means that the reflected signal is only 1% of the transmitted signal in (excellent). Return loss is commonly expressed as a positive value, so a return loss of −20 dB is often expressed simply as "20 dB" (Eqn. 19.15).

Fig. 19.15 Range of components/devices tested using network analyzers

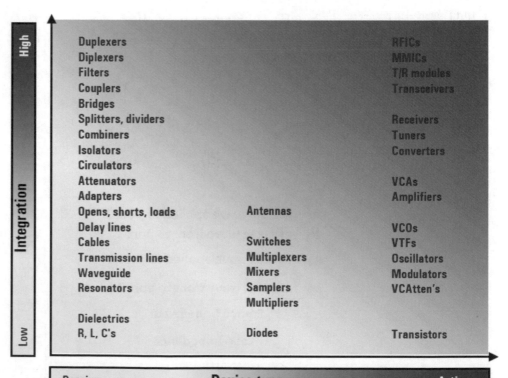

Table 19.2 Return loss versus reflected power and mismatch

Return loss (dB)	Reflected power (%)	Forward power (%)	Mismatch loss (dB)	VSWR	Reflection coefficient
0.00	100.00	0.00	∞	∞	1.00
1.00	79.43	20.57	6.87	17.39	0.89
2.00	63.10	36.90	4.33	8.72	0.79
3.00	50.12	49.88	3.02	5.85	0.71
4.00	39.81	60.19	2.20	4.42	0.63
5.00	31.62	68.38	1.65	3.57	0.56
6.00	25.12	74.88	1.26	3.01	0.50
7.00	19.95	80.05	0.97	2.61	0.45
8.00	15.85	84.15	0.75	2.32	0.40
9.11	12.59	87.41	0.58	2.10	0.35
10.00	10.00	90.00	0.46	1.92	0.32
12.00	6.31	93.69	0.28	1.67	0.25
15.00	3.16	96.84	0.14	1.43	0.18
20.00	1.00	99.00	0.04	1.22	0.10
30.00	0.10	99.90	0.00	1.07	0.03
∞	0.00	100.00	0.00	1.00	0.00

The magnitude element of S_{21} is known as gain or insertion loss, depending on whether the DUT is active or passive—these are listed in Table 19.2. If it is an active device such as an amplifier, then it has gain since it increases the magnitude of an input signal. A passive component like a filter does not have a gain but has insertion loss, and output signal is smaller in magnitude than the input signal.

Many parameters related to complex impedance are shown in Table 19.2 to offer a sense of understanding: how much power is sent; how much is reflected and how much power is lost due to mismatch [20]. The last two columns indicate VSWR and the Reflection Coefficient Γ which is a ratio of reflected wave/Incident wave expressed as phasors. The reflection coefficient corresponds directly to a specific impedance as seen at the point it is measured. Characteristic impedance is Z_0 and load impedance is Z_L are related to the reflection coefficient by: $\Gamma = \frac{Z_L - Z_0}{Z_L + Z_0}$; if that load Z_L is not directly measured through the transmission line then the magnitude is identical, but the phase will have shifted according to the formula: $\Gamma' = \Gamma e^{-i2\varnothing}$, where $\varnothing$ is the electrical length (expressed as phase) of the transmission line at the frequency considered.

Γ is often listed in data sheets of devices. For quick reference, formulas used are provided below

$$\text{Return Loss} = 10 \log \frac{\text{Forward Power}}{\text{Reflected Power}} \quad (19.19)$$

$$\text{Return Loss(dB)} = -20 \log |\Gamma| \quad (19.20)$$

$$\text{Mismatch Loss(dB)} = 10 \log [1 - \Gamma^2] \quad (19.21)$$

$$\text{Reflected Power(\%)} = 100 * \Gamma^2 \quad (19.22)$$

$$\text{Forward Power(\%)} = 100 [1 - \Gamma^2] \quad (19.23)$$

$$\text{VSWR} = \frac{\{1 + |\Gamma|)}{\{1 - |\Gamma|)} \quad (19.24)$$

$$\text{Reflection Coefficient } \Gamma = 10^{(\frac{-\text{Return Loss}}{20})} \quad (19.25)$$

Calculation of complex impedances in vector plane (Cartesian coordinates) used by Smith Chart is shown in Fig. 19.16. The Smith chart bends the right-hand side of Cartesian quadrants 1 and 4 into the lower and upper half of a circle [21]. Since resistance is always positive, the 2nd and 3rd quadrants are not used (where resistance would be negative). Figure 19.16 shows this chart in the form of a circle. Smith chart was originally invented by Philip Hagar Smith of Bell Labs, whose work was published in Jan 1939. He described many ways in which this chart could be used. It has remained popular even today, although sophisticated alternatives such as the Vector Network Analyzer are available now which incorporate Smith Chart within the instrument. However, it is helpful to understand how the smith chart is used for calculating and matching impedance.

In Smith chart, top half of the circle represents the inductive region (positive part of imaginary) and bottom half represents the capacitive region (negative part of imaginary). The center line that separates the upper and lower halves

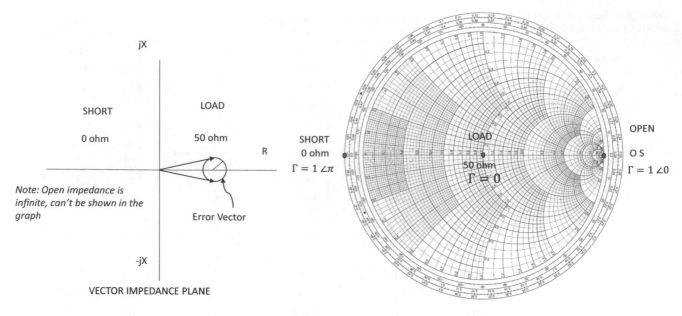

Fig. 19.16 Impedance values represented in vector plane and smith chart

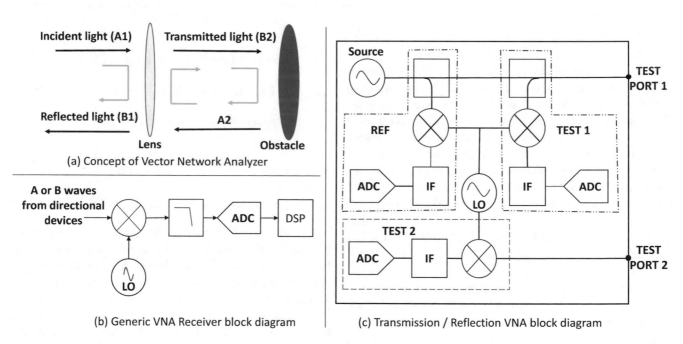

Fig. 19.17 Block diagram of vector network analyzer (VNA)

indicates Resistance. The Smith chart is still very useful in visualizing complex impedances. It is also used for tuning and matching radio networks since it shows Load impedance Z_L with reference to Source Impedance Z_0. An example of how Smith Chart is used is described well in a video with illustrations [21]. There are many sources (video and books) that explain Smith chart in detail.

In Fig. 19.16 impedance values for the reference terminations are represented in two ways. With vector impedance plane (left) and with a Smith chart (right). The Smith chart contains

almost all possible impedances, real or imaginary, within one circle. All imaginary impedances from −infinity to + infinity get represented. Since Smith chart was invented prior to existence of computer, it was an important tool. Such calculations are computed in the Network Analyzer, using a built-in microprocessor. This is illustrated in Fig. 19.17 where a simplified diagram of modern Vector Network Analyzer is presented using a three-part diagram [22].

In the top left diagram (a) the concept of Vector Network Analyzer (VNA) is introduced using light (instead of RF) as an example. Light is also an electromagnetic wave and acts similar to RF. Part of the incident light A1 gets reflected by the lens (as B1), and part of it gets transmitted through the lens. If there is no obstacle after lens the light will pass though smoothly. But if there is obstacle after lens (shown in blue), then part of light is reflected back, (shown as A2) towards lens and this gets split into two parts where one-part merges with original reflection (B1) and the other part merges with transmitted light (B2). This was explained earlier using the S-parameter (or Scattering parameter) equations Eqs. 19.13 and 19.14. followed by four Eqs. 19.16, 19.17, and 19.18 where S-parameters using dB scale were indicated.

VNA usually measures incident and reflected waves through series of bridges or couplers. These are referred to as signal separation devices in VNA. These are also referred as directional devices. Imperfections in VNA measurements lies on coupling factor as well as directivity of directional devices. Third part figure (c) depicts VNA block diagram depicting transmitter and receiver parts of vector network analyzer [23].

The other component is source which is the main stimulus used to characterize the response of the DUT (Device Under Test). This source is tunable which tunes both frequency and power. The output can be configured as single tone or multi-tone for different type of measurements. The source shown in (c) will sweep across the range of frequencies to collect the frequency response of the DUT. The source can also sweep across range of power values to obtain gain compression results of the DUT at fixed continuous wave frequency.

Figure 19.17b shows the simplified block diagram of VNA as a receiver. VNA receiver part measures phase and magnitude of traveling wave. This is done by converting analog form into digital form (using ADC). The figure indicates single stage down converter which is used to convert RF to IF (using Local Oscillator or LO) before processing incident and reflected waves.

There are two architectures of VNA (Vector Network Analyzer) viz. T/R set or the Transmission/eflection test set) and full S-parameter test set.

19.4 Noise and Interference

Any unwanted signal is considered "noise". In all radio systems, the term "interference" also refers to useful signals in the vicinity, but they are being operated by other users/systems and therefore causes "interference". In all communication systems, noise is an important topic that must be addressed to design any RF network with clear reference to known standards of Noise [23]. Yes, there are primary reference noise sources available, that are used to

calibrate the LNA (Low Noise Amplifier) for example [24]. Classical definition of noise power as defined in Eq. 13.5 is $P_N = kTB$ (Watts). Note that for a given bandwidth B, the only option to reduce noise is to lower the temperature T (in Kelvin), since k—Boltzmann's constant and B—given bandwidth are both fixed. In broadband systems, B is quite large which increases noise power. The classic treatment indicated in Chap. 13, showing how satellite systems are designed, apply to the radio link of the system. In all cases the generic term to describe the condition is "Signal to Noise Ratio" commonly known as SNR. But this parameter does not include harmonic distortion. Since input power is concerned with thermal noise, noise power P_N or N_P whichever term is used is expressed in watts and deals with system noise. It is not noise voltage $V = \sqrt{4kTBR}$.

19.4.1 Noise and Its Source—Electrical Power

Since noise is fundamental to all electronic circuits, it is important to begin from the fundamental source of noise—the electrical power from utility. Whether it is a base station, radar transmitter, or satellite station, for example, all of them have this fundamental feature. The only exception to this would be wireless devices that operate on battery power, lately there are many of them that were reviewed in earlier chapters. Following are the basic sources of noise that start at any utility provided power, which unintentionally gets into the circuit:

- Disturbances associated with mains distribution (220 V/50 Hz or 120 V/60 Hz)
- Disturbances arising from DC power supplies
- Noise generated by mechanical contact switching
- Noise emitted by digital circuits
- Transformer noise
- Pulse noise due to electrostatic discharge

This power distribution network involves a large array of wires connecting various outlets to the primary system [24]. For the electronic or wireless designer, it is important not to transform it into a noise transport vehicle and take care to minimize or isolate noise sources.

The generic configuration for supplying electronic equipment is shown in Fig. 19.18. The conductor labeled

Fig. 19.18 Power source for electronic equipment

earth (green or yellow-green wire) provides a current path for activating a circuit breaker—intent is to limit damage due to a ground fault. During normal operating conditions, this conductor carries no current. The conductor labeled neutral is separately connected to supply ground and acts as a return path for current flowing through phase.

All conductors distributing mains power act as receiving antennas and are sensitive to many different types of surges (lightning discharges, power switching, etc.) but also to electromagnetic waves emitted by broadcast services (especially LW, MW, and SW stations—at higher frequencies such as VHF, UHF, and cellular bands the supply lines act as a low-pass filter and do not work well as antennas). At the same time, these conductors are contaminated by low-frequency disturbing signals (f <1 MHz) resulting from transients and switching-on and -off of electrical equipment, since all of them are eventually supplied by the same line [23].

There are two basic rules for the designer to follow:

- **Rule 1.** Minimize the frequency spectrum of the system of interest by controlling rise- and fall-times, as well as the pulse repetition rate. This can substantially reduce noise.

  This rule applies mainly to logic circuits or digital communication systems which we will address later in the form of ADC and DAC. Improving performance by increasing clock speed has its direct consequence—a significant increase in frequency spectrum. While this is the classic approach there are some alternatives that will be discussed in the next sub-section. In general, if a particular application can operate at a lower clock rate (without affecting overall performance), the level of emitted noise can be decreased. It is important to remember that there is a close relationship between the clock rate and the risk of interference.

- **Rule 2.** In any electronic system, power conversion and control circuits are the principal sources of interfering signals. Think of methods to isolate the noise source.

  For example, DC/DC converters and switching-mode power supplies are well known as powerful sources of disturbances. Consequently, whenever possible, selecting a traditional DC power supply instead of a switching-mode supply will considerably reduce the associated noise. Also, whenever possible, it is advisable to replace the thyristors and "TRIAC" (which has an exceptional dv/dt or di/dt performance, but that exacerbates the noise problem) with power MOS (Metal Oxide Semiconductor) devices. In the same category are fluorescent lamps whose choke and switching circuit are the source of interference to portable radio units, for example.

19.4.1.1 Methods to Consider Noise During Design

The equations and circuit consideration for analysis of noise make it laborious to calculate many circuit related noise sources. Since good simulation programs like SPICE are available, it is widely used to perform fast and accurate noise analysis. In practice, noise analysis is either available as an option in most electrical simulators, or dedicated software packages (such as NOF) meant to assist in design of microwave communication networks.

For wireless communication systems operating the higher millimeter wave bands, there are multiple "Primary Noise Standards" available over different frequency bands [24]. Noise power Primary standards have precision waveguide horn with known insertion loss and monitor a blackbody embedded in a cryogenic, liquid nitrogen bath. The benefits of this construction are better accuracy, lower noise temperature, and improved repeatability. Such standards are used for noise temperature and noise source calibrations and Satellite communication earth station conformance verifications, for example.

In wireless communication systems, the "Noise Figure (NF)" or the related "noise factor (F)" is a number used to specify the performance of a radio receiver. Lower the value of noise figure, better will be the performance. Noise Figure is expressed in terms of the four noise parameters $Gamma_{pot}$, R_n, and NF_{min} are included. $Gamma_{pot}$(Go) is a complex number [25]. Source pull involves varying the source impedance presented to a device under test and monitoring a single or set of performance parameters. When used in conjunction with Noise Figure Analyzer (NFA), source pull is often used to measure noise figure (or noise power) as a function of source impedance presented to the DUT (Device Under Test).

The first question is "what is noise figure?" Noise figure is a value which explains the level of noise produced when a signal passes through an active device. Noise added by such an active device, causes degradation in the original signal, to the extent that the content within the original signal might become "unrecoverable". The lower the noise figure, the cleaner the resulting output signal (i.e., a NF of 0.3 dB is lower and more advantageous than a NF of 1 dB or 5 dB). Amplifiers with the associated noise figures are normally listed in their data sheets. Amplifiers not only amplify the signal but will also amplify noise. This is an important reason why NF must be chosen low, particularly when amplifier gain is very high, such as 40 dB which is an amplification of 10,000 (quite common in satellite systems).

The second question is "why do we need to measure noise figure under non-50 Ω conditions?" The term noise figure normally refers to the 50 Ω noise figure of an amplifier which is at or near the minimum possible noise figure value for that amplifier if it was designed well. However, in the case of unmatched transistors, the minimum noise figure will not likely occur at 50 Ω but at some other impedance. It is therefore critical for low noise amplifier designers to understand how to best match their transistors to obtain the least amount of noise possible.

The value of the least amount of noise is referred to as the minimum noise figure, or NF_{min} "NFmin" usually refers to the decibel value of "Fmin" which is a linear ratio. This minimum noise figure occurs at some specific impedance referred to as $Gamma_{opt}$ in real and imaginary values. A fourth value, noise resistance or Rn, represents the rate of change of the level of noise when varying the source impedance presented to the DUT. Lower Rn means a more desirable, shallower slope where noise figure does not rapidly increase as impedance moves away from $Gamma_{opt}$. The four noise parameters ($Gamma_{opt}$, NF_{min}. $Gamma_{opt}$, and n) are graphically represented by the image shown in Fig. 19.19a.

Figure 19.19b shows NF, collapsed onto the two dimensions of a Smith chart, where the contours of constant noise figure (in linear or decibel format) form perfect circles [25].

$$F = F_{min} + \left(\frac{R_n}{G_s}\right)|Y_s - Y_{opt}|^2$$
$$= F_{min} + \frac{4R_n}{Z_0}\frac{|\Gamma_{opt} - \Gamma_s|^2}{|1 + \Gamma_{opt}|^2(1 - |\Gamma_s|)^2} \quad (19.26)$$

where

F = Noise Figure (linear ratio)
F_{min} = minimum Noise Figure (linear ratio)
Γ_{opt} = Optimum complex reflection coefficient
R_n = Noise resistance
Γ_s = Complex source reflection coefficient

Noise figure varies as a function of "source impedance", which was discussed in Sect. 19.3.4, using Fig. 19.14 where source impedance Z_S was explained using transmission line theory.

Figure 19.19a, b represents the noise parameters and how Noise figure is related to them as indicated in Eq. 19.26. Although both represent the Noise figure, the second figure (b) has been used to provide a 3D perspective of Noise figure as it relates to the Smith chart. It is important to realize that Smith chart is "normalized" for a given impedance, which is 50 Ω in most RF networks [26]. Smith chart was discussed earlier in Sect. 19.3.4 showing $\Gamma = 0$ at the center of Smith chart in Fig. 19.16.

It is beyond the scope of this book to discuss details of noise consideration in all two-port and four-port signal networks. Those interested may pursue the references provided [24–26].

19.4.2 Noise in High-Speed Data Converters

In digital systems, all signals must be converted from analog to digital prior to onward transmission. Underlying layer of every digital system whether based on wireless or wired network, always involves ADC and DAC. Let us consider the ADC (Analog to Digital Converter) where the sources of noise begin (electronic circuits shown in Fig. 19.18). At the receiving end, digital signal is converted back to analog using DAC (Digital to Analog Converter). These two units (ADC and DAC) are better understood by considerations of noise. ADC performance is better understood using SNR (Signal to Noise Ratio), while DAC performance is better understood using NSD (Noise Spectral Density). It is possible to derive SNR from NSD [27]. Let us begin with noise considerations in ADC, using SNR.

SNR in an ADC is contributed by three individual noise sources:

SNR_{QUANT} = SNR due to Quantization
SNR_{JITTER} = SNR due to Jitter
SNR_{THERM} = SNR due to Thermal and Transistor noise

The total SNR can be calculated by the sum of these individual noise sources:

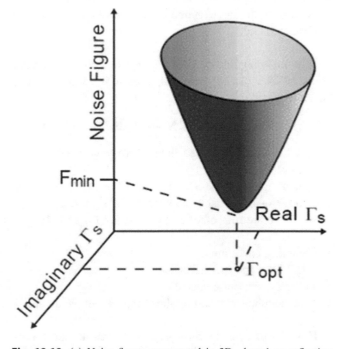

Fig. 19.19 (a) Noise figure represented in 3D—based on reflection coefficient Γ_s, (b) Noise parameters represented on Smith Chart

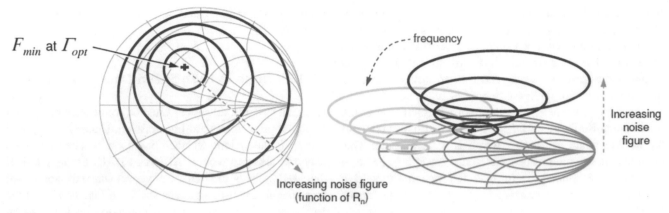

F_{min} at Γ_{opt}

Increasing noise figure
(function of R$_n$)

frequency

Increasing
noise
figure

Fig. 19.19 (continued)

$$SNR_{Total} = 10\log\left(\frac{1}{10^{-\left\{\frac{SNR_{QUANT}}{10}\right\}} + 10^{-\left(\frac{SNR_{JITTER}}{10}\right)} + 10^{-\left(\frac{1}{SNR_{THERM}}\right)}}\right)$$
(19.27)

- Quantization is a process inherent to the way analog voltage is converted to digital. The input approaches the different quantization level values with equal probability.
- The quantization error is not correlated with the input.
- The quantizer has a large number of quantization levels (such as when a high-resolution ADC is used—let us consider 12-bit ADC).
- The quantization steps are uniform (unlike the data converters used in telephony that have a logarithmic characteristic).

Without going into detailed analysis to arrive at the result in Eq. 19.28, it can be stated that "if input is a Sine wave and the ADC has N-bits", then:

$$SNR_{QUANT} = 6.02\,N + 1.76\,\text{dB}$$
(19.28)

If a 12-bit ADC is used then $SNR_{QUANT} = 6.02\cdot(12) + 1.76$ dB = 74.00 dB.

Which represents the SNR for the 12-bit ADC. Note that each additional bit adds another 6.2 dB of quantization noise. There are four other factors that limit the SNR:

1. Sampling clock jitter.
2. ADC jitter and thermal noise.
3. Other system noise sources.
4. Oversampling rate and application channel bandwidth.

Next, let us consider SNR_{JITTER} and what are the contributions that make up this term. First, $Clock_{JITTER}$ is the random variation of the clock edge compared to its ideal point in time and is mathematically expressed by the equation:

$$\text{SNR}_\text{J}(\text{dBc}) = -20\log(2\pi f_{in} \cdot \tau_j)$$
(19.29)

where f_{in} is the analog input frequency and τ_j is the clock Jitter. Total jitter is the RMS (Root Mean Square) sum of the individual jitter contributions. For all A to D Converters, this is a combination of the external clock jitter and aperture jitter. Noise due to Jitter is dominant at higher frequencies.

$$\tau_{\text{Total}} = \sqrt{\tau_{\text{external}}^2 + \tau_{\text{aperture}}^2}$$
(19.30)

The external clock jitter can be measured using a phase noise analyzer, but the aperture jitter cannot be measured by phase noise analyzer. Aperture jitter is internal to the ADC circuit with internal clock buffers. Aperture jitter is the sample-to-sample variation during the encoding process. It has three distinct effects on system performance. First, it can increase system noise. Second, it can contribute to the uncertainty in the actual phase of the sampled signal itself, giving rise to increases in Error Vector Magnitude (EVM). Third, it can increase Inter-Symbol Interference (ISI). Calculation of jitter using phase noise is possible. In Eq. 19.28 the term clock jitter τ_j is due to phase noise. By integrating the clock phase noise over a specified bandwidth and converting to seconds, it can be mathematically expressed as

$$\tau_j = \frac{\sqrt{2\cdot 10^{\frac{\phi_N}{10}}}}{2\pi f_{clock}}$$
(19.31)

where ϕ_N is the phase noise power (dBc/Hz), f_0, f_1 are frequency limits of integration (not indicated here) and τ_j is the clock jitter. By substituting jitter equation into the SNR equation, we get

$$\text{SNR}_J(\text{dBc}) = -20\log(2\pi f_m \frac{\sqrt{2 \cdot 10^{\frac{\phi_N}{10}}}}{2\pi f_{clock}}) \qquad (19.32)$$

After simplification this becomes

$$\text{SNR}_J(\text{dBc}) = -20\log\left(\sqrt{2 \cdot (10)^{\frac{\phi_N}{10}}}\right) + 20\log\left(\frac{f_{clock}}{f_{in}}\right) \qquad (19.33)$$

Equation 19.33 is a general equation that gives a better estimate of SNR, offering a clue on how to improve SNR. The first term is based on phase noise, while the second term indicates a correction term related to "clock". Following Rule 1, described earlier, this second term allows SNR to be improved using "oversampling". Oversampling is the process of sampling a signal at a sampling frequency significantly higher than the Nyquist rate. This may seem contrary to Rule 1 that directed reduction in clock rate to reduce noise. Won't oversampling "increase noise"? The answer is yes, but another point to consider is that broadband digital systems use extensive filtering, retaining only signals of interest within the sub-band. Bearing in mind this point, jitter won't be measured over the entire band of clock frequency but will be measured over to that narrow sub-band [27]. The traditional approach would have used the following equation —a function of analog input frequency and jitter performance over the entire band, indicated as

$$SNRj[dbc] = -20 \cdot log_{10} 2\pi f_m \cdot \tau_j \qquad (19.34)$$

In contrast, the new approach now focuses on specific part or sub-channel which is limited to the frequency range of interest, rather than considering the entire band. When the entire band is considered, noise would have been significant, which is not done here. It turns out the SNR equation is also a function of clock frequency as well. By choosing specific part of the sub-channel, it is possible to control SNR which will be discussed now, as illustrated in Fig. 19.20

The new approach uses over-sampling and considers the second term in Eq. 19.33 carefully with the objective of

reducing SNR_j in an ADC. Figure 19.20 shows frequency spectrum where the "wanted signal" occupies only part of the band (narrow red block). Noise performance over this "wanted signal" portion is important for the system. ADC clock usually has better noise behavior as the offset frequency increases, and the clock is often well filtered.

Researchers performed a simple experiment to validate their assumptions regarding general ADC equation for SNR [27]. A high-speed DAC was used to generate a clock with known but exaggerated level of noise, around the frequency of interest, by oversampling it at 250 MHz rate. This was used to drive the ADC sampling clock, and these two different input signals (tones) were sent into the ADC. One tone was at 10 MHz and the other tone was at 100 MHz. The resulting signals captured by the ADC are overlaid and shown together in Fig. 19.21.

As expected, the sampling clock phase noise is coupled to the input signals at 10 MHz and at 100 MHz. The oversampling correction factor improves the phase noise at 10 MHz by "over- sampling ratio of 25 times", resulting in a 28 dB reduction in noise power (shown on the left). At 100 MHz the oversampling is only 2.5 times (250/100) resulting in a reduction in noise power of just 8 dB (shown on the right).

Observe the results in Fig. 19.21 on the left end, which was oversampled at 250 MHz. Since that is 25 times 10 MHz it shows that the noise reduction is $20\log(250/10) = 28 dBc$ in the first case. But in the second case, 250 MHz is only 2.5 times 100 MHz. Hence, reduction is only marginal at $20\log(250/100) = 8 dBc$ in the second case.

Note that dBc indicated here means relative to the carrier. The term dBc represents the ratio of input signal (modulated) power vs the carrier signal power. It is known as the "decibels relative to the carrier". If the value of ratio is positive then the power of input signal (modulated) is greater than the carrier signal. If the value is negative then the carrier signal has more power.

$X_{dBc} = 20log_{10}\frac{P_i}{P_C}$ where P_i is the input signal power (modulated) and P_C is the carrier signal power. dBc is the

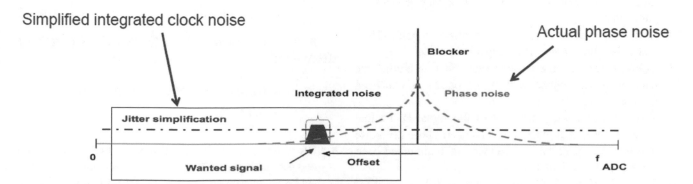

Fig. 19.20 Simplification of integrated clock—noise in a specific sub-channel

Fig. 19.21 ADC test result of sampling at 10 and 100 MHz

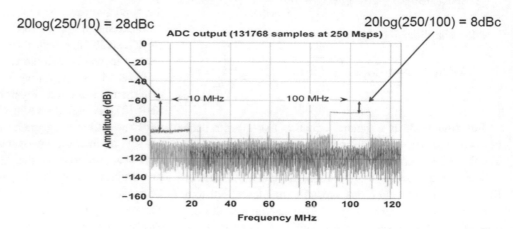

unit used to measure the level of phase noise and harmonics/spurious signals with respect to carrier power level. The term *dBc* can also be used as a measurement of Spurious-Free Dynamic Range (SFDR) between the desired signal and unwanted spurious outputs resulting from the use of signal converters such as a digital-to-analog converter or a frequency mixer.

In summary, the noise floor is made up of contributions from quantization, clock and aperture jitter, and thermal and transistor noise. The jitter limit for SNR is also treated the same way. Till now we addressed noise in the ADC.

The next section will discuss the concept of noise floor in the DAC output. For DAC, the Noise Spectral Density (NSD) is the preferred specification rather than SNR. The estimate of the SNR for DAC is the same as for ADC but is not very useful as we will observe.

For DACs, the NSD specification is generally more important than SNR, due to the following:

1. The shape of the NSD around the carrier must generally meet some transmission mask.
2. Usually, when a specific SNR is required, customers will limit the bandwidth of the signal with a bandpass or low-pass filter.
3. DAC output noise consists of–Quantization noise–Thermal noise–Jitter noise–Data dependent noise.
4. In real systems, there is often tight filtering around the band of interest, where all the noise outside of that band is filtered out.
5. Rather than showing the SNR of the signal in the first Nyquist zone, it is more convenient to show the noise power so that the total noise power in the unfiltered band can be readily calculated.
6. The SNR of the DAC can be calculated from the NSD spec.

7. The SNR was traditionally defined as the ratio of the power of the fundamental to the power of the noise integrated over the first Nyquist zone, relationship is shown below.

$$SNR_{dBc} = P_{dBm.fundamental} - \left(NSD_{\frac{dBm}{Hz}} + 10\log\left(\frac{F_s}{2}\right) \right)$$

(19.35)

SNR can also be calculated directly in dBFS (dB Full Scale) from the NSD in *dBc*/Hz, using the following equation:

$$SNR_{dBFS} = 0 \; dBFS - \left(- NSD_{\frac{dBc}{Hz}} + 10\log\left(\frac{F_s}{2}\right) \right) \quad (19.36)$$

We can use example of two filters to see how it affects a real "DAC 3484" (from Texas Instruments) operating at 1228.8 MSPS with a band of interest of 100 MHz and the following filters.

–A 614.4-MHz low-pass filter (passing full first Nyquist zone):

$$\begin{aligned} SNR_{dBFS} &= 0 \; dBFS - \left(- 160_{\frac{dBc}{Hz}} + 10\log(614.4) \right) \\ &= 72.12 \; dBFS \end{aligned}$$

(19.37)

–A 100-MHz low-pass filter: 23 SNR dBFS (*dBC* Hz (MHz)) dBFS

$$\begin{aligned} SNR_{dBFS} &= 0 \; dBFS - \left(- 160_{\frac{dBc}{Hz}} + 10\log(100) \right) \\ &= 80 \; dBFS \end{aligned}$$

(19.38)

• SNR estimates based on dBFS in these two equations are good estimates for SNR for the entire Nyquist band—but may be too pessimistic for bandwidth limited application.

Table 19.3 Transistor and thermal noise—SNR

Noise mechanism	Operator	Spectral profile	Source	Cause
Shot	I_{DC}	White noise	P-N junctions	DC bias current is not constant
Flicker	$\frac{1}{f}$	$\frac{1}{f}$	Active devices	Carriers are "trapped" and released in a semiconductor
Thermal	T	White noise $\frac{kT}{C}$	Resistors	Thermal excitation of carriers in a conductor

- SNR estimates based on NSD (typically measured at some MHz offset) do not account for close-in phase noise that may affect in band EVM (Error Vector Magnitude)— Useful for out-of-band estimates like ACPR (Adjacent channel Power Ratio)—Also useful for transmitting mask requirements.
- Using the clock NSD curve and bandwidth limited noise calculations would be the ideal solution for in-band and out-of-band measurements.
- Noise floor can then be estimated with the pass band of the filter using NSD.
- This is one of the main reasons most new data sheets of   DAC   will show NSD, bandwidth limited versus SNR, first Nyquist. This is because.
  - SNR can also be calculated directly in dBFS from the NSD in *dBC*/Hz.
  - NSD is useful for out-of-band estimates, such as noise limited ACPR (Adjacent channel Power Ratio).
  - Finally, the third SNR term, SNR_{THERM}. Thermal noise is dominant at lower frequencies. This is a combination of three sources of noise. These three are indicated in Table 19.3. "Track and hold" is a dominant source of noise in ADC, while Resistors cause $\frac{kT}{C}$ noise.

The resultant noise of the ADC can be thought of as a combination of Jitter and thermal noise.

Although noise and its implications are spread throughout communication systems, this explanation regarding ADC and DAC can be viewed as a brief overview of how they are identified and corrected to the extent possible, using clever techniques.

19.5   Transmitter and Receiver Measurements

Transmitter and Receiver often exist together in bi-directional systems such as cellular, land mobile, and WiFi. Some of them such as AM, FM broadcast, GPS, etc., exist as transmit only or receive only systems. Measurements to assess all of them are considered here.

For low power units that contain transmit and receive segments also use the term "modem" quite often. Independent of the modes of how each system is operated, there are

certain measurements specific to the transmitter and some that are specific to receiver. These parameters and their implications on the quality will be reviewed here.

19.5.1   Transmitter Measurements

Types of transmitters vary considerably, but there are nine essential tests to make sure transmitter meets specifications. These tests are.

1. Output Power
2. Power in a Band
3. Unwanted Signals
4. Phase Noise
5. Modulation Quality
6. Gain
7. Gain Flatness (linearity)
8. dB Compression Point
9. dB Compression Point

10. Noise Figure

Output power

Transmitter output can be in many forms—unmodulated CW (Continuous Wave) or modulated by a variety of modulation techniques:

- Pulsed (Radar)—high power for very short duration
- Complex IQ techniques – such as different PSK, or QAM schemes
- OFDM where output power resides only in parts of the signal

Traditionally most common instrument used was a power meter, which was ideal for analog, modulated, power. The power meter is a scalar instrument and cannot measure phase [28]. Since power meters are broadband instruments, they may measure power in other "unwanted" bands. The output therefore must be carefully filtered to allow "power in band" measurement or "Occupied Band Width (OCW)", since regulators will specify limits based on this. Accuracy is limited to about ±2% but more often ±5%. Intermodulation products in the output of transmitter do occur and it is important these are kept to a minimum. Despite such

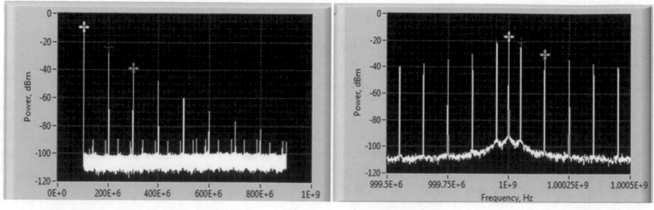

(a) Power Level of output products   (b) Frequency of output products

Fig. 19.22 Spectrum analyzer measures Power and Phase of transmitter output

limitations, power meter is still in use due to its simplicity. Spectrum analyzer is used for laboratory measurements.

Power in band: Fig. 19.22a shows spectrum analyzer display with example of relative power levels of the main output indicated in yellow (in-band power), followed by unwanted signals which are the first, second harmonics in red and blue respectively. The same set of signals are represented on the right in Fig. 19.22b but using the wanted signal placed in the center and spurious (unwanted signals) on either side indicating if these are intermodulation products which often occur in pairs. For example, if a frequency F_c is modulated by a signal F_m then intermodulation products are $F_c + F_m$ and $F_c - F_m$ followed by the next pair $F_c + 2F_m$ and $F_c -2F_m$ and the next pair will be $F_c + 3F_m$ and $F_c -3F_m$ and so on. This is indicated in Fig. 19.22b on the right which shows such pairs, which is a quick method to identify them.

Unwanted signals: Unwanted spurious and harmonics are typically spread over the lower and higher bands (left and right of the main frequency). Intermodulation products are signals generated by nonlinear interactions in the transmitter components. The difference between spurious and harmonics is that harmonics are a multiple of the transmitter (carrier) frequency. Spurious occur at other frequencies within the band but are not multiple of the main RF carrier. But both of these are unwanted signals.

Phase noise: Next, let us consider phase noise, inherent in RF transmitters. All RF transmitters and receivers typically have several frequency conversion stages. Instability in the frequency sources (such as LO—Local Oscillators) used for such conversion, result in phase noise. Earlier, in Sect. 19.1.2 while discussing the topic of crystal oscillator,

its short-term stability was discussed. In any oscillator, short-term instability can be minimized, but not eliminated since it is inherent to the crystal. In Fig. 19.23 upper part shows typical test set up where on the left is the device under test—transmitter in this case. After a preselector and band-pass filter, only the transmitter output frequency band to be measured is mixed.

A reference source (such as crystal oscillator) in the measuring unit is also connected to the mixer, whose output is sent to the display. It is expected that the reference source is considerably more stable and accurate (possibly a TCXO —Temperature Controlled Crystal Oscillator) than the transmitter and mixer. The output presented on the display (lower part of the figure) shows intermodulation components which are displayed as phase noise. The phase noise plot measured with a vector signal analyzer shows the frequency offset from the carrier on the horizontal axis and the relative amplitude on the vertical axis. The red plot represents the raw data and the white trace in the middle is the mean value. Note the low-level spurious signals (in red) are in the range between 1.0 and 10 MHz offset.

Transmitter's "in band" frequency is usually kept small (by filtering) to avoid the second harmonic. Such gradual reduction in phase noise is consistent with the relative power level of harmonics were shown earlier in Fig. 19.23a. The difference is that during "measurement of phase noise" it is important to make sure that it is being measured against a very stable reference source that is clean (with much better short-term stability). The measurement principle is based on the standard principle of measurement that "reference source used to measure must be at least 10 times better in than the Device Under Test". Usually, such reference frequency sources are TCXO that are enclosed in an oven in which they use a closed loop control to keep the short-term stability within tight bounds. Such stable sources are also frequently

Fig. 19.23 Phase noise measurement set up and display result

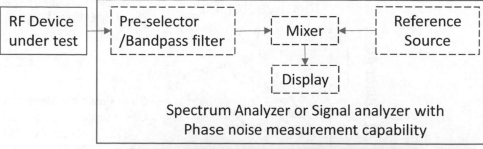

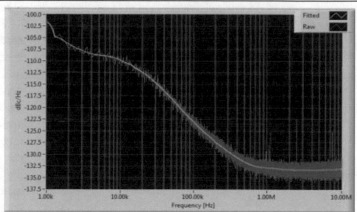

Phase noise as shown on the display

calibrated against secondary or primary standards that were described in 19.1.1.

Mixer shown in upper part of Fig. 19.23 is an example of a circuit component with nonlinear behavior, causing spurious. Phase noise over the entire band is shown in the display at the lower part of Fig. 19.23. Also, harmonic products occur in pairs as indicated earlier in the display in Fig. 19.22. Spectrum analyzer displays RF power and frequency with accuracies ranging from ±0.5 to ±2dB. There are other sophisticated instruments such as Vector signal/network analyzers that measure magnitude and phase, that help in error correction with improved accuracy. Such instruments are particularly useful for pulsed applications. Complex waveform measurement follows methods similar to noise, requiring detailed measurements like peak-to-average ratio (PAR), and Complementary Cumulative Distribution Function (CCDF). These measurements so far addressed the first four measurements viz., output power, power in band, unwanted signals, and phase noise.

Modulation quality: Quality of modulation is termed as "modulation index" in analog systems which is measured by markings of signal levels using an oscilloscope.

For example, in AM schemes, the modulation index refers to the amplitude ratio of the modulating signal to the carrier signal. With the help of Fast-Fourier-Transforms (FFT), the modulation index can be obtained by measuring the sideband amplitude and the carrier amplitude. A carrier with amplitude modulation (AM) can be represented as:

$$V_t = U_c[1 + ma(t)]\cos(2\pi f_c t)$$

where

V(t)—the amplitude Modulated Signal
U (t)—the amplitude of the Carrier Signal
"m"—the modulation index
"a (t)—the normalized modulation Signal
f_c—the carrier frequency

Modern oscilloscopes provide FFT as part of its application software package that allows "modulation index" to be displayed as part of the measurement functions.

In systems that use digital modulation, modulation quality is expressed by two terms known as EVM and Origin Offset. In such systems "In Phase and Quadrature vector components that are orthogonal" must be measured. Hence, the term "Error Vector Magnitude" is used (EVM for short). EVM has two important components (magnitude and phase) whose measurements are needed to determine modulation quality. Figure 19.24 shows the "Vector" and possible errors to it that defines modulation quality.

The ideal vector P1 is shown by the "green" arrow which extends from the origin to center of the "Error Vector Circle", where two errors are possible. One is increase/decrease

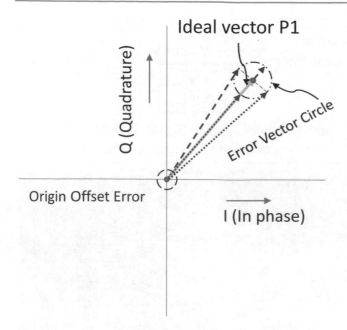

Fig. 19.24 Error vector magnitude—definitions

in magnitude of the vector. An increase in the magnitude results in the "red" dotted arrow, with extra magnitude (error) taking it to outer edge of error circle. A decrease in magnitude results in a red vector that extends from the origin but only up to edge of the error circle. The other error that can occur is the phase error shown by "violet" arrow. The "violet" vector has shifted in phase, compared to the ideal "green" vector that can occur on either side of the error circle.

Since various combination of errors in magnitude and phase are possible, the EV circle represents the limit of error of any vector as percentage of the ideal (center of error circle). For example, the EV circle is usually specified as 10% or 6% which supports a dynamic combination of magnitude and phase. Smaller this percentage, more stringent is the specification of EVM.

The second error that can occur is shift in the origin itself, and this is the "Origin Offset Error". The vector instead of starting at (0,0) may start at (±0.07, ±0.04) for example. It therefore has error offset at the origin itself (0.07 in I and 0.04 in Q), shown by dotted circle at origin in Fig. 19.24. This origin offset is specified in dB, for example, if −20 dB is specified, it will mean that offset error allowed can be no more than 1/100 at the origin, or ±0.01 is limit to zero on the I and Q axes (0 ± 0.01, 0 ± 0.01). Error circles represent the limit of error—smaller the circle, more stringent is the specification. Origin offset is derived as part of the modulation accuracy measurement. This is a measure of the DC offset in the I and Q paths of the transmitter and is expressed in dB (as a ratio of nominal signal vector magnitude). Frequency error is also derived from this measurement.

Wireless standards specify errors for EVM and Origin Offset. The basic EVM equation is

$$EVM = \frac{\sqrt{(I2 - I1)^2 + (Q2 - Q1)^2}}{|P1|}$$

Some examples: GSM specifies ±4% for EVM and Origin Offset suppression of ±1.5 dB

ETSI specifies for TETRA an EVM <10%, and Origin Offset of −30 dBc.

Figure 19.25 shows a simulated example of 16-QAM modulation with considerable noise, typically encountered in wireless networks [29]. Transmitter design is improved after making simulations such as these and improvement can be done using various error correction schemes. Depending on the standard, unit circles around each of the 16 symbols are verified by the measuring instruments. This process of setting the framework of unit circles for each standard imposes complexity that increases with modulation schemes involving higher modulation. 256 QAM is commonly used by many standards such as 802.11AC for WiFi, or any digital microwave link. Its scatter plot will have 256 such points (instead of 16 shown here in Fig. 19.25) and each would have a limit circle around it.

Gain and gain flatness: These measurements are related to transmitters that have a Power Amplifier (PA) stage. Any PA will not be able to provide uniform gain across the entire

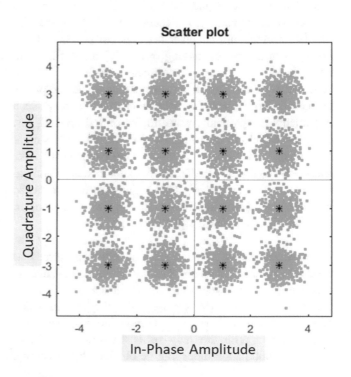

Fig. 19.25 Showing a simulation of 16-QAM modulation

frequency band. Therefore, gain flatness is the other parameter that is often measured. The measurement can be made using a spectrum analyzer, checking responses at specific frequency points across the entire band or across OCW (Operating Channel Width), depending on the standard specified.

1 dB compression point: Specific measurements such as are compression point are used to verify dynamic range of the PA. It is also known as gain compression point, which indicates the output power till DUT or Device Under Test (PA) operates in the linear range and does not go into saturation condition. The input power is varied, and the output power is measured. During this measurement the 1 dB compression point is the level of input power for which the gain is 1 dB less than what it would be.

Noise Figure: Noise Figure is a term used to indicate how much extra noise is added by the device, to the system. It is the measure of the amount of noise generated by amplifier/DUT. It is related to noise temperature as mentioned by the equation:

$$NF = 10\log(1 + \frac{T}{290}) \qquad (19.39)$$

where T is the noise temperature in Kelvins and NF is in dB.

19.5.2 Cellular Receiver Measurements

In all duplex systems, receiver works in conjunction with the transmitter. User devices such as 4G "Modem" or "Hotspot" typically have "transmitter and receiver" within the same physical unit. But there are specific systems such as television used in homes which have only receivers.

In RF receivers, "sensitivity" is an important and key specification that defines the quality of the receiver. This is true of digital receivers too, but the term sensitivity is expanded to multiple levels. Modern wireless standards give different names to BER (Bit Error Rate) measurements such as Minimum Input Power Sensitivity, Minimum Input Level Sensitivity, Adjacent Channel Rejection, Adjacent Channel Selectivity, Reference Sensitivity Level, Dynamic Range, Blocking, and Intermodulation. All these are BER measurements under different conditions.

A receiver's performance is determined by its ability to receive and demodulate a wanted signal in the presence of noise and/or other interfering signals [30]. Sensitivity in a receiver is normally taken as the minimum input signal (S_{min}) required to produce a specified output signal having a specified signal-to-noise (S/N) ratio. Signal level is directly linked to throughput in digital radio. Sensitivity is defined as

the minimum signal-to-noise ratio that the receiver can detect, *times* the mean noise power (kTB) and noise figure of the receiver, leading to the expression:

$$S_{\min} = \left(\frac{S}{N}\right)_{\min} \cdot k \cdot T_o B(NF) \qquad (19.40)$$

which is receiver sensitivity ("black box" performance parameter). This term S_{min} doesn't include antenna. If antenna is included, then its gain must be also be considered (system sensitivity). The term Minimum Operating Sensitivity (MOS) specifies the absolute minimum signal the receiver can tolerate.

$$Min\ Op\ Sensitivity = \left(\frac{S}{N}\right)_{\min} \cdot k \cdot T_o B\left(\frac{NF}{G}\right) \qquad (19.41)$$

Equation 19.40 is modified to provide system sensitivity where the receiver is connected to an antenna (transmission line loss included with antenna gain). In Eq. 19.41 a MOS (Minimum Operating Sensitivity) for a chip is considered since that includes antenna $(\frac{S}{N})$ min or the minimum SNR needed to process received signal, k = Boltzmann's constant, T_o is the temperature of the receiver which is 290 °K (at room temperature), B is receiver Bandwidth, and G is antenna gain or system gain. Based on this equation, the measured quantity is BER, which indicates the probability that a transmitted bit will be received but has detected some error. Of course, better receivers have a lower BER, since considerable effort is put in to correcting such errors using FEC (Forward Error Correction) schemes implemented in DSP. How is this BER detected and estimated? That will be answered now.

When a wireless link is established, the goal is to ensure that sufficient power reaches the receiver (using link budget). The amount of power received drives the design of the receiver itself. Receiver sensitivity gives the lowest received power that a particular receiver needs to successfully detect the signal and demodulate the data. It is known that thermal noise is always present and can dramatically impact receiver's ability to detect a signal. It is important that the signal-to-noise ratio (SNR or S/N) be large enough (good margin) to enable receiver detection and demodulation (i.e., meet minimum receiver sensitivity) under normal conditions.

In RF communications, carrier signals are used to transmit the data. The data is modulated on a carrier wave using methods such as Phase Shift Keying (PSK), Frequency Shift Keying (FSK), or Amplitude Shift Keying (ASK). Receiver detects the carrier wave as well as noise. Therefore, designers are concerned with the amount of power with carrier to noise ratio (C/N) that reaches the receiver. Let us use C/N instead of SNR. Let us revisit typical link equation in decibel form:

Fig. 19.26 Typical RF communication system receiver block diagram

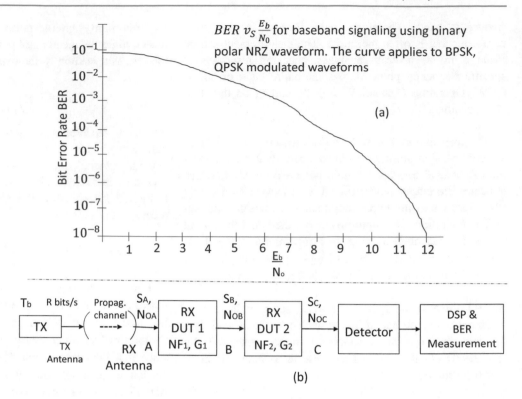

$BER\ vs\ \frac{E_b}{N_0}$ for baseband signaling using binary polar NRZ waveform. The curve applies to BPSK, QPSK modulated waveforms

(a)

(b)

$$[P_r] = [P_t] + [G_t] + [G_r] - [FSL] \qquad (19.42)$$

where P_r = receive power, P_t = transmit power, G_t= transmit antenna gain, G_r = receive antenna gain, FSL = Free Space Loss.

In the above equation, we are concerned with the amount of power that finally reaches the receiver after the receive antenna gain. The received power, P_r, is the power of the carrier signal received (C), or P_r = C. We also know that there will be thermal noise associated with any carrier signal.

Therefore, $$\frac{c}{N_o} = \frac{P_r}{kT} \qquad (19.43)$$

N_o = Noise density (noise power in 1 Hz), k = Boltzmann's constant $-1.38E -23$ J/K, and T = temperature in Kelvin.

As an example, let us assume that designer performed link calculation and determined that the carrier power received will be C = -41dBm. Designer also determined that the noise density of the system is No = -96dBm. Therefore, in decibels, $[C/N_o] = -41-[-96] = +55$ dBm.

Now that we know what the C/N$_o$ is and provided the receiver's sensitivity is low enough to detect the signal, we can then calculate an achievable data rate based on the bit error rate (BER) desired. This is where the ratio of energy per bit to noise power density (E_b/N_o) becomes relevant. Here, E_b is the Error/bit and N$_o$ is the spectral noise density. The upper part of Fig. 19.26a shows a plot of BER versus E_b/N_o. The curve in this graph is specific to the type of RF modulation method used. Let's assume that BER needed is at least 1×10^{-5}(or 1bit error out of 100,000 bits received) to establish a reliable telecommunications link. Following the curve, an E_b/N_o of approximately 10 dB is needed.

The following equation is used to determine the data rate possible given C/N$_o$ and E_b/N_o:

$$\left[\frac{c}{N_o}\right] = \frac{E_b}{N_o} + [R], \text{ in decibels,} \qquad (19.44)$$

where R = data rate

In Fig. 19.26b or lower part, initial two blocks represent transmitted signal and propagation channel between the transmit and receive antennas. The transmitted signal contains data with bit time T_b with bit rate R bits/sec. The propagation channel includes significant attenuation and propagation effects (phase, amplitude, multi-path fading, etc.).

- A is the receiver antenna output.
- B is a mid-point within the receiver system.
- C is the receiver system pre-detection point.

- RX DUT 1 is the receiver's "RF front-end" and contains any lossy lines before other parts of receiver such as front-end amplifiers, filters, and mixers. For this discussion, let us assume it has gain in dB (G1) and Noise Figure in dB (NF1).
- RX DUT 2 is the receiver backend which includes devices before detection. This includes amplifiers, filters, matched filters, and samplers. For this discussion let us assume it has Gain in dB (G2) and Noise Figure in dB (NF2).
- Now BER can be measured, with suitable DSP algorithms, that are part of the FEC (Forward Error Correction) algorithm. Such algorithms are designed to account for number of frames and detailed methods to assess BER, as a statistical average over multiple frames.

At each point in the system A, B, and C, there is a measurable value for the signal (S_A, S_B, S_C) and noise density (N_{OA}, N_{OB}, N_{OC}), where the signal is measured in Watts (W) and noise density is measured in Watts/Hz (W/Hz). Detailed analysis is provided and the following conclusion is summarized in [30]:

- Local S/N_o and E_b/N_o values are not the same at all points of measurement in the receiver.
- System S/N_o and E_b/N_o values are only measurable at the receiver pre-detection point (point C in Fig. 19.26b may be inferred at the other receiver points (points A and B) and have only one value defined for the system.
- Always specify if the S/N_o or Eb/N_o used is the Local or the System value.

In typical Wireless Test Bench data displays, the E_b/N_o value displayed happens to be the local Eb/N_o value at the receiver input, equivalent to point A in Fig. 19.26b. In receiver measurements, the sensitivity measurements are based on different standards and care must be taken to observe at which point in the system, are they defined. In generic terms, E_b/N_o allows comparison of any two digital systems without considering bandwidth (since this is SNR per bit).

19.6 Field Measurements at the Site

Consider the field technician who must make measurements in the field, during installation, system integration set up and finally turn on the system to verify that it operates as designed. An important consideration for field measurement is that a portable measuring equipment or device is needed. It must operate on battery power for extended periods of

time. In wireless sites, often the site is not near mains or AC power sources. Another reason is due to the limitation indicated in Fig. 19.18, which showed that when mains power source is used for electronic equipment, there is possibility of noise and interference. It is important to avoid an instrument directly connected to mains AC power, to reduce noise and interference. For instruments operated on a battery, the measurements parameters are the same as were reviewed so far: Frequency/time, RF power, Impedance, Noise and interference, Transmitter, and receiver measurements.

Frequency/time and RF power are measured by many handheld field instruments—their accuracy may be lower than the instruments used in laboratory. For impedance measurement, it is common to see reflection loss measurement. Noise measurements are somewhat difficult, but interference may take different forms including PIM (Passive Inter-Modulation) that will be reviewed soon in 19.6.3. Transmitter and receiver measurements also offer their challenges since in the field a wide variety of signals are encountered. Let us briefly consider each of them now.

19.6.1 Cell Site Measurements

Cell planning: Since 5G is currently being deployed, complex field instruments used in such planning must make sure the network operates properly. This is accomplished by three field measuring instruments, as examples. The first instrument is 5G site testing unit with full RF capability, the second is 5G "Qualipoc" Android smartphone-based site tester, which has special software to support detailed measurements. The third is a generic cable and antenna tester used at any RF site such as cellular, radar, microwave, or any other technology. Before describing use of these instruments, a brief recollection of cell planning is appropriate. Recall from Chap. 11 that 4G LTE is widely deployed throughout the global cellular network. Now, 5G NR (New Radio) is being deployed, which has cell sites with a Physical Cell ID (PCI). This PCI is also used in LTE to distinguish cells with each base station [31]. PCI planning for 5G NR is an important part of deployment. Inaccurate planning can affect the synchronization procedure, demodulation, and handover signaling. Without careful planning, it can also degrade the network performance.

In 5G NR, there are altogether 1008 unique PCIs that identify cell site. It is based on the formula:

$$N_{IC}^{cell} = 3N_{ID}^{(1)} + N_{ID}^{(2)} \tag{19.45}$$

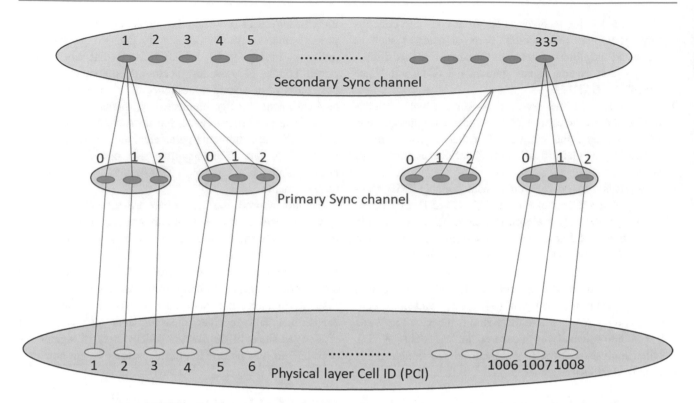

Fig. 19.27 Physical layer segmentation of PCI in 5G NR

where,

- $N_{ID}^{(1)}$ = Secondary Synchronization Signal (SSS) which has range {0, 1…0.335}
- $N_{ID}^{(2)}$ = Primary Synchronization Signal (PSS) which has range {0, 1, 2}

Segmentation of physical layer PCI or Physical Cell ID is represented in terms of the primary and secondary synchronization signals (channels).

Figure 19.27 shows segmentation of PCI in 5G NR standard. With 1008 distinct PCI it is expected that each cell can avoid using the same PCI twice. By keeping the identity of PCI distinct the cell site will clearly recognize data stream from users. To achieve this clarity there are well laid out rules or principles [31]. In terms of planning the following principles apply:

- Avoid PCI Collision: Important principle of network planning–neighboring cells cannot be allocated the same PCI. If neighboring cells are allocated the same PCI, only one of the neighboring cells can be synchronized during the initial cell searching in the overlapping area. However, the cell may not be the most appropriate one. This conflict is known as collision. Physical separation between cell using the same PCI is adequate to make sure

the UE (User Equipment such as smartphone) never receives same PCI from more than one cell. This can be implemented by maximizing reuse distance for PCI.

Consequence of improper cell planning can result in the following:

- PCI collision can result delay in DL (Down Link) synchronization in overlapping zone.
- High BLER (Block Error Rate) and decoding failure of physical channels scrambled using PCI.
- Handover failures.

The instruments used must address these and provide help in proper planning of PCI. There is another level of complexity in 5G—multiple beams within one cell sector. In a typical 5G NR FR1 site (4.1–7.125 GHz), there are three sectors (PCI), with seven beams per sector [31]. Note that FR2 (>24.250 GHz) is not being deployed currently.

Figure 19.28 shows a single sector in 5G NR FR1 site, which has seven separate beams SSB0 to SSB6. Synchronization Signal Block (SSB) is used by each cell site in 5G. SSB beams are static, or semi-static, always pointing to the same direction. They form a grid of beams covering the whole cell area. The UE (smartphone) searches for and measures the beams, maintaining a set of candidate beams that it can go to. Typical track taken by UE while making

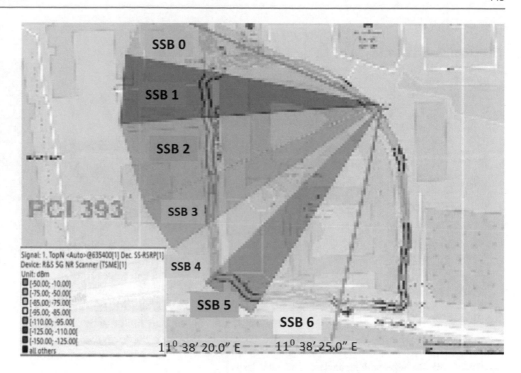

Fig. 19.28 One cell sector (PCI 393) 5G NR FR1 site with seven beams SSB 0–SSB 6

measurements is shown in Fig. 19.28. The candidate set of beams may contain beams from multiple cells. This is similar to the concept of neighbor cells described earlier in Chap. 11.

The field instrument must accurately measure signal from each beam within the PCI. This is important to make sure the site is properly aligned within the network. For many of the abbreviations used here, please refer to the details provided at the end of the section in this chapter. The number of beams emanating from each cell site can vary depending on down tilt (how the antenna is oriented electrically). Note that these colors are used only to show separate beams, but all the beams radiate RF and not these colors as light. 5G standard of 3GPP allows up to 64 such beams but that level of complexity is expected to be deployed only in the coming years. General expectation is that 5G deployment will continue over this decade of 2020 and practical systems of 6G are expected towards later part of this decade.

During installation the 5G site testing unit will perform the following functions during site installation:

- Check VSWR reflections via OSS (Operations Support System) counters.
- Ensure the PCI and beams are visible in the expected location (SS-RSRP, SS-SINR).
- Conduct RF power measurements on the allocated LTE anchor cell (RSRP, SINR, etc.).
- Run functional tests to ensure the correct integration of 5G cell into the network.

Installation may not be always perfect—it could have errors. When errors are encountered, the 5G site testing unit will perform the following verifications:

- If reflection counters show problems, measure the reflection/DTF (Distance-To-Fault) with a cable and antenna tester (cable and antenna tester is unit on the right).
- If beams/PCI are not visible, check and correct the 5G site configuration file.
- Check cable and antenna (using cable and antenna tester).
- If the 5G cell is not utilized or the throughput is too low, check if 5G cell is advertised on the system information Block (SIB) which will be LTE- SIB 2 in this case [32].
- If the SINR is not as expected, check for internal interference (e.g., sidelobes) and external interference.

Some of the tests or verifications may/may not be needed —but test instruments are indicated as illustrations.

Figure 19.29 is a typical 5G site tester with "Qualipoc Android" on a smartphone, with traditional cable and antenna tester [33]. "SwissQual Qualipoc" Android is a high-performance, smartphone-based optimization test product for mobile radio networks. These three units were briefly mentioned earlier, as part of the required instruments needed at a cell site for planning. Note that a traditional cable and antenna tester is required as well. That type of tester is used at every cell site (whether 4G, 5G, or any other technology) to test coaxial cable and antenna, to make sure

5G QualiPoc Android smartphone

5G Site tester with smartphone

Cable and Antenna tester

5G site testing unit with full RF capability

Site Tester unit in bag and Tablet with QualiPoc Software shown in the foreground

50Ω termination   **Power Sensor**

Fig. 19.29 5G site tester with smartphone and cable tester

the return loss (impedance, VSWR) is aligned properly. Figure also shows the cable tester (right side) with 50 Ω termination and a power sensor in front of the tester. Sections 19.2, 19.3, 19.4, and 19.5 described the role of cable and antenna test equipment. The 5G site tester is expected to support the following measurements:

Automatic channel detection

- SS/PS/PBCH/DM-RS -RSRP over PCI and beam (note: PBCH DMRS is a special type of physical layer signal which functions as a reference signal for decoding PBCH. In LTE (at least in TM1, 2, 3, 4), this type of special DMRS for PBCH is not needed, since it is possible to use CRS (Cell Specific Reference Signal) for PBCH decoding. However, in 5G/NR there is no CRS.
- RSSI over PCI and beam
- SS/PS/PBCH/DM-RS-SINR over PCI and beam
- SS/PS/PBCH/DM-RS-RSRQ over PCI and beam
- Parallel measurements for LTE and 5G NR
- FR1 and FR2 (FR2 using downconverter)—these are separate units and not part of the site tester

The smartphone (Android platform including "Qualipoc" software) is expected to perform the following:

- Layer one parameters of anchor cell (LTE).
- Layer one parameters of secondary cell (5G).
- Scheduled and net throughput (LTE and 5G cell).

- Block error rate, BLER (LTE and 5G cell).
- Layer 3 signaling, MIB, and SIB. The very first step for UE to gain access to the LTE network is to read the Master Information Block (MIB) transmitted every 40ms. MIB arrives over the air on BCCH (Logical channel), BCH (Transport channel or broadcast channel) and PBCH (Physical Broadcast channel). The MIB contains three important information elements: Down Link (DL) bandwidth, System Frame Number (SFN) and PHICH. The PHICH is used for uplink re-transmission in LTE, but not needed in 5G NR. Earlier 5G NR was discussed earlier in Chap. 11, and Fig. 11.25 indicated the uplink and downlink.
- Functional tests (ping, data DL/UL)—alternatively any other tester with RFC 2544.
- Dropbox transfer (or some other major data base download for data transfer).
- Facebook test—could be any other social site with multiple video streams.
- Traceroute (UDP only).

From many of these tests, it is obvious that 5G NR test sites are expected to become the primary means of Internet access and cellular mobility functions. Such cell sites would become an alternative to the traditional wired connections (such as fiber from traditional wireline carriers, CATV or other cable operators, DSL, or other copper wire).

19.6.2 Generic Receiver Measurements

With modern wireless technology spread over a variety of applications, the generic version of receiver that can detect and measure multiple signals in the air has become popular. Let us consider such a receiver and its measuring methods [33].

Portable monitoring receiver units also incorporate spectrum analyzer functions, to accommodate complex spectrum—since measuring just power and frequency does not provide full information. Some of the important functions of such units include:

- Detection, analysis, and location of RF signals from 8 kHz to 8 GHz; extendable to 18 GHz with high frequency directional antenna with a downconverter.
- Wide range of tools for frequency and time domain analysis with up to 40 MHz real-time bandwidth.
- High RF performance optimized for use in dense spectrum environments thanks to sub-octave preselection and automatic overload protection.
- Optimized for demanding field operation with minimal size, weight, and power consumption.
- Convenient, simple, and intuitive to operate with application-oriented user interface which has become possible due to many Android applications.

Unlike spectrum analyzers used in the laboratory environment, such portable receivers are optimized for use with an antenna in dense spectrum environments. They typically use gap-free, real-time fast Fourier transform (FFT) signal processing based on a powerful FPGA (Field Programmable Gate Array which is an Integrated Circuit that can be programmed to perform a customized operation for a specific application).

FFT allows such receivers to detect even signals that transmit for a period as short as 1.5 µs (short pulse) with 100% Probability of Intercept (POI).

Figure 19.30 shows typical portable monitoring receiver, which can receive a wide range of signals. With the help of DDC or Digital Down Converter that uses DSP (Digital Signal Processing) this instrument can preselect signals over a wide range of frequencies and which have very different signal formats. DDC not only detects the type of signal based on its header and frame information, it can also align the receiver to the proper channel.

In Fig. 19.30 on the right are typical spectrum available from FM radio station on the receiver display [34]. The top portion display shows scan results of up to three configurable traces in F-Scan (Full Scan) mode tuned from 88

to108 MHz in 120 kHz steps. An external High gain Vertically Polarized (VP), wide band antenna was used in this measurement. The bottom display shows an event that was missed but revisited using the history function to look back in time and examine that signal. Due to history of trace, it is possible to see two subcarriers one at 25 kHz and another in the upper 600 MHz band. The yellow indicates "max hold" meaning snapshot of signal. These can be accumulated over a period of time. The green signal is an accumulated stream of historic monitoring of signals is also known as "waterfall" display. To locate the presence of 5G network it is important to locate the SIB (System Information Block) signal which is the only "always on" signal.

Details of methodology used to set up a cell site for 5G network is described in [35]. Although Chap. 11 described cellular systems, this video presentation describes the latest release 15 related to 5G deployment with standalone version showing how user equipment can differentiate between a 4G LTE signal and 5G LTE signal. It is important to remember that LTE (Long-Term Evolution) has the concept of backward compatibility to 4G. This concept came about due to constant changes in the entire technology of cellular systems where 3G would not work with 4G or 2G, etc. This non-compatibility resulted in considerable hardship to end user, which is now avoided by using the concept of "backward compatibility" where the user cell phone will use whatever generation of cellular service is available in the area. Starting with LTE in 4G all future systems will be backward compatible.

19.6.3 Passive Inter-Modulation (PIM) Measurements

Passive Inter-Modulation (PIM) is a form of intermodulation distortion that occurs in components normally thought of as linear, such as cables, connectors, and antennas. However, when subject to the high RF powers normally emitted by cellular systems, these devices can generate intermodulation signals at –80 dBm or higher [36].

With the rollout of spread-spectrum modulation techniques, such as W-CDMA, and OFDM technologies such as LTE and WiMAX, it has become essential to test both PIM and impedance parameters accurately. PIM lowers the reliability, decreases capacity and data rate of cellular systems. It does this by limiting the receive sensitivity by increasing the level of interference. Up to 2G, RF engineers could select channel frequencies that would not produce PIM in the desired receive bands. From 3G and beyond cellular usage increased rapidly, and licensed spectrum became

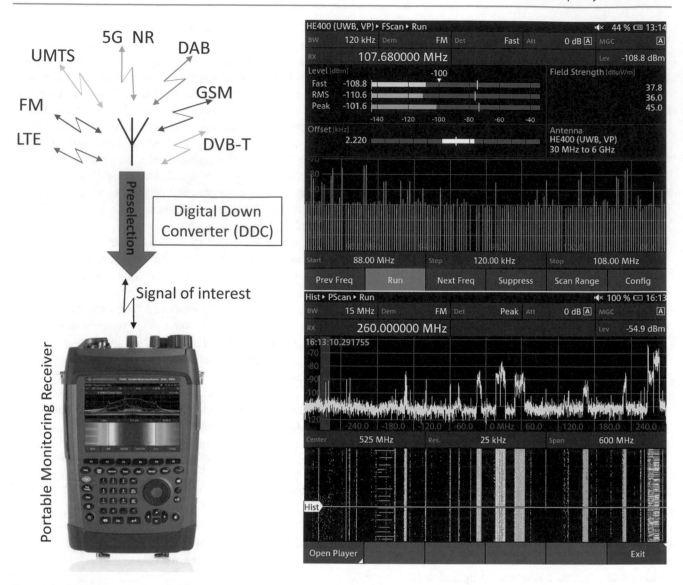

Fig. 19.30 Portable monitoring receiver and its spectrum capability

crowded—leading to using bands but must use methods of avoiding PIM interference. Now let us understand what PIM is and why it could occur anywhere.

Let us consider two frequencies F1 and F2 as shown in Fig. 19.31. Observe that PIM gets generated as harmonics, there are two equations that indicate how PIM occurs.

$$nF1 \; - mF2 \qquad\qquad (19.46)$$

and

$$nF2 \; - mF1 \qquad\qquad (19.47)$$

where F1 and F2 are fundamental carrier frequencies; "n" and "m" are positive integers. When referring to PIM products, "n + m" is termed as product order. For example,

if m = 2, n = 1 then 2 + 1 = 3 is the third-order product. Similarly, 5th and 7th order products are also possible. From the signal spectrum on the left side of Fig. 19.31, observe that as the order increases, the band width of the harmonic increases too, but amplitude reduces.

Intermodulation created from modulated signals occupies more bandwidth than the fundamentals. For example, if both fundamentals are 1 MHz wide, the third-order product will have a 3 MHz bandwidth, the fifth-order product, a 5 MHz bandwidth, and so forth. PIM products can be very wide-band, covering wide swaths of frequencies. Let us consider the PCS and AWS band and the lower cellular band to show how PIM affects the receiver [35].

Figure 19.32 shows two examples—Fig. 19.32a in the PCS/AWS band where PIM gets generated at 1750 MHz which is within the 1710–1755 MHz of the AWS-1 band.

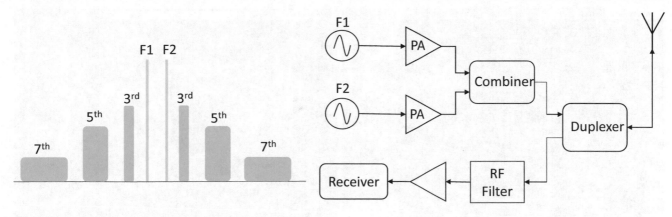

Fig. 19.31 Generation of PIM from two frequencies F1 and F2

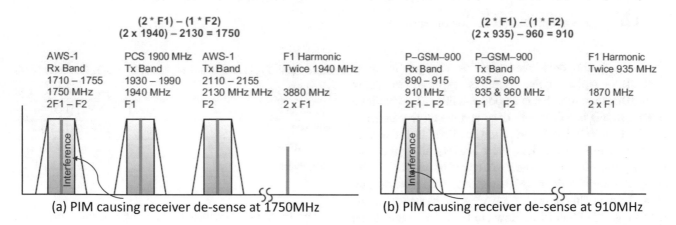

Fig. 19.32 Harmonics due to PIM affect different bands–PCS/AWS and cellular

Similarly, in the lower band of cellular Fig. 19.32b shows PIM generated at 910 MHz which is within the 890–915 MHz of the P-GSM-900 band. These are provided as examples to show how PIM affects the receiver in every band. Since these can be calculated, it is best to use components that meet PIM standards. There are other precautions that can help avoid PIM—usually this involves tests in the field where actual system is set up. A typical PIM tester indicates proximity of PIM source (how close).

Causes of PIM

There are many causes of PIM—some of these are listed here. Detailed explanation of each of these causes is provided in [36]. But it is provided here to indicate a wide range of phenomenon that can cause PIM.

- Mechanical considerations (metal joints, fixtures, etc.)
- Metallic contact
- Tunneling effect
- Rusty bolt effect
- Fritting
- Ferromagnetic materials
- Surface effects
- Time dependency on PIM sources
- Components
- Connectors
- Cables
- Antennas
- Nearby corrosion
- Lightning arrestors

In the field, when the network is being set up, a systematic approach to test for PIM is first conducted. If the interference is high, then there are specific methods to locate the source of PIM [36]. Initial sweeping of the frequency band (cellular 800/900 MHz or the PCS 1900 MHz or the AWS 1700/2100 MHz are common) provides an overview of "Inter-Modulation" products in the area, which result in raising the noise floor. This is followed by locating the source of PIM.

Figure 19.33 shows (a) measurement of typical noise floor in the cellular 900 MHz band where there is no interference.

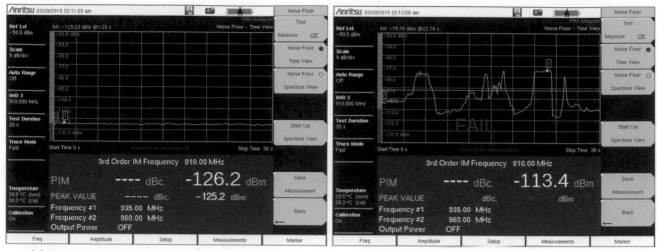

(a) Typical Noise floor – no external interference   (b) Noise floor with external interference

Fig. 19.33 PIM measurements—with and without interference

Observe that the noise floor in this example is below −125 dBm—typical noise floor is expected to below −120 dBm, therefore this is a good condition. Typical limits for PIM in this band are −97 dBm and the display indicates that PIM is well below the allowed limit. In the second measurement (b) on the right, there is PIM, raising the noise floor beyond −97 dBm. The peak levels of interference are being seen at −79.8 dBm which is about 17 dB more than the allowed limit. This is the reason why the second measurement shows "fail" on PIM. How are these noise floor limits set? Noise floor depends on usage of that band and typical bandwidth used for the system. The general thumb rule continues to be, SNR (Signal to Noise Ratio) must be at least 12 dB for all audio systems. Given that limitation −120 dBm is sufficiently low since cellular systems can operate well in the −30–110 dBm. Handoff to a better channel is set around −100 dBm based on actual measurement reported from the handset to the base station. Therefore, 20 dB margin (−120 to −100) is provided as a safe measure. The −97 dBm floor indicated above is a PIM limit placed after allowing an additional 3dB margin. At −97 dBm there is a good possibility of cellular performance being normal.

PIM noise is within the 900 MHz band—it will cause degradation in performance. It is necessary to perform the Distance-To-Fault (DTF) indicated earlier in Sect. 19.6.1. In this case such location changes to DTP which is Distance to PIM source. Resolution of how close to the source the instrument can detect depends on the frequency band. Table 19.4 shows typical bands of operation and the resolution. Other than the wavelength in that band the wave propagation velocity factor needs to be accounted for—in this case it is taken as 0.88.

Table 19.4 Typical calculated resolution of sample frequency bands

Frequency band	Standard resolution
850 MHz	5.5 m (18.0 Ft)
900 MHz	5.3 m (17.3 Ft)
1800 MHz	2.4 m (7.9 Ft)
1900 MHz	3.3 m (10.8 Ft)

The resolution calculated and shown in the table is the best possible distance for resolving two PIM sources located close to each other. Distance limitation to such PIM sources is not due to both measurement limitation but also due to how obstructions around that area are located. It also depends on whether the PIM source sends strong or weak signals, and also whether it has intermittent signal due to some loose connector, etc [36].

Overall, the aspect of PIM and its impact on wireless systems will continue to be a growth area for concern and scope for further improvements, since higher throughput from wireless links will need cleaner noise floors [37].

19.7 RF Components—Their Functions

There are many components that help build the RF network—most are passive components. They are used on the lab bench as well as in field installations. Many of them act as "pass through", others may provide "branch off" and some are terminators. All of them have important functions, which is why they are popular and have been used over many decades. We will briefly review each of them with their functions.

Some of them are:

(1) Lightning Arrestors
(2) Directional Couplers
(3) Duplexers, Diplexers, Splitters
(4) Balun transformers
(5) Circulators/Isolators
(6) Attenuators

19.7.1 Lightning Arrestor (Surge Arrester)

Earliest experiments to study lightning date back to Benjamin Franklin's kite experiments in 1792. However in 1859, Yale University described the use of the first spark gap protection solution based on his experiments. Lightning arrestors are important units that provide a safe path for the enormous energy discharged to earth due to lighting that last between 10 and 100 mS. Since many antennas are mounted on top of tall structures, probability of lightning striking them is high. RF coaxial cables are connected to the antennas—they have metal conductors allowing lightning voltage to pass on to the radio equipment. A surge arrester is a protective device for limiting voltage on equipment by discharging or bypassing surge current. An arrester does not absorb lightning or stop lightning. It diverts the lightning current to ground, limits the voltage build up, and protects the equipment installed in parallel. To avoid damage, its design uses alternate or easier path for this energy to reach earth, using an arrester unit and grounding rod [38]. To do so three approaches are in common use.

(a) The Gas Discharge Tube (GDT, Gas Capsule).
(b) High Pass type.
(c) Band Pass type.

Figure 19.34 shows the three approaches to implement the concept of "alternate path" for lighting to get discharged to ground—RF equipment shown on right by blue box [37]. Since each of these techniques use "Thruline or feed-through connectors" it is important that the coaxial line has a direct grounding to earth "outside" the RF shelter, such that the current does not enter the shelter or damage other equipment. This means that the lightning arrestor while allowing the coaxial path for RF, must make it easier for discharging current by providing "path of least resistance to ground" known as the "alternate path". Surge immunity must be tested with a current test pulse, to make sure surge protector operates, but won't through (not extremely high current). A gas capsule with inert gas filled in is used to provide this "alternate path".

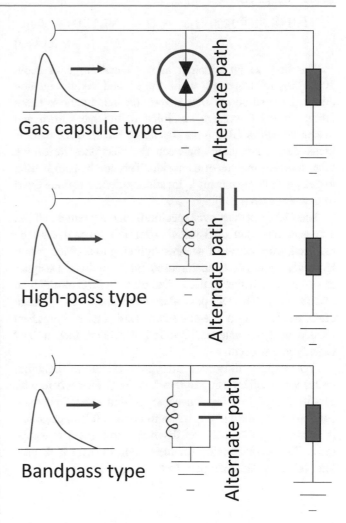

Fig. 19.34 Three approaches to lightning arrestor discharge design

The international standard IEC 61,000-4-5 defines a combined current test pulse of 1.2/50 μs voltage and 8/20 μs for surge protection devices to determine their protection performance. Despite its relevance for general induction and power switching interferences, this pulse is used to describe protection quality of lightning protectors worldwide. Protection performance data shows residual pulse values because of a 1.2/50 μs; 8/20 μs combination is comparable.

The gas capsule (GDT) type (topmost figure) takes the approach of using a small bulb within which an inert gas or air is filled and has "two electrodes placed within arching distance". The design carefully selects two voltages. Value of DC Break Down voltage VDCBD, and the Impulse voltage carried per microsecond are considered during design. For example, one GDT design uses the following:

$$\text{VIMPULSE } @100V/\mu s = (1.4 \times \text{VDCBD}) + 140$$

$$(19.48)$$

$$\text{VIMPULSE } @1000\text{V}/\mu s = (1.6 \text{ x VDCBD}) + 300 \tag{19.49}$$

The first design equation allows small surge of about 100 V per microsecond, while the second design equation allows large surge of 1000 V over the same period. These terms show DC voltage sustained before breakdown and discharge within GDT (Gas Discharge Tube). Once there is break down, arc jumps between the electrodes inside the tube, ionizing the gas or air inside. This mechanism is used to design multiple electrode locations offering various types of discharge rates [38].

The GDT operates over specific frequency range and has a typical life span of 5 years. After this time, it must be replaced, irrespective of whether lightning took place or not. However, type of connector used offers options. Designer can consider whether to use a "DC pass" or "DC block" type of connector. The DC pass allows the use of DC voltage often needed for a pre-amplifier (Low Noise Amplifier) located near the antenna. If it is not needed, then a "DC block" type is used.

Figure 19.34 illustrates different options to lightning protection. It offers using different types of filter where with careful design low inductance to ground offers "alternate path" to move current discharge to earth, while capacitance is increased to reduce the discharge towards RF equipment. The second option (in the middle) uses a high pass filter allowing alternate path to those signals.

The third option of Fig. 19.34 (bandpass filter) is elaborated further in Fig. 19.35 where (a) shows a shorting stub based on the "Through line or feed through" type of transmission. A coaxial shorting stub is used to short circuit the line at its end, and its length is matched to the mid-band frequency of the operation band. It thereby forms a bandpass filter which was indicated earlier in Fig. 19.34 (bottom figure). Its bandwidth can be adjusted by up to ±50% of the center frequency. The lightning discharge path is shown by the red line and arrow points to its path towards grounding rod. Let us now consider Fig. 19.35b—in schematic on the right; it shows a quarter wave lightning protector.

During normal operation, the RF signal reaches the entry of the shorting stub (shown as point 1). The signal then runs along the shorting stub up to the short (point 2). This corresponds to a 90° phase shift. At the short, the signal is reflected (point 2')—a sudden phase shift of 180° is created and signal flows back to the start of the shorting stub (point 1'), where it arrives after another 90° phase shift. As a result, the reflected signal is again in phase with the arriving signal. This is indicated by the waveforms at the top left of (b). Therefore, the RF signal does not ≪detect≫ the short [38].

Both the second and third options consider other factors such as PIM [38], described earlier in the previous section. Of the three approaches—the first one using GDT is in widespread use and costs less. It has frequency band limitation but offers "DC by-pass" option if there is no LNA. But its limitations are lifetime is typically about 5 years after which it must be replaced.

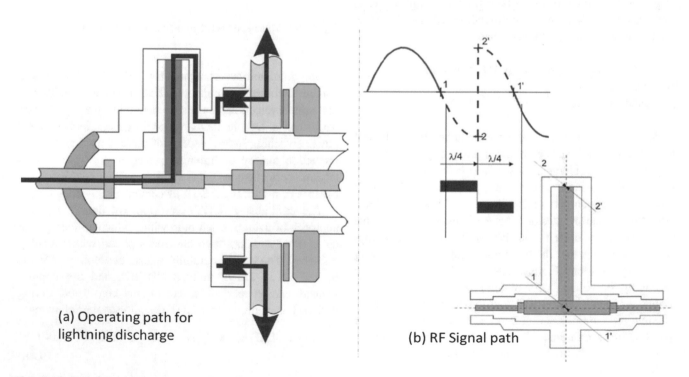

(a) Operating path for lightning discharge

(b) RF Signal path

Fig. 19.35 Principle of RF quarter wave lightning protector

The second option of high pass filter is not widely used. The third option of bandpass is popular since it does not have limitations in terms of lifetime. It supports wide range of frequencies but does not provide the option for using DC if there is LNA next to antenna. It only offers "DC by-pass" since there is a short in the transmission line.

19.7.2 Directional Couplers

The term "directional coupler" inherently indicates the direction of signal travel is predefined. Its major function is to direct RF power in one direction.

Figure 19.36 shows the schematic diagram of directional coupler. Two representations are shown in (a) and (b), both of which are commonly used in texts and diagrams. The signal is allowed to go from Input port 1 (Incident port) to transmitted port 2 (Output port). However, the coupling is such that in the forward direction, coupled port 3 (forward coupled port) will allow only a portion of the forward power to pass (usually -20 dB or $\frac{1}{100}$ of the input power). Similarly, the isolated port 4 (reverse coupled port) will allow only a part of the reverse power (again -20 dB of return power). It is important to note that this is not a resistive splitter or an isolator. It does not use resistive elements and the main objective is to move power in one direction (input to output port) allowing other circuits to use the power.

Therefore, coupled ports are usually kept terminated (normally with 50 Ω), and used to test forward power without interrupting transmission. Similarly, the isolated port is used to check returned power level. Directional couplers are implemented using a variety of techniques including strip line, coaxial feeder, and lumped or discrete elements. They may also be implemented in a variety of packages from blocks with RF connectors, or solder pins, or they may be contained on a substrate carrier, or they may be constructed as part of a larger unit containing other functions. When used as an individual unit, directional coupler is useful as a device that samples a small amount of radio power for measurement purposes. Such power measurements include incident power, reflected power, VSWR values, etc.

19.7.3 Duplexer, Diplexer, and Splitter

Although Duplexer and Diplexer are both three-port devices, their functions are different. Let us first start with the Duplexer.

Duplexer is a key element that is not part of the antenna but essential when the antenna serves to "transmit and receive" signals, quite common in cellular and all other bi-directional systems.

Figure 19.37 shows a "Duplexer" which is a three-port device—Transmit, Receive, and Antenna ports. Many antennas on cell towers do transmit and receive signals. Such systems are based on FDD—Frequency Division Duplex, which was explained in Chap. 11, Sects. 11.4 and 11.8. Since there is a specific transmit frequency band and an associated receive frequency band 45 MHz away, they are known as "Tx-Rx channel pair". Similarly, at the smartphone the same type of duplexer is needed. It will be much smaller in size but performs the same function. Such a duplexer is absolutely essential since the cell phone also uses a transmitter and receiver connecting to the common antenna port.

It is straight forward to separate the Transmit/Receive signals using "band pass filters" that allow the entire transmit band to pass through. Similarly, there would be a matched receiver filter that allows its entire associated receive band to pass through. Such design allows the use of single cable up the tower all the way to the antenna. Similarly, the cell phone also uses a single port to connect to its antenna. In modern cell phone the antenna is internal, which was discussed in Chap. 17. Considerable isolation between the transmit and receive bands must be maintained in filter design—this separation determines quality of service. Efficiency of the antenna becomes an important parameter in such systems.

Diplexer and Splitter

In contrast to the duplexer, "Diplexer" is a three-port device, but it is used for "combining" two independent frequencies from two distinct signal sources and sending it to the same antenna. The two frequencies must be on separate

Fig. 19.36 Directional coupler —schematic

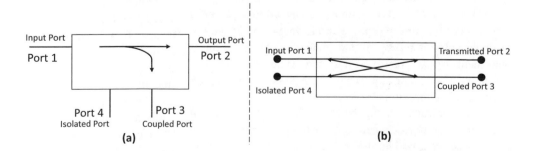

Input Port
Port 1

Output Port
Port 2

Port 4
Isolated Port

Port 3
Coupled Port

(a)

Input Port 1

Isolated Port 4

Transmitted Port 2

Coupled Port 3

(b)

Fig. 19.37 Duplexer used to connect antenna for transmit and receive

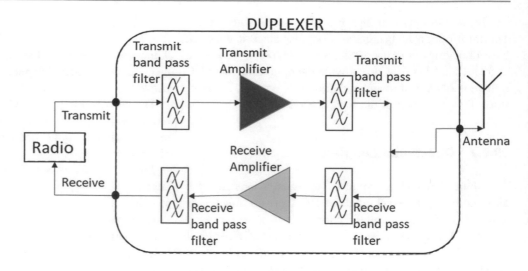

Table 19.5 Splitter division reduction loss

Number of output ports	Theoretical division reduction (dB)
2	3.0
3	4.8
4	6.0
5	7.0
6	7.8
8	9.0
10	10.0

bands so that they remain distinct. In order to achieve this, two distinct filters are used to keep signal from the sources [39].

Figure 19.38a shows block diagram of a diplexer—observe that both signals are in the same direction. Also, the two filters are in distinct frequency bands A and B which are clearly separated. Diplexer does not provide isolation between frequency bands.

In contrast, the splitter as the term indicates, only splits signals. The input signal P1 is split into two paths P2 and P3 therefore each path provides only half the power. Splitter can be designed either using resistors or hybrid elements but operates over a specified frequency band. It is possible to split multiple times, but each split results in reducing the signal levels. This is used as an advantage when a powerful signal must be uniformly spread as a lower power signal. For example, splitting the signal into smaller signal allows linear placement of antennas carrying the same level of signal along a long corridor. Table 19.5 shows how splitter loses signal as the number of branches increase.

Losses indicated in the table are based on theoretical calculations. In practice, each splitter has connector loss and cables used, etc., making the actual loss slightly more than

what is shown on the right-hand column. But splitters are very useful in serving multiple TV connections or extending cellular signal along a corridor, etc.

19.7.4 Balun Transformers

The word "Balun" is shortened form of "Balance-Unbalance". It is an important element used in television and other RF circuits. What is balance and unbalance—in every circuit the source and the load must be in balance or 1:1 so that energy is not wasted. Quite often this is not the case since the load and source may have different impedances. Balun transformer is primarily used to bring about balance in an unbalanced circuit. Its secondary use is isolating the source from the load. A common example in TV is the 300 Ω ribbon cable (balanced) from an antenna, to be attached to coaxial cable of 75 Ω (unbalanced or single ended). Balun transformer in this case uses 4:1 ratio in coil winding. This application is illustrated in Fig. 19.39a.

In Fig. 19.39 three types of Balun transformers are illustrated as examples of impedance matching [40]. The first one shown in (a) is often used in TV circuits. The second type shown in (b) is a center tapped "Ruthroff Balun". The center tap can be used for grounding—this type of Baluns provide impedance matching as well as isolation. There are situations where such isolation may not be necessary. Then "Guanella" type of transmission line balun is shown in (c) which can be used to operate in higher frequency bands of about 3 GHz. Most wideband Balun transformers operate at lower power of about 250 mW and also lower currents of no more than 250 mA. This is generally not a limitation since many cellular applications in the cellular, PCS, and AWS bands operate at powers lower than this.

Fig. 19.38 Diplexer and splitter
—differences

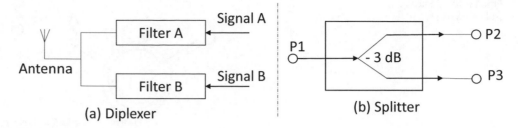

(a) Diplexer   (b) Splitter

Fig. 19.39 Three examples of
Balun transformers

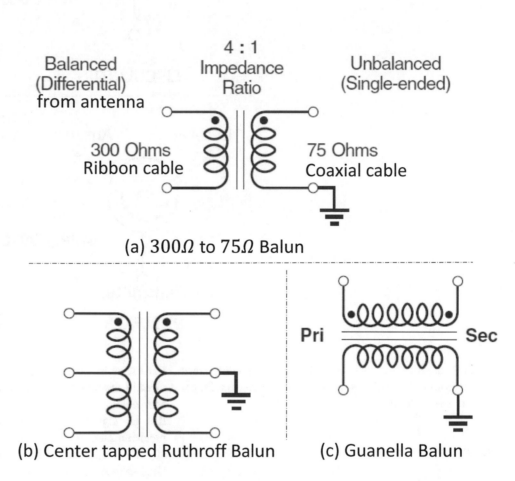

(a) 300Ω to 75Ω Balun

(b) Center tapped Ruthroff Balun   (c) Guanella Balun

19.7.5   Circulators/Isolators

RF Circulator is a non-reciprocal ferrite device which consists of 3 or more ports. Due to ferrite material placed inside circulator or an isolator, its strong magnetic interacts with the incoming signal and moves the signal to the adjacent port. So long as this port has little mismatch and good impedance with VSWR closer to 1.0, these devices operate well [41].

Figure 19.40 shows Circulator and an Isolator—both have similar features except that in Isolator port C terminated. The circulator symbol indicates that signal arriving at any port leaves at the subsequent port. In theory in a circulator if signal enters at port "N" it leaves at port "N + 1".

RF circulator has many applications. It can be used as an isolator (as shown), or as a diplexer, as a duplexer, or as protection for the equipment that provides signal generation and analysis often used in test and measurement applications.

The upper part of figure shows the symbols while lower part shows typical applications. For example, Duplexer function can be accomplished using a circulator, so long as the transmit and receive frequencies are considerably apart and well isolated. Circulator can be used to reduce the cost of expensive, traditional duplexer shown earlier in Fig. 19.37, which would have filters, amplifiers, and associated connections. Similarly, in measurement scenarios discussed throughout this chapter, signal source at one port

Fig. 19.40 Circulator and
Isolator symbols

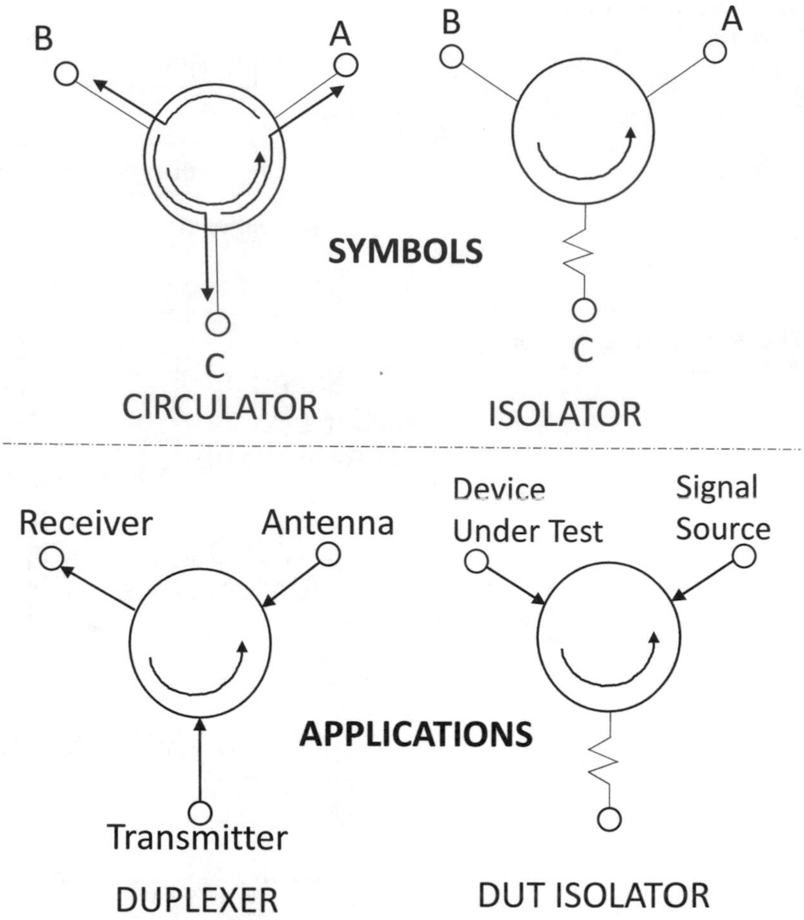

SYMBOLS

CIRCULATOR   ISOLATOR

APPLICATIONS

Receiver   Antenna

Transmitter

DUPLEXER

Device Under Test   Signal Source

DUT ISOLATOR

and DUT at the other offer isolation to the source, since reaction from DUT can be isolated. In such cases, traditional duplexer may not be needed.

19.7.6 Attenuators

An attenuator is a two-port resistive network designed to weaken or "attenuate" (as the name suggests) the power being supplied by a source to a level that is suitable for the connected load.

A passive attenuator reduces the amount of power being delivered to the connected load by either a single unit of fixed value, or a variable unit or in a series of known switchable steps [41]. All such steps are part of RF design in laboratory and attenuators are very useful during design implementation and also to trouble shoot networks in the field. When receiver levels are not known, it is always a good practice to insert an attenuator at the receiving instrument.

Attenuators are generally used in radio, communication, and transmission line applications to weaken a strong signal.

For example, power output from a transmitter shall not be directly measured on a precision instruments such as spectrum analyzer or a network analyzer. It must be attenuated so that the signal reduces to within the range acceptable to precision measuring instruments. In fact, the golden rule of all RF measurement laboratories is—first insert a "high-power variable attenuator" to "protect the measuring instrument". This is practiced as a "precautionary approach". Once the actual output is determined, a fixed value attenuator may replace the variable unit, but such fixed attenuator should be capable of dissipating the power level. If power levels are high, then attenuators placed in cooling oil baths are used.

Figure 19.41 shows the logical connection of how an attenuator is inserted between source and load. In most RF circuits the impedance used is 50 Ω. Since R1 and R2 determine the amount of attenuation, various configurations are used. Attenuator is a 4-port passive device which should be designed to minimize any distortion of the signal. There are even active attenuators that use transistors and integrated circuits to adjust any phase shift and avoid wire wound resistors to achieve "distortion-less" attenuators.

Fig. 19.41 Typical connection for an attenuator

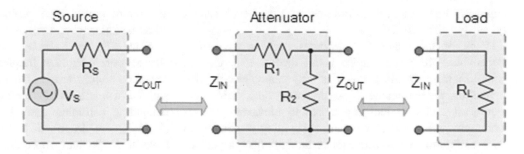

Source   Attenuator   Load

Table 19.6 Attenuator loss table

$\frac{V_{out}}{V_{in}}$	1	0.7071	0.5	0.25	0.125	0.0625	0.03125	0.01563	0.00781
Log value	20log (1)	20log (0.7071)	20log (0.5)	20log (0.25)	20log (0.125)	20log (0.0625)	20log (0.03125)	20log (0.01563)	20log (0.00781)
in dB's	0	−3 dB	−6 dB	−12 dB	−18 dB	−24 dB	−30 dB	−36 dB	−42 dB

Fig. 19.42 Three common configurations used to form attenuators

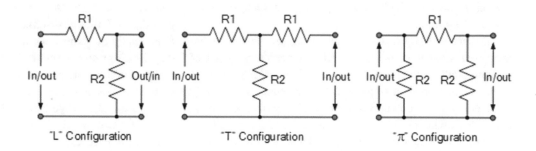

"L" Configuration   "T" Configuration   "π" Configuration

Attenuation normally takes into account the ratio of V_{in} and V_{out}. Since attenuation is based on V_{out} as the desired outcome:

$$dB_v = 20 \log_{10} \frac{V_{out}}{V_{in}} \ (dB) \qquad (19.50)$$

For all passive circuits, the convention used is to take 0 dB as the reference (input to attenuator taken as reference) and then subtract using the -dB notation. For example, if an attenuator of 6 dB is used, we can derive the ratio starting with a known. Since 3 dB is represented by output/input = 0.707, then 20 log (0.707) = −3 dB. Instead, if this ratio is 0.5 then 20 log (0.5) = −6 dB.

Using Eq. 19.7.6.1 it is possible to provide ratios of output/input, that correspond to standard value in dB. Table 19.6 shows standard values in terms of dB. For the sake of convenience, the well-regarded 0 dB reference point is taken, followed by higher attenuations (smaller ratios).

Since dB is a logarithmic scale, all calculations use dB allowing simple addition or subtraction. There is difference between dB in terms of power and dB in terms of Voltage (described here). These differences are clearly explained in chapter 2 under Sect. 2.1.1, where the concept of using dBm scale was reviewed.

Figure 19.42 shows three commonly used configurations in attenuators [42]. The ratio of input/output is a key design factor that provides value of the attenuator. The most common unit for all RF measurements is in dB, therefore attenuators follow this concept indicating attenuation in dB. There are other combinations of attenuators such as the switched attenuator, variable attenuator that are well described in [42].

19.8 Conclusion

This was indeed one of the most important aspects of wireless communication. Measurement and how accurately RF parameters are quantified establishes the criteria whether a system works and how well it will function. Quantifying also provides a better vision in terms of planning.

For a very long time RF measurements are deeply embedded in fundamentals of Physics which were pointed out often during this chapter. Many of the specific manufacturer equipment were brought as examples and are not intended as an endorsement of their products. The extensive knowledge brought about by application notes by vendors is often not available in textbooks. Such knowledge learnt by engineers with practical measurements at the bench cannot be easily substituted.

Each of the major RF attributes such as frequency, impedance, RF power, noise is all deeply embedded in concepts from Physics, which is why many Nobel prize inventions were referenced. At the same time, field measurements and how they are linked to fundamental measurements were used as revelation providing reasons as to why calibration of instruments is an important aspect of wireless communication.

Finally, some RF components were reviewed with the intent of how these are different from other electronic parts. At higher frequencies, every aspect, whether it is PIM, attenuator or using Balun for impedance matching, take an important role. Referencing all RF measurements to DC is correct and was evaluated theoretically. In practice, components designed for the frequency band in use must be carefully chosen, which is why parameters such as return loss are important.

Readers are urged to pay attention to what is being measured, how is it measured and whether the very act of measurement causes errors, why and how. Many of these were indicated and in retrospect, it is essential to carefully evaluate the connections before turning the power on. It may not seem apparent, but an expensive equipment can get permanently damaged if the input limits are not carefully verified (some of them can tolerate no more than ½ watt of power).

Summary

The chapter started with Frequency and time—which have become the very fundamental units of measurement with both length and mass specified in terms of time. The accuracy of frequency and time far exceeds any other quantity that is measurable today. Another important anchor in this area is the use of crystal oscillator whose stability and connection to other primary standard phenomenon such as the Cesium beam were explained.

This was followed by RF power which is very difficult to quantify accurately. The general approach taking DC as reference and moving to micro-calorimeter as a reference standard provides another important steppingstone in primary standards. RF power with its ability to steer all wireless systems both in capacity and coverage distance plays a major role. Power meters with thermistor mount play a role both as secondary standard and working standard. Other network issues such as impedance matching and how it affects accuracy of RF power were discussed, leading to the measurement of impedance. In Chap. 17 on Antennas, a separate Sect. 17.1.4 was devoted to matching network. Here this aspect was discussed using a Thruline power meter, underling its simplicity and ability to read forward and return power were discussed showing how a simple "field instrument" can remain versatile over many decades.

Since RF impedance is inherently a vector quantity that changes considerably as frequency varies, physicists took an important step in terms of using the Smith chart (uses first and fourth quadrants of cartesian system) to using a complex computing instrument known as "Vector Network Analyzer (VNA)" were described. Since series and parallel circuits have their popular representation of resistance and its reciprocal admittance and similar reciprocity terms—these were discussed, showing why in some case admittance seems to offer a better understanding.

RF-IV method of voltage ratio measurements using a fixed value resistor were described. Such ratio-based methods provide better accuracy by referencing to a stable value of resistor.

Scattering parameters or S-parameters are still the major approach used by vector and signal analyzers to characterize impedance. This was discussed in detail, indicating how they functionally evaluate impedance and establish the nature of impedance in any network. Network analyzers test a wide range of active and passive devices—these were described with examples establishing how Smith chart has paved the way to automated impedance calculations for both capacitive and inductive loads—showing how matching can be achieved in each case.

Noise is inherent in all RF systems—it is a fundamental quantity that must be dealt with and addressed seriously to make sure the system works. Starting with the power line from utility, noise begins its systematic journey into all networks. Both precautionary measures and aggressive methods to reduce noise were described. There are many software packages available that simulate noise models well. They provide a better view of this complex, and random quantity that every communication system must address.

Starting with design simulation indicating noise source and reduction techniques, other sources of noise in data converter were discussed to bring an important insight into the world of digital radio, where data converters and sources dominate baseband. A detailed investigation in terms of providing techniques to overcome noise in such areas is an important pointer to impress upon the reader as to how this area continues to grow in addressing noise within ADC and DAC. Noise Factor which is an important measurement was discussed in terms of how it affects the first unit and subsequent network units.

Subsequent topic in measurements was that of transmitter and receiver parameters—it laid the foundation for how phase noise and modulation parameters and their measurement play a key role in establishing quality of a transmitter. Similarly, modern digital receivers place significant importance on $\frac{E_b}{N_o}$ for bit error rate or frame error rate measurements. The most important parameter of a receiver being "sensitivity". It was described in detail. Details went on to show how it is measured with the help of Digital Signal Processing (DSP).

Field measurements and instruments are in the process of a major transformation given the pace with which cellular systems are changing. Starting with how a cell site is established often in harsh environments, measurements that provide an insight into their network deployment particularly in 5G. Their complexity is significant and detailed analyzers are required to set up a cell site today. With the use of broadband systems, it has become necessary to address PIM —the Passive Inter-Modulation sources that are a major source of noise in cellular systems. This was described briefly showing how PIM sources are identified in the field and eliminated.

Many RF components are unique in their approach to design and operation. They are in widespread use in the lab and in the field and are an integral part of all measurement set ups. A major sample of such components were reviewed. Although majority of them are passive in nature, they contribute significantly in terms of deployment planning, performance, and network behavior. Starting with lightning arrestors which are mandatory in all cell sites, were described—along with how lightening current can be diverted by careful design. Other components such as couplers, duplexers, circulators, and attenuators were described indicating their design.

Emphasis on participation

1. What are the primary and secondary standard bodies in your area? How do they trace standards for RF power, impedance, and frequency? Survey at least three primary standard bodies and compare their accuracy levels.
2. What is the noise floor level expected in a typical cellular channel in the 800 MHz band? What is the level in the 800 MHz public safety band? Why is there a considerable difference? How does it compare with noise in satellite networks?

Knowledge gained

- Relationships between primary, secondary, and working standards
- Challenges of measuring RF power accurately—alternatives. Working units of RF power measurements—units that need no battery and flexible
- Crystal oscillator as a major frequency source—its accuracy and traceability
- Methods of impedance measurement at different frequency bands in RF

- Nature of noise, its sources in the RF and digital converter networks. Compensation methods and measurement of noise
- Practical measuring instruments in the field—complexity of a typical 5G base station measurement and alignment
- RF components—their simplicity and precautions needed for various circuits
- Difference between duplexer, diplexer, and splitter
- Circulator and its use as a simple duplexer and isolator
- Attenuators—different types and their use in circuits— precautions while measuring transmit power.

Home work problems

(1) Print out a Smith chart from any online site or use a hard copy (if available). A load of 100-j50 is to be compensated at 500 MHz, on a standard RF network with characteristic impedance of 50 Ω. How would you proceed?
(2) Which method would you choose for high accuracy impedance measurement at 450 MHz? Use two different methods to calculate the impedance of a load—one with 54-j8 and another with 48 + j9. Can there be a compensation circuit to accommodate both? Provide a proposal of such a network.
(3) A source is connected to an antenna with "Thruline" power meter inserted in the line. The power meter measures 8.5 Watts in the forward direction and 1.5 W in the reverse direction. What is the reflection coefficient? Hint: Table 19.2.
(4) An LTE network operates on the 1850–1910 MHz band for transmit and 1930–1990 MHz band for receive. Using duplexer in the diagram shown in Fig. 19.38, draw a scheme to connect the transmitter, receiver, and antenna. Using the circulator shown in Fig. 19.41 draw a similar scheme to connect transmitter, receiver, and antenna. What is the difference between the two? Why are duplexers used more often in industry?

Project ideas

Using laboratory equipment—device a method to measure short-term and long-term accuracy of a crystal oscillator. Place the oscillator in an oven or any enclosure where temperature can be kept within ±5 °C. Record readings and propose design of an enclosure to maintain it within ±2 °C

Out of the box thinking

Equation 19.33 discusses the jitter noise in an ADC. Oversampling is indicated as one of the methods. Are there different methods that can reduce noise in an ADC, particularly for those using 12 bits or more? How much of oversampling is possible? What is the point of compromise when oversampling is no longer helpful?

Using reference [29] or any equivalent Math Lab software, generate 16-QAM data streams. Using Forward Error Correction (FEC) scheme what would be the data rate show how much is the error improvement. Provide scatter plots with and without FEC.

References

1. Explorable, Indian Astronomy in the first millennium, Martyn Shuttleworth, https://explorable.com/indian-astronomy

2. How are atomic clocks so accurate at keeping time? World Time Server. https://www.worldtimeserver.com/learn/how-are-atomic-clocks-so-accurate-at-keeping-time/current_time_in_US-NY/

3. Jefferts SR Primary frequency standards at NIST. NIST—time and frequency division.https://www.aps.org/units/maspg/meetings/upload/jefferts-111517.pdf

4. Scientists Are About to Redefine the Kilogram and Shake Up Our System of Measures, By Jay Bennett, SMITHSONIANMAG. COM, NOVEMBER 14, 2018. https://www.smithsonianmag.com/science-nature/redefine-kilogram-180970798/

5. Understanding Frequency Accuracy in Crystal Controlled Instruments, Application Note, https://www.testworld.com/wp-content/uploads/understanding-frequency-accuracy-in-crystal-controlled-instruments.pdf

6. Time realization and distribution, GPS data archive, NIST, Boulder. Colorado. https://www.nist.gov/pml/time-and-frequency-division/services/gps-data-archive

7. New Crystal Oscillators transform the market, Electronic pages, 21 Nov 2019, By Nnamdi Anyadike, https://www.electropages.com/blog/2019/11/new-crystal-oscillators-transform-market

8. John Vig, Quartz Crystal Resonators and Oscillators, 2016, (Slide 167, section 4–73). http://www.beckelec.com/john-vigcrystal-tutorial-2.html

9. MEMS oscillators: Enabling Smaller, Lower Power IoT & Wearables, SI Time, https://www.sitime.com/company/news/media-coverage/mems-oscillators-enabling-smaller-lower-power-iot-wearables-0

10. Clock works configurator – MEMS single output and multioutput oscillator. https://clockworks.microchip.com/microchip/

11. Traceability of RF measurement quantities to National standards, Rohde and Schwarz. https://www.rohde-schwarz.com/file/N199_28-33_PTB-Traceability-of-RF-measurement_e.pdf

12. Module 16 – Introduction to test equipment, Navy electricity and electronics training series (NEETS), Chapter 5, SPECIAL-APPLICATION TEST EQUIPMENT, https://www.rfcafe.com/references/electrical/NEETS-Modules/NEETS-Module-16-5-1-5-10.htm

13. Bird RPM-16, Average reading RF Power meter, Manual. https://birdrf.com/Products/Test%20and%20Measurement/RF-Power-Meters/Wattmeters-Line-Sections/RF-Wattmeters/APM-16_Average-Reading-Power-Meter.aspx

14. Farad and Impedance metrology, National Institute of Science and Technology, Projects and Programs. https://www.nist.gov/programs-projects/farad-and-impedance-metrology

15. What are Josephson Junctions? How do they work, Nov 24, 1997, Richard Newrock, Scientific American, https://www.scientificamerican.com/article/what-are-josephson-juncti/

16. A primary voltage standard for the whole world, April 4, 2013. https://www.nist.gov/news-events/news/2013/04/primary-voltage-standard-whole-world

17. Impedance measurement handbook, A guide to measurement technology and techniques. 6th edition, Keysight Inc. https://www.cmc.ca/wp-content/uploads/2019/07/Keysight-Technologies-impedance-measurement-handbook.pdf, https://www.cmc.ca/wp-content/uploads/2019/07/Keysight-Technologies-impedance-measurement-handbook.pdf

18. How to accurately evaluate Low ESR, High Q, RF chip devices, Application note 1369–6, Agilent Technologies, http://literature.cdn.keysight.com/litweb/pdf/5989-0258EN.pdf

19. Brushing up on network analyzer fundamentals, Chris Demartino, June 27, 2018. https://www.mwrf.com/print/content/21849241

20. Network Analyzer application notes, http://na.support.keysight.com/faq/Document_list.html

21. Understanding the smith chart by Paul Denisowski, Rohde and Schwarz, https://www.youtube.com/watch?v=rUDMo7hwihs

22. Vector Network Analyzer tutorial: https://www.rfwireless-world.com/Tutorials/Vector-Network-Analyzcr-VNA-tutorial.html

23. Electronic Noise and Interfering Signals, Principles and applications, by Gabriel Vasilescu, Laboratoire des Instruments et Systèmes d'Ile de France (LISIF) Université Pierre & Marie Curie Paris Cedex 05 France. December 18, 2010, Springer-Verlag Berlin Heidelberg 2005. Print ISBN :978–3–540–40741–6, Online ISBN: 978–3–540–26510–8

24. Primary Noise sources, NBS series. https://www.noisecom.com/Portals/0/Datasheets/PrimNoiseNBS4pages.pdf

25. Noise parameters a practical example, Microwaves101. Com https://www.microwaves101.com/encyclopedias/noise-parameters-a-practical-example

26. New ultra-fast noise parameter system, David Balo and Gary Simson, April 2009, https://www.maurymw.com/Support/pres/ultra-fast_noise_para.pdf

27. Understanding signal to noise ratio and noise spectral density in high-speed data converters, by Ken Chen, Texas Instruments, TIPL 4703. July 26, 2017 https://training.ti.com/understanding-signal-noise-ratio-snr-and-noise-spectral-density-nsd-high-speed-data-converters

28. Five RF transmitter measurements every engineer should know, National Instruments, July 23, 2019. https://www.ni.com/en-us/innovations/white-papers/14/5-rf-transmitter-measurements-every-engineer-should-know.html

29. Examine 16-QAM using MATLAB, Mathworks Help Center, R2020b. https://www.mathworks.com/help/comm/gs/examine-16-qam-using-matlab.html

30. Wireless Test bench simulation, Chapter 5, Keysight Technologies, https://literature.cdn.keysight.com/litweb/pdf/ads2004a/pdf/adswtbsim.pdf

31. 5G NR physical cell ID PCI planning, http://www.techplayon.com/5g-nr-physical-cell-id-pci-planning/

32. LTE quick reference, SIB - System Information Block, https://www.sharetechnote.com/html/Handbook_LTE_SIB.html

33. Site testing and trouble shooting in 5G mobile networks, R & S Application brochure version 01.01. https://www.rohde-schwarz.com/us/product/5g-sts-productstartpage_63493-766529.html

34. PR 200 Portable monitoring receiver, product brochure 4.01, R & S, https://www.rohde-schwarz.com/us/product/pr200-productstartpage_63493-594881.html

35. Webinar on successful 5G site testing by Peter Busch and Simon Allemon, R & S https://www.rohde-schwarz.com/us/knowledge-center/webinars/webinar-successful-5g-site-testing-and-acceptance_252686.html#media-gallery-4

36. Understanding PIM – site maintenance, Anritsu application note. https://dl.cdn-anritsu.com/en-us/test-measurement/files/Application-Notes/Application-Note/11410-00629F.pdf

37. Anritsu Company, Measurement guide MW82119B. Part Number: 10580–00402 Revision: H, Published: April 2019. https://dl.cdn-anritsu.com/en-us/test-measurement/files/Manuals/Measurement-Guide/10580-00402L.pdf

38. Lightening protectors—Richardson RFPD, Huber-Suhner. Switzerland, Edition 2007, https://www.richardsonrfpd.com/docs/rfpd/LP_catalogue.pdf

39. What is the difference between diplexer and duplexer? Editorial team, Everything RF, Nov 28, 2016. https://www.everythingrf.com/community/what-is-the-difference-between-a-diplexer-and-duplexer

40. Using Baluns and RF components for impedance matching, Coilcraft, Document 1077–1 Revised 03/27/13. https://www.coilcraft.com/pdfs/Doc1077_Baluns_and_Impedance_Matching.pdf

41. RF Microwave blog—Isolator circulator basics, http://www.e-meca.com/rf-microwave-blog/isolator-circulator-basics

42. Electronic tutorials—passive attenuators. https://www.electronics-tutorials.ws/attenuators/attenuator.html

Printed in the United States
by Baker & Taylor Publisher Services